Modern Dictionary
of
ENVIRONMENTAL SCIENCE AND ENGINEERING

Modern Dictionary of ENVIRONMENTAL SCIENCE AND ENGINEERING

Comprehensive and Illustrated Encyclopaedic Dictionary

Edited by

DR. GAUHAR MEHMOOD

Professor, Department of Civil Engineering
Jamia Millia Islamia, New Delhi

NuTech Books
from

DEEP & DEEP PUBLICATIONS PVT. LTD.
F-159, Rajouri Garden, New Delhi - 110 027

MODERN DICTIONARY OF
ENVIRONMENTAL SCIENCE AND ENGINEERING

ISBN 81-7629-936-7

Printed in India at MAYUR ENTERPRISES
WZ Plot No. 3, Gujjar Market, Tihar Village, New Delhi - 110 018

NuTech Books are Published by DEEP & DEEP PUBLICATIONS PVT. LTD.,
F-159, Rajouri Garden, New Delhi - 110 027 • Phone : 25435369, 25440916
E-mail : ddpubs@gmail.com • ddpbooks@yahoo.co.in
Showroom :
2/13, Ansari Road, Daryaganj, New Delhi - 110 002 • Telefax : 23245122

Contents

Preface

Environmental Science and Engineering is the study of the interactions among the physical, chemical and biological components of the environment; with a focus on pollution and degradation of the environment related to human activities; and the impact on biodiversity and sustainability from local and global development. It is inherently an interdisciplinary field that draws upon not only its core scientific areas, but also applies knowledge from other non-scientific studies such as economics, law and social sciences. Physics is used to understand the flux of material and energy interaction and construct mathematical models of environmental phenomena. Chemistry is applied to understand the molecular interactions in natural systems. Biology is fundamental to describing the effects within the animal kingdoms.

The concept of environmental science has existed for centuries. It came alive as a substantive, active field of scientific investigation in the 1960s and 1970s driven by (a) the need for a large multi-disciplined team to analyze complex environmental problems, (b) the arrival of substantive environmental laws requiring specific environmental protocols of investigation and (c) the growing public awareness of a need for action in addressing environmental problems.

Environmental science encompasses issues such as climate change, conservation, biodiversity, groundwater and soil contamination, use of natural resources, waste management, sustainable development, air pollution and noise pollution. Due to the inherent interdisciplinary nature of environmental science, teams of professionals commonly work together to conduct environmental research or to produce Environmental Impact Statements. Atmospheric sciences examines the phenomenology of the Earth's gaseous outer layer with emphasis upon interrelation to other systems. Atmospheric sciences comprises meteorological studies, greenhouse gas phenomena, atmospheric dispersion modeling of airborne contaminants, sound propagation phenomena related to noise pollution, and even light pollution.

Physicists create computer models of atmospheric circulation and infra-red radiation transmission, chemists examine the inventory of atmospheric chemicals and their reactions, biologists analyze the plant and animal contributions to carbon dioxide fluxes, and specialists such as meteorologists and oceanographers add additional breadth in understanding the atmospheric dynamics. Ecology studies typically analyse the dynamics among an interrelated set of populations, or a population and some aspect of its environment. These studies could address endangered species, predator/prey interactions, habitat integrity, effects upon populations by environmental contaminants, or impact analysis of proposed land development upon species viability.

An interdisciplinary analysis of an ecological system which is being impacted by one or more stressors might include several related environmental science fields. For example one might examine an estuarine setting where a proposed industrial development could impact certain species by water pollution and air pollution. For this study biologists would describe the flora and fauna, chemists would analyze the transport of water pollutants to the marsh, physicists would calculate air pollution emissions and geologists would assist in understanding the marsh soils and bay muds.

Environmental chemistry is the study of chemical alterations in the environment. Principal areas of study include soil contamination and water pollution. The topics of analysis involve chemical degradation in the environment, multi-phase transport of chemicals (for example, evaporation of a solvent containing lake to yield solvent as an air pollutant) and chemical effects upon biota.

Physicists would develop a computer model to understand the extent of soil contamination and sub-surface transport of solvent, chemists would analyze the molecular bonding of the solvent to the specific soil type and biologists would study the impacts upon soil arthropods, plants and ultimately pond dwelling copepods who are the food of the endangered amphibian.

Geosciences includes environmental geology, environmental soil science, volcanic phenomena and evolution of the earth's crust. In some classification systems it can also embrace hydrology including oceanography. As an example study of soils erosion, calculations would be made of surface runoff by soil scientists. Hydrologists would assist in examining sediment transport in overland flow. Physicists would contribute by assessing the changes in light transmission in the receiving waters. Biologists would analyze subsequent impacts to aquatic flora and fauna from increases in water turbidity.

This environmental dictionary includes over 6000 terms, with heavy emphasis on Agriculture Biology Environment, Atmospheric Chemistry, Carbon Dioxide and Climate, Civic, Coastal Terminology, Ecosystem, Ecosystem Management Terms, Environment Health Terminology, Forest Genetics, Hurricane and Tropical, Meteorology, Mountain Meteorology, Pollution Control, Recycling Terminology, Snow Glossary, Toxic Substances, Water Quality and Waste Management, Water Quality Terms, Weather Climate Geology Meteorology, Weather Glossary Words and phrases such as "rad," "greenhouse effect," "ozone layer," and "nuclear winter" are all here. A useful list of acronyms such as EPA, EIS, and GRAS is also included.

The need to consider the environmental impacts for any industrial activity, is now taken for granted. Environmental science has become an interdisciplinary subject dealing with elements of ecology, chemistry, economics, engineering, biochemistry, politics and sociology. This clearly presented and thoroughly comprehensive Dictionary, is indispensable to all those concerned with the world's environment.

I am grateful to all who have cooperated in finalising this book. I acknowledge with thanks the help of various librarians and authors of various books which have been referred to in this book. I would also like to thank the publisher for bringing this book out in time.

DR. GAUHAR MEHMOOD

Agricultural Biology Environment

A

❒ activator

Activators are additives to the compost pile which contain a nitrogen source or sugars. Their purpose is to increase microbial activity. Generally, adequate nitrogen organic waste is the only activator needed. If you have insufficient nitrogen, a substance like cottonseed meal may be added to encourage decomposition.

❒ aerate, aeration

To aerate soil is to introduce air into the soil. This can be accomplished by many methods including plowing, roto-tilling, using a pitch fork or pole to punch holes into the earth.

❒ aerated static pile

A compost pile that is not turned (static), but is aerated through ventilation pipes that run through the pile. These may be PVC pipes with holes drilled into them.

❒ aerobic

Containing air, containing oxygen.

❒ aggregation

A mass or body of individual units or particles. Healthy soil has good aggregation. As microorganisms and worms feed, they form polysaccharides which act like glue to hold individual soil particles together, creating groups, or aggregates, of particles. This loose formation allows soil to hold both water and air, and does not restrict the growth of roots.

❒ ambient temperature

Air temperature, i.e., the temperature of the air around the pile and not affected by the heat of the pile.

❒ anaerobic

Lacking air, lacking oxygen.

❒ appropriate land use 1: minimum tillage

- Conventional tillage with tractors and ploughs is a major cause of severe soil loss in many developing countries.

- Today's conventional plowing methods cause severe soil loss and desertification in many developing countries.
- The Food and Agricultural Organization (FAO) estimates that some 40 percent of land degradation around the world is caused by soil erosion.

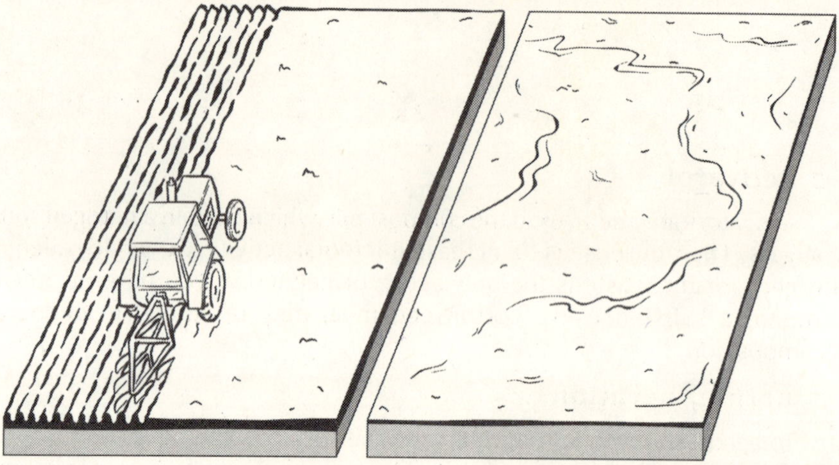

Intensive tillage leaves the soil bare and unprotected

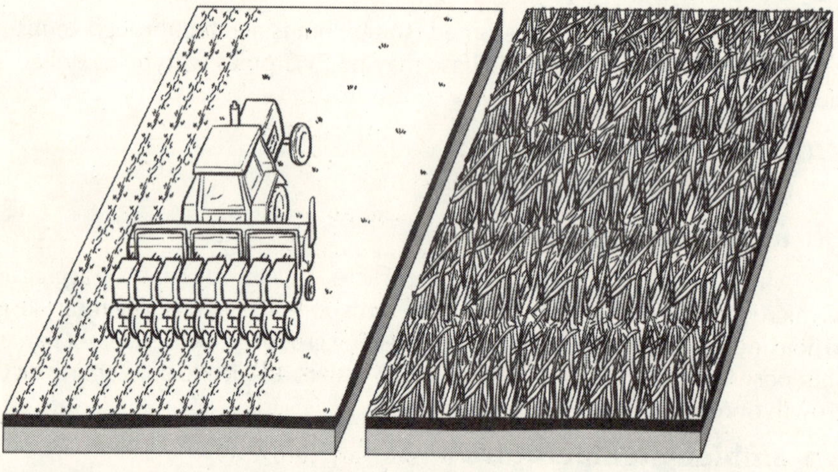

Minimum tillage protects the soil against erosion
Special no-till planters cut through crop residues and plant into soil without plowing or disking.

- Intensive tillage causes more erosion and soil degradation, especially in warmer areas where the topsoil layer is thinner.
- Minimum tillage systems offer numerous benefits that intensive tillage systems cannot match.
- These include: reduced labor requirements, time saving, reduced machinery wear, fuel savings, improved long-term productivity, improved surface water

quality, reduced soil erosion, greater soil moisture retention, improved water infiltration, and reduced release of carbon gases.

❑ appropriate land use 2: intercropping

- Intercropping is the growing of two or more crops in proximity to promote interaction between them.

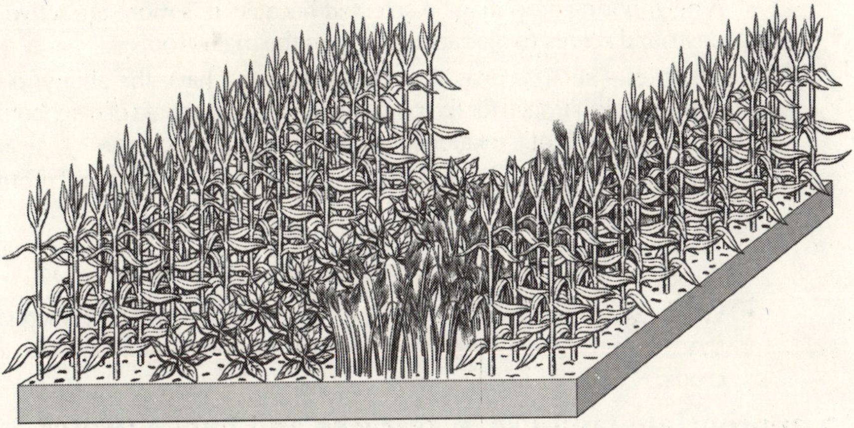

Corn, soybeans, and wheat strip-cropped on a US farm
There is overwhelming evidence that plant mixtures support lower numbers of pests than do pure stands.

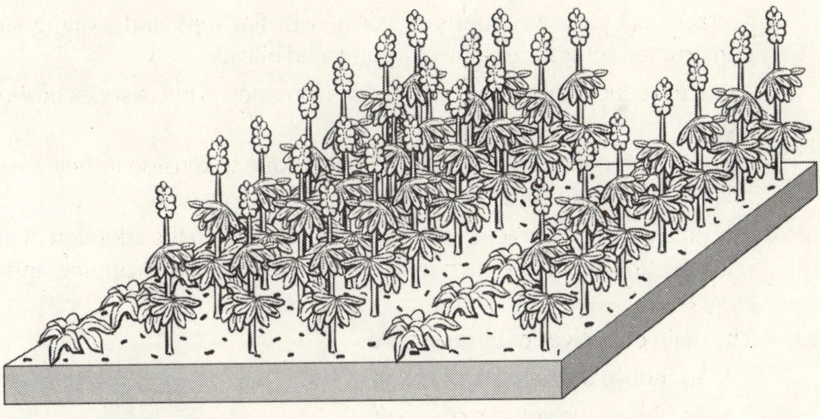

Young cotton planted into lupines
Lupines fix nitrogen and provide a beneficial insect habitat, so interplanting reduces inputs of fertilizer and pesticides for the cotton.

- There are four basic types of intercropping:
  - Row intercropping – growing two or more crops at the same time with at least one crop planted in rows.
  - Strip intercropping – growing two or more crops together in strips wide enough to permit separate crop production using machines but close enough for the crops to interact.

- Mixed intercropping – growing two or more crops together in no distinct row arrangement.
- Relay intercropping – planting a second crop into a standing crop at a time when the standing crop is at its reproductive stage but before harvesting.
- Intercropping has various advantages for the farmer.
- A neighboring crop may be selected because it is more attractive to pests and serves to distract them from the main crop.
- Legumes – such as peas, beans, and clover – have the ability to fix atmospheric nitrogen for their own use and for the benefit of neighboring plants via symbiotic relationship with Rhizobium bacteria.
- Some plants exude chemicals that suppress or repel pests and protect neighboring plants.
- Tall or dense-canopied plants may protect more vulnerable species through shading or by providing a windbreak.
- When two or more crops are growing together, each must have adequate space to maximize cooperation and minimize competition between the crops.

❑ appropriate land use 3: terraces and bunds (water and soil conservation)

Terraces

- Terraces are raised mounds of earth with flat tops and sloping sides, constructed across the slope of a cultivated hillside.
- Terraces reduce soil erosion by breaking long slopes into a series of shorter slopes.
- Terraces can considerably reduce soil loss due to erosion if they are well planned, correctly constructed and properly maintained.
- The efficiency of a terrace system will also depend on the adoption of other conservation practices such as contour sowing, strip cropping and soil cover.
- The main objectives of terracing are:
  - To reduce the velocity of runoff;
  - To reduce the volume of runoff;
  - To reduce the losses of soil, seed and fertilizer;
  - To increase soil moisture content through improved infiltration;
  - To reduce peak discharge rates of rivers.

Bunds

- Contour bunding is one of the simplest methods of soil and water conservation.
- Bunds are embankments of earth or stones which act as a barrier to water flow, reducing soil erosion and keeping the soil moist.

- Bunding plays an important role in soil and water conservation in fields with a medium slope.
- Trees which fix nitrogen in to the soil can be planted along bunds.

Terracing of shallow slope and steep slope

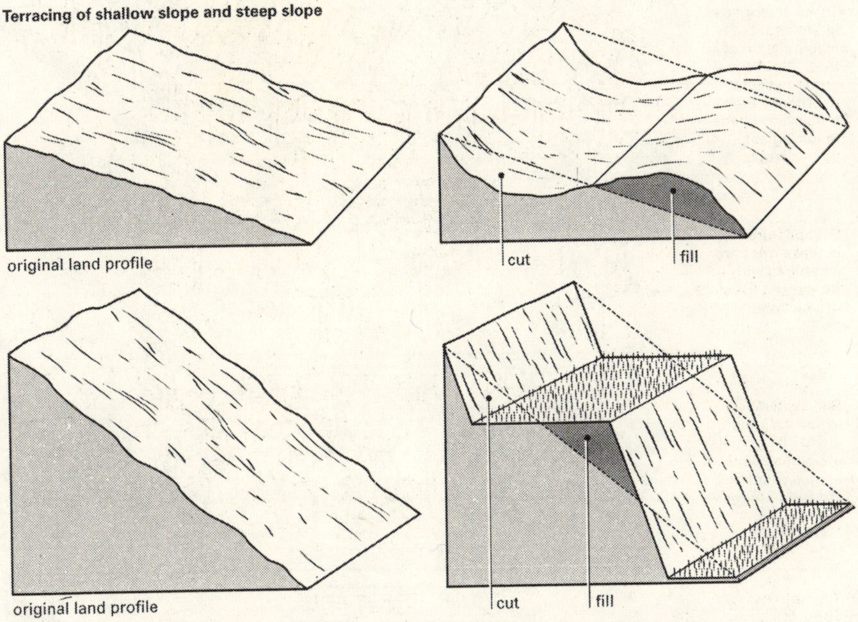

Contour bunding on fields with moderate slopes helps in soil and water conservation.

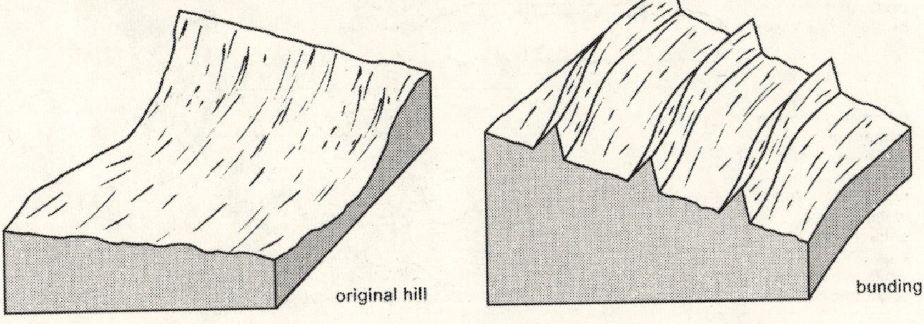

❐ appropriate land use 4: agroforestry

- Agroforestry is a land management approach that deliberately combines the production of trees with other crops and/or livestock.
- By blending agriculture and forestry with conservation practices, agroforestry strives to optimize economic, environmental and social benefits.
- Intensive management of trees, non-timber forest crops, agricultural crops and animals on traditional forest and agricultural lands is the key to successful agroforestry.
- There are five basic types of agroforestry:

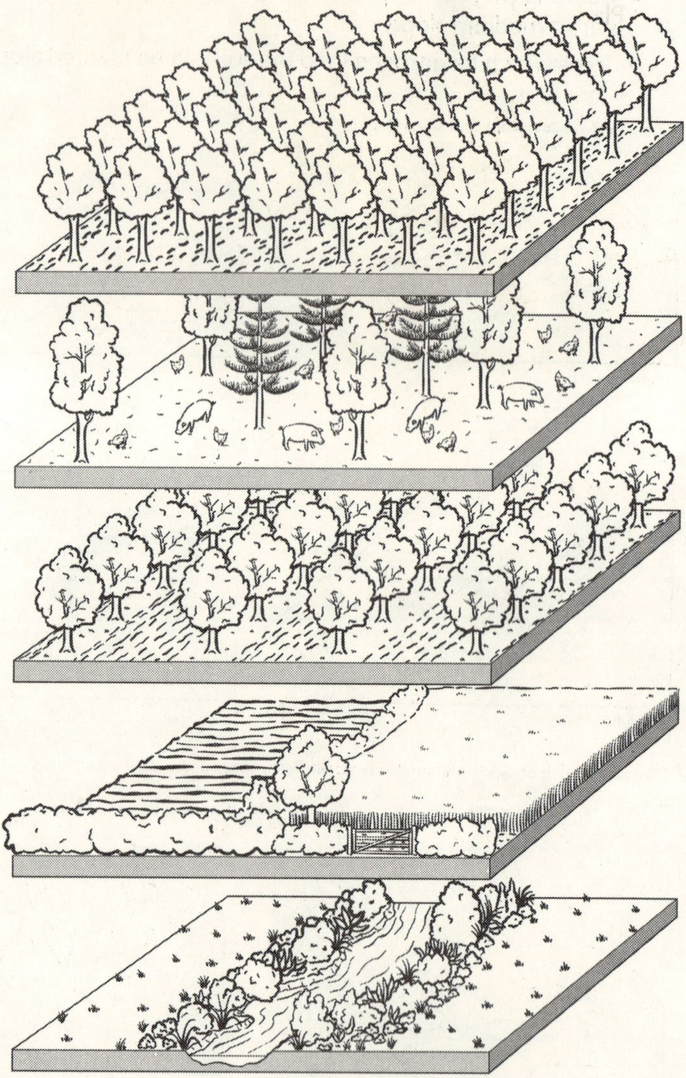

Shade systems
These comprise the intentional manipulation of trees, shrubs and ground vegetation to produce both timber and shade-requiring non-timber crops.

Silvopasture
In these trees are combined with forage and livestock production.

Sun systems
In sun systems woody perennials are planted with annual or other crops that require full sun.

Timberbelts
These consist of multiple rows of trees planted for both environmental protection and production of crops.

Integrated riparian management
The management of areas bordering waterbodies and watercourses to enhance and protect aquatic resources while generating economic benefits.

- Shade systems: shade agroforests and forest farming
- Silvopasture
- Sun systems: sun agroforests and intercropping
- Timberbelts.
- Integrated riparian management

❑ appropriate land use 5: soil protection/rejuvenation

How soil fertility is protected and restored

a. Reforestation to give soil stability and control runoff from the watershed

b. Reduce number of grazing animals to prevent overgrazing.

c. Plant shelter belts to reduce the effect of wind.

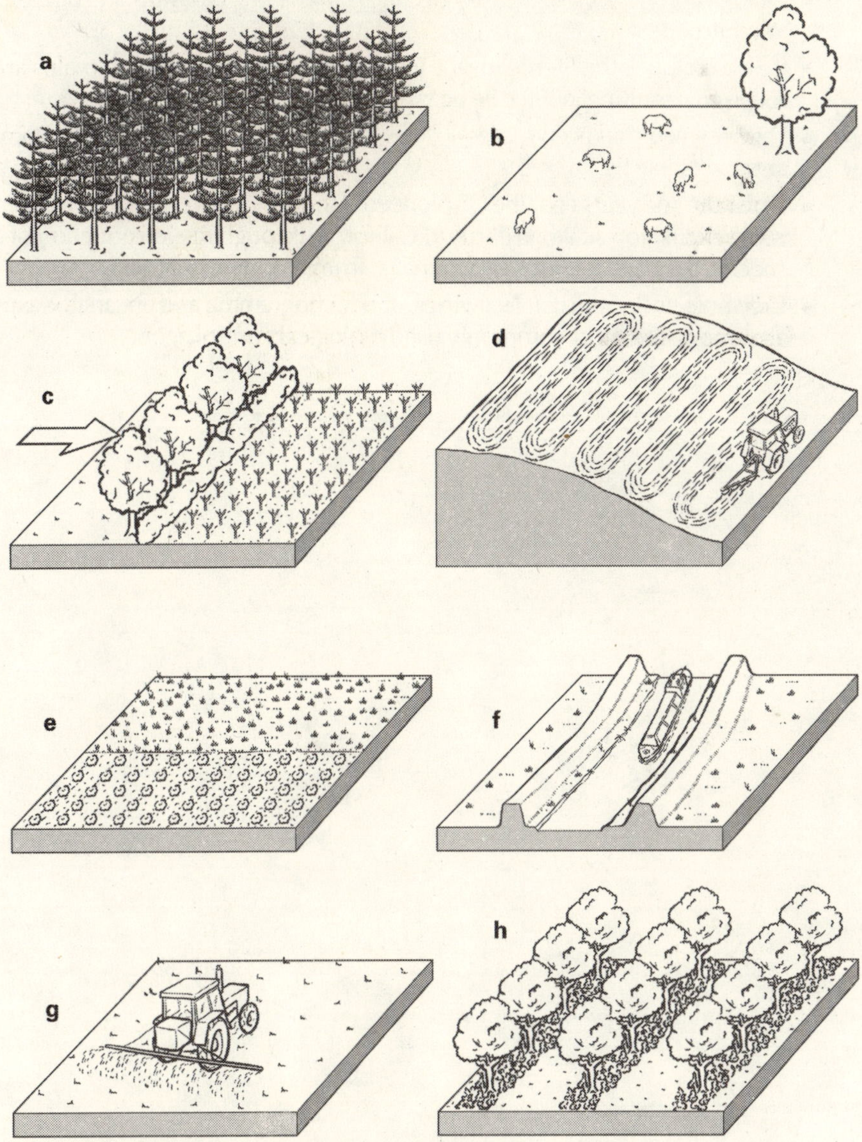

d. Contour plow to prevent gully formation.

e. Rotate crops and use fallow periods to prevent excessive use of soil nutrients.

f. Reinforce riverbanks to prevent removal of topsoil by flooding.

g. Apply crop residues and organic manures to restore fertility.

h. Plant nitrogen-fixing plants and trees as intercrops or green manure.

❐ appropriate land use 6: biological pest control

Biological control

- Biological control uses naturally occurring predators, parasites and diseases to control pests.
- It also includes the Sterile Insect Technique, in which sterilized males are released, resulting in infertile eggs.
- This technique has been successfully used against the New World screwworm and the tsetse fly.
- One hundred years ago, the US pioneered the biological control of the alien cottony cushion scale on citrus in California through the introduction of a specific lady beetle, Rodolia cardinalis, from its native Australia.
- Lady beetles (ladybirds), lacewings, and trichogramma and encarsia wasps are among the most commonly used biological control agents.

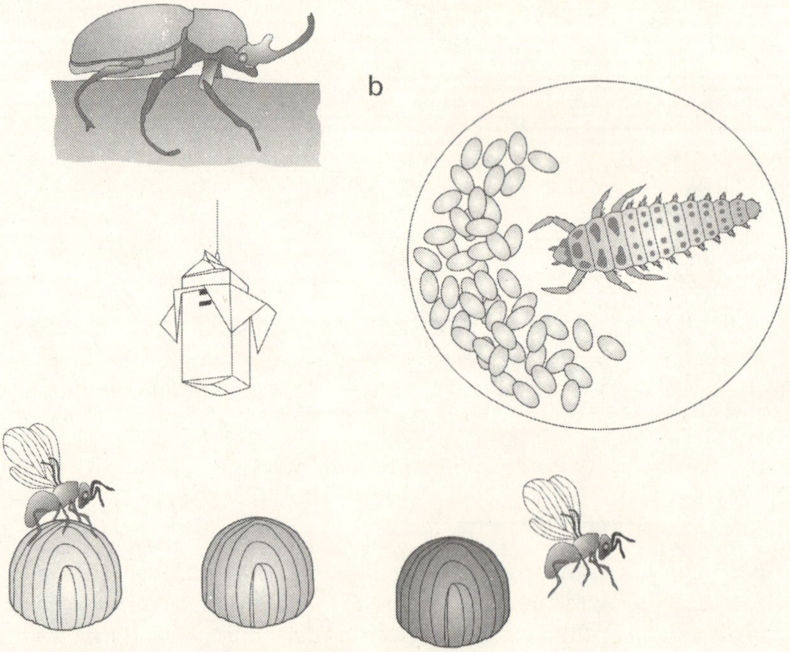

a Rhinoceros beetle, a pest of coconut palms, was successfully controlled by a virus.
b The larva of the lady beetle can eat hundreds of aphids every day, and the adult beetle also eats them.
c The Gypsy Moth, introduced to the United States from Europe, can defoliate trees and vines, but is controlled by pheromone traps.
d The boll weevil, a pest of cotton growers in the United States, can be controlled by a parasitic wasp, *Trichogramma*. Diagram shows wasp laying egg in bollworm egg; its larva feeds on the egg, and emerges as an adult wasp 8 days later.

Bacterial control

- Bacteria or viruses, such as Bacillus thuringiensis, are also successfully used to control pests.
- A baculovirus was successfully used against the rhinoceros palm beetle (Oryctes rhinoceros), when it devastated Pacific island coconut palms.
- Sex pheromones are also used in sticky traps to attract male insects.
- The myxoma virus was used to control rabbits in Australia, with partial success.

Results

- To date, of the over 6,000 introductions of alien natural enemies to control insects, weeds and other pests, only a proportion, perhaps 25–30 percent have led to substantial success.
- But because the control is permanent, successes yield enormous returns on investment to governments and communities.
- Few if any introductions, successful or unsuccessful, have led to environmental problems because the alien natural enemy is becoming a pest itself.

❒ auxin

substances found in places such as plant sprouts and human urine which stimulates growth in plant tissues promoting root formation.

B

❒ bacteria

Microorganisms that break down organic materials in the first stages of composting. It is bacteria that generate the heat associated with hot composting. The three types of bacteria are psychrophilic, mesophyllic, and thermophilic.

❒ bat guano

Manure, i.e., excretions, of bats.

❒ batch process

. An entire cubic-yard pile is built at the same time, and finishes decomposing at the same time. This is the opposite of the "add as you go" method. Because the entire pile is built at one time, factors such as moisture, C:N ratio, variety of textures and sizes, etc. can be more closely controlled for fast decomposition.

❒ biodegradable

Able to be broken into simpler chemical compounds by microorganisms. Organic materials are biodegradable.

❒ biodiversity

In an environments created solely by nature, there is a variety of plant and animal life, ranging from the very small to the very large. Nature has created a natural system for post and disease control. However, when we only incorporate limited variety in our landscapes, the system of checks and balances breaks down. In general, the more diverse we can make our gardens, the healthier they will be.

❑ browns

The term "browns" is used to denote organic materials high in carbon, more specifically, materials whose carbon to nitrogen ratio is higher than 30:1. (Materials high in nitrogen are referred to as "greens"). Achieving a carbon-to-nitrogen ratio of about 30:1 is one factor in creating favorable conditions for backyard pile composting.

C

❑ carbon-to-nitrogen ratio (c:n)

The relative amount of carbon to nitrogen, e.g., a 2:1 ratio means that there is twice as much carbon as nitrogen. Bacteria, like all living organisms, require quite a bit of carbon and comparatively less nitrogen. By providing them with materials that provide these elements in the correct proportion, they thrive, grow, and multiply. Therefore, they can decompose your compost pile at their highest speed. Achieving a carbon-to-nitrogen ratio of about 30:1 is one factor in creating favorable conditions for backyard pile composting.

❑ carnivores

Macroorganisms that feed on other animals. Includes ant, beetle, centipede, enchytraed, fly, mite, mole, scorpion, slug, snail, spider, springtail.

❑ castings

Manure, i.e., excretion, of earthworms. Earthworm castings are high in nutrients for plants and microorganisms.

❑ chemical

In science, chemicals are elementary substances such as oxygen, hydrogen, nitrogen, etc. In the context of home composting, however, the word "chemical" is often used to describe a philosophy considered to be in opposition to the organic philosophy. In general, the chemical philosophy encourages people to force nature to do what they want by applying synthetic pesticides and fertilizers which may get the temporary results they want, but may harm or not enhance the general soil condition and environment.

❑ clay soils

Soils with clay particles and small air pores. Water retention is high creating poor drainage conditions.

❑ clippings

Blades of grass cut off during the mowing process.

▢ cold composting, cold pile

When less attention is given to providing and maintaining optimum conditions for compost piles, the resulting environment that will attract psychrophilic bacteria, possibly mesophyllic bacteria, but not thermophilic bacteria. As the psychrophilic bacteria work, the compost pile will reach about 55 degrees F. This is the slow method of creating compost from a backyard pile, and can take as long as 6 months to 2 years to create compost. However, there is little maintenance other than occassionally turning the pile. This type of compost piles requires the least effort.

▢ compaction (of soil)

Compaction of soil is a lack of air or oxygen. Particles of soil are pressed together so tightly that there is insufficient air space. The obvious way this may occur is when a great weight is present, i.e., during construction when large trucks are daily rolled over the land. However, chemical overuse and poor irrigation are more common causes. In healthy soil, natural processes provide aeration, notably the presence of earthworms burrowing their way through the soil.

▢ compost

Completely decayed organic matter. It is dark, odorless, and rich in nutrients.

▢ cytokinins

Organic compounds acting like hormones in plants. Stimulates or alters cellular RNA resulting in a modification to plant development. Generally acts by affecting cellular division and differentiation into roots and shoots. This inhibits aging in the plant.

D

▢ daily food intake, by region and food constituent 1

- Diagrams show the total calorie intake by region, as well as the total weight of protein and fat per capita intake in total and by basic foods.
- They demonstrate the wide divergence of food intake across the developing and industrialized world.

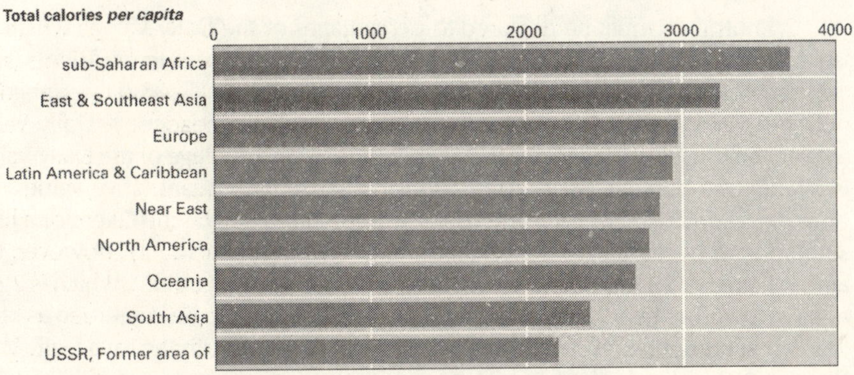

- The average total calorie intake worldwide is around 2,800.
- The World Health Organization (WHO)- recommended minimum daily per capita energy requirement is 2,100 calories, and the average safe protein intake per person per day is 46 grams of mixed-diet protein.
- These are average figures, and adolescents and pregnant and lactating women need considerably more protein.

Total protein *per capita* (grams)

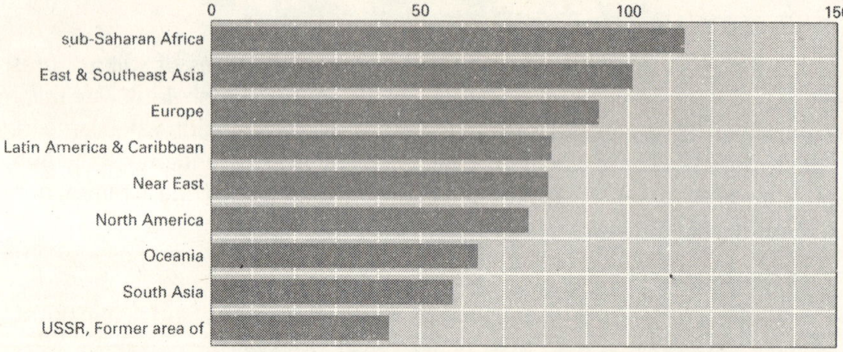

Total fat *per capita* (grams)

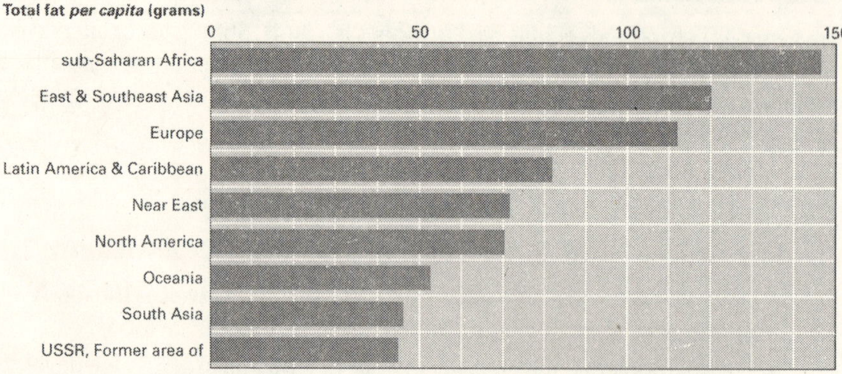

❑ daily food intake, by region and food constituent 2

All nutrients must be declared as percentages of the Daily Values which are label reference values. The amount, in grams or milligrams, of macronutrients (such as fat, cholesterol, sodium, carbohydrates, and protein) are still listed to the immediate right of these nutrients. But, for the first time, a column headed "% Daily Value" appears on the far right side. Declaring nutrients as a percentage of the Daily Values is intended to prevent misinterpretations that arise with quantitative values. For example, a food with 140 milligrams (mg) of sodium could be mistaken for a high-sodium food because 140 is a relatively large number. In actuality, however, that amount represents less than 6 percent of the Daily Value for sodium, which is 2,400 mg. On the other hand, a food with 5 g of saturated fat could be construed as being low in that nutrient. In fact, that food would provide one-fourth the total Daily Value because 20 g is the Daily Value for saturated fat.

A Total calories per capita
B Total protein per capita (grams)
C Total fat per capita (grams)

highest calorie intake
highest protein intake

Eggs and milk

	sub-Saharan Africa	East & Southeast Asia	Europe	Latin America and Caribbean	Near East	North America	Oceania	South Asia	USSR, Former area of
A	49.1	31.2	221.2	177.3	117.7	240.2	154.5	103.0	258.6
B	2.6	2.0	12.8	10.0	6.2	14.1	8.8	5.1	14.5
C	2.9	2.0	12.9	9.7	7.5	13.8	8.7	6.5	14.4

Fish & Meat

	sub-Saharan Africa	East & Southeast Asia	Europe	Latin America and Caribbean	Near East	North America	Oceania	South Asia	USSR, Former area of
A	66.7	169.5	375.1	285.3	117.8	473.8	431.0	33.0	218.5
B	6.4	14.0	29.2	22.4	10.2	45.1	35.7	3.6	17.2
C	4.4	12.0	27.6	20.9	8.2	31.1	30.7	2.0	16.0

Sugars (Calories only)

	sub-Saharan Africa	East & Southeast Asia	Europe	Latin America and Caribbean	Near East	North America	Oceania	South Asia	USSR, Former area of
A	94.2	201.7	370.8	477.8	251.4	662.4	367.4	230.5	324.3
B									
C									

Pulses

	sub-Saharan Africa	East & Southeast Asia	Europe	Latin America and Caribbean	Near East	North America	Oceania	South Asia	USSR, Former area of
A	80.8	35.0	25.5	101.9	61.5	42.1	61.5	106.7	3.6
B	5.3	2.2	1.7	6.5	4.0	2.8	4.1	6.3	0.2
C	0.4	0.2	0.1	0.5	0.4	0.2	0.4	0.9	0.0

Vegetables

	sub-Saharan Africa	East & Southeast Asia	Europe	Latin America and Caribbean	Near East	North America	Oceania	South Asia	USSR, Former area of
A	22.4	45.9	75.6	34.1	83.2	83.9	67.1	33.6	58.3
B	1.2	2.4	3.6	1.4	3.8	3.8	2.9	1.8	2.6
C	0.2	0.4	0.6	0.3	0.7	0.8	0.6	0.3	0.4

Fruits

	sub-Saharan Africa	East & Southeast Asia	Europe	Latin America and Caribbean	Near East	North America	Oceania	South Asia	USSR, Former area of
A	93.3	66.1	90.1	118.6	134.8	122.3	143.6	48.9	35.5
B	1.0	0.8	1.0	1.5	1.6	1.4	1.6	0.6	0.3
C	0.4	0.4	0.6	0.8	0.7	0.7	1.0	0.3	0.2

Roots and tubers

	sub-Saharan Africa	East & Southeast Asia	Europe	Latin America and Caribbean	Near East	North America	Oceania	South Asia	USSR, Former area of
A	444.7	108.6	168.6	120.4	63.7	103.0	195.3	44.6	186.3
B	4.4	0.8	4.0	2.0	1.2	2.7	3.5	0.8	4.4
C	0.7	0.2	0.2	0.3	0.1	0.2	0.4	0.1	0.3

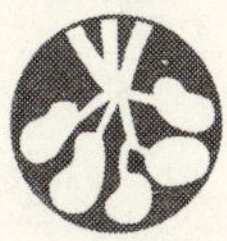

Cereals

	sub-Saharan Africa	East & Southeast Asia	Europe	Latin America and Caribbean	Near East	North America	Oceania	South Asia	USSR, Former area of
A	1,015.8	1,628.9	1,001.6	1,033.6	1,662.7	844.0	738.9	1,462.4	1,252.9
B	26.4	33.1	30	25.5	47.1	24.8	22.6	34.0	36.4
C	7.5	5.8	3.9	5.4	9.4	3.5	2.8	6.1	4.5

❐ daily per capita calorie intake, by country

- There is currently an average of 2,790 calories of food available each day for every human on the planet–23 percent more than in 1961 and enough to feed everyone.

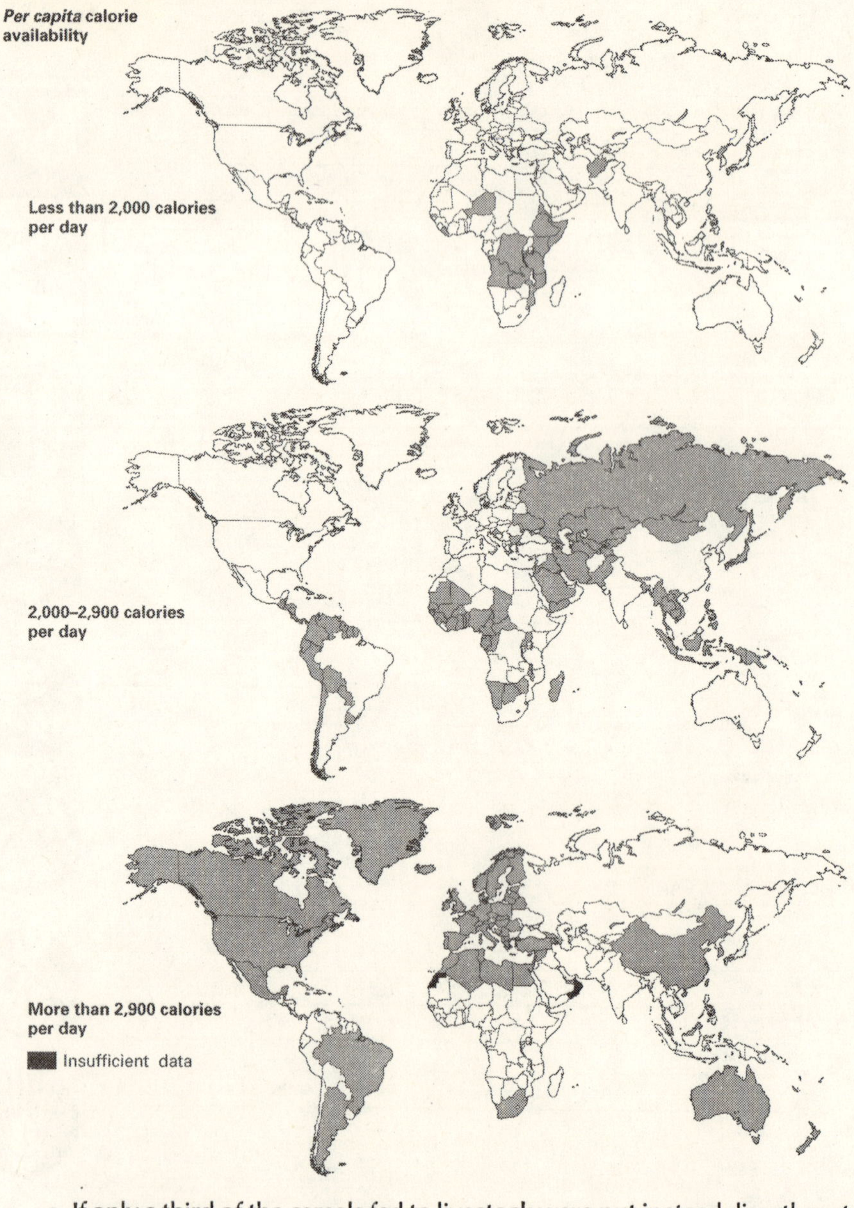

- If only a third of the cereals fed to livestock were put instead directly onto human plates, the per-capita calories available daily would rise to 3,000.

- But the average sub-Saharan African consumes only 2,200 calories a day, significantly less than the global average.
- Recent Food and Agriculture Organization (FAO) figures estimate that 316 million Africans, or approximately 35 percent of the continent's total population, are chronically undernourished.
- In 1998, there were 78 low-income countries that neither grew enough food to feed their populations, nor had the resources to make up the deficit with imports. Of these, more than half were in Africa.

❐ decompose, decomposition

Decay. Rot. The breaking down of organic materials into smaller particles until the original material is no longer recognizable.

❐ detritivores

Macroorganisms that eat decaying matter. Includes ant, beetle, centipede, cricket, earthworm, earwig, enchytraeid worm, millipede, mite, scorpion, slug, snail, spider, springtail, termite, woodlice.

❐ dripline

The ground that lies directly below the outermost reaches of a trees branches.

E

❐ earthworm castings

Manure, i.e., excretion, of earthworms. Earthworm castings are high in nutrients for plants and microorganisms.

F

❐ fertilizer use worldwide

Increasing Use

- Major elements required by most plants are: nitrogen, potassium, phosphorous, calcium, magnesium and sulfur.
- Fertilizer use was scant until the 1950s but since then has increased exponentially.
- In 1950 the global fertilizer use was 20 kg per hectare. In 1990 this had increased globally to 91 kg per hectare.
- One half of all the commercial fertilizer ever produced has been applied since 1984.
- There is considerable potential for increasing world food supply by increasing fertilizer use in low-production countries.
- It has been estimated that the developing world could at least triple its crop production by raising fertilizer use to the world average.

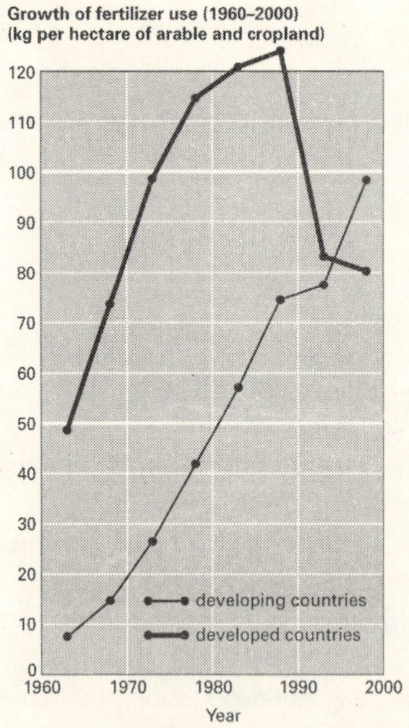

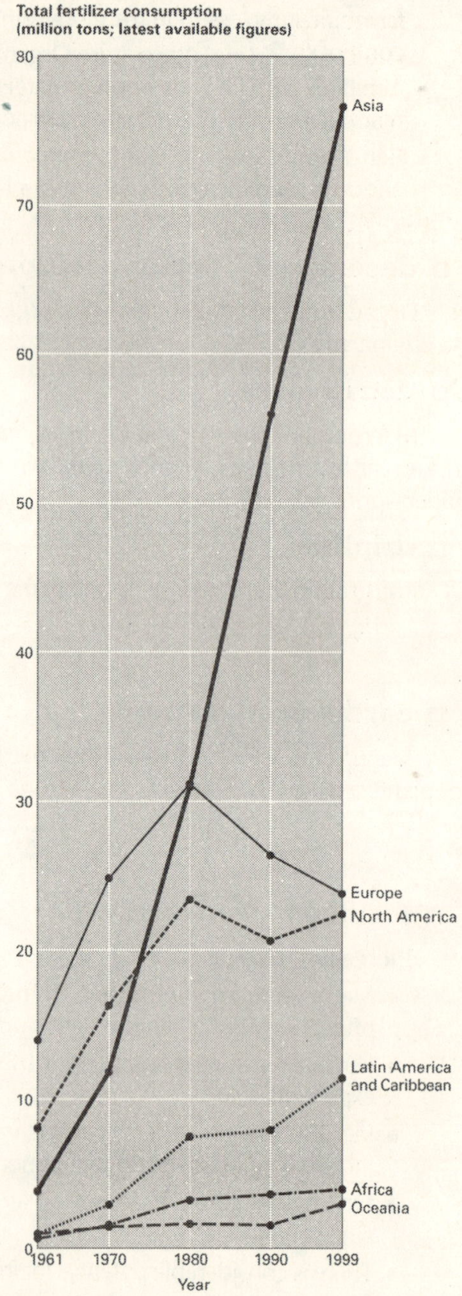

Problems

- Agriculture accounts for by far the largest amount of human-generated nitrogen–some 86 percent.

- Overfertilization is a problem in the developed world where affluence means farmers tend to overfertilize because they are unaware of the specific nutrient content of soils or the needs of the crops.
- About one half of every ton of fertilizer applied to fields does not reach plant tissue but evaporates or is washed into local watercourses.
- Better timing of fertilizer applications, plus more exact calculation of doses, and more accurate delivery could cut this waste substantially.

❐ friable

Easily crumbled. Healthy soil is friable, so if you hold up a handful of soil and wiggle your fingers the particles of soil should fall out of your hand.

G

❐ genetic homeland agriculture livestock

- Of the 4,000 species of wild mammals in the world, only 14 were domesticated to become the valuable livestock of the modern agrarian world.
- Of these, 13 are species found in Europe and Asia.
- Only two species were domesticated outside of Eurasia: the llama and the turkey.

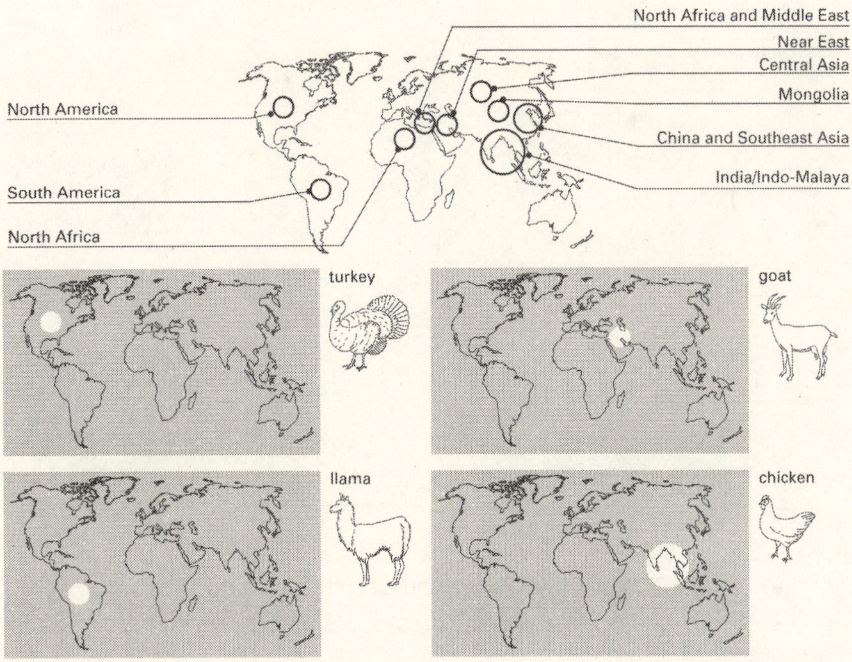

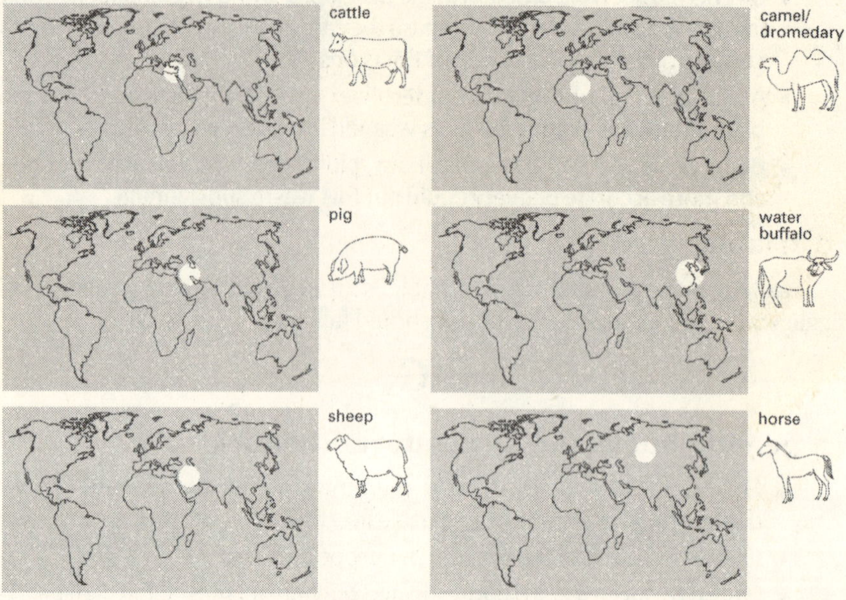

❑ genetic homeland of major food crops: 1

- Of 270,000 plants known to exist, only 7,000 have ever been cultivated for food.

Areas of agricultural origins and regions of cultivated plant diversity

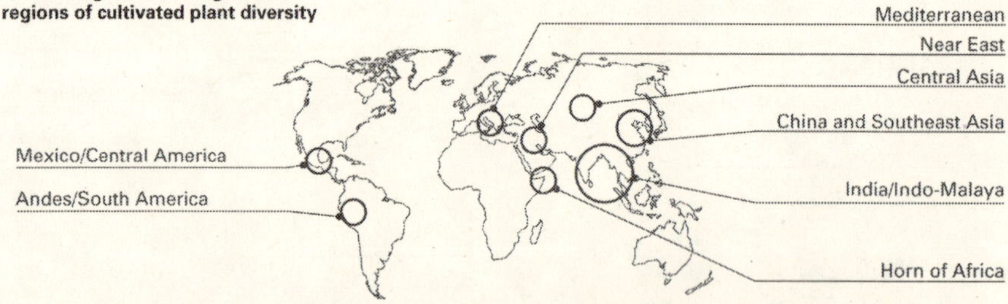

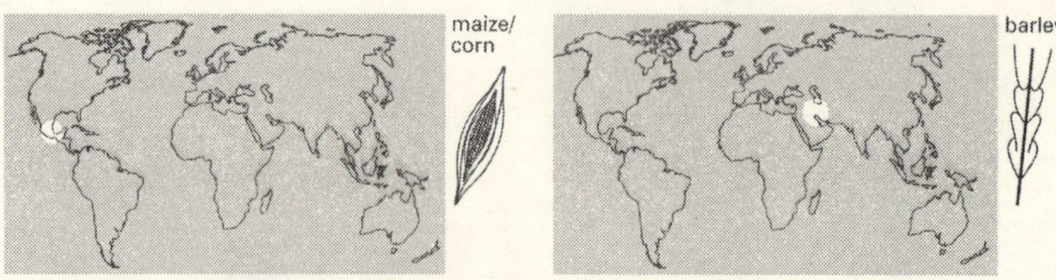

- Of those 7,000, only 120 are widely cultivated today.
- Of those 120, nine species provide 75 percent of human food and 21 species provide 20 percent.
- Crop genetic resources are disappearing from fields at an estimated 1 or 2 percent a year.

- Many traditional plant varieties can now be found only in gene banks.
- The abilities of certain plant varieties to withstand drought, grow on poor soil, resist insects and disease, prosper in local environments or under global warming, or simply taste better, are encoded in their genes.

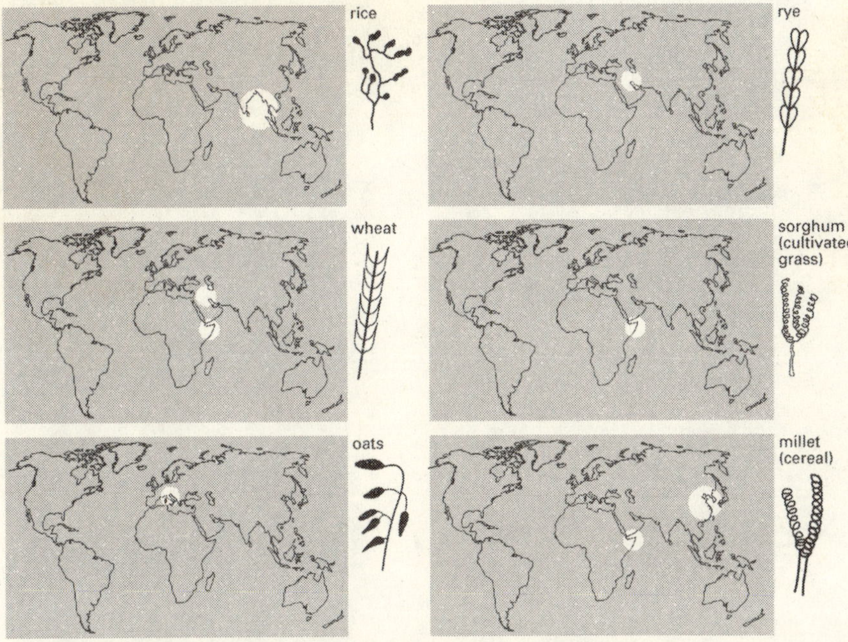

❑ genetic homeland of major food crops: 2

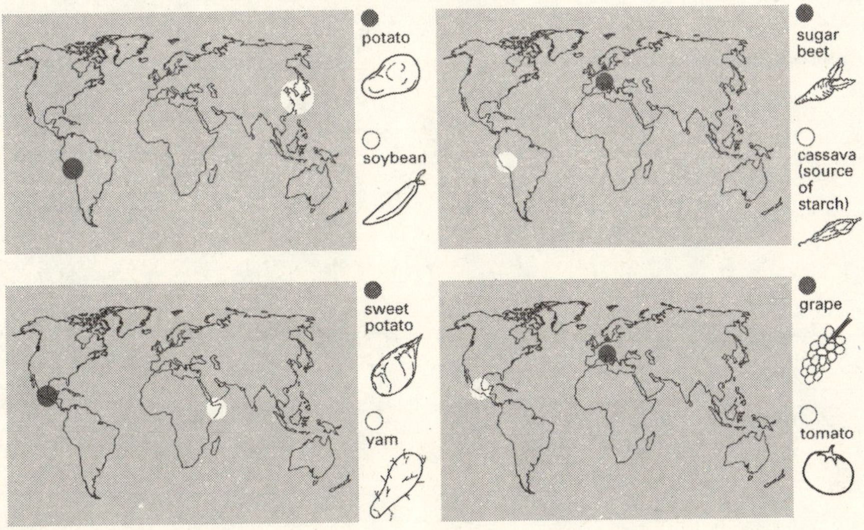

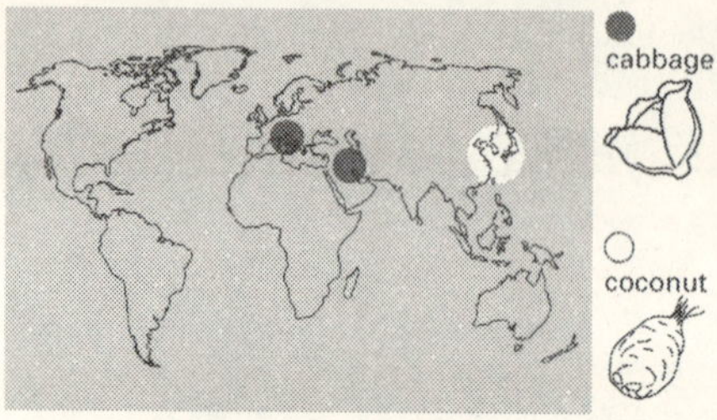

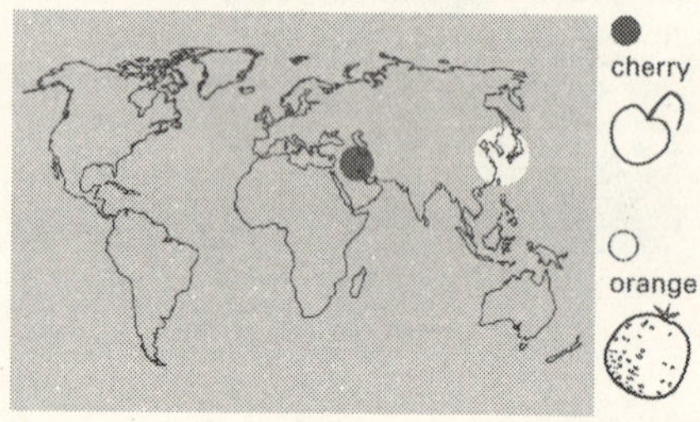

❐ genetic homeland of major non-food crops

Genetic homelands of major non-food crops

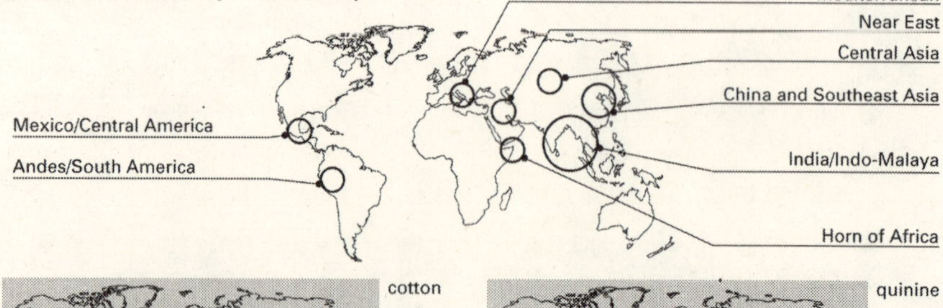

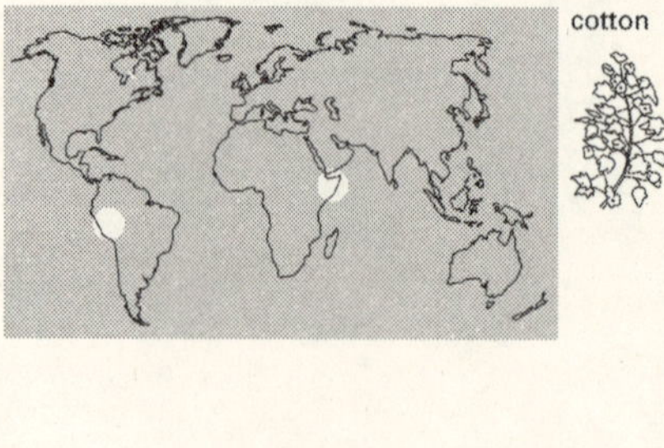

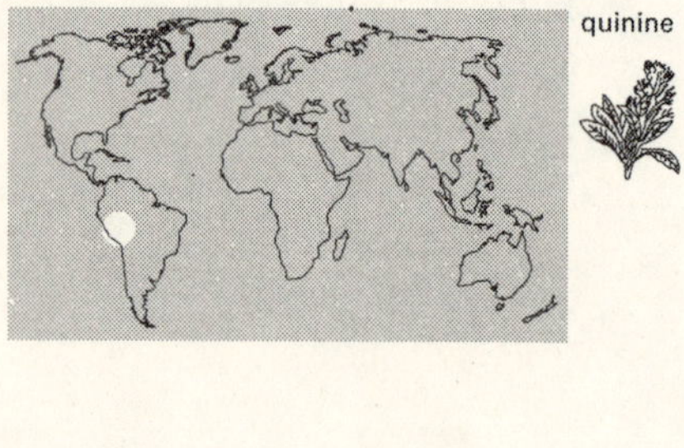

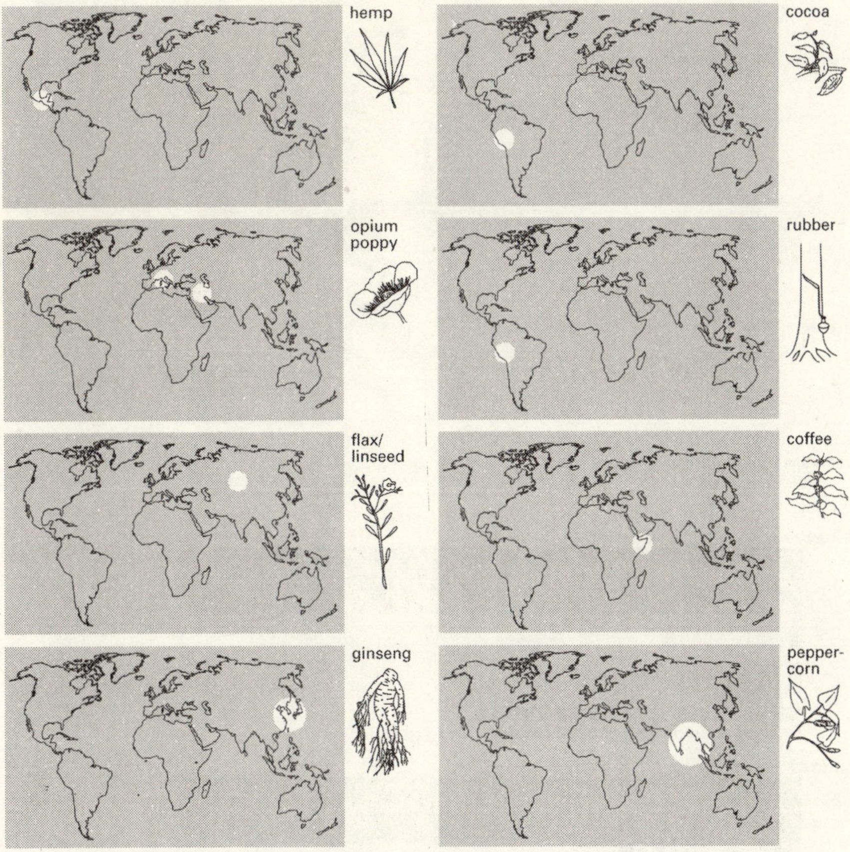

▫ global water use and per capita availability

- Freshwater sources are dwindling or becoming contaminated throughout the world.
- Chronic or acute water shortage is increasingly common in many countries with fast-growing populations, becoming a potential source of conflict.
- The distribution of water resources around the globe is highly unequal, even at the continental level.
- Asia has more than 60 percent of the world population but only 36 percent of river runoff.
- South America has just 6 percent of the global population but 26 percent of runoff.
- Canada has more than 30 times as much water available to each of its citizens as China.
- More than 60 percent of the water used in the world each year is diverted for irrigating crops.

- In Asia, which has two thirds of the world's irrigated land, 85 percent of water goes for irrigation.

Countries which are water-scarce

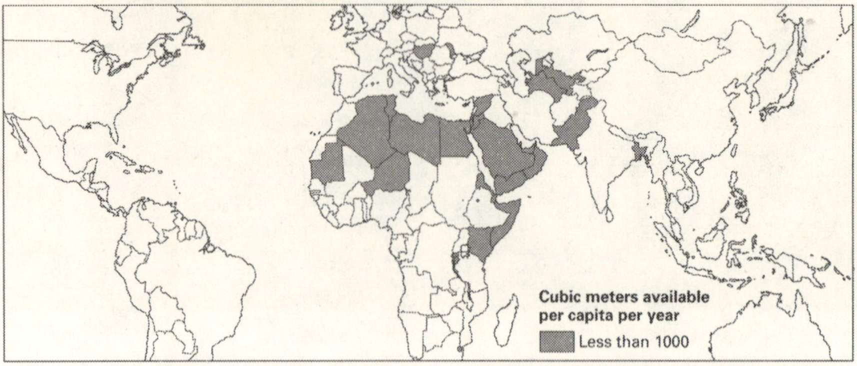

Water use by sector in selected countries

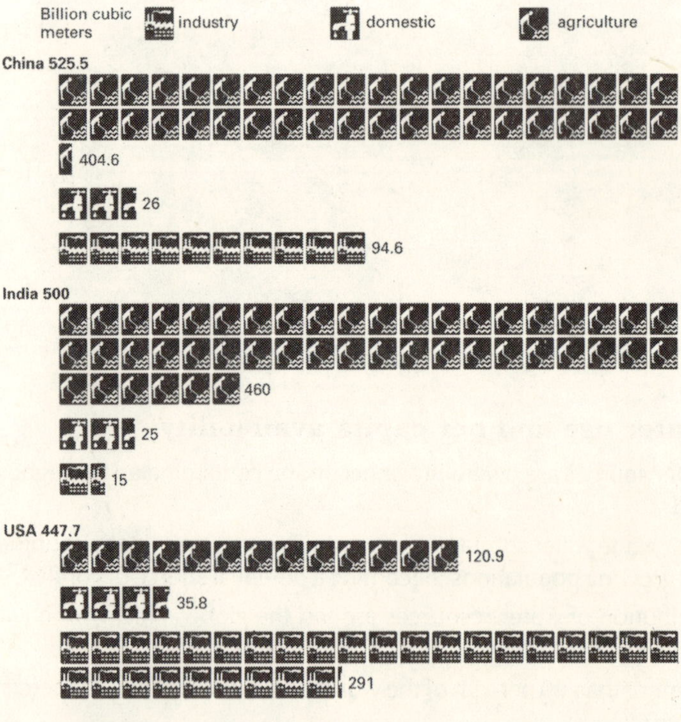

❐ grasscycling

Leaving grassclippings on your lawn to break down and add nutrients. You must mow a little more often so only short clippings are left on the lawn.

❐ greens

The term "greens" is used to denote organic materials high in nitrogen, more specifically, materials whose carbon to nitrogen ratio is lower than 30:1. (Materials high in carbon are referred to as "browns"). Achieving a carbon-to-nitrogen ratio of about 30:1 is one factor in creating favorable conditions for backyard pile composting.

❐ growing number of resistant pests

- The use of pesticides increased more than 30 times between 1950 and the end of the 1980s, but pests still cost the world billions of dollars annually in lost agricultural production.
- More species of weeds, diseases and insects are becoming resistant, up from under 100 in the 1950s to around 700 today.
- At least 520 species of insects and mites, 50 plant diseases and 113 weeds have become resistant to pesticides meant to control them.
- Resistance or tolerance arises as a product of natural selection within a population which multiplies rapidly.
- Individuals with natural immunity are produced by chance.
- They quickly spread their immune genes within the depleted population, and before long a new race appears with a built-in resistance or tolerance to the chemical.

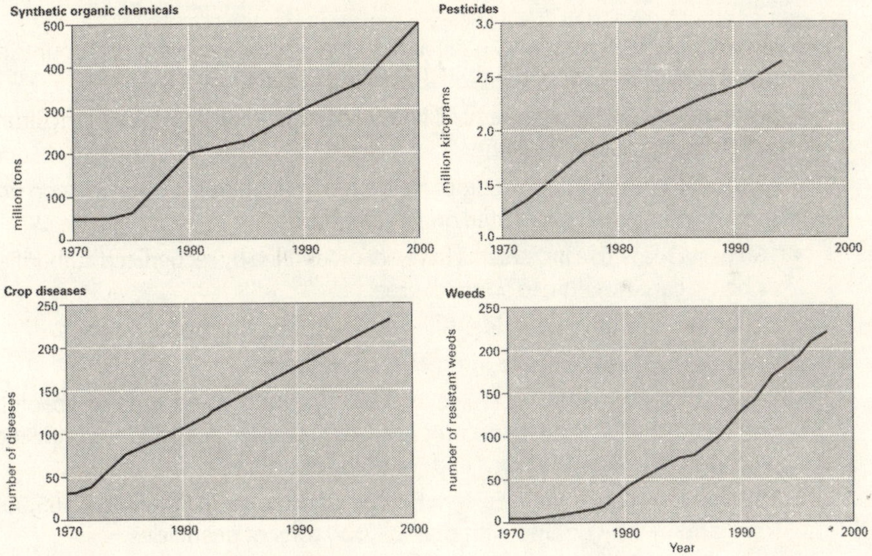

❐ guano

Manure, i.e., excretions, of bats and birds. Can be purchased after being dried and composted.

H

❐ harvesting

Removing compost from the composting environment for use in the landscape, lawn or garden.

❑ herbivores

Macroorganisms that eat plants. Includes insect and beetle larva, grub, mouse, slug, snail, termite, woodchuck.

❑ hot composting, hot pile

Optimum conditions for compost piles, including 30:1 Carbon-to-Nitrogen ratio, 1" or smaller particles of various sizes and textures, moisture, air, volume of 3 cubit feet, produce an environment that will attract psychrophilic, mesophyllic, and then thermophilic bacteria. As the thermophilic bacteria work, the compost pile will reach as high as 170 degrees F. This is the fastest method of creating compost from a backyard pile, and can take as little as 3 weeks if the pile is monitored and turned each time the temperature starts to fall.

❑ human casualties of pesticide use

- The World Health Organization (WHO) estimates that every year 3 million people suffer acute, severe pesticide poisoning. Over 20,000 may die.
- The 1984 gas leak in Union Carbide's pesticide plant in Bhopal killed over 3,500 people at the time: registered deaths had reached 22,149 by December 1999.
- Women and men working in the agricultural sector in developing countries make up 59 percent of the global working population.
- Pesticides cause 14 percent of all known occupational injuries in agriculture and 10 percent of all fatalities.
- Around 30 percent of pesticides marketed in developing countries do not meet internationally accepted quality standards.
- 160 pesticide active ingredients have been classified by recognized authorities as being carcinogenic to some degree.
- 118 pesticides have been identified as disrupting hormonal balance.
- There are 39 pesticides of international concern.
- It is estimated that there are several hundred thousand tons of obsolete pesticides–that is, banned, decomposed, or unwanted pesticides–worldwide, with well over 100,000 tons in developing countries.
- The problem also exists in Eastern Europe, particularly in Poland with 65,000 tons, and in the Ukraine with over 23,000 tons of pesticides.
- The obsolete pesticides in developing countries include Aldrin, DDT, Dieldrin, Endrin, Fenitrothion, HCH, Lindane, Malathion, and Parathion.
- Dieldrin is said to be five times as toxic as DDT when swallowed, and 40 times as toxic when absorbed through the skin.
- Endrin, no longer sold, is 15 times as toxic as DDT to mammals, 30 times as toxic to fish, and up to 300 times as toxic to some birds.
- Some 20,000 tons of obsolete pesticides are to be found in Africa alone.
- The Food & Agricultural Organization (FAO) has estimated the total cost for cleaning up Africa at $80 to $100 million.

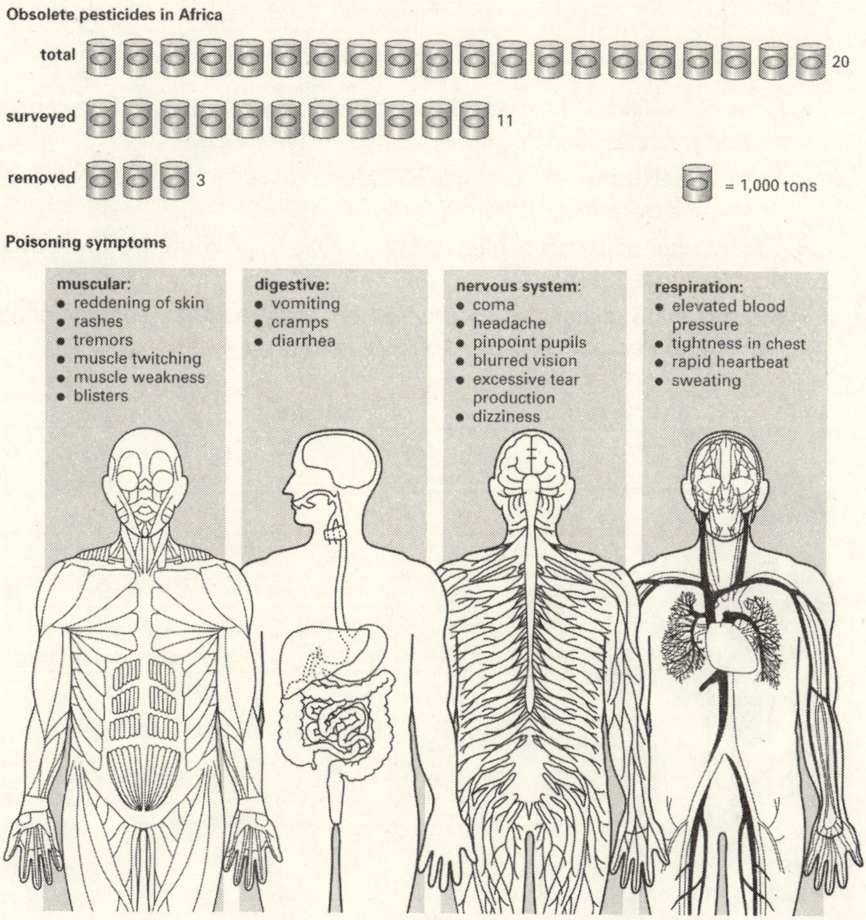

❐ humus

Decomposed organic matter. Healthy soil will consist of about 3.5 to 5% of this organic matter. Humus is soft, sweet-smelling, shapeless dark and crumbly and smells like the forest floor (more correctly, the forest floor smells like humus because that is what is naturally there).

It is this stage of the decomposition process which provides nutrients for plant life. It contains about 30% each of lignin, protein, and complex sufars. It contains 3 - 5% nitrogen and 55 - 60% carbon. Humus is the slow-release food source for microorganism development. It is constantly being transformed into acids, enzymes and minerals and, therefore, must be constantly replenished for proper vegetative nutrition.

I

❑ industrial and artisanal fisheries

- Of the 3.5 million fishing vessels worldwide, about 38,000, or 1 percent by number, are classified as large-scale, industrial vessels.
- These vessels, however, constitute 50–60 percent of the total vessel 'capacity' (measured by tonnage) of the world's fishing fleet.
- The number of people fishing and practicing aquaculture worldwide has doubled since 1970.
- More than 25 million are full-time fishers and around 200 million of the world's poorest people depend on fishing for their livelihood.

		industrial	artisanal
	total fishing vessels worldwide	38,000	3.4 million
	percentage of all ships	1 percent	99 percent
	percentage of fish catch capacity	50–60 percent	40–50 percent
	total individuals depending on fishing	200,000	200 million

Distribution of fish catchers worldwide

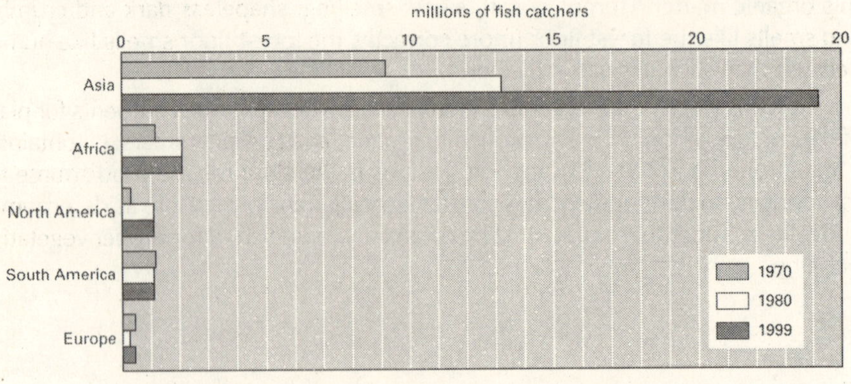

- In contrast large-scale fishing employs only about 200,000 people.
- 24,000 people engaged in fishing, fish farming and processing are killed at work every year.
- In the US the fatality rate for the fishing industry is 16 times higher than for fire-fighting or police work and 40 times the national average.
- In developing countries, the small-scale sector produces more economic and social benefits than the industrial one.
- It uses less capital and fuel, generally more selective gear, is based in rural areas and usually produces more food for the domestic market than for export.
- Fish for export are caught with less waste than in the over-subsidised industrial sector.
- There is potential conflict between the artisanal fishery which is coastal and inshore, and the commercial mechanized fleets exploiting the offshore resources.

❑ inoculants

Dominant microorganisms which may be added to a compost pile. Generally, these are not necessary as there are microorganisms living on all organic matter, so your pile already has these in it.

L

❑ land productivity: developing regions

- Of the world's agricultural land, only around a third is used to grow crops, with the remaining two-thirds dedicated to livestock pasture.
- The world has doubled food production in the past 35 years, more than keeping pace with population growth.
- This has largely been made possible by new crop varieties, increased chemical inputs to fields and the extension of land under irrigation.
- But the rate of increase has slowed in the past 10 years, and per capita productivity has declined in some regions.

In economics, Productivity is the amount of output created (in terms of goods produced or services rendered) produced per unit input of used. For instance, labor productivity is typically measured as output per worker or output per labor-hour.

With respect to land, the "yield" is equivalent to "land productivity". Some economists write of "capital productivity" (output per unit of capital goods employed), the inverse of the capital/output ratio. "Total factor productivity," sometimes called multifactor productivity, also includes both labor and capital goods in the denominator (weighted by their incomes). Unlike labor productivity, the calculation of both capital productivity and total factor productivity are dependent on a number of doubtful assumptions and is subject to the Cambridge critique.

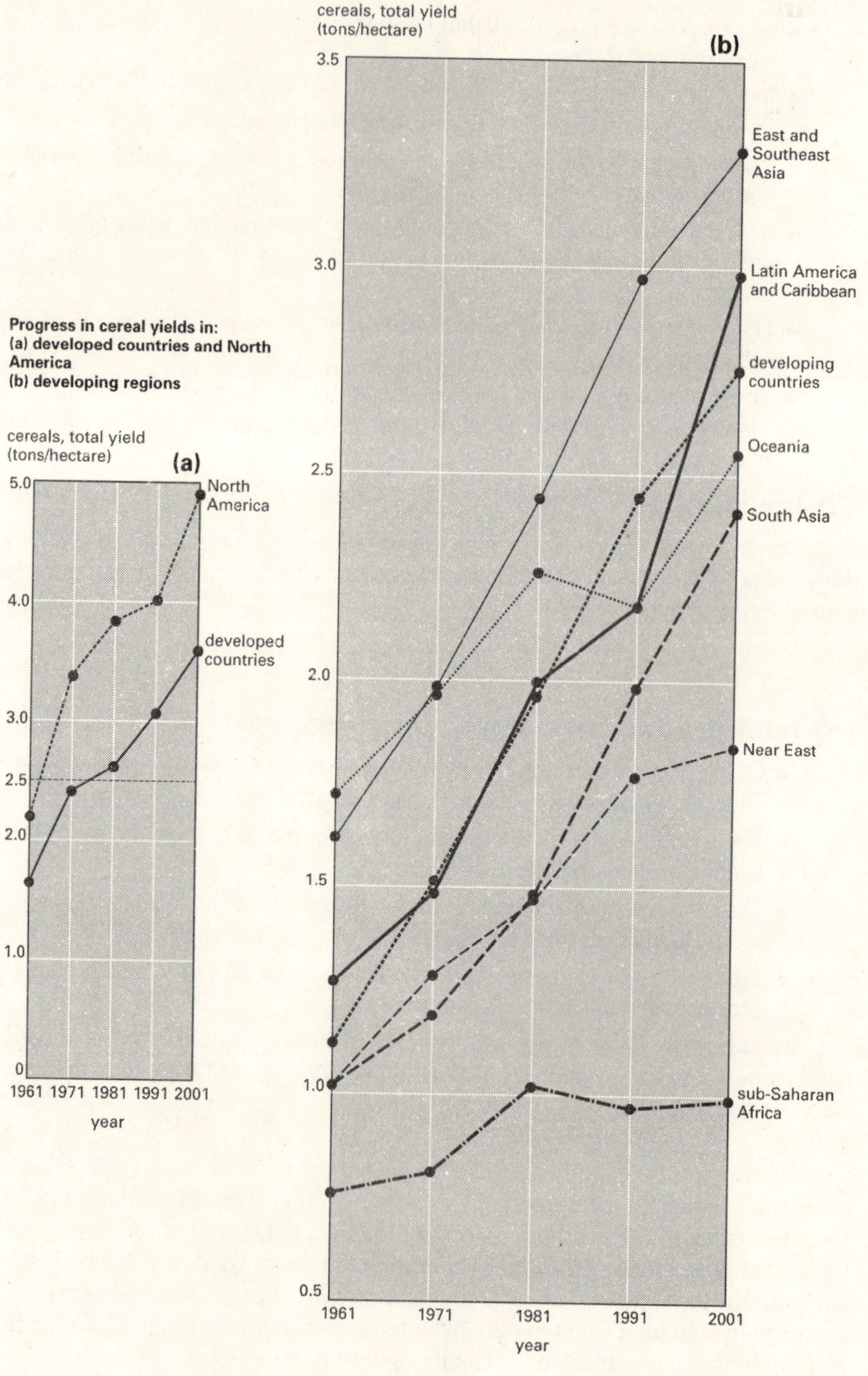

Progress in cereal yields in:
(a) developed countries and North America
(b) developing regions

❑ land use and population

- As world population and the global economy have grown, ever more land has been cleared, drained, or irrigated to plant cash crops for export, such as sugar and palm oil, coffee and rubber, or to grow food crops for livestock.

Road network density, around 2000, measured in km of route per 100 sq km of land area

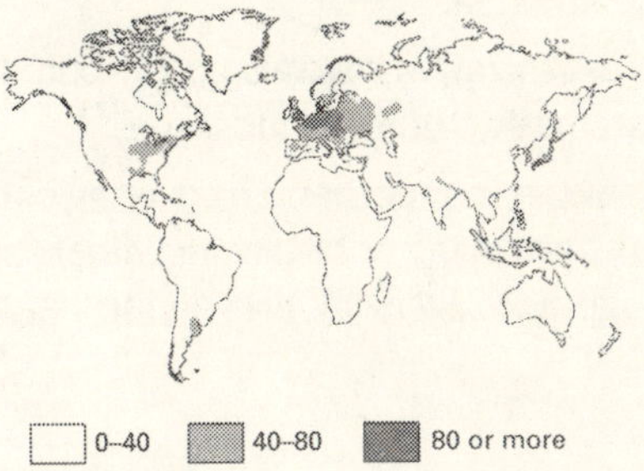

Level of human disturbance by region 2000

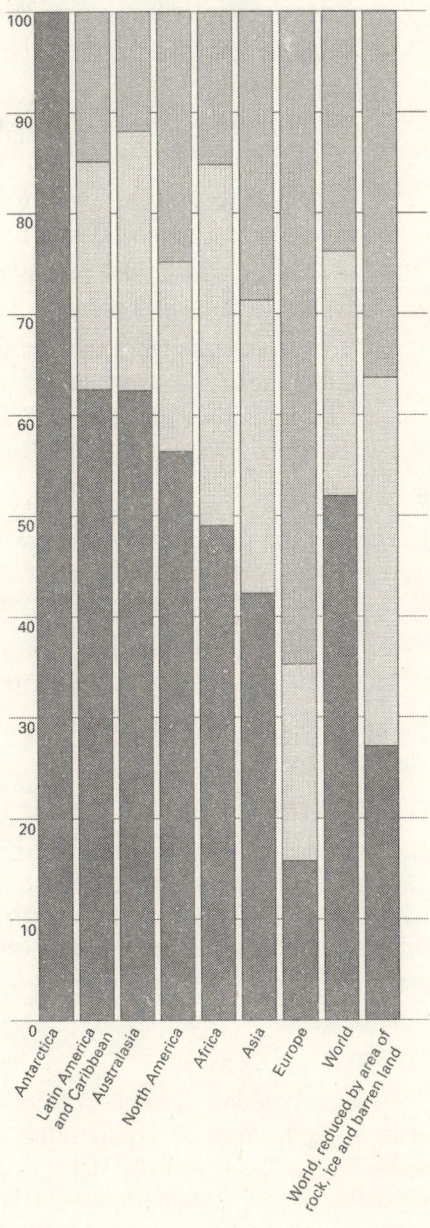

- Extensification – the expansion of arable land – has overwhelmingly been a response to fastgrowing populations, with subsistence in food the main driving force.
- The 60 years from 1860 to 1920 saw 440 million hectares (over 1 billion acres) of land brought under cultivation (an area larger than India).
- More than half of this took place in the temperate lands of North America and in the region that became the Soviet Union.
- A similar scale of transformation took place in the subsequent 60 years, from 1920 to 1980.
- By then, most of the potentially productive temperate lands of the northern hemisphere were occupied and the rate of population growth was slackening.
- The new 'frontier lands', where population growth rates remained high, were in Africa, South Asia and South America.
- Mining, industrial development and urbanization have also contributed to the transformation of natural ecosystems into human landscapes.
- The second half of the 20th century saw an unprecedented covering of the landscape with concrete and tarmac, destroying or displacingwildlife and causing major disruption to drainage and rivers by preventing natural seepage.

❑ landfill

Pleasant term for a garbage dump which is located in a cavity in the ground so that, when full, it may be covered up and look like part of the land. Today's landfills are sanitary and require special technology to eliminate methane gas and toxic leachate produced by the garbage.

❑ landfill mining and reclamation

An operation to excavate and process solid wastes which were previously land filled.

❑ leachate

Liquid "run-off". Leachate from the compost pile contains nutrients generated in the composting process. In contrast, as groundwater and rain flow through a landfill, they pick up weak acids created by decaying organic matter. As these acids react with other garbage, the leachate can become toxic which may contaminate streams and groundwater unless the landfill is properly constructed to contain the run-off.

❑ lime

Lime is sometimes added to compost piles to increase pH. However, unless you are seeking a high-pH compost it is unlikely you need to add lime. Compost piles become acidic in the initial stages as organic acids are formed. However, as the composting process continues, the pH returns to a balanced state. If you add lime, an odor may occur because of the formation of ammonia gas.

M

❑ major fishing grounds and fishing countries

Major fishing grounds and estimates of the year in which they became fully fished

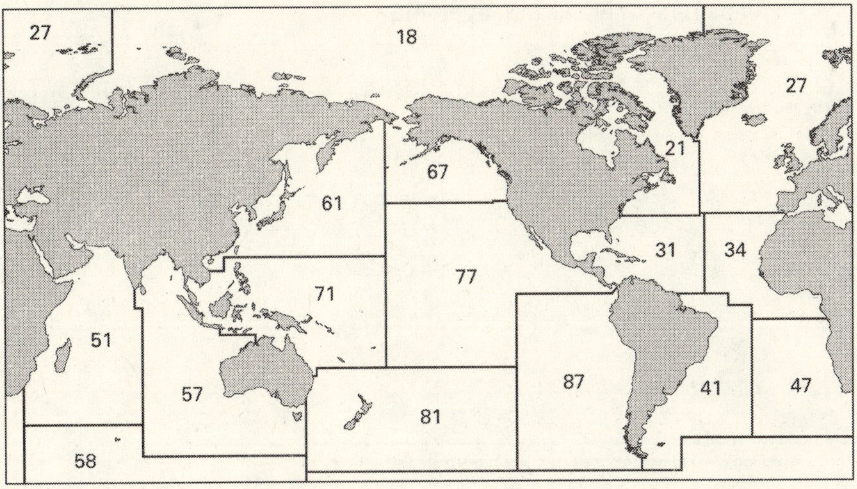

Comparison of catches by region (latest available figures)

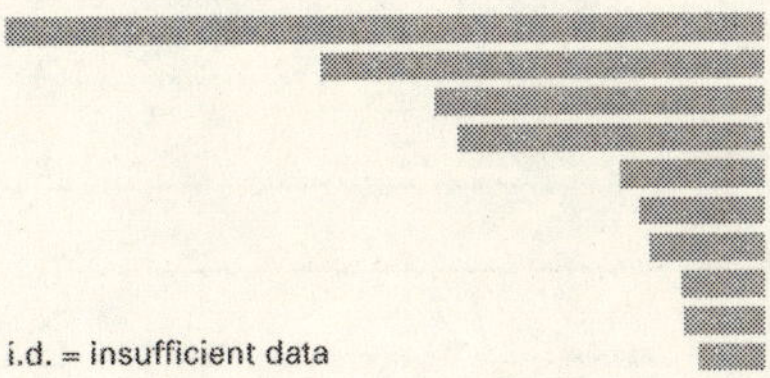

code*	fishing ground	1999 catch (millions of tons)	year fully fished
[61]	Northwest Pacific	24,121,100	1998
[87]	Southeast Pacific	14,171,000	2001
[27]	Northeast Atlantic	10,489,900	1983
[71]	West Central Pacific	9,754,300	2003
[57]	Eastern Indian Ocean	4,553,800	i.d.
[51]	Western Indian Ocean	3,894,600	i.d.
[34]	Eastern Central Atlantic	3,591,000	1984
[67]	Northeast Pacific	2,590,400	1990
[41]	Southwest Atlantic	2,497,100	i.d.
[21]	Northwest Atlantic	2,031,000	1971

i.d. = insufficient data

[*fishing ground codes allocated by FAO]

The top ten fishing countries (latest available figures)

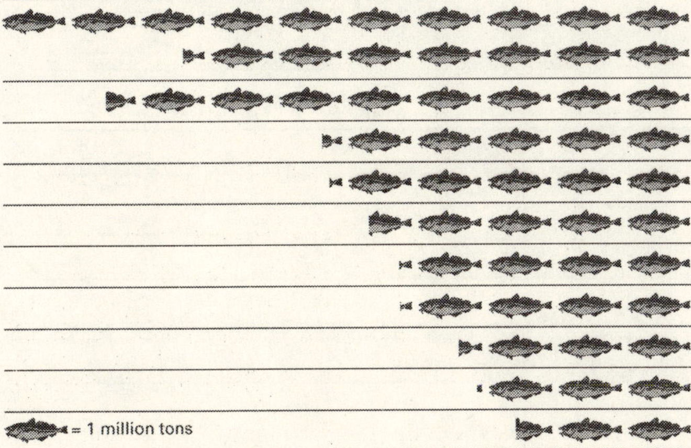

Country	Catch
China	17,240,032
Peru	8,429,290
Japan	5,176,460
Chile	5,050,528
USA	4,749,645
Indonesia	4,149,420
Russian Fed.	4,141,157
India	3,316,815
Thailand	3,004,900
Norway	2,620,073

❐ major land use categories, by continent

- New land for arable farming has generally been obtained through the annexation of grazing pastures, deforestation and the drainage of wetlands.
- The largest areas of pasture 'lost' to cropland were in the United States Great Plains, the South African veldt, the Russian steppe and the campos and pampas of Brazil and Argentina.

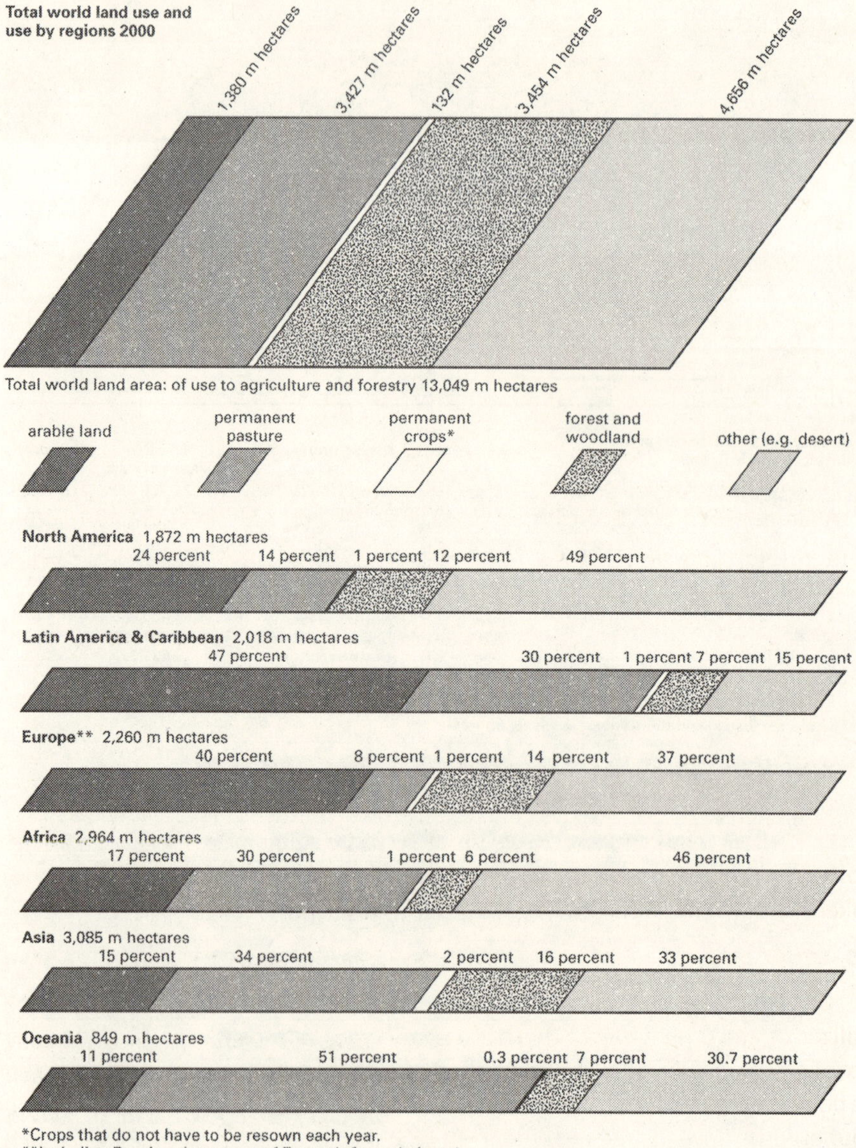

*Crops that do not have to be resown each year.
**Including Russia, where most of Europe's forest is found.

- With the potential for new colonization reduced in much of the world, farming has increasingly invested in intensification by purchases of fertilizers, high-yield seeds and machinery.

❒ mesophyllic bacteria

This group of bacteria species work to break down organic matter under "warm" conditions of 40 degrees up to 110 degrees. The ideal temperature of their environment is 70 to 90 degrees.

❒ methane gas

Explosive (when highly concentrated) gas that is formed when organic materials decompose in anaerobic conditions which exist in landfills. Landfill operators must have a method of venting methane gas before it becomes volatile.

❒ microbe

Used interchangeably with "microorganism".

❒ micronutrients

Nutrients that plants require in small doses, primarily to enhance the ability to absorb macronutrients. They include iron, manganese, copper, boron, molybdenum,chlorine, cobalt and zinc.

❒ microorganism

These are microscopic plants and animals. They exist in soil for the purpose of breaking down organic matter into basic mineral elements. They include bacteria, fungi, actinomycetes, algae, protozoa, yeast, germs, ground pearls, and nematodes.

❒ mineralization

Organic matter is broken down into humus, then acids, enzymes and compounds, and finally into basic mineral elements.

❒ minerals

Supply food and nutrients for plants and microorganisms. Webster defines minerals as "any naturally occurring substance that is neither vegetable nor animal". In other words, these are the most basic form into which organic matter can be broken. At the mineral stage, the particles are inorganic.

❒ mulch

Covering for soil. Mulch should not generally be mixed into the soil, it is not a fertilizer or soil amendment. There are many types of mulch, including partially decomposed compost, bark, wood chips, hay, nut shells, pine needles, and others. The point is to cover bare ground so that top soil is not washed away, soil temperature is buffered, and weeds are reduced from lack of light. A good organic mulch will also supply nutrients to the earth as it decomposes.

N

❑ nitrogen draft

Incorporating high-carbon matter like wood chips into the soil (i.e., mixed into soil, not placed on top) can cause deficiencies in the nitrogen available to plant roots. Organic matter composts and, in order to compost, the high-carbon material requires the nitrogen from the soil to create the desired diet for microbial action.

❑ n-p-k

N-P-K is an abbreviation for nitrogen (N), phosphorus (P), and potassium (K). In the chemical philosophy, these three elements are considered important to force crop production (as opposed to the organic philosophy goal of improving the biodiversity of the soil). U.S. law requires that the ratio of these three elements be specified on every bag of commercially-available fertilizer. A ratio of 3-1-2 or 4-1-2 is considered good.

O

❑ organic

Literally refers to something derived from plant or animal matter. Includes anything that is or was living, made from something living, excreted from something living. The term "organic" is used to describe a philosophy of working within the laws and systems existing in nature to achieve a healthy environment that is bountiful long-term. Healthy soil is the foundation of this philosophy, therefore, the term comes up frequently in discussions of home composting.

P

❑ per capita food production indices by country and region 1

Comparison of countries where food production fell and rose 1979–2000

Countries where food production fell

Country	1979–81	1998–2000
Bulgaria	105.5	72.2
Cuba	90.1	58.2
Finland	93.5	89.7
Haiti	101.3	95.4
Iraq	78.0	77.7
Japan	94.0	92.4
Mongolia	88.1	87.8
Namibia	107.2	97.3
Puerto Rico	99.7	81.7
Romania	113.0	93.4
Singapore	154.3	41.5

Increases in international dollar-earning value of food production 1979–2000 by countries' average income levels

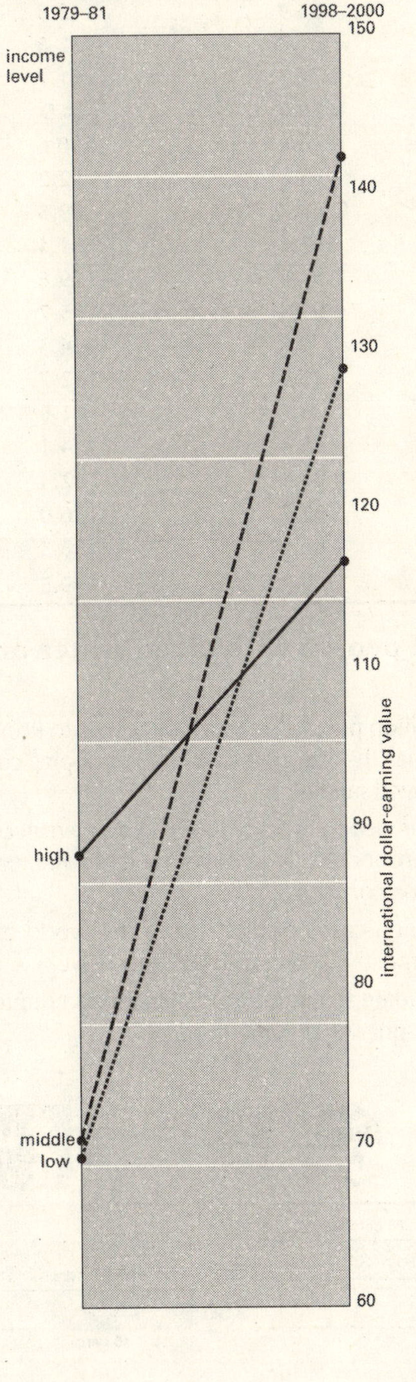

Countries that more than doubled food production

Country	1979–81	1998–2000
Nigeria	57.2	152.2
Pakistan	66.4	143.3
Paraguay	61.0	132.8
Peru	77.3	161.7
Saudi Arabia	26.7	88.5
United Arab Emirates	48.8	250.6
Vietnam	63.8	152.2
Ghana	68.7	163.5
Iran	61.1	151.4
Jordan	57.5	139.6
Kuwait	91.4	185.7
Laos	70.3	144.3
Lebanon	59.2	142.7
Libya	78.7	161.6
Malaysia	55.4	134.1
Benin	63.1	152.4
Burkina Faso	62.7	136.0
China	60.9	168.5
Egypt	68.2	149.5

❑ percentage of people with clean water and sanitation, by region

- Globally, 1.1 billion people are without access to improved water supply and 2.4 billion (half the total population of developing countries) are without access to improved sanitation.
- Of the 4.9 billion people worldwide who do have access to water supply services, only around 3 billion have the convenience of access through house connections or yard taps.
- There are 4 billion cases of diarrhea in the world every year, with 2.2 million deaths, mostly among children under five.
- Safe water, adequate sanitation, and hygiene can reduce diarrheal disease by between one-quarter and one-third.

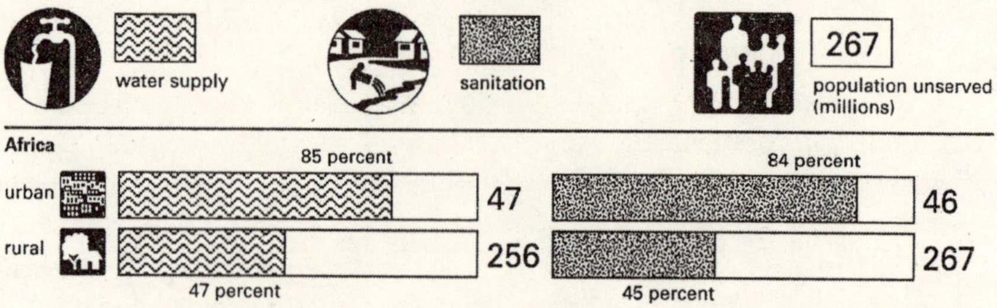

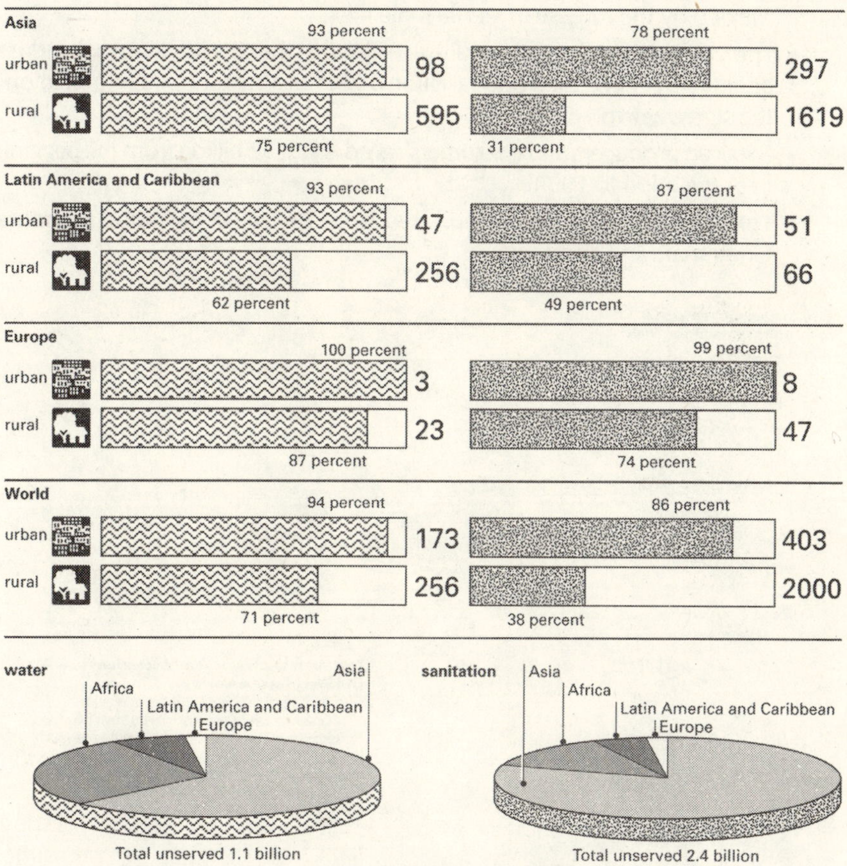

❐ persistence (i.e., chemical persistence)

The length of time a chemical will remain essentially unchanged in the environment.

❐ pest hazard study 1: the new world screwworm

- The New World screwworm (Cochliomyia hominivorax) is a serious pest of domestic animals and man.
- Female screwworms lay their eggs at the edges of wounds on living mammals or on natural body openings such as the nostrils, eyes, mouth, ears and vagina. Virtually any wound is attractive.
- Within 24 hours, larvae emerge and immediately begin to feed on the underlying tissues, burrowing head-downward into the wound.
- As they feed, the wound is enlarged and deepened, resulting in extensive tissue destruction.
- The larvae reach maturity about five to seven daysafter hatching and leave the wound, falling to the ground, into which they burrow to pupate.

- The screwworm was successfully eradicated from the United States and Mexico by the release of sterile male flies.
- The US Department of Agriculture estimates that the US livestock industry benefits by more than $900 million a year as a result of the eradication of the screwworm.
- Mexican producers and consumers saved about $2 billion from the beginning of eradication to 1991.
- The pest has now been eradicated from Belize, El Salvador, Honduras, Guatemala and Nicaragua.

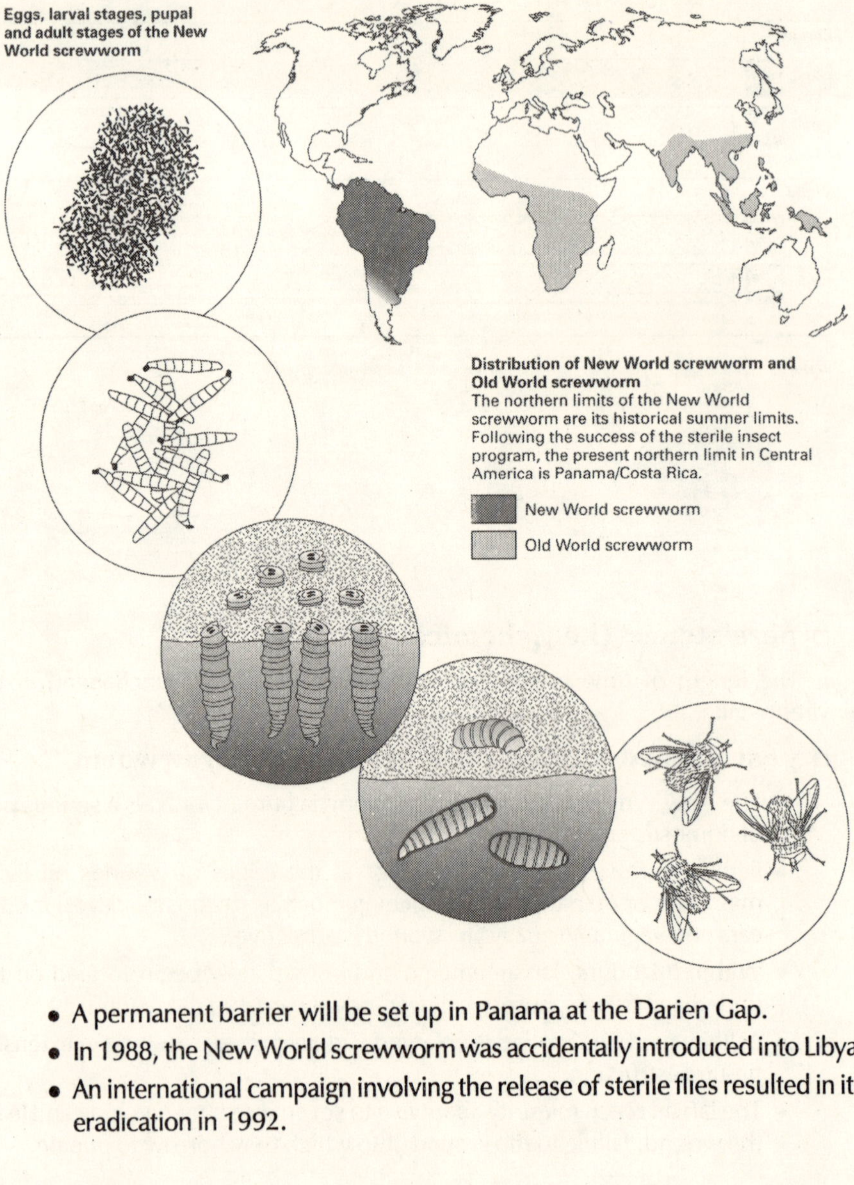

Eggs, larval stages, pupal and adult stages of the New World screwworm

Distribution of New World screwworm and Old World screwworm
The northern limits of the New World screwworm are its historical summer limits. Following the success of the sterile insect program, the present northern limit in Central America is Panama/Costa Rica.

- A permanent barrier will be set up in Panama at the Darien Gap.
- In 1988, the New World screwworm was accidentally introduced into Libya.
- An international campaign involving the release of sterile flies resulted in its eradication in 1992.

- The Old World screwworm (Chrysomya bezziana) is a similar plague of Africa and Asia.

❐ pest hazard study 2: spread of the cassava mealy-bug

Cassava

- Cassava (also called manioc or tapioca) was first introduced to Africa from South America in the 16th century by Portuguese travelers, and has become an important staple food crop.
- Its starchy, thickened roots are a source of cheap calories for 200 million Africans, providing up to 70 percent of their daily energy intake.
- The leaves and tender shoots are also eaten in many parts of Africa.

cassava

cassava root

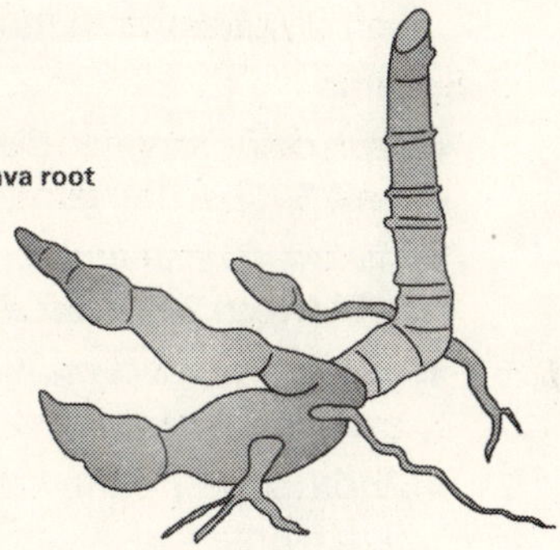

mealybug

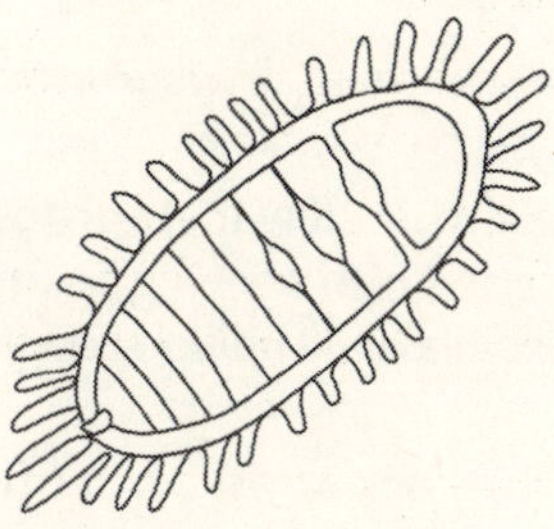

parasitic wasp

Mealybug

- The cassava mealybug, Phenacoccus manihoti, was inadvertently introduced into Africa in the early 1970s on infected planting materials from South America.
- It quickly spread from field to field and from country to country to become the most important pest of cassava during the 1970s and 1980s.
- It ravaged plots, drastically reducing leaf and root yields, sometimes by as much as 80 percent.

Parasitic wasp

- Scientists at the International Institute of Tropical Agriculture discovered that the mealybug has a natural enemy, a parasitic wasp called *Apoanagyrus lopezi*, which uses the mealybug as a host for laying its eggs. The developing larvae then kill the mealybug.
- The wasp was brought to Africa, large numbers were reared, and it was released in some 30 countries, covering the cassava belt of Africa.
- Reduction in mealybug damage was seen in the first season following release, and full control was typically achieved within 2–4 years.
- The parasitic wasp continues to control the mealybug across the cassava belt of sub-Saharan Africa.

Benefits

- Each dollar spent on the mealybug control project brought returns worth at least $150 to the farmer.
- The overall economic benefit of controlling the mealybug has been estimated at between $9 billion and $20 billion.
- Cassava production has since increased in many countries; Nigeria, for example, is now the world's highest cassava producer.
- Another pest – the cassava green mite, more recently introduced from South America – has also been controlled by use of a predatory insect from Brazil.

▫ pest hazard study 3: tsetse fly areas

- The tsetse fly (Glossina) spreads a parasitic disease, trypanosomosis, to both cattle (nagana) and humans (sleeping sickness).
- The World Health Organization (WHO) estimates that there are between 300,000 and 500,000 people currently infected with sleeping sickness, the human form of trypanosomosis, and an additional 60 million people are at risk of contracting the disease.
- The pest prevents the widespread use of livestock across a belt of over 10 million square km of the African continent.
- There are 22 different species of tsetse fly, slightly larger than a horsefly, and they live only in Africa.
- They breed along rivers and streams, are active during the day and feed exclusively on blood.

- Tsetse flies transmit sleeping sickness through their saliva, after first biting an animal or person that is already infected with a tiny microorganism called a trypanosome.
- The tsetse fly has been eradicated from the island of Zanzibar by use of the sterile male technique.
- In 2001, a Pan-African campaign involving the UN Food and Agriculture Organization (FAO) and the International Atomic Energy Agency (IAEA) was launched to eradicate the tsetse fly using sterile insects.

Tsetse-infested areas and cattle distribution in Africa

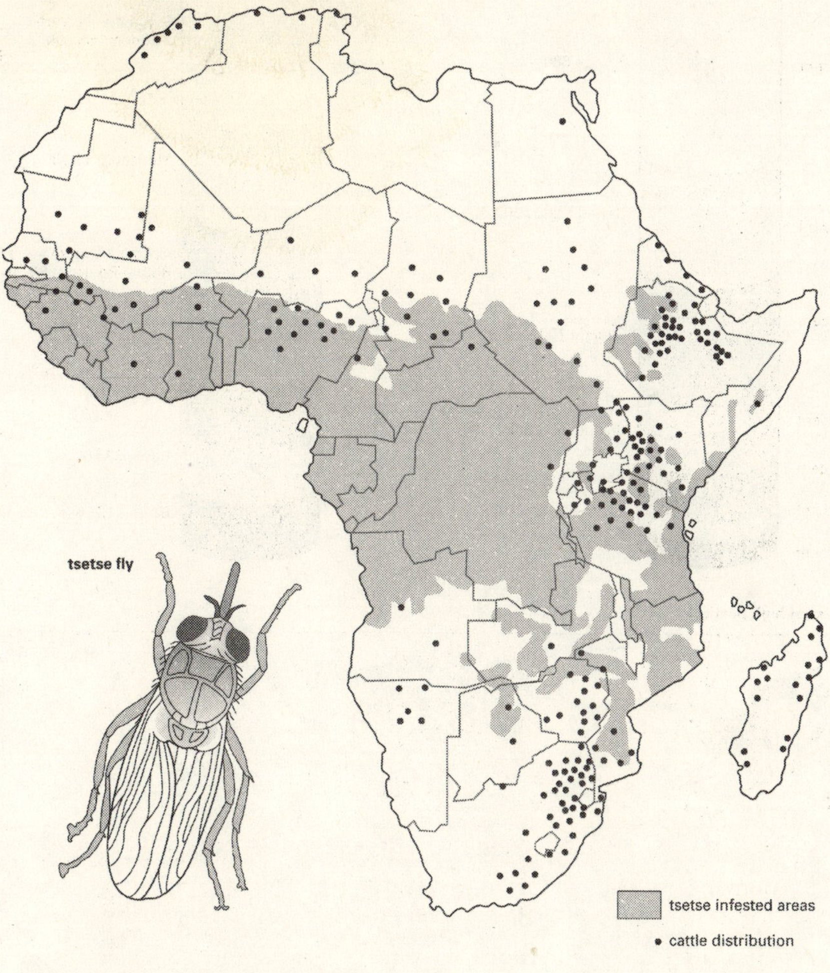

❑ pesticide use worldwide

- In 1995, world pesticide consumption peaked at 2.6 million metric tons of active ingredients with a market value of $38 billion.
- Only seven companies now supply some 90 percent of the pesticide market.

- Use of pesticides in the developed world is now decreasing, in part as a result of the substitution of new, more powerful chemicals which are used in much smaller amounts.
- However, it is still increasing in developing countries.

Annual use by region (thousands of metric tons)

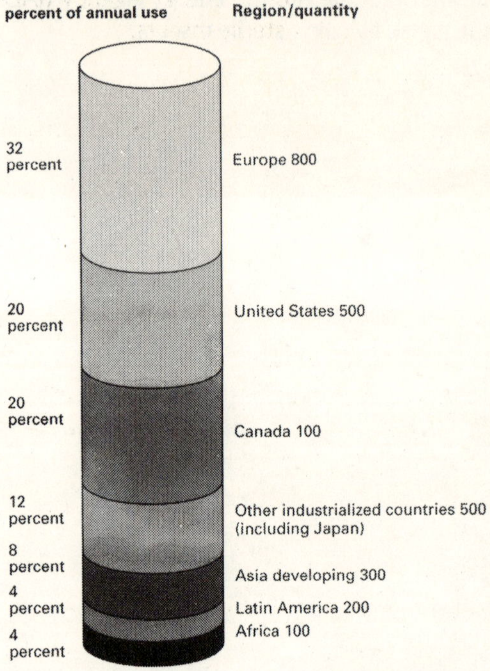

Annual sales of seven companies $22.7 billion

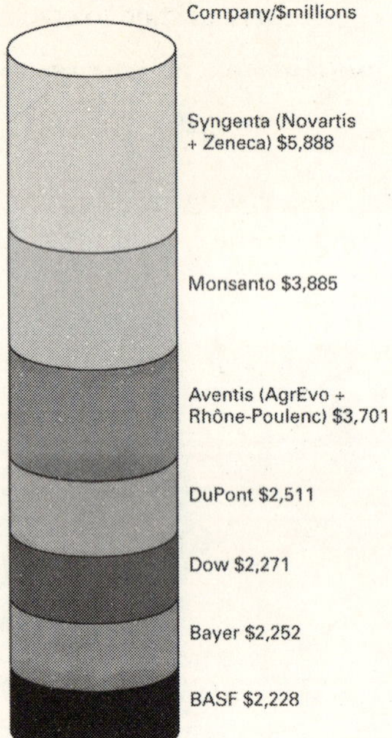

Areas of high pesticide use

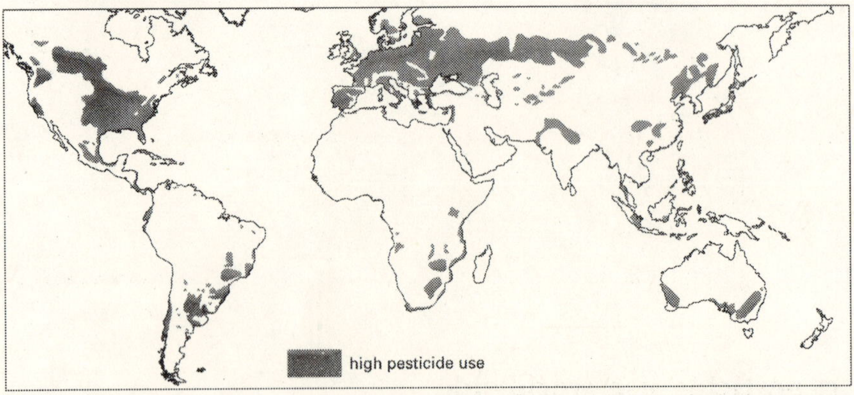

- Applications of pesticides inevitably lead to residues in soils which may evaporate into the air or be washed into watercourses, causing contamination of food and the environment, and endangering human health.

❐ ph

For our purposes, pH is a scale from 1 to 14 which expresses the relative acidity or alkalinity of the water in soil. A pH of 7 is neutral, i.e., neither more alkaline or more acidic. Values below 7 are acidic, increasingly acidic toward 1. Values above 7 are alkaline, increasingly alkaline as the values increase toward 14. pH is the standard abbreviation for "potential hydrogen" which denotes the hydrogen ion concentration of a solution.

❐ phytotoxic

Toxic to plants. Partially composted organic matter may have acids or alchohols present that will harm young or sensitive plants. Partially-decomposed compost is therefore referred to as phytotoxic.

❐ plant species used for food

- The world currently uses only a tiny fraction of the genetic resources available for food.
- Of 270,000 plants known to science only around 120 are widely cultivated today and nine provide 75 percent of our food.
- The most important food crops are rice, wheat, maize, sorghum, millet, potatoes, soybean, yams and cassava.
- For thousands of years, farmers have bred new crop varieties and tailored their farming methods to maintain both food supply and their land's fertility.
- The result was a huge diversity of plant varieties and farming methods.

Uncultivated species of plant germplasm

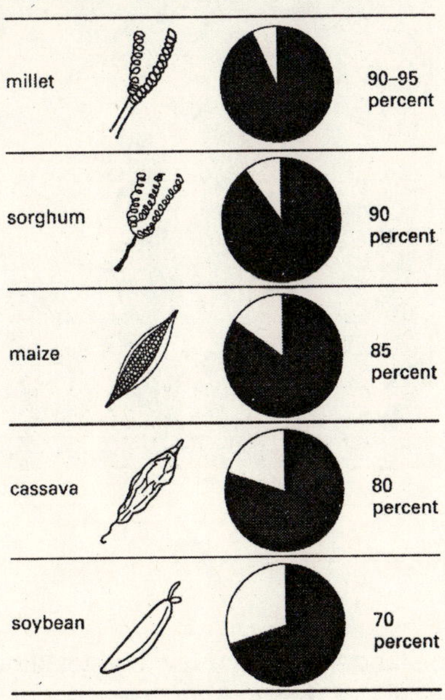

Germplasm (reproductive tissue) of nutritionally important crops held by international centers

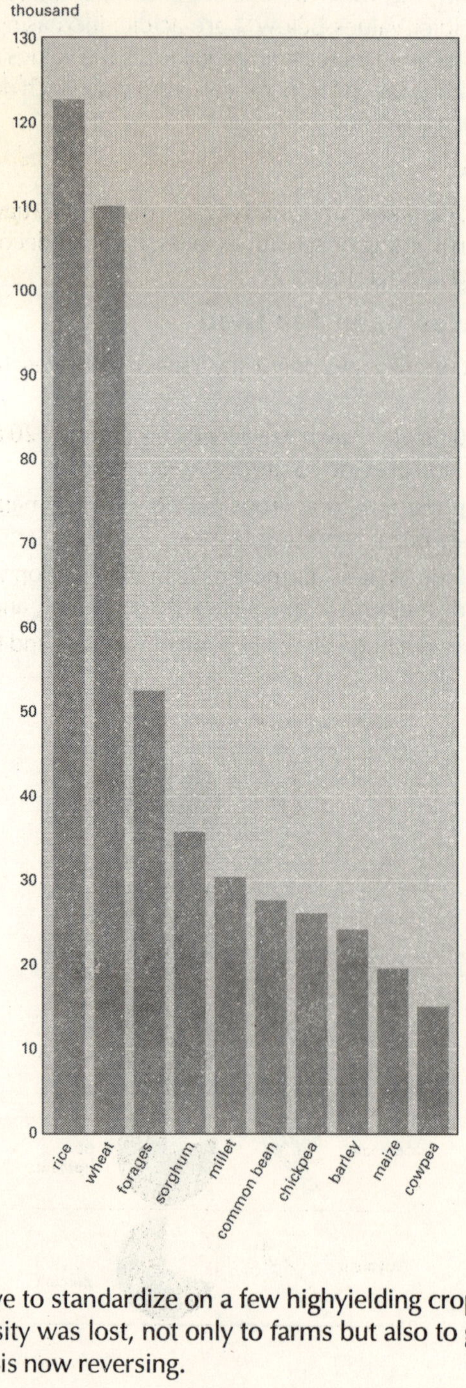

- In the drive to standardize on a few highyielding crop varieties, much of this diversity was lost, not only to farms but also to gene banks, though this trend is now reversing.

- Some 70 percent of wild rice species, and 90 percent of wild millet and sorghum, have yet to be collected.
- Human populations can be placed in three main dietary groups:
    - grain-eaters (the vast majority)
    - root crop or plantain eaters
    - meat and dairy produce eaters

❑ polysaccharides

Glue-like material formed by microorganisms in the process of mineralization, or breaking down organic matter. They are complex sugars which hold individual soil particles together. This creates larger particles known as aggregates.

❑ post-harvest losses

- Estimates of the post-harvest losses of food grains in the developing world from mishandling, spoilage and pest infestation are put at 25 percent.
- This means that one quarter of grain produced never reaches the consumer, and the effort and money required to produce it are lost forever.
- Fruit, vegetables and root crops perish quickly, and if care is not taken in their harvesting, handling and transport, they will soon decay and become unfit for human consumption.

A Physiological deterioration can be caused by extremes of temperature, atmospheric modification or contamination. This may cause unpalatable flavors, failure to ripen or other changes in the living processes of the produce, making it unfit for use.

B Mechanical damage can be caused by careless handling of fresh produce, causing internal bruising, splitting or skin breaks, thus rapidly increasing water loss and the rate of normal physiological breakdown. Skin breaks also provide sites for infection by disease organisms and fungi causing decay.

C Diseases and pests. All living material is subject to attack by parasites. Grain is often eaten by rodents. Fresh produce can become infected before or after harvest by diseases widespread in the air, soil and water. Some diseases are able to penetrate the unbroken skin of produce; others require an injury in order to cause infection. Damage so produced is probably the major cause of loss of fresh produce.

- Some estimates put losses of sweet potatoes, plantain, tomatoes, bananas, and citrus fruit in developing countries as high as 50 percent, or half of what is grown.

D Transport can often speed deterioration through vibration and exposure to heat or cold.

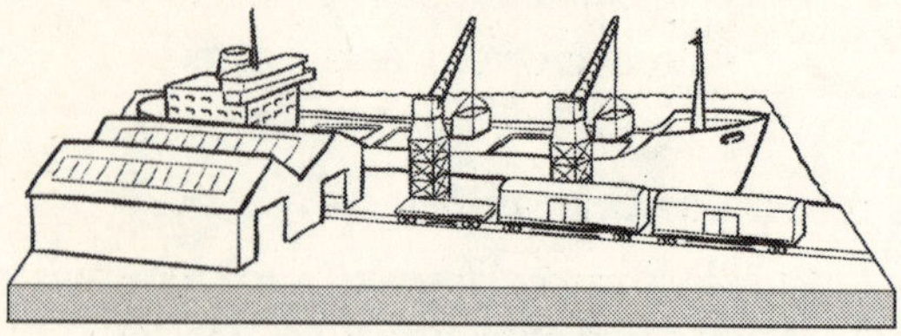

E Treatment. Careless harvesting and handling can add to loss. Some crops need to be stored at a constant temperature to 'cure' them or sprayed with fungicide to prolong life.

❒ psychrophilic bacteria

This group of bacteria species work to break down organic matter under "cold" conditions of 0 degrees up to over 55 degrees. They generate low levels of heat.

R

❒ reclamation

Processing of excavated landfill waste.

❒ recycle

Systems which may be run by private enterprise or local government to gather recyclable materials and remake them into similar or dissimilar products for market. Common examples are newspapers, glass, plastic, steel, and aluminum.

❒ reduction, source reduction, reduce (as part of "reduce, reuse, recycle")

Practices which result in the reduction of wastes without additional energy expended for recycling, composting, disposal, etc. Examples are: minimal packaging, lowering demand for disposable products, leaving grass clippings and leaves on the lawn, and learning to read/write/review data online without printing it onto paper.

❒ reuse

Practices which find alternate uses or alternate avenues for use of an item rather than expending energy to dispose it or alter its form by recycling or composting. Examples would be donating used books and magazines to a nursing home, using the fronts of Christmas cards to make new cards, using plastic margarine tubs to hold leftovers or craft supplies, donating clothing to charity.

S

❐ sandy soils

Low total quantity of pore space even though individual pores are large. Water retention is low.

❐ soil amendment

Matter than, when added to the land, will make the soil healthier by such means as balancing and adding nutrients, balancing the pH, encouraging the presence of microorganisms. From a legal standpoint, this is different than "fertilizer" and is not governed by the laws which regulate fertilizers.

❐ source separation

In homes or commercial operations, waste is separated into categories for recycling, composting or land filling. This is a fancy name for separating your newspapers, glass, yard wastes, plastic bottles, etc. into separate containers or piles for waste processing.

❐ sustainable development

- In 1987, the United Nations' World Commission on Environment and Development released its report Our Common Future (or the Brundtland Report), which brought the terms 'sustainability' and 'sustainable development' into widespread use.
- Our Common Future defined sustainable development as 'development that meets the needs of the present without endangering the needs of future generations.'
- At the 1992 Earth Summit in Rio de Janeiro, more than 100 nations signed Agenda 21–an international approach towards sustainable development.
- Fundamental to the philosophy of sustainable development is the fact that human communities and economies are not separate from the environment.
- An interdependence exists between the way humans manage the environment, the level of cohesion in a community and economic performance.
- The concept of sustainable development promotes not only protection of the environment and creation of wealth but equitable distribution of that wealth within society.
- Sustainable development seeks to ensure the needs of the present are met without compromising the ability of future generations to meet their needs.
- Sustainable development has four primary objectives:
  - minimized risk of environmental damage ari sing from incomplete knowledge.
  - ecological sustainability and environmental protection.
  - socio/cultural sustainability recognizing the needs of all.

- economic sustainability maintaining high and stable levels of economic growth and employment.

Conservation agriculture

- One form of sustainable development, 'conservation agriculture' aims to conserve, improve and make more efficient use of natural resources through integrated management of available soil, water and biological resources combined with external inputs.
- Conservation agriculture contributes to environmental conservation as well as to enhanced and sustained agricultural production.
- Conservation agriculture maintains a permanent or semi-permanent organic soil cover.
- This can be a growing crop or a dead mulch. Its function is to protect the soil physically from sun, rain and wind, and to feed soil biota.
- The soil micro-organisms and soil fauna take over the tillage function and soil nutrient balancing.
- Mechanical tillage disturbs this process. Therefore, zero or minimum tillage and direct seeding are important elements of conservation agriculture.
- A varied crop rotation is also important to avoid disease and pest problems.
- Livestock production can be fully integrated into conservation agriculture, by making use of the recycling of nutrients.
- This reduces the environmental problems caused by concentrated intensive livestock production.
- Globally, conservation agriculture is currently practiced on about 58 million hectares (140 million acres) of land, from the tropics almost to the Arctic Circle: mostly in the United States, Brazil, Argentina, Canada and Paraguay.

❑ thatch

A layer of organic matter that begins to grow between the base of a plant and the soil. Usually discussed as a problem with lawns as dead roots and stems accumulate. Proponents of the organic philosophy say that in an organic program where soil has achieved a healthy balance, microorganisms are present to feed on mulched grass clippings and prevent thatch. However, if long grass (generally meaning growth of more than a week) is cut and mulched, there may be an excess of clippings which is too great for the existing population of microbes to handle in a timely way. Use your own judgement in determining how much is too much, then remove excess clippings from the yard and put them into the compost bin. Clippings of any length that are wet or matted should be removed from the lawn and placed in the compost bin.

❒ the environmental costs of modern agriculture

- The environmental costs of modern agriculture include massive soil erosion, deforestation, pollution and loss of plant and animal genetic diversity.
- Supplies of land, water and firewood are shrinking in many regions.
- By 2020, Southeast Asia will have less than one hectare of arable land per person.
- Within a few years, 300 million people, mainly in North Africa, will face critical water scarcities.

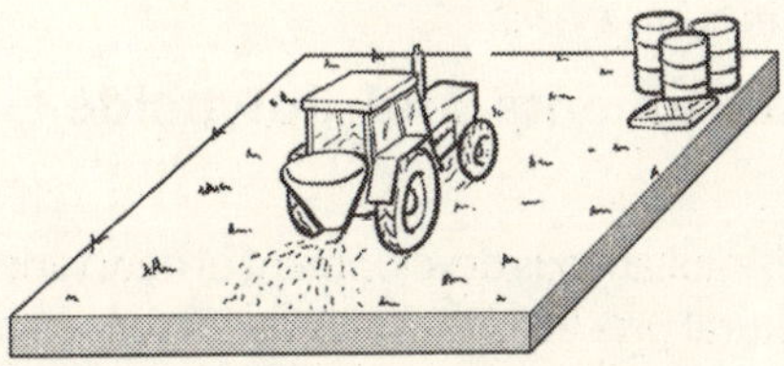

Mechanized farming with heavy use of fertilizer, using finite fossil fuels and generating pollution and CO_2

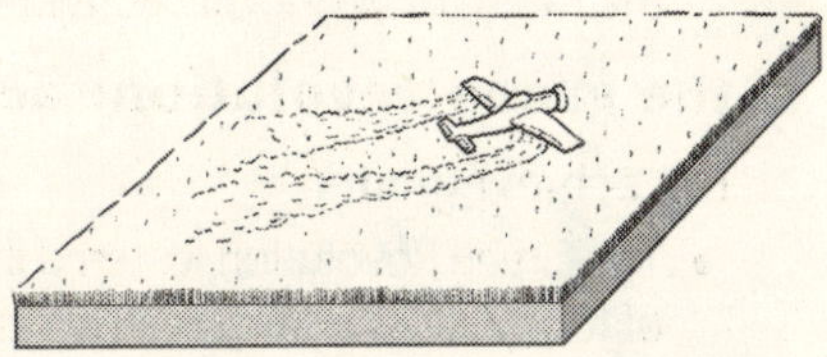

Intensive use of chemicals to control resistant pests

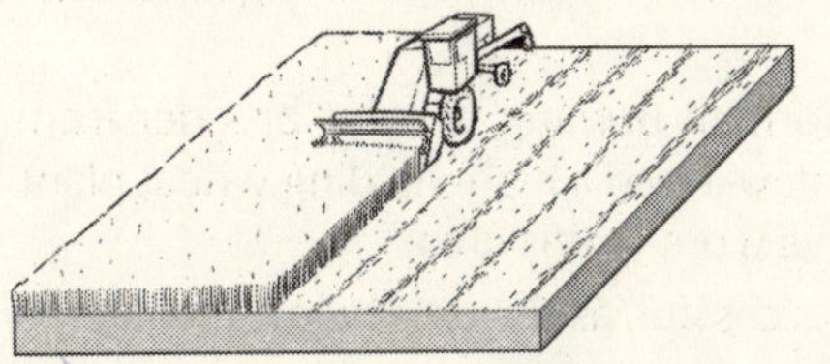

Monocrop cultivation leading to loss of diversity

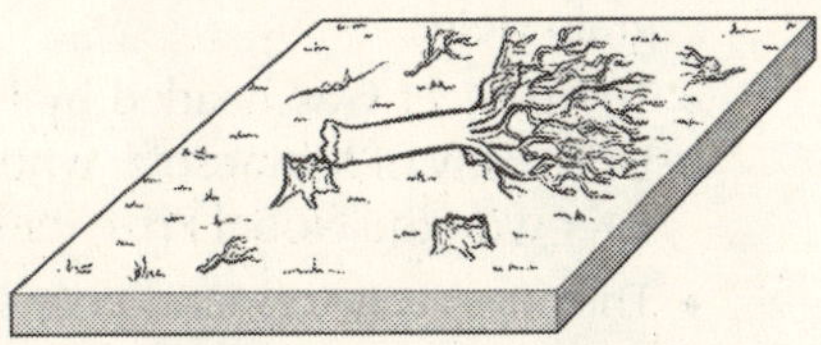

Deforestation leading to shortages of wood for fuel

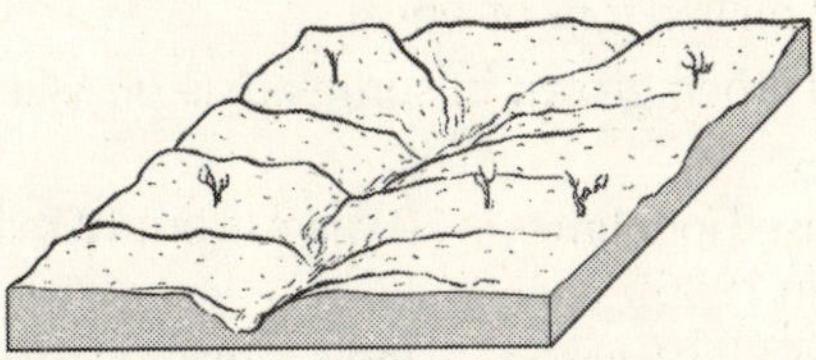

Overcropping, overgrazing leading to soil erosion

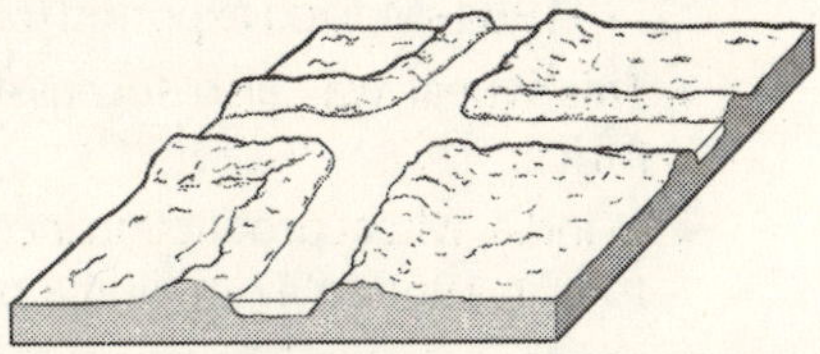

Demands on water resources for irrigation, which can destroy soil by salinization

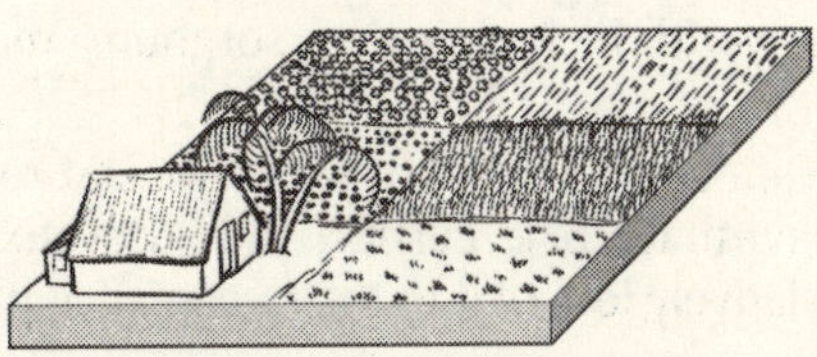

Population growth leading to smaller plots and more intensive farming

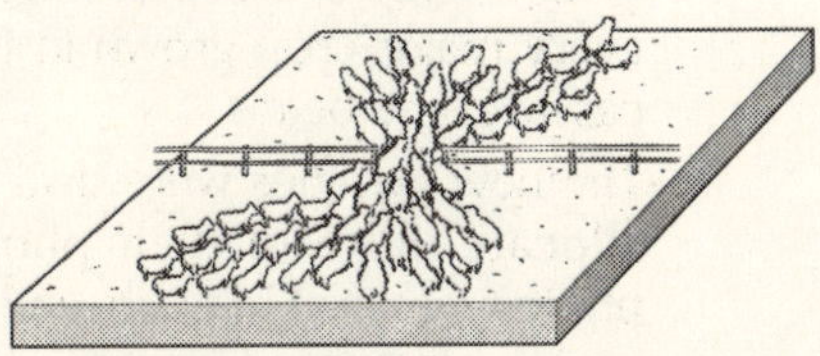

Loss of plant and animal genetic diversity caused by large-scale farming

- More than 130 million Africans already live in areas where firewood consumption outpaces the regenerative capacity of their forests.
- Increasing food production while protecting the environment – particularly in marginal areas – requires production systems that increase productivity while reducing pollution and resource degradation.
- Many ecologically sound technologies – such as biological control of pests, agroforestry, biogas digesters and improved irrigation management – are already in use or being developed, but are not widespread.
- Sustainable use of natural resources must also be socially and economically viable.
- The availability of energy is a key factor in achieving food security with environmental protection – it increases labor efficiency and helps diversify the range of economic activities in rural areas.

❑ the green revolution: achievements and downside

The achievements

- The Green Revolution fostered the simultaneous development of new varieties of food plants and altered agricultural practices that greatly increased crop yields.
- It began in 1944, when the Rockefeller foundation and Mexican government established a plantbreeding station in NW Mexico, with a goal of boosting grain yields.
- The project was headed by Norman Borlaug, a plant breeder from the University of Minnesota, who developed a high-yielding wheat plant and later won the Nobel Prize for his work in this area.
- The effort was tremendously successful, as the data below indicate:
  - 1944–Mexico was importing half its wheat
  - 1956–Mexico was self sufficient in wheat production
  - 1964–Mexico exported 0.52 million tons of wheat
- This wheat was also successful when grown in some areas of Asia and Africa.
- In India, wheat production increased four times in 20 years (from 12 million tons in 1966 to 47 million tons in 1986).
- At present, over 75 percent of the wheat acreage in India is planted to these new high-yielding varieties.
- Over the years, breeders have produced high yielding varieties of virtually every major crop grown in the world, including rice, sorghum, maize, cassava, and beans.
- The new varieties were dwarf varieties because they have been bred to allocate more of their photosynthate (the carbon that they fix via photosynthesis) to grain, and relatively less of it to their vegetative parts, such as stems and leaves.
- These varieties are also highly responsive to fertilizer and to water, and need heavy inputs of fertilizers and irrigation.

- Corn yields in the US have more than quadrupled from 20 bushels/hectare to 100–250 bu/há in the last 60 years.
- Since the advent of the Green Revolution, there have been increases in cultivated acreage of about 24 percent between 1950–1981.
- On a per capita basis, world grain production increased about 40 percent between 1950–1984.
- Gains in food availability have been greatest in the developing world, where the green revolution enabled a rise of 38 percent between 1961 and 1998 to 2,660 calories per person daily.
- Life expectancy in the developing countries increased by 10 years in two decades (from less than 43 years in the early 1950s to over 53 years in the early 1970s), with a major portion of the increase attributable to improved nutrition.

The downside

- The Green Revolution spread a farming method that relies heavily on chemical fertilizers, monocultural cropping practices and decreased fallow times, coupled with more intensive plowing and irrigation.
- Local water supplies became degraded and crop diversity decreased, leading to land degradation and erosion.
- Concern has increased over the implications of pests that thrive on now dominant strains of grain and maize.
- Chemical applications to fields have soared. More than half of all the commercial fertilizer ever produced has been applied to fields since 1984.
- The doubling of agricultural production during the past 35 years has required a 600 percent increase in nitrogen fertilizer and a 250 percent increase in phosphate fertilizer.
- This has been accompanied, over the same period, by a 70 percent increase in irrigated cropland to more than 260 million hectares.
- Almost 40 percent of the global food harvest now comes from the 17 percent of the world's croplands that are made productive in this way
- Human activity has taken over from nature as the dominant source of fixed nitrogen in the environment.
- Natural sources from soil bacteria, algae and lightning release 140 million tons of fixed nitrogen a year; human sources now total 210 million tons per year, of which 86 percent percent comes from agricultural activity, with fertilizer responsible for most of it.
- The result has been a nitrogen overload of natural ecosystems, particularly in Western Europe and East Asia, where average annual applications on arable land are highest.
- Between 1950–2000, total global grain production increased on average about 2.6 percent per year, and grainland productivity (yields per ha) rose more than 2 percent per year–comfortably above the rate of population growth for most of that time.

- But between 1990–1996, the harvest increased only at 0.7 percent per year, no longer keeping ahead of population growth.
- There were similar increases for rice and other staple cereals.

Total grain production 1950–2000 (A) and total grain yield per hectare (1950–2000) (B)

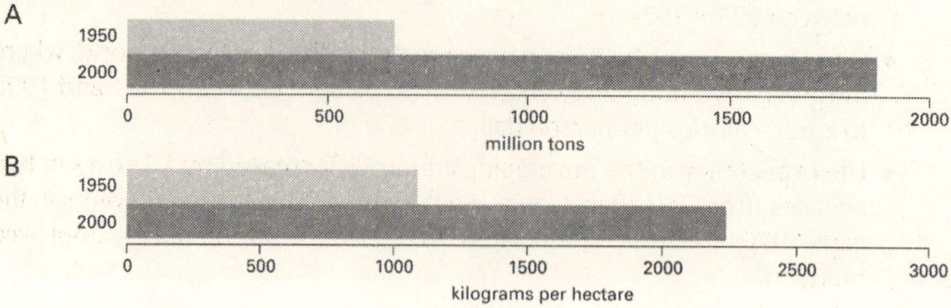

❐ the problems of overfishing

Scale of overfishing

- The world marine catch stands at nearly 100 million tons per year.
- However, 25 per cent of this is 'cheap fish' – such as pollack, jack mackerel and pilchards – most of which ends up as animal feed.
- A further 27 million tons of 'by-catch' – nearly a third of the global catch – is thrown back dead into the sea because of the widespread use of unselective fishing gear.
- In the Gulf of Mexico shrimpers throw back 9 lb of dead fish for every lb of shrimp they actually catch.
- On a wider scale this death toll includes thousands of dolphins and porpoises as well as vulnerable species such as turtles, billfish and sharks.
- According to the FAO, 15 of the world's 17 major ocean fisheries are already depleted or overexploited.
- The collapse of the cod stocks off the east coast of Newfoundland in 1992 forced the Canadian government to close the fishery. Over 40,000 people lost their jobs.
- These trends are even more troubling when population growth is considered.
- World population, now at six billion, continues to grow by over 60 million people per year with nearly half this growth in areas within 100 km of a coastline.
- Over one billion people in Asia already depend on ocean fish for their entire supply of protein as does 1 out of every 5 Africans.
- When the US National Academy of Sciences recently brought together many of world's leading marine biologists, they concluded that overfishing – not global warming or pollution – was the greatest single threat to the diversity of life in the world's oceans.
- Government subsidies have allowed national fisheries in the North to swell to an unsustainable size.

- In 1997 the fishing industry in Europe received subsidies totaling $1.4 billion – about $14,000 per boat.
- This cash funds overcapacity – the World Wide Fund for Nature (WWF) estimates that Europe has 40 percent more boats than needed to catch fish on a sustainable basis.

Modern fishing techniques

- The technologies of modern fishing–radar, sonar, global positioning systems–allow great efficiency in tracking down and catching every last stock.
- Modern net designs practically vacuum the sea and sea bottom clean of living organisms.
- There is a very high level of by-catch of unwanted species, or juveniles that would be illegal to land.
- Distant Water Fleets (DWF) use mechanized fishing methods that net fish at two and a half times the rate needed to ensure that fish stocks are able to naturally replenish themselves.
- Stern factory-trawlers, or draggers, haul enormous, baglike nets, as long as a football field, held open by a combination of huge steel plates or 'doors' and heavy chains and rollers that plow and scrape the ocean bottom.
- They drag up as many as 400 tons of fish in one scoop, catching whole schools of fish and anything else in the way, and inflicting immense damage to immature target and non-target fish and the benthic (bottom-dwelling) community.
- Some fishing gear is particularly deadly for certain animals in some situations.
- Driftnets indiscriminately killed millions of marine creatures, while targeting for just one or two commercially valuable species.
- Marine mammals are frequently killed in great numbers in trawls, set nets and purse seine nets.
- Fishing practices like bottom trawling physically disturb marine habitats such as the ocean floor, sea grass beds or coral reefs.

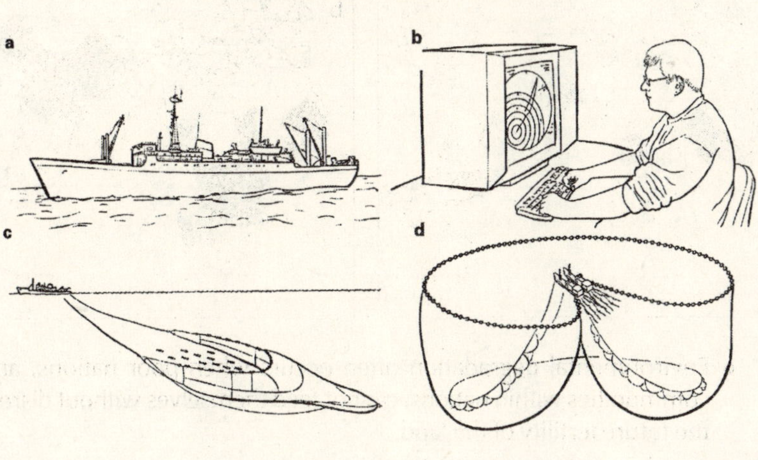

Vast factory trawlers (a), using sophisticated location technology (b), use bottom trawl nets (c) or purse seine nets (d) to trap hundreds of tons of fish.

❐ the world's undernourished people

- In 1998, there were 78 low-income countries that neither grew enough food to feed their populations, nor had the resources to make up the deficit with imports.
- Of these, more than half were in Africa, where population growth rates are highest and poverty is greatest, soils are generally most vulnerable to degradation and modern advances in agricultural technology have had the least impact.
- The World Food Summit pledged in 1996 to halve malnutrition within 20 years. But the UN Food and Agriculture Organization (FAO) predicts that in some regions of the world chronic undernourishment is likely to persist, rising to 30 percent of the population of sub-Saharan Africa in 2010.
- Poverty and hunger frequently cause a cycle of environmental decline that further undermines food security.

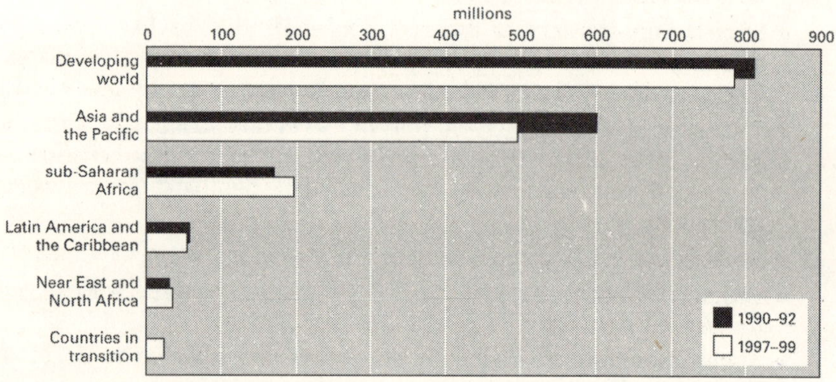

Prevalence of malnutrition
- Over 200 million children in developing countries under the age of five are malnourished.
- Malnutrition contributes to more than half of the nearly 12 million under-five deaths in developing countries each year.
- Map shows percentage of children significantly underweight.

Percentage significantly underweight
a 40 percent or more
b 20–39 percent

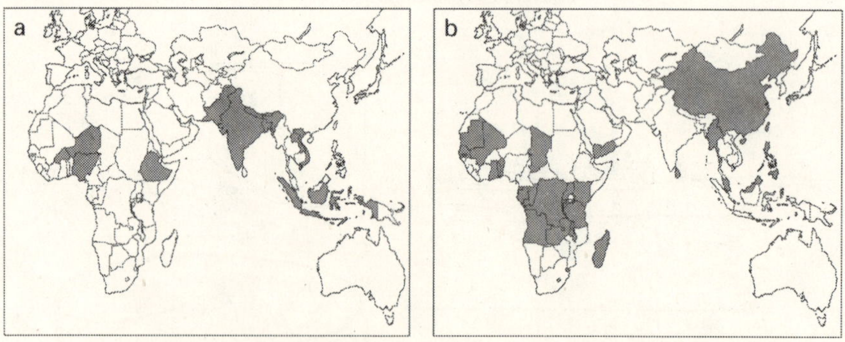

- Environmental degradation often occurs when poor nations, and poor communities within nations, cannot feed themselves without disregarding the future fertility of the land.

- They overcultivate or overgraze land to meet immediate needs, or annexe inappropriate land with steep slopes, and shallow, infertile, stony, toxic or poorly drained soils.
- In 1997–99 there were 815 million undernourished people in the world.

Undernourished population by region in 1990–92 and 1997–99 (latest available figures)

- The highest number of undernourished people is found in Asia.
- The greatest reduction in undernourished people since 1990–92 is also in Asia.
- Increases can be observed for sub-Saharan Africa and for the Near East and North Africa.

❏ thermophilic bacteria

This group of bacteria species work to break down organic matter under "hot" conditions of 104 degrees up to 170 degrees. This type of bacteria can perform the greatest decomposition in the shortest amount of time.

❏ tilth

Tilth is land which has been prepared for growing crops by plowing and fertilization. In discussions of composting and organic gardening, the phrase "the tilth of the soil" is often used to describe the general health of the soil including a balance of nutrients, water, and air.

U

❏ using hill slopes

- Soil erosion increases rapidly with slope, and is worst between 10 and 25 percent slopes.
- Sustainable cropland is not steeper than 5 percent.
- Many agricultural practices in common use can dramatically reduce erosion from hillslopes.
- Terracing uses the runoff from the hill to increase soil retentiveness and arability and is often part of a larger irrigation system that includes canals.
- It is widely used in the humid tropics to grow rainfed rice.
- Contour farming is the practice of preparing land, planting crops, and cultivating them on a level or nearly level contour around a slope. Each crop row serves as a small dam to hold water on the slope.
- Compared to up-and-downhill farming, contour farming can cut soil losses by as much as 50 percent on long, gentle slopes.
- Tree planting on higher slopes provides deep roots that hold the soil, cycles deep nutrients and lost nutrients, provides organic matter and more.
- Stripcropping reduces the downslope length of the field considerably.
- It is most useful on gentle slopes, where it may reduce erosion to acceptable levels without any banks or drains.

- Stone lines, or stop wash lines, retain some rainwater and eroded soil on gentle slopes.
- Grass strips can also be used as stop-wash lines.

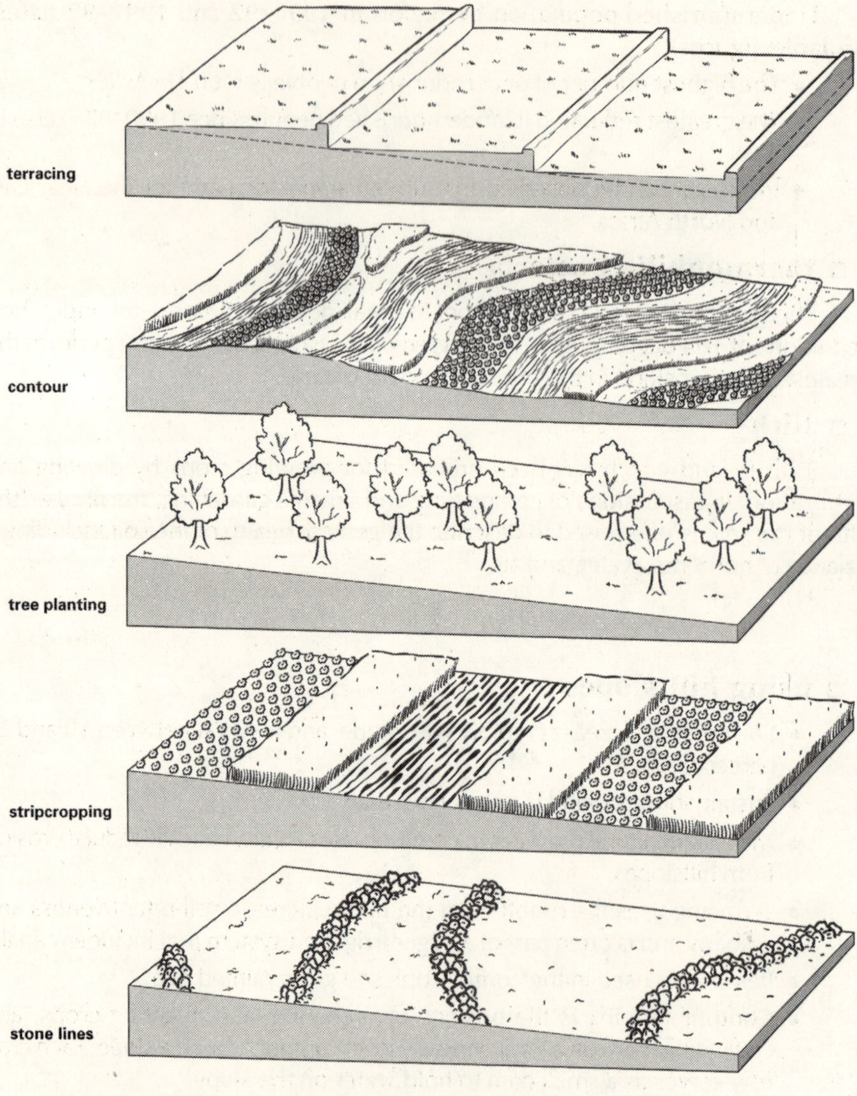

❑ vavilov centers of genetic crop diversity

- Several regions – known as Vavilov Centers of Diversity after Nikolai I. Vavilov, the Russian botanist who first described the pattern in the 1930s– have been identified as locations of highly diverse crop genetic resources.
- The centers of crop genetic diversity are characterized by a long agricultural history, ecological diversity, mountainous terrain, cultural diversity, and a lack of heavy forest cover.

- These centers may or may not be located where the crop was first domesticated.
- Wheat and barley were domesticated in southwest Asia, but a current center of their varietal diversity is in Ethiopia.

Vavilov Centers of high crop diversity and origins of food crops

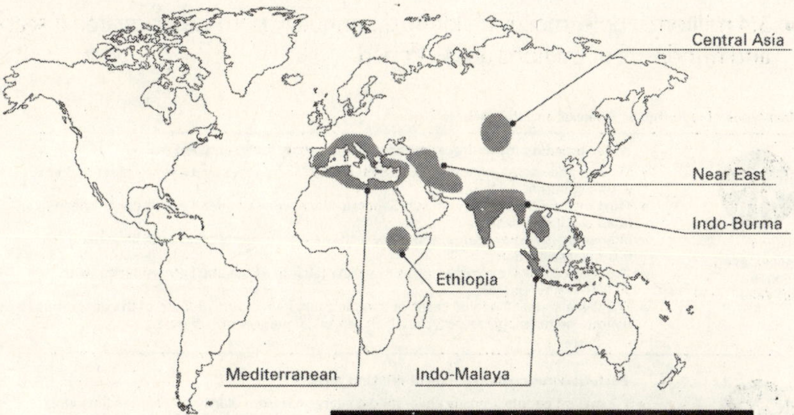

region	crop
Ethiopia	barley, coffee, sorghum
Mediterranean	oats, olives, wheat
Near East	barley, lentil, oats, wheat
Central Asia	apple, chickpea, lentil
Indo-Burma	eggplant, rice, yam
Indo-Malaya	banana, coconut, sugar cane

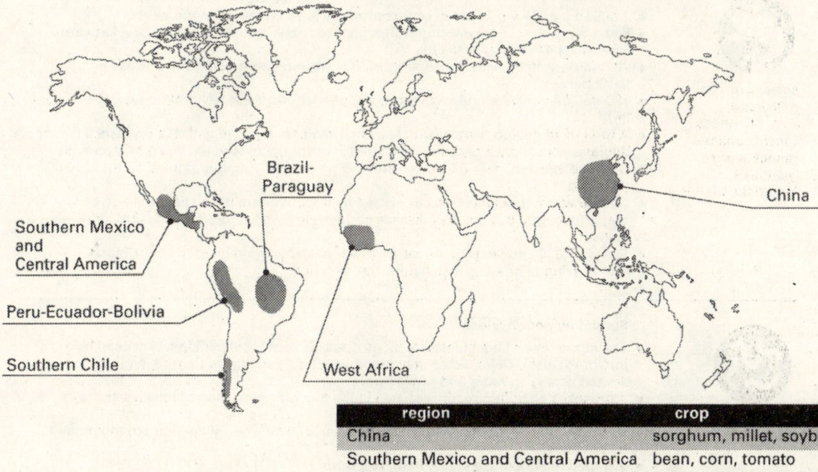

region	crop
China	sorghum, millet, soybean
Southern Mexico and Central America	bean, corn, tomato
Peru-Ecuador-Bolivia	bean, potato, squash
Southern Chile	potato
Brazil-Paraguay	peanut
West Africa	millet, sorghum

❑ vermicomposting, vermiculture

Using redworms to compost food scraps, newspapers, and cardboard, yielding nutrient-rich castings.

W

❑ waterborne diseases

- 3.4 million people, mostly children, die annually from water-related diseases, and hundreds of millions are sickened.

Main types of waterborne diseases and their effects

malaria, yellow fever, dengue, Rift Valley fever

Water-breeding mosquitoes spread malaria, yellow fever, and dengue.

- Malaria affects 300 million and kills over one million people every year – most of them young children.
- Most malaria deaths occur in sub-Saharan Africa, where malaria accounts for one in five of all childhood deaths.
- Intensified irrigation, dams, and other water-related projects contribute importantly to this disease burden.
- There are 200,000 estimated cases of yellow fever in Africa and Latin America (with 30,000 deaths) per year.
- There are some 50 million cases of dengue every year, some of them of the dangerous dengue hemorrhagic fever type, which can kill 20 percent of sufferers.

cholera, typhoid, dysentery, diarrhea

Bacteria/virus contamination of drinking water

- 2.2 million people (mostly children) die every year from diarrheal diseases (including cholera) associated with inadequate water supply, sanitation and hygiene.
- Other major diarrheal diseases include typhoid fever and rotavirus.
- Diarrheal diseases impose a heavy burden on developing countries – accounting for 1.5 billion bouts of illness a year in children under five.
- The burden is highest in deprived areas where there is poor sanitation, inadequate hygiene and unsafe drinking water.
- The polio virus used to be a major water-borne scourge, but has now been virtually eradicated.

schistosomiasis (bilharzia), river blindness (onchocerciasis), guinea worm, giardiasis, lymphatic filariasis

Parasites in polluted water

- 2 billion people are at risk of schistosomiasis in 74 countries of the world.
- 200 million people are infected with schistosomiasis, which is caused by a flatworm spread via a freshwater snail.
- 20 million suffer severe consequences. Basic sanitation reduces the disease by up to 77 percent.
- 120 million people world-wide are at risk of onchocerciasis (river blindness), 96 percent in Africa.
- A total of 18 million people (almost all in Africa) are infected with the disease, a type of filariasis caused by a parasitic worm that lives for up to 14 years in the human body.
- Of those infected, over 6.5 million suffer from severe itching or dermatitis and 270,000 are blind.
- Guinea worm (*dracunculiasis*) is caused by a roundworm which can grow up to one meter in length. It is the only disease exclusively associated with unhealthy drinking water.
- Some 85,000 people, all in Africa, contract the debilitating disease every year.
- The filtering of drinking water stops transmission.

trachoma, hookworm

Spread by poor hygiene

- Trachoma, one of the oldest infectious diseases known to mankind, is caused by a microorganism, *Chlamydia trachomatis*, which spreads through poor hygiene and limited access to water and sanitation.
- It provokes an inflammatory reaction in the eye with formation of follicles in the conjunctiva, which can cause blindness.
- 6 million people are irreversibly blinded by trachoma and 146 million are threatened by blindness.
- 500 million people are at risk from trachoma.
- Hookworm infections occur mostly in tropical and subtropical climates and are estimated to infect about 1 billion people.

arsenic

Poisoned drinking water

- Arsenic contamination of ground water is a global problem. It has been found in many countries, including Argentina, Bangladesh, Chile, China, India, Mexico, Thailand, and the United States.
- 35–77 million from a total of 125 million in Bangladesh are estimated to be at risk of drinking contaminated water.

- Directly associated with poor sanitation services and unclean drinking water are the severe waterborne diseases: malaria, dysentery, cholera, and the many parasitic diseases found in Africa and Asia, such as schistosomiasis and guinea worm.

❐ windrow system

Rather than making a square or round compost pile, some people make a long row. This is especially true of commercial operations. The compost pile is about as tall as it is wide, but may be as long as space allows. This row of compost is called a windrow.

❐ world fish catch, tonnage and target species

- Worldwide, humanity obtains 16 percent of its animal protein from marine sources.

Species of which over 500,000 tons were caught worldwide in 1999 (latest available figures)
Not all are destined for human consumption: the species with the highest tonnage, the anchoveta, is used for the production of fishmeal and fish oil.

Species	Tons
Anchoveta (= Peruvian anchovy)	8,723,265
Alaska pollock	3,362,473
Atlantic herring	2,403,543
Skipjack tuna	1,976,479
Chub mackerel	1,955,053
Japanese anchovy	1,820,259
Chilean jack mackerel	1,423,447
Largehead hairtail	1,418,944
Blue whiting	1,323,167
Yellowfin tuna	1,258,386
Atlantic cod	1,092,859
Argentine shortfin squid	1,091,299
Capelin	904,840
European pilchard (=Sardine)	901,427
Araucanian herring	782,142
Gulf menhaden	694,242
European sprat	684,164
Atlantic mackerel	610,947
Akiami paste shrimp	598,602
European anchovy	598,438
Japanese Spanish mackerel	595,103
Japanese pilchard	515,477

= 500,000 tons

- Around half a billion people gain their livelihoods from harvesting the oceans.
- Of some 3 million fishing vessels known to be at sea worldwide, more than a million are large 'industrial' vessels that travel the globe.
- The world's marine fish catch increased fivefold in the later 20th century, reaching around 90 million tons a year.
- Despite increasing investment in ships, nets and tracking equipment, the catch stagnated since 1990.
- The impact of overfishing has resulted in reduced catch of several major species, including the Atlantic cod, haddock and Alaska pollock.
- According to the UN Food and Agriculture Organization (FAO), nine of the world's 17 major international fishing grounds are at or beyond the point at which yields will decline.
- In the North Atlantic, cod stocks are at half their former levels.
- Asian fleets fished out the North Atlantic squid stocks in the 1980s.
- Atlantic mackerel, redfish, and herring catches are all less than half their size of 30 years ago.
- The total world catch (marine and inland waters) for1999 was 92,866,553 tons, according to the FAO.

❐ world food output

- Over the past four decades, the rate of worldwide food production has increased faster than the rate of world population.
- Gains in food availability have been greatest in the developing world, where the green revolution enabled a rise of 38 percent between 1961 and 1998 to 2,660 calories per person daily.
- But the developed world, with a quarter of the world's population, still takes some 49 percent of the world's agricultural products, partly because it converts more grain to meat.

World food production relative to population growth

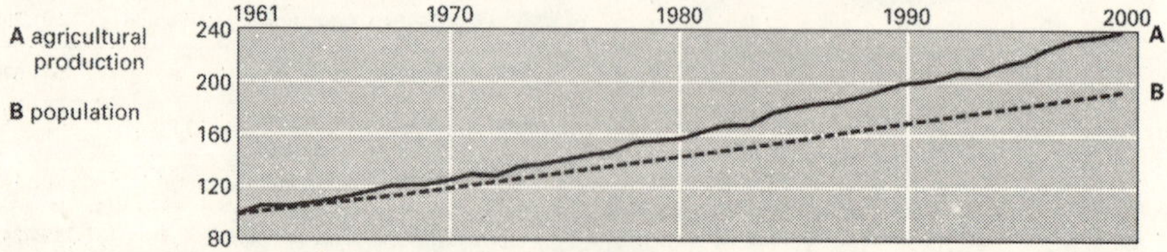

meat (million tons) 2000

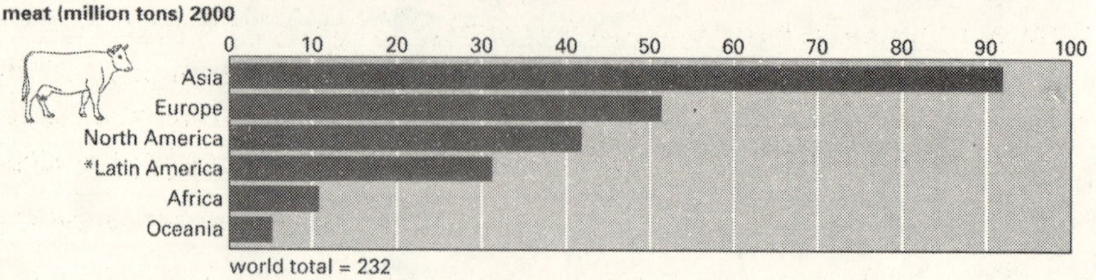

cereals (million tons) 2000

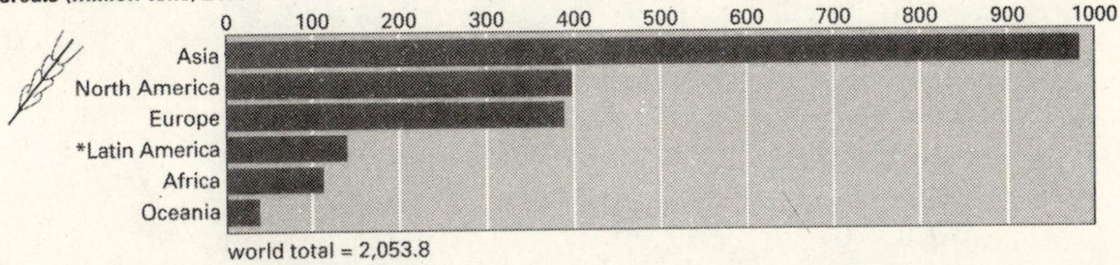

pulses (million tons) 2000

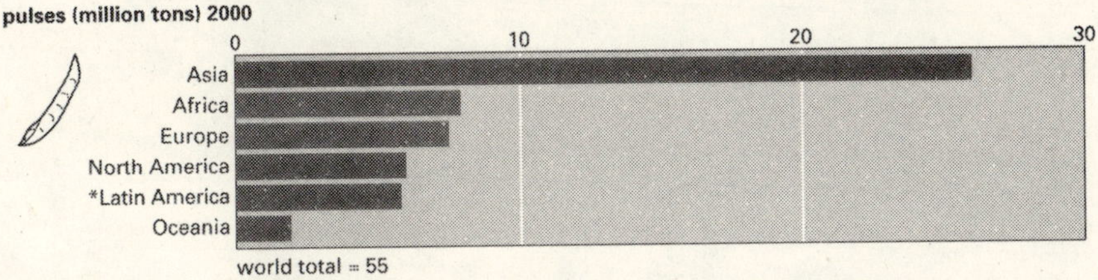

roots and tubers (million tons) 2000

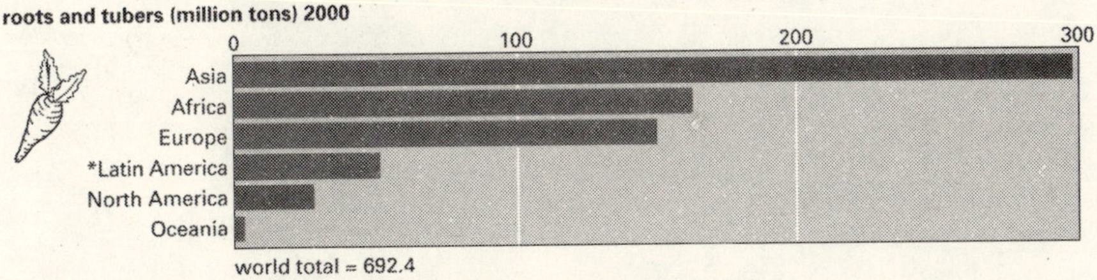

- Even so, differences in food availability within the developing world are now greater than between typical developed and developing countries.

Atmospheric Chemistry

❐ 1,3-butadiene

is produced during a petrochemical combustion. It is also leaked into the atmosphere from the storage of petrochemicals. Typical half-life in the atmosphere is around 2 hours. It is also a known human carcinogen.

❐ 2-hexanone

$C_6H_{12}O$ is a volatile organic compound (VOC) which are a large family of carbon-containing compounds which are emitted into the atmosphere from a variety of industrial processes. 2-hexanone may have the potential to contribute directly to global warming by absorbing infrared radiation from the earth's surface. In general the more complex a VOC, the greater its ability to absorb infrared radiation. 2-hexane may also contribute indirectly to global warming through the change in concentration of ozone, which is a potent greenhouse gas.

A

❐ abatement

To lower, depreciate, or reduce. Often refers to emission reductions in air quality control.

❐ ablation

The weathering of a glacier by surface melting, or rock weathering by hydraulic erosion.

❐ absorption cross section

A measurement of an atom or molecule's ability to absorb light at a specified wavelength, measured in square cm/particle.

❐ abstraction reaction

a reaction that takes any atom away from another chemical species. Classical examples in atmospheric chemistry are the gas phase removal of hydrogen from methane by hydroxyl radical or the following solution phase reaction:

$$HSO_3\text{-} + H2O_2 \rightarrow HSO_4\text{-} + H_2O$$

❐ accretion

The addition of air particles to hydrated drops (snow, rain, sleet, etc.) by coagulation as the drops fall through the sky.

❐ accumulation mode particles

(Also known as secondary particles) These are particles that are formed in the atmosphere due to both the chemical and physical processes that take place with the interactions of primary gaseous emissions. The primary gaseous emissions are injected into the atmosphere by combustion processes such as from a car or from a coal burning plant.

❑ acetaldehyde (C_2H_4O)

This is a fairly simple aldehyde (second in the analogous series after formaldehyde) that is found in the atmosphere as a result of emissions from the manufacture of acetic acid, plastics, raw materials, and as a product in some polluted air oxidation reactions.

❑ acetic acid ($C_2H_4O_2$)

is a carbonyl compound that is emitted into the troposphere by both natural and anthropogenic processes. In the troposphere, acetic acid is present in the gas phase and is highly water soluble. Since acetic acid is highly water soluble, it is found in high concentration as acidic precipitation, such as in fog water and cloud droplets in urban areas.

❑ acetone (C_3H_6O)

is a carbonyl compound that is found in the atmosphere as a reactive gas. Acetone is considered to be a volatile organic compound (VOC), which is emitted into the atmosphere by industrial processes. Acetone has been linked to the formation of ozone in the troposphere due to the fact that it is a source of free radicals.

❑ acid rain

Acidified particulate matter in the atmosphere that is deposited by precipitation onto a surface, often eroding the surface away. This precipitation generally has a pH less than 5 and sometimes much lower depending on the concentration of acidic components.

❑ acid-pulse (dry deposition)

deposit of powder-like substance over the ground surface; especially effecting plant leaves; that when contacted by water has a very low pH.

❑ activation energy

The energy barrier that must be overcome during a collision of two potential reactants in order for a reaction to occur.

❑ adiabatic lapse rate

The rate of decrease of temperature with increasing altitude in the atmosphere. If heat is neither gained nor lost from the air parcel under consideration, then the lapse rate is said to be adiabatic.

❑ advection

To transport or carry. In air quality, the rate at which particulate matter is transported.

❑ aerobic

Oxygen-dependent or requiring oxygen.

❑ aerobic respiration

Organisms utilize oxygen to break down components, derive energy, and generate needed biomolecules. Carbohydrates are cycled into water and carbon dioxide.

❑ aerosol particles

One of the components of an atmospheric air parcel, comprised of minute solids particles part of which is almost certainly water.

❑ aerosol scavengers

Cloud droplets which attract an aerosol into snow, rain or some other water precipitate by coagulation and accretion.

❑ afforestation

The process or act of changing land into forest by planting trees, seeding, etc. on land formerly used for something other than forestry. This can obviously be contrasted with deforestation.

❑ air parcel

a volume of air, the component makeup and concentrations of which remain relatively static, that tends to travel around the earth changing geographic location and position above earth as an intact unit. Air parcels can be tracked.

❑ air pollution

The contamination of the atmosphere by any toxic or radioactive gases and particulate matter as a result of human activity.

❑ air quality assessment

A prescribed level of atmospheric pollution allowed for a certain compound during a specific time in a specific geographical area. Standards are set by some regulating body, office or agency.

❑ albedo

Refers to reflection power or the portion of solar radiation that reaches a body (earth) surface and is immediately reflected back into the atmosphere. The uncertainty in the reflective power of clouds unfortunately increases the complexity of determining clouds' effects on modeling global warming.

❑ ammonia

The molecular formula for ammonia is NH_3. Ammonia is one of the most important inorganic nitrogen compounds in atmospheric water droplets. It reacts with strong acids and is one of the only known basic, gas phase atmospheric components. Atmospheric ammonia can also enhance the nucleation rate and the production of new particles in the atmosphere. These new particles can be activated to become condensation nuclei and then, through various processes, grow to a particle size of 0.05 micrometer or larger which can then be effective as cloud condensation nuclei. This process can, therefore, affect the global radiation budget.

The major sources of ammonia are decaying natural organic matter, livestock wastes, fertilizers, and industrial activity.

❐ anaerobic respiration

Living or acting in the absence of oxygen. Cellular respiration in the absence of oxygen.

❐ anoxic

The lack of oxygen such as the inadequate oxygenation of the blood (anoxia). In aquatic environmental chemistry it refers to water that has become oxygen poor due to the bacterial decay of organic matter.

❐ antarctic ozone hole

Recent data suggest that due to the release into the atmosphere of CFCs by human activities, the ozone hole can as large as North America during a given astral spring. The CFCs are broken apart by UV light--to form free radicals of chlorine--after they have diffused into the upper stratosphere from the troposphere. This long term movement from the troposphere to the stratosphere is possible for these chlorine containing chemicals because of there long atmospheric lifetimes; however, this type of movement is not important for more reactive species such as tropospheric ozone because of their reactivity and therefore short atmospheric lifetime. In the Antarctic stratosphere, the reaction that converts reservoir species of chlorine into an active form--which destroys ozone--takes place on the surface of particles in polar stratospheric clouds as the temperature drops below about200K. This is possible because of the unique isolation of the south polar vortex during the austral winter. The surfaces of these (nitric acid/water) clouds act as catalysts for reactions that release molecular chlorine which quickly photolyzes to chlorine's (radical) active state. It is this radical which destroys ozone.

❐ antarctic vortex

the combination of a drastic temperature and corresponding pressure drop along with the rotation of the Earth on its axis produces a spinning/rotating volume of air. The rotational speed of the winds commonly reaches as high as 180 mph. The motion of these winds form an impenetrable barrier such that the trapped air inside is unmixed, as it is separated from the air outside, and remains quite cold (temperatures drop below ?80 Celsius) until October. Inside the whirling volume of freezing air, the cold temperatures facilitate the condensation of gases into particles that eventually form polar stratospheric clouds.

❐ anthropogenic

Refers to something originating from humans and the impact of human activities on nature.

❐ anthropogenic forcing

Influence exerted on a habitat or chemical environment by humans. This obliquely refers to the idea of managed environment as opposed to a wild or non-disturbed environment. The variability of assumptions about radiative forcing can cause significant changes in the results from computer models that require an estimate of those values.

❐ anticyclonic flow

The air flow produced about a high pressure center by the combination of two forces: the pressure gradient accelerating the air away from the center and the Coriolos force acting inward. Anticyclonic flows are clockwise in the northern hemisphere and counterclockwise in the southern hemisphere.

❐ argon

An element that is a member of the 8A group (Noble Gases) of the periodic table. Argon is represented by the atomic symbol Ar, has an atomic number of 18, and an atomic weight of 39.948. It is colorless, odorless, and a very inert gas. It comprises about 1% of the Earth's atmosphere.

❐ aromatic hydrocarbons

Hydrocarbon compounds in which the carbon atoms are connected by a ring structure that is planar and joined by sigma and pie bonds between the carbon atoms. An example of an aromatic compound is benzene, C_6H_6.

❐ asbestos

A naturally occurring fibrous mineral found in certain types of rock formations. The primary use of asbestos was as a fireproofing material, applied to structural steel members in multistory commercial buildings to attain proper fire protection. With its resistance to heat, asbestos was able to keep the building structure from bending and warping if exposed to fire. After widespread use of asbestos, over a 15 to 25 year time span, a pattern of illness gradually began to occur in asbestos workers. Three diseases linked to asbestos exposure are asbestosis, a fibrous scarring of the lungs, lung cancer, and mesothelioma, a cancer of the lining of the chest or abdominal cavity.

❐ assimilation cycle

A natural process by which a water body uses microbes, which convert nonliving substances into protoplasm or cells, to purify itself from pollutants.

❐ atmosphere

The sum total of all the gases surrounding the Earth, extending several hundred kilometers above the surface in a mechanical mixture of various gases in fluid-like motion. The permanent constituents are molecular nitrogen; 78.1%, molecular oxygen; 20.9%, argon; 0.934%, and approximately 0.036% carbon dioxide. Various other components exist in trace amounts. Not to be under emphasized, these trace components are where the interesting atmospheric chemistry occurs. The atmosphere can also be artificially divided into layers. Example: the troposphere (the layer closest to the earth) and the stratosphere (the layer above the troposphere).

❐ atmospheric convection

Process in which heat energy is transported through a medium, usually a gas or liquid. An example, in the atmosphere occurs when warm air with a lower density experiences an upward force until it cools and its density matches the surrounding air, generally termed convection cells.

❐ austral spring

Related to the Antarctic ozone depletion, south polar austral spring denotes the season of spring in the southern hemisphere when the greatest amount of ozone is lost, generally beginning in September. Astral relates to the hemisphere that the observer is currently in and the season that relates to the observer, while austral refers to the south.

❐ autumnal equinox

Point at which on the celestial sphere that the equator and the Ecliptic intersect. Generally the autumnal equinox occurs on or about September 23 in the northern hemisphere; this also signifies spring in the southern hemisphere.

B

❐ backscattering

Process by which up to 25% of radiant energy from the sun is reflected or scattered away from the surface by clouds. Serves the greatest importance in the Atmospheric Heat Budget. Large errors in the assumed value of this variable may have important effects on computer models of the atmosphere.

❐ barometric pressure

The downward pressure, at any given point in the atmosphere, of the gases directly above that point. Average pressure globally at sea level is 1,013,000 dynes per centimeter squared or 760 torr. This is defined as one atmosphere.

❐ beer's law (beer-lambert law)

For monochromatic radiation, absorbance (A) is determined by the relationship: $A = abc$, with a = absorptivity, b = path length through the medium, and c = concentration of the absorbing species. The intensity of a ray of light which has gone through a medium.

❐ benthic

Term denoting the lower section of a surface or bottom dwelling. The largest part of all biological life are benthic organisms in the greater part of the atmosphere.

❐ benzene

is an aromatic hydrocarbon. It can be found in the air by several different ways of transmission. It can be produced for use with plastic or produced through the burning of fossil fuels. Benzene can also be found in the soil as well as some areas of groundwater pollution. Exposure to benzene has been linked to leukemia.

❐ beta particle decay

A radioactive process in which the atomic number, Z, of the radioactive nuclide changes but the mass number, A, does not. In electron emission, Z increases. In positron emission or in electron capture, Z decreases.

❑ biogeochemical carbon

The biological and geographic study of the properties of carbon's chemical properties in relation to other reactive gases in the atmosphere.

❑ biological amplification or bioaccumulation

Increase in concentration of DDT, PCB, and other slowly degradable, fat-soluble chemicals in organisms at successively higher levels.

❑ biomass

The complete dry weight of organic material found in the biosphere or less strictly, the matter in the biosphere that is contained in living organisms.

❑ biomass burning

The process of oxidizing living material. This process produces atmospheric particulates as well as the production of greenhouse and reactive tropospheric gases. These gases include CO_2, CO, NOx, CH_4, CH_3Cl along with the addition of black carbon. All of these chemical species can be lofted relatively high in the atmosphere due to the convective heating of a fire.

❑ biosphere

A volume including the lower part of the troposphere (as high as living organisms can fly or be lofted) and the surface of the earth including the oceans. This region, by definition, encompasses all the living matter of the earth. Some very important atmospheric chemicals are produced in this region and pass into the atmosphere. This region exchanges chemicals and particulate matter with the atmosphere and soils and waters of the earth.

❑ blackbody radiation

Any physical body absorbs and emits electromagnetic radiation when its temperature is above absolute zero. Planck's law determines the radiant flux of a body at a specific wavelength. In atmospheric chemistry, the calculation involving the earth's blackbody radiation shows that the earth's surface temperature would be below the freezing point of water if it did not have an atmosphere which absorbed some of the outgoing radiation.

❑ bolide impacts

Asteroids or comets striking the earth; a possible cause of major climate changes and mass extinctions in the Earth's history.

❑ bond stretching vibrations

The oscillatory motion of two bonded atoms relative to each other. This motion consist of the two bonded atoms stretching passed their equilibrium position, then returning to their equilibrium position, and finally contracting passed their equilibrium position. The result is an increase in frequency of the bond which allows it to absorb light at that same frequency.

❑ boundary layer

An area in the troposphere which is affected by the solar heating, radioactive cooling, and surface friction of the earth. The height of the boundary layer can range anywhere from 100m to 300 km and coincides with the height at which pollutants are mixed.

❑ bromochlorodifluoromethane

Chemical formula $CBrClF_2$ Synonyms: Halon 1211, Freon 12B1, chlorodifluorobromomethane. This chemical belongs to the freon family. Halon 1211 was introduced in the 1960s as an effective gaseous fire suppression agent for application in the protection of computer control rooms, museums, telecommunication switches and other areas containing highly valuable materials. Results of studies done in the late 1980s indicated that the agent was an ozone depleting chemical. Currently FM-200 and Inergen have been approved as alternatives for its use by the EPA. This chemical is stable in the troposphere but photodissociates to yield halogen radicals, which can catalytically destroy stratospheric ozone.

❑ bromomethane

a volatile compound with a high mixing rate in the atmosphere. Methyl bromide escapes easily into the atmosphere where it contributes to the depletion of ozone. It works similarly to CFCs in its reaction with ozone molecules (O3). In the stratosphere, with the Sunís UV light as a catalyst, methyl bromide breaks down and exchanges a bromide ion for oxygen from ozone. By breaking down the ozone molecules in the stratosphere, methyl bromide thins the ozone layer and lets more UV light pass through.

❑ bromotrifluoromethane

Also known as Halon-1301, this chemical with formula CBrF3 is one of the most ozone destructive substances known to man. It is listed as a Class I ozone depleting chemical in the United States Clean Air Act amendments of 1990. This chemical is commonly used in fire fighting equipment used around sensitive electronic equipment such as computer rooms, telecommunication centers, and aviation equipment. It is know to cause headache and unconsciousness in humans.

❑ brownian motion

Three-dimensional, random movement of particles in a gas or liquid.

C

❑ carbon 14

An isotope of carbon 12 (containing two more neutrons) that is radioactive and used in carbon dating.

❑ carbon cycle

A complex cycle that circulates carbon through the atmosphere, oceans, and land which includes vegetation and soil and carbon is in various forms and oxidation states throughout the cycle.

❐ carbon dioxide

CO_2, a compound consisting of one carbon and two oxygens. It is a reactant in photosynthesis and necessary for plant life. Abundant in the atmosphere due to anthropogenic and natural activities. It is a greenhouse gas.

❐ carbon disulfide

chemical formula CS_2; a compound used to manufacture products such as rayon and cellophane. Carbon disulfide is produced naturally by microbial activity in marshes and in volcanic ash. Since carbon disulfide does not adhere well to sediments, if it comes into contact with soil, it could percolate into groundwater, where it is very soluble. CS_2 can also add to photochemical smog development when it reacts with other organic substances in the atmosphere, such as methane or oxides of nitrogen.

❐ carbon monoxide

CO, a toxic, odorless, colorless gas produced during fossil fuel or biomass burning. Compound consisting of one carbon and one oxygen. Except for carbon dioxide, it is one of the longest lived naturally occurring atmospheric carbon compounds (this wording is meant to exclude chlorofluorocarbons). The recent change in tropospheric CO content may portend a change in the balance between oxidants and reductants in the atmosphere.

❐ carbon tetrachloride

CCl_4, a compound consisting of a carbon and 4 chlorines that is active in ozone depletion when the compound is broken down and releases chlorine atoms (radicals). Chlorine reacts with the ozone creating diatomic oxygen and chlorine monoxide which cycles back to chlorine radicals.

❐ carbon tetrafluoride (CF_4)

Known as Freon 14, carbon tetrafluoride is a stable, non-flammable, colorless gas. It absorbs light at 8 mm (in the infrared), is very stable, and does not react with water. This compound is theoretically considered a likely candidate for warming the earth during the next ice age.

❐ carboniferous period

The time period between 280-345 Myr BP of Earth's geologic history. Characterized by glacial onsets and melting and massive migration and extinctions of species during this period.

❐ carbonyl sulfide

COS, a gas that is very stable and unreactive in the troposphere, but, it is thought, photolyzes to form carbon monoxide, CO, and sulfur, S, in the stratosphere. Through stratospheric chemical reactions, the sulfur atoms are converted to SO_2 and H_2SO_4 which form sulfate aerosol and cloud condensation nuclei, but which eventually settle into the troposphere and reacts to form sulfuric acid, a component in acid rain. The major biospheric sources of COS are thought to be biological.

❑ catalytic converter

An air pollution control device using the exhaust system of cars. The converter helps complete combustion of any fuel that was not burned in the engine and reduce the presence of other harmful emission concentrations. The converter changes the unburned hydrocarbons and carbon monoxide in the exhaust into carbon dioxide and water vapor. The converter use chemical catalysis to create this change.

❑ chappuis band of ozone absorption

The wavelengths of light absorbed by ozone in the visible region, approximately 380 to 750 nm. Compare to ozone's ultraviolet absorption region, the Hartley-Huggins band.

❑ chemical lifetime

The length of time a chemical species can survive without reacting, photolyzing, dissociating, or otherwise changing into another chemical species. Highly reactive chemicals have short lifetimes.

❑ chloracne

A painful, disfiguring condition similar to common acne that it caused by people being exposed to high concentrations of PCB's (polychlorinated biphenyls). It is a biological response to the exposure of different types of organochlorine compounds.

❑ chlorine

Cl_2, in the stratosphere, atomic (radical) chlorine is very destructive because it depletes the greatly needed ozone layer which protects the earth from ultraviolet radiation from the sun. in the Antarctic stratosphere, molecular chlorine along with nitric acid are formed by the reaction of hydrogen chloride and chlorine nitrate--both stratospheric chlorine reservoir species. This process occurs on polar stratospheric clouds which serve as the reaction sites. Once formed, Cl_2 vaporizes into the surrounding air as nitric acid--also formed in that process--binds with the ice matrix. Cl_2 is then photodissociated in sunlight (lamda $<=$ 450 nm) into chlorine radicals. These chlorine radicals then catalytic destroy ozone.

❑ chlorine atoms

Cl, the seventeenth element in the periodic table of elements. It has a atomic weight of 35.453 grams per mole. It has 17 protons in its nucleus and 7 electrons in its outer shell and is a very effective catalyst in the reaction that breaks down ozone in the stratosphere over Antarctica.

❑ chlorine dioxide

ClO_2, a radical, undergoes photodecomposition in the stratosphere where the products of this reaction react with ozone. Since this is a photochemical reaction it only takes place while the sun is up. Experiments over Antarctica have shown a direct relation between polar ozone loss and the increase in halocarbon chemistry, which comes from anthropogenic sources. Scientist are currently looking at the molecular behavior of chlorine dioxide in the atmosphere in order to understand its role in depletion of ozone more thoroughly.

❐ chlorine monoxide

ClO, a radical species which plays an important role in the breakdown of stratospheric ozone over Antarctica. Formed by the photolysis of CFCs in the stratosphere and the subsequent destruction of an ozone molecule, these radicals can act as a catalyst in the destruction of ozone while not being destroyed themselves. ClO, reacting with a oxygen atom (present from the Chapman Mechanism), releases a free chlorine radical once again. As a result, one Cl atom can destroy thousands of ozone molecules before being sequestered as HCl or another reservoir species.

❐ chlorine nitrate

$ClONO_2$, this is a stratospheric reservoir species for chlorine and nitrogen, two of the catalysts in the breakdown of ozone. Frankly, it is named in a confusing manner; it is formed from the reaction of chlorine monoxide and nitrogen dioxide (not chlorine atoms with nitrate). It reacts with HCl at low temperatures on the surfaces of polar stratospheric clouds (PSCs over Antarctica and probably also in the stratosphere over the Arctic). That normally slow reaction heterogeneously produces molecular chlorine and nitric acid. The former outgases from the PSC surface and is quickly photolyzed by 450 nm or shorter wavelength light to form chlorine radicals which rapidly catalyze the breakdown of ozone.

❐ chlorobenzene

C_6H_5Cl, a colorless liquid that is manufactured for use as a solvent. It quickly evaporates in the air and is degraded by hydroxyl radicals that are produced photochemically. The gas acts as a source of ClOx, which helps in the breakdown of stratospheric ozone.

❐ chlorodifluoromethane

Formula $CHClF_2$, this chemical is released into the atmosphere directly from aerosol cans. The use of this chemical has been banned in aerosol cans in the United States, but not in many parts of the world. There are research efforts to find chemical replacements for these compounds because they are know to reach the stratosphere and destroy ozone.

❐ chloroethane

This manmade VOC is highly reactive in the atmosphere. It is a gas at room temperature and when released, it readily reacts with oxidizing agents. This releases the chlorine atoms to circulate and break down ozone in the troposphere. The carbon and hydrogen atoms are released to be further oxidized to create CO_2 and HO. Chloroethane is often used in the manufacturing and production of insecticides, dyes and drugs.

❐ chlorofluorocarbons-cfcs

Very stable chemical compound, used in refrigerants, solvent, and (in the past in the U.S.) aerosols, which release chlorine (important) and fluorine (less important) into the upper atmosphere. In the stratosphere, CFCs are photolyzed (by incoming solar UV) to form carbon dioxide, CO_2, hydrogen fluoride, HF, and ultimately (after multiple UV absorption events) chlorine radicals. These chlorine species are crucial in the destruction of the ozone layer over Antarctica and probably elsewhere.

❐ chloroform

$CHCl_3$, a colorless liquid that evaporates easily into the air. The compound is released into the air by direct and indirect sources and breaks down in the lower atmosphere into carbon dioxide, phosgene (carbonic dichloride), and hydrogen chloride. The degradation occurs in the troposphere by the reaction of the compound with hydroxyl radicals.

❐ cirrus clouds

High clouds that are formed entirely from ice crystals. They appear delicate and wispy and can reach a height of 35,000 feet (10,700 meters). Other types of cirrus clouds include cirrostratus and cirrocumulus. Cirrostratus is a thin sheet of cloud that often causes a halo to appear around the sun or moon. Cirrocumulus look like many small tufts of cotton; however, these clouds rarely form.

❐ clathrates

Also called gas hydrates, formed by or having molecules which are interlaced in a lattice-like geometrical pattern.

❐ clean air act

The Clean Air Act passed in 1970 and later in November of 1990 made into law established nationwide levels of acceptable air pollution from automobiles, individuals, and industry. The Environmental Protection Agency is responsible for enforcement of standards and regulations of the Clean Air Act.

❐ climate

Determined by the daily weather interactions over many years. Characteristics used in determining climate are temperature, precipitation, humidity, sunshine, and cloudiness, wind, and air pressure. Climatologists describe climate in terms of average temperature and precipitation amounts.

❐ cloud

Condensed water vapor floating in air. They can take many different shapes due to wind patterns and moisture content. They play an important part in the world's weather because of the water they bring and because of their radiative properties vis a vis global warming.

❐ cloud condensation nuclei (ccn)

Condensed water vapor that is so small that it can only be seen through a microscope. CCN are actually the center of the droplet. Many of these nuclei are tiny salt particles, sulfate or nitrate aerosol, or small particles present in smoke.

❐ column ozone

The total amount of ozone that is found in a column of air. The majority of this amount of O_3 is typically found in the stratosphere.

❒ condensation

(Gaseous) water vapor that begins to change to tiny water droplets (a liquid state) or ice crystals when the air gets cold enough. This process begins at the dew point. Energy required to vaporize the water is released, about 585 cal/g of water at 20 degrees Celsius.

❒ continental drift

A term applied to early theories supporting the possibility that the continents are in motion over the Earth's surface.

❒ convection

Process by which, in a fluid being heated, the warmer part of the mass will rise and the cooler portions sink. This is also a component of the theory for continental drift, in that the circulating movements of crustal materials push the continents apart.

❒ coriolis force

A fictional source associated with the earth's rotation. It results in the deflection of all objects not at the equator to the right in the direction of motion in the northern hemisphere and to the left in the southern hemisphere.

❒ cryosphere

The portion of the earth which consists of the ice masses and snow deposits which include continental ice sheets, mountainous glaciers, sea ice, surface snow cover and lake/river ice. Alterations in the extent of snow cover are due to seasonal fluctuations and are interrelated with atmospheric circulation. Sea level and hydrologic cycle variations can affect the volume of water tied up in the glaciers and ice sheets.

❒ crystallization

Physical or chemical process or action which results in the formation of regularly-shaped, -sized, and -patterned solid forms known as crystals.

❒ cumulus clouds

Clouds forming in the troposphere which are vertically formed with flat bases and fluffy, rounded tops. They have often been described as cauliflower-like in structure. They occur at heights of 500-6000 meters in elevation from the earth and most often occur scattered or in dense heaped packs. They are formed due to buoyant upward convection during warm, anti-cyclonic summer weather.

❒ cycle

In the atmosphere or biosphere a sequence of events in repetitive motion in which the final output feeds back into the initial input. Examples of this include biogeochemical cycles, including the nitrogen, carbon, and sulfur cycles in which these species are chemically processed in gas, solid, and solution phase by physical and biological processes which change their form, oxidation state, and physical state.

D

❑ ddt

Para-dichlorodiphenyltrichloroethane was commonly used during the second World War. It was used to kill mosquitoes that carried malaria and yellow fever and it was also used to kill body lice that can transmit typhus. Many problems have been associated with the use of DDT. Some of these problems are the bioaccumulation in the food chain, decline of smaller animals, and softening of birdsí eggs. In 1973, the EPA banned all use except those essential to public health. DDT is still being used extinsively in developing countries.

❑ deforestation

Type of "forestry practice" which involves the permanent removal of forests and their undergrowth so that the land can serve another purpose. This practice has had a profound effect on global environmental problems (air pollution, global warming), soil erosion, desertification, sedimentation of water courses, alteration of climate and hydrological cycles, alteration of the atmospheric oxygen and carbon dioxide balance and has caused many species to become extinct, reducing worldwide biodiversity. A common deforestation practice is to simply burn the forest while clearing the land for a seasonal crop. This also creates a large-scale release of carbon dioxide further compounding the global warming effect. Since most of the available nutrients are contained in the forest biomass the newly cleared land does not support a crop for more than a few seasons before adjacent land must be deforested.

❑ denitrification

Step in the nitrogen cycle which involves the reduction of nitrates into nitrite, nitrous oxide, ammonia or elemental nitrogen. It is carried out by certain forms of denitrifying bacteria in the soil and serves as an important part of the breakdown of dead organism. It is responsible for the loss of much of the soil's natural and synthetic fertilizers. This process is favored most in warm, anaerobic conditions.

❑ dewpoint

The temperature at which air is cooled enough to reach its saturation point (100% humidity) for a given volume of water vapor in the air. If the air cooling occurs because of adiabatic expansion, clouds are formed; if it is due to contact of the air with a colder object, condensation (known as dew) is formed on the object; and if it is due to mixing of a warmer, moist air parcel with a cooler, dry parcel, fogs and condensation trails are created. Unlike humidity, the dew point is not based upon the temperature or pressure of the air, but solely upon the water vapor content. It can be measured using a dew point hygrometer, from the known temperature and relative humidity and by measuring the difference in wet and dry bulb temperatures. It is of importance in predicting fogs, frost, dew and overnight minimum temperatures.

❑ dibromotetrafluoroethane

A chemical also know as Halon-2402 which has the chemical formula $C_2F_4Br_2$. It is a known ozone depleting chemical and is listed as a Class I ozone depleting chemical in the United States Clean Air Act amendments of 1990. It is commonly used as a fire extinguishing agent. The reason it has so much ozone depleting potential is because it contains bromine, which has many times the ozone depleting potential of chlorine.

❑ dichlorodifluoromethane CCL_2F_2,

This compound, also known as Freon 12, is the primary CFC that has been linked to the greenhouse effect. Even though the compound is found in very small concentrations, its properties make it a significant factor in stratospheric chemistry. CCl_2F_2 absorbs UV light in the upper stratosphere where it arrives decades after being released in the troposphere. CFCs, which were originally created by Dupont corporation, were used as refrigerants for everything from air conditioning to home refrigerators because they are non-toxic and extremely unreactive (in the troposphere). It is this unreactivity that makes them such a problem for the atmosphere, because they are able to last long enough to make it to the stratosphere, where they cause so many problems. CCl_2F_2 production, along with all CFCs, has been banned since the mid 90s, but some scientists believe that the damage has already been done.

❑ dichloromethane (CH_2CL_2)

Dichloromethane is a clear, colorless organic solvent that has a sweet but mild odor and does not occur naturally in the environment. Its principle use is as a solvent in paint removers and as an aerosol propellant in paints and automotive spray products. In addition, it is used as a blowing and cleaning agent of metal, as a solvent in the production of polycarbonate resins, and in film processing, as well as many other industrial applications. In the food industry, it is used as an extraction solvent for caffeine, spices, and hops. All of these applications result in the release of dichloromethane into the environment, with the primary sources being the industries that manufacture the chemical or use it in production, such as the plastic and synthetics manufacturers, as well as the electronics industry. Upon release of the liquid chemical into the environment, dichloromethane quickly evaporates to a gas, where it is then degraded by photochemically produced products in the air. Because of its reactivity in the troposphere, it is not believed to react with ozone in the upper atmosphere. Despite its solubility in water, it has low acute toxicity to aquatic animals. In animals (as well as humans), dichloromethane is metabolized into carbon monoxide, which has a higher affinity for the heme groups in hemoglobin than does oxygen and results in deprivation of oxygen. Interestingly, dichloromethane has been found to be an antisickling agent in sickle cell hemoglobin.

❑ dimethyl sulfide

Released by bacteria on the continents and in the oceans. Oxidized in the marine atmosphere to partially form cloud condensation nuclei and this may effect the formation of clouds over the oceans.

❑ dinitrogen pentoxide

A compound that contributes to ozone depletion during springtime in the Antarctic. This occurs because ozone converts some nitrogen dioxide radical to nitrogen trioxide radical, which combines with other dinitrogen radicals to produce dinitrogen pentoxide. The reaction is reversible except in the presents of stratospheric liquid droplets where nitric acid occurs. This reaction ties up the nitrogen dioxide radical allowing more chlorine atoms to be present in the active form.

❐ dioxins

Family of 75 different toxic chlorinated hydrocarbons formed as by-products in chemical reactions involving chlorine and hydrocarbons, usually at high temperatures.

❐ disproportionation

A reaction in which an element is oxidized and reduced. This is how stable molecules are made from radicals.

❐ dissolution

The transfer of gas to the surface of a water droplet from the bulk atmosphere, and the gas mixing within the droplet.

❐ diurnal ozone concentration

The ozone concentration in the troposphere is directly related to the pollutants that are in the air and the time of day. For example, during the early hours of the day many pollutants are released into the air. As the day progresses, the sun becomes more overhead and the concentration of the sunlight increases. This solar radiation causes chemical reactions to occur and the primary pollutants decrease while the concentration of ozone increases. Eventually the ozone breaks back down and the process is ready to proceed the next day.

❐ dobson units

Measurement unit for determining the total amount of ozone present in a vertical column of air above the surface of the earth. An air layer at atmospheric pressure of 1013 hPa and temperature of 298 K which measures 1 mm in thickness and is equivalent to 100 dobson units.

❐ doldrums

Calm, light winds, or squalls lying along the length of the equator. Winds which are part of the general circulation of the atmosphere, which are driven by the input of solar energy and modified by the rotation of the Earth. It is here that 17th and 18th century sailing ships had the most trouble making way and lack of fresh water or food sometimes spelled disaster.

❐ dry adiabatic lapse rate

Rate at which unsaturated air cools as it travels vertically, provided that all temperature change is adiabatic (without heat exchange), and no condensation occurs. In dry air it can be approximated as 1 degree Celsius per 100 meters of rise. It can be used as a basis of comparison for actual temperature profiles of air and can help predict smoke stack gas dispersion characteristics.

❐ dry deposition

The transfer of trace species (gases or particles) from the atmosphere to surfaces as a consequence of molecular diffusion, brownian diffusion, or gravitation settling, in the absence of active precipitation. The term refers to the transfer process and not to the surface themselves, which may be moist.

E

❐ ecliptic

The plane of orbit of the earth around the sun. An animation showing this process is available here.

❐ ecology

Branch of biology which focuses on the relationships among various species and among the species and their environments. The basic units for study are the species (all the organisms which are capable of interbreeding), population (all of the members of a species occupying a certain geographical area) and community (number of populations interacting within a certain area). The emphasis in environmental science, however, is on the processes of energy flows and nutrient cycles within these communities and ecosystems and involves such ideas as species abundance, biodiversity, complexity, trophic (feeding status) levels, ecological niches, and structure of communities.

❐ eddy diffusion

The exchange of gaseous components of the atmosphere in a turbulent flow by rapid mixing of fluid eddies.

❐ efflorescence

The reverse process of deliquescence; the drying of a salt solution when the vapor pressure of the saturated solution of a substance is greater than that of the ambient air.

❐ effusion

The movement of gas molecules through an opening that has relatively large holes.

❐ el nino

A condition caused by the decrease in atmospheric pressure over the Eastern Pacific Ocean, weakening the prevailing westerly winds and resulting in warm waters and less nutrient replacement from cold, deeper waters into the Eastern Pacific along the coast of South America. El Nino creates strange weather patterns worldwide sometimes causing flooding in California and droughts in South America simultaneously. The effect gets it name from the Spanish for the Christ Child--the little one--because it is experienced in December by the local fisherman off the western coast of South America as a poor fishing season.

❐ el nino-southern oscillation (enso)

The relationship of the El Nino conditions and the Southern Oscillation pattern (The Walker Circulation). A high pressure over the Pacific Ocean and low pressure over the East Indies causes heavy rainfall across the Pacific, rise in the sea level and increased surface temperatures.

❐ electron capture detector

One of the most sensitive gas chromatographic detector for halogen-containing compounds like chlorofluorocarbons. James Lovelock's early work with this detector led to the discovery that anthropogenic chlorofluorocarbons used most heavily in the Northern hemisphere were very long-lived and therefore well-mixed throughout the troposphere. Later work by Rowland and Molina suggested that CFCs also difussed into the stratosphere where their chlorine atoms could by freed by photolysis-CFC bond breaking by UV light. The result of this is the destruction of ozone molecules in the important ozone layer of the stratosphere.

❐ electrostatic precipitator

A device that reduces pollution from smokestacks. It gives the particles in smoke a electrical charge. The charged particles are then collected on filters that have the opposite charge and thereby removed from the emission stream.

❐ elliptic

An object having the shape of an ellipse: a squashed or elongated circle. The sun follows an elliptic path across the sky.

❐ emission

The release or giving out of a gas, light or heat. An automobile releases uncombusted or partially oxidized hydrocarbon emissions from its tailpipe when the engine is running. Atoms in an excited state often emit light in order to reach a lower energy state. These emissions can be used to determine the presence of the atoms in a sample.

❐ emission spectrum

a spectrum of electromagnetic radiation containing narrow spectral lines from an element.

❐ endothermic

A chemical reaction in which heat is absorbed from the surrounding environment.

❐ enthalpy

A thermodynamic variable that is a system's internal energy plus the product of its pressure and volume: $H = E + PV$.

❐ equatorial rain forests

An area located near the equator worldwide, covering seven percent of the earth's surface, characterized by high precipitation, absent or short dry season, high tree and vegetative growth rates, high humidity, high photosynthetic rates and extreme biodiversity. The equatorial rain forests produce much of the oxygen in the Earth's atmosphere and absorbs large quantities of carbon dioxide from the air.

❐ equilibrium-constant expressions

These are algebraic equations that describe the concentration relationships that exist among reactants and products at equilibrium. Among other things, equilibrium-constant expressions permit calculation of the error resulting from the quantity of unreacted analyte that remains when a steady state has be reached.

❐ equinox

Twice during the year, September 21 and March 21, the length of day and night are equal because the tilt of the Earth's axis (in relationship to the sun) is nullified and both the Northern and Southern Hemispheres receive equal quantities of sunlight.

❐ ethane

A hydrocarbon found in the atmosphere whose primary fate is to react with free radicals such as chlorine and nitrous oxide. This reaction would prevent these radicals from reacting with ozone which is needed in the stratosphere to block ultraviolet light from reaching the surface of the earth.

❐ ethanol

Also known as ethyl alcohol or alcohol, (C_2H_5OH) is an oxygenated hydrocarbon that can be burned as a fuel, or blended into gasoline. Thought of as an alternative fuel and clean fuel, ethanol is a renewable energy source that is made from corn or other grains. If blended into gasoline, ethanol helps with the combustion process and therefore less unburned hydrocarbons form. Ethanol blended fuels competed with methyltertbutylether (MTBE), but because of the cost and the fact MTBE can be shipped through existing pipelines, MTBE was chosen. MTBE is now being banned because of its ability to travel in ground water and the bad taste it imparts to drinking water, because of this ethanol is once again being put back into the gas.

❐ eustatic changes

A consistent change in the sea level that affect oceans, throughout the world.

❐ evaporation

The process of the change in the state of a liquid or solid to a gas or vapor. Vanishing of the surface of a liquid to the atmosphere.

❐ evapotranspiration

The total water loss due to the transpiration of vegetation plus the evaporation from the soil; higher climatic temperatures result in a higher evapotranspiration rate.

❐ exothermic

A chemical reaction associated with energy, or heat that is released into the surroundings.

❐ extinction coefficient

A measure of the rate of the reduction of transmitted light through a substance.

F

❐ fauna

Animals or animal life of a particular region or a particular time.

❑ feedback cycle

The increase of the input into a system, which produces a negative or a positive output in a system of interlocking components; the return of a part or the effects of the output to the input. A proposed positive feedback cycle involves the outgassing of carbon dioxide from the planet's oceans: Since CO_2 solubility in water DECREASES with temperature and since CO_2 is a greenhouse gas, a positive (and possibly runaway) feedback cycle might involve the following: increased greenhouse warming of the atmosphere by IR absorbing CO_2 in the gas phase and the subsequent heating of the oceans could increase the amount of carbon dioxide released into the atmosphere from the oceans. This might thereby increase greenhouse warming, which would, in turn, again heat the atmosphere and the ocean, etc. Though the process is no doubt more complicated than this, a runaway greenhouse effect is usually given as the reason for Venus' high surface temperatures.

❑ firn

Granular snow that has endured a summer without melting and will transform into glacial ice.

❑ flora

A plant or plant life of a specific region or particular period.

❑ flux

The rate of continuous change, flow or movement of liquid, particles or energy. The rate of discharge of a liquid, removal of energy or particle depositing from one body to another.

❑ foehn

A warm, dry wind on the lee side of a mountain range that owes its relatively high temperatures largely to adiabatic heating during descent down mountain slopes. In Zurich Switzerland, this means a warm but clear view of the Alps.

❑ fog

The state in which particles of very small water droplets are in contact or close to the earth's surface, generally reducing horizontal visibility.

❑ formaldehyde

CH_2O, A colorless, pungent, carcinogenic, and toxic gas which can irritate membranes. It is used as formalin in its aqueous state (37 %). Formaldehyde is reactive due to its carbonyl functional group.

❑ formic acid

CH_2O_2, A weak acid that photolytically reacts in the atmosphere forming COOH radical and H radical. This compound often reacts with the hydroxyl radical and atomic oxygen to form formic acid and a hydroperoxy radical. This is a sink for the hydroxyl radical and it also helps drive other reactions with the addition of hydroperoxy radicals to the atmosphere.

❒ fossil fuels

There are coal, crude oils, oil shales, tar sands and natural gases such as butane, ethane, methane which occur naturally from the decomposition of plant and sea and land organisms over millions of years. These natural resources contain stored energy from the sun which is released upon combustion. These fuels also release various types of pollutants such as sulfur dioxide, carbon dioxide and nitrogen dioxide when burned.

❒ free radicals

Highly reactive atoms or molecules with incomplete (electronic) octets and therefore uneven numbers of electrons. (In the case of hydrogen radical this is an incomplete duet.) Free radicals species are very electrophilic, will abstract atoms from other molecules to complete their octets, and will, in the process, generate new radicals. In the atmosphere, most free radical species have short life times; however, they can promote the conversion of oxygen to ozone and also take part in the catalytic cycle of ozone destruction.

❒ freons

Stable liquids or gases usually produced for solvents, aerosol propellants, refrigeration, air conditioner or styrofoam purposes. These high vapor pressure liquids/gases are chemically stable and have long atmospheric lifetimes; therefore, they can eventually become well-mixed in the troposphere and ultimately diffuse into the stratosphere. Upon encountering high energy UV light in the upper stratosphere these species are photo dissociated and ultimately release all of their atoms there. The free radical chlorines (or bromines in the case of some halons) catalytically destroy ozone (see chlorine) in the stratosphere.

❒ frontal wedging

The lifting of air resulting when cool air acts as a barrier over which warmer, lighter air will rise.

G

❒ gaia hypothesis

An hypothesis proposed during the early 1970s by James Lovelock which states that all living organisms have the ability to affect their surroundings such as the atmosphere, lithosphere, and climate to maximize its biological success. The hypothesis connects the evolution and survival of a species to the evolution and conditions of its environment.

❒ general circulation models

Three dimensional computer simulations of climate and weather which, through calculations, are used to investigate gravity wave fields. The models may help to predict atmospheric chemistry distribution and meteorological trends for the future.

❒ geologic time scale

A time scale in which the earth and its atmosphere are dated. It is broken into two division; Relative Time: Determines the age relative to other surrounding objects; Absolute Time: Obtained with radiometric dating of the object.

❒ gigayears

Time measured in billions of years; a common measure of geologic time.

❒ glaciation

The process of forming glaciers.

❒ glacier

A slow-moving mass of ice formed in higher latitudes and elevations. When snowfall is greater than melting and the increasing amounts of snow become compacted and pressurized it forms firn (see above) and ultimately a glacier. As the glacier moves, it carries rocks and soil, and can form u-shaped valleys over geologic time. Glaciers can move at variable rates.

❒ global climate model

A computer program that is used by atmospheric scientists to study and predict world-wide trends in weather patterns and the forces and physical laws that affect climate. The formulae are complicated and require skilled mathematicians, atmospheric scientists, and powerful computers for execution. Five regimes that influence climate are often taken into consideration: atmosphere, hydrosphere, biosphere, cryosphere, and pedosphere. Atmospheric and environmental physical laws as well as principles of thermodynamics and hydrodynamics are all used as part of the extensive mathematical parameters for the program. The results of these models are used as reference for other models, policy decisions, and by meteorologists.

❒ global warming

The overall increase of atmospheric temperature due to a buildup of greenhouse gases.

❒ glycolic aldehyde

($C_2H_4O_2$), also called hydroxyacetaldehyde, is formed during the oxidation of ethane and isoprene, and also by the emission of biomass fires. It is the simplest form of sugar. Glycol aldehydes have also been found in trace amounts in dust clouds near the Milky Way of outer space.

❒ gondwanaland

A hypothetical super continent comprising approximately the present continents of the southern hemisphere.

❒ gravity

A natural force exerted on a body toward the earth's surface or between two bodies. Gravity has a strong effect on how matter interacts. Gravitational attraction depends on the masses of the bodies involved and the distance between them.

Centrifugal force caused by the earth's rotation on its axis tends to lessen the force of gravity.

❐ greenhouse effect

The phenomenon in which outgoing infrared radiation that would normally exit from a planet's atmosphere but instead, is trapped or reflected because of the presence of the atmosphere and its components is called the greenhouse effect. It has been calculated that this effect is necessary to maintain the earth's climate and surface temperature and, more importantly, the liquid state of water in the majority of the earth's biosphere; however, the best scientific estimates to date suggest that increasing amounts of greenhouse gases are resulting in higher temperatures worldwide. This could result in melting of icecaps that would raise the sea level and cause devastating floods in coastal areas, more extremes in rainfall and intensity, and the distribution of species in the biosphere.

❐ greenhouse gases

Those atmospheric components that absorb strongly in the infrared region of the spectrum. Infrared radiation is reflected and emitted by the earth's surface as heat and causes a fairly large warming effect when trapped by these gases in the atmosphere. In order of abundance and importance as greenhouse gases are water vapor, carbon dioxide, ozone, nitrous oxide, methane, and chlorofluorocarbons (CFCs). Absorption by water vapor, the most common greenhouse gas, explains why many humid or cloudy days feel much hotter than dry, clear days of the same air temperature. Because of these gases, only about 5% of the radiation escapes from the atmosphere while more than 90% is radiated back to the surface of the earth.

❐ gulf stream

The Gulf Stream is a powerful current in the Atlantic Ocean. Its movement is related to density distribution system of waters and currents. Beginning in the Gulf of Mexico, it flows along the east coast of North America and across the Atlantic to Europe. The Gulf Stream carries warm water from the tropics and warms the climate of the land near where it passes.

H

❐ hadley circulation

proposed by George Hadley in 1735, it is the tropical convection system that describes the atmospheric thermal circulation of air from the equator (0° latitude) poleward to 30° both north and south of the equator. When saturated air at the intertropical convergence zone (ITCZ), located approximately at the equator, is heated, convergence and convection causes it to rise into the upper atmosphere, creating an area of low pressure at the earth's surface. As the pocket of saturated air at the ITCZ rises, it cools, releasing large amounts of precipitation. Rather than remaining in the tropics, the air then flows horizontally towards the higher latitudes. At approximately 30° both north and south of the equator, the pocket of air in the upper atmosphere has sufficiently mixed and cooled enough to return to the surface. To complete the circle, the area of low pressure at the equator pulls air in from the higher latitudes.

❐ halocline

In the oceans, a well-defined vertical gradient of salinity.

hartley-huggins band of ozone absorption

This spectral band in the UV is responsible for the absorption and filtering of solar ultraviolet radiation. Photons with a wavelength longer than 210 nm are weakly absorbed by oxygen in the atmosphere. Ozone then becomes the major UV absorber, with a band of 210-380 nm. Compare to ozone's Chappuis band in the visible wavelength region.

❐ haze

An atmospheric aerosol of sufficient concentration to be visible. The particles are so small that they cannot be seen individually but are still effective in visual range restriction. The Blue Ridge Mountains of the Eastern United States have a haze caused by the natural emissions of hydrocarbons which are oxidized and form aerosols.

❐ heat flux

The amount of heat that is transferred across a surface of unit area in a unit of time.

❐ henry's law

The relationship that defines the partition of a soluble or partially soluble species between the gas and solution phases. It is often represented as Hc. It is the relationship of species dissolving but not reacting in solution to those in equilibrium in the gas phase. The equation used for this is: [Caq] = [Hc]*[Cg]. Hc is the Henry's Law Constant. This may be regarded as a measure of the molecules in a gas to those dissolve in solution, and this relationship is temperature dependent. Cg and Caq are the concentrations or partial pressure of species C in gas and aqueous solution, respectively.

❐ hexafluoroethane (C_2F_6)

An odorless halocarbon that meets all requirements of a greenhouse gas: unreactive to OH· and ozone, does not photodissociate from either UV or visible light, is insoluble in water, and absorbs in the infrared spectrum.

❐ horse latitudes

This is the area between 30° and 35° latitude where the dense cool air descends and the winds become weak. According to some traditions sailors relying solely on wind power threw their horses overboard in hopes that speed would increase due to the resulting lighter load of the ship. Other traditions state that the horses were thrown overboard to avoid depleting their provisions and still another states that this is simply the region where the horses perished.

❐ hplc

High-performance liquid chromatography (HPLC) is used to determine the amount of a compound in various media such as the air, water or blood. Samples of the desired media (air, water, etc.) are dissolved in solution and then the components are separated by injecting the solution through a mobile phase solution and then a stationary phase solution. As the solution passes through the stationary phase, the components are separated out due to different partitioning behaviors between the components in the mobile phase and the stationary phase. This allows each component to be analyzed separately as they pass through the detector.

❒ humic matter

is part of humus, and is a mixture of amorphous, polydispersed colloidal, yellow to brown-black organic substances. They are hydrophyllic, and high in molecular weights, which range from a few hundreds to thousands of daltons. They have originated from the decomposed, non-humic, organic fraction by "new formation," called humification.

❒ humidity

The moisture content of air. Relative humidity is the ratio of the amount of water vapor actually present in the air to the greatest amount possible at that same temperature.

❒ hydrocarbons

Chemicals containing only carbon and hydrogen. These are of prime economic importance because they encompass the constituents of the major fossil fuels, petroleum and natural gas, as well as plastics, waxes, and oils. In urban pollution, these components--along with NOx and sunlight--contribute to the formation of tropospheric ozone.

❒ hydrochlorofluorocarbons

HCFCs, chemical species slated to replace CFCs in the near future in most Western nations. When the normal chlorofluorocarbons (see above) reach the stratosphere and are photo-decomposed, their released chlorine radicals destroy the natural ozone that acts as our umbrella to shield the Earth from ultraviolet radiation (see chlorine and ozone). With one or more hydrogen-carbon bonds, HCFCs are still useful as replacement for CFCs in most applications. However, this bonding structure makes HCFCs much more chemicallyy unstable--as compared to CFCs--and therefore subject to hydroxyl radical and ozone attack early in their gas phase career in the atmosphere. In fact they react in the troposphere instead of the stratosphere. Their atmospheric lifetime is shorter than CFCs and they, therefore, have a smaller chance of reaching the stratosphere where their chlorine could be released by destructive photolysis and enter the catalytic ozone destruction cycle.

❒ hydrofluoric acid (hf)

A clear, corrosive liquid that has an extremely pungent odor and forms dense white vapor clouds if released. Both the liquid and the vapor can cause severe burns to all parts of the body, and medical treatment is required for all exposures. In fact, HF burns covering only 2% of the body can be fatal. If this chemical comes in contact with glass, concrete and other silicon-containing materials, it yields silicon tetrafluoride gas. If it comes into contact with common metals, it will produce hydrogen fluoride, which poses a fire and explosion hazard. In general, this chemical is used to produce environmentally safer fluorocarbon products, chemical derivatives, and fluoropolymers. Environmentally speaking, hydrofluoric acid is much too reactive to ever reach the upper atmosphere, and would not interfere with the ozone layer.

❐ hydrogen bromide

HBr, in the stratosphere, ozone is continuously being formed via the Chapman mechanism. However, halogen radicals such as bromine and chlorine (released from anthropogenic and some natural sources) catalyze the breakdown of ozone which results in a net ozone loss:

$O_3 + BR \rightarrow BRO + O_2$

A reservoir species is a compound whose formation breaks the destruction cycle by taking the halogen radical out of the cycle. Two examples are HCl and HBr. Unfortunately, the success of HBr as a reservoir species is very limited due to the fact that it photodissociates quickly and easily. Thus, the ozone depleting effects of bromine are, atom for atom, more severe than those of chlorine.

❐ hydrogen chloride

HCl, anhydrous hydrochloric acid. Its molecular weight is 36.47 g/mol; it is a colorless, corrosive gas. Hydrogen chloride is a covalent bonded nonflammable gas and ionizes almost completely when dissolved in water. When dissolved in water, hydrogen chloride forms a strong acid, hydrochloric acid. In the stratosphere, HCl acts as a reservoir species temporarily removing chlorine radicals from a catalytic ozone destruction cycle.

❐ hydrogen peroxide

H2O2, this compounds molecular weight is 34.02g/mol; it is a colorless, rather unstable oxidant with a bitter taste and caustic to the skin. Hydrogen peroxide will decompose, liberating oxygen. Pure hydrogen peroxide is stable, but the slightest impurity will enhance decomposition, often violently. Concentrated solutions of hydrogen peroxide are highly corrosive and toxic. H2O2 is used as bleach, deodorizer, and in the manufacturing of rocket fuel. The hydrogen peroxide in your bathroom is approximately 3 percent in water. In the atmosphere this is probably one of the oxidizers for sulfur dioxide in cloud water droplets that produces sulfuric acid, a major component in acid rain.

❐ hydrogen sulfide

H_2S, hydrogen sulfide is a reducing species, which is relatively unstable but survives for rather a long time owing to its slow reaction with atmospheric oxygen. This sulfur-containing gas is a major participant in gas to particle conversion in the atmosphere. Many sulfur-containing gases are reactive and thus are rapidly converted to sulfuric acid.

❐ hydrologic cycle

The recycling of water on earth. There are five steps: Condensation occurs when water vapor in the atmosphere turns into a liquid and forms clouds. The clouds release precipitation in the form of rain, snow, or sleet when the moisture in the clouds becomes too heavy. Infiltration occurs when the water seeps into the ground, The infiltration rate depends on the permeability of the ground. When there is too much water on the surface to infiltrate into the ground, it becomes runoff and ends up in lakes, streams, or oceans. Evapotranspiration is a two part process driven by the sun. Evaporation occurs when water is turned into a vapor from the ground and transpiration is water loss by plants, The water vapor rises and starts the process over again.

❑ **hydrometeors**

Any condensed form of water--larger than a single water molecule--that is falling or suspended in the atmosphere; some examples of hydrometeors are snow, fog, cloud or rain. When these particles reach the size of 10-20 micrometers, available water is depleted and a stable cloud is formed.

❑ **hydroperoxy radical**

HO_2, this radical readily reacts with nitrogen oxides and hydrocarbons in the atmosphere; in urban atmospheres, the major source of the hydroperoxy radical is formaldehyde photolysis.

❑ **hydrophobic**

A substance having a strong tendency to reject association with water.

❑ **hydrosol**

A dispersion of solid particles is a liquid system.

❑ **hydrosphere**

That part of the atmosphere which contains water in the liquid, solid, or gaseous phase.

❑ **hydrothermal chemistry**

Aqueous chemistry under high temperature conditions. In biospheric chemistry, this refers to water whose chemical composition has been altered by deep, hot rock and which tends to come to the surface at fault lines. This creates boiling springs and geysers.

❑ **hydroxyl radical**

OH, a radical consisting of one hydrogen atom and one oxygen atom which does not normally exist in a stable form; this radical readily reacts with methane and carbon monoxide. The source of the hydroxyl radical in the atmosphere occurs primarily through the 1) photolysis of hydrogen peroxide, heat, and light and 2) the attack on water of an excited oxygen radical (created by the photolysis of ozone).

❑ **hygroscopic**

Having the characteristic of drawing moisture from the atmosphere.

❑ **hypsometric diagram**

A graph that shows in any way the relative amounts of Earth's surface at different elevations with regard to sea level.

❒ hypsometry

Hypsometry is a technique of mapping the elevation contour of both land and the ocean floor using tinting to illustrate a change in contour:

I

❒ ice core

Often used by humans to determine the historic chemical composition of the atmosphere; the source of the ice core is the snow which falls on the glacier which consolidates upon freezing and in which gas bubbles which are trapped upon freezing and can be used to determine the atmospheric quality of a number of constituents at that time.

❒ indoor air quality

A measure of the value or comfort of people with the air they breathe inside buildings and homes. "Suitable" indoor air quality can conferred on air in which there are no known contaminants at harmful levels as determined by appropriate authorities or--more likely--air with which 80% or more of the people exposed do not express dissatisfaction.

❒ infrared radiation (ir)

Energy that is emitted in the form of electromagnetic waves at a wavelength greater than about 750 up to approximately 30,000 nanometers. Although the earth absorbs almost all the IR, UV, and visible radiation the hits it, the surface of the earth emits only the longest wavelength radiation of these three, IR, in any significant amounts. It is this re-emission of IR towards space--and its subsequent absorbance, re-emission and scattering--that contributes to the process that heats the atmosphere. This re-emission and subsequent heating is a part of the greenhouse effect.

❒ insolation

The solar radiation incident on a unit horizontal surface at the top of the atmosphere. It is sometimes referred to as solar irradiance. The latitudinal variation of insolation supplies the energy for the general circulation of the atmosphere. Insolation depends on the angle of incidence of the solar beam and on the solar constant.

❒ intertropical convergence zone (itcz)

The location in the atmosphere along both sides of the equator at which the strong upward motion of air (Hadley cell) is characterized by heavy precipitation in convective thunderstorms and by relatively low surface pressure.

❒ inversion

An anomaly in the normal positive atmospheric lapse rate (change of temperature with increasing altitude). This usually refers to a thermal inversion, in which temperature of the atmosphere increases rather than decreases with height.

❐ irradiance

The total radiant flux received on a unit area of a given real or imaginary surface. Also called the radiant flux density.

❐ isoprene

C_5H_5, a colorless volatile liquid which is insoluble in water, soluble in alcohol, and very reactive because of its low vapor pressure and double bonds. From an atmospheric chemistry point of view, isoprene is emitted from trees and plants, globally on the order of teragrams (1012 g) per year. This reactive, natural hydrocarbon may influence the oxidative/reductive balance in the biosphere where its concentrations are significant, in regions such as rainforests and large temporal forests. Most isoprene is directly emitted into the atmosphere from plant surfaces and some no doubt condenses on particulate matter. Emission rates from trees are positively correlated with leaf temperature and therefore, to a degree, time of day. Ultimately the carbon in isoprene is oxidized to carbon monoxide and finally carbon dioxide.

❐ isotopes

Two or more forms of an element that have the same atomic number but different numbers of neutrons in the atomic nucleus and therefore a different atomic weight. Most elements occur naturally as mixtures of isotopes. The symbol for an isotope usually consist of the mass number placed as a leading superscript before the elemental symbol. In atmospheric chemistry, long lived isotopes such as carbon-14 (14C) are used to determine the age of objects that contained living matter. Other isotopes are used to track the routes of air parcels in the atmosphere.

J

❐ jet stream

The major atmospheric air currents centers over the Pacific and Atlantic Oceans, North America and over the Middle East. These systems are fairly robust from period to period, and across different observational data sets. The most important jet stream from our perspective at Sam Houston State University, in Huntsville Texas, USA, is the Pacific Jet, which helps set the storm tracks over the west coast of North America--like most northern hemispheric weather--moving towards the east.

❐ joule (j)

Defined as a SI unit which expresses a unit of energy. 1 J = 1 kg*m2/s2. 3

K

❐ katabatic winds

Strong winds associated with mountains and valleys. They occur when cold, dense air is produced by radiational cooling flows downhill due to the pull of gravity to a lower, warmer region such as depressions and valleys.

❐ kelvin scale

Scale for measuring temperature that sets zero degrees at the point of which molecular motion desists, that is, the point at which there is no heat. It is also known as the absolute scale.

❐ kinetics

The study of the rates at which chemical reactions occur and the influence of physical and chemical conditions of these rates.

❐ krakatua winds

The winds produced as a result of the Krakatoa eruption on the island of Krakatua in 1883. This volcanic blast is the largest in recent times whose effects caused several atmospheric changes such as lowering the global temperature by 1.2 degrees Celsius and changing tides and wind conditions and providing beautiful red sunsets as far away as the United States.

L

❐ lapse rate

The theoretical rate of decrease of temperature with increasing height in the atmosphere. If heat is neither gained nor lost from the air parcel under consideration, then the lapse rate is said to be adiabatic. For air parcels of differing water content this value is different.

❐ latent heat

If a change of state occurs from gas to liquid or liquid to solid, internal energy in the form of heat is released. If a change of state occurs from solid to liquid or liquid to gas, heat is required. Different compounds absorb and release different amounts of latent heat. Water releases nearly 600 kilocalories for each kilogram of water condensed.

❐ lithology

The branch of mineralogy that deals with the study of rocks and their composition.

❐ lithosphere

The outer, rigid shell of the Earth, situated above the asthenosphere and containing the crust, continents and plates.

❐ lpg (liquid petroleum gas)

is a mixture of short hydrocarbons with most of the volume being propane and butane. LPG is considered an alternative fuel that burns cleaner than gasoline because of the very short hydrocarbon chains that tend to not produce a high volume of CO_2, NOx radicals. LPG is now being found in the air in the unburned state most likely from leaking pipes and tanks, in this physical state LPG is now being seen a contributor to urban air pollution.

M

❐ mesopause

The upper limit and the coldest portion of the mesosphere. The transition zone between the mesosphere and the thermosphere.

❐ mesosphere

In the atmosphere, the region immediately above the stratosphere and innemdiately below the thermosphere. The mesosphere begins about 50 kilometers high at the stratopause and ends about 80 kilometers high at the mesopause. The temperature in the mesosphere decreases sharply with increased altitude.

❐ mesozoic period

Period spanning from 225 to 65 million years before present (Myr BP), this period was marked by widespread warmth of the planet's surface, especially in the higher latitudes. During this time the continents were joined in a single super continent. This situation most likely facilitated the meridional transport of heat by ocean currents, thus resulting in an even distribution of global heat.

❐ metastable state

A object that returns to its original position after force is applied.

❐ meteorology

The science of the atmosphere and its direct effects upon the earth's surface. Meteorology is especially concerned with how atmospheric conditions effect the weather.

❐ methane

CH_4, a colorless, odorless, flammable, greenhouse gas. It is the simplest of all hydrocarbons with a formula of CH_4. Methane is released naturally into the air from anaerobic environments such as marshes, swamps, and rice fields, and from symbiotic microbes in the guts of ruminant animals (such as cattle, sheep, and camels), and sewage sludge. Methane is released from methane producing bacteria (methanogens) that live in these anaerobic places. Methanogens in termite guts are the source of methane released by termites. The discovery that termites may be a significant source of atmospheric methane is attributed to work by Patrick Zimmerman and his research group members at the National Center for Atmospheric Research in Boulder Colorado (see reference immediately below and papers in the scientific literature).

❐ methane flux

The (continuous) flow of methane from one place in the atmosphere to another. It is usually measured as a rate of flow across a given area where methane is produced; for instance, the average amount of methane being released from a rice field of a specific area every hour would be considered the methane flux rate for that field.

❑ methanesulfonic acid (msa)

This may be a significant component of atmospheric aerosol particles that form from the atmospheric oxidation of dimethylsulfide (DMS), when this DMS is produced by phytoplankton. These aerosol particles can lead to the formation of cloud condensation nuclei, water droplet formation, and hence clouds and storm formation. This therefore may contribute to the turbulent mixing of oceanic surface waters and subsequent upwelling of nutrients that feed those microorganisms (a positive feedback cycle). It is thought that increased emissions of dimethylsulfide, and subsequent methanesulfonic acid over the oceans from an increase in sea water temperature may have a negative feedback on global warming by increasing cloud formation; although the net effect of clouds on global warming is very unclear at this point.

❑ methyl bromide

CH3Br, this halocarbon is released, to a degree, naturally from the oceans, but is more commonly released from its anthropogenic use as a soil fumigant or pesticide. Methyl bromide is persistent enough to reach the stratosphere where it photochemically decomposes to yield atomic bromine (radical) and proceeds to destroy stratospheric ozone in the same manner as the atomic chlorine radical. On an atom-for-atom basis, stratospheric bromine is more efficient at destroying ozone than is chlorine because the HBr reservoir species is more photochemically active than HCl; however, there is much less of hydrogen bromide in the stratosphere.

❑ methyl chloride

CH3Cl, this compound supplies chlorine to the stratosphere by occasional volcanic eruptions and by tropospheric to stratospheric transport. Methyl chloride is also produced by seaweed. The natural chlorine content of the stratosphere as a consequence of these sources is about 0.6 ppbv.

❑ methyl chloroform

CH_3CCL_3, a synthetic organic compound that has been used as a substitute for earlier solvents that contributed to air pollution. Methyl chloroform is now known to cause destruction to stratospheric ozone and is scheduled to be phased out by the year 2005.

❑ methyl radical

CH_3. is formed, along with water, from the reaction of methane (CH_4) and the hydroxyl radical (OH.). This begins a series of reactions that eventually produces ozone(O_3).

❑ midocean ridge

A major, elevated, linear feature of the seafloor, consisting of many small, slightly offset segments, with a total length of 200 to 200,000 km. This type of plate boundary occurs in a divergence zone, which is a site where two plates are being pulled apart and new oceanic lithosphere is being created.

❐ mie scattering

Processes by which particles of similar size and electrical characteristics separate or disperse different wavelengths (read colors) of light. First described by Gustav Mie in 1808. Since the sun's visible spectrum contains a mixture of (traditionally ordered) red, orange, yellow, green, blue, indigo, and violet colors, these wavelengths are differentially scattered by particles as they travel through the atmosphere. Red--longer wavelength--light is scattered not much and blue--shorted wavelength--light is scattered much more. This is why the sky appears blue: the sun's blue light is scattered back towards your eyes from atmospheric particles so when you look up--that is, away from the sun, the light you see is light scattered to your eye from atmospheric particles. (You might ask yourself why the sky--viewed at an off angle from the sun--isn't black instead of blue.) Mie scattering is also the reason why sunsets appear red: the sun's red light is NOT scattered as much as blue light by atmospheric particles and so solar blue light is scattered away from your eyes on its way from the sun and red light is scattered less. The result is that more solar red light hits your eyes than blue and sunsets appear red. Notice also that the redness of sunsets increases at the amount of atmospheric particles between you and the sun increase, that is as the sun "goes down" mie scattering increases as the amount of particles between you and the sun increases.

❐ mixing ratio

The fixed proportions in which two or more substances may become combined, such as the amount of nitrogen in the air compared to the rest of the air. Atmospheric scientists routinely discuss the gas phase concentrations of trace components in mixing ratios expressed in ratios of VOLUMES, as in "the present tropospheric mixing ratio for methane is approximately 1.7 ppmv."

❐ mobile source

(also call non-point source) consist of emissions from on-road and off-road sources. Cars, light, medium and heavy-duty trucks are considered on-road mobile sources, off-road mobile sources; are tractors, lawn mowers, boats, and airplanes. Although the off road sources usually put out more pollution per hour of work than the on-road, the great number of on-road source put out a larger overall volume into the atmosphere.

❐ molecular oxygen

A molecule that is composed of two oxygen atoms, O_2, and has no color, odor, or taste. It is present in both the atmosphere and dissolved in the oceans and freshwater sources exposed to the atmosphere. Solar radiation with wavelengths less then 242 nm can break it back into oxygen atoms, $O_2 \rightarrow O + O$. One of these oxygen radicals in turn can combine with O_2 to form ozone, $O2 + O \rightarrow O_3$.

❐ monsoon

Heavy winds characterized by a pronounced seasonal change in direction. Winds usually blow from land to sea in the winter, while in the summer, the flow reverses and precipitation is more common. Monsoons are most typical in India and southern Asia.

❐ montreal protocol

A international environmental agreement (one of the first) to prevent the use of substances that are harmful to ozone, such as chlorofluorocarbons and halons, in order to protect the ozone layer in a global manner. It was agreed upon in 1987 and has been amended repeatedly since that time.

❐ mtbe

the compound methyl tertiary butyl ether (MTBE) was blended in gasoline originally in 1979 to serve as an oxygenate. This would theoretically cause the combustion process in internally combustion engines to be more complete. This effort to improve air quality by producing a fuel that was more readily oxidized and decrease the resulting troposheric ozone production was somewhat effective. A problem arises, however, in the fact that MTBE moves extremely well in water (it may actually move upstream because of its high affinity for water molecules), causes fish kills due to its toxicity to fish, and even at low concentrations, it can render large surface or groundwater sources to be useless. Its use has since been banned in California.

N

❐ nacreous clouds

Clouds that occur in the stratosphere at altitudes above 20 km and are usually iridescent and luminous in color. They may also be called mother of pearl clouds.

❐ naphthalene

a naturally occurring volatile organic compound, is a component of petroleum and coal and is released by the burning of wood or tobacco. It is also used in moth repellants, dyes and pharmaceuticals. Like other nonmethane hydrocarbons, it reacts with NO in the atmosphere to produce NO_2. The NO_2 is photolzed to produce NO and O. The O then reacts with O_2 to produce ozone.

❐ national ambient air quality standards (naaqs)

Sets the levels of air quality for the United States, in the Code of Federal Regulations (40 CFR §50.2), to protect the population's health. These are the minimum and might be more stringent from state to state.

❐ national center for atmospheric research (ncar)

A non-profit organization dedicated to furthering understanding of the Earth's atmosphere. Located in Boulder, Co., NCAR is operated by the University Corporation for Atmospheric Research (UCAR) and sponsored by the National Science Foundation (NSF).

❐ national oceanic and atmospheric administration (noaa)

A United States Government agency created in 1970 as part of the United States Department of Commerce to determine how our oceans and atmosphere should be developed, regulated, analyzed, formed, and mined. NOAA tries to achieve these goals with the help of eight major federal services, The National Oceanic Survey, The National Weather Service, The National Fisheries Service, The Environmental Research Laboratories, The Environmental Data Service, The Environmental Satellite Service, The Office of Sea Grant, and The Office of Coastal Zone Management. NOAA has been threatened with extinction in the spring of 1995 with the possible dissolution of the Department of Commerce under the guise of "necessary" budget cuts.

❒ nimbostratus clouds

Cloud formation consisting of dark-gray layers of clouds and occurring at altitudes which produce prolonged rain and snow.

❒ nimbus clouds

Nimbus is a term used to describe a cloud that is precipitating. [Physics Today; v 47; pages 36-42; 1994.] [Journal of the Atmospheric Sciences; v 51; pages 3183-3193; 1994.]

❒ nitric acid

HNO3, this is a corrosive, non-volatile, and inorganic acid. It is a strong acid (dissociates completely in aqueous solution) and is also an oxidizer. In the atmosphere it is formed by the conversion of nitrogen monoxide into nitrogen dioxide, and ultimately into nitric acid: $2NO + O_2 \rightarrow 2NO_2$, $3NO_2 + H_2O \rightarrow 2HNO_3 + NO$ Nitrogen monoxide in this process most often come from (fossil fuel-based) combustion processes that use atmospheric air (containing 78.1% N_2) which combines with atmospheric oxygen in those high temperature combustion process (see nitrogen monoxide and nitrogen dioxide entries below). Nitric acid is highly water soluble. This solubility with water allows easy removal of nitric acid from the troposphere by atmospheric precipitation. Commonly, this is referred to as acid rain or snow. Nitric acid has a relatively low concentration in the atmosphere but provides an important role in the production of sulfuric acid. It acts as a catalyst in the conversion of sulfur dioxide to sulfuric acid. Sulfuric acid is usually formed within cloud droplets by oxidation of dissolved sulfur dioxide again most foten released by fossil fuel combustion.

❒ nitric oxide radical

This radical is a primary pollutant from many sources, the most common and abundant is automobile exhaust. Its importance in the atmosphere is great because it is a catalyst in the destruction of ozone.

❒ nitrification

The conversion of nitrogen organic compounds to inorganic compounds of nitrogen. This conversion is accomplished, in the main, by bacteria.

❒ nitrogen

N_2, a colorless, tasteless, odorless gas which makes up 78.1% of the atmosphere. Atmospheric nitrogen is converted by nitrogen fixation and nitrification into compounds used by plants and animals. In the far upper atmosphere, N_2 is broken down when large numbers of energetic secondary electrons are produced and available to react with the N_2. This leads to the eventual production of NO in that part of the atmosphere and is not--by definition--anthropogenic in nature.

❐ nitrogen dioxide

NO_2, is the lesser of the two emitted NOx gases from high temperature combustion in air. It is an important species in the atmosphere. Since it absorbs in the visible wavelength region--creating the Brown Cloud see over Denver, LA, Mexico City, Beijing, etc.--and can be photolyzed and yield oxygen atoms that can react with molecular oxygen to create ozone, NO_2 and the NO/NO_2 ratio is important in tropospheric chemistry.

❐ nitrogen monoxide (nitric oxide)

NO is the principal emitted NOx gas from high temperature combustion in air. This gas can acts a catalyst in the reactions that cause the destruction of ozone. Reacting with ozone and tropospheric radicals, NO is inextricably linked with the polluted urban atmospheric production of NO_2, ozone, and other secondary pollutants such as peroxyacetyl nitrate.

❐ nitrogen oxides

NOx (pronounced "nox") are produced from high temperature combustion in air. They are nitrogen oxide and nitrogen dioxide.

❐ nitrous oxide

N_2O, this is a by-product of biological activity of a symbiotic bacteria living in leguminous plant roots. It is a principal greenhouse gas that absorbs in the infrared wavelength region and unfortunately falls in an IR "window" between IR absorbing features of water and carbon dioxide (a characteristic of all the "trace" greenhouse gases with significant radiative forcing). It is also laughing gas used in medicine as a gentle general anesthetic.

❐ noctulucent clouds

Relatively unusual wavy, thin, bluish-white clouds that form at altitudes of about 80 to 90 km.

❐ nonmethane hydrocarbons (nmhcs)

These are hydrocarbons such as ethylene, butane, hexane, propane and, by definition, exclude the first member of that analogous series, CH_4. Large quantities of NMHCs are emitted from vegetation, the vast majority as isoprene, C_5H_8. The natural emission of isoprene is significant compared to that of anthropogenic NMHC.

❐ nuclear winter

In the event of a significant nuclear war, researchers (Paul Crutzen and John Birks initially) predicted that a barrier of smoke (from incinerated urban centers and forests) would fill tropospheric skies, and eventually surround the earth, reducing the tropospheric temperature and causing damage to ecosystems and atmospheric components such as stratospheric ozone.

❐ nucleation

A process in which water condenses on a small atmospheric particle. Nucleation aides in the process of condensation.

❐ nucleation scavenging

Term for water vapor that removes particulate matter from the atmosphere.

O

❐ ocean

The oceans cover approximately 70% of the Earth's surface and are apparently unique in the solar system (although keep an eye on some of the giant planets' moons...). The oceans are important for the regulation of tropospheric heat and the storage and production of oxygen and carbon dioxide.

❐ one dimensional model

An atmospheric modelling technique in which a concentration or another unknown can be calculated relative to certain parameters, such as altitude or temperature.

❐ opacity

A measure of light blocking by atmospheric particles.

❐ outgassing

The release of "young" gases into the earth's atmosphere by volcanic activity.

❐ oxidants

Chemical species that readily accept electrons. The electron accepting species is reduced in a chemical reaction.

❐ oxidation

a process that results in a species having a higher oxidation state, that is, the net loss of electrons from an atom. Quite often oxygen is involved in the oxidation of another atom therefore it is an oxidant.

❐ oxygen

O_2, is found on Earth as a gas and constitutes about 20.8% of the air we breathe. Elemental molecular oxygen consists of two oxygen atoms bonded together. A photochemical reaction of oxygen is (ultimately) responsible for the production of ozone in the stratosphere. Oxygen concentrations found in ice core samples (using isotopic 16O/18O rations) have been used to determine past atmospheric levels of oxygen and have helped in determining past climates.

❐ oxygen radical

O. is the result of the oxygen molecule (O_2) being photo dissociated at a wavelength greater than 242 nm. These two oxygen atoms react with O_2 and create ozone (O_3).

❐ ozone

O_3, is a molecule that consists of three oxygen atoms bonded together. The ozone layer in the stratosphere absorbs UV radiation and creates a warm layer of air in the stratosphere and is therefore responsible for the thermal structure of the stratosphere. Ozone that is present in the troposphere is mostly a result of anthropogenic pollution and therefore higher concentrations are found in urban areas. Ozone is involved with NOx in the photochemical production of many of the constituents of pollution environments.

❐❐ ozone depletion potential(odp)

The is the ratio of a particular ozone depleting compound environmental effect compared to the depleting ability of a standard compound, CFC-11, which arbitrarily defined as 1.0. An example would be the ODP of halon-1301 which is about 13.

❐ ozone destruction

In addition to photolytic destruction by infrared light (wavelengths of 1140 nanometers or less), stratospheric ozone is destroyed by interaction with a variety of chemical substances in the stratosphere. Some of these chemicals, such as the chlorine free radical--liberated by the photolysis of chlorofluorocarbons, are not normally produced there in nature at such high concentrations. This leads to the conclusion that human activities, and not natural chemical cycles, are responsible for the current rate of stratospheric ozone destruction. Ozone destruction is most noticeable over the South Pole, where seasonal plummeting in stratospheric ozone concentrations have been measured since the 1970s and are apparently still increasing. That said, estimates based on declining CFC use (because of the Montreal Protocol) suggest that the damage to the global ozone layer will max out soon in the 21st century and that that ozone damage will decrease within the next 75 to 100 years.

❐ ozone distribution

Nearly all of the planet's natural ozone in the atmosphere is found in the stratosphere. The thickness and distribution of this stratospheric ozone vary seasonally and by location. Ozone also occurs in the troposphere as a pollutant, often in photochemical smog.

❐ ozone formation

Molecular oxygen in the stratosphere is broken into a pair of oxygen radicals by light with a wavelength of 240 nanometers or less. If one of these O radicals encounters an oxygen molecule, it can bond to produce ozone. This reaction is only stable if another molecule is present to absorb the excess energy released as the oxygen radical and molecule bond. This is a called a three body reaction, and the third body exhibits its removal of the excess energy by whizzing off at a higher energy and thereby increasing the temperature of the atmosphere where this reaction occurs.

❐ ozone hole

The ozone hole refers to the total amount of ozone in the atmosphere over Antarctica that was discovered to be decreasing at an alarming rate over an observation site in Antarctica. The hole was discovered by J.C. Farman, B.G. Gardiner and J.D. Shanklin.

P

❐ pangea

A huge protocontinent thought to have existed 200 million years ago and from which all present continents are thought to have formed.

❐ parent elements

The element that underwent radioactive decay to produce a radioactive isotope and eventually produce a more stable product. Can be by the emission of an alpha particle, two protons and two neutrons. or by the emission of a beta particle, emission of an electron which changes a neutron to a proton. Is used in the dating of objects because the decay happens in a predictable known pattern.

❐ parts per million (ppm) and parts per billion (ppb)

These terms give scientists a way to describe how much of a substance is contained in a sample: parts of analyte per million parts of sample, for instance. In atmospheric chemistry these become volume parts of analyte per volume parts of atmosphere: ppmv, ppbv, etc. At low analyte gas phase concentrations the analyte is assumed to act as an ideal gas. For instance, a 1 ppmv concentration of formaldehyde would represent 1 liter of formaldehyde per every 1,000,000 liters of air; also equivalent to 1 microliter of H_2CO per 1 L air. In gas phase concentrations these units are also called gas phase mixing ratios.

❐ pcv systems

A PCV system is an air pollution control device in automobile engines. PCV stands for positive crankcase ventilation. Automobile makers began installing PCV systems in all new cars in 1963. It helps to recycle hydrocarbon contaminated gases inside the engine (blowby) back into the engine's combustion process.

❐ pedosphere

This is the area of the Earth comprised of the solid plates of the continental crust, loose rocks and soil.

❐ permafrost

Soil, silt and rock located in perpetually cold areas which remains frozen year-round. Though a thin layer may thaw during summer months, the majority of the permafrost in a given location will remain frozen until the local climate dramatically changes, as from continental drift toward the equator or global warming.

❐ peroxyacetylnitrate (pans)

A class of chemical substances found as a pollutant in the troposphere, formed by photolysis from natural and manufactured organic chemicals. These chemicals act as irritants and mutagens in mammals and are toxic to many plants.

❐ ph

The negative logarithm of the hydrogen ion concentration of a solution. It is the quantitative expression of the acidity and alkalinity of a solution and has a scale that ranges from about 0 to 14. pH_7 is neutral; <7 is acidic; and >7 is alkaline.

❐ phenol

1) A corrosive poisonous crystalline acidic compound present in coal tar and wood tar that in dilute solutions is used as a disinfectant and 2) any of various acidic compounds analogous to phenols and regarded as hydroxyl derivatives of aromatic hydrocarbons. When coal and wood are burned for energy, phenols are released into the atmosphere.

❐ photic zone

In the ocean (or in lakes) this is the area near the surface with enough light to support photosynthesis. In this zone, both the chemical and biological activity can be enhanced by penetrating solar radiation. This zone is about 20 meters in depth in the ocean. In the upper surface of this zone, the marine photochemistry is most vigorous. Changes in the ultraviolet light content penetrating this region because of changes in the UV-blocking ability of the ozone layer may create unexpected biological results there.

❐ photochemical smog

Natural and artificially emitted hydrocarbons in the presence of oxides of nitrogen undergo photochemical reactions which produce a cloud of toxic chemicals including ozone and a variety of caustic agents. This process is powered by sunlight and some of the products, such as ozone, reach a peak soon after photon flux from the sun reaches a maximum, around midday. The thermal inversions often associated with some cities can lead to a dangerous buildup of smog in urban areas. Human deaths have been attributed to photochemical smog since the Industrial Revolution in cities such as London and New York.

❐ photodissociation

This is a specific type of photolysis reaction, in which the destruction of a molecule is initiated strictly by solar energy.

❐ photolysis

The destruction of a molecule by electromagnetic radiation, which provides the energy required for a constituent atom to break the chemical bonds between it and the other atoms comprising the molecule.

❐ photon

A particle that has no rest mass, no electrical charge and whose energy is determined by its electromagnetic wavelength. A photon is considered to be an amount of electromagnetic radiation energy that is proportional to the frequency of that radiation. It may also be considered as a particle of light with energy, hv, where h is Planck's constant and v is the frequency of light.

❐ photostationary state

This describes a condition in which there is equilibrium in a photochemical reaction. For this condition to take place the rate of dissociation of the reactants has to be equal to the rate of their recombination.

❐ photosynthesis

Photosynthesis is a biological process which plays a vital role in cycling the atmosphere's carbon dioxide, while simultaneously producing oxygen and other complex substances. Green plants, with chlorophyll, utilize the sun's energy to convert carbon dioxide and water into the plant biomass while oxygen that is released into the atmosphere.

❐ planck's law

This is a derived formula, from the German physicist Max Planck, that portrays the amount of radiation emitted by a blackbody as theoretically determined by its temperature. It is an equation that produces a curve, termed Planck's blackbody radiation curve, which illustrates that the warmer a body is, the greater is its blackbody emission at each wavelength and the shorter is the wavelength at which emissions peak.

❐ plankton

Plankton are forms of marine, organic life that gather vital energy for life through photosynthesis. Since they are very low on the food chain, they are vital to all marine life. They are also vital to the atmosphere in that they consume large amounts of carbon dioxide, which contribute to enhanced global warming when excess amounts are present. Plankton are considered to be one of the more important contributors of oxygen to the atmosphere.

❐ plate tectonics

The process of planetary surface plate formation, movement, interaction, and destruction. The consequence of this process is that many regions of the planet are situated near interacting--and slowly moving--plates. The process causes earthquakes such as those along the West Coast of the United States.

❐ pleistocene

An epoch in Earth history from about 2-5 million years to 10,000 years ago. Also refers to the rocks and sediment deposited in that epoch.

❐ plume

A plume is a visible smoke-like structure, which may contain pollutants emitted from an exhaust or smoke stack and released into the atmosphere. This elongated band of smoke has changing characteristics that vary with its local environmental conditions. These conditions may include the physical and chemical nature of the pollutant, weather conditions and downwind topography.

❐ point source

Pollutants which are put into the atmosphere by stationary objects, such as refineries, power plants, mills, and the like. In Texas any source that emits more than one ton in a calendar year is considered a point source.

❐ polar vortex

This is a phenomenon that occurs during the polar winter in which stratospheric air moves in a circular motion, with an area of relatively still air in its center. The temperature in the vortex is approximately -130 degrees F (-80 degrees C), which assists in the formation of polar stratospheric clouds. Though usually more prolonged and colder over Antarctica, the Arctic polar vortex does form to a degree, and when the temperatures there are coldest and the vortex persists, Arctic stratospheric ozone destruction on Arctic polar stratospheric clouds has also been observed.

❐ pollutants

A pollutant may be considered as any substance, usually an unwanted by-product or waste, that is released into the environment as a result of (human) activities that alter the chemical, physical and biological characteristics of the environment. These substances may be found in any of the solid, liquid or gas phases. Their long-term effects are difficult to predict and depend upon future human or cultural judgments.

❐ polynuclear aromatic hydrocarbons

(PAHs) These compounds are large ring aromatic hydrocarbons that are molecules containing carbon and hydrogen. It could also be any large ring that consists of five or more cyclic compounds of carbons and any elements that may be bonded to the ring. They usually exist in nature in low amounts. Many times these hydrocarbons exist within the atmosphere in the aerosol phase from combustion processes.

❐ precambrian

All geologic time from the beginning of Earth history to 570 million years ago. Also refers to the rocks that formed in that epoch.

❐ precipitation

What often occurs when clouds form upon reaching 100% relative humidity. If condensation nuclei are present, liquid or solid particles of water will form. When this material becomes heavy enough to fall towards the ground, it is known as the common phenomenon snow, rain, sleet, or hail.

❐ primary pollutant

This refers to the types of pollutants that are either found in the atmosphere or on the ground and are emitted directly from some sources such as automobile tailpipes or smoke stacks. This is contrasted with secondary pollutants which in and of themselves are not directly emitted from a source. Instead they are created from chemical and photochemical reactions of primary pollutants. Nitrogen monoxide is an example of a primary pollutant and ozone is an example of a secondary pollutant.

❑ propagation

In chemical terms, a propagation step is any step but the first, initial step, in a chain reaction (a series of chemical reactions in which the first reaction causes the second which causes the third, which causes the fourth, etc).

❑ propionaldehyde

(C_3H_6O), a colorless liquid that enters the atmosphere by emissions from the manufacture of plastics, synthesis of rubber, and as a disinfectant and a preservative.

❑ pyrodenitrification

This is best defined as way that nitrogen is lost or emitted by the use of heat or radiation, mainly, from the sun. This can occur in the air or on the ground around vegetative regions. It is less likely to occur in arid regions, but it is possible.

Q

❑ quantum yield

A measure of how efficient an emission process is. The quantum yield is a ratio of absorbed photons and photons used up in the chemical process. Quantum yield is usually dependent on temperature, pressure, and wavelength.

❑ quasi-biennial oscillation (use buttons at top in this link)

Periodic variation in the direction, either easterly or westerly, of tropical lower stratospheric winds. The direction changes every 26 months.

❑ quaternary

The period of Earth's history from about 2 million years ago to the present; also, the rocks and deposits of that age.

R

❑ radiation flux

This is defined as the way heat from the sun is absorbed, scattered, returned, or deposited around the earth and can be expressed as some radiation arrival rate per the unit of area of the surface. The radiation flux can be the total amount of radiation that the earth collects, emits, and disperses.

❑ radiative forcing

Reduction of radiation reflecting from earth per units of increase of a chemical component that causes this change in the atmosphere. The result is often used to compare the relative effect of greenhouse gases which can alter climate on earth. This means that components that are present in the atmosphere at low concentrations (e.g. methane at 1.7 ppmv) but with high radiative forcing have a greater effect on global warming than higher concentration components with a lower radiative forcing.

❐ radical (free radical)

A highly reactive molecule or atom with an unpaired electron. The species is often represented by a formula with a single dot as the unpaired electron.

❐ radical chain reaction

Free radicals are molecules that, through photolysis or chemical reaction, have an unpaired electron in their outer valance shell. These radicals are very reactive and thus have a short life. When a free radical reacts with a more stable molecule, the radical often pulls an atom from it and becomes a stable molecule itself. The original molecule then becomes a free radical and will react with other species of atoms and molecules in a long series (or chain) of reactions until the process reaches the termination phase. In this phase two free radicals combine, sharing the pair of electrons and breaking the chain.

❐ radical propagation

Free radicals react with molecules, atoms, other radicals in long series of chain reactions. A radical is formed in the atmosphere through photolysis or chemical reaction. Radicals are very reactive and will pull atoms or electrons from other molecules to fill their outer valance shell. New radicals may formed from the original molecule which then react with another molecule, etc. This reaction propagates from one species to another until a termination reaction occurs. The termination reaction usually involves the reaction of two free radicals to form a stable product.

❐ radioactive decay

The spontaneous breakdown of certain kinds of atomic nuclei into one or more nuclei of different elements, involving the release of energy and subatomic particles.

❐ radiocarbon dating

Establishing the relative age of various materials with the use of carbon-14. This involves measuring the amount of 14C and of 12C and comparing the measured ratio to the one established by the production of 14C in the upper atmosphere by cosmic rays. When an organism is alive the 14C/ 12C ratio in its biomass is constant (because of constant atmospheric 14C production, difussion to the lower atmsophere, absorption by organisms, and because of constant 14C radioactive decay) but when the organism dies the ratio begins to change--at a very predictable rate. Therefore knowing the carbon-14/carbon-12 ratio now in some artifact gives a very good measure of how long the carbon has been "dead." The object however must obviously have organic material either in it or on it for this method to work. Also the lenth of time one can "look back" is limited because the amount of carbon-14 must be detected with some certainty for the age to be known with confidence, and the longer the sample has been dead the less 14C is present because it is continually decaying.

❐ rainout

One item necessary for a rain drop to form is called a condensation nucleus. Pollutants in the air such as nitric acid and sulfuric acid form nucleation sites (particles) where water vapor often condenses and forms a droplet. As these droplets grow they collect more moisture and may ultimately fall out as precipitation. This accounts for some of the acid rain deposition on the planet's surface.

❐ rayleigh scattering

The scattering of light by a body with a particle diameter (Dp) less than 0.03 micrometers is termed Rayleigh Scattering. The wavelength of light scattered is dependent on the Dp, and the amount of light scattered is dependent on the number of particles present per unit volume. Shorter visible wavelengths, such as blue, are scattered by smaller particles than are the longer wavelengths like red.

❐ reaction rate

The speed of chemical reactions is expressed as a rate of change in the concentration of a participating species with respect to time.

❐ recombination

In chemical terms, a recombination is one of the ways in which free radicals are lost (the other being disproportionation). In this case, two free radicals join to create one product. This could be from a molecule that broke apart (ionized) and reconstituted, or two free radicals from different sources that form a new molecule.

❐ relative humidity

This is the ratio between the actual water vapor content of the atmosphere and the maximum water vapor content possible at that given temperature. If the temperature of a given parcel of air rises, the amount of moisture it can hold increases and as such its relative humidity decreases. If there is no change in temperature but the moisture content decreases then the relative humidity will again decrease because the ratio of actual water vapor present is less than the maximum amount the air could hold. Conversely, if the water content does not change but the temperature falls, the relative humidity increases until saturation and possible precipitation occurs.

❐ return stroke

A short-lived, high amplitude discharge of electricity that accompanies a cloud-to-ground lightning event. The bright lightning flash that is visible to the human eye is actually composed of extremely rapid electric discharges called strokes. The return stoke follows a downward extending leader, or conductive path of ionized air. The electrons in the cloud flow down this conductive pathway toward the surface of the earth. As electrons continually migrate down the path, electrons remaining higher up on the path in the cloud begin to consecutively move down the channel to the surface. Since the path of electron flow is progressively lengthened upward, the discharge of the elevated electrons high in the cloud to a lower place in the cloud and then down the pathway to the ground is called the return stroke.

❐ rule of 90

The systematic numbering system originally created by the Du Pont Chemical Company in the early 1930s, which is now the standard method in the naming of chlorofluorocarbons (CFCs). Chlorofluorocarbons, such as dichlorodifluoromethane

(CCl_2F_2), and trichlorofluorocarbon (CCl_3F), were given the nicknames CFC-12 and CFC-11, respectively, by Du Pont so that they could be remembered more easily. In order to derive the chemical structure of a chlorofluorocarbon from its common name, the rule of 90 states that you would add 90 to the number at the end of the name. In the case of CFC-12, add 90+12 to get 102. The hundreds place, the one in this example, represents the number of carbon atoms in the molecule. The tens place, the 0, represents the number of hydrogen atoms in the molecule, and the ones place represents the number of fluorine atoms in the molecule. According to the rule of 90, CFC-12 has 1 carbon atom, 0 hydrogen atoms, and 2 fluorine atoms. The final aspect of the rule of 90 pertains to the necessity of carbon to form four bonds. If, after assigning hydrogen and fluorine atoms to it, the carbon atom still requires more bonds to reach the required four, those bonds will be to chlorine atoms. So far, the one carbon atom of CFC-12 has only formed two bonds, both with fluorine atoms. Thus the remaining two bonds go to chlorine atoms.

S

❑ satellite

Satellites are electronic devices placed into orbit to monitor and examine regions which we may not be able observe from earth. These are used routinely to monitor atmospheric ozone, ultraviolet radiation, and many other constituents. Satellite altimetry and radar interferometry are used to measure magnetic field intensity and track Antarctic ice flow.

❑ sea level

This is the average level of the ocean over the entire earth. Tidal fluctuation is taken into account when determining sea level. Mean Sea Level (MSL) is used to reference the height of structures above ground level and depth below the ocean's surface. Atmospheric pressure is also referenced from sea level. At sea level, 760mm (29.92 inches) of mercury is normally displaced by the pressure of the air column above that point on the earth.

❑ sediment

Particles carried by water or air that are deposited in other areas or elevation are called sediment. Usually associated with water flow, sediment will accumulate at the mouth of a river or stream as it empties into a larger, slower moving body of water. Sediment deposition is also effected by particle size.

❑ sink

This is a natural removal process for pollutants in the troposphere such as dissolution and removal via rain.

❑ smog

This is a term used to describe a mixture of smoke and fog. Smog occurs when high concentrations of moisture is combined with smoke (often containing oxides of sulfur and nitrogen) in the presence of high temperatures or thermal inversions and the absence of wind. These conditions cause polluted air to stagnate over industrial areas and can create a respiratory health hazard. Large coastal industrial centers with surrounding high ground are more prone to smog. There is often a diurnal (over a day) variation in the process of smog formation because one of the necessary components for its formation is sunlight.

❒ smoke

It is the product of incomplete combustion and consisting mostly of carbon and small liquid and solid particles. Smoke is primarily caused by the burning of wood, coal, gasoline, or other biomass. Smoke particles are usually less than 1.0 micrometers in size. Wind intensity, particle size, air temperature, and atmospheric pressure all effect how smoke behaves in the environment.

❒ solar radiation

This is electromagnetic radiation (light energy) emitted by the sun. This energy is transmitted through space in the units of electromagnetic energy called photons. The strength of the solar energy that reaches our outer atmosphere is called the solar constant and has a value of approximately 2.0 calories per minute per square centimeter. Roughly one-half (1.0 cal min-1 cm-2) of all the solar radiation that enters the upper atmosphere is available at the earth's surface. Of the fifty percent loss, approximately 32 percent is reflected back into space by either clouds, dust, or aerosols; the remaining 18 percent is absorbed by clouds, dust, water, and atmospheric gases. The short high-energy wavelengths that are absorbed by the Earth are re-radiated at longer, lower energy wavelengths mostly in the IR wavelength region. These lower-energy wavelengths are, in part, prevented from escaping into space by the greenhouse gases in the atmosphere. Solar radiation is the fuel source that provides a mechanism for all biological processes on earth. Without solar radiation, life on earth would not exist. (This statement, of course, neglects the contribution to the earth's surface temperature from radioactive decay in the earth.)

❒ solar wind

The outflow of charged particles from the solar corona into space. Because of the high temperature of the particles of the corona (mostly protons and electrons), they are moving at speeds higher than the solar escape velocity. At the orbit of the Earth, these particles are moving at about 500 km/sec. Some of these particles are captured by the magnetic fields of the planets, forming their magnetospheres.

❒ sorption

A class of processes by which one material is taken up by another. Absorption is refers to the process of the penetration of one material into another; adsorption to the action of one material being collected on another's surface.

❒ spring equinox

In the Northern hemisphere, March 21 or 22; on these dates, the rays of the sun are striking vertically at noon at the equator (0 degrees latitude), for the earth is in such a position in its orbit that the axis is tilted neither toward nor away from the sun. For latitudes above the equator, the dates are earlier then these dates.

❒ stratocumulus clouds

Low altitude gray colored clouds composed of water droplets that have a patchy appearance. Each cloud patch consists of a rounded mass. This cloud has a somewhat uniform base and normally covers the entire sky. Between the patches blue sky can be seen.

❒ stratopause

The boundary between the stratosphere and the mesosphere. It occurs at a height in the atmosphere of approximately 50 kilometers; however this depends on latitude. The atmosphere is characterized by a decrease in pressure with respect to increased altitude. More importantly, regions within the atmosphere like the troposphere, stratosphere, and mesosphere are distinguishable because of distinct temperature gradients with relatively well-defined starting and ending points. The stratopause is the highest portion of the stratosphere, with a temperature of approximately 0 degrees Celsius; the stratopause can also be described as the warmest region between the mesosphere and the stratosphere.

❒ stratosphere

The thermal atmospheric region of the atmosphere between the troposphere and the mesosphere. The lower boundary of the stratospheric region is marked by the tropopause and begins at approximately 13 kilometers; however, this altitude of the troposphere depends on latitude. The upper limit of the stratosphere is marked by the stratopause at approximately 50 kilometers. The stratosphere is characterized by relatively stable temperatures (between -80 and -50 degrees Celsius) in the lower regions, and begins warming near 20 kilometers, reaching its maximum temperature of approximately 0 degree Celsius at the stratopause. Stratospheric chemistry is of particular interest to scientists because ozone, the principal substance that shields the earth from incoming solar ultraviolet radiation, is found in the stratosphere. It should also be noted that wind currents in the stratosphere are primarily horizontal in nature.

❒ stratosphere/troposphere exchange

The exchange between the stratosphere and the troposphere is extremely important. It is caused by upwelling in the troposphere and when troposphere species are pushed into the stratosphere. It starts in the tropics and moves toward the poles. If certain ozone destroying species are brought into the stratosphere from the troposphere when they mix damage to the ozone layer and global warming can result.

❒ stratospheric clouds

These are clouds found in the stratospheric region of the atmosphere. Unlike the moisture laden troposphere, the stratosphere is almost moisture free. A lack of moisture means that cloud formation in the stratosphere is a relatively uncommon occurrence. Because of the very low temperatures, nacreous clouds, a type of stratospheric cloud, occur during winter conditions at the higher latitudes. A second type of stratospheric cloud, called a polar stratospheric cloud (PSC), occurs in the polar regions during winter and early spring. Polar stratospheric clouds are of particular interest to scientists studying the Antarctic Ozone Hole. Polar stratospheric clouds form under extremely cold conditions, whereby nitric acid, water vapor and other trace chemicals freeze to form ice crystals. PSCs provide a solid surface on which chlorine-containing reservoir compounds can gather, and on this surface the chemical reactions involved in the depletion of ozone are greatly increased.

❏ stratospheric ozone

This is also called the ozone layer, where ozone concentrations are as high as 10 parts per million, and is a vitally important region of the atmosphere. This layer of ozone is located approximately 20-50 kilometers above the earth's surface. Stratospheric ozone is important because it prevents most of the high-energy ultraviolet solar radiation from reaching the earth's surface. Photo dissociation, a photochemical process, is responsible for the formation of the protective ozone layer in the stratosphere. In the upper atmosphere, diatomic oxygen absorbs high-energy ultraviolet radiation. The absorption of radiation causes the diatomic oxygen molecule to break forming two oxygen radicals. The oxygen radical can then recombine with other diatomic oxygen molecules to form triatomic oxygen, or ozone. In the middle regions of the stratosphere, ozone is found in concentrations as high as 10 parts per million. Ozone can also form in the lower portions of the troposphere, due to anthropogenic activity and by a completely different mechanism. Without the protective stratospheric ozone layer in the upper atmosphere, life (as we know it) on earth would not be possible.

❏ stratus cloud

In general, clouds are classified and named according to their appearance and altitude at which they occur, being either high, medium, or low. Stratus clouds are low clouds, and as the root word suggests they are layered in appearance. They occur in broad sheets across most or all of the sky. Generally their formation is due to condensation within layered air that is not subjected to strong vertical movement.

❏ styrene

chemical formula C_8H_8, also commonly called vinyl benzene. Styrene is an aromatic hydrocarbon originating from petroleum and natural gas by-products. It is used to manufacture resilient, lightweight packaging materials and fiberglass. Styrene is extremely volatile in air and evaporates quickly; as a result, it does not remain in the atmosphere for long periods of time.

❏ sulfate anion (SO_{42}-)

The sulfate anion in the atmosphere is the result of the oxidation of sulfur dioxide (SO_2). The sulfate anion is water soluble and is removed from the atmosphere through precipitation.

❏ sulfur dioxide

This is a colorless gas consisting of a single sulfur atom and two oxygen atoms. Sulfur dioxide is a major primary pollutant in the atmosphere originating mostly from coal fired power plants and other fossil fuels combustion. In the atmosphere, sulfur dioxide is usually oxidized by ozone and hydrogen peroxide to form sulfur trioxide, a secondary pollutant. Sulfur trioxide, similar to sulfur dioxide, is extremely soluble in water. If these sulfur oxides are present in the atmosphere when condensation occurs, droplets of sulfuric acid (acid rain) are formed. Volcanic eruptions provide a natural source of sulfur dioxide in the atmosphere. However the real problem associated with the production of sulfuric acid in the atmosphere is not with the sulfur dioxide expelled by volcanoes. Anthropogenic production of sulfur dioxide, caused by the burning of fossil fuels, is largely responsible for damage caused by acid rain.

❐ sulfuric acid

In the atmosphere sulfur oxides (see sulfur dioxide) are converted to sulfuric acid. Oxides of sulfur and nitrogen combine with atmospheric moisture to produce acid rain. Although some sulfur oxides are introduced into the atmosphere by natural means, such as volcanic eruptions, the majority of the sulfur oxides responsible for the damaging effects of acid rain come from anthropogenic sources, mainly the burning of fossil fuels. Areas in the northeastern United States, eastern Canada, and northern Europe have suffered damage due to the effects of acid rain. In many areas damage to forests, crops, lakes, and streams are so severe that they are completely devoid of any life forms. Steps are now being taken in many parts of the world to reduce the amount of sulfur dioxide introduced into the atmosphere. In 1990, amendments where made to the Clean Air Act that places restrictions on the release of sulfur dioxides by power plants. The amendment calls for the reduction of sulfur emissions from a 1990 level of nearly 20 million tons per year to approximately 10 million tons per year by January 1, 2000.

❐ summer solstice

In the Northern Hemisphere occurs on about June 21 of each year. The summer solstice is the longest day of the year and marks the beginning of the summer season. If the earth's axis were not tilted in relation to the plane of earth's orbit around the sun (the ecliptic) there would be no seasonal changes on earth. But because the earth's axis is tilted in relationship to its plane of orbit around the sun (~23.5 degrees) and assumes a constant direction--with the North Pole pointing towards the North Star, seasons occur, and daylight hours can fluctuate in length. Each day prior to the summer solstice the sun appears farther north in the sky; but at the summer solstice, the sun reaches its most northerly position in the sky, directly above the Tropic of Cancer, 23.5 degrees north latitude. Following the summer solstice the sun moves farther south in the sky, and on about September 23 it is directly overhead at the equator (the fall equinox).

❐ sunspot cycle

This cycle, lasting an average period of 11 years, involves recurring increases and decreases in the number of sunspots that occur on the sun's surface.

❐ sunspots

A dark area, which is strongly magnetic, that occurs in the area of the photosphere on the disk of the sun.

T

❒ tectonic activity

The process of the formation, movement, interaction, and destruction of parts of the earth's crust on a generally large scale. This can be used to explain seismicity, volcanism, and mountain building.

❒ tectonics

The structural arrangement of rocks in any crust of a planet. Used to understand rock structures.

❒ temperature inversion

The abnormal reversal of temperature in the troposphere caused by the entrapment of urban air pollution. Under normal circumstances, air in the troposphere is cold at high altitudes and warm at low altitudes, or near the earth's surface. When air pollutants, such as SO_2 or NOx, in urban areas get too concentrated, smog is produced. Since cold air underlies the warmer air at high altitudes, the air does not mix well which can lead to human health disturbances in people with respiratory illnesses such as asthma.

❒ termolecular reaction

A reaction containing three particles. In the atmosphere an example of a termolecular reaction is one in which a molecule acquires the excess energy released by a reaction so that the products created do not convert back to the original state of the reactants. The reaction of molecular oxygen with an oxygen atom (radical) to produce ozone is a good example of an important atmospheric termolecular reaction. In the atmosphere this third body is most often molecular nitrogen because statistically this is the most likely next collision since atmospheric N2 = 78% by volume.

❒ terpene

A naturally occurring organic compound, of the general empirical formula, $C_{10}H_{16}$, biologically built from a naturally occurring "monomer" called isoprene, C_5H_8, which is found as a volatile oil in plants. A clearly important--and olifactorily obvious--example is alpha-terpinene found in lemon oil.

❒ tertbutyl alcohol

a highly flammable, volatile, colorless liquid which has a camphor-like odor. It is miscible in esters, and aromatic and aliphatic hydrocarbons, and soluble in water. Based on its vapor pressure, tert-butyl alcohol will exist mainly in the vapor phase in the atmosphere. It is most likely to react with the hydroxyl (OH) radical at this state. The reaction products are formaldehyde and acetone.

❒ tertiary period

This extended from about 65 million years ago to about 2.5 million years ago.

❒ tetradecane

a light, oily hydrocarbon with a melting point of 5.89°C and a boiling point of 253.7°C. As a gas-phase pollutant it is inactive as a human contact allergen.

❐ thermodynamics

The science of heat and temperature and of the laws governing the conversion of heat into mechanical, electrical, or chemical energy.

❐ thermosphere

The outermost shell of the atmosphere, between the mesosphere and outer space; where temperatures increase steadily with altitude.

❐ toluene

A highly volatile and water-soluble benzene derivative that is highly toxic to most forms of life. Its estimated lifespan in air is 60 hours in the case of a photochemical reaction with hydroxyl radical and is eventually oxidized to carbon dioxide. Toluene is used in a variety of industries including the manufacture of TNT, pharmaceuticals, rubbers, and other benzene derivatives, and is also a fuel additive and a paint solvent. Toluene also reaches the atmosphere from natural sources such as coal tar, mineral oils, and forest fires.

❐ toms (total ozone mapping spectrometer)

An instrument that measures the total ozone in a vertical column of atmosphere. Measurements are made by comparing the amount of backscattered ultraviolet radiation with the amount of incoming solar energy. These changes in this ratio, the earth's albedo, at specific wavelengths can be used to determine the amount of ozone present. 4 satellites create global maps daily of the earth's ozone measurements.

❐ total column ozone

The total amount of ozone that is found in a column of air above the earth from the surface to the top of the atmosphere. The majority of this amount is typically found in the stratosphere.

❐ trace gases

These are gases in the atmosphere that do not occur in large quantities but are significant to life on Earth or are important constituents of the chemical cycles in the atmosphere.

❐ trade winds

Wind systems occupying most of the tropics, which blow from 30 degrees North and South toward the equator; winds are northeasterly in the Northern Hemisphere and southeasterly in the Southern Hemisphere.

❐ trichloroethylene (tce)

C_2HC_{13}, this is an industrial solvent used primarily for vapor degreasing and cold cleaning. It is an extraction solvent for greases, oils, fats, waxes, and tars, and is used by the textile processing industry to sour cotton, wool, and other fabrics. Exposure to moderate amounts of trichloroethylene may cause various health effects such as headaches, vertigo, visual disturbance, tremors, and dizziness. Death may occur from inhalation of large amounts. People located near or downwind of sites where TCE may be in operation might be exposed to higher levels. It is a carcinogen.

❐ trichlorofluoromethane

One of the principal greenhouse gases, a gas with absorption bands in the infrared portion of the spectrum. There is extensive evidence showing that a class of synthetic compounds, the chlorofluorocarbons (CFCs), are responsible for the destruction of the ozone layer. CFCs are molecules that contain one or more atoms of both chlorine and fluorine. In September of l992, the ozone hole over Antarctica was the largest ever recorded and was almost THREE times as large as the area of the U.S. Because CFCs are so unreactive, they do not break down when released into the air in the troposphere where they are spilled. In time, air currents and diffusion carry them into the stratosphere, where, under the influence of UV radiation, they release chlorine radicals that initiate the destruction of ozone. Data collected by NASA have shown conclusively, that there is an inverse relationship between ozone concentration and the chlorine monoxide radical in the stratosphere; ClO is formed by chlorine atom attack on O3. CFCs are very useful inert, nontoxic, nonflammable compounds that had been used for years as coolants and as spray can propellants for aerosol forms of hair sprays and deodorants. They had been unsurpassed as solvents for cleaning electronic microcircuits. Commercially, the most important CFCs are the halogenated methanes, Freon-11 (trichlorofluoromethane) and Freon-12 (dichlorodifluoromethane). Over 50% of asthma inhalers contain chlorofluorocarbns as the solvent and gaseous propellent. These CFCs have relatively recently been prohibited in all products except in those medicinal inhaler dispensers for asthmatics and a few other limited exceptions.

❐ tropopause

The area where the temperature in the troposphere no longer decreases, indicating the boundary between the troposphere and the stratosphere.

❐ troposphere

The lowest layer of the Earth's atmosphere, ranging from the ground to the base of the stratosphere with a altitude of 10-15 kilometers depending on the latitude. This is where all weather occurs. The word troposphere is derived from the Greek words tropos--turning and spharia--sphere.

U

❐ uars (upper atmosphere research satellite)

A satellite deployed by NASA to study the chemistry in the upper atmosphere of Earth.

❐ uranium-thorium dating

An absolute dating technique which uses the properties of the radioactive half-life of Uranium-238 and Thorium-230. The half-life of uranium-238 is 4.5 billion years, that is, in that many years half of the original amount is still uranium-238 while the other half has lost nucleons to form a different element which is more stable.

❐ urban air pollution

Pollution that occurs in cities from the burning of fossil fuels and the emissions of hydrocarbons. Urban air pollution can be in the form of ozone, smog, acid rain, carbon monoxide, particulate matter (PM-10), and nitrogen oxides.

❐ uv radiation

energy that is emitted in the form of electromagnetic waves with a wavelength of 1-380nm, which is composed of UV-A, UV-B, and UV-C light; UV-A light being the longest wavelength and having the smallest energy and UV-C light being the shortest wavelength and having the highest energy. All UV-C light and most UV-B light is filtered out of the atmosphere via the ozone layer. However, UV-A light travels to the surface of the earth.

❐ uv/vis spectrophotometry

Ultraviolet/visible spectrophotometry is used to measure the absorption capabilities of certain compounds with reference to wavelengths of light on the ultraviolet and visible light ranges.

❐ uv-a

Ultraviolet light that has wavelengths of 320nm to 400nm. The wavelengths of this light are shorter than visible light and are not absorbed in the stratosphere by ozone.

❐ uv-b

An ultraviolet wavelength of light between 280 nm and 320 nm whose intensity is increasing at the earth's surface (probably causing increases in skin cancer), This UV increase is because of decreases in statospheric ozone.

❐ uv-c

Light that spans the spectrum with wavelengths of 10nm to 290nm. These wavelengths of ultraviolet light are extremely dangerous to human and animal tissue and is totally absorbed in the stratosphere by ozone and molecular oxygen.

V

❐ vadose zone

The unsaturated zone in a soil where chemical processes are at their most active. Its extent is determined partly by the content of the soil water, but it cannot extend beyond the water table, below which voids are completely filled with water.

❐ valence electrons

The outermost electrons of an atom that are in the orbital of highest principal quantum number of all electrons in the atom and don't fill that orbital completely.

❐ vapor

A substance in the gas state that is below its critical temperature but still suspended in air. It is possible for a vapor to be liquefied by increased pressure. Vapors are usually not visible.

❐ venus

Second planet from the sun; mass 4.872 x 1027g; radius of cloud surface about 6100 Km; radius of solid surface about 6056 Km; surface temperature 737K; temperature of cloud tops 250K; cloud composition more than 75% H2SO4 and about 25% water; atmosphere (by volume) 90-95% CO2 with traces of SO2, water, CO, H2SO4, Ar, Ne, HF, HCl. Atmospheric pressure 92-95 atm.

❐ vernal equinox

A time during the spring season when the sun is directly above the equator which results in an equal amount of time for day and night.

❐ vertical flux

In dry deposition this refers to the rate at which a gas particle is deposited onto another surface without the aid of precipitation. Vertical flux equals concentration of the species multiplied by the deposition velocity.

❐ vertical mixing

The movement of air in a vertical direction usually caused by differences in temperature and density.

❐ vinyl bromide

a toxic compound that was established as a federal hazardous air pollutant in 1993. Sources of vinyl bromide may include rubber substitution, flame-retarding treatment of acrylics, laminating, and/or preparing films. It is also important to point out that vinyl bromide may be the result of the atmospheric degradation of 1,2-dibromoethane. Based on a typical atmospheric gas phase reaction with a hydroxyl radical, the lifetime of vinyl bromide in the atmosphere is between 1.5 and 2.1 days resulting in the products formaldehyde and formyl bromide.

❐ vinyl chloride

Chemical Formula C_2H_3Cl: Synonyms: Ethylene, chloro-; Chloroethene; Chloroethylene; Monochloroethylene; Vinyl chloride; Vinyl chloride monomer; Vinyl C monomer; C_2H_3Cl; Ethylene monochloride; Monochloroethene; Chlorethene; Chlorethylene; Chlorure de vinyle; Cloruro di vinile; Rcra waste number U043; Trovidur; UN 1086; VC; VCM; Vinile; Vinylchlorid; Vinyl chloride, inhibited; Vinyle(chlorure de); Winylu chlorek. This colorless gas is used mainly to produce polyvinylchloride. As there are no natural sources known for this compound, all sources areanthropogenic. The compound reacts with hydroxyl radicals in the atmosphere to yield chlorine radicals, which catalytically destroy ozone.

❐ visibility

A measurement of the ability to see and identify objects at different distances.

❐ volcanoes

Openings in the earth's crust that release molten rock, steam, reduced and oxidized compounds, and ash from within the earth's interior into the atmosphere.

W

❑ walker cell

A zonal circulation of the atmosphere confined to equatorial regions and driven principally by the oceanic temperature gradient. In the Pacific, air flows westward from the colder, eastern area to the warm, western ocean, where it acquires warmth and moisture and subsequently rises. A return flow aloft and subsidence over the eastern ocean complete the cell.

❑ water

Water consists of two hydrogen atom and one oxygen atom (H_2O). Water covers three-fourths of the Earth's surface, 60-70% of the world's weight, regenerates and redistributes through evaporation and other atmospheric processes. Water is involved in electrical charge separation because it has two types and positions of atoms giving it a net dipole movement. Water vapor also absorbs 17% of solar radiation in the troposphere, thus making it one of the two principal greenhouse gases. Of the solar energy absorbed by the Earth's surface a little more than half goes into latent heat, which is heat absorbed by water because of its transformation from a liquid to a gas. When these molecules condense back into a liquid, usually higher in the atmosphere, they released that energy back into the atmosphere as local warming.

❑ water cycle

The process by which water is transpired and evaporated from the land and water, condensed in the clouds, and precipitated out onto the earth once again to replenish the water in the bodies of water on the earth.

❑ water vapor

Water present in the atmosphere in gaseous form; the source of all forms of condensation and precipitation. Water vapor, clouds, and carbon dioxide are the main atmospheric components in the exchange of terrestrial radiation in the troposphere, serving as a regulator of planetary temperatures via the greenhouse effect. Approximately 50 percent of the atmosphere's moisture lies within about 1.84 km of the earth's surface, and only a minute fraction of the total occurs above the tropopause.

❑ weak sun paradox

This refers to the fact that our sun has probably increased its luminosity during its lifetime (approximately 3.5 billion years). This assumption is based on astrophysical observations of many other stars. The best scientific conclusions are that when our sun was young it had a luminosity (total energy output) of 70-80% of what it is today; however, there has clearly been (based on geological records and long term evidence for the presence of liquid water) approximately the SAME long term planetary temperature on earth (about 15 degrees C, averaged planet-wide) instead of colder earlier surface temperatures followed later by warmer surface temperatures. This paradox suggests that the planet's surface temperature may have been controlled by atmospheric gas concentrations that varied over time while keeping (via the greenhouse effect) the surface temperature relatively constant.

❒ wet deposition

The deposition of dissolved substances such as airborne particles and gaseous molecules by precipitation. Airborne particles/gaseous molecules in the atmosphere are trapped in water droplets due to falling precipitation (i.e. rain and snow). These substances are then dissolve in the water droplets and are deposited on the surface in which the water droplets come into contact with.

❒ wien's displacement law

The relationship between wavelength and temperature in relation to blackbody radiation. The wavelength decreases as the temperature increases. The common formula to determine this relationship is Plank's Formula.

X

❒ xylene

An aromatic hydrocarbon used as a common industrial solvent, also named dimethylbenzene; exists as 3 isomers: m-xylene, o-xylene, and p-xylene. Xylene is usually obtained from petroleum and natural gas distillates. It is used in paint, rubber and leather industries, and also as a cleaning agent or degreaser.

Z

❒ zeolite

Microporus silicate or aluminosilicate structured minerals that can act as an absorbing filter or sieve on a molecular level. Mainly used by the petroleum industry for the cracking of petroleum or use as a filter against various compounds.

Carbon Dioxide and Climate

A

❐ ablation (glacial)

All processes, which include melting, evaporation (sublimation), wind erosion, and calving (breaking off of ice masses), that remove snow or ice from a glacier or snowfield. The term also refers to the amount of snow or ice removed by these processes.

❐ abscission

Shedding by a plant of its parts, such as leaves, flowers, fruits, or seeds. The process is regulated by the plant hormone abscisic acid.

❐ absorption coefficient

A measure of the amount of radiant energy, incident normal to a planar surface, that is absorbed per unit distance or unit mass of a substance.

❐ acclimation (acclimatization)

Change that occurs in an organism to allow it to tolerate a new environment.

❐ accumulation (glacial)

All processes, which include snowfall, condensation, avalanching, snow transport by wind, and freezing of liquid water, that add snow or ice to a glacier, floating ice, or snow cover. The term also includes the amount of snow or other solid precipitation added to a glacier or snowfield by these processes.

❐ acidity profile

The acid concentration in ice core layers as a function of depth as determined from electrical measurements. The magnitudes of some volcanic eruptions in the Northern Hemisphere have been estimated from the acidity of annual layers in ice cores taken in Greenland. This methodology is sometimes referred to as acidity signal or acidity record.

❐ adaptation

The adjustment of an organism or population to a new or altered environment through genetic changes brought about by natural selection.

❐ adiabatic process

A thermodynamic change of state of a system such that no heat or mass is transferred across the boundaries of the system. In an adiabatic process, expansion always results in cooling, and compression in warming.

❐ advection

The predominately horizontal large-scale movement of air that causes changes in temperature or other physical properties. In oceanography, advection is the horizontal or vertical flow of sea water as a current.

❐ aerosol

Particulate material, other than water or ice, in the atmosphere ranging in size from approximately 10x-3 to larger than 10x2 m in radius. Aerosols are important in the atmosphere as nuclei for the condensation of water droplets and ice crystals, as participants in various chemical cycles, and as absorbers and scatterers of solar radiation, thereby influencing the radiation budget of the earth-atmosphere system, which in turn influences the climate on the surface of the Earth.

❐ agglomeration

In meteorology, the process by which precipitation particles grow larger by collision or contact with cloud particles or other precipitation particles.

❐ agung

Active volcano 10,380 ft (3,141 m) high in Bali, Indonesia.

❐ airborne fraction

The portion of CO_2 released from all energy consumption and land use activities that remains in the atmosphere as opposed to the amounts absorbed by plants and oceans. How the world's total carbon is partitioned among the oceanic, terrestrial, and atmospheric pools is determined by complex biogeochemical and climatological interactions.

❐ airborne particulates

Total suspended matter found in the atmosphere as solid pieces or liquid droplets. Airborne particulates include windblown dust, emissions from industrial processes, smoke from the burning of wood and coal, and the exhaust of motor vehicles.

❐ airmass

A widespread body of in the atmosphere that gains certain meteorological or polluted characteristics while set in one location. The characteristics can change as it moves away.

❐ albedo

The fraction of the total solar radiation incident on a body that is reflected by it.

❐ alfisols

An order of soils with a medium-to-high base supply, horizons of clay accumulation, and gray-brown surface horizons.

❐ algae

Simple rootless plants that grow in sunlit waters in relative proportion to the amounts of nutrients available. They are food for fish and small aquatic animals.

❐ algal blooms

Sudden spurts of algal growth that can indicate potentially hazardous changes in local water chemistry.

❐ alkalinity

A pressure- and temperature-independent property of seawater that determines in part the carbon content of seawater. Carbonate alkalinity is the sum of the concentration of bicarbonate plus two times the concentration of the carbonate ions. Total alkalinity is the amount of acid required to bring seawater to a pH at which all dissolved inorganic carbon becomes freely exchangeable. The alkalinity of the oceans is determined with potentiometric or normal titration techniques that detect and measure the presence of bicarbonate, carbonate, and borate ions.

❐ altithermal period

A period of high temperature, particularly the one from 8000 to 4000 B.P. (before the present era), which was apparently warmer in summers, as compared with the present, and with the precipitation zones shifted poleward.

❐ anadromous

Fish that spend their adult lives in the sea but swim upriver to freshwater spawning grounds to reproduce.

❐ analog (climate)

A large-scale weather pattern of the past that is similar to a current situation in its essential characteristics.

❐ anthropogenic

Man made. Usually used in the context of emissions that are produced as the result of human activities.

❐ anticyclone (high-pressure area)

An atmospheric high-pressure closed circulation with clockwise rotation in the Northern Hemisphere, counterclockwise in the Southern Hemisphere, and undefined at the Equator.

❐ aragonite

A mineral species of calcium carbonate ($CaCO_3$) with a crystal structure different from that of vaterite and calcite, which are the other two forms of $CaCO_3$. It is precipitated from ocean surface waters mainly by organisms (e.g., coral) that use it to make their shells and skeletons.

❐ arctic haze

A persistent winter diffuse layer in the Arctic atmosphere whose origin may be related to long-range transport of mid latitude continental man-made pollutants.

❐ atmosphere (an)

A standard unit of pressure representing the pressure exerted by a 29.92-in. column of mercury at sea level at 45 degrees latitude and equal to 1000 g/cm2.

❑ atmosphere (the)

The envelope of air surrounding the Earth and bound to it by the Earth's gravitational attraction. Studies of the chemical properties, dynamic motions, and physical processes of this system constitute the field of meteorology.

❑ atmospheric turbulence

A state of the flow of air in which apparently random irregularities occur in the air's instantaneous velocities, often producing major deformations of the flow.

❑ atmospheric window

The spectral region between 8.5 and 11.0 microns where the atmosphere is essentially transparent to long wave radiation.

❑ autotrophic

An organism that produces food from inorganic substances.

B

❑ baroclinic model

A model of atmospheric circulation that, in contrast with barotropic models, does not constrain constant-pressure surfaces to coincide with constant-density surfaces.

❑ basal sliding (basal slip)

The movement or speed of movement of a glacier on its bed.

❑ bathymetry

The science of measuring ocean depths to determine the topography of the sea floor.

❑ benthic organism (benthos)

A form of aquatic plant or animal life that is found on or near the bottom of a stream, lake, or ocean.

❑ benthic region

The bottom layer of a body of water.

❑ biogeochemical cycle

The chemical interactions among the atmosphere, biosphere, hydrosphere, and lithosphere.

❑ biological productivity

The amount of organic matter, carbon, or energy content that is accumulated during a given time period.

❑ biomass

The total dry organic matter or stored energy content of living organisms that is present at a specific time in a defined unit (community, ecosystem, crop, etc.) of the Earth's surface.

❒ biosphere

The portion of Earth and its atmosphere that can support life. The part (reservoir) of the global carbon cycle that includes living organisms (plants and animals) and life- derived organic matter (litter, detritus). The terrestrial biosphere includes the living biota (plants and animals) and the litter and soil organic matter on land, and the marine biosphere includes the biota and detritus in the oceans.

❒ biota

The animal and plant (fauna and flora) life of a given area.

❒ bitumen

Dark, naturally occurring solid or semisolid substances composed mainly of a mixture of hydrocarbons with little oxygen, nitrogen, or sulfur.

❒ buffer factor (revelle factor)

The ratio of the instantaneous fractional change in the partial pressure of CO_2 (pCO_2) exerted by seawater to the fractional change in total CO_2 dissolved in the ocean waters. The buffer factor relates the partial pressure of CO_2 in the ocean to the total ocean CO_2 concentration at constant temperature, alkalinity and salinity. The Revelle factor is a useful parameter for examining the distribution of CO_2 between the atmosphere and the ocean, and measures in part the amount of CO_2 that can be dissolved in the mixed surface layer.

C

❒ C_3 plants

Plants (e.g., soybean, wheat, and cotton) whose carbon- fixation products have three carbon atoms per molecule. Compared with C_4 plants, C_3 plants show a greater increase in photosynthesis with a doubling of CO_2 concentration and less decrease in stomatal conductance, which results in an increase in leaf-level water-use efficiency.

❒ C_4 plants

Plants (e.g., maize and sorghum) whose carbon fixation products have four carbon atoms per molecule. Compared with C_3 plants, C_4 plants show little photosynthetic response to increased CO_2 concentrations above 340 ppm but show a decrease in stomatal conductance, which results in an increase in photosynthetic water-use efficiency.

❒ calcrete

A surficial gravel and sand conglomerate cemented by calcium carbonate.

❒ caliche

Also called hardpan; an opaque, reddish-brown-to-white calcareous material, which occurs in layers near the surface of stony soils in arid and semiarid areas.

❒ calvin cycle

The incorporation of CO_2 into glucose by enzymatic reactions.

❒ cam plants (crassulacean acid metabolism)

Plants (e.g, cactus and other succulents) that, unlike the C_3 and C_4 plants, temporarily separate the processes of carbon dioxide uptake and fixation when grown under arid conditions. They take up gaseous carbon dioxide at night when the stomata are open and water loss is minimal. During the day when the stomata are closed, the stored CO_2 is released and chemically processed. When CAM plants are not under water stress, they then follow C_3 photosynthesis.

❒ canopy

The branches and leaves of woody plants that are formed some distance above the ground.

❒ carbon budget

The balance of the exchanges (incomes and losses) of carbon between the carbon reservoirs or between one specific loop (e.g., atmosphere - biosphere) of the carbon cycle. An examination of the carbon budget of a pool or reservoir can provide information about whether the pool or reservoir is functioning as a source or sink for CO_2.

❒ carbon cycle

All parts (reservoirs) and fluxes of carbon; usually thought of as a series of the four main reservoirs of carbon interconnected by pathways of exchange. The four reservoirs, regions of the Earth in which carbon behaves in a systematic manner, are the atmosphere, terrestrial biosphere (usually includes fresh water systems), oceans, and sediments (includes fossil fuels). Each of these global reservoirs may be subdivided into smaller pools ranging in size from individual communities or ecosystems to the total of all living organisms (biota). Carbon exchanges from reservoir to reservoir by various chemical, physical, geological, and biological processes.

❒ carbon density

The amount of carbon per unit area for a given ecosystem or vegetation type, based on climatic conditions, topography, vegetative-cover type and amount, soils, and maturity of the vegetative stands.

❒ carbon dioxide fertilization

Enhancement of plant growth or of the net primary production by CO_2 enrichment that could occur in natural or agricultural systems as a result of an increase in the atmospheric concentration of CO_2.

❒ carbon dioxide reference gas

A mixture of a known quantity of CO_2-in air or CO_2-in-N_2 used to calibrate carbon dioxide analyzers.

❑ carbon flux

The rate of exchange of carbon between pools (reservoirs).

❑ carbon isotope ratio

Ratio of carbon-12 to either of the other, less common, carbon isotopes, carbon-13 or carbon-14.

❑ carbon pool

The reservoir containing carbon as a principal element in the geochemical cycle.

❑ carbon sink

A pool (reservoir) that absorbs or takes up released carbon from another part of the carbon cycle. For example, if the net exchange between the biosphere and the atmosphere is toward the atmosphere, the biosphere is the source, and the atmosphere is the sink.

❑ carbon source

A pool (reservoir) that releases carbon to another part of the carbon cycle.

❑ carbon-based resources

The recoverable fossil fuel (coal, gas, crude oils, oil shale, and tar sands) and biomass that can be used in fuel production and consumption.

❑ catanadromous

Fish that swim downstream to spawn.

❑ chernozem (tchernozem)

A major group of dark-colored zonal soils with a rich and deep humus horizon occurring in temperate-to-cool, subhumid climates.

❑ chlorofluorocarbons

A family of inert nontoxic and easily liquified chemicals used in refrigeration, air conditioning, packaging, and insulation or as solvents or aerosol propellants. Because they are not destroyed in the lower atmosphere, they drift into the upper atmosphere where their chlorine components destroy ozone.

❑ chloroplast

An organelle in the cells of green plants. It contains chlorophyll and functions in photosynthesis and protein synthesis.

❑ clear cutting

A forest-management technique that involves harvesting all the trees in one area at one time.

❑ climate

The statistical collection and representation of the weather conditions for a specified area during a specified time interval, usually decades, together with a description of the state of the external system or boundary conditions. The properties that characterize the climate are thermal (temperatures of the surface air, water, land, and ice), kinetic (wind and ocean currents, together with associated vertical motions and the motions of air masses, aqueous humidity, cloudiness and cloud water content, groundwater, lake lands, and water content of snow on land and sea ice), nd static (pressure and density of the atmosphere and ocean, composition of the dry ir, salinity of the oceans, and the geometric boundaries and physical constants of the system). These properties are interconnected by the various physical processes such as precipitation, evaporation, infrared radiation, convection, advection, and turbulence.

❑ climate change

The long-term fluctuations in temperature, precipitation, wind, and all other aspects of the Earth's climate. External processes, such as solar-irradiance variations, variations of the Earth's orbital parameters (eccentricity, precession, and inclination), lithosphere motions, and volcanic activity, are factors in climatic variation. Internal variations of the climate system also produce fluctuations of sufficient magnitude and variability to explain observed climate change through the feedback processes interrelating the components of the climate system.

❑ climate sensitivity

The magnitude of a climatic response to a perturbing influence. In mathematical modeling of the climate, the difference between simulations as a function of change in a given parameter.

❑ climate signal

A statistically significant difference between the control and disturbed simulations of a climate model.

❑ climate system

The five physical components (atmosphere, hydrosphere, cryosphere, lithosphere, and biosphere) that are responsible for the climate and its variations.

❑ climate variation

The change in one or more climatic variables over a specified time.

❑ climatic analog

A past climate situation in which changes similar to the present occurred. Used in making climatic projections.

❑ climatic anomaly

The deviation of a particular climatic variable from the mean or normal over a specified time.

❑ climatic optimum

The period in history from about 5000 to about 2500 B.C. during which surface air temperatures were warmer than at present in nearly all regions of the world. In the Arctic region, the temperature rose many degrees, and in temperate regions, the increase was 1.0 degrees - 1.7 degrees C. In this period, glaciers and ice sheets receded greatly, and the melt-water raised sea level by about 3 meters.

❑ cloud

A visible mass of condensed water vapor particles or ice suspended above the Earth's surface. Clouds may be classified on their visible appearance, height, or form.

❑ cloud albedo

Reflectivity that varies from less than 10 to more than 90% of the insolation and depends on drop sizes, liquid water content, water vapor content, thickness of the cloud, and the sun's zenith angle. The smaller the drops and the greater the liquid water content, the greater the cloud albedo, if all other factors are the same.

❑ cloud feedback

The coupling between cloudiness and surface air temperature in which a change in surface temperature could lead to a change in clouds, which could then amplify or diminish the initial temperature perturbation. For example, an increase in surface air temperature could increase the evaporation; this in turn might increase the extent of cloud cover. Increased cloud cover would reduce the solar radiation reaching the Earth's surface, thereby lowering the surface temperature. This is an example of negative feedback and does not include the effects of longwave radiation or the advection in the oceans and the atmosphere, which must also be considered in the overall relationship of the climate system.

❑ coastal zone

Lands and waters adjacent to the coast that exert an influence on the uses of the sea and its ecology or whose uses and ecology are affected by the sea.

❑ composting

The natural biological decomposition of organic material in the presence of air to form a humus-like material.

❑ continental crust

The layer of the Earth that lies under continents and the continental shelves. It ranges in thickness from 35 to 60 km. Its upper layer has a density of 2.7 g/cm3 and is composed of rocks that are rich in silica and alumina.

❑ continental plate

A thick continental crust.

❑ continental shelves

Those parts of the continent that are covered by water. They are several to more than 322 km wide and about 122 m deep. At the edges of the shelves, the continental slopes drop rapidly from 100 to 200 m to 3000 to 3700 m.

❐ convection

Atmospheric or oceanic motions that are predominately vertical and that result in vertical transport and mixing of atmospheric or oceanic properties. Because the most striking meteorological features result if atmospheric convective motion occurs in conjunction with the rising current of air (i.e., updrafts), convection is sometimes used to imply only upward vertical motion.

❐ convective adjustment

A numerical procedure applied in many atmospheric models to approximate the vertical nonradiative heat transport. This procedure adjusts the lapse rate whenever necessary so that some prescribed critical lapse rate is never exceeded.

❐ convergence

The quasi-horizontal flow of a fluid toward a common destination from different directions. When waters of different origins come together at a point or along a line (convergence line), the denser water from one side sinks under the lighter water from other side. The ocean convergence lines are the polar, subtropical, tropical, and equatorial.

❐ coriolis effect

The tendency for an object moving above the Earth to turn to the right in the Northern Hemisphere and to the left in the Southern Hemisphere relative to the Earth's surface. The effect arises because the Earth rotates and is not, therefore, an inertial reference frame.

❐ cotyledon

A leaf or leaves of the embryos of seed plants. They can function in food storage and can become photosynthetic when the seed germinates.

❐ crop water-use efficiency

A measure at the ecosystem level of how well plants use available water in growth. The grams of dry weight gained by plants during the growing season per unit land area are divided by the millimeters of water lost (including evaporation directly from the soil).

❐ cryosphere

The portion of the climate system consisting of the world's ice masses and snow deposits, which includes the continental ice sheets, mountain glaciers, sea ice, surface snow cover, and lake and river ice. Changes in snow cover on the land surfaces are by and large seasonal and closely tied to the mechanics of atmospheric circulation. The glaciers and ice sheets are closely related to the global hydrologic cycle and to variations of sea level and change in volume and extent over periods ranging from hundreds to millions of years.

D

❐ decomposers

Heterotrophic organisms that break down dead protoplasm and use some of the products and release others for use by consumer organisms.

❒ decomposition

The breakdown of matter by bacteria and fungi. It changes the chemical makeup and physical appearance of materials.

❒ deep water

That part of the ocean below the main thermocline.

❒ deforestation

The removal of forest stands by cutting and burning to provide land for agricultural purposes, residential or industrial building sites, roads, etc. or by harvesting the trees for building materials or fuel. Oxidation of organic matter releases CO_2 to the atmosphere, and regional and global impacts may result.

❒ dendrochronology

The dating of past events and variations in the environment and the climate by studying the annual growth rings of trees. The approximate age of a temperate forest tree can be determined by counting the annual growth rings in the lower part of the trunk. The width of these annual rings is indicative of the climatic conditions during the period of growth; wide annual rings signify favorable growing conditions, absence of diseases and pests, and favorable climatic conditions, while narrow rings indicate unfavorable growing conditions or climate.

❒ dendroclimatology

The use of tree growth rings as proxy climate indicators. Tree rings record responses to a wider range of climatic variables over a larger part of the Earth than any other type of annually dated proxy record.

❒ desertification

The progressive destruction or degradation of vegetative cover especially in arid or semiarid regions bordering existing deserts. Overgrazing of rangelands, large-scale cutting of forests and woodlands, drought, and burning of extensive areas all serve to destroy or degrade the land cover. The climatic impacts of this destruction include increased albedo leading to decreased precipitation, which in turn leads to less vegetative cover; increased atmospheric dust loading could lead to decreased monsoon rainfall and greater wind erosion and/or atmospheric pollution.

❒ differentiation

The process by which single cells grow into particular forms of specialized tissue (e.g., root, stem, or leaf).

❒ divergence

A horizontal flow of water, in different directions, from a common center or zone; it is often associated with upwelling.

❒ downwelling

The process of accumulation and sinking of warm surface waters along a coastline. A change of air flow of the atmosphere can result in the sinking or downwelling of warm surface water. The resulting reduced nutrient supply near the surface affects the ocean productivity and meteorological conditions of the coastal regions in the downwelling area.

❐ dust veil index

A quantitative method developed by H. H. Lamb for comparing the magnitude of volcanic eruptions. The formulae use observations either of the depletion of the solar beam, temperature lowering in middle latitudes, or the quantity of solid matter dispersed as dust. The reference dust veil index is 1000, assigned to the Krakatoa 1883 eruption, and the index is calculated using all three methods, where the information is available, for statistical comparison purposes.

E

❐ ecosystem

The interacting system of a biological community and its nonliving environmental surroundings.

❐ eddy

A circular movement of water or air that is formed where currents pass obstructions or between two adjacent currents that are flowing counter to each other.

❐ el chicon

Active volcano 7300 ft (2225 m) high in Mexico. The last eruption was in 1983.

❐ el nino

An irregular variation of ocean current that from January to March flows off the west coast of South America, carrying warm, low-salinity, nutrient-poor water to the south. It does not usually extend farther than a few degrees south of the equator, but occasionally it does penetrate beyond 12 degrees S, displacing the relatively cold Peru Current. The effects of this phenomenon are generally short-lived, and fishing is only slightly disrupted. Occasionally (in 1891, 1925, 1941, 1957 - 58, 1965, 1972 - 73, 1976, and 1982 - 83), the effects are major and prolonged. Under these conditions, sea surface temperatures rise along the coast of Peru and in the equatorial eastern Pacific Ocean and may remain high for more than a year, having disastrous effects on marine life and fishing. Excessive rainfall and flooding occur in the normally dry coastal area of western tropical South America during these events. Some oceanographers and meteorologists consider only the major, prolonged events as El Nino phenomena rather than the annually occurring weaker and short-lived ones. The name was originally applied to the latter events because of their occurrence at Christmas time.

❐ emissions

Materials (gases, particles, vapors, chemical compounds, etc.) that come out of smokestacks, chimneys, and tailpipes.

❐ emissivity

The ratio of the radiation emitted by a surface to that emitted by a black body at the same temperature.

❐ energy balance models

An analytical technique to study the solar radiation incident on the Earth in which explicit calculations of atmospheric motions are omitted. In the zero-dimensional models, only the incoming and outgoing radiation is considered. The outgoing infrared radiation is a linear function of global mean surface air temperature, and the reflected solar radiation is dependent on the surface albedo.

The albedo is a step function of the global mean surface air temperatures, and equilibrium temperatures are computed for a range of values of the solar constant. The one-dimensional models have surface air temperature as a function of latitude. At each latitude, a balance between incoming and outgoing radiation and horizontal transport of heat is computed.

❐ environment

The sum of all external conditions affecting the life, development, and survival of an organism.

❐ epidemiology

The study of diseases as they affect populations, including the distribution of disease or other health-related states and events in human populations, the factors (e.g., age, sex, occupation, and economic status) that influence this distribution, and the application of this study to control health problems.

❐ equilibrium line

The level on a glacier where accumulation equals ablation and the net balance equals zero.

❐ estuary

Regions of interaction between rivers and near-shore ocean waters, where tidal action and river flow create a mixing of fresh and salt water.

❐ euphotic zone

The layer of a body of water that receives sufficient sunlight for photosynthesis. The depth of this layer, which is about 80 m, is determined by the water's extinction coefficient, the cloudiness, and the sunlight's angle of incidence.

❐ evapotranspiration

Discharge of water from the Earth's surface to the atmosphere by evaporation from bodies of water, or other surfaces, and by transpiration from plants.

F

❐ fen

A type of wetland that accumulates peat deposits; they are less acidic than bogs, deriving most of their water from groundwater rich in calcium and magnesium.

❐ firn

Material that is transitional between snow and glacier ice. It is formed from snow after passing through one summer melt season and becomes glacier ice after its permeability to liquid water falls to zero.

❐ first detection

Identification of a precursor signal, detectable above the noise of natural climatic variability, of a significant change in a climate parameter and attribution of this change to an increase in atmospheric carbon dioxide concentration. The signal may be estimated by numeric modeling of the climate, and the noise can be estimated using instrumental data. For any modeled signal that is estimated, the corresponding noise can be estimated from observational data, and a signal-to-noise ratio can be calculated to provide a quantitative measure of detectability.

❐ flaring

The burning of waste gases through a flare stack or other device before releasing them to the air.

❐ flow law

In glaciology, a constitutive relation for the analysis of three-dimensional deformation states of ice subjected to stress.

❐ fog

Liquid particles less than 40 microns in diameter that are formed by condensation of vapor in air.

❐ food chain

A sequence of organisms, each of which uses the next lower member of the sequence as a food source.

❐ fossil fuel

Any hydrocarbon deposit that can be burned for heat or power, such as petroleum, coal, and natural gas.

❐ fungi

Molds, mildews, yeasts, mushrooms, and puffballs, a group of organisms that lack chlorophyll and therefore are not photosynthetic. They are usually nonmobile, filamentous, and multicellular.

G

❐ general circulation models

Hydrodynamic models of the atmosphere on a grid or spectral resolution that determine the surface pressure and the vertical distributions of velocity, temperature, density, and water vapor as functions of time from the mass conservation and hydrostatic laws, the first law of thermodynamics, Newton's second law of motion, the equation of state, and the conservation law for water vapor. Atmospheric general circulation models are abbreviated AGCM, while oceanic general circulation models are abbreviated OGCM.

❐ geomorphology

The study of present-day landforms, including their classification, description, nature, origin, development, and relationships to underlying structures. Also the history of geologic changes as recorded by these surface features. The term is sometimes restricted to features produced only by erosion and deposition.

❐ geosphere

The solid mass (lithosphere) of the Earth as distinct from the atmosphere and hydrosphere or all three of these layers combined.

❐ geostrophic flow

A type of movement where the Coriolis force balances exactly the horizontal pressure force.

❐ glacial maximum

The position or time of the greatest advance of a glacier (e.g., the greatest equatorward advance of Pleistocene glaciation).

❐ glacial rebound

The isostatic adjustment of previously glaciated areas after glacial retreat (e.g., the uplift of Scandinavia after the most recent glaciation).

❐ glacier

A mass of land ice that is formed by the cumulative recrystallization of firm. A glacier flows slowly (at present or in the past) from an accumulation area to an ablation area. Some well-known glaciers are: the Zermatt, Stechelberg, Grindelwald, Trient, Les Diablerets, and Rhone in Switzerland; the Nigards, Gaupne, Fanarak, Lom, and Bover in Norway; the Wright, Taylor, and Wilson Piedmont glaciers in Antarctica; the Bossons Glacier in France; the Emmons and Nisqually glaciers on Mt. Ranier, Washington; Grinnell glacier in Glacier National Park, Montana; the Dinwoody glacier in the Wind River Mountains and the Teton glacier in Teton National Park, both in Wyoming; and many glaciers in the Canadian Rockies.

❐ glacier flow (ice flow)

The slow downward or outward movement of ice in a glacier caused by gravity.

❑ greenhouse effect

A popular term used to describe the roles of water vapor, carbon dioxide, and other trace gases in keeping the Earth's surface warmer than it would be otherwise. These " radiatively active " gases are relatively transparent to incoming short wave radiation, but are relatively opaque to outgoing longwave radiation. The latter radiation, which would otherwise escape to space, is trapped by these gases within the lower levels of the atmosphere. The subsequent reradiation of some of the energy back to the surface maintains surface temperatures higher than they would be if the gases were absent. There is concern that increasing concentrations of greenhouse gases, including carbon dioxide, methane, and manmade chlorofluorocarbons, may enhance the greenhouse effect and cause global warming.

❑ greenhouse gases

Those gases, such as water vapor, carbon dioxide, tropospheric ozone, nitrous oxide, and methane, that are transparent to solar radiation but opaque to longwave radiation. Their action is similar to that of glass in a greenhouse.

❑ grenz

A soil horizon, which is frequently marked by a bed of clay, that results from a temporary halt in the accumulation of vegetal material.

❑ gross primary production

The total amount or weight of organic matter created by photosynthesis over a defined time period (total product of photosynthesis).

❑ ground cover

Plants grown to keep soil from eroding.

❑ grounding line

The boundary between the area where an ice shelf or a glacier is floating on water and where it is in contact with the shore or underlying earth (grounded).

❑ groundwater

The supply of fresh water found beneath the surface of the Earth (usually in aquifers) that often supplies wells and springs.

❑ growth water-use efficiency

A measure at the individual plant level of how well plants use available water in growth. The units of dry matter synthesized are divided by the units of water lost.

❑ gulf stream meander

A transient winding bend in the Gulf Stream. These bends intensify as the Gulf Stream merges into the North Atlantic and can break up into detached eddies at about 40 degrees S.

❑ gyres

Major circular flow patterns in the oceans. The wind- driven eastward- and westward-flowing equatorial currents are blocked by the continents and rotate slowly in a clockwise direction in the North Atlantic and Pacific Oceans and in a counter-clockwise direction in the South Atlantic, South Pacific, and Indian Oceans.

H

- **hadley cell**

A direct thermally-driven and zonally symmetric large- scale atmospheric circulation first proposed by George Hadley in 1735 as an explanation for the trade winds. It carries momentum, sensible heat, and potential heat from the tropics to the mid-latitudes (30 degrees). The poleward transport aloft is complemented by subsidence in the subtropical high pressure ridge and a surface return flow. The variability of this cell and the Walker cell is hypothesized to be a major factor in short-term climatic change.

- **halocline**

In the oceans, a well-defined vertical gradient of salinity.

- **heat flux (thermal flux)**

The amount of heat that is transferred across a surface of unit area in a unit of time.

- **heat island effect**

A dome of elevated temperatures over an urban area caused by the heat absorbed by structures and pavement.

- **herbivore**

An animal that feeds on plants.

- **heterotrophs**

Organisms that break down and use organic matter.

- **histosol**

Wet organic soils, such as peats and mucks.

- **holdridge life zone**

A climate category defined by three weighted climatic indexes, namely, mean annual heat, precipitation, and atmospheric moisture.

- **holocene**

The most recent epoch of the Quaternary period , covering approximately the last 10,000 years.

- **humus**

Decomposed organic material.

- **hydrologic budget**

A quantitative accounting of all water volumes and their changes with time for a basin or area.

❑ hydrologic cycle

The process of evaporation, vertical and horizontal transport of vapor, condensation, precipitation, and the flow of water from continents to oceans. It is a major factor in determining climate through its influence on surface vegetation, the clouds, snow and ice, and soil moisture. The hydrologic cycle is responsible for 25 to 30 percent of the mid-latitudes' heat transport from the equatorial to polar regions.

❑ hydrology

The science dealing with the properties, distribution, and circulation of water.

❑ hydrosphere

The aqueous envelope of the Earth, including the oceans, freshwater lakes, rivers, saline lakes and inland seas, soil moisture and vadose water, ground waters, and atmospheric vapor.

❑ hydrostatic equation

In the vector equation of motion, the form assumed by the vertical component when all Coriolis, earth-curvature, frictional, and vertical-acceleration terms are considered negligible compared with those involving the vertical pressure force and the force of gravity. The error in applying the hydrostatic equation to the atmosphere for cyclonic-scale motions is less than 0.01%. In extreme situations, the strong vertical accelerations in thunderstorms and mountain waves can be 1% of gravity.

❑ hypsithermal period

The period about 4000 to 8000 years ago when the Earth was apparently several degrees warmer than it is now. More rainfall occurred in most of the subtropical desert regions and less in the central midwest United States and Scandinavia. It is also called the altithermal period and can serve as a past climate analog for predicting the regional pattern of climate change should the mean Earth surface temperature increase from an increase in atmospheric carbon dioxide concentration.

I

❑ ice age

A glacial epoch or time of extensive glacial activity. Also, as Ice Age, which refers to the latest glacial epoch, the Pleistocene Epoch.

❑ ice and snow albedo

The reflectivity of ice and snow-covered surfaces. The albedo of freshly fallen snow may be as much as 90%, while older snow may have values of 75% or less. The larger the a real extent of snow and ice cover, the higher the albedo value. The surface albedo will also increase as a function of the depth of snow cover up to 13 cm and be unaffected by increased snow cover after reaching that depth.

❑ ice and snow-albedo-temperature feedback

Interactions that can be described as a theoretical concept of a feedback mechanism in which the interacting elements are the a real extent of polar ice and snow cover, the albedo of the polar region (dependent on a real extent of ice and snow), absorption of solar radiation (dependent on the albedo), temperature (dependent on the absorption of solar radiation) and the area of ice and snow cover (dependent on temperature). Less snowfall would mean more absorption of solar radiation, therefore a surface warming would occur. Climate modeling studies indicate an amplification effect (i.e., positive feedback) of the ice and snow-albedo feedback on increased surface air temperatures caused by increases in the atmospheric concentration of carbon dioxide.

❑ ice cover

During the present time, the extent, especially the thickness, of glacier ice on a land surface. Also the same as ice concentration, which is the ratio of an area of sea ice to the total area of sea surface within some large geographic area.

❑ ice front

The floating vertical cliff that forms the seaward face or edge of a glacier or an ice shelf that enters water. It can vary from 2 to 50 m in height.

❑ ice sheet (continental glacier)

A glacier of considerable thickness and more than 50,000 sq km in area. It forms a continuous cover of ice and snow over a land surface. An ice sheet is not confined by the underlying topography but spreads outward in all directions. During the Pleistocene Epoch, ice sheets covered large parts of North America and northern Europe but they are now confined to polar regions (e.g., Greenland and Antarctica).

❑ ice shelf

A sheet of very thick ice with a level or gently undulating surface. It is attached to the land on one side, but most of it is floating. On the seaward side, it is bounded by a steep cliff (ice front) 2 to 50 m or more above sea level. Ice shelves have formed along polar coasts (e.g., Antarctica and Greenland); they are very wide with some extending several hundreds of kilometers toward the sea from the coastline. They increase in size from annual snow accumulation and seaward extension of land glaciers. They decrease in size from warming, melting, and calving.

❑ infiltration (soil)

Movement of water from the ground surface into the soil.

❑ infrared radiation

Electromagnetic radiation lying in the wavelength interval from 0.7 micrometers to 1000 micrometersm. Its lower limit is bounded by visible radiation, and its upper limit by microwave radiation. Most of the energy emitted by the Earth and its atmosphere is at infrared wavelength. Infrared radiation is generated almost entirely by large-scale intra- molecular processes. The tri-atomic gases, such as water vapor, carbon dioxide, and ozone, absorb infrared radiation and play important roles in the propagation of infra- red radiation in the atmosphere.

❑ insolation

The solar radiation incident on a unit horizontal surface at the top of the atmosphere. It is sometimes referred to as solar irradiance. The latitudinal variation of insolation supplies the energy for the general circulation of the atmosphere. Insolation depends on the angle of incidence of the solar beam and on the solar constant.

❑ instantaneous transpiration efficiency

A measure at the physiologic level of how well plants use available water in photosynthesis. The assimilation rate is divided by the transpiration rate; the moles of CO_7 taken up are divided by the moles of water lost through transpiration in a unit of time.

❑ inversion

An anomaly in the normal positive lapse rate; usually refers to a thermal inversion, in which temperature increases rather than decreases with height.

❑ irradiance

The total radiant flux received on a unit area of a given real or imaginary surface. Also called the radiant flux density.

❑ isopynic

A line on a chart that connects all points of equal or constant density.

isostatic adjustment (isostatic compensation)

The process whereby lateral transport at the Earth's surface from erosion or deposition is compensated for by movements in a subcrustal layer to maintain equilibrium among units of varying masses and densities.

❑ isotherm

A line on a chart that connects all points of equal or constant temperature.

❑ isotope

One of two or more atoms that have the same atomic number (i.e., the same number of protons in their nuclei) but have different mass numbers.

K

❑ krakatoa (krakatau)

Active volcano 2667 ft (813 m) high in West Indonesia. It forms an island between Java and Sumatra. Its eruption in 1883, which was one of the most violent in modern times, scattered debris and darkened skies over vast areas. Additional eruptions occurred in late 1927 and in the l960s.

L

❑ lapse rate

The rapidity with which temperature decreases with altitude. The normal lapse rate is defined to be 3.6 degrees F per 1000 feet change in altitude. The dry adiabatic lapse rate is about 5.5 degrees F per 1000 feet, and the wet adiabatic lapse rate varies between 2 and 5 degrees F per 1000 feet.

❑ latent heat

Energy transferred from the earth's surface to the atmosphere through the evaporation and condensation processes.

❑ le chatelier 's principle

When an external force is applied to an equilibrium system, the system adjusts to minimize the effect of the force.

❑ life zone

A climatically-defined class that can be associated with regions of soil and biota with a high uniformity in species composition and environmental adaptation.

❑ lithosphere

The component of the Earth's surface comprising the rock, soil, and sediments. It is a relatively passive component of the climate system, and its physical characteristics are treated as fixed elements in the determination of climate.

❑ litter

Undecomposed plant residues on the soil surface.

❑ little ice age

A cold period that lasted from about A.D. 1550 to about A.D. 1850 in Europe, North America, and Asia. This period was marked by rapid expansion of mountain glaciers, especially in the Alps, Norway, Ireland, and Alaska. There were three maxima, beginning about 1650, about 1770, and 1850, each separated by slight warming intervals.

❑ loess

A buff-colored, wind-blown deposit of fine silt, which is frequently exposed in bluffs with steep faces. The thickness can range from 6 to 30 m. The loess of the USA and Europe is thought to be the fine materials first transported and deposited by the waters of melting ice sheets during the glacial period. It was later blown consider- able distances with, in some cases, deposition in lakes. The origin of Asiatic loess, however, is apparently wind-blown dust from central Asian deserts.

❑ longwave radiation

The radiation emitted in the spectral wavelength greater than 4 micrometers corresponding to the radiation emitted from the Earth and atmosphere. It is sometimes referred to as terrestrial radiation or infrared radiation, although somewhat imprecisely.

M

❑ marsh

A type of wetland that does not accumulate appreciable peat deposits and is dominated by herbacious vegetation. Marshes may be fresh- or saltwater, tidal or nontidal.

❑ mass balance

The application of the principle of the conservation of matter. For example, the mass of a glacier is not destroyed or created; the mass of a glacier and all its constitutive components remains the same despite alterations in their physical states. The mass balance of a glacier is calculated with the input/output relationships of ice, firn, and snow, usually measured in water equivalent. Output includes all ablative processes of surface melting, basal melting, evaporation, wind deflation, calving, and internal melting. Input includes direct precipitation, avalanching, and the growth of superimposed ice.

❑ mauna loa

An intermittently active volcano 13,680 ft (4,170 m) high in Hawaii. Last eruption was in 1984.

❑ mauna loa record

The record of measurements of atmospheric carbon dioxide concentrations taken at the Mauna Loa Observatory, Mauna Loa, Hawaii, since March 1958. The Mauna Loa record is the longest reliable daily record of atmospheric carbon dioxide measurements in the world.

❑ maunder minimum

The period from 1654 to 1714 when it was believed that there were no sunspots. It is now thought that there were some sunspots during that time but less than the numbers counted after 1800.

❑ mean sea level

The average height of the sea surface, based upon hourly observation of the tide height on the open coast or in adjacent waters that have free access to the sea. In the United States, it is defined as the average height of the sea surface for all stages of the tide over a nineteen year period. Mean sea level, commonly abbreviated as MSL and referred to simply as sea level, serves as the reference surface for all altitudes in upper atmospheric studies.

❑ mesic environment

A habitat with a moderate amount of water.

❑ mesoscale eddies (mode eddies)

In the ocean, dense and irregularly-oval high- and low- pressure centers about 400 km in diameter. The intensities of currents in these centers are about 10 times greater than the local means.

❑ milankovitch theory

An astronomical theory formulated by the Yugoslav mathematician Milutin Milankovitch that associates climate change with fluctuations in the seasonal and geographic distribution of insolation determined by periodic variations of the Earth's eccentricity and obliquity and the longitude of the perihelion.

❐ mist

Liquid particles 40 to 500 microns in diameter that are formed by condensation of vapor in air.

❐ modeling

An investigative technique that uses a mathematical or physical representation of a system or theory that accounts for all or some of its known properties. Models are often used to test the effects of changes of system components on the overall performance of the system.

❐ monsoon

A name for seasonal winds, first applied to the winds over the Arabian Sea that blow for six months from the northeast and for six months from the southwest. The term has been extended to similar winds in other parts of the world (i.e., the prevailing west to northwest winds of summer in Europe have been called the European monsoon). The primary cause for these seasonal winds is the much greater annual variation of temperature over large land areas compared with neighboring ocean surfaces, causing an excess of pressure over the continents in winter and a deficit in summer, but other factors, such as topography of the land, also have an effect. The monsoons are strongest in the southern and eastern sides of Asia, but also occur along the coasts of tropical regions wherever the planetary circulation is not strong enough to inhibit them. The monsoon climate can be described as a long winter-spring dry season, which includes a cold season followed by a short hot season just preceding the rains; a summer and early autumn rainy season, which is generally very wet but may vary greatly from year to year; and a secondary warming immediately after the rainy season.

N

❐ natural selection

The process of survival of the fittest by which organisms that adapt to their environment survive while those that do not adapt disappear.

❐ negative feedback

An interaction that reduces or dampens the response of the system in which it is incorporated.

❐ net primary production

The part of the gross primary production that remains stored in the producer organism (primarily green plants) after deducting the amount used during the process of respiration.

❐ nutrient

Any substance assimilated by living things that promotes growth.

O

❐ ocean-mixing

Processes that involve rates of advection, upwelling/ downwelling, and eddy diffusion and that determine how rapidly excess atmospheric carbon dioxide can be taken up by the oceans.

❐ opacity

The degree of obscuration of light; for example, a glass window has almost 0% opacity, whereas a concrete wall has 100% opacity.

❐ optical thickness (optical depth)

In calculating the transfer of radiant energy, the mass of an absorbing or emitting material lying in a vertical column of unit cross-sectional area and extending between two specified levels. Also, the degree to which a cloud prevents light from passing through it; the optical thickness then depends on the physical constitution (crystals, drops, and/or droplets), the form, the concentration, and the vertical extent of the cloud.

❐ ozone

A molecule made up of three atoms of oxygen. In the statosphere, it occurs naturally and it provides a protective layer shielding the Earth from ultraviolet radiation and subsequent harmful health effects on humans and the environment. In the troposphere, it is a chemical oxidant and major component of photochemical smog.

P

❐ paleosol

An ancient soil or soil horizon that formed on the surface during the geologic past.

❐ paludification

The expansion of a bog caused by the gradual rising of the water table as accumulation of peat impedes water drainage.

❐ palynology

The science of reconstructing the past flora and past climate from pollen data obtained from lake and bog sediments. The fossil pollen record is a function of the regional flora and vegetation at a given time and location.

❐ particulate matter

Very small pieces of solid or liquid matter, such as particles of soot, dust, aerosols, fumes, or mists.

❐ past climate analogs

The reconstructing of past climates at a given locality from modern climatic conditions in a different elevation or latitudinal zone to infer past climatic conditions.

❐ pco2

The partial pressure of CO_2 in the atmosphere and the ocean. In the atmosphere, the partial pressure of CO_2 is defined as the pressure the CO_2 would exert if all other gases were removed. The sum of the partial pressure of all the atmospheric gases will equal the atmospheric pressure. The partial pressure of CO_2 in the atmosphere is determined by the atmospheric CO_2 concentration and atmospheric temperature. In the ocean, the pCO_2 is determined by the amount of dissolved CO_2 and H_2CO_3. It varies with alkalinity, latitude, depth, and temperature. Biological processes in the ocean also exert an influence on the pCO_2 in the ocean.

❐ percolation

The movement of water downward and radially through the subsurface soil layers, usually continuing downward to the groundwater.

❐ permafrost

Perennially frozen ground that occurs wherever the temperature remains below 0 degrees C for several years.

❐ phenology

The study of periodic biological phenomena with relation to climate, particularly seasonal changes. These phenomena can be used to interpret local seasons and the climatic zones.

❐ photochemical smog

Air pollution caused by chemical reactions among various substances and pollutants in the atmosphere.

❐ photoelectric

Of or relating to the electrical effects of light, including the emission of electrons, the generation of a voltage, or a change in resistance.

❐ photosynthesis

The manufacture by plants of carbohydrates and oxygen from carbon dioxide and water in the presence of chlorophyll with sunlight as the energy source. Oxygen and water vapor are released in the process. Photosynthesis is dependent on favorable temperature and moisture conditions as well as on the atmospheric carbon dioxide concentration. Increased levels of carbon dioxide can increase net photosynthesis in many plants.

❐ phytoplankton

That portion of the plankton community comprised of tiny plants (e.g., algae and diatoms).

❐ planetary albedo

The fraction (approximately 30%) of incident solar radiation that is reflected by the earth-atmosphere system and returned to space, mostly by backscatter from clouds in the atmosphere.

❑ planetary boundary layer

The transition region between the turbulent surface layer and the normally nonturbulent free atmosphere. This region is about 1 km in thickness and is characterized by a well-developed mixing generated by frictional drag as the air masses move over the Earth's surface. This layer contains approximately 10% of the mass of the atmosphere. Also called the atmospheric boundary layer or frictional layer.

❑ plankton

Passively floating or weakly motile aquatic plants (phytoplankton) and animals (zooplankton).

❑ pleistocene

The earlier of the two epochs of the Quaternary period, starting 2 to 3 million years before the present and ending about 10,000 years ago. It was a time of glacial activity.

❑ positive feedback

An interaction that amplifies the response of the system in which it is incorporated.

❑ precipitation

Any or all forms of liquid or solid water particles that fall from the atmosphere and reach the Earth's surface. It includes drizzle, rain, snow, snow pellets, snow grains, ice crystals, ice pellets, and hail. The ratio of precipitation to evaporation is the most important factor in the distribution of vegetation zones. Precipitation is also defined as a measure of the quantity, expressed in centimeters or milliliters of liquid water depth, of the water substance that has fallen at a given location in a specified amount of time.

❑ primary succession

The natural development of vegetation and soil on a site that had not previously borne vegetation (e.g., a sand dune or lava flow), which vegetation will be replaced by other, successive plant communities.

❑ primitive equations

The Eulerian equations of fluid motion in which the primary dependent variables are the velocity components of the fluid. In meteorology, they can be specialized to apply directly to the cylonic-scale motions.

❑ proxy climate indicators

Dateable evidence of a biological or geological phenomenon whose condition, at least in part, is attributable to climatic conditions at the time of its formation. Proxy data are any material that provides an indirect measure of climate and include documentary evidence of crop yields, harvest dates, glacier movements, tree rings, varves, glaciers and snow lines, insect remains, pollen remains, marine microfauna, isotope measurements: 18O, in ice sheets, 18O, 2H, and 13C in tree rings; $CaCO_3$

in sediments; and speleothems. There are three main problems in using proxy data: (1) dating, (2) lag and response time, and (3) meteorological interpretation. Tree rings, pollen deposits from varved lakes, and ice cores are the most promising proxy data sources for reconstructing the climate of the last five millennia because the dating are precise on an annual basis while other proxy data sources may only be precise to +/- 100 years.

❐ pycnocline

In the ocean, a region where the water density increases rapidly with depth.

❐ pyrgeometer

An instrument that measures radiation from the earth's surface into space.

Q

❐ quaternary period

The latest period of geologic time, covering the most- recent 2,000,000 years of the Earth's history. It is divided into two epochs: the Pleistocene - 2 million years ago to approximately 10,000 years ago - and the Holocene - the period from approximately 10,000 years ago to the present. The Quaternary period is the artificial division of time separating prehuman and human periods. It contains five ice ages and four interglacial ages, and temperature indicators seem to show sharp and abrupt changes by several degrees.

R

❐ radiant flux density

The total flow of radiation received on a unit area of a given real or imaginary surface.

❐ radiation balance

The difference between the absorbed solar radiation and the net infrared radiation. Experimental data show that radiation from the earth's natural surfaces is rather close to the radiation from a black body at the corresponding temperature; the ratio of the observed values of radiation to black body radiation is generally 0.90 - 1.0.

❐ radiative-convective models

Thermodynamic models that determine the equilibrium temperature distribution for an atmospheric column and the underlying surface, subject to prescribed solar radiation at the top of the atmosphere and prescribed atmospheric composition and surface albedo. Submodels for the transfer of solar and terrestrial radiation, the heat exchange between the earth's surface and atmosphere, the vertical redistribution of heat within the atmosphere, the atmospheric water vapor content and clouds are included in these one-dimensional models. Abbreviated as RCM.

▫ radiatively active gases

Gases that absorb incoming solar radiation or outgoing infrared radiation, thus affecting the vertical temperature profile of the atmosphere. Most frequently being cited as being radiatively active gases are water vapor, CO_2, methane, nitrous oxide, chlorofluorocarbons, and ozone.

▫ radiosonde

A balloon-borne instrument for the simultaneous measurement and transmission of meteorological data up to a height of approximately 30,000 meters (100,000 feet). The height of each pressure level of the observation is computed from data received via radio signals.

▫ recharge

The process by which water is added to a reservoir or zone of saturation, often by runoff or percolation from the soil surface.

▫ reflectivity

The ratio of the energy carried by a wave that is reflected from a surface to the energy of a wave incident on the surface.

▫ relative sea level

The height of the boundary between sea and air as measured in relationship to a fixed reference point on land.

▫ reservoir

Any natural or artificial holding area used to store, regulate, or control a substance.

▫ residence time

The size of any specific reservoir or pool of mass (e.g., carbon) divided by the total flux of mass into or out of that pool.

▫ respiration

A biochemical process by which living organisms take up oxygen from the environment and consume organic matter, releasing both carbon dioxide and heat. In plants, the organic matter in photosynthate produced during daylight hours.

▫ revelle factor

The ratio of the instantaneous fractional change in the partial pressure of CO_2 (pCO_2) exerted by seawater to the fractional change in total CO_2 dissolved in the ocean waters. The buffer factor relates the partial pressure of CO_2 in the ocean to the total ocean CO_2 concentration at constant temperature, alkalinity and salinity. The Revelle factor is a useful parameter for examining the distribution of CO_2 between the atmosphere and the ocean, and measures in part the amount of CO_2 that can be dissolved in the mixed surface layer.

❐ rocketsonde

A rocket-borne instrument for measurement and transmission of upper-air meteorological data in the lower 76,000 meters (250,000 feet) of the atmosphere, especially that portion inaccessible to radiosonde techniques.

❐ runoff

That part of precipitation, snowmelt, or irrigation water that flows from the land to streams or other surface waters.

S

❐ salinity

The degree of salt in water.

❐ salt water intrusion

The invasion of fresh, surface, or groundwater by salt water.

❐ sea surface temperature

The temperature of the layer of seawater (approximately 0.5 m deep) nearest the atmosphere.

❐ sea surface temperature anomalies

Temperature of emitted energy from the sea surface. SST anomaly = (SST - SST mean).

❐ seasonal variation

The change in a set of meteorological parameters averaged over three months. Seasonal variation is the largest climatic variation, and temperature is the most frequently observed meteorological parameter. Often, monthly averaged data are grouped into seasons, according to the prescribed definition.

❐ secular carbon dioxide trend

The fairly uniform and accelerating increase of carbon dioxide concentration in the atmosphere, as illustrated by the Mauna Loa record. The secular trend reflects the increase in global atmospheric carbon dioxide concentrations produced by combustion of fossil fuels, kilning of limestone, and possibly a net biospheric release of carbon dioxide resulting from deforestation.

❐ sensible heat

The excess radiative energy that has passed from the Earth's surface to the atmosphere through advection, conduction, and convection processes.

❐ shortwave radiation

The radiation received from the sun and emitted in the spectral wavelengths less than 4 m. It is also called solar radiation.

❑ signal-to-noise ratio

A quantitative measure of the statistical detectability of a signal, expressed as a ratio of the magnitude of the signal relative to the variability. For first detection of a CO_2-induced climate change, the model signal is the mean change or anomaly in some climatic variable, usually surface air temperature, attributed by a numerical model to increased concentrations of carbon dioxide. Observed noise is the standard deviation or natural variability computed from observations of that variable and adjusted for sample size, autocorrelation, and time averaging.

❑ silviculture

Management of forest land for timber.

❑ smog

Air pollution associated with oxidants.

❑ smoke

Particles suspended in air after incomplete combustion of materials.

❑ soil carbon

A major component of the terrestrial biosphere pool in the carbon cycle. Organic soil carbon estimates, rather than total soil carbon, are generally quoted. The amount of carbon in the soil is a function of historical vegetative cover and productivity, which in turn is dependent upon climatic variables.

❑ solar constant

The rate at which solar energy is received just outside the Earth's atmosphere on a surface that is normal to the incident radiation and at the mean distance of the Earth from the sun. The current value is 0.140 watt/cm2.

❑ solar cycle

The periodic change in sunspot numbers. It is the interval between successive minima and is about 11.1 years.

❑ southern oscillation

A large-scale atmospheric and hydrospheric fluctuation centered in the equatorial Pacific Ocean. It exhibits a nearly annual pressure anomaly, alternatively high over the Indian Ocean and high over the South Pacific. Its period is slightly variable, averaging 2.33 years. The variation in pressure is accompanied by variations in wind strengths, ocean currents, sea-surface temperatures, and precipitation in the surrounding areas.

❑ southern oscillation index

An indicator based on the pressure gradient between the quasi-stationary low pressure region and the center of a subtropical high pressure cell. A positive index corresponds to an anomalously high pressure difference between the two centers of action.

❐ statistical-dynamical models

Computer programs that calculate simplified climate models based on versions of the conservation equations that have been averaged over longitude, with the effects of the synoptic eddies parameterized statistically in the meridional plane.

❐ steric height

The mean dynamic depth (or height) of the ocean for the month minus the annual mean dynamic depth for the same isobaric reference level.

❐ stoma, plant stomata

A minute pore in the epidermis of plant leaves or stems. Stoma, which are bordered by guard cells that regulate the size of the opening, function in gas exchange between the plant and the external environment. The stomatal apparatus or stomate consists of the stoma plus guard cells.

❐ stratification

Separating into layers.

❐ stratosphere

The region of the upper atmosphere extending from the tropopause (8 to 15 km altitude) to about 50 km. The thermal structure is determined by its radiation balance and is generally very stable with low humidity.

❐ suess effect

The relative change in the 14C/C or 13C/C ratio of any carbon pool or reservoir caused by the addition of fossil- fuel CO_2 to the atmosphere. Fossil fuels are devoid of 14C because of the radioactive decay of 14C to 14N during long underground storage and are depleted in 13C because of isotopic fractionation eons ago during photosynthesis by the plants that were the precursors of the fossil fuels. Carbon dioxide produced by the combustion of fossil fuels is thus virtually free of 14C and depleted in 13C. The term Suess effect originally referred to the dilution of the 14C/C ratio in atmospheric CO_2 by the admixture of fossil-fuel produced CO_2, but the definition has been extended to both the 14C and 13C ratios in any pool or reservoir of the carbon cycle resulting from human disturbances.

❐ sunspot

A relatively dark, sharply defined region on the solar disk, marked by an umbra approximately 2000K cooler than the effective photospheric temperature, surrounded by a less dark but also sharply bounded penumbra. The average spot diameter is about 3700 km, but can range up to 245,000 km. Most sunspots are found in groups of two or more, but they can occur singly. Sunspots are cyclic, with a period of approximately 11 years. The quantitative description of sunspot activity is called the Wolf sunspot number, denoted R. The Wolf sunspot number is also referred to as Wolfer sunspot number, Zurich relative sunspot number, or relative sunspot number.

❐ surface air temperature

The temperature of the air near the surface of the Earth, usually determined by a thermometer in an instrument shelter about 2 m above the ground. The true daily mean, obtained from a thermograph, is approximated by the mean of 24 hourly readings and may differ by 1.0 degrees C from the average based on minimum and maximum readings. The global average surface air temperature is 15 degrees C.

❐ surface albedo

The fraction of solar radiation incident on the Earth's surface that is reflected by it. Reflectivity varies with ground cover, and during the winter months it varies greatly with the amount of snow cover (depth and areal extent). Roughness of terrain, moisture content, solar angle, and angular and spectral distribution of ground-level irradiations are other factors affecting surface albedo.

❐ surface water

All water naturally open to the atmosphere.

❐ swamp

A type of wetland that is dominated by woody vegetation and does not accumulate appreciable peat deposits; it may be fresh- or saltwater, and tidal or nontidal.

T

❐ tephra

Any rock material produced by a volcano.

❐ terrestrial radiation

The total infrared radiation emitted by the Earth and its atmosphere in the temperature range of approximately 200-300K. Because the Earth is nearly a perfect radiator, the radiation from its surface varies as the fourth power of the surface's absolute temperature. Terrestrial radiation provides a major part of the potential energy changes necessary to drive the atmospheric wind system and is responsible for maintaining the surface air temperature within limits for livability.

❐ thermocline

A transition layer of water in the ocean, with a steeper vertical temperature gradient than that found in the layers of ocean above and below. The permanent thermocline separates the warm mixed surface layer of the ocean from the cold deep ocean water, and is found between 100- and 1000-m depths. The thermocline first appears at the 55 - 60 degree N and S latitudes, where it forms a horizontal separation between temperate and polar waters. The thermocline reaches its maximum depth at mid-latitudes and is shallowest at the equator and at its northern and southern limits. The thermocline is stably stratified, and transfer of water and carbon dioxide across this zone occurs very slowly. Thus, the thermocline acts as a barrier to the downward mixing of carbon dioxide.

❐ thermohaline

Refers to the combined effects of temperature and salinity that contribute to density variations in the oceans.

❒ tidal marsh

Low, flat marshlands traversed by channels and tidal hollows and subject to tidal innundation; normally, the only vegetation present are salt-tolerant bushes and grasses.

❒ trace gas

A minor constituent of the atmosphere. The most important trace gases contributing to the greenhouse effect are water vapor, carbon dioxide, ozone, methane, ammonia, nitric acid, nitrous oxide, ethylene, sulfur dioxide, nitric oxide, dichlorofluoromethane or Freon 12, trichlorofluoromethane or Freon 11, methyl chloride, carbon monoxide, and carbon tetrachloride.

❒ transient tracers

Chemical elements (often radioactive) or compounds that have finite lifetimes. Atmospheric testing of nuclear weapons from the mid-1950s to the early 1960s released large quantities of radionuclides to the atmosphere. Atmosphere-ocean exchange processes have transferred some of these elements to the oceans. Studying the behavior and distribution of these specific isotopes and other chemical tracers in the ocean will provide information on: residence times of the water and its dissolved components in gyres, basins, etc.; the mode and rate of formation and the subsequent spreading rates of specific water types, such as the polar water of the Norwegian and Greenland Seas; deep-ocean circulation and ocean- mixing processes, such as advection and upwelling; and the flux of anthropogenic carbon dioxide into the ocean through its correlation with several different transient tracers.

❒ transpiration

The process in plants by which water is taken up by the roots and released as water vapor by the leaves. The term can also be applied to the quantity of water thus dissipated.

❒ tree rings

Annual growth increments of trees that indicate, among other factors, the climatic conditions that enhance or limit growth. Tree ring widths and indexes have been used to search for solar-terrestrial relationships and climatic cycles and to reconstruct past climates.

❒ trophic level

A segment of the food chain in which all organisms obtain food and energy in, basically, the same manner (e.g., photosynthesis, herbivory, or carnivory) and in which all organisms are the same number of links from the photosynthetic segment.

❒ tropopause

The boundary between the troposphere and the stratosphere (about 8 km in polar regions and about 15 km in tropical regions), usually characterized by an abrupt change of lapse rate. The regions above the troposphere have increased atmospheric stability than those below. The tropopause marks the vertical limit of most clouds and storms.

❑ troposphere

The inner layer of the atmosphere below about 15 km, within which there is normally a steady decrease of temperature with increasing altitude. Nearly all clouds form and weather conditions manifest themselves within this region, and its thermal structure is caused primarily by the heating of the Earth's surface by solar radiation, followed by heat transfer by turbulent mixing and convection.

❑ tundra

A type of ecosystem dominated by lichens, mosses, grasses, and woody plants. It is found at high latitudes (arctic tundra) and high altitudes (alpine tundra). Arctic tundra is underlain by permafrost and usually very wet.

❑ turnover rate

The fraction of the total amount of mass (e.g., carbon) in a given pool or reservoir that is released from or that enters the pool in a given length of time. The turnover rate of carbon is often expressed as GtC/year.

U

❑ upwelling

The vertical motion of water in the ocean by which subsurface water of lower temperature and greater density moves toward the surface of the ocean. Upwelling occurs most commonly among the western coastlines of continents, but may occur anywhere in the ocean. Upwelling results when winds blowing nearly parallel to a continental coastline transport the light surface water away from the coast. Subsurface water of greater density and lower temperature replaces the surface water, and exerts a considerable influence on the weather of coastal regions. Carbon dioxide is transferred to the atmosphere in regions of upwelling. This is especially important in the Pacific equatorial regions, where 1 - 2 GtC/year may be released to the atmosphere. Upwelling also results in increased ocean productivity by transporting nutrient-rich waters to the surface layer of the ocean.

V

❑ vapor

The gaseous phase of substances that are liquid or solid at atmospheric pressure (e.g., steam).

❑ varve

A layer of sediment deposited in lakes during one year. Each layer consists of two parts, which are deposited at different seasons and which differ in color and texture; thus, the layers can be counted and measured. In a complete series, the number of layers gives the date on which the ground was vacated by the retreating ice.

W

❑ walker cell

A zonal circulation of the atmosphere confined to equatorial regions and driven principally by the oceanic temperature gradient. In the Pacific, air flows westward from the colder, eastern area to the warm, western ocean, where it acquires warmth and moisture and subsequently rises. A return flow aloft and subsidence over the eastern ocean complete the cell.

❑ water stress effect

The closing of the stomata by a plant in response to excessive water loss through transpiration or in response to drought conditions. The stomatal closing reduces CO_2 uptake as well as water loss, thus decreasing the photosynthetic rate. However, under conditions of elevated CO_2 concentration, the CO_2 gradient between the atmosphere and the leaf is higher than under ambient conditions, and CO_2 can pass through partially closed stomates at a rate similar to that under conditions of lower CO_2 and open stomates. The humidity gradient remains the same at higher CO_2, and transpiration is impeded. The net result is improved water-use efficiency by some plants.

❑ water table

The level of groundwater.

❑ water vapor

Water present in the atmosphere in gaseous form; the source of all forms of condensation and precipitation. Water vapor, clouds, and carbon dioxide are the main atmospheric components in the exchange of terrestrial radiation in the troposphere, serving as a regulator of planetary temperatures via the greenhouse effect. Approximately 50 percent of the atmosphere's moisture lies within about 1.84 km of the earth's surface, and only a minute fraction of the total occurs above the tropopause.

❑ water vapor feedback

A process in which an increase in the amount of water vapor increases the atmosphere's absorption of longwave radiation, thereby contributing to a warming of the atmosphere. Warming, in turn, may result in increased evaporation and an increase in the initial water vapor anomaly. This feedback, along with carbon dioxide, is responsible for the greenhouse effect and operates virtually continuously in the atmosphere.

❑ water-use efficiency

A measure of the amount of water used by plants per unit of plant material produced. The term can be applied at the leaf, whole-plant, and ecosystem levels. At the leaf level, it is more precisely referred to as the instantaneous transpiration efficiency, the CO_2 assimilation rate (photosynthesis) divided by the transpiration rate (the moles of CO_2 taken up divided by the moles of water lost through transpiration in a unit of time per unit leaf area). At the whole-plant level, it is more precisely referred to as the growth water-use efficiency, the units of dry matter synthesized divided by the units of water lost. At the ecosystem level, it is more precisely referred to as the crop water-use efficiency, the grams of dry weight gained by plants during the growing season per unit land area divided by the millimeters of water lost (including evaporation directly from the soil).

❏ weather

The instantaneous state of the global atmosphere-ocean- cryosphere system.

❏ wetlands

An area that is regularly saturated by surface water or groundwater and subsequently is characterized by a prevalence of vegetation that is adapted for life in saturated-soil conditions.

Z

❏ zonally-averaged models

Statistical-dynamical or energy-balance models in which only the latitudinally averaged quantities are determined and the effects of the longitudinally varying transports are determined parametrically.

❏ zooplankton

That portion of the plankton community comprised of tiny aquatic animals eaten by fish.

Civic Environment Disaster

A

❐ absorbed dose

The amount of radiation energy absorbed in a unit of mass of an organ or living thing from ionizing radiation.

❐ absorbing capacity

The buffering ability that enables a society to dissipate the effects of an event. It is a function of the level of preparedness of the society to respond to the event plus the resilience of the population and environment. Absorbing capacity = resilience + preparedness.

❐ acceleration

A force having the units of gravity that denotes the rate of change of the back and forth movement of the ground during an earthquake. Velocity (the rate of the ground motion at a given instant of time with units of cm/s) and displacement (the distance the ground has moved from its rest position with units of cm) are derived from an accelerogram.

❐ accelerogram

The record or time history obtained from an instrument called an accelerometer showing acceleration of a point on the ground or a point in a building as a function of time. The peak acceleration, the largest value of acceleration on the record is typically used in design criteria. The velocity and displacement time histories and the response spectrum are derived analytically from the time history of acceleration.

❐ acceptable risk

The probability of occurrences of physical, social, or economic consequences of an earthquake that is considered by authorities to be sufficiently low in comparison with the risks from other natural or technological hazards that these occurrences are accepted as realistic reference points for determining design requirements for structures, or for taking social, political, legal, and economic actions in the community to protect people and property.

❐ acid rain

Sulphuric acid (H_2SO_4) in the atmosphere, formed by the combination of sulphur trioxide with water, resulting in a relatively stable mist of acid droplets. In excessive

excessive concentrations in the air it increases the acidity of the soil and disturbs the ph causing agricultural and ecological damage.

❐ acquired immunodeficiency syndrome (aids)

A highly infectious disease of pandemic proportions, caused by the HIV virus. A person who has the virus is a carrier and can infect others. Spread is by sexual intercourse, by contaminated needles and syringes, transfusions of infected blood, by an infected mother to her unborn child. Spread is unlikely through daily social contact, such as shaking hands. Note: Non-infective immune deficiency can also be acquired from radiation.

❐ active fault

A fault is active if it exhibits physical characteristics such as historic earthquake activity, surface fault rupture, geologically recent displacement of stratigraphy or topography, or physical association with another fault system judged to be active. When these characteristics are suspected or proven, it is classed as active and judged to be able to undergo movement. See Fault.

❐ acute exposure

An exposure, often intense, over a relatively short period of time.

❐ advance planning activities

Collectively called disaster preparedness. Includes 1) strategic planning, 2) contingency planning, and 3) forward planning

❐ advanced life support

Those invasive measures (such as intubation and ventilation, infundation and thoracic drainage) as to preserve life of ABC-unstable patients.

❐ affected people

People requiring immediate assistance during a period of emergency, i.e. Requiring basic survival needs such as food, water, shelter, sanitation and immediate medical assistance (definition considered in EM-DAT; included in the field " total affected "); Appearance of a significant number of cases of an infectious disease introduced in a region or a population that is usually free from that disease.

❐ aftershocks

The long, exponentially decaying sequence of smaller earthquakes that follow a large-magnitude earthquake for months to years, exacerbating the damage.

❐ agency on toxic substances and disease registry (atsdr)

Federal agency that conducts public health and safety studies.

❐ aid

Free material or financial assistance or other support given to an organization, community or country.

❐ air pollution

Presence of considerable quantities of gaseous, liquid or solid contaminants in the atmosphere and liable to be harmful to animal, vegetable and human life.

❐ air scent canine

A search dog that will detect airborne human scent.

❐ alara

As Low As Reasonably Achievable; concept utilized in relation to intervention levels following the release of dangerous chemical or nuclear materials.

❐ alarm procedure

Alerting every party concerned precedes repressive disaster management. Various optical and acoustical means of alarm are possible: flags, lights, sirens, radio and telephone.

❐ alert

Term used for physical signs a canine or horse gives when they are interested in something. Canines may alert by barking, horses may alert by placing their ears forward, distending their nostrils, focusing their eyes, stopping forward movement, etc; Automated Local Evaluation in Real Time.

❐ alpha particle

An electrically charged (+) particle emitted from the nucleus of some radioactive chemicals, e.g. Plutonium. It contains two protons and two neutrons, and is the hardest of the atomic particles emitted by radioactive chemicals. It can cause ionization. Alpha particles are relatively heavy and large and cannot easily penetrate clothing or skin, but they are highly dangerous if inhaled or ingested. See 'Ionizing radiation'.

❐ alpha radiation

The least penetrating type of nuclear radiation; not considered dangerous unless alpha-contaminated particles enter the body.

❐ ambient

Any unconfined portion of the atmosphere; open air; outside surrounding air.

❐ ambient temperature

Temperature of the surroundings.

❐ ambulance support

When disaster strikes a certain (ambulance) region, ambulance support is needed from surrounding regions according to a preplanned scheme.

❐ american chemical council

The chemical industry's trade association.

❐ american red cross

The American Red Cross is a humanitarian organization, led by volunteers, that provides relief to victims of disasters and helps people prevent, prepare for, and respond to emergencies. It does this through services that are consistent with its

Congressional Charter and the Principles of the International Red Cross Movement.

❑ ascender

Mechanical device (clamps) utilized by high angle rescue teams, that allows the flow of rope to be controlled in a specific direction (usually upward).

❑ asphyxiation

One of the six types of harm (see TRACEM) that can be encountered at a terrorist incident. Asphyxiants interfere with oxygen flow during normal breathing. There are two types of asphyxiants: simple and chemical.

❑ assessment

Survey of a real or potential disaster to estimate the actual or expected damages and to make recommendations for preparedness, mitigation and relief action.

❑ atom

The smallest particle of matter that can take part in a chemical reaction. Some atoms may combine to form larger but still tiny molecules.

❑ attack

A hostile action taken against the United States by foreign forces or terrorists, resulting in the destruction of or damage to military targets, injury or death to the civilian population or damage or destruction to public and private property.

❑ attenuation

A decrease in the strength of seismic waves and seismic energy with distance from the point where the fault rupture originated. Also referred to as Seismic Wave Attenuation Function.

❑ attraction search

Search based on the idea that the victim is alive, willing to be found and will, upon hearing or seeing signs of rescue, make them self and their position known.

❑ avalanche

Sudden slide of a huge mass of snow and ice, usually carrying with it earth, rocks, trees and other debris; Rapid and sudden sliding and flowage of masses of usually unsorted mixtures of snow/ice/rock material.

B

❑ background radiation

This includes emissions from radioactive chemicals which occur naturally and those which result from the nuclear fission process. The meaning of this term is vague. In a licensing process it includes radiation from all sources other than the particular nuclear facility located on the same site (US regulation). Radioactive chemicals released from a nuclear power plant are called 'background' after one year.

❒ bacteria

Single-celled organisms that multiply by cell division and can cause disease in humans, plants, or animals. Examples include anthrax, cholera, plague, tularemia, and Q fever.

❒ basic life support

Those non-invasive measures (such as elimination of airway obstruction, cardo-pulmonary resuscitation, hemorrhage control, woundcare, and immobilization of fractures) as to preserve life of ABC-unstable patients.

❒ bastard search

A search in which, for one reason or another and unknown to the searchers, there is no subject, as the subject is elsewhere rather than in the search area.

❒ batf

Bureau of Alcohol, Tobacco and Firearms

❒ becquerel (bq)

Unit of nuclear activity. 1 bq represents the amount of radioactive substance where on disintegration occurs per second.

❒ belay

Term used by high/low angle rescue teams, for the act of one person controlling the descent of another person or object.

❒ beta emitter

A nuclide which releases beta radiation.

❒ beta particle

An electrically charged (-) particle emitted from some radioactive chemicals. It has the mass of an electron. Krypton-85, emitted from nuclear power plants, is a strong beta emitter. Beta particles can penetrate a centimetre or more of a living tissue but can be stopped by thin sheets of denser materials. They can also cause ionization. See 'Ionizing radiation'

❒ beta radiation

A type of nuclear radiation that is more penetrating than alpha radiation and can damage skin tissue and harm internal organs.

❒ bhopal

A city in India which, in 1984, suffered the worst chemical accident in history, with a gas cloud from a Union Carbide plant that killed 3,000 people overnight and severely injuring 200,000.

❒ bilateral cooperation

Technical cooperation or assistance given by a donor country to a recipient country, through direct adreement between the two governments without UN or other intermediary

❑ biological agent

(A) Living organisms, or the materials derived from them, that cause disease in, or harm, humans, animals, or plants, or cause deterioration of material. Biological agents may be found as liquid droplets, aerosols, or dry-y powders. A biological agent can be adapted and used as a terrorist weapon, such as anthrax, tularemia, cholera, encephalitis, plague, and botulism. There are three different types of biological agents: bacteria, viruses, and toxins;

(B) Microorganisms or toxins from living organisms that have infectious or non-infectious properties which produce lethal or serious effects in plants and animals (FBI).

❑ biological incident

An event in which a biological agent is used as a terrorist weapon.

❑ biological safety

A specialized area within the field of Occupational Safety and Health, Biological Safety or "biosafety" has as its goal the protection of workers from hazards created by the manipulation of biological life forms. Containment, decontamination, and protective equipment procedures are used to reduce the hazards from agents of disease such as bacteria and viruses.

❑ biological terrorism/bioterrorism

The use and dissemination of various kinds of microbes or toxins with the intent to intimidate or coerce a government or civilian population to further political or social objectives; humans, animals, and plants are often targets

❑ biological warfare

The intentional spread of disease in warfare through the dispersal of infective bacteria, rickettsiae, viruses or toxins which cause diseases, such as anthrax, plague, typhoid, brucellosis. There is a un convention against biological weapons. Biological and chemical weapons are usually considered together (cbw). Synonyms: bacteriological warfare, biological weapon, bw.

❑ blind fault

A fault system that is not visible at the surface of the ground and can only be detected by using geophysical techniques such as drilling, seismic reflection profiles, gravity profiles, or magnetic profiles.

❑ blister agent

A chemical agent, also called a vesicant, which causes severe blistering and bums to eyes, skin, and tissues of the respiratory tract. Exposure is through liquid or vapor contact. Also referred to as mustard agents; examples include mustard and lewisite.

❑ blood agent

A chemical agent that interferes with the ability of blood to transport oxygen and causes asphyxiation. These substances injure a person by interfering with cell respiration (the exchange of oxygen and carbon dioxide between blood and tissues). Common examples are hydrogen cyanide and cyanogen chloride.

❑ bloodhound

A search dog that is remarkable for its keen sense of smell and often used as tracking/trailing canines.

❑ blue plains

The name of a sewage treatment plant in Washington, DC which switched from liquid chlorine to sodium hypochlorite for wastewater treatment in October 2001.

❑ b-nice

Five categories of terrorist incidents: Biological, Nuclear, Incendiary, Chemical and Explosives; The acronym for identifying the five categories of terrorist incidents: Biological, Nuclear, Incendiary, Chemical, and Explosives.

❑ bowline

One of the more common fixed loop knots utilized in rescue situations.

❑ briefing

A meeting before a search in which information is disclosed to the search teams in order to initiate and properly execute the search.

❑ buffer zones

The amount of land set aside between chemical storage, use or production facilities and surrounding land uses.

❑ built environment

The buildings and lifelines (or infrastructure) of the community.

❑ burnout

Mental and emotional problems that occur soon after or even during a stressful incident or series of incidents.

C

❑ c/b-rrt

Chemical/biological-rapid response team

❑ cal/osha

Under the California Department of Labor, Cal/OSHA is the primary state agency that is charged with implementation and enforcement of worker safety and health standards as set forth by the California Code of Regulations and federal Occupational Safety and Health Administration.

❑ cap

Civil air patrol

❑ capture technologies

Equipment added to chemical facilities to impede a release of chemicals from reaching the external environment.

❑ carabiner

A device utilized extensively by high/low angle rescue teams. It is usually a "D" shaped type of metal link that opens on one side to insert rope or webbing.

❑ casualty

Any person suffering physical and/or psychological damage by outside violence leading to either death, injuries, or material losses only

❑ casualty clearing station

Collecting point for T1 and T2 victims in the immediate vicinity of the disaster site where further triage and basic and advanced life support can be provided.

❑ caustic material

A substance capable of destroying or eating away living tissue.

❑ caving

The art of being able to navigate through caves usually for recreational purposes, but may be for rescue purposes.

❑ central holding area

An ambulance assembly location in the filter area from where ambulances either leave to pick-up patients from the casualty clearing station or to leave for one of the neighbouring hospitals according to a victim distribution plan.

❑ centre for research on the epidemiology of disasters

A pioneering facility at the School of Public Health, Catholic University of Louvain, for research, study and training in the epidemiology and medical aspects of disasters. A WHO Collaborating Centre.

❑ cert

Community emergency response team

❑ chain of medical care

The chain of medical care from the disaster site to the hospital bed, along which the patient is medically handled and treated, can be divided into 3 phases: the medical organization at the site, the distribution of patients among neighbouring hospitals, and the organization in the hospitals.

❑ checklist

Written (or computerized) enumeration of actions to be taken by an individual or organization, meant to aid memory rather than provide detailed instruction.

▫ chemical agent

Solids, liquids, or gases that have chemical properties that produce lethal or serious effects in plants and animals (FBI). There are five classes of chemical agents, all of which produce incapacitation, serious injury, or death: (1) nerve agents, (2) blister agents, (3) blood agents, (4) choking agents, and (5) irritating agents. A chemical substance used in military operations is intended to kill, seriously injure, or incapacitate people through its physiological effects.

▫ chemical asphyxiant

Referred to as blood poisons, these are compounds that interrupt the flow of oxygen in the blood or the tissues in three ways: (1) They react more readily than oxygen with the blood. Carbon monoxide is the best-known example. (2) They liberate the hemoglobin from red blood cells, resulting in a lack of transport for oxygen. Hydrazine is one such Asphyxiant. (3) They cause a malfunction in the oxygen carrying ability of the red blood cells. Benzene and toluene are two of these.

▫ chemical harm

One of the six types of harm (see TRACEM) that can be encountered at a terrorist incident. There are two broad types of chemical agents that can cause harm: toxic and corrosive materials.

▫ chemical hygiene plan

Required by Federal Regulation 29 CFR 1910 and California Regulation Title 8, Section 5191, the Chemical Hygiene Plan establishes safety standards for workers exposed to hazardous chemicals in laboratories. A written plan to implement control measures, training, and other protective measures is required of laboratories meeting specified criteria.

▫ chemical incident

An event in which a chemical agent is used as a terrorist weapon.

▫ chemical lesions

Body lesions caused by chemical substances either through external (skin and mucous membranes) or internal (inhalation and ingestion) contamination, or both, leading to a variety of reactions from skin irritation and ventilatory problems to systemic effects and even death.

▫ chemical release 1: seveso, italy, 1976

- On Saturday, July 10, 1976 a major accident at a chemical plant in the small town of Meda in Northern Italy led to the release of a toxic cloud containing about 3 tons of pollutants, including the deadly chemical dioxin.
- The ICMESA (Industrie Chimiche Meda Società) plant, owned by Givaudan S.A. in Genevá, was producing intermediate compounds for the cosmetics, pharmaceutical and agricultural industries. This included 2,4,5 trichlorophenol (TCP), a toxic compound used to make herbicides.
- The plant had been shut down for the weekend, part way through production of a batch of TCP. The accident occurred when an unexpected heatreleasing

reaction raised the temperature in the TCP reactor vessel from the normal 300°F to 850°F. This high temperature caused dioxin to be produced and led to a pressure build-up, rupturing the plate on a safety valve. The hot chemicals burst through the roof into the air, forming a huge toxic cloud in the 20 minutes before a worker noticed and stopped the release.

- A brisk 15 foot/second wind carried the cloud southeastward from ICMESA, contaminating over 270 acres, including parts of the large town of Seveso from which the incident takes its name.
- Dioxin (2,3,7,8 tetrachlorodibenzo-para-dioxin) is an extremely toxic compound, which is very stable, only breaking down slowly once released into the environment. It builds up in the fatty tissues of exposed animals, becoming more concentrated as it passes up the food chain. The dangers of dioxin were poorly understood at the time.
- Estimates of the amount of dioxin released vary widely from just 12 oz to nearly 300 lb. There were probably several pounds of dioxin released. The cloud also contained TCP, caustic soda, ethylene glycol, solvents, chlorinated phenols and other toxic compounds.

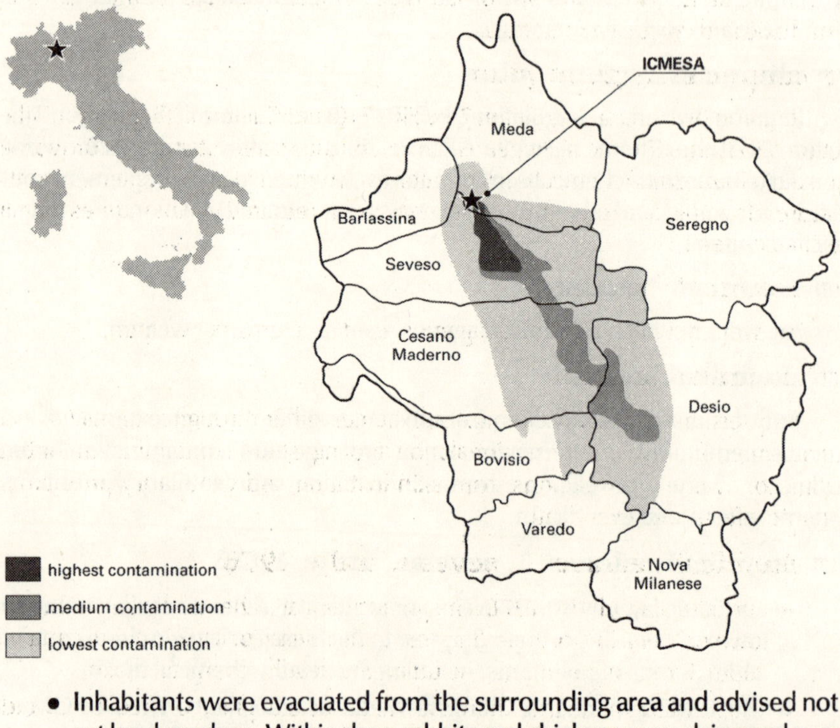

- Inhabitants were evacuated from the surrounding area and advised not to eat local produce. Within days, rabbits and chickens were dying and people began reporting skin rashes. Later, all 80,000 farm animals in the worst affected area were slaughtered to reduce the risk of humans ingesting dioxin. In all, an estimated 37,000 people were exposed to some level of contamination.

- Although only one person died from liver cancer attributed to dioxin, special measures remained in place for several years while a huge decontamination program was carried out. Life in Seveso eventually returned to normal, although there are lingering concerns over the health effects of the accident.
- 200 cases of skin disease (chloracne) were eventually reported, and some long-term studies have suggested increased mortality from liver disease, cardiovascular disease, respiratory disease, diabetes and some cancers as a result of the accident, although the findings are inconclusive. Young men who wre exposed to dioxin during the accident were less likely to father male children subsequently, possibly due to hormonal disruption.
- The scale of the accident prompted the European Union to review legislation concerning industrial risk. The 1982 Seveso Directive aimed to reduce the chances of such an accident happening again, and was updated as the Seveso II Directive on the Control of Major Accident Hazards in 1996.

❑ chemical release 2: bhopal, india, 1984

- Around midnight on December 2, 1984 44 tons of highly poisonous methyl isocyanate (MIC) gas escaped from the Union Carbide Corporation pesticide production facility in Bhopal, India.
- The gas, because it is denser than air, stayed close to the ground as it swept over hundreds of shanties and huts before spreading 5 miles downwind over the sleeping city of around 900,000 people.
- MIC is an organic chemical used in the production of Sevin, a commercial insecticide, and various other pesticides. The Indian government encouraged their use as part of a program of industrialization and a move towards the intensive farming methods widely used in developed countries.
- MIC is extremely toxic to humans and even brief exposure may cause death or serious health effects including pulmonary edema, bronchitis, bronchial pneumonia, brain damage and reproductive problems.
- At least 2,000 people died that night. By 1991 the official death toll had risen to 3,800 plus approximately 11,000 seriously disabled, and now stands at more than 5,000 dead. Other sources suggest much higher figures of up to 20,000 dead and 300,000 injured, 70,000 permanently. Suffering, illness and death continue to the present day.
- The precise cause of the release is disputed. A major Union Carbide investigation concluded that the plant was sabotaged by a disgruntled employee, who deliberately added water to a tank of MIC in an attempt to spoil it. Others accuse the owners of negligence, pointing to design flaws, poor maintenance, ineffective safety procedures and inadequate staffing resulting from cost-cutting.
- The addition of water to the tank of MIC, however it occurred, lead to a vigorous reaction which quickly raised the temperature and pressure inside the tank. MIC escaped for nearly two hours through the reliefvalve system into the night air.
- Workers first detected the leak at 11:30 p.m., experiencing irritation of their eyes and throats. They reported their concerns to supervisors, who

decided to have their tea before investigating the source of the leak. By the time they took action at around 12:15 a.m., it was too late.

- Union Carbide (now owned by Dow Chemical) agreed to an out-of-court settlement with the Indian government of $470 million in 1989. This move effectively blocked claims being pursued through the US courts, which would undoubtedly have been far more costly.
- Much of the money has never reached the victims and their families. The average payout to those who have received anything has been about $580.

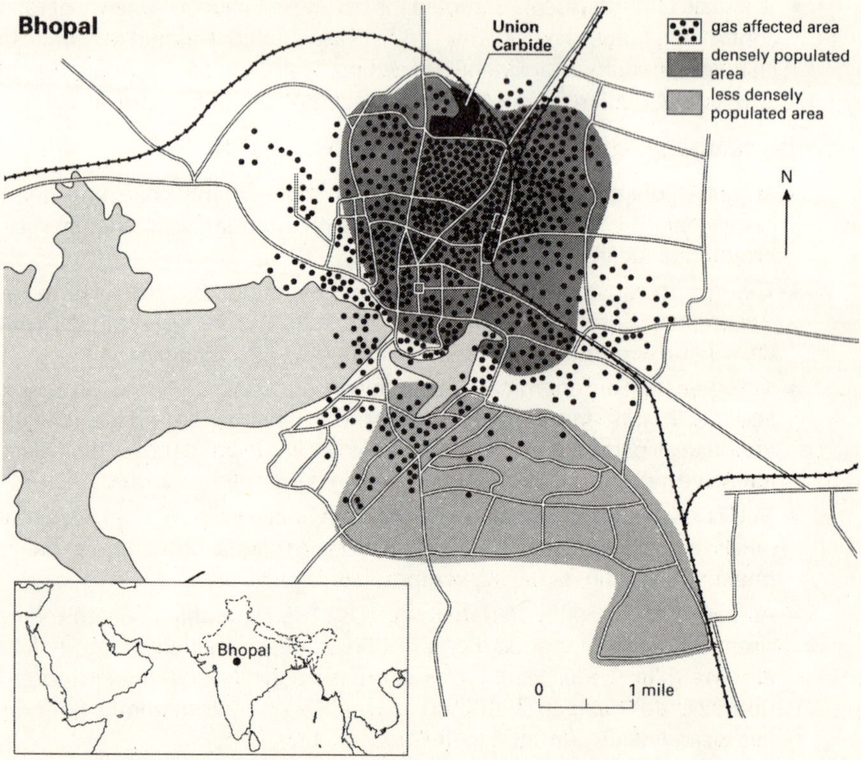

❐ chemical release 3: basel, switzerland, 1986

- On November 1, 1986 a fire broke out in a large storehouse at the Sandoz chemical factory near Basel, Switzerland. It contained around 1,400 tons of toxic chemicals including fungicides, pesticides and dyes.
- The fire destroyed most of the chemicals, but over 30 tons were washed into the nearby river Rhine by the 4 million gallons of water used to combat the fire. Pollution also entered the atmosphere and soaked into groundwater through soil near the site.
- Pollution from the accident, including over 400 lbs of mercury, poisoned life in the Rhine for hundreds of miles downstream. The water turned red s,

in just two hours, more pollutants entered the river than it would normally receive in a whole year.

- The Rhine stretches for 820 miles. 50 million people live in the Rhine valley, which passes through nine European countries: Germany, Switzerland, France, Holland, Austria, Luxembourg, Italy, Liechtenstein and Belgium.
- High levels of atrazine were found, a chemical not listed as being stored at the Sandoz site. It was then disclosed that another chemical company, Ciba- Geigy, had been responsible for an accidental discharge into the Rhine only the day before.

Chemical plants on the River Rhine

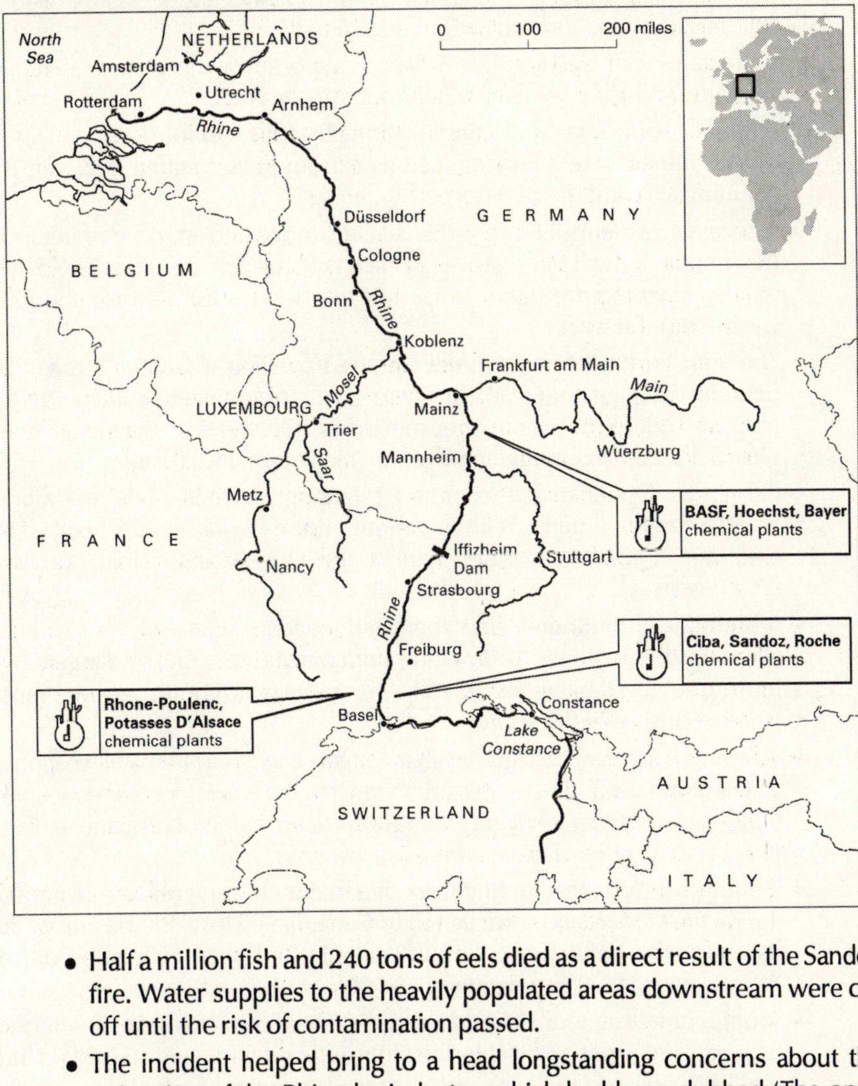

- Half a million fish and 240 tons of eels died as a direct result of the Sandoz fire. Water supplies to the heavily populated areas downstream were cut off until the risk of contamination passed.
- The incident helped bring to a head longstanding concerns about the poisoning of the Rhine by industry, which had been dubbed 'The open sewer of Europe'. Concerted cross-border efforts to clean up the river have

now been largely successful, although there are still fears over the risks posed by persistent chemical pollutants eventually leaching through the riverbed into the groundwater.

- The cleanup has been one of the most successful reversals of humanly caused environmental damage.

❑ chemical release 4: mississauga, canada, 1979-412 (ci)

The 'Mississauga Miracle'

- Just before midnight on Saturday, November 10, 1979 a 106-car freight train carrying poisonous chemicals and explosive gases was derailed in Mississauga near Toronto in Ontario, Canada.
- Canadian Pacific Railway Train 54 was transporting a mixed cargo including propane, chlorine, styrene, toluene and caustic soda.
- 25 miles from Toronto, friction resulting from lack of lubricating oil in an old-fashioned wheel bearing lead to extreme overheating between the spinning axle and the car above.
- Trackside residents later reported seeing smoke and sparks coming from the middle of the train. The intense heat split the axle, sending its glowing wheels crashing through a fence into the backyardof a house about 15 yards from the track.
- The train continued on for three minutes, travelling about two more miles until the dangling undercarriage of the 33rd car, containing toluene, left the track at 11:53 p.m. Twenty three more cars followed the toluene tanker off the tracks, some containing propane which burst into flames.
- Other derailed tanker cars ruptured, releasing a volatile mixture of vapors and flammable liquids. Within a minute there was a massive explosion, creating a fireball rising 5,000 feet into the night air and visible from over 60 miles away.
- A member of the train's crew managed to close a cock on the 32nd car, allowing the engineer to drive the undamaged cars out of danger. Two more propane tankers exploded in the following minutes as emergency services arrived at the scene.
- Chlorine is a poisonous heavier-than-air gas, used as a chemical weapon in World War I. As soon as it became known that a leaking tanker containing 90 tons of chlorine was lying amongst more unexploded propane tankers, the first of a series of evacuations was ordered.
- Shifting winds and mounting fears of a major chlorine release prompted the biggest emergency evacuation in Canadian history. By the end of the day after the derailment, 218,000 people had been evacuated out of Mississauga's total population of 284,000.
- A long, complicated, and hazardous operation followed to make the wrecked tankers safe. Most of the chlorine, it turned out, had been sucked up into the flames of the initial fire and dispersed harmlessly over Lake Ontario. Only luck had prevented a Bhopal-scale disaster.

- By Friday evening, all the evacuees were allowed back to their homes. Miraculously, no casualties resulted from the incident.

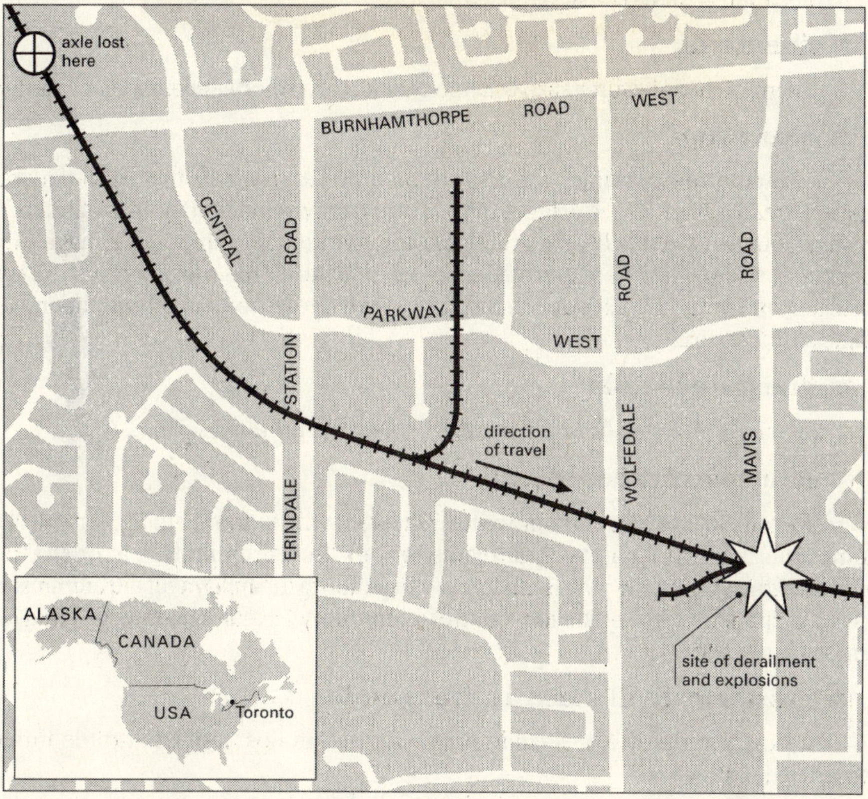

❐ chemical spill

Accident release occurring during the production, transportation or handling of hazardous chemical substances.

❐ chemical terrorism

Attacks meant to cause mass devastation in which terrorist organizations release toxins; chemical attacks meant primarily to terrorize, blackmail, or cause economic damage; a specific attack in a particular product, particularly a food product

❐ chemical warfare

War in which harmful chemical substances are used with the intention to kill, injure, or otherwise incapacitate humans or to destroy the environment and national economies. The many chemical weapons are grouped in seven main categories in terms of their toxic properties: nerve agents (lethal); pulmonary toxics (lethal); cyanide (lethal); tissue damaging vesicants (mustards, lewisites, halogenated oximes); psychomimetics; riot control agents (incapacitating); and defoliants. Chemical weapons are internationally outlawed by the 1925 geneva protocol.

❐ chemical warfare agents

Highly toxic chemicals that can be disseminated as vapors, gases, liquids, or aerosols or adsorbed to dust particles

❐ chemtrec

Chemical manufacturers' association chemical transportation emergency center.

❐ chernobyl

A town in the Ukraine, U.S.S.R, site of a nuclear reactor that exploded and burned on 26 April 1987, causing extensive radioactive contamination locally and in distant countries, with deaths and agricultural and environmental damage. The exploded reactor has been permanently out of action. The most serious nuclear disaster up to that date, resulting also in extensive revision and strengthening of supervision by IAEA.

❐ chernobyl aids

Radiation-induced Aids which breaks down the immune system.

❐ chief executive official

The official of the community who is charged with authority to implement and administer laws, ordinances, and regulations for the community. He or she The official of the community who is charged with authority to implement and administer laws, ordinances, and regulations for the community. He or she may be a mayor, city manager, etc.

❐ child survival - lost in the woods

A program that teaches children how not to get lost and what to do if they should get lost.

❐ choking agent

A chemical agent that causes physical injury to the lungs. In extreme cases, membranes swell and lungs become filled with liquid, which can result in asphyxiation resembling drowning. Death results from lack of oxygen; hence, the victim is "choked." Common examples are chlorine and phosgene.

❐ chronic

An exposure, often mild, over a long period of time.

❐ clean air act amendments of 1990

Amendments enacted by Congress to the nation's key air pollution law, which added requirements for prevention of chemical catastrophes through risk management planning and a facility owner or operator's general duty to prevent chemical accidents.

❐ clean production

Industrial design strategies which seek to minimize the damage of an industrial process on environment throughout the production lifecycle, including materials extraction and disposal, as well as waste and energy consumption issues at the production facility.

❒ cold wave

Long lasting period with extremely low surface temperature.

❒ collapse

Accident involving the collapse of building or structure. Can either involve industrial structures or domestic / non-industrial structures.

❒ combustible

Substance that catches fire and burns easily.

❒ commission of european communities

The institutional arrangements bringing together the European Communities with the aim of gradually integrating their economies and moving towards a political unity, with a European Parliament in Brussels. It encompasses the European Economic Community (EEC or Common Market). Has an active programme for disaster relief and assistance to developing countries.

❒ communicable disease

An infectious condition that can be transmitted from one living person or animal to another through a variety of channels, according to the nature of the disease

❒ community

A political entity which has the authority to adopt and enforce laws and ordinances for the area under its jurisdiction. In most cases, the community is an incorporated town, city, township, village, or unincorporated area of a county. However, each State defines its own political subdivisions and forms of government.

❒ community labor coalitions

Alliances of chemical facility workers, unions, local residents, and other organizations representing members of the community or the public interest.

❒ community lifelines

Structures and facilities that provide the essential functions of supply, disposal, communication, and transportation in a community. They are also called infrastructure.

❒ community profile

Characteristics of the local environment prone to a chemical or nuclear accident: population density, age distribution, roads, railways, waterways, types of dwellings and buildings, and the relief agencies locally available.

❒ compass

A device that indicates northerly direction based on the needle pointing at the nearest magnetic core.

❑ confinement search

Also a containment measure. Search in which the object is to confine the subject to a specific area.

❑ consequence management

As described in PDD-39, consequence management is the response to the disaster, and focuses on alleviating damage, loss, hardship, or suffering. The Federal Emergency Management Agency (FEMA) has the lead in consequence management.

❑ contamination

Accidental release of hazardous chemical or nuclear materials leads to pollution of the environment in which man could be contaminated by these materials, either externally (skin and mucous membranes) or internally (by inhalation or ingestion) or both.

1. Invasion of a person or animal by pathogenic germs (contaminants).
2. Presence of an infectious agent on inanimate articles such as clothes, surgical instruments, dressings, water, milk, food.
3. Transfer and propagation of a contaminant.

❑ contingency plan

A document setting out an organized, planned, and coordinated course of action to be followed in case of fire, explosion, or other accident that releases toxic chemicals, hazardous waste, or radioactive materials that threaten human health or the environment; An anticipatory emergency plan to be followed in an expected or eventual disaster, based on risk assessment, availability of human and material resources, community preparedness, local and international response capability, etc.

❑ contingency planning

Site-specific and recognizes that a disaster could occur at any time.

❑ conventional weapons

More frequently used weapons of terrorism such as bombs and guns

❑ corrosive

Substance that causes visible destruction of living tissue.

❑ corrosive materials

One type of chemical agent that can cause chemical harm at an incident scene. They are liquids or solids causing visible destruction or irreversible alterations in human skin tissue at the site of contact.

❑ cosmic radiation

Beams of very high energy particles (protons, alpha particles and certain heavier nuclei) of solar, galactic or extragalactic origin.

❑ council of europe

Organization bringing together, at Strasbourg, 21 States of Europe, to "achieve a greater unity for the purpose of safeguarding and realizing the ideals and principles which are their common heritage and facilitating their economic and social progress." Promotes the European Centre for Disaster Medicine in San Marino.

❑ cpc

Chemical protective clothing.

❑ crash team

Team comprising a doctor and a nurse specialized in advanced trauma life support and meant for stabilizing seriously wounded victims.

❑ creeping disaster

A disaster of insidious onset and slow progress, such as famine, drought, desertification, health deterioration or epidemic, that does not become manifest until damage and suffering reach extensive proportions and need massive emergency response.

❑ crisis management

Measures to identify, acquire, and plan the use of resources needed to anticipate, prevent, and/or resolve a threat or act; As described in PDD-39, crisis management is the law enforcement response, and focuses on the criminal aspects of the incident. The Federal Bureau of Investigation (FBI) has the lead in crisis management.

❑ critical incident stress

Problems that can arise as a result of an over taxing situation, such as a major disaster event.

❑ critique

A formal review of any SAR incident, no matter how large or small. The critique is a process intended to identify the lessons learned from an exercise or mission and is not for placing blame, pointing the finger or allowing adversaries to embarrass each other.

❑ crop failure

Abnormal reductions in crop yield such that is insufficient to meet the nutritional or economic needs of the community.

❑ csepp

The EPZ is divided into three concentric circular zones

1) Immediate Response Zone (IRZ),
2) Protective Action Zone (PAZ), and
3) Precautionary Zone (PZ).

❑ cu

Weapons of Mass Destruction Countermeasures Unit (FBI)

❑ cyclone

Large-scale closed circulation system in the atmosphere above the Indian Ocean and South Pacific with low barometric pressure and strong winds that rotate clockwise. Maximum wind speed of 64 knots or more.

D

❑ d.o.

Doctor of Osteopathic Medicine

❑ dam

A barrier built across a watercourse for the purpose of impounding, controlling, or diverting the flow of water.

❑ damage assessment

The process used to appraise or determine the number of injuries and deaths, damage to public and private property, and the status of key facilities and services such as hospitals and other health care facilities, fire and police stations, communications networks, water and sanitation systems, utilities, and transportation networks resulting from a man-made or natural disaster.

❑ dart

Downed aircraft rescue transmitter also disaster assistance response team.

❑ dead

Persons confirmed as dead and persons missing and presumed dead.

❑ debriefing

A meeting after a search or after a team has been called in, in which all information found is disclosed and where suggestions can be made for the future.

❑ decontamination

The removal of hazardous chemical or nuclear substances from the skin and/ or mucous membranes by showering or washing with water, or out of wounds by rinsing with sterile solutions; The reduction or removal of a chemical, biological, or radiological material from the surface of a structure, area, object, or person.

❑ definition. classification. and scoring of disaster

From a medical point of view, a disaster needs only two criteria: victims and a discrepancy between the number and treatment capacity. Disasters can then be classified utilizing various parameters; man-made versus God-made, the radius of the disaster site, the number of dead, the number of wounded, the average severity of the injuries sustained, the impact time, and the rescue time. By attributing 0, 1 or 2 to each of them, increasing with intensity, number or time a scale can be produced varying between 0 and 13, which is called the Disaster Severity Scale.

❐ delayed onset disaster

Includes events that have a prolonged duration.

❐ department of transportation

The federal agency responsible for policies and procedures governing the transport of materials, including hazardous wastes.

❐ df

Direction Finding-as in equipment described under "ELT".

❐ dioxin

The chemical compound 2, 3, 7, 8, tetra chlorodibenzo-p-dioxin, an extremely toxic substance used in manufacturing some herbicides. The harmful effects are very persistent and capable of causing severe illness and chromosomal malformations. A major accident occurred in seveso.

❐ dirty bomb

See radiological dispersion bomb

❐ disaster

An occurrence, either natural or man made, that causes human suffering and creates human needs that victims cannot alleviate without assistance; There are many definitions of a disaster, and these definitions have been discussed in detail by Al-Mahari and Keller in a recent issue of Prehospital and Disaster Medicine. The definitions used seem dependent upon the discipline using the term. Thus, no definition of "disaster" is accepted universally. In the course on Disaster Management published in Prehospital and Disaster Medicine, Frederick C. Cuny defined a disaster as "a situation resulting from an environmental phenomenon or armed conflict that produced stress, personal injury, physical damage, and economic disruption of great magnitude."

(1) Perez and Thompson in their series on Natural Disasters, define a disaster as "the occurrence of widespread, severe damage, injury, or loss of life or property, with which the community cannot cope, and during which the affected society undergoes severe disruption."

(2) These definitions note that a disaster disrupts the infrastructure of the society stricken by the event. Furthermore, Cuny stresses that the event resulting in a disaster does not comprise the disaster: it is what results from the event, not the precipitating event itself.

The result of a vast ecological breakdown in the relations between man and his environment, a serious and sudden event (or slow, as in drought) on such a scale that the stricken community need extraordinary efforts to cope with it, often with outside help or international aid. Synonym: catastrophe

An occurrence of a natural catastrophe, technological accident, or human-caused event that has resulted in severe property damage, deaths, and/or multiple injuries. As used in this Guide, a "large-scale disaster" is one that exceeds the response capability of the local jurisdiction and requires State, and potentially Federal, involvement. As used in the Stafford Act, a "major disaster" is "any natural catastrophe or, regardless of cause, any fire, flood, or explosion, in any part of the United States, which in the determination of the President causes damage of sufficient severity and magnitude to warrant major disaster assistance under [the] Act to supplement the efforts and available resources or States, local governments, and disaster relief organizations in alleviating the damage, loss, hardship, or suffering caused thereby."

Situation or event, which overwhelms local capacity, necessitating a request to national or international level for external assistance (definition considered in EM-DAT); An unforeseen and often sudden event that causes great damage, destruction and human suffering. Though often caused by nature, disasters can have human origins. Wars and civil disturbances that destroy homelands and displace people are included among the causes of disasters. Other causes can be: building collapse, blizzard, drought, epidemic, earthquake, explosion, fire, flood, hazardous material or transportation incident (such as a chemical spill), hurricane, nuclear incident, tornado or volcano.

❑ disaster field office

The office established in or near the designated area of a Presidentially declared major disaster to support Federal and State response and recovery operations. The DFO houses the FCO and ERT, and where possible, the SCO and support staff.

❑ disaster medicine

The combination of medical and medico-organizational measures undertaken in case of disaster covering the entire range of medical care from the scene of the disaster to the hospital bed; The study and collaborative application of various health disciplines--e.g., paediatrics, epidemiology, communicable diseases, nutrition, public health, emergency surgery, social medicine, community care, international health--to the prevention, immediate response and rehabilitation of the health problems arising from disaster, in cooperation with other disciplines involved in comprehensive disaster management.

❑ disaster mitigation

Separate and aggregate measures taken prior to or following a disaster to reduce the severity of the human and material damage caused by it.

❑ disaster preparedness

The aggregate of measures to be taken in view of disasters, consisting of plans and action programmes designed to minimize loss of life and damage, to organize and facilitate effective rescue and relief, and to rehabilitate after disaster. Preparedness requires the necessary legislation and means to cope with disaster or similar emergency situations. It is also concerned with forecasting and warning, the education and training of the public, organization and management, including plans, training of personnel, the stockpiling of supplies and ensuring the needed funds and other resources.

❑ disaster prevention

The aggregate of approaches and measures to ensure that human action or natural phenomena do not cause or result in disaster or similar emergency. It implies the formulation and implementation of long range policies and programmes to eliminate or prevent the occurrence of disasters. Based on vulnerability analysis of risks, it also includes legislation and regulatory measures in the field of town planning, public works and environmental development.

❐ disaster recovery center

Places established in the area of a Presidentially declared major disaster, as soon as practicable, to provide victims the opportunity to apply in person for assistance and/or obtain information relating to that assistance. Drcs are staffed by local, State, and Federal agency representatives, as well as staff from volunteer organizations (e.g., the ARC).

❐ disaster site

Area where the immediate impact of the disaster took place. The first duty of the police is to seal down this area by an inner cordon. Outside this cordon a second one can be laid: the outer cordon. The area in between both cordons is then called the filter area, through which at one or two points the disaster site can be reached by rescuers.

❐ disinfestation

Technique or process used to destroy parasites, insects and other undesirable small animal species such as arthropods or rodents present on the person, on clothing, domestic animals or in the environment. Delousing is disinfestation against body lice.

❐ distance

One of the three components of the time, distance, and shielding (TDS) response; refers to the recommendation that one maintain distance from a hazard if at all possible.

❐ dog team

One handler and one canine.

❐ domestic terrorism

Terrorist acts by individuals or groups within a given country, without foreign direction or involvement

❐ donor

A country, organization, or agency that provides relief or, in different ways, comes to the assistance of a population in disaster.

❐ dose

Energy imparted to matter by nuclear transformations (radioactivity). This is the general term for the quantity of radiation absorbed by the body.

❐ dose (radiation)

A general term indicating the quantity (total or accumulated) of ionizing radiation or energy absorbed by a person or animal.

▫ dose rate

The amount of ionizing radiation which an individual would absorb per unit of time.

▫ dosemeter

A device used to measure doses of ionizing radiation.

▫ dosimeter

An instrument for measuring and registering total accumulated exposure to ionizing radiation.

▫ dot

Department of Transportation.

▫ drought

Period of deficiency of moisture in the soil such that there is inadequate water required for plants, animals and human beings.

▫ duration

A measure of the length of time the ground shaking exceeds a given threshold of shaking, such as 5 % of the force of gravity. Also, a description of the length of time between the onset and the departure of a natural hazard.

E

▫ earth flow

Mass of water-logged earth, sliding by gravity along a slope at a relatively slow speed of a few kilometers per hour

▫ earthquake

The violent shaking of the ground produced by deep seismic waves, beneath the epicentre, generated by a sudden decrease or release in a volume of rock of elastic strain accumulated over a long time in regions of seismic activity (tectonic earthquake). The magnitude of an earthquake is represented by the richter scale; the intensity by the mercalli scale. The sudden motion or trembling of the ground produced by abrupt displacement of rock masses, usually within the upper 10 to 20 miles of the earth's surface. Sudden break within the upper layers of the earth, sometimes breaking the surface, resulting in the vibration of the ground, which where strong enough will cause the collapse of buildings and destruction of life and property.

▫ earthquake cycle

The notion of an earthquake cycle dates back to the hypothesis formulated by H. F. Reid on the basis of his observations in the 1906 San Francisco, CA earthquake. The concept holds that two comparable sized earthquakes rupturing the same segment of a fault will be separated by a period of time long enough to reaccumulate strain in the amount equal to the elastic strain drop in the first earthquake. The stages in the cycle are: 1)a long period of seismic quiescence, except for the aftershocks, following a major earthquake, 2)a shorter period of increased seismicity as elastic strain accumulation approaches the critical strain level, and 3) the next major earthquake as the critical strain level is suddenly exceeded.

❑ earthquake hazards

The physical effects generated in an earthquake (e.g., ground shaking, ground failure, surface fault rupture, regional tectonic deformation, tsunami run up, seiches, and aftershocks).

❑ earthquake resistant buildings

Buildings that are sited, designed and constructed in such a way that they are able to resist the ground shaking from large-magnitude earthquakes without collapsing and from moderate-magnitude earthquakes without significant loss of function and with damage that is repairable.

❑ effect(s)

The consequences of an event. The effects may be single, but most often are multiple and involve multiple basic functions. Any or all of the basic functions of the society affected may become impaired as part of the effects of the impact: 1) medical services; 2) public health; 3) sanitation and water supplies; 4) food; 5) shelter/clothing; 6) energy supply; 8) public works; 9) environment; 10) logistics/transportation; 11) security; 12) communications; and/or 13) economy. In addition, the structure that provides for coordination and control of these functions may become impaired or inoperative. Effects may be acute and have a short time course or may have several phases usually described as 1) alert/notification; 2) response and recovery; and 3) reconstruction and rehabilitation. It is important to recognize that there may be primary and secondary effects of an event. Primary effects are those that are a direct result of the event. Secondary effects are those that result from the primary effects or from the responses to the event. Although described as acute, some effects may be ongoing and stretch over long periods of time (e.g., famine, drought, epidemics, complex human emergencies). These effects are functions of the vulnerability of the population and the environment and to the human responses to the impact of the event. Effects = vtotal +a1z1 +a2z2 where vtotal is the total, aggregate vulnerability of the population, a1z1 are the positive responses, and a2z2 are the negative responses. Positive responses produce a positive result (outcome) and negative responses result in a negative outcome.

❑ ehs

Extremely hazardous substance.

❑ el nino

Anormalous warming of ocean water resulting from the oscillation of current in the South Pacific, usually accompanied by heavy rainfall in the coastal region of Peru and Chile, and reduction of rainfall in equatorial Africa and Australia.

❑ electromagnetic pulse

A sharp pulse of energy radiated instantaneously by a nuclear detonation which may affect or damage electronic components and equipment.

❑ electron

A negatively charged electronic particle that is lighter than protons and neutrons. All atoms are made up in part of electrons. A free electron is one which is not associated with a nucleus.

❑ element

A substance which consists only of atoms of the same atomic number and which cannot be split up into simpler substances by chemical means. There are over 100 different elements, of which 92 occur naturally.

❑ elements at risk

People, ecosystems, natural resources, the environment, buildings and infrastructure, essential facilities, and critical facilities that are voluntarily or involuntarily exposed to natural and technological hazards.

❑ elt

Emergency Locator Transmitter-A low-power transmitting device carried on light aircraft as required by federal law, and on most military aircraft. It is impact activated and transmits a distinctive sweeping tone that can be detected by aircraft mounted or hand-held direction finding (DF) equipment.

❑ emergency

A sudden and usually unforeseen event that must be countered immediately to minimize the consequences. The term is often used for disaster. With rational planning, emergencies can be tackled more effectively. Any occasion or instance--such as a hurricane, tornado, storm-n, flood, tidal wave, tsunami, earthquake, volcanic eruption landslide, mud slide, snowstorm, fire, explosion, nuclear accident, or any other natural or man-made catastrophe--that warrants action to save lives and to protect property, public health, and safety.

❑ emergency response team

An interagency team consisting of the lead representative from each Federal department or agency assigned primary responsibility for an ESF and key members of the FCO 's staff, formed to assist the FCO in carrying out his/her coordination responsibilities. The ERT may be expanded by the FCO to include designated representatives of other Federal departments and agencies as needed. The ERT usually consists of regional-level staff.

❑ emergency response team advance element

For Federal disaster response and recovery activities under the Stafford Act, the portion of the ERT that is first deployed to the field to respond to a disaster incident. The ERT-A is the nucleus of the full ERT.

❑ emergency alert system (eas)

A digital technology (voice/text) communications system consisting of broadcast stations and interconnecting facilities authorized by the Federal Communication Commission. The system provides the President and other national, State, and local officials the means to broadcast emergency information to the public before, during, and after disasters.

❒ emergency and hazardous chemical inventory

An annual report by facilities having one or more extremely hazardous substances or hazardous chemicals above certain weight limits, as specified in Section 311 and 312 of EPCRA.

❒ emergency disaster plan

A plan prepared by organizations to provide a swift, efficient, and cost-effective response to medical, fire, care and shelter, and communications needs after disasters such as earthquakes, explosions, or fires.

❒ emergency environmental health services

Services required to correct or improve damaging environmental health effects on humans, including inspection for food contamination, inspection for water contamination, and vector control; providing for sewage and solid waste inspection and disposal; clean-up and disposal of hazardous materials; and sanitation inspection for emergency shelter facilities.

❒ emergency health kit

Basic drugs and medical equipment calculated for the emergency needs of a population of 10,000 people over three months. One prepackaged kit contains 10 identical smaller kits, each for 1,000 persons.

❒ emergency health services

Services required to prevent and treat the damaging health effects of an emergency, including communicable disease control, immunization, laboratory services, dental and nutritional services; providing first aid for treatment of ambulatory patients and those with minor injuries; providing public health information on emergency treatment, prevention and control; and providing administrative support including maintenance of vital records and providing for a conduit of emergency health funds from State and Federal governments.

❒ emergency medical services

Services, including personnel, facilities, and equipment required to ensure proper medical care for the sick and injured from the time of injury to the time of final disposition, including medical disposition within a hospital, temporary medical facility, or special care facility, release from site, or declared dead. Further, emergency medical services specifically include those services immediately required to ensure proper medical care and specialized treatment for patients in a hospital and coordination of related hospital services.

❒ emergency mortuary services

Services required to assure adequate death investigation, identification, and disposition of bodies, removal, temporary storage, and transportation of bodies to temporary morgue Facilities; notification of next of kin; and coordination of mortuary services and burial of unclaimed bodies.

❐ emergency operating center

The protected site from which State and local civil government officials coordinate, monitor, and direct emergency response activities during an emergency.

❐ emergency operations plan

A document that: describes how people and property win be protected in disaster and disaster threat situations; details who is responsible for carrying out specific actions; identifies the personnel, equipment, facilities, supplies, and other resources available for use in the disaster; and outlines how all actions will be coordinated.

❐ emergency operations plan (eop)

An EOP is a document that

(1) assigns responsibility to organizations and individuals for carrying out specific actions at projected times and places in an emergency that exceeds the capability or routine responsibility of any one agency;

(2) sets forth lines of authority and organizational relationships, and shows how all actions will be coordinated;

(3) describes how people and property will be protected in emergencies and disasters;

(4) identifies personnel, equipment, facilities, supplies, and other resources available for use during response and recovery operations; and

(5) identifies steps to address mitigation concerns during response and recovery activities.

❐ emergency planning zones (epz)

Areas around a facility for which planning is needed to ensure prompt and effective actions are taken to protect the health and safety of the public if an accident occurs. The REP Program and CSEPP use the EPZ concept.

❐ emergency preparedness

Planning and actions undertaken in advance of a possible or probable disaster. Preparation of an Emergency Disaster Plan is one component of Emergency Preparedness, but construction design, food and sanitation equipment storage, and conducting drills are also included.

❐ emergency preparedness coordinator

The local government official designated to be notified immediately of chemical emergencies (e.g., spills, chemical releases, explosions, or fires) under EPCRA.

❐ emergency responders

Firefighters, police, health care workers and hazmat teams that respond to chemical incidents.

❑ emergency response plans

The plans prepared by emergency responders and LEPC's prescribing how various individuals and institutions will take action in the event of a future chemical accident or other emergency in the community.

❑ emergency response team national

An ERT that has been established and rostered for deployment to catastrophic disasters where the resources of the FEMA Region have been, or are expected to be, overwhelmed. Three ERT-Ns have been established.

❑ emergency stage

Involves the immediate response to the effects of the disaster, immediate community provides assistance or aid because outside sources of aid have not yet arrived

❑ emergency support function (esf)

In the FRP, a functional area of response activity established to facilitate the delivery of Federal assistance required during the immediate response phase of a disaster to save fives, protect property and public health, and to maintain public safety. Esfs represent those types of Federal assistance which the State will most likely need because of the impact of a catastrophic or significant disaster on its own resources and response capabilities, or because of the specialized or unique nature of the assistance required. ESF missions are designed to supplement State and local response efforts. The Federal Response Plan (FRP) details 12 esfs in place to coordinate operations during Federal involvement in an incident: transportation, communications, public works and engineering, firefighting, information and planning, mass care, resource support, health and medical services, urban search and rescue, hazardous materials, food, and energy.

❑ emergency support team

An interagency group operating from FEMA headquarters. The EST oversees the national-level response support effort under the FRP and coordinates activities with the ESF primary and support agencies in supporting Federal requirements in the field.

❑ employee exposure records

Information, results, or records concerning employee exposures to toxic or harmful substances or agents in the workplace. Examples are work area sampling results, biological monitoring results (blood tests, etc.), inventories of chemicals, and Material Safety Data Sheets.

❑ employee medical records

Records concerning the health status of employees that are made by physicians, nurses, or other health professionals. Examples include results of medical examinations, first aid records, medical complaints and diagnoses, opinions, and treatments recommended by a physician.

❑ end -of -pipe pollution control technologies

Equipment added to facilities to capture waste or pollutants generated by processes within the operation of the facility. The opposite of end-ofpipe control technologies is pollution prevention or source reduction, which seek to change processes to avoid the creation of the wastes in the first place.

❑ environment

The aggregate, at any given time, of the physical, chemical and biological agents and social factors that can have a direct or indirect, immediate or late effect on living organisms and on human activities

❑ epa

Environmental protection agency.

❑ epcra

Emergency Planning and Community Right to Know Act Also know as SARA Titile III, this law is intended to improve local community access to information about chemical hazards and to improve state and local emergency response capabilities. Some features of EPCRA include community Right-to-Know provisions and the TRI reporting requirement. See Also:CERCLA, SARA, LEPC, MSDS, SERC, TRI.

❑ epicenter

The point on the earth's surface vertically above the subsurface point where the fault rupture originated; The point or area on the earth's surface immediately above the focus of an earthquake; Point on the earth's surface directly above the place of origin (i.e. Focus, or hypocenter) of an earthquake.

❑ epidemic

Either an unusual increase in the number of cases of an infectious disease, which already exists in the region or population concerned; or the appearance of an infection previously absent from a region.

❑ epidemiology

The medical discipline that studies the influence of such factors as the life style, biological constitution and other personal or social determinants on the incidence and distribution of disease.

❑ epidemiology of disaster

Only with a uniform and standard definition, classification and scoring system for (the medical aspects of) disasters is epidemiologic research feasible.

❑ epirb

Emergency Position Indicating Radio Beacon-The marine version of the "ELT". It will float and is manually activated or activated when submerged in the water.

❑ estimated damage

The economic impact of a disaster usually consists of direct (e.g. Damage to infrastructure, crops, housing) and indirect (e.g. Loss of revenues, unemployment, market destabilisation) consequences on the local economy. In EM-DAT estimated damage are (if available) given in 2 different currencies (in thousand): Dam US$ and Dam Euros. If cost damage is given in the local currency, it will be directly converted in US$ and in EURO for European countries. For each disaster, the registered figure corresponds to the damage value at the moment of the event, i.e. The figures are shown true to the year of the event.

❑ etiological harm

One of the six types of harm (see TRACEM) that can be encountered at a terrorist incident. Involves exposure to a living microorganism, or its toxins, which causes, or may cause, human disease. Biological agents are the most obvious examples of etiological agents.

❑ european centre for disaster medicine

Intergovernmental centre established in San Marino under the aegis of the Council of Europe, to promote prevention and mitigation of the effects of natural and technological disasters through research, training programmes and international collaboration, in particular among European countries..

❑ evacuation

Organized, phased, and supervised dispersal of people from dangerous or potentially dangerous areas.

❑ evacuees

All persons removed or moving from areas threatened or struck by a disaster.

❑ event

Occurs when the hazard is realized or becomes manifest. For the current discussion, it means an occurrence that negatively affects living beings and/or their environment. Such occurrences have a magnitude and a temporal component. Temporally, they may be either sudden onset or delayed onset and may be acute (short term) or prolonged.

❑ event intensity

A measure of the power or force (magnitude) of the event in relation to the time over which the event occurs and the area over or in which it occurs (Dorland 843).

❑ event magnitude

The overall size of the event (e.g., area of drought, number of persons potentially or actually affected, etc.). Magnitude usually expressed as a mathematical quantity.(ox 820) (i.e., richter scale for earthquakes, safir-simpson scale for tropical cyclones).

❑ exceedance probablilty

A term used in probabilistic ground shaking maps. The probability (for example, 10%) that an earthquake will generate a level of ground motion that exceeds a specified reference level during a given exposure time.

- **explosions**

Explosions involving buildings or structures. Can either involve industrial structures or domestic/non-industrial structures.

- **explosive**

As defined by the U.S. Department of Transportation, "a substance fitting into one of these two categories: (1) any substance or article, including a device, designed to function by explosion; or (2) any substance or article, including a device, which, by chemical reaction within itself, can function in a similar manner even if not designed to function by explosion.

- **explosive incident**

An event in which an explosives device is used as a terrorist weapon.

- **exposure (radiological)**

A quantitative measure of gamma or x-ray radiation at a certain place based on its ability to produce ionization in air.

- **exposure rate (radiological)**

The amount of ionizing radiation to which an individual would be exposed or which he or she would receive per unit of time.

- **exposure time**

A term used in probabilistic ground shaking maps. The period of time (for example, 50 years) that a structure or a community is exposed to potential earthquake ground shaking and other earthquake hazards.

- **extreme temperature**

Disaster type term used in EM-DAT comprising the two disaster subsets " heat wave " and " cold wave ".

- **extremely hazardous substances**

Any of 406 chemicals identified by EPA as toxic, and listed under SARA Title III. The list is subject to periodic revision.

F

- **facility emergency coordinator**

Representative of a facility covered by environmental law (e.g., a chemical plant) who participates in the emergency reporting process with the Local Emergency Planning Committee (LEPC).

- **fallout**

Radioactive materials which descend from the atmosphere onto the surface of the earth. Fallout may be caused by nuclear explosions, by the release of radioactive substances after a nuclear accident, and by the routine operation of nuclear facilities.

❐ famine

Catastrophic food shortage affecting large numbers of people due to climatic, environmental and socio-economic reasons (in EM-DAT, famine can have four disaster subsets: " crop failure "," food shortage " and " drought ").

❐ far-field

Following a nuclear accident on-site, e.g. A nuclear plant, the immediate vicinity is called near-field with a diameter varying between 2 and 20 kilometers, depending on the source strength. The area outside the near-field is called the far-field, where effects are still noticeable after an accident.

❐ fault

A fracture or a zone of fractures in the earth along which displacement of the two sides relative to one another has occurred as a consequence of compression, tension, or shearing stresses. A fault may rupture the ground surface during an earthquake, especially if the magnitude is greater than M 5.5. The length of the fault is related to the maximum magnitude, with long faults able to generate larger-magnitude earthquakes than short faults.

❐ federal coordinating officer

The person appointed by the President to coordinate Federal assistance in a Presidentially declared emergency or major disaster.

❐ federal response plan (frp)

Developed to help expedite Federal support to disasters. Generally, the FRP is activated when the State's resources are not sufficient to cope with a disaster, and the governor has requested Federal assistance.

❐ fema

Federal emergency management agency.

❐ field assessment team

A small team of pre-identified technical experts that conduct an assessment of response needs (not a PDA) immediately following a disaster. The experts are drawn from FEMA, other agencies and organizations--such as the U.S. Public Health Service, U.S. Army Corps of Engineers, U.S. Environmental Protection Agency, and the American Red Cross--and the affected State(s). All FAST operations are joint Federal/State efforts.

❐ figure 8

One of the more common knots utilized in search and rescue operations, especially within the high angle/rappelling unit.

❐ filter area

The area between the inner and outer cordon around the disaster site. See disaster site.

❐ fire and life safety program

A plan implemented by businesses and other organizations to protect constituents (employees, building occupants, etc.) From fire and other hazards that are immediately hazardous to life. In the context used here, this entails such elements as fire prevention, pre-fire planning, assessment of building design for compliance with fire and safety codes, ensuring proper emergency egress, avoidance of electrical hazards or oxygen-deficient atmospheres, etc.

❐ firewall

A protective element in software or hardware to prevented unwanted intrusions to computers via internet or networks.

❐ first acts

The first acts of doctors and nurses at the disaster site are Anticipation, Control, Triage, Treatment and Transport (ACTTT).

❐ first-aid post

Collecting point for T3 victims in the immediate vicinity of the disaster site, however separate from the casualty clearing station, as to divide the T1 and T2 flow of patients from the T3 one.

❐ flash flood

Follows a situation in which rainfall is so intense and severe and runoff so rapid that it precludes recording and relating it to stream stages and other information in time to forecast a flood condition.

❐ flir

Forward Looking Infrared-A thermal imaging device designed to be aircraft mounted, usually on a helicopter, that can detect heat sources at night. Useful in locating persons on foot, in concealed vehicles, aircraft wreckage, and other heat sources.

❐ flood

A general and temporary condition of partial or complete inundation of normally dry land areas from overflow of inland or tidal waters, unusual or rapid accumulation or runoff of surface waters, or mudslides/mudflows caused by accumulation of water; Significant rise of water level in a stream, lake, reservoir or coastal region.

❐ focal depth

The vertical distance between the point where the fault rupture originated and the earth's surface.

❐ food shortage

Lack of alimentation bases (in EM-DAT, " food shortage " is a disaster subset of disaster type " famine ").

❐ footprint

The vulnerable area indicated by a map, showing how large an area is vulnerable to injury from to chemical releases.

❐ forest fire

Fires in forest that cover extensive damage. They may start by natural causes such as volcanic eruptions or lightning, or they may be caused by arsonists or careless smokers, by those burning wood, or by clearing a forest area.

❐ forward control point

The point next to the disaster site where the first ambulances to arrive are those to function as a command, coordination and communication post.

❐ forward planning

Occurs when a disaster is imminent and some details regarding the threat are known to the crisis manager.(cuny1)

❐ friedrich's time

4-6 hours following sustainment of mechanical injuries, T2 victims may become ABC unstable when untreated. It is therefore important to provide first-aid measures within this period of time.

❐ fumigation

The process of dispersion of fine gaseous particles of chemical agents used to kill harmful animal species, such as insects.

G

❐ gamma radiation

Gamma rays are high-energy, ionizing radiation that travel at the speed of light and have great penetrating power. They can cause skin burns, severely injure internal organs, and have long-term, physiological effects.

❐ gamma rays

Electromagnetic radiation of high energy and penetration released by some nuclear transformations. There are similar to X-rays and will penetrate through the human body. Iodine-131 emits gamma rays. Both gamma rays and X-rays cause ionization. See 'Ionizing radiation'

❐ gar

Govemor 's Authorized Representative

❐ gedaper

An acronym used to describe an incident analysis process. The steps include

(1) Gathering information,

(2) Estimating course and harm,

(3) Determining strategic goals,

(4) Assessing tactical options and resources,

(5) Planning and implementing actions,

(6) Evaluating, and

(7) Reviewing.

❐ gene

A unit of heredity composed of DNA, occupying a fixed position on a chromosome. A gene may determine a characteristic of an individual or regulate or repress the operation of other genes.

❐ general duty clause

Provisions of law requiring measures to prevent known hazards.

❐ genetic effects of radiation

Those effects which are experienced by the children of the individual receiving the radiation or by later generations - in contrast to somatic effects, which occur in the individual receiving the radiation, and teratogenic effects, which occur in the embryo or foetus inside the mother's womb.

❐ gibbs

A specific type of rescue ascender.

❐ gis

Geographic information system

❐ glawars commission report

The extensive investigation carried out by the Greater London Area War Risk Study Commission on the likely effects of modern warfare on a major metropolitan centre like London. Published under the title "London Under Attack," its conclusions have been found applicable to most modern urban centres. Translated into Italian under the title of "Attacco alla Citta."

❐ goal

A defined aim towards which to strive, and the actions taken to achieve it. Example: the goal may be to have an environment that is conducive to health, or to have primary health care available to everybody in a refugee settlement.

❐ golden hour

ABC-unstable victims (T1) should be stabilized as soon as possible, at least within one hour following injury, otherwise they will die.

❐ good neighbor agreement

A binding agreement negotiated between community organizations and site owners, sometimes including others, detailing the owners' commitments to environmental or safety improvements and other issues of concern to the community.

❐ governor 's authorized representative

The person empowered by the Governor to execute, on behalf of the State, all necessary documents for disaster assistance.

❐ gps

Global (or Ground) Positioning System-Based upon satellites, this small computer will give exact locations using latitude and longitude.

❐ gpsar

Greater Philadelphia Search and Rescue - The premier search and rescue organization in Pennsylvania, New Jersey and Delaware.

❐ graphite

A soft form of carbon used as a moderator in nuclear reactors.

❐ grid search

All searchers move forward on a line. They are spaced closely together. This search is slow and methodical and takes a lot of time.

❐ ground failure

A term referring to the permanent, inelastic deformation of the soil and/or rock triggered by ground shaking.

❐ ground shaking

The dynamic, elastic, vibration of the ground in response to the arrival and propagation of the elastic P, S, Love, and Rayleigh seismic waves. Ground shaking, of primary interest to the engineer performing design, is characterized in terms of amplitude, frequency composition, and duration. All structures are vulnerable at some amplitude, period, and duration of ground shaking. Depending upon the available data, ground shaking is quantified in terms of Modified Mercalli Intensity (the least precise) or in terms of ground acceleration, ground velocity, ground displacement, and spectral response (the most precise). Ground shaking can be increased by soil amplification, source directivity, topography, a shallow focal depth, surface fault rupture, and the fling of the fault which is thought to be the cause of the "killer pulse," a long-duration acceleration pulse.

H

❐ half-life, biological

The time required for the body to eliminate one half of an administered dose of a radioactive chemical. It is influenced by health and diet. The main methods of elimination are via urine, faeces, exhalation and perspiration.

❐ half-life, physical or radioactive

The time taken for the activity of a radionuclide to lose half its value by decay. The chemical resulting from the transformations may be either radioactive or non-radioactive. The process of radioactive decay is independent of temperature, pressure or chemical condition. Half-life ranges from less than a millionth of a second to millions of years. The half-life is a characteristic constant for each particular nuclide. An individual nuclide may decay before or after the half-life. Radiation is released every time a radioactive material changes to the next material in its decay series. Radioactivity per unit weight is inversely proportional to the half-life. For example, a specified quantity of caesium, -137 (half-life 30.04 years) is about 76,000 times more radioactive than the same quantity of caesium - 135 (half-life 3 million years).

Iodine -131 is a radio-isotope with a half-life of eight days. In traditional nuclear physics there is a rule of thumb that after ten half-lives a substance is considered to have decayed to a 'safe' level. However, this rule does not consider the size of original quantity of the radioactive nuclide, nor is there a universally accepted definition of what is 'safe'.

❑ handler

The person who controls and directs a canine during a search. This person is usually, but does not have to be, the owner or trainer of the canine.

❑ hasty search

A search whose purpose is to cover the most obvious places a subject might be in the least time possible. Usually the first kind of search tactic to be utilized.

❑ hazard

The probability of the occurrence of a disaster caused by a natural phenomenon (earthquake, cyclone), or by failure of man-made sources of energy (nuclear reactor, industrial explosion) or by uncontrolled human activity (overgrazing, heavy traffic, conflicts).-undrosome authors use the term in a broader sense, including vulnerability, elements at risk and the consequences of risk; Threatening event, or probability of occurrence of a potentially damaging phenomenon within a given time period and area.

❑ hazard assessment

An estimate of the range of the threat from natural and technological hazards to humans and their welfare. The physical parameters used to characterize the earthquake threat include: magnitude, frequency, duration, two-dimensional areal extent, speed of onset, three-dimensional spatial dispersion, and temporal spacing (e.g., the tendency of large-magnitude earthquakes to cluster in time).

❑ hazard awareness and communication program

Originally mandated by Federal Regulation CFR 1910.1200 and later codified into Title 8 of the California Code of Regulations, this program requires evaluation of possible health effects from hazardous chemicals in the workplace and communication of this information to affected employees. A written program including worker training, chemical labeling and Material Safety Data Sheet distribution is required and has been implemented at SDSU.

❑ hazard communication standard

An OSHA regulation that requires chemical manufacturers, suppliers, and importers to assess the hazards of the chemicals they make, supply, or import, and to inform employers, customers, and workers of these hazards through a Material Safety Data Sheet.

❑ hazard control procedures

Specific guidelines or step-by-step instructions that must be followed to control a particular hazard or to comply with health and safety regulations that apply to the hazard.

▫ hazard control program

A formal written program that has been prepared and implemented to control one or more types of occupational health and safety hazards or to prevent degradation of the environment. These programs are prepared to provide protection against specific types of hazards (e.g., Ionizing Radiation) or in response to regulatory requirements (e.g., Hazard Awareness and Communication Program). Hazard Control Programs describe requirements and assign specific responsibilities for meeting those requirements. Hazard Control Programs provide guidance in the form of Hazard Control Procedures that are specific instructions that must be followed in managing certain hazards.

▫ hazard environment

The geologic, geophysical, and geotechnical setting of the community that controls where, why, and how frequently earthquakes occur, how big they are, and their severity.

▫ hazard mitigation

Any action taken to reduce or eliminate the long-ten-n risk to human life and property from hazards. The term is sometimes used in a stricter sense to mean cost-effective measures to reduce the potential for damage to a facility or facilities from a disaster event.

▫ hazard priority classification

A system for categorizing the severity of consequence and probability of injury or damage for identified hazards. This information is used to determine the level and speed of corrective action to be taken after a hazard has been identified.

▫ hazardous chemical

EPA's designation for any hazardous material that requires a Material Safety Data Sheet. Such substances are capable of producing adverse physical effects (fire, explosion, etc.) Or adverse health effects (cancer, dermatitis, etc.).

Any chemical as defined by SARA Title III Section 311.(e), which reads: "Substances as defined within the meaning of 29 CFR 1910.1200(c), except that the term does not include the following:

(1) Any food, food additive, color additive, drug or cosmetic regulated by the FDA;

(2) Any substance present as a solid in any manufactured item to the extent that exposure to the substance does not occur under normal circumstances;

(3) Any substance to the extent that it is used for personal, family or household purposes or is present in the same form and concentration as a product packaged for distribution and use by the general public;

(4) Any substance to the extent that is is used on a research laboratory or a hospital or other medical facility under the direct supervision of a technically qualified individual;

(5) Any substance to the extent that it is used in routine agricultural operations or is a fertilizer held for sale by a retailer to the ultimate consumer."

❐ hazardous material

Any substance or material that when involved in an accident and released in sufficient quantities, poses a risk to people's health, safety, and/or property. These substances and materials include explosives, radioactive materials, flammable liquids or solids, combustible liquids or solids, poisons, oxidizers, toxins, and corrosive materials.

❐ hazardous materials

Substances that, because of their chemical, biological, or physical nature, pose a potential risk to life, health, or property if released

❐ hazmat

Hazardous material.

❐ health

1. The state of complete physical, mental and social well being, and not merely the absence of disease or infirmity. -WHO
2. The state of an individual or a community free from disabilitating conditions, demonstrating a reasonable resistance to diseases and living in a salubrious environment.

❐ heat exhaustion

A heat related condition in which the victim shows elevated temperature, poor heart action, clammy skin, cramping and dizziness. It is caused by prolonged exposure to high temperature.

❐ heat stroke

A heat related condition which is serious and life threatening. The victim will show rapid pulse, no sweating, and hot skin. It is caused by prolonged exposure to high temperature.

❐ heat wave

Long lasting period with extremely high surface temperature (in EM-DAT, " heat wave " is a disaster subset of disaster type " extreme temperature ").

❐ helibase

A location within the general incident area for parking, fueling, maintenance and loading of helicopters.

❐ helispot

A location where a helicopter can take off and land.

❐ high angle

A term used by rappelling teams. Its specific indication is usually anything over 45 degrees.

❑ high-hazard areas

Geographic locations that for planning purposes have been determined through historical experience and vulnerability analysis to be likely to experience the effects of a specific hazard (e.g., hurricane, earthquake, hazardous materials accident, etc.) Resulting in vast property damage and loss of life.

❑ hmrt

Hazardous materials response team

❑ hmru

Hazardous materials response unit (fbi)

❑ hmta

Hazardous Materials Transportation Act The HMTA provides for the safe transportation of hazardous materials. Regulations developed from the HMTA cover shipment preparation and labeling, handling, routing, emergency and security planning, incident notifications, and liability insurance. See Also:DOT, RCRA.

❑ homeless

People needing immediate assistance in the form of shelter (included in the field " total affected ").

❑ hospital treatment capacity

The number of T1 and T2 victims, which can be treated in a hospital per hour, according to current medical standards.

❑ hot spot

A local geographic area where radioactive environmental contamination levels are higher than average for the surrounding area. Hot spots may be a tiny point containing only a few radioactive fragments, or they may encompass many square kilometres, such as when an air mass or cloud containing a high concentration of radioactive dust becomes part of rain or snow. Areas of body tissue where a much higher than average concentration of radioactivity occurs are also referred to as hot spots.

❑ hug-a-tree

A class taught by certified teachers to young children. This course teaches children and their parents how not to get lost and what to do if in fact they should get lost.

❑ human immunodeficiency virus (hiv)

The causal organism of AIDS. Synonym: HIV

❑ hurricane

A tropical cyclone, formed in the atmosphere over warm ocean areas, in which wind speeds reach 74 miles per hour or more and blow in a large spiral around a relatively calm center or "eye". Circulation is counter-clockwise in the Northern Hemisphere and clockwise in the Southern Hemisphere.

❐ hygiene

Science that deals with the principles, methods and practical aspects of disease prevention, sanitation and improvement of health. It is usually divided into such fields as personal hygiene, domestic hygiene, food hygiene, industrial hygiene.

❐ hygienic measures

Those measures as to prevent diseases following a major disaster, because the infrastructure of the stricken area is non- or malfunctioning.

❐ hypocentre

The point on the ground vertically beneath an air explosion of a nuclear bomb. Synonym: ground zero.

❐ hypothermia

A generalized cooling that may lower the body temperature below normal.

I

❐ identification of dead

Disasters with "unknown" dead necessitate identification of the bodies or their remains. This is important for the bereaved, not only from an emotional point of view, but also for judicial and insurance purposes. Various medical disciplines are involved in matching ante-and post-mortem findings.

❐ immediate response zone (irz) see also csepp

A circular zone ranging from 10 to 15 km (6 to 9 miles) from the potential chemical event source, depending on the stockpile location on-post. Emergency response plans developed for the IRZ must provide for the most rapid and effective protective actions possible, since the IRZ will have the highest concentration of agent and the least amount of warning time.

❐ immunodeficiency

Defective or deficient immunological mechanisms of the body due to insufficiency in one of the components of the immune process or to a defect in the B-lymphocyte or T-lymphocyte systems. Immunilogical deficit may result from infection, as in AIDS, or excessive radiation, as in nuclear war, or toxic substances. Synonyms: immune deficiency, immunological deficit.

❐ impact stage

A time when the disaster event has occurred and the community experiences the immediate effects; the community is rapidly assessed for damage, types and extent of injuries suffered, and immediate needs of the community

❐ incendiary device

Any mechanical, electrical, or chemical device used intentionally to initiate combustion and start a fire.

❑ incendiary incident

An event in which an incendiary device is used as a terrorist weapon.

❑ incident

An occurrence or event, either human caused or natural phenomena, that requires action by emergency service personnel to prevent or minimize loss of life or damage to property and or natural resources.

❑ incident command post

That location at which the primary command functions are executed.

❑ incident command system

A standardized organizational structure used to conunand, control, and coordinate the use of resources and personnel that have responded to the scene of an emergency. The concepts and principles for ICS include common terminology, modular organization, integrated communication, unified command structure, consolidated action plan, manageable span of control, designated incident facilities, and comprehensive resource management.

❑ incident command system (ics)

An organizational scheme wherein one person, normally the Fire Chief, takes charge of an integrated, comprehensive emergency response. This commander is backed by an Emergency Operations Center which provides support, resources, communications, and advice.

❑ incident commander

The individual responsible for the management of all incident operations.

❑ industrial accident

Disaster type term used in EM-DAT to describe technological accidents of an industrial nature/involving industrial buildings (e.g. Factories). It comprises of a number of disaster subsets:Chemical spill/leak " Ind:Chemical spill "; Explosions " Ind:Explosion "; Radiation leakages " Ind:Radiation "; Collapses " Ind:Collapse "; Gas leaks from industrial sites " Ind:Gas leak "; Poisoning " Ind:Poisoning "; Fires " Ind:Fire "; and other technological accidents involving industrial sites " Ind:Other ".

❑ industrial hygiene

A specialized area within the field of Occupational Health and Safety, industrial hygiene has as its goal the recognition, evaluation, and control of worker exposures to harmful physical or chemical agents or conditions. Workplace monitoring of noise levels, ventilation rates, airborne contaminants, heat exposures, and radiation dose may be conducted by industrial hygienists. Various engineering, administrative, and other methods are employed to control or reduce worker exposures.

❑ industrial safety procedures

Specific guidelines for implementing safety regulations and practices in industrial settings. Safety procedures are designed to prevent accidents or acute illnesses, as opposed to many industrial hygiene procedures that reduce chronic (longer-term) exposures.

❐ infection

The entry and development or multiplication of an infectious agent (virus, bacteria, fungus, parasite) in the body of man or animal.

❐ infestation

The penetration and development of arthropods and parasites on the body or in clothing.

❐ infrastructure

Structures and facilities that provide the essential functions of supply, disposal, communication, and transportation in a community. They are also called community lifelines.

❐ ingeslion pathway (50-mile epz) rep program

A circular geographic zone (with a 50-n-dle radius centered at the nuclear power plant) for which plans are developed to protect the public from the ingestion of water or foods contaminated as the result of a nuclear power plant accident.

❐ inherent safety

Elements in the design or redesign of a facility which eliminate the potential for catastrophic chemical releases or incidents.

❐ injured

People suffering from physical injuries, trauma or an illness requiring medical treatment as a direct result of a disaster.

❐ injury and illness log

Required by federal and state regulations, this log documents injuries and illnesses caused by work-related activities that result in lost work time, fatalities, offsite treatment, or restricted work activity. Employers are required to maintain and post this information at the worksite.

❐ injury and illness prevention program (iipp)

Mandated by California Code of Regulations, Title 8, Section 3203, this program is required of all employers in the state. Its purpose is to prevent job-related accidents and illnesses through implementation of a written plan to identify, evaluate, and correct unsafe workplace hazards. Included are requirements to provide safety training to employees, investigate accidents, and maintain safety records. At SDSU, the Injury and Illness Prevention Program serves as the umbrella program for many of the university's health and safety efforts.

❐ insect infestation

Pervasive influx and development of insects or parasites affecting humans, animals, crops and materials.

❑ inspections

Periodic audits of the workplace environment, including equipment, chemicals, building structure, documented procedures, records, and employee knowledge of job requirements and hazards. Inspections may be undertaken for the purpose of departmental self-evaluation or by Environmental Health and Safety or outside agencies for analysis of compliance with health and safety regulations.

❑ insult

The damage (negative result) resulting from the impact of the event.

❑ intensity

A numerical index denoted by Roman numerals from I to XII describing the physical effects of an earthquake at a specific location on the earth's surface, man, or on structures built by man. These values are determined subjectively by individuals performing post earthquake investigations to determine the nature and spatial extent of the damage distribution. There are no instrumental readings. The most commonly used scales throughout the world are Modified Mercalli Intensity (MMI), developed in the 1930's and named for the Italian author, and the MSK scale, developed in the 1960's and denoted by the first initial of the last name of two scientists of the former Soviet Union and one scientist of the former Czechoslavakia who created the scale. Intensity VI denotes the threshold for potential ground failure such as liquefaction. Intensity VII denotes the threshold for architectural damage. Intensity VIII denotes the threshold for structural damage. Intensity IX denotes intense structural damage. Intensities X to XII denote various levels of destruction up to total destruction. The MMI and MSK scales are essentially equivalent for intensities VII to X. An earthquake has many intensities, but only one magnitude. See Magnitude.

❑ intensity (seismic)

The degree of shaking or of vibrations, signifying the intensity of an earthquake as measured numerically on the Mercalli scale

❑ international assistance

Assistance provided by one or more countries or international and voluntary organizations to a country in need, usually for development or for an emergency. The four main elements of assistance within the international community are:

(a) the intergovernmental agencies--united nations, common market.

(b) non-governmental organizations.

(c) the red cross, and

(d) bilateral agreements.

❑ international terrorism

Terrorist acts directed by foreign groups who transcend national boundaries, affecting people in several countries

▫ intervention levels

Levels of radiation or concentrations of chemicals in the environment. These levels determine the measures to be taken in the measure zone and can be regarded as aids to decision-making.

▫ iodine prophylaxis

Radioactive iodine is often an important component of radioactive isotopes to be discharged into the atmosphere after a nuclear accident. Stable iodine prevents the absorption of radioactive iodine in the thyroid gland provided it is administered beforehand.

▫ ion

An originally neutral atom which has become electrically charged by losing or acquiring electrons. Loss of an electron results in a positive ion (cation), and acquisition in a negative ion (anion).

▫ ionization

This occurs when sufficient energy is deposited in a neutral molecule to displace an electron, thus replacing the neutral molecule with positive and negative ions.

▫ ionizing radiation

Particulate or electromagnetic radiation that has sufficient energy to remove electrons from the atoms of material with which it interacts. Ionizing radiation may be emitted from radioactive materials or from radiation-producing machines. The most important forms of ionizing radiation include alpha particles, beta particles, neutrons, gamma rays, and X-rays.

Radiation with enough energy to ionize atoms, i.e. To remove electrons from them. It includes X-rays and gamma rays and the particles that are emitted from radioactive substances. High levels of ionizing radiation can have noticeable, severe and lethal effects on the human body. At low levels, none of our senses can detect ionizing radiation. We cannot hear it, see it, feel it or smell it. However, although we are unable to detect it with our senses, damage to the human body can occur which only becomes apparent years later. Radioactive nuclides pose the greatest threat to human health when they are inhaled or ingested. However, radiation-emitting fragments can be so small that they fasten in the many sweat pores and hair follicles all over the body. There are three main types of ionizing radiation: Alpha. Beta and Gamma. Alpha emitters are the most harmful to living cells if ingested or inhaled, but the distance which alpha particles can travel is only a few centimetres in air and mere tens of microns in soft tissue. Beta particles can travel up to about 80 centimetres in air and they cannot go through steel or wood thicker than about one centimetre. The difference between alpha and beta particles is like that between cannon-balls and bullets. Alpha particles, like cannon-balls, have less penetrating power but impact. For this reason the biological damage of alpha radiation is considered to be about 20 times that of the same absorbed dose of beta or gamma radiation. Gamma rays are a type of electromagnetic radiation unlike alpha and beta radiation, which are forms of particle radiation. Gamma radiation can be as harmful as beta radiation and can travel great distances. The majority of gamma radiation is stopped by a few centimetres of concrete. When discussing the distance which different forms of radiation are capable of travelling, it is important to distinguish between radioactive particles and radioactive emitters. Alpha and beta particles travel a very short distance, due to radioactive decay. In contrast minute dust fragments containing alpha, beta and gamma emitters can be transported great distances by wind and water. Sometimes the false impression is given that locations only centimetres away from a radioactive source are isolated from exposure. See also 'Alpha particle', 'Beta particle' and 'Gamma ray'. Causes the long-term destructive force of the dirty bomb, an ion's electrical charge may lead to unnatural chemical reactions inside the cells, break DNA chains, either causing DNA strand death or DNA mutations; unlikely ill effects in very small doses

❐ ionosphere

The zone of the atmosphere, from about 70 km to 500 km, in which charged particles, ions and electrons are formed by photo-ionization under the effect of the sun's radiation.

❐ irritant

A substance that causes a reversible inflammatory response.

❐ irritating agent

A chemical agent, also known as riot control agents or tear gas, which causes respiratory distress and tearing designed to incapacitate. Common examples include chloropicrin, MACE, tear gas, pepper spray, and dibenzoxazepine.

❐ isar

Inland Search and Rescue - A search management class taught by the U.S. Coast Guard and U.S. Air Force OR Introduction to Search And Rescue-A basic search and rescue course taught by the National Association for Search and Rescue.

❐ isotope

One of two or more forms of an element that have differing atomic weights, i.e. They have the same number of protons but different numbers of neutrons. It is considered that after ten half-lives a radioactive isotope has decayed into another radioactive isotope (a daughter product) or has become stable. See 'Radioactivity'.

J

❐ joint information center

A central point of contact for all news media near the scene of a large-scale disaster. News media representatives are kept informed of activities and events by public information officials who represent all participating Federal, State, and local agencies that are collocated at the JIC.

❐ joint information system

Under the FRP, connection of public affairs personnel, decision-makers, and news centers by electronic mail, fay, and telephone when a single Federal-State-local JIC is not a viable option.

K

❐ kerma

Acronymic term for Kinetic Energy Released in Matter, the measure of intensity of ionizing radiation at a given place. The dose is expressed in grays.

❐ kuwaiti oil spill 1991

- The biggest oil spill to date occurred during the 1991 Persian Gulf war when the Iraqi Army occupying Kuwait began destroying tankers, oil terminals, and oil wells late in January 1991.

Oil slick spreading in the Persian Gulf January–March 1991

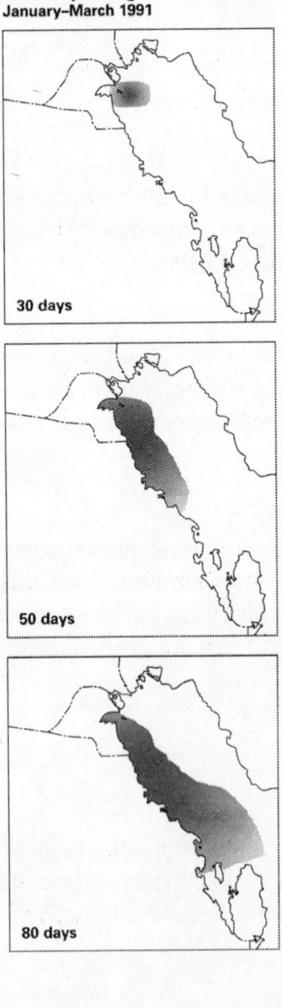

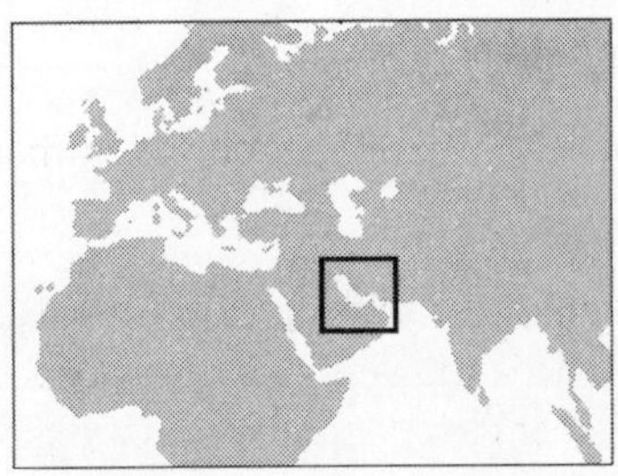

Smoke and soot plume in Gulf region on February 2, 1991

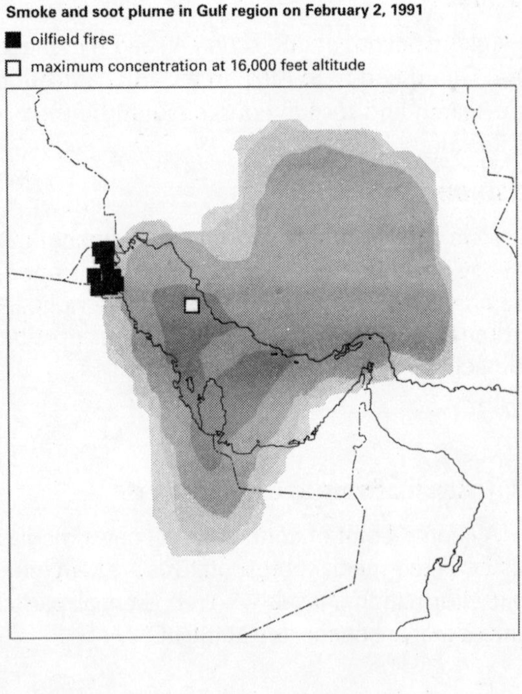

- Approximately 1 million barrels of oil spilled onto the land and some 8 million barrels spilled directly into the Arabian Gulf forming a 600-square-mile oil slick.
- Four hundred miles of the western shoreline of the Gulf was polluted with oil, and between 15,000 and 30,000 birds are thought to have been killed.
- An estimated third of the total amount of oil released evaporated.
- Some 74 million tons of oil were burned, which produced about 2.3 million tons of soot and 2.2 million tons of sulfur.
- The smoke from the burning oil created darkness during daylight hours, and the World Health Organization (WHO) estimated that death rates in Kuwait rose by 10 percent over the following year because of associated breathing difficulties and skin problems.
- Recovery operations using skimmers, vacuum trucks, and booms recovered 1,400,000 barrels of oil from the Gulf.
- Desalinization, industry, and power plant intakes were protected by booms and skimming operations. Sensitive natural areas were also boomed to prevent oiling. la nina (" little girl " in spanish)

It is essentially the opposite of El Nino. The ocean becomes much cooler than normal. Although, La Nina is not as well understood as El Nino, it is thought to occur due to an increase in the strength of the trade winds. This increases the amount of cooler water that upwells toward the West Coast of South American and reduces water temperatures.

L

❑ landslide

A massive and more or less rapid sliding down of soil and rock, causing damage in its path; In general, all varieties of slope movement, under the influence of gravity. More strictly refers to down-slope movement of rock and/or earth masses along one or several slide surfaces.

❑ landslides

The most common and wide spread type of ground failure; consists of falls, topples, slides, spreads, and flows of soil and/or rock on unstable slopes.

❑ legal responsible authority

The official and/or agency having legal responsibility for the emergency response to a SAR or disaster incident.

❑ lepc

Local Emergency Planning Committee. In Bucks County, the Bucks County Local Emergency Planning Committee.

❑ liquefaction

Occurs mainly in young, shallow, loosely compacted, water saturated sand and gravel deposits when subjected to ground shaking; results in a temporary loss of bearing strength. See ground failure.

- **litter**

A stretcher like device utilized in rescue to transport a victim.

- **lkp**

Last Known Position-The last known location for the missing subject as determined by physical evidence such as a discarded object or a footprint. In the case of an aircraft, the LKP may be the last reported point or the last observed radar position.

- **lng**

Liquefied natural gas.

- **local emergency planning committee (lepc)**

The body appointed by the State Emergency Response Commission (SERC), as required by EPCRA, which develops comprehensive emergency plans for Local Emergency Planning Districts, collects MSDS forms and chemical release reports, and provides this information to the public. Each county and some large city governments participate in an LEPC.

- **local eop**

The local EOP focuses on essential measures for protecting the public, to include warning, emergency public information, evacuation, and shelter. To be included in a local EOP should be a mechanism for emergency responders and managers to notify and activate State resources.

- **local sar coordinator**

The person responsible for coordination of SAR operations within a given jurisdiction.

- **local sar coordinator**

The person responsible for coordination of SAR operations within a given jurisdiction.

- **loss**

A range of adverse consequences impacting communities and individuals (e.g., damage, loss of economic value, loss of function, loss of natural resources, loss of ecological systems, environmental impact, health deterioration, mortality, morbidity).

- **low angle**

Anything under 45 degrees.

- **low-level radiation**

This includes background radiation and man-made radiation from low-level nuclear waste. The International Commission of Radiological Protection assesses radiation damage on the premise that radiation is always potentially harmful, however small the dose.

❐ lpg

Liquefied petroleum gas.

M

❐ magnitude

A numerical quantity, devised by the late American Professor, Charles F. Richter, in the 1930's and denoted by Arabic integers with one decimal place accuracy (for example 7.8) to characterize earthquakes in terms of the total energy released after adjusting for difference in epicentral distance and focal depth. Magnitude differs from intensity in that magnitude is determined on the basis of instrumental records; whereas, intensity is determined on the basis of subjective observations of the damage. Measured on a logarithmic scale, magnitude is open ended theoretically, with the two largest magnitude earthquakes to date being the M 9.5 Chile earthquake of 1960 and the M 9.2 Alaska earthquake of 1964. Moderate-magnitude earthquakes have magnitudes of 5.5 to 6.9; large-magnitude earthquakes have magnitudes of 7.0 to7.9; and great-magnitude earthquakes have magnitudes of 8.0 and greater. The energy increases exponentially with magnitude. For example, a magnitude 6.0 earthquake releases 31.5 times more energy than a magnitude 5.0 earthquake, but (31.5) (31.5) or approximately 1,000 times more energy than a magnitude 4.0 earthquake; The overall size of the event (e.g., area of drought, number of persons potentially or actually affected, etc.). Magnitude usually expressed as a mathematical quantity.(ox 820) (i.e., richter scale for earthquakes, safir-simpson scale for tropical cyclones).

❐ magnitude of earthquake

The "size" of an earthquake, expressing the amount of energy released in the form of elastic waves as measured by a seismograph, on a scale such as Richter's.

❐ major consequences

Exceed State and local capabilities, requiring a Federal response.

❐ management of change

OSHA requirements for regulated facilities to anticipate changes in technologies or operations, and to assess and prepare for the safety of the changes before they are undertaken.

❐ mandatory or directed evacuation

This is a warning to persons within the designated area that an imminent threat to life and property exists and individuals must evacuate in accordance with the instructions of local officials.

❐ man-made disaster

A disaster caused not by natural phenomena, but by man's or society's action, involuntary or voluntary, sudden or slow, directly or indirectly, with grave consequences to the population and the environment. Examples: technological disaster, toxicological disaster, desertification, environmental pollution, conflict, epidemics, fires.

❐ mass care

The actions that are taken to protect evacuees and other disaster victims from the effects of the disaster. Activities include providing temporary shelter, food, medical care, clothing, and other essential life support needs to those people that have been displaced from their homes because of a disaster or threatened disaster.

❐ mass casualty

The definition of disaster implies a discrepancy between number of victims and its treatment capacity. This does not necessarily means a mass casualty situation, in which case the number of victims is overwhelming.

❐ mass casualty event

May have the same magnitude in terms of human life and suffering, but does not destroy the infrastructure of the society. Examples of mass casualty events include epidemics, complex human emergencies, etc. The impact of such events is close to or even may exceed that of disasters, but the infrastructure remains intact and mechanisms can be developed within the infrastructure to cope with the circumstances.

❐ mass casualty incident

A situation with 100 or more casualties, significantly overwhelms available emergency medical services, facilities, and resources

❐ material safety data sheet (msds)

Printed material concerning a hazardous chemical, or Extremely Hazardous Substance, including its physical properties, hazards to personnel, fire and explosion potential, safe handling recommendations, health effects, fire fighting techniques, reactivity, and proper disposal. Originally established for employee safety by OSHA.

Literature prepared by a chemical or chemical product manufacturer that contains hazard and safety information about the product. Information typically includes a list of hazardous ingredients, safety precautions for handling, spill or release response procedures, and first aid instructions. Environmental Health and Safety files these sheets as part of the Hazard Awareness and Communication Program.

❐ mcl

Maximum contaminant level. The maximum level of contamination allowed under EPA standards.

❐ measure zone

The zone where measures are to be taken in case of nuclear or chemical accidents. The community profile and the source strength determine these zones.

❐ mechanical harm

One of the six types of harm (see TRACEM) that can be encountered at a terrorist incident. Causes trauma from contact with mechanical or physical hazards.

One form of mechanical injury can result from an explosive device. Other types include routine slip, trip, and fall hazards.

❑ mechanical lesions

Mechanical impact on the human body creates injuries like wounds, lacerations, fractures, bleedings (internal and external) and concussions. Mechanical lesions also include burns.

❑ medical coordination

In the chain of medical care, coordination between its phases and in each phase between doctors, nurses and paramedics, is of paramount importance. Simplification and standardization of materials and methods utilized is therefore a prerequisite.

❑ medical disaster

Definition adopted by the World Health Organization and the United Nations as established by Gunn: "the result of a vast ecological breakdown in the relationships between man and his environment, a serious and sudden (or slow, as in drought) disruption on such a scale that the stricken community needs extraordinary efforts to cope with it, often with outside help or international aid."(3) This definition also indicates that it is the impact on society that constitutes the disaster, not the event that is the disaster; The most common medical definition of a disaster used is an event that results in casualties that overwhelm the health-care system in which the event occurs. A health disaster often is considered a medical disaster. However, a health disaster infers impaired public health while a medical disaster is related to the health care or break in health care to individuals resulting from the event. The Task Force broadened this definition to: Any event that results in a precipitous or gradual decline in the overall health status of a community with which it is unable to cope adequately.

❑ medical disaster preparedness

The medical preparedness in the chain of medical care is determined by personnel, materials, and methods. With the aid of this basic concept medical disaster preparedness can be expressed in a figure ranging from 1 to 5.

❑ medical rescue capacity

The number of victims that could be rescued and stabilized at the disaster site per hour by doctors, nurses and paramedics.

❑ medical severity index

The ratio between the number of victims times the average severity of the injuries sustained and the treatment capacity in the chain of medical care. When this ratio is larger than 1 the event can be considered a disaster.

❑ medical surveillance

A program whereby the health of workers is monitored through regularly scheduled and post-incident medical examinations. Medical surveillance may be required to determine the suitability of workers to perform a certain type of job (for example, to wear a respirator), to establish a baseline medical status of a worker, or to identify potential changes in the overall health status of workers.

❑ **medical transport capacity**

The number of victims that could be transported to and distributed between the hospitals surrounding the disaster site, per hour.

❑ **mic**

Methyl isocyanate, the chemical involved in the 1984 Bhopal chemical disaster in which 3,000 people were killed in a single night.

❑ **military medicine**

The art and science of medicine, including in particular, critical care, emergency surgery and traumatology as applied to mass casualty situations, battlefront conditions and the needs of soldiers.

❑ **miscellaneous accident**

Disaster type term used in EM-DAT to describe technological accidents of a non-industrial or transport nature (e.g. Houses). It comprises of a number of disaster subsets:Explosions " Misc:Explosion "; Collapses " Misc:Collapse "; Fires " Misc:Fire "; and other miscellaneous accidents involving domestic/non-industrial sites " Misc:Other ".

❑ **mission**

A specific incident such as a search and rescue operation.

❑ **mission coordinator**

The person designated by the legal responsible authority to coordinate and manage on-scene operations during a specific SAR incident. Depending upon the management system used and the organizations involved, this person may be called the Mission Coordinator, On-Scene Commander or Incident Commander.

❑ **mitigate**

To lessen or decrease the seriousness of the process to which the word is applied. Mitigate is the action verb and mitigation is the result of this action.

❑ **mitigation**

Action taken to prevent or reduce the harmful effects of a disaster on human health or property, involves future-oriented activities to prevent subsequent disasters or to minimize their effects; A range of policy, legislative mandates, professional practices, and social adjustments that are designed to reduce or minimize the effects of earthquakes and other natural hazards on a community. Mitigation measures implemented over the last 20 years have included:

1) Land use planning and management,
2) engineering codes, standards and practices,
3) control and protection works,

4) prediction, forecasts, warning, and planning,
5) recovery, reconstruction, and planning, and
6) insurance.

▫ mobile medical teams

Instead of bringing the patient to the hospital, the hospital comes to the patient for whom mobile medical teams are created in order to stabilize the patient on the spot. This could shorten the treatment delay.

▫ mock search

Staged training that is set up to approximate as closely as possible, a real search effort.

▫ modification

The aggregate of all approaches and measures to modify the intensity of the event that would have occurred without human intervention. Thus, through human activities, the intensity and/or magnitude and/or the time course of the resulting event either may be augmented (increased) or attenuated (decreased).

▫ mountaineering

The art of mountain climbing usually for recreational purposes, but may be for rescue purposes.

▫ mounted team

One rider and one horse.

▫ multiple casualty incident

Disaster in which there are more than 2 but fewer than 100 persons injured

▫ mutual aid search

A search that is too large scale to be handled by one organization. One or more organizations or groups outside the initial group are called in to assist in the effort.

N

▫ nasar

National Association for Search and Rescue.

▫ national response center (nrc)

The primary communications center operated by the U.S. Coast Guard to receive reports of major chemical and oil spills and other hazardous substances into the environment. The NRC immediately relays reports to a predesignated federal On-Scene Coordinator.

▫ national response team (nrt)

Representatives from 15 federal agencies with interests and expertise in various aspects of emergency response to pollution incidents. EPA serves as chair and the

U.S. Coast Guard serves as vicechair. The NRT is primarily a national planning, policy, and coordinating body and does not respond directly to incidents. The NRT provides policy guidance prior to an incident and assistance as requested by a federal On-Scene Coordinator via a Regional Response Team during an incident. NRT assistance usually takes the form of technical advice, access to additional resources or equipment, or coordination with other rrts.

❑ national strike force (nsf)

Operated by the U.S. Coast Guard, the NSF is composed of three strategically located teams (Atlantic, Pacific, and Gulf coasts) who back up the federal On-Scene Coordinator. These teams are extensively trained and equipped to respond to major oil spills and chemical releases. These capabilities are especially suited to incidents in a marine environment but also include site assessment, safety, action plan development, and documentation for both inland and coastal zone incidents. The NSF Coordination Center is at Elizabeth City, NC.

❑ natural background radiation

Low-level radiation from space and from radioisotopes in rock, soil and building materials. It includes emissions from radioactive chemicals which are not man made. These chemicals include uranium, radon, potassium and other trace elements. They are made more hazardous through human activities such as mining and milling, since this makes them more available for uptake in food, air and water.

❑ natural disaster

A sudden major upheaval of nature, causing extensive destruction, death and suffering among the stricken community, and which is not due to man's action. However, (a) some natural disasters can be of slow origin, e.g. Drought, and (b) a seemingly natural disaster can be caused or aggravated by man's action, e.g. Desertification through excessive land use and deforestation.

❑ natural hazard

A potential threat to humans and their welfare caused by rapid and slow onset events having atmospheric, geologic, and hydrologic origins on solar, global, regional, national, and local scales (e.g., floods, severe storms, earthquakes, landslides, volcanic eruptions, wild fires, tsunamis, droughts, winter storms, coastal erosion, and space weather); The probability of occurrence, within a specific period of time in a given area, of a potentially damaging phenomenon of nature.

❑ nbc

Nuclear, biological, chemical

❑ nerrtc

National Emergency Response and Rescue Training Center

❑ nerve agent

A substance that interferes with the central nervous system. Exposure is primarily through contact with the liquid (skin and eyes) and secondarily through inhalation of the vapor. Three distinct symptoms associated with nerve agents are pinpoint pupils, an extreme headache, and severe tightness in the chest. Examples of nerve agents are sarin, Soman, tabun, and VX agent.

❐ neutron

A neutron elementary particle. In the nucleus of an atom it is stable, but when free it decays.

❐ night vision

A monocular or binocular piece of equipment that allows you to see in near total darkness by intensifying existing light. It may also be utilized in total darkness with an infrared attachment. Night vision is very useful for search and rescue missions.

❐ noaa

National Oceanic and Atmospheric Administration-A mobile weather data collection and forecasting facility including personnel, which can be utilized within the incident area.

❐ nonconventional weapons

Less frequently used weapons of terrorism such as chemical, nuclear, and biological categories

❐ nondisaster stage

Also known as the interdisaster stage, the time for prevention, planning, preparedness, and mitigation activities; the threat of a disaster is still in the future

❐ non-governmental organization

A private, international, not governmental organization (as distinct from an inter-governmental organization), constituted as a single association or as a federation of various national organizations, without governmental or state ties. The most important ngos are given consultative status with the united nations or its specialized agencies.and are active in disaster work.

❐ nonionizing radiation

Radiation of electromagnetic form that does not have a high enough energy level to ionize (i.e., remove electrons from) material with which it comes in contact. Examples of nonionizing radiation include lasers, microwaves, extremely low frequency (ELF) electromagnetic fields, infrared and ultraviolet light, ultrasound, etc.

❐ nsars

National Search and Rescue School-A national school which conducts search management programs administered by the U.S. Coast Guard and the U.S. Air Force.

❐ nuclear activity

The number of spontaneous nuclear disintegrations within a radionucleotide at any given time. The old unit of activity, the curie (ci) has been replaced by the becquerel.

▫ nuclear detonation

An explosion resulting from fission and/or fusion reactions in nuclear material, such as that from a nuclear weapon.

▫ nuclear disaster 1: three mile island, united states, 1979

- The Three Mile Island pressurized water reactor (PWR) power station is near Harrisburg, Pennsylvania.
- Its original reactor entered service in 1974 and is still working today.
- A second reactor was destroyed in March 1979 when it had just entered service. It was the first major accident at a commercial nuclear power plant, and the worst in United States history.
- A combination of human error, mechanical failure, and poor control room design led to a loss-of-coolant accident involving partial reactor core meltdown and release of radioactivity into the environment.

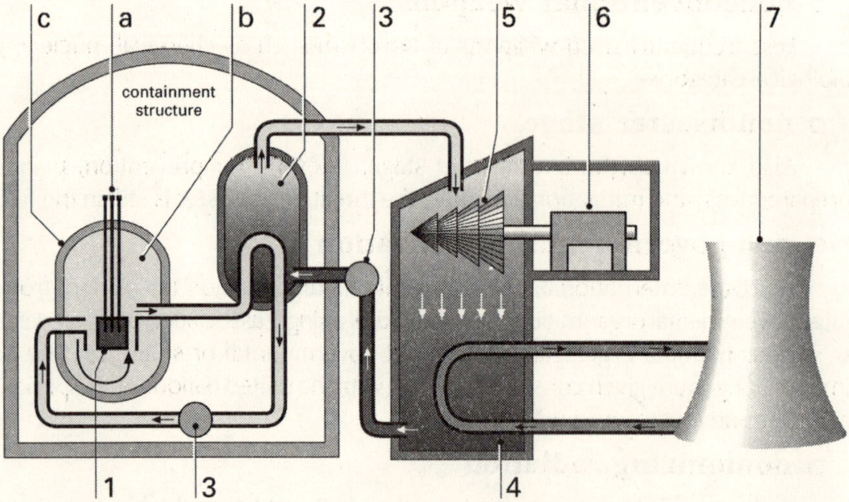

1 reactor core
2 steam generator
3 pump
4 condenser cooling water
5 turbine
6 generator
7 cooling tower
a control rods
b reactor
c reactor vessel

- No one was killed, and no adverse health effects caused by radiation have been found in the population around Three Mile Island.
- The average radiation dose to people within ten miles of the plant was 8 millirem, about the same as having a chest X-ray.
- The clean-up took 12 years and cost around 1 billion dollars.
- The accident caused widespread alarm and dealt a serious blow to public confidence in nuclear energy, leading to a sharp decline in nuclear construction through the 1980s and 1990s.

How a PWR works

A pressurized water reactor (PWR) generates heat by promoting nuclear fission in the reactor core.

Control rods slow the reaction by absorbing neutrons, so they are withdrawn to allow the reactor to heat up to operating temperature.

Water, kept under pressure to prevent it boiling, circulates through the core and carries heat from the reactor to a heat exchanger.

The heat exchanger boils separate water in the steam generator.

The steam spins a turbine connected to a generator, which produces electricity.

1 reactor core

2 steam generator

3 pump.

4 condenser cooling water

5 turbine

6 generator

7 cooling tower

a control rods

b reactor

c reactor vessel

What went wrong:

- At 4:00 a.m. on March 28, 1979 a minor malfunction stopped water flowing round the secondary cooling circuit (b). This allowed the temperature of the water in the primary cooling circuit (a) to rise, increasing its pressure.
- A safety valve opened to relieve the pressure building up in the reactor vessel.
- The reactor 'scrammed' – the control rods (c) automatically dropped into the reactor to shut it down.
- The relief valve should then have shut, but became stuck open, allowing more coolant to escape. The operators did not realize this was happening.
- Emergency pumps automatically pushed replacement water into the reactor system, but the operators shut these down, believing that the reactor vessel (c) was already full of coolant.
- As coolant boiled away, the reactor (b) became hotter and the core was partly exposed, damaging the fuel rods (a) and releasing radioactive material, some of which escaped into the environment.
- A chemical reaction in the exposed part of the reactor produced hydrogen, some of which ignited, raising fears of a major explosion.
- It took a month to achieve a safe 'cold shutdown', with the reactor cooled by convection of unpumped coolant.

❑ nuclear disaster 2: chernobyl, ukraine, 1986

- The accident at the Chernobyl nuclear power plant on April 26, 1986 was the most serious in the history of the nuclear industry.

- The accident was the result of a flawed reactor design, operated by inadequately trained personnel with insufficient awareness of safety.
- The resulting steam explosion and fire released about 5 percent of the radioactive reactor core into the atmosphere.
- Most radioactive material was deposited over a wide area around the plant, but a significant amount was carried downwind.
- The fallout contaminated about 58,000 square miles of Belarus, the Russian Federation and Ukraine, with a combined population of about five million people.
- Traces of radioactive fallout from Chernobyl have been measured throughout the northern hemisphere.
- 30 power plant employees and firemen were killed within days or weeks, including 28 with acute radiation syndrome.
- 700 cases of thyroid cancer, including 10 deaths, have been linked to radiation exposure resulting from the accident.
- During 1986, about 116,000 people were evacuated from around the reactor. About 220,000 people weresubsequently relocated from the contaminated areas.

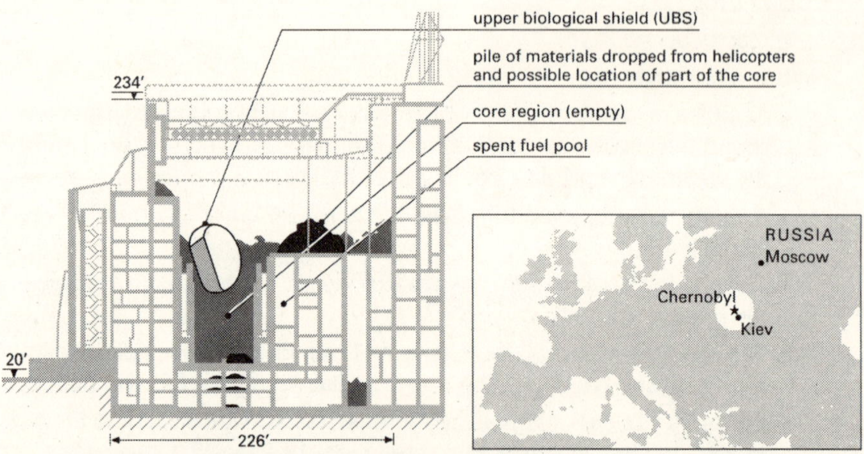

- A UN report in 2000 found that there is no firm scientific evidence of significant radiation-related health effects to most of the estimated 1 million people exposed, although lack of information prior to 1986 makes comparisons difficult.
- Psychosocial effects among those affected are emerging as a serious problem, similar to those arising from other major disasters such as earthquakes, floods and fires.

The accident

- Prior to a routine shutdown, the crew at Chernobyl Reactor Number 4 began preparing for a test to find out how long the turbines would spin and supply electricity following a loss of power from the reactor.

- They carried out a series of actions, including disabling automatic shutdown mechanisms and withdrawing control rods in contravention of safety protocols.
- Similar tests had been carried out at Chernobyl and other plants, despite the fact that these reactors were known to be very unstable at low power.
- As the flow of coolant water was reduced, the reactor became hotter. When the operator tried to shut down the unstable reactor, there was a catastrophic power surge. This was a know design problem, resulting from the use of graphite as the moderator in combination with water as the coolant.
- The fuel elements ruptured and the explosive force of superheated steam blew the massive steel cover plate off the reactor, releasing fission products to the atmosphere.
- A second explosion threw out fragments of burning fuel and graphite from the core and allowed air to rush in, causing the graphite moderator to burst into flames.
- The graphite burned for nine days, causing the main release of radioactivity into the environment.

The clean-up

- 600 emergency workers fought to contain the fire on the night of the accident, receiving high doses of radiation.
- Around 5,500 tons of boron, dolomite, sand, clay, and lead were dropped by helicopter onto the burning core in an attempt to extinguish the blaze and stop the release of radioactive particles.
- About 600,000 people (civilian and military) took part in recovery operations to decontaminate the reactor block, reactor site and roads, as well as construction of the sarcophagus and of a new town for reactor personnel. These tasks were completed by 1990.
- The remaining three reactors at Chernobyl resumed operation for several years, although they have now been decommissioned (the last one in December 2000).
- The original concrete sarcophagus sealing in the wrecked reactor is unsafe, and remedial work will cost approximately $715 million.

❒ nuclear disaster 2: spread of chernobyl gases, ukraine, 1986

Material released from the Chernobyl explosion included xenon gas, iodine, caesium, and approximately 5 percent of any radioactive material remaining in the reactor after the explosion. The bulk of the material fell as dust and debris in the immediate vicinity of the reactor. Lighter materials were carried by wind over southern Russia. Some circulated over Northern Euope and the North Atlantic.

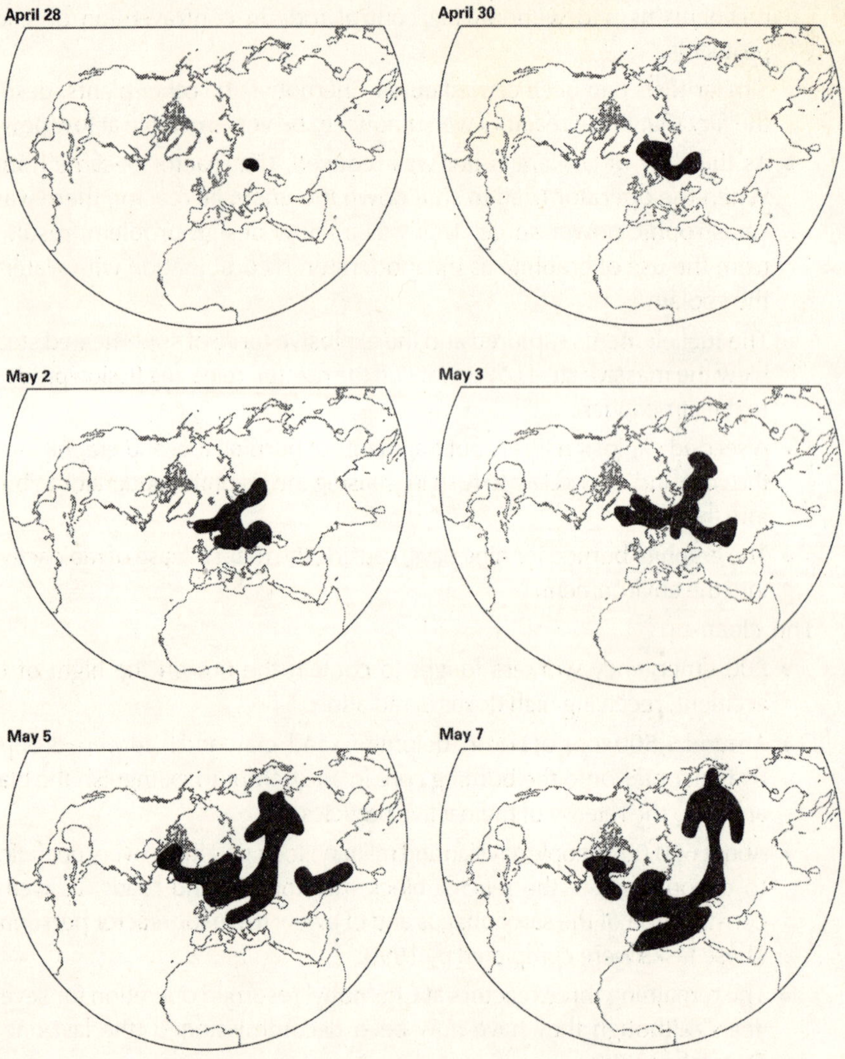

❐ nuclear disaster 3: tokaimura, japan, 1999

- On September 30, 1999 the worst nuclear accident in the history of the Japanese nuclear power programme occurred in Tokaimura, 87 miles northwest of Tokyo.
- The plant, owned by JCO Co., was preparing a small batch of nuclear fuel for use in the experimental JOYO fast breeder reactor, using uranium enriched to 18.8 percent U-235.
- Three employees poured 35 lb of a solution containing the enriched uranium (a) into a precipitation tank containing nitric acid (b), instead of the 5.2 lb normally used.

- Bringing together so much enriched uraniumtriggered a limited uncontrolled nuclear chain reaction, known as a 'criticality'.
- There was no explosion, but a flash of blue light followed by alarms as the nuclear reaction emitted intense neutron and gamma radiation.
- The three workers received massive doses of radiation. Two of them died as a result.
- The criticality continued intermittently for 20 hours. It was sustained by the water used in the process, which acted as a moderator, slowing the neutrons sufficiently to keep the chain reaction going. A sleeve containing cooling water was also acting as a neutron reflector, bouncing neutrons back into the reaction.

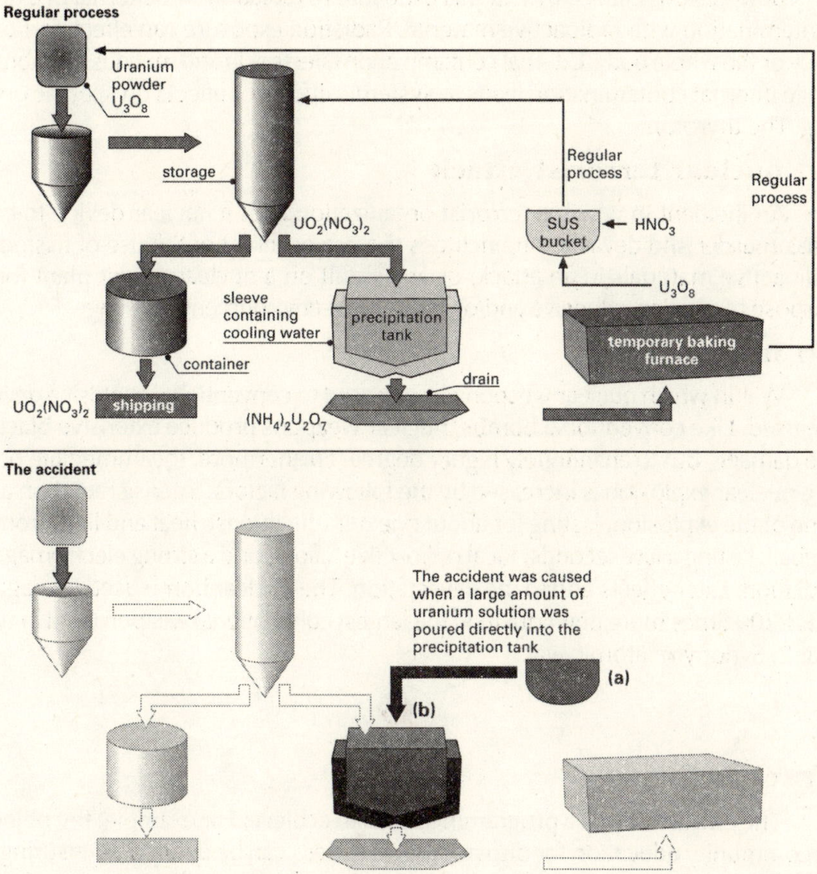

- 161 people were evacuated from 39 houses within a 380 yard radius 5 hours after the criticality began.
- The chain reaction was finally stopped when workers crushed the pipes supplying water to the cooling sleeve surrounding the precipitation tank, allowing it to drain. Boric acid solution, a neutron absorber, was then added to the tank to ensure that it stayed subcritical.

- These operations exposed a further 27 employees to some radiation.
- The International Atomic Energy Agency rated the accident as the worst since Chernobyl, blaming it on 'human error and serious breaches of safety principles.'

❑ nuclear incident

An event in which a nuclear agent is used as a terrorist weapon. There are two fundamentally different threats in the area of nuclear terrorism: (1) the use, or threatened use, of a nuclear bomb; and (2) the detonation of a conventional explosive incorporating nuclear materials.

❑ nuclear lesions

Body lesions caused by external exposure to radiation and internal of external contamination with radioactive material. Radiation exposure can effect part of the body or the whole body. External contamination effects skin and mucous membranes, while internal contamination leads to systemic effects or effects on specific organs (e.g. The thyroid).

❑ nuclear terrorist attack

An incident in which a terrorist organization uses a nuclear device to cause mass murder and devastation, includes the use or threat of the use of fissionable radioactive materials in an attack, or an assault on a nuclear power plant for the purpose of causing extensive and/or irreversible environmental damage

❑ nuclear war

War in which nuclear weapons-as opposed to conventional explosive devices-are used. Like conventional bombs, nuclear weapons produce extensive blast and fire damage, but to an infinitely higher degree. Furthermore, the immediate power of a nuclear explosion is increased by the following factors: intense radiation at the time of the explosion, lasting for about one minute; intense heat and light from the fireball, lasting a few seconds; local radioactive fallout; and a strong electromagnetic radiation. Later effects add to the devestation. The nuclear bomb used on nagasaki was 2200 times more powerful than the largest conventional weapon used in world war 2. Synonym: atomic war.

O

❑ objective

The end result that a programme seeks to achieve. For example, the objective of community education for disaster preparedness can be defined as ensuring that people in risk areas will want to be less vulnerable, know how to act in case of disaster, do what they can individually and collectively at the time of the emergency, and do the necessary before the emergency so that they can be prepared for it.

❑ occupational health and safety

The spectrum of endeavors, encompassing many technical disciplines, aimed at protecting workers from injury or illness associated with exposure to hazards encountered in the workplace. These hazards include both hazardous materials (such as poisons or reactive chemicals) and hazardous conditions (such as oxygen-deficient environments or unshored trenches).

❑ occupational safety and health administration (osha)

Agency within the U.S. Department of Labor that is responsible for the implementation of the Occupational Safety and Health Act. Process Hazard Analysis (PHA) The analysis of a facility's hazards conducted by the facility or its consultants as required by EPA regulations on risk management planning.

❑ offsite consequence analysis

An analysis in a facility's risk management plan, and risk management plan summary, that indicates how many people are placed in danger of injury or death in the event of a chemical release incident at the facility.

❑ oil spills

Number and frequency

- In the United States alone,14,000 oil spills are reported to the Environmental Protection Agency each year. The vast majority of these involve only small amounts of oil.

Numbers of spills over 770 tons

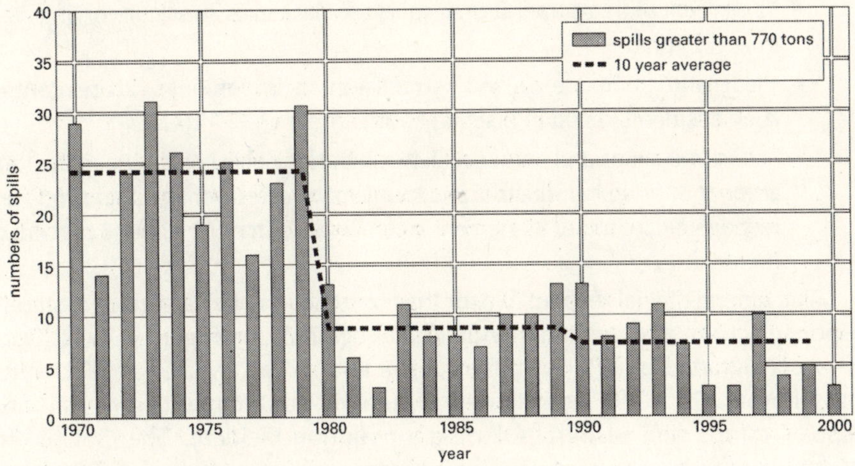

Percentage causes of large spills (>770 tons) from tankers 1974–2000

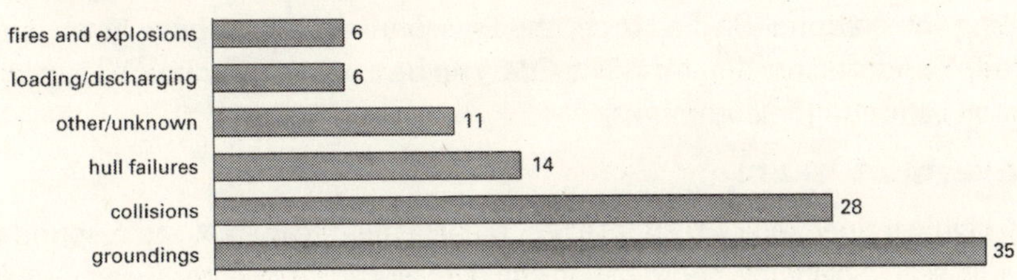

- Large spills (more than 771 tons) from tankers are rare, and their frequency has decreased significantly over the last 30 years. Although few in number, these large spills can have devastating consequences for the environment.
- The average number of large spills per year during the 1990s was about 33 percent the number that occurred during the 1970s.
- Although rare, large spills account for a high percentage of the total oil spilt. In the ten year period 1990–1999, 1 percent of incidents were responsible for 75 percent of the total oil spilt.

Quantity spilled

- Over the last 30 years, the total amount of oil spilt per year has varied widely from 9,918 tons (1995) to 670,000 tons (1979).
- 85 percent of spills are classed as small (less than 7.7 tons).
- 1979 was the worst year for oil spills. Of the 670,000 tons spilt, 316,000 tons was from a single incident (the Atlantic Empress spill in 1979). on-scene coordinator (osc)

Causes

- Most tanker spills happen during routine operations in ports or at oil terminals. Such operations include loading and unloading the oil cargo, and refuelling the ship (known as bunkering).
- 92 percent of these operational spills are classed as small, involving less than 7.7 tons.
- Most large spills are caused by collisions or groundings. 20 percent of such accidents result in discharges of more than 771 tons.
- Natural oil-seeps and human activity on the land also contribute to the total amount of oil spilt. Industrial and municipal waste discharge and runoff are responsible for about 37 percent and natural sources for about 8 percent of the total.

The federal official responsible for the coordination of a hazardous materials response action, as specified in individual Regional Contingency Plans. Oscs are predesignated by EPA for inland areas and by the U.S. Coast Guard for coastal areas. The OSC coordinates all federal containment, removal, and disposal efforts and resources during a pollution incident. The OSC is the point of contact for the coordination of federal efforts with those of the local response community. The OSC has access to extensive federal resources, including the National Strike Force, the Environmental Response Team, and Scientific Support Coordinators. The OSC can be a source of valuable support and information to the community.

☐ overhead team

A highly trained, quick response SAR management team that can respond to assist with search planning, coordination and operations. The overhead team usually consists of a search manager and one or two assistants.

P

❑ paleoseismology

A field of study based on geologic techniques developed in the 1980's to extend the historical record of earthquake activity on a specific fault system or areas exhibiting prehistoric liquefaction through trenching and age dating to chronicle either the periods of deformation of the rocks or liquefaction episodes in past earthquakes. The technique can extend the record back in time several thousand years, much longer than the record of instrumental seismicity.

❑ panic

Acute and overwhelming sense of fear and dread, usually of sudden onset and most often self-limiting and of short duration, from a few seconds to hours, the accompanying restlessness resulting in an urge to escape. A frequent but not lasting phenomenon following disasters and major emergencies.

❑ parasitic diseases

Infections, infestations and other disease states caused by parasites of animal origin. Some examples common in disaster situations are amoebiasis, intestinal worms, schistosomiasis, malaria, trypanosomiasis, scabies, pediculosis.

❑ pema

Pennsylvania emergency management agency.

❑ perimeter search

A search of an outside line that is believed to be across the direction of travel of the subject. The object is to cut the trail and find tracks or other evidence of the subject being there.

❑ perimeter search

A search of an outside line that is believed to be across the direction of travel of the subject. The object is to cut the trail and find tracks or other evidence of the subject being there.

❑ permissible exposure limit (pel)

The maximum amount or concentration of a chemical to which a worker may be exposed.

❑ personal protective equipment (ppe)

Equipment worn or used by workers to protect themselves from exposure to hazardous materials or conditions. The major types of PPE include respirators, eye protection, ear protection, gloves, hard hats, protective suits, etc.

❑ pest management

Practices that are aimed at controlling pests or vectors (e.g., insects, rodents, etc.) And ensuring the safe use of pesticides and herbicides.

❐ pesticide

Chemical compound used for killing organisms that are dangerous, undesirable, or a nuisance to man, animals or plants. They are named according to their action, such as: fungicide, herbicide, insecticide, molluscicide, nematocide, ovicide, rodenticide, virucide.

❐ plan

A pre-established course of action which, when implemented, is expected to lead to the attainment of the expected ends and objectives. An orderly set of decisions on the ways and means of achieving the impact and objectives sought.

❐ plan of action

A written document that consolidates all of the operational actions to be taken by various personnel in order to stabilize the incident.

❐ planning case scenario

The potential situation of a chemical release upon which a company builds its risk management plan under EPA rules. See also worse case scenario.

❐ plans, procedures and protocols

A well thought-out and fixed way of acting in order to reach a certain goal. Written down as plans, procedures, and protocols.

❐ plb

Personal Locator Beacon -The personal version of the "ELT" designed to be carried by a person on foot. It is manually activated.

❐ pls

Point (or Place) Last Seen - The location where the missing subject was actually seen by another person.

❐ plume exposure pathway (10-mile epz) rep program

A circular geographic zone (with a 10-mile radius centered at the nuclear power plant) for which plans are developed to protect the public against exposure to radiation emanating from a radioactive plume caused as a result of an accident at the nuclear power plant.

❐ plutonium

A man made highly toxic synthetic metallic element with a half-life of 24,400 years. Traces of plutonium occur in uranium ore and it is produced in a nuclear reactor by neutron bombardment of uranium-238. The isotope plutonium-239 readily undergoes fission and is used as a reactor fuel in nuclear power stations and in nuclear weapons. Plutonium is an extremely dangerous substance because of its radioactivity and the fact that when ingested as an oxide or other compound it deposits in the bone and is excreted very slowly. Metallic plutonium is not absorbed by digestive organs. Inhalation of only a few thousandths of a gram may lead to death within a few years and much smaller quantities can cause lung cancer after a latent period of about 20 years. Plutonium should be handled by remote control using extreme caution to avoid the release of dusts to the atmosphere. Plutonium metal is highly reactive and thus must be stored at low temperatures in dry air to avoid corrosion. It was first identified in experiments at the University of California in 1940, and plutonium-239 was isolated a year later. There are 16 isotopes of plutonium, of which only five are produced in significant quantities: plutonium-238.-239,-240,-241,-242.

❒ poa or pa

Probability of Area (chances that the subject, or clues, are in the area being searched).

❒ pod or pd

Probability of Detection (chances of finding the subject or clues given that they are in the area being searched).

❒ poisoning

Poisoning of atmosphere or water courses due to industrial sources (in EM-DAT referred to as " Ind:Poisoning " and forms a disaster subset of disaster type " industrial accident ").

❒ policy environment

The forum and the process for making decisions about plans, laws, and practices to reduce unacceptable risk to people, property, and infrastructure in the community and strategic plans to implement them over time.

❒ pos or ps

Probability of Success (chances of being successful with a particular search in a given area).

❒ post traumatic stress/shock

Emotional problems that can arise after a critical incident that can be long lasting. Search Managers, Team Leaders, etc., should know the signs and symptoms and be alert to the well-being of every team member.

❒ post-traumatic stress disorder (ptsd)

An anxiety disorder that can develop after exposure to a terrifying event or ordeal in which grave physical harm occurs or is threatened

❒ post-traumatic stress syndrome

Following a period of intense stress (like in disaster) a person may encounter short or long term psychic disorders varying among other things from anxiety, insomnia, feelings of guilt, irritability, and concentration problems.

❒ ppa

Performance partnership agreement

❐ ppe

Personal protective equipment.

❐ precautionary zone (pz) see also csepp

The outermost portion of the EPZ for CSEPP, extending from the PAZ outer boundary to a distance where the risk of adverse impacts to humans is negligible. Because of the increased warning and response time available for implementation of response actions in the PZ, detailed local emergency planning is not required, although consequence management planning may be appropriate.

❐ predisaster stage

Occurs when there is knowledge about an impending disaster; activities include warning, preimpact mobilization, and evacuation if appropriate

❐ preliminary damage assessment

A mechanism used to determine the impact and magnitude of damage and the resulting unmet needs of individuals, businesses, the public sector, and the community as a whole. Information collected is used by the State as a basis for the Governor's request for a Presidential declaration, and by FEMA to document the recommendation made to the President in response to the Governor 's request. Pdas are made by at least one State and one Federal representative. A local government representative familiar with the extent and location of damage in the community often participates; other State and Federal agencies and voluntary relief organizations also may be asked to participate, as needed.

❐ preparedness

The aggregate of all measures and policies taken by humans before the event occurs that allows mitigation of the impact caused by the event through responses to the impact of the event. Preparedness includes warning systems, evacuation, relocation of dwellings (e.g., for floods), stores of food and water, temporary shelter, energy, management strategies, disaster drills and exercises, etc. Contingency plans and responses are included in the preparedness in the sense used in this document. As preparedness increases, the ability of the society to absorb the event and mitigate the impact (damage) is augmented as a dependent variable of the level of preparedness; A range of policies, legislative mandates, professional practices, and social adjustments that are used by individuals, businesses, organizations, communities, and Nations to plan for emergency response, recovery, and reconstruction after a damaging earthquake.

❐ presidential decision directive 39 (pdd-39

Issued in June 1995, PDD-39, United States Policy on Counter terrorism, directed a number of measures to reduce) the Nation's vulnerability to terrorism, to deter and respond to terrorist acts, and to strengthen capabilities to prevent and manage the consequences of terrorist use of nuclear, biological, and chemical weapons.

❐ prevention

The aggregate of approaches and measures taken to ensure that human actions or natural phenomena DO NOT cause or result in the occurrence of an event related to the identified or unidentified hazard.(Gunn) It does not mean decreasing the severity or intensity of the event. See also disaster prevention; Primary prevention of disaster is possible through technical, organizational and judicial means. Secondary prevention implies the optimal management of disaster itself. Tertiary prevention combats the complications of disaster. The better secondary prevention, the less tertiary prevention is needed.

❐ primary health care

Essential health care made universally accessible to individuals and families in the community by means acceptable to them, through their full participation and at a cost that the community and country can afford. It forms an integral part both of the country's health system, of which it is the nucleus, and of the overall social and economic development of the community.

❐ primary prevention

Front-end solutions to issues of facility hazards, which involve designing the process so that it eliminates vulnerabilities. Also known as inherent safety.

❐ principal investigators

Individuals (usually faculty members) who are responsible for directing or supervising research on specific projects or in specific laboratories.

❐ process safety management (psm)

The OSHA PSM rule requires employers subject to the rule to engage in various activities to prevent or minimize the consequences of catastrophic releases of toxic, reactive, flammable and explosive chemicals.

❐ protective action zone (paz) see also csepp

An area that extends beyond the area to approximately 16 to 50 km (10 to 30 miles) from the stockpile location. The PAZ is that area where public protective actions may still be necessary in case of an accidental release of chemical agent, but where the available warning and response time is such that most people could evacuate. However, other responses (e.g., sheltering) may be appropriate for institutions and special populations that could not evacuate within the available time.

❐ proton

A stable, positively charged particle in an atomic nucleus.

❐ prusik

A friction or ascending knot commonly used in belaying and climbing. Originally known as a Magnus Hitch.

❐ ptsd

Post Traumatic Stress Disorder (see Post Traumatic Stress).

❑ public health

The discipline in health sciences that, at the level of the community or the public, aims at promoting prevention of disease, sanitary living, laws, practices and a healthier environment.

❑ public information officer

A Federal, State, or local government official responsible for preparing and coordinating the dissemination of emergency public information.

❑ public policy

A plan, a rule, professional practice, or a way of acting that has the force of law. Public policy is designed to manage the unacceptable risk that is the result of being exposed over time to the occurrence of one or more of the earthquake hazards. When implemented over periods of 20-30 years or more, seismic safety policy is expected to reduce unacceptable risk. The policy options encompass: a) stop increasing the risk to elements that will be exposed in the future to natural and technological hazards, b) start decreasing the risk to existing elements already at risk from natural and technological hazards, and c) continue to plan for the inevitable event. The measures to implement these policies include mitigation and preparedness.

R

❑ races

Radio amateur civil emergency services.

❑ rad

The old unit of absorbed dose of ionizing radiation. One rad is equal to 1 rem for most forms of radiation. The rad was first defined in 1953. In 1956 the rad replaced the roentgen (a unit of radiation exposure) for clinical work involving X-rays or radioactive sources. In practice the rad and the roentgen both represent about the same amount of energy, since 1,000 rad equals 1,0550-1,150 roetgens, but unlike the roetgen, the rad is applicable to all types of radiation. Historically, the rad can be traced back to 1918, when it was suggested as a name for the unit of X-ray dose sufficient to kill a mouse.

❑ radiation

The emission and propagation of energy through space or tissue in the form of waves. This term usually refers to electromagnetic radiation, classified by its frequency: radio, infrared, visible, ultraviolet, X-rays, gamma rays and cosmic rays; In this self-study program, refers to nuclear radiation, not radiation as a type of heat transfer. There are three types of nuclear radiation:

(1) alpha,

(2) beta, and

(3) gamma. Radiation is the cause of one of the six types of harm (see TRACEM) that can be encountered at a terrorist incident.

❐ radiation injury

Somatic and genetic damage to living organisms caused by ionizing radiation

❐ radiation safety

Practices that are aimed at minimizing exposure to the potentially harmful effects of radiation. At SDSU, there are two separate Hazard Control Programs for Radiation Safety: Ionizing Radiation, which is aimed at protection against sources of ionizing radiation (e.g., alpha or beta particles, neutrons, gamma rays or X-rays), and Nonionizing Radiation, which is aimed at controlling exposures to sources such as laser rays or electromagnetic fields.

❐ radiation sickness

One of the results of DNA mutation inside cells exposed to ionizing radiation; can be deadly, but survivable with bone marrow transplantation;The symptoms characterizing the sickness known as radiation injury, resulting from excessive exposure of the whole body to ionizing radiation.

❐ radioactive contamination

The undesireable presence of radioactive material in the air, in man, or on any substance.

❐ radioactivity

The emission of radiation from atomic nuclei. Nuclear radiation includes alpha and beta particles, gamma rays, X-rays and neutrons; The phenomenon of spontaneous disintigration in a nuclide accompanied by the emission of ionizing radiation.

❐ radiological dispersal devices (rdd)

A conventional explosive incorporating nuclear materials.

❐ radiological dispersion bomb

The most accessible nuclear device to be used by a terrorist, consists of a conventional explosive such as trinitrotoluene (TNT) packed with radioactive waste by-products from nuclear reactors, discharges deadly radioactive particles into the environment; also called a dirty bomb

❐ radiological monitoring

The process of locating and measuring radiation by means of survey instruments that can detect and measure (as exposure rates) ionizing radiation.

❐ rappel

To descend a rope at a safe controlled speed.

❐ react

Radio emergency associated communications teams

❒ reconstruction stage

Restoration, reconstitution, and mitigation take place, rebuilding, replacing lost or damaged property, returning to school and work; when life returns to some semblance of "normal," another name for recovery stage

❒ recovery

The long-ten-n activities beyond the initial crisis period and emergency response phase of disaster operations that focus on returning all systems in the community to a normal status or to reconstitute these systems to a new condition that is less vulnerable.

❒ red cross

Red Cross, or International Red Cross, general terms used for one or all the components of the worldwide organization active in humanitarian work. The official overall name is the International Red Cross and Red Crescent Movement, which has 3 components. 1. International Committee of the Red Cross (ICRC): acts mainly in conflict disasters as neutral intermediary in hostilities and for the protection of war victims. Guardian of the Geneva Conventions. 2. League of the Red Cross and Red Crescent Societies (LRCS): international federation of the National Societies, active in non-conflict disasters and natural calamities. 3. The individual National Red Cross or Red Crescent Society of every country.

❒ regional operating center

The temporary operations facility for the coordination of Federal response and recovery activities, located at the FENU Regional Office (or Federal Regional Center) and led by the FEMA Regional Director or Deputy Director until the DFO becomes operational. Once the ERT-A is deployed, the ROC performs a support role for Federal staff at the disaster scene.

❒ regional tectonic deformation

Changes in elevation over regional distances; is a feature of earthquakes having magnitudes of 8 or greater. A type of ground failure.

❒ relief

Assistance in material facilities, personal needs and services given to needy persons or communities, without which they would suffer.

❒ rem

The acronym for roentgen-equivalent-man. This unit of dose equivalent was replaced by the sievert.

❒ rep program

The two epzs are Plume Exposure Pathway (10-mile EPZ) and Ingeslion Pathway (50-mile EPZ)

❒ resiliency

Pliability, flexibility, or elasticity to absorb the event. Resiliency is offered by types of construction, barriers, composition of the land, (geological base), geography, bomb shelters, location of dwelling, etc. As resiliency increases, so does the absorbing capacity of the society and/or the environment. Resilience is the inverse of vulnerability.

❒ resource conservation and recovery act (rcra)

The federal law on hazardous and solid waste management.

❒ resource management

Those actions taken by a government to: identify sources and obtain resources needed to support disaster response activities; coordinate the supply, allocation, distribution, and delivery of resources so that they arrive where and when most needed; and maintain accountability for the resources used.

❒ resources

All personnel and major items of equipment available, or potentially available, for assignment to incident tasks on which status is maintained.

❒ response spectrum

A graph of the output of a mathematical model which shows how an idealized ensemble of lightly damped, simple harmonic oscillators representing a complete spectrum of short (short periods of vibration) to tall (long periods of vibration) buildings will respond to a particular dynamic ground motion accelerogram. The accelerogram is used to excite the ensemble into vibration in the 0.05-10 seconds period range, the range of interest to engineers. The concept of the response spectrum is used in building codes and the design of essential and critical structures.

❒ responsible person

The individual at a place of employment who has the responsibility and authority to implement the IIPP on behalf of the business or institution. At SDSU, the President has delegated the Vice President for Business and Financial Affairs the responsibility for implementing the IIPP.

❒ richter scale

Devised by C.F. Richter in 1935, an index of the seismic energy released by an earthquake (as contrasted to intensity that describes its effects at a particular place) expressed in terms of the motion that would be measured by a specific type of seismograph located 100 km from the epicentre of an earthquake.

❒ right-to-know

The public's rights to information regarding chemical sites, established under various laws.

❒ right-to-know act

1986 Emergency Planning and Community Right-To-Know Act RMP-Risk Management Planning. Planning that is required under the Air Toxics regulations of the Clean Air Act. Risk Management Plans- due by June 21, 1999-are intended to detect and prevent or minimize accidental airborne release of a set of "extremely hazardous substances" and to provide a prompt emergency response to any such release.

▫ risk

The objective (mathematical) or subjective (inductive) probability that the hazard will become an event. Factors (risk factors) can be identified that modify this probability. Such risk factors are constituted by personal behaviors, life-styles, cultures, environmental factors, and inherited characteristics that are known to be associated with health-related questions; The expected number of lives lost, persons injured, damage to property and disruption of economic activity due to a particular natural phenomenon, and consequently the product of specific risk and elements at risk.

▫ risk assessment

An objective scientific assessment of the chance of loss or adverse consequences when physical and social elements are exposed to potentially harmful natural and technological hazards. The endpoints or consequences depend on the hazard and include: damage, loss of economic value, loss of function, loss of natural resources, loss of ecological systems, environmental impact, deterioration of health, mortality, and morbidity. Risk assessments integrate hazard assessments with the vulnerability of the exposed elements at risk to seek reliable answers to the following questions1. What can happen? 2. How likely are each of the possible outcomes? 3. When the possible outcomes happen, what are the likely consequences and losses?

▫ risk indicator

Descriptor that briefly denotes a risk that may cause a disaster

▫ risk management

The public process of deciding what to do when risk assessments indicate that risk, or the chance of loss, exists. Risk management encompasses choices and actions for communities and individuals (i.e., prevention, mitigation, preparedness, recovery) which are designed to: a) stop increasing the risk to future elements that will be placed at risk to natural and technological hazards, b) start decreasing the risk to existing elements already at risk, and c) continue planning ways to respond to and recover from the inevitable natural and technological hazard, including the imponderable extreme situation or catastrophic event.

▫ risk map

Cartographic representation of the types and degrees of hazards and of natural phenomena that may cause or contribute to a disaster

▫ risk marker

An attribute of the hazard that is associated with an increased probability that an event may occur and can be used as an indicator of an increased or increasing risk that the event will occur.

▫ robert t. stafford disaster relief and emergency assistance act, public law 93-288

Authorizes the Federal government to respond to disasters and emergencies in order to help State and local governments save lives, and to protect public health, safety, and property.

❐ **rural sar**

A search and rescue mission conducted in an area often given to farming or to woodland and is usually sparsely to moderately populated.

S

❐ **safety communication**

The collective means by which safety information is disseminated to employees. Classroom and departmental safety meetings, and written communications such as posters, newsletters, and posting of regulatory agency inspection findings are included.

❐ **safety coordinators**

Individuals within departments or colleges who have been appointed by their supervisors or administrators to assist in implementing the IIPP in their respective areas.

❐ **sanitation**

A comprehensive term referring to the development and application of measures designed to protect public health. Sanitation procedures attempt to prevent the spread of disease through the proper disposal of sewage and other waste, correct handling of food and water supplies, and control of the carriers of disease (mosquitoes, rats, etc.).

❐ **sar**

Search and Rescue.

❐ **sar emergency**

Any SAR incident, whether related to any other type of incident or not, that requires the utilization of resources to resolve, due to the threat or potential threat to human life.

❐ **sara title iii**

The third part of the Superfund Amendments and Reauthorization Act of 1986, also known as EPCRA.

❐ **scrub fire**

Fires in scrub or bush that cover extensive damage. They may start by natural causes such as volcanic eruptions or lightning, or they may be caused by arsonists or careless smokers, by those burning wood, or by clearing a forest area.

❐ **search manager**

The individual "managing" the search operations. Also known as "MSO" for Managing Search Operations or "MLPI" for Managing the Lost Person Incident or "ISAR" for Inland Search and Rescue Specialist or "NSARS" for National Search and Rescue School; Also known as "MSO" for Manager of Search Operations or "MSF" for Manager of the Search Function or "MSI" for Managing the Search Incident.

❐ secondary hazard

A threat whose potential would be realized as the result of a triggering event that of itself would constitute an emergency. For example, dam faffure n-dght be a secondary hazard associated with earthquakes.

❐ secondary prevention measures

Hazard reduction measures that accept the existence of a design hazard and seek to reduce the consequences of a potential accident by adding on technologies (e.g. Measures to capture releases) and operational practices (e.g. Security guards). Opposite of inherent safety or primary prevention.

❐ seiches

Standing waves induced in lakes and harbors by earthquake ground shaking. A type of ground failure.

❐ seismic zonation

A public policy tool to link earthquake risk assessment and earthquake risk management, with the objective being to identify, delineate, and highlight those geographic areas of a community where investments in expanded risk assessment and specific mitigation measures and regulations are needed to mitigate, prevent, or reduce the community's perceived unacceptable risk. When fully implemented, seismic zonation is the link between earthquake risk assessment and earthquake risk management. It requires three kinds of activities that build upon the current state-of-the-art and the continually evolving understanding of science, technology, and public policy. They are: 1) development of hazards maps for use in a risk assessment. These maps characterize aspects of the hazard environment that contribute to the risk. They are constructed by integrating information and databases on the characteristics of the earthquake source, regional seismic wave propagation paths, and the local site. The maps are typically based on probabilistic concepts and depict the hazards of ground shaking, ground failure, surface fault rupture, regional tectonic deformation, and tsunami wave run up, and the aftershock sequence during a specified exposure time. Simultaneously with the development of hazards maps, information on the perceived vulnerabilities and expected performance of the community's built environment in a damaging earthquake is integrated and formatted for use in a risk assessment. 2) applications of information management techniques (i.e., gis) and analytical models (i.e., analytical models such as hazus) to assess the risk to elements of the built environment exposed to the earthquake hazards of ground shaking, ground failure, surface fault rupture, regional tectonic deformation, tsunami wave run up, and the aftershock sequence. The products of a risk assessment include statements on the risk, which can be classified by the community policy makers and stake holders through a consensus process into two categories: acceptable risk and unacceptable risk. 3) enactment, adoption, enforcement, and implementation of community-specific public policies and professional practices that will increase a community's earthquake resistance and mitigate, prevent, or reduce unacceptable risk

❐ seismicity

Earthquake activity, as measured in terms of number of events, their magnitude, distribution, and frequency.

❐ serc

State Emergency Response Commission. One of the public agencies that must be notified under the term of EPCRA regarding the presence of certain hazardous substances on a facility property. The LEPC must also be notified if there is a release of a hazardous substance into the environment See Also:EPCRA, LEPC.

❐ seveso

A village near Milan, Italy, site of a chemical plant which, in July 1976, accidentally discharged the toxic compound Dioxin, causing severe illness and toxic manifestations among the surrounding population, with extensive environmental damage.

❐ shelter in place

An emergency response strategy in which people living or working near a facility where a chemical incident occurs, in which they are advised to stay indoors and to seal off ventilation against external air sources.

❐ sheltering

The extent to which a shelter can protect potential victims from exposure to ionizing radiation and contamination with radioactive material depends on its location and type. Exposure is at a maximum level in the open air and at a minimum in a cellar of a concrete building with a ventilation filter; Action that consists of providing asylum or provisional lodgings to an individual or group.

❐ shielding

One of the three components of TDS; refers to maintaining significant physical barriers between you and the hazard. Examples include vehicles, buildings, walls, and PPE.

❐ sievert

The international unit of dose equivilant. One sievert is equal to 100 rems.

❐ significant threat

The confirmed presence of an explosive device or WMD capable of causing a significant destructive event, prior to actual injury or property loss (FBI).

❐ simple asphyxiant

Generally, an inert gas that displaces the oxygen necessary for breathing, and dilutes the oxygen concentration below the level that is useful for the human body.

❐ simplification

Simplification of medical procedures saves time so that more attention can be paid to the seriously wounded victims; e.g. Large wounds should be disinfected and covered and in a later stage closed with plastic and reconstructive surgery.

❐ sizeup

The rapid mental evaluation of the factors that influence an incident. Sizeup is the first step in determining a course of action.

❐ slide

Disaster type term used in EM-DAT comprising the two disaster subsets " avalanche " and " landslide ".

❐ slow disaster

Disaster, usually natural, the beginnings of which are slow, sometimes imperceptible until the full effect is felt, as in poor crops leading to drought and famine. Synonym: creeping disaster

❐ soil amplification

Soils have a period-dependent effect on the ground motion, increasing the level of shaking for certain periods of vibration and decreasing it for others as a function of the "softness" and thickness of the soil relative to the underlying rock and the three-dimensional properties of the soil/rock column.

❐ soil/structure resonance

A physical phenomenon increasing the potential for destructiveness that results when the input seismic waves causes the underlying soil and the structure to resonate, or vibrate at the same period.

❐ somatic effects of radiation

Those effects which are experienced by the individual receiving the radiation - in contrast to genetic effects, which occur in offspring or future generations, and teratogenic effects, which occur in the embryo or foetus inside the mother's womb.

❐ sound sweep search

An attraction search in which all searchers call out, whistle or make some other sound at a prescribed time and then are all quiet and listening for an answer during a second prescribed time.

❐ source directivity

A phenomenon that increases ground shaking at a site. It results from the directional aspects of the fault rupture that cause most of the energy to be released in a particular direction instead of in all directions.

❐ space blanket

Emergency blanket that looks a lot like foil. This thin, light blanket keeps the body heat in.

❐ spontaneous evacuation

Residents or citizens in the threatened areas observe an emergency event or receive unofficial word of an actual or perceived threat and without receiving instructions to do so, elect to evacuate the area. Their movement, means, and direction of travel is unorganized and unsupervised.

❐ stafford act

See Robert T. Stafford Disaster Relief and Emergency Assistance Act.

❐ staging area

That location where incident personnel and equipment gather and are assigned.

❐ standard operating procedure

A set of instructions constituting a directive, covering those features of operations which lénd themselves to a depth, step-by-step process of accomplishment. Sops supplement eops by detailing and specifying how tasks assigned in the EOP are to be carried out.

❐ standardization

Standardization of medical procedures, like the administration of drugs, antibiotics, analgesics and anticoagulants in the chain of medical care avoids errors, simplifies the transfer of medical information in this chain and is more economic.

❐ state coordinating officer

The person appointed by the Governor to coordinate State, Commonwealth, or Territorial response and recovery activities with FRP-related activities of the Federal Government, in cooperation with the FCO.

❐ state eop

The State EOP is the framework within which local eops are created and through which the Federal government becomes involved. The States play three roles:

(1) they assist local jurisdictions whose capabilities are overwhelmed by an emergency;

(2) they themselves respond first to certain emergencies; and

(3) they work with the Federal government when Federal assistance is necessary.

❐ state liaison

A FEMA official assigned to a particular State, who handles initial coordination with the State in the early stages of an emergency.

❐ stockpile

A place or storehouse where material, medicines and other supplies needed in disaster are kept for emergency relief. Examples: undro warehouse in pisa, unipac in copenhagen.

❐ storm

Wind with a speed between 48 and 55 knots; (in EM-DAT, " storm " is a disaster subset of the disaster type " wind storm ").

❐ storm surge

A dome of sea water created by the strong winds and low barometric pressure in a hurricane that causes severe coastal flooding as the hurricane strikes land.

❑ **strategic**

Goals Strategic goals are broad, general statements of intent.

❑ **strategic planning**

Consists of preparing the organization to respond to disaster threats in locations that are not specified and not immediately threatened.

❑ **stress**

Any strain, anxiety, psychological shock or excessive pressure that disturbs the smooth functioning of a person or organism (and by extension, a group). Disasters are stressful events.

❑ **strike team**

Specified combinations of the same kind and type of resources, with common communication and a leader.

❑ **strontium-90**

A radio-isotope with a half-life of 28.2 years. It is one on the hazards of fallout.

❑ **supervisors**

Employees who have authority to direct the tasks of other employees and are therefore responsible for the job-related environments to which their workers are exposed. Where required, supervisors are to develop local area procedures, train their workers regarding health and safety precautions, enforce rules, and maintain records.

❑ **surface fault rupture**

A phenomenon that increases ground shaking at a site. Refers to the physical phenomenon of the rupturing fault breaking the surface of the ground, instead of stopping beneath the ground surface, and releasing more energy on the side of the fault that is moving than on the stationary block.

❑ **susceptibility**

The degree of ease by which a person or a population is affected by a given phenomenon. In the context of the guidelines, susceptibility and vulnerability will be used interchangeably.

❑ **sweeping triage**

The first triage at the disaster site in order to locate the most seriously wounded T1 victims.

T

❑ **task force**

A group of resources with common communications and a leader temporarily assembled for a specific mission.

❐ tds

Time, Distance and Shielding. Three types of protective measures commonly associated with hazardous materials training.

❐ technical assistance

The system of providing assistance, on a bilateral or multilateral basis, through technicians, experts, teachers or equipment, to a developing country. Synonym: technical cooperation.

❐ technical operations

Include operations to identify, assess, dismantle, transfer, dispose, and decontaminate personnel and property exposed to explosive ordnance or NBC/WMD material.

❐ technological disaster

Man-made disaster due to a sudden or slow break-down, technical fault, error or involuntary or voluntary human act that causes destruction, death, pollution and environmental damage.

❐ technological hazard

A potential threat to humans and their welfare caused by technological factors (e.g., chemical release, nuclear accident, dam failure). Earthquakes and other natural hazards can trigger technological hazards.

❐ technology options analysis

Analysis of opportunities for enhancing the inherent safety of a facility.

❐ teratogenic effects of radiation

Those effects which are experienced by the embryo or foetus inside the mothers womb - in contrast to genetic effects, which occur in children of the individual receiving the radiation or in later generations, and somatic effects, which occur in the individual receiving the radiation.

❐ terrorism

The unlawful use of force against persons or property to intimidate or coerçe a government, the civilian population, or any segment thereof, in the furtherance of political or social objectives; The systematic use of terror, the deliberate creation and exploitation of fear for bringing about political change; As defined by the FBI, "the unlawful use of force against persons or property to intimidate or coerce a government, the civilian population, or any segment thereof, in the furtherance of political or social objectives." This definition includes three elements:

(1) Terrorist activities are illegal and involve the use of force.

(2) The actions are intended to intimidate or coerce.

(3) The actions are committed in support of political or social objectives. Emergency management is typically concerned with the consequences of

terrorist acts directed against large numbers of people (as opposed to political assassination or hijacking attacking, which may also be considered "terrorism").

❑ terrorism incident annex

The annex to the FRP that describes the Federal concept of operations to implement PDD-39 when necessary to respond to terrorist incidents within the U.S. Please see Appendix A for a copy of the annex.

❑ terrorist incident

A violent act, or an act dangerous to human life, in violation of the criminal laws of the United States or of any State, to intimidate or coerce a government, the civilian population, or any segment thereof, in furtherance of political or social objectives (FBI).

❑ that others may live

Motto of the search and rescue community.

❑ the world's most environmentally damaging oil tanker Sills-403 (ci)

- The most notorious oil spills are not necessarily the largest. Spills make headlines when there is widespread and visible damage to the environment that occurs when the oil is washed ashore.
- The Exxon Valdez oil spill in 1989 is very well known because of the extensive damage to the Alaskan coastline that resulted, yet it is only ranked as the 34th biggest spill of all time.
- Large spills out to sea attract relatively little publicity. The ABT Summer, for example, spilt 280,000 tons of oil 700 nautical miles off the coast of Angola in 1991 but, because the coastline was not immediately and obviously polluted, escaped widespread attention in the media.

rank	ship's name	year	location	oil lost (tons)
1	Atlantic Empress	1979	off Tobago, West Indies	316,000
2	Castillo de Bellver	1983	off Saldanha Bay, South Africa	277,000
3	Amoco Cadiz	1978	off Brittany, France	246,000
4	Haven	1991	Genoa, Italy	160,000
5	Torrey Canyon	1967	Scilly Isles, United Kingdom	131,000
6	Urquiola	1976	La Coruña, Spain	110,000
7	Independenta	1979	Bosphorus, Turkey	105,000
8	Jakob Maersk	1975	Oporto, Portugal	97,000
9	Braer	1993	Shetland Islands, United Kingdom	94,000
10	Aegean Sea	1992	La C`oruña, Spain	82,000

The world's ten biggest oil tanker spills resulting in pollution of coastlines

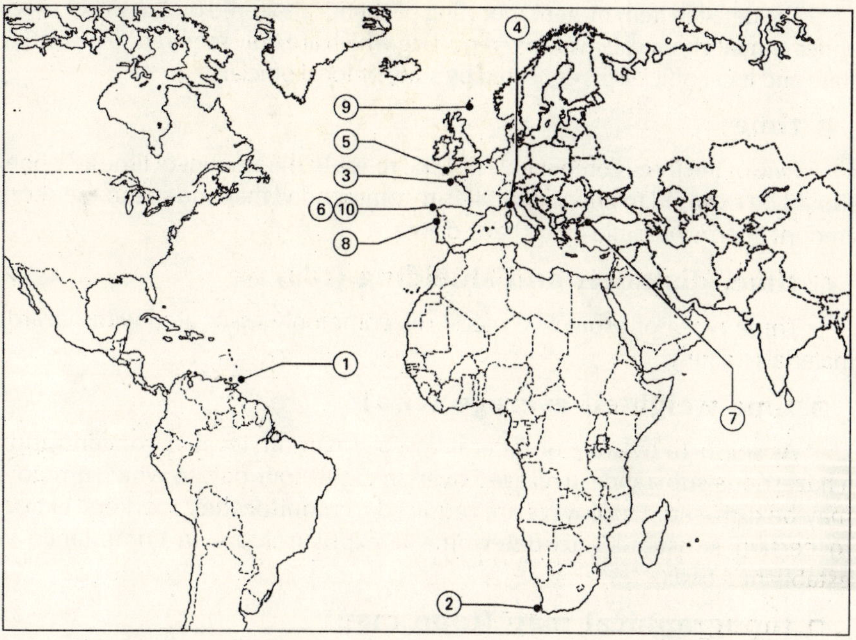

▫ thermal harm

One of the six types of harm (see TRACEM) that can be encountered at a terrorist incident. Thermal harm is the result of exposure to the extremes of heat and cold.

▫ threshold limit value (tlv)

The concentration of an airborne substance to which an average person can be repeatedly exposed without adverse effects.

▫ threshold planning quantity (tpq)

The amount of an extremely hazardous substance present at a facility above which the facility's owner/operator must give emergency planning notification to the LEPC and PEMC.

▫ tidal wave

Abrupt rise of tidal water (caused by atmospheric activities) moving rapidly inland from the mouth of an estuary or from the coast.

▫ tier 1

The categorical reporting of chemicals stored on the site of a facility under annual chemical inventory reporting. Required to automatically be filed with state and local officials under federal law.

❐ tier 2

The detailed item by item reporting of chemicals stored on the site of a facility under annual chemical inventory reporting which are only required to be filed with state and local officials if requested by state or local officials.

❐ time

One of the three components of TDS; refers to the amount of time a responder should be exposed to an incident. It is recommended that one spend the shortest amount of time possible in the hazard area.

❐ time, distance, and shielding (tds)

Three types of protective measures commonly associated with hazardous materials training.

❐ time-weighted average (twa)

As set forth by state or federal law, a maximum level or concentration of a hazardous substance, averaged over an eight-hour day, to which employees may be exposed. Employers are required to monitor their workers' exposure to certain substances to determine if exposures are in compliance with established twas.

❐ topographical map (topo map)

A map used for search (usually 1:24,000) which shows physical features of a region and allows a certified search manager to deploy resources accordingly.

❐ tornado

A local atmospheric storm, generally of short duration, formed by winds rotating at very high speeds, usually in a counter-clock-Mse direction. The vortex, up to several hundred yards wide, is visible to the observer as a wild pool-like column of winds rotating about a hollow cavity or funnel. Winds may reach 3 00 miles per hour or higher; Violently rotating storm diameter; the most violent weather phenomenon. It is produced in a very severe thunderstorm and appears as a funnel cloud extending from the base of a cumulonimbus to the ground.

❐ total affected

In EM-DAT, people that have been injured, affected and left homeless after a disaster are included in this category.

❐ toxic cloud

An airborne mass of gases, vapors, fumes, or aerosols of toxic materials.

❐ toxic materials

A type of chemical that can cause chemical harm at an incident scene. They produce harmful effects depending on the concentration of the materials and the length of exposure to them. An individual can have chronic or acute exposures to toxic materials.

❐ toxic release inventory (tri)

A database of annual toxic releases from certain manufacturers compiled from EPCRA Section 313 reports. Manufacturers must report annually to EPA and the states the amounts of almost 350 toxic chemicals and 22 chemical categories that they release directly to air, water, or land, inject underground, or transfer to off-site facilities. EPA compiles these reports and makes the information available to the public under the "Community Right-to-Know" portion of the law.

❐ toxic substance

A chemical or mixture that can cause illness, death, disease, or birth defects. The quantities and exposures necessary to cause these effects can vary widely. Many toxic substances are pollutants and contaminants in the environment.

❐ toxic-by-inhalation (tih)

Materials which present the danger of forming a toxic cloud which can cause injury or death.

❐ toxicological disaster

Serious environmental pollution and illness caused by the massive accidental escape of toxic substances into the air, soil or water, and to man, animals or plants.

❐ toxin

Substance secreted by certain living organisms, capable of causing harmful (toxic) effects in the receiving organism.

❐ toxins

Toxic substances of natural origin produced by an animal, plant, or microbe. They differ from chemical substances in that they are not manmade. Toxins may include botulism, ricin, and mycotoxins.

❐ tracem

The acronym used to identify the six types of harm one may encounter at a terrorist incident: Thermal, Radioactive, Asphyxiation, Chemical, Etiological, and Mechanical.

❐ tracker

An individual who can detect and follow signs of human passage.

❐ tracking canine

A search dog that will follow the ground scent of a person who has passed through an area in which the dog is searching; A search dog that will follow the scent trail of a specific individual, after the dog has been allowed to smell an article or object that has been in contact with that individual. The trailing dog is scent discriminating (the ability to distinguish the scent of one person and follow it).

❐ training

Classroom instruction, job-site safety meetings, on-the-job training, and written materials provided to employees to make them aware of workplace hazards and how to prevent accidents and illnesses.

❐ transboundary pollution

Pollution and pollutants that have been produced in one country and that have passed international boundaries through water or air to other countries, causing pollution. The effects can be mitigated only through international agreements as the damage is caused outside the boundaries of the victim country.

❐ transport accident

Disaster type term used in EM-DAT to describe technological transport accidents involving mechanised modes of transport. It comprises of four disaster subsets: accidents involving aeroplanes, helicopters, airships and balloons " Transport:Air "; accidents involving sailing boats, ferries, cruise ships, other boats " Transport:Boat "; accidents involving trains " Transport:Rail "; and accidents involving motor vehicles on roads and tracks " Transport:Road ".

❐ trauma

Physical injury caused by violent or disruptive action, or by the introduction into the body of a toxic substance, or a psychic injury resulting from a severe emotional shock

❐ tri

Toxic Release Inventory. An an annual inventory that must be reported to EPA and the state of releases of listed "toxic chemicals" and transfers of toxic chemicals that exceed specified threshold amounts. This submittal must also include information about the environmental media where releases take place. See Also: EPCRA; Toxic Chemical Release Inventory. A federal program described in SARA Title III that requires certain facilities to report releases and transfers of toxic chemicals to the U.S. EPA

❐ triage

Sorting out the wounded at a major incident so that the most serious cases are treated first; Selection and categorization of the victims of a disaster with the view to appropriate treatment according to the degree of severity of illness or injury, and the availability of medical and transport facilities.

❐ tropical storm

Generic term for a non-frontal synoptic scale cyclone originating over tropical or sub-tropical waters with organised convection and definite cyclonic surface wind circulation.

❐ tsunami

An oceanic tidal wave generated by an under-water upheaval such as earthquake or volcanic eruption. The waves move out in all directions over 100 miles, causing great destruction; Sea waves produced by an undersea earthquake. Such sea waves can reach a height of 80 feet and can devastate coastal cities and low-lying coastal areas.

❐ tsunami (" wave in the port " in japanese)

Series of large waves generated by sudden displacement of seawater (caused by earthquake, volcanic eruption or submarine landslide); capable of propagation over large distances and causing a destructive surge on reaching land. The Japanese term for this phenomenon, which is observed mainly in the Pacific, has been adopted for general usage.

❐ tsunami run up

A type of ground failure that results when the long period ocean waves generated by the sudden, impulsive, vertical displacement of a submarine earthquake reaches low lying areas along the coast.

❐ typhoon

Large-scale closed circulation system in the atmosphere above the western Pacific with low barometric pressure and strong winds that rotate clockwise in ther southern hemisphere and counter-clockwise in the northern hemisphere. Maximum wind speed of 64 knots or more.

U

❐ u.s. epa

United States Environmental Protection Agency. The U.S. government agency responsible for developing and administering environmental regulations.

❐ unified command

A method for all agencies or individuals who have jurisdictional responsibility at the incident, to contribute to determining overall objectives for the incident and selection of a strategy to achieve the objectives.

❐ uniform fire code

The nationally developed code of fire prevention which has been adopted by many communities through local regulations or ordinances.

❐ united nations

The supreme intergovernmental world body established in 1945 with the purposes of 1. Maintaining international peace and security, 2. Developing friendly relations among nations, 3. Solving international problems through international cooperation, and 4. Harmonizing the actions of all nations for these common ends. The un acts through various mechanisms, such as specialized agencies, e.g. Who; centres, e.g. Human rights; other constituted bodies, e.g. Unhcr; committees, e.g. Disarmament; funds, e.g. Unicef; major programmes, e.g. Undp; peace keeping forces, e.g. Unifil; institutes, e.g. Unitar, etc. Undro is responsible for the direction and coordination of the un response and capability in natural and other disasters. The general assembly has designated the 1990s as the international decade for natural disaster reduction.

❐ uranium

The heaviest naturally occurring element. This dark grey, radioactive, metallic element was discovered by the German chemist H.M. Klaproth in 1789. Uranium is both radiologically and chemically toxic and poses a health hazard as a heavy metal as well as a radioisotope. Uranium -235 is used as a source of nuclear energy by fission.

❒ urban fire

Urban fire involving buildings or structures. Can either involve industrial structures (in EM-DAT referred to as " Ind:Fire " and forms a disaster subset of disaster type " industrial accident ") or domestic / non-industrial structures.

❒ urban sar

A search and rescue mission conducted in moderately to heavily populated areas given primarily to business and residential development.

❒ urban search and rescue

A search and rescue mission conducted in moderately to heavily populated areas given primarily to business and residential development.

❒ urgency profile

A way to determine the urgency of each individual search and rescue mission. The combination of factors affecting the urgency will help determine not only how quickly to respond, but the nature and level of response as well. On the Urgency Profile chart, the lower the numerical value, the more urgent the response should be.

V

❒ vesicants

Chemical agents, also called blister agents, which cause severe bums to eyes, skin, and tissues of the respiratory tract. Also referred to as mustard agents, examples include mustard and lewisite.

❒ victim

Casualty with sustained lesions of mechanical, chemical or nuclear nature or combinations.

❒ victim distribution

Victims should be transported to and distributed among neighbouring hospitals according to their hospital treatment capacity, while the nearest hospital should be avoided, since walking T3 victims will overcrowd this one. For this a preplanned victim distribution plan is required.

❒ virus

The simplest type of microorganisms, lacking a system for their own metabolism. They depend on living cells to multiply and cannot live long outside of a host. Types of viruses are smallpox, Ebola, Marburg, and Lassa fever.

❒ volcanic eruption (in em-dat, volcano)

Discharge (aerially explosive) of fragmentary ejecta, lava and gases from a volcanic vent.

❑ voluntary agency

A non-profit, non-governmental, private association, maintained and supported by voluntary contributions. Among its activities, assistance in emergencies and disasters is notable. Icva, the international council of voluntary agencies, represents their federation.

❑ voluntary evacuation

This is a warning to persons within a designated area that a threat to life and property exists or is likely to exists in the immediate future. Individuals issued this type of waning order are NOT required to evacuate, however it would be to their advantage to do so.

❑ vulnerability

The susceptibility of the population to the type (nature) of the event. In this way, vulnerability represents the susceptibility of an individual or population to injury or contagion.(dorland 1849) or from gothenberg group 1: the degree of possible/potential loss to a given element at risk resulting from a given hazard at a given intensity. Thus, it is the inverse of resiliency; The potential loss in value of an element at risk from the occurrence and consequences of natural and technological hazards. The factors that influence vulnerability include: demographics, the age and resilience of the built environment, technology, social differentiation and diversity, regional and global economies, and political arrangements. Vulnerability is a result of flaws in planning, siting, design, and construction.;The degree of loss to a given element at risk, or set of such elements, resulting from the occurrence of a natural phenomenon of a given magnitude and expressed on a scale from 0 (=no damage) to 1 (=total loss). -undro.

❑ vulnerability study

Study and investigation of all the risks and the hazards susceptible to cause a disaster.

❑ vulnerability zone

The area mapped by a facility in risk management planning which reflects the area in which people could suffer injury or death as a result of a chemical incident.

W

❑ warning

The alerting of emergency response personnel and the public to the threat of extraordinary danger and the related effects that specific hazards may cause. A warning issued by the NWS (e.g., severe storm warning, tornado warning, tropical storm warning) for a defined area indicates that the particular type of severe weather is imminent in that area.

❑ watch

Indication by the NWS that, in a defined area, conditions are favorable for the specified type of severe weather (e.g., flash flood watch, severe thunderstorm watch, tornado watch, tropical storm watch).

❐ water curtain

A technology which sprays water to block the passage of toxic gases from a facility that is experiencing a chemical release.

❐ wave/surge

Disaster type term used in EM-DAT comprising the two disaster subsets " tsunami " and " tidal wave ".

❐ weapon of mass destruction (wmd)

(A) Any destructive device as defined in section 921 of this title, (which reads) any explosive, incendiary, or poison gas, bomb, grenade, rocket having a propellant charge of more than four ounces, missile having an explosive or incendiary charge of more than one quarter ounce, mine or device similar to the above; (B) poison gas; (C) any weapon involving a disease organism; or (D) any weapon that is designed to release radiation or radioactivity at a level dangerous to human life.

❐ webbing

Flat nylon strap used for various rescue situations, especially in high and low angle rescue missions.

❐ whopax report

Abridged designation for the Report of the WHO Management Group on the Role of Physicians and Other Health Workers in the Preservation and Promotion of Peace, published under the title "Effects of Nuclear War on Health and Health Services." It concludes that "the only approach to the treatment of health effects is the prevention of nuclear war."

❐ wild fire

Disaster type term used in EM-DAT comprising the two disaster subsets " forest fire " and " scrub fire ".

❐ wilderness medicine

The administration of first aid and medical protocols for emergency's where support is not immediately forthcoming. This type of "first aid" may differ significantly from standard first aid.

❐ wilderness sar

A search and rescue mission conducted in an area generally uncultivated and uninhabited and often inaccessible by roads.

❐ wind chill

The temperature when the temperature is lowered by the speed or gusts of wind to create an actual temperature that one would feel against the body.

❐ wind storm

Disaster type term used in EM-DAT comprising the following disaster subsets " cyclone ", " hurricane ", " storm ", " tornado ", " tropical storm ", " typhoon ", " winter torm ".

❒ **winter storm**

Snow (blizzard), ice or sleet storm.

❒ **wmd**

Weapons of Mass Destruction

❒ **wmdou**

Weapons of Mass Destruction Operations Unit (FBI)

❒ **world health organization**

The health arm of the United Nations, aiming at "the attainment by all peoples of the highest possible level of health". Coordinates efforts to raise health levels worldwide and promotes the development of primary health. Besides multiple public health programmes and actions, it is engaged in disaster preparedness and relief both at headquarters and at six Regional Offices, and coordinates the health sector of any UN involvement in major emergencies.

❒ **worst case scenario**

The worst chemical release that can happen at a facility, based on EPA's guidelines for calculating this in risk management plans. This scenario involves the release of all of the contents of the largest tank holding extremely hazardous materials at a site.

Z

❒ **zaction level**

A predesignated level of employee exposure to an airborne contaminant or other hazard which, if equaled or exceeded, requires that personal monitoring, medical evaluation, or other regulatory provisions be applied. Action levels are specified for a number of substances by state and federal law.

Coastal Terminology

A

❐ abrasion

Frictional erosion by material transported by wind and waves.

❐ accelerometer

A device used in wave buoys for measuring acceleration.

❐ accretion

The accumulation of (beach) sediment, deposited by natural fluid flow processes.

❐ active margin

A margin consisting of a CONTINENTAL SHELF, a CONTINENTAL SLOPE, and an oceanic trench.

❐ aeolian deposits

Wind-deposited sediments, such as sand DUNES.

❐ aggradation

The geologic process by which various parts of the surface of the earth are raised in ELEVATION or built up by the deposition of material transported by water or wind.

❐ allochtonous

A term applied shelves that presently experience deposition of river-derived sediments.

❐ alluvial deposits

Detrital material which is transported by a river and deposited - usually temporarily - at points along the FLOODPLAIN of a river. Commonly composed of sands and GRAVELS.

❐ alongshore

Parallel to and near the shoreline; same as LONGSHORE.

❐ altimeter

An instrument that determines its distance above a particular surface.

❐ altimeter, laser

An instrument that determines altitude by measuring the length of time needed for a pulse of coherent light to travel from the instrument to the surface and back, and multiplies half this time by the speed of light to get the straight-line distance to the surface.

- **amplitude**

Half of the peak-to-trough range (or height) of a wave.

- **angle of repose**

The maximum slope (measured from the HORIZON) at which soils and loose materials on the banks of CANALS, rivers or embankments stay stable.

- **aquifer**

A geologic formation that is water-bearing, and which transmits water from one point to another.

- **aseismic ridge**

A submarine ridge with which no earthquakes are associated.

- **astronomical tide**

The tidal levels and character which would result from gravitational effects, e.g. of the Earth, Sun and Moon, without any atmospheric influences.

- **attenuation**

The loss or dissipation of wave energy, resulting in a reduction of wave height (AMPLITUDE).

- **autochthonous**

A term applied to shelves on which older shelf sediments are primarily being reworked by modern shelf processes.

- **automatic tide gage**

An instrument that automatically registers the rise and fall of the tide. In some instruments, the registration is accomplished by printing the heights at regular intervals, in others by a continuous graph in which the height of the tide is represented by the ordinates of the curve and the corresponding time by the abscissae.

- **avulsion**

(1) Rapid EROSION of the shoreland by waves during a storm.

(2) A sudden cutting off of land by flood, currents or change in course of a body of water.

B

- **backshore**

(1) The upper part of the active BEACH above the normal reach of the tides (high water), but affected by large waves occurring during a high.

(2) (SMP) The accretion or erosion zone, located landward of ordinary high tide, which is normally wetted only by storm tides.

❐ backwash

(1) The seaward return of the water following the uprush of the waves. Also called backrush or run down. (2) Water of waves thrown back by an obstruction such as a ship, BREAKWATER, CLIFF, etc.

❐ backwash ripples

Low amplitude ripple marks formed on fine sand beaches by the BACKWASH of the waves.

❐ bank

The rising ground bordering a lake, river or sea.

❐ bar

An offshore ridge or mound of sand, GRAVEL, or other unconsolidated material which is submerged (at least at high tide), especially at the mouth of a river or ESTUARY, or lying parallel to, and a short distance from, the BEACH.

❐ barrier beach

A BAR essentially parallel to the shore, which has been built up so that its crest rises above the normal high water level. Also called BARRIER ISLAND and OFFSHORE BARRIER.

❐ barrier island

A detached portion of a BARRIER BEACH between two inlets.

❐ barrier spit

Similar to a BARRIER ISLAND, only connected to the mainland.

❐ basin

A large submarine DEPRESSION of a generally circular, elliptical or oval shape.

❐ bathymetry

The measurement of DEPTHS of water in oceans, seas and lakes; also the information derived from such measurements.

❐ bay

A recess or INLET in the shore of a sea or lake between two capes or headlands, not as large as a GULF but larger than a COVE.

❐ baymouth bar

A BAR extending partly or entirely across the mouth of a bay.

❐ beach

(1) A deposit of non-cohesive material (e.g. sand, GRAVEL) situated on the interface between dry land and the sea (or other large expanse of water) and actively "worked" by present-day hydrodynamics processes (i.e. waves, tides and CURRENTS) and sometimes by winds. (2) The zone of unconsolidated material that extends

landward from the LOW WATER LINE to the place where there is marked change in material or physiographic form, or to the line of permanent vegetation. The seaward limit of a BEACH - unless otherwise specified - is the MEAN LOW WATER LINE. A BEACH includes FORESHORE and BACKSHORE. (3) (SMP) The zone of unconsolidated material that is moved by WAVES, wind and TIDAL CURRENTS, extending landward to the COASTLINE.

❐ beach crest

The point representing the limit of high tide storm wave run-up.

❐ beach erosion

The carrying away of beach materials by wave action, tidal currents, littoral currents or wind.

❐ beach face

The section of the BEACH normally exposed to the action of wave uprush.

❐ beach head

The CLIFF, dune or sea wall looming the landward limit of the active beach.

❐ beach nourishment

The process of replenishing a BEACH by artificial means; e.g., by the deposition of dredged materials, also called beach replenishment or beach feeding.

❐ beach profile

A cross-section taken perpendicular to a given beach contour; the profile may include the face of a dune or sea wall, extend over the BACKSHORE, across the foreshore, and seaward underwater into the near shore zone.

❐ beach ridge

A low extensive ridge of beach material piled up by storm waves landward of the BERM. Usually consists of very coarse sand, GRAVEL or shells. Occurs singly or as a series of more or less parallel ridges.

❐ beach scarp

(1) An almost perpendicular slope along the beach FORESHORE; an erosional feature due to wave action, it may vary in height from a few centimeters to several meters, depending on wave action and the nature and composition of the BEACH.

❐ beach width

The horizontal dimension of the BEACH measured normal to the shoreline.

❐ beaufort scale

The relationship between sea state and wind speed. The Beaufort Scale can be used to estimate wind speed at sea, but is valid only for waves generated within the local weather system, and assumes that there has been sufficient time for a fully developed sea to have become established.

❑ bed

The bottom of a watercourse, or any body of water.

❑ bed load

Heavy or large sediment particles that travel near or on the BED.

❑ bedding plane

A surface parallel to the surface of deposition, which may or may not have a physical expression. The original attitude of a bedding plane should not be assumed to have been horizontal.

❑ bench

(1) A level or gently sloping erosion plane inclined seaward. (2) A nearly horizontal area at about the level of maximum HIGH WATER on the sea side of a DIKE.

❑ bench mark

A mark affixed to a permanent object in tidal observations, or in a survey, to furnish a DATUM level.

❑ bench mark, tidal

A BENCH MARK whose ELEVATION has been determined with respect to MEAN SEA LEVEL at a nearby tide gauge; the tidal bench mark is used as reference for that tide gauge.

❑ benefits

The economic value of a scheme, usually measured in terms of the cost of damages avoided by the scheme, or the valuation of perceived amenity or environmental improvements.

❑ benthic

Pertaining to the sub-aquatic bottom.

❑ benthos

Those animals who live on the sediments of the sea floor, including both mobile and non-mobile forms.

❑ berm

(1) On a BEACH: a nearly horizontal plateau on the BEACH FACE or BACKSHORE, formed by the deposition of beach material by wave action or by means of a mechanical plant as part of a beach recharge scheme.

(2) On a structure: a nearly horizontal area, often built to support or key-in an armour layer.

(3) (SMP) A linear mound or series of mounds of SAND and/or GRAVEL generally paralleling the water at or landward of the line of ordinary high tide.

❑ berm breakwater

Rubble mound with horizontal BERM of armour stones at about sea-side water level, which is allowed to be (re)shaped by the waves.

❑ berm crest

The seaward limit of the BERM, or the minimum DEPTH of a submerged BERM; also called berm edge.

❑ bifurcation

Location where a river separates in two or more reaches or branches (the opposite of a CONFLUENCE).

❑ bight

A slight indentation in a COAST forming an open BAY, usually crescent shaped.

❑ blowout

A DEPRESSION on the land surface caused by wind EROSION.

❑ bog

(SMP) A wet, spongy, poorly drained area which is usually rich in very specialized plants, contains a high percentage of organic remnants and residues and frequently is associated with a spring, seepage area, or other subsurface water source. A bog sometimes represents the final stage of the natural processes of eutrophication by which lakes and other bodies of water are very slowly transformed into land areas.

❑ boil

An upward flow of water in a sandy formation due to an unbalanced hydrostatic pressure resulting from a rise in a nearby stream, or from removing the overburden in making excavations.

❑ bottom boundary layer

The lower portion of the water flow that experiences frictional retardation based on its proximity to the BED.

❑ boulder

A rounded rock on a BEACH, greater than 256 mm in diameter, larger than a cobble.

❑ box gage

A tide gage that is operated by a float in a long vertical box to which the tide is admitted through an opening in the bottom. In the original type of box gage the float supported a graduated rod which rose and fell with the tide.

❑ breaching

Failure of the BEACH HEAD or a DIKE allowing flooding by tidal action.

❑ breaker

A wave that has become so steep that the crest of the wave topples forward, moving faster than the main body of the wave. Breakers may be roughly classified into four kinds, although there is much overlap: 1. Spilling - bubbles and turbulent water spill down the front face of wave. The upper 25 percent of the front face may become vertical before breaking. Breaking generally across over quite a distance. 2. Plunging - a crest curls over air pocket; breaking is usually with a crash. Smooth splash-up usually follows. 3. Collapsing - breaking occurs over lower half of wave. Minimal air pocket and usually no splash-up. Bubbles and foam present. 4. Surging - wave peaks up, but bottom rushes forward from under wave, and wave slides up BEACH FACE with little or no bubble production. Water surface remains almost plane except where ripples may be produced on the BEACHFACE during BACKWASH.

❑ breaker index

Maximum ratio of wave height to water DEPTH in the surf zone, typically 0.78 for SPILLING WAVES, ranging from about 0.6 to 1.5.

❑ breaker zone

The zone within which waves approaching the COASTLINE commence breaking, typically in water DEPTHS of between 5 m and 10 m.

❑ breaking depth

The still-water DEPTH at the point where the wave breaks.

❑ breakwater

(1) A structure protecting a HARBOR, anchorage, or BASIN from waves.

(2) (SMP) Offshore structure aligned parallel to the SHORE, sometimes shore-connected, that provides protection from waves.

❑ buffer area

A parcel or strip of land that is designed and designated to permanently remain vegetated in an undisturbed and natural condition to protect an adjacent aquatic or wetland site from upland impacts, to provide habitat for wildlife and to afford limited public access.

❑ bulkhead

(1) A structure separating land and water areas, primarily designed to resist earth pressures.

(2) A structure or partition to retain or prevent sliding of the land. A secondary purpose is to protect the upland against damage from wave action.

❑ buoy

A float; especially a floating object moored to the bottom, to mark a CHANNEL, anchor, shoal rock, etc. Some common types include: a nun or nut buoy is conical in shape; a can buoy is squat and cylindrical above water and conical below water; a spar buoy is a vertical, slender spar anchored at one end; a bell buoy, bearing a bell, runs mechanically or by the action of waves, usually marks shoals or rocks; a whistling buoy, similarly operated, marks shoals or channel entrances; a dan buoy carries a pole with a flag or light on it.

❐ buoyancy

The resultant upward forces, exerted by the water on a submerged or floating body, equal to the weight of the water displaced by this body.

❐ bypassing, sand

Hydraulic or mechanical movement of sand from the accreting updrift side to the eroding DOWNDRIFT side of an INLET or HARBOR ENTRANCE. The hydraulic movement may include natural as well as movement caused by man.

C

❐ california current

A deep-ocean boundary current that flows south-southeasterly along the U.S. west coast. The current is shallow, broad and slow moving carrying cold, nutrient poor waters toward the equator.

❐ calm

The condition of the water surface when there is no WIND WAVES or SWELL.

❐ camera, aerial

A camera especially designed for photographing the Earth's surface from above the ground; usually carried in aircraft and Earth-orbiting satellites.

❐ camera, metric

A camera designed particularly for photogrammetry, constructed so that the image is distorted geometrically as little as possible and the camera characteristics do not change from photograph to photograph.

❐ canal

An artificial watercourse cut through a land area for such uses as navigation and irrigation.

❐ cape

A relatively extensive land area jutting seaward from a continent or large island which prominently marks a change in, or interrupts notably, the coastal trend; a prominent feature.

❐ capillary wave

A wave whose velocity or propagation is controlled primarily by the surface tension of the liquid in which the wave is traveling. A water wave in which the wave length is less than 2.5 cm is considered to be a capillary wave, while waves longer than 2.5 cm and shorter than 5cm are in an indeterminate zone between CAPILLARY and GRAVITY WAVES.

❒ carrying capacity

The maximum number of species that any particular area can support over an extended period of time.

❒ cartography

The science and art of making maps.

❒ celerity

The magnitude of WAVE VELOCITY.

❒ change of tide

The change of one tide condition (rising or falling) for the other (falling or rising), or of one TIDAL CURRENT direction flow for the other.

❒ channel

(1) A natural or artificial waterway of perceptible extent which either periodically or continuously contains moving water, or which forms a connecting link between two bodies of water.

(2) The part of a body of water deep enough to be used for navigation through an area otherwise too shallow for navigation.

(3) The deepest portion of a stream, BAY, or strait through which the main volume of CURRENT of water flows.

(4) (SMP) An open conduit for water either naturally or artificially created, but does not include artificially created irrigation, return flow or stock watering channels.

❒ chart datum

The plane or level to which soundings, tidal levels or water DEPTHS are referenced, usually LOW WATER DATUM.

❒ choppy sea

Short, rough waves tumbling with a short and quick motion. Short-crested waves that may spring up quickly in a moderate breeze, and break easily at the crest.

❒ clastic rocks

Rocks built up of fragments which have been produced by the processes of weathering and EROSION, and in general transported to a point of deposition.

❒ clay

A fine grained sediment with a typical grain size less than 0.004 mm. Possesses electromagnetic properties which bind the grains together to give a bulk strength or cohesion.

❒ cliff

A high steep face of rock.

❐ climate change

Refers to any long-term trend in MEAN SEA LEVEL, wave height, wind speed, drift rate etc.

❐ coast

A strip of land of indefinite length and width (may be tens of kilometers) that extends from the SEASHORE inland to the first major change in terrain features.

❐ coastal currents

(1) Those CURRENTS which flow roughly parallel to the shore and constitute a relatively uniform drift in the deeper water adjacent to the surf zone. These currents may be tidal currents, transient, wind-driven currents, or currents associated with the distribution of mass in local waters.

(2) For navigational purposes, the term is used to designate a current in coastwise shipping lanes where the tidal current is frequently rotary.

❐ coastal defense

General term used to encompass both coast protection against EROSION and sea defense against flooding.

❐ coastal forcing

The natural processes that drive coastal hydro- and morphodynamics (e.g. winds, waves, tides, etc).

❐ coastal management

The development of a strategic, long-term and sustainable land use policy, sometimes also called shoreline management.

❐ coastal plain

The plain composed of horizontal or gently sloping strata of clastic material fronting the COAST and generally representing a strip of recently emerged sea bottom that has emerged from the sea in recent geologic times.

❐ coastal processes

Collective term covering the action of natural forces on the shoreline, and the near shore seabed.

❐ coastal zone

The land-sea-air interface zone around continents and islands extending from the landward edge of a BARRIER BEACH or SHORELINE of coastal bay to the outer extent of the CONTINENTAL SHELF.

❐ coastline

(1) Technically, the line that forms the boundary between the COAST and the SHORE.

(2) Commonly, the line that forms the boundary between land and the water.

(3) (SMP) The line where terrestrial processes give way to marine processes, TIDAL CURRENTS, wind waves, etc.

❐ cobble

Rounded rocks ranging in diameter from approximately 64 to 256 mm.

❐ colloid

As a size term refers to particles smaller than 0.00024 mm, smaller than CLAY size.

❐ comber

(1) A deepwater wave whose crest is pushed forward by a strong wind; much larger than a whitecap. (2) A long-period BREAKER.

❐ compass, surveyor's

A compass for determining the magnetic azimuth of a line of sight by means of a sighting device, a graduated horizontal circle, and a pivoted magnetic needle.

❐ confluence

The junction of two or more river reaches or branches (the opposite of a BIFURCATION).

❐ conservation

The protection of an area, or particular element within an area, accepting the dynamic nature of the environment and therefore allowing change.

❐ continental shelf

(1) The zone bordering a continent extending from the line of permanent immersion to the DEPTH, usually about 100 m to 200 m, where there is a marked or rather steep descent toward the great depths.
(2) The area under active LITTORAL processes during the Holocene period.
(3) The region of the oceanic bottom that extends outward from the shóreline with an average slope of less than 1:100, to a line where the GRADIENT begins to exceed 1:40 (the CONTINENTAL SLOPE).

❐ continental slope

The declivity from the offshore border of the CONTINENTAL SHELF to oceanic depths. It is characterized by a marked increase in slope.

❐ contour current

A bottom current that flows parallel to the slopes of the continental margin (along the contour rather than down the slope).

❐ contour line

A line connecting points, on a land surface or sea bottom, which have equal ELEVATION. It is called an ISOBATH when connecting points of equal DEPTH below a DATUM.

❐ control network

Geodetic control together with the measured or adjusted values of the distances, angels, directions, or heights used in determining the coordinates of the control.

❐ control, geodetic

A set of control stations established by geodetic methods.

❐ control, ground

A point or set of points, the coordinates of which have been determined by survey, used for fixing the scale and position of a photogrammetrically determined network.

❐ control, horizontal

The geometric data relating to the horizontal coordinates of a control station.

❐ control, photogrammetric

Geodetic or other control established to provide scale, location, and orientation for photogrammetric network.

❐ control, vertical

The elevations (or approximations thereto) associated with control points.

❐ controlling depth

The least DEPTH in the navigable parts of a waterway, governing the maximum draft of vessels that can enter.

❐ coordinate system

A set of rules for specifying how coordinates are to be assigned to points.

❐ core

(1) A cylindrical sample extracted from a BEACH or seabed to investigate the types and DEPTHS of sediment layers.

(2) An inner, often much less permeable portion of a BREAKWATER, or BARRIER BEACH.

❐ coriolis effect

Force due to the Earth's rotation, capable of generating currents. It causes moving bodies to be deflected to the right in the Northern Hemisphere and to the left in the Southern Hemisphere. The "force" is proportional to the speed and latitude of the moving object. It is zero at the equator and maximum at the poles.

❐ coset

A group of units of CROSS-BEDDING which shows a uniform direction of current flow.

❐ co-tidal lines

Lines which link all the points where the tide is at the same stage (or phase) of its cycle.

❐ countercurrent

A secondary current usually setting in a direction opposite to that of a main current.

❐ cove

A small sheltered recess in a shore or COAST, generally inside a larger EMBAYMENT.

❐ creek

(1) A stream, less predominant than a river, and generally tributary to a river. (2) A small tidal CHANNEL through a coastal MARSH.

❐ creep

Very slow, continuous down slope movement of soil or debris.

❐ crenulate

An indented or wavy shoreline beach form, with the regular seaward- pointing parts rounded rather than sharp, as in the CUSPATE type.

❐ cross sea

Confused, irregular state of the sea due to different groups of waves from different directions raised by local winds.

❐ cross-bedding

An arrangement of relatively thin layers of rock inclined at an angle to the more nearly horizontal BEDDING PLANES of the larger rock unit. Also referred to as cross-stratification.

❐ cross-shore

Perpendicular to the shoreline.

❐ current

(1) The flowing of water, or other liquid or gas.

(2) That portion of a stream of water which is moving with a velocity much greater than the average or in which the progress of the water is principally concentrated.

(3) Ocean currents can be classified in a number of different ways. Some important types include the following:

1. Periodic - due to the effect of the tides; such CURRENTS may be rotating rather than having a simple back and forth motion. The CURRENTS accompanying tides are known as TIDAL CURRENTS.
2. Temporary - due to seasonal winds.
3. Permanent or ocean - constitute a part of the general ocean circulation. The term DRIFT CURRENT is often applied to a slow broad movement of the oceanic water.

4. Nearshore - caused principally by waves breaking along a shore. Also, COASTAL CURRENTS such as CALIFORNIA and DAVIDSON CURRENTS that run parallel to the COAST.

❐ current meter

An instrument for measuring the velocity of a CURRENT. It is traditionally operated by a wheel equipped with vanes of cups which is rotated by the action of the impinging CURRENT. A recording device is provided to indicate the speed of rotation which is correlated with the velocity of the CURRENT.

❐ cusp

One of a series of short ridges on the FORESHORE separated by crescent-shaped troughs spaced at more or less regular intervals. Between these cusps are hollows. The cusps are spaced at somewhat uniform distances along beaches. They represent a combination of constructive and destructive processes.

❐ cuspate

Form of BEACH shoreline involving sharp seaward-pointing CUSPS (normally at regular intervals) between which the shoreline follows a smooth arc.

❐ cuspate foreland

A large, sandy cusp-shaped projection of the COAST.

❐ cuspate sandkey

A cusp-shaped sand island.

❐ cuspate spit

A sandy cusp-shaped projection of the shoreline, found on both sides of some lagoons.

D

❐ daily retardation of tides

The amount of time by which corresponding tidal phases grow later day by day (about 50 minutes).

❐ dam

Structure built in rivers or estuaries, basically to separate water at both sides and/or to retain water at one side.

❐ datum

Any position or element in relation to which others are determined, as datum point, datum line, DATUM PLANE.

❐ datum plane

A horizontal plane used as a reference from which to determine heights or DEPTHS. The plane is called a TIDAL DATUM when defined by a certain phase of the tide. Datum planes are referenced to fixed points known as BENCH MARKS, so that they can be recovered when needed.

❐ davidson current

Deep-ocean boundary current off the west coast of the U.S. which brings warmer, saltier, low oxygen, high phosphate equatorial type water from low to high latitudes.

❐ debris line

A line near the limit of storm wave UPRUSH marking the landward limit of debris deposits.

❐ decay area

Area of relative CALM through which waves travel after emerging from the GENERATING AREA.

❐ decay distance

The distance through which waves travel after leaving the GENERATING AREA.

❐ decay of waves

The change which occurs in waves when they leave a GENERATING AREA and pass through a CALM (or region of lighter winds). In the process of decay the significant wave length increases and the SIGNIFICANT WAVE HEIGHT decreases.

❐ deep water

In regard to waves, where DEPTH is greater than one-half the WAVE LENGTH. Deep-water conditions are said to exist when the surf waves are not affected by conditions on the bottom.

❐ deep water waves

A wave in water the DEPTH of which is greater than one-half the WAVE LENGTH.

❐ deflation

The removal of loose material from a beach or other land surface by wind action.

❐ degradation

The geologic process by means of which various parts of the surface of the earth are worn away and their general level lowered, by the action of wind and water.

❐ delta

(1) An ALLUVIAL DEPOSIT, usually triangular, at the mouth of a river of other stream. It is normally built up only where there is no tidal or CURRENT action capable of removing the sediment as fast as it is deposited, and hence the DELTA builds forward from the COASTLINE.

(2) A TIDAL DELTA is a similar deposit at the mouth of a tidal INLET, put there by TIDAL CURRENTS.

(3) A WAVE DELTA is a deposit made by large waves which run over the top of a SPIT or BAR BEACH and down the landward side.

❐ density stratification

The lateral expansion of a sediment plume as it moves out of the distributary mouth, where salt and fresh water mix. This is most likely to occur where the speed of the river flow is moderate to low and the distributary mouth is relatively deep.

❐ density-driven circulation

Variations in salinity create variations in density in estuaries. These variations in density create horizontal pressure gradients, which drive estuarine circulation.

❐ depression

A general term signifying any depressed or lower area in the ocean floor.

❐ depth

Vertical distance from still-water level (or DATUM as specified) to the bottom.

❐ design storm

Coastal protection structures will often be designed to withstand wave attack by the extreme DESIGN STORM. The severity of the storm (i.e. return period) is chosen in view of the acceptable level of risk of damage or failure. A DESIGN STORM consists of a DESIGN WAVE condition, a design water level and a DURATION.

❐ design wave

In the design of HARBORS, harbor works, etc., the type or types of waves selected as having the characteristics against which protection is desired.

❐ detached breakwater

A BREAKWATER without any constructed connection to the shore.

❐ detritus

Small fragments of rock which have been worn or broken away from a mass by the action of water or waves.

❐ diapositive

A positive photograph on a transparent medium.

❐ differential erosion / weathering

These features develop in rocks which have varying resistance to the agencies of EROSION and/or weathering so that parts of the rock are removed at greater rates than others. A typical example is the removal of soft beds from between harder beds in a series of sedimentary rocks. The term may be applied to any size of feature, from small-scale 'etching' to the regional development of hills and valleys controlled by hard and soft rocks.

❒ diffraction

The phenomenon occurring when water waves are propagated into a sheltered region formed by a BREAKWATER or similar barrier that interrupts a portion of the otherwise regular train of waves, resulting in the multi-directional spreading of the waves.

❒ dike

Sometimes written as dyke; earth structure along a sea or river in order to protect LITTORAL lands from flooding by high water; DIKES along rivers are sometimes called levees.

❒ direction of current

Direction toward which CURRENT is flowing.

❒ direction of waves

Direction from which waves are coming.

❒ direction of wind

Direction from which wind is blowing.

❒ disperse

(1) To spread or distribute from a fixed or constant source. (2) To cause to become widely separated.

❒ dispersion

(1) Act of dispersing, or state of being dispersed. (2) The separation of waves by virtue of their differing rates of travel.

❒ diurnal

Literally of the day, but here meaning having a period or a tidal day, i.e. about 24.84 hours.

❒ diurnal current

The type of tidal CURRENT having only one flood and one EBB period in the tidal day. A ROTARY CURRENT is DIURNAL if it changes its direction through all points of the compass once each tidal day.

❒ diurnal inequality

The difference in height of the two high waters or of the two low waters of each day. Also, the difference in velocity between the two daily flood or EBB CURRENTS of each day.

❒ dock

The slip or waterway between two piers, or cut into the land, for the reception of ships.

❑ double ebb (tidal)

An EBB CURRENT having two maxima of velocity separated by a smaller EBB velocity.

❑ double flood (tidal)

A flood CURRENT having two maxima of velocity separated by a smaller flood velocity.

❑ double tide

A double-headed tide; that is, a high water consisting of two maxima of nearly the same height separated by a relatively small DEPRESSION, or a LOW WATER consisting of two minima separated by a relatively small ELEVATION.

❑ downdrift

The direction of predominant movement of littoral materials.

❑ downwelling

A downward movement (sinking) of surface water caused by onshore Ekman transport, converging CURRENTS or when a water mass becomes more dense than the surrounding water.

❑ drainage basin

Total area drained by a stream and its tributaries.

❑ dredging

(SMP) Excavation or displacement of the bottom or SHORELINE of a water body. Dredging can be accomplished with mechanical or hydraulic machines. Most is done to maintain channel depths or berths for navigational purposes; other dredging is for shellfish harvesting or for cleanup of polluted sediments.

❑ drift current

A broad, shallow, slow-moving ocean or lake CURRENT. Opposite of STREAM CURRENT.

❑ drift sector

(SMP) A particular reach of marine shore in which LITTORAL DRIFT may occur without significant interruption, and which contais any and all natural sources of such drift, and also any accretion shoreforms accreted by such drift.

❑ drying beach

That part of the BEACH uncovered by water (e.g. at LOW TIDE). Sometimes referred to as subaerial beach.

❑ dunes

(1) Accumulations of windblown sand on the BACKSHORE, usually in the form of small hills or ridges, stabilized by vegetation or control structures. (2) A type of bed form indicating significant sediment transport over a sandy seabed.

❑ duration

In forecasting waves, the length of time the wind blows in essentially the same direction over the FETCH (GENERATING AREA).

❑ duration of ebb

The interval of time in which a tidal current is ebbing, determined from the middle of the slack waters.

❑ duration of fall

The interval from high water to low water.

❑ duration of flood

The interval of time in which a tidal current is flooding, determined from the middle of slack waters.

❑ duration of rise

The interval from LOW WATER to high water.

❑ duration, minimum

The time necessary for steady-state wave conditions to develop for a given wind velocity over a FETCH.

❑ dynamic equilibrium

Short term morphological changes that do not affect the morphology over a long period.

E

❑ ebb

Period when tide level is falling; often taken to mean the EBB CURRENT which occurs during this period.

❑ ebb current

The movement of a tidal current away from shore or down a tidal stream. In the semidiurnal type of reversing current, the terms GREATER EBB and LESSER EBB are applied respectively to the ebb currents of greater and lesser velocity of each day. The terms of MAXIMUM EBB and MINIMUM EBB are applied to the maximum and minimum velocities of a continuously running ebb current, the velocity alternately increasing and decreasing without coming to a slack or reversing. The expression MAXIMUM EBB is also applicable to any EBB CURRENT at the time of greatest velocity.

❑ ebb interval

The interval between the transit of the moon over the meridian of a place and the time of the following strength of EBB.

❑ ebb strength

The EBB CURRENT at the time of maximum velocity.

❑ ebb tidal delta

The bulge of sand formed at the SEAWARD mouth of TIDAL INLETS as a result of interaction between tidal currents and waves. Also called inlet-associated bars and estuary entrance shoals.

❑ ebb tide

A nontechnical term used for falling tide or EBB CURRENT. The portion of the tidal cycle between high water and the following low water.

❑ echo sounder

An instrument for determining the DEPTH of water by measuring the time of travel of a sound-pulse from the surface of a body of water to the bottom and back.

❑ ecosystem

The living organisms and the nonliving environment interacting in a given area.

❑ eddy

A current of air, water, or any fluid, forming on the side of the main current, especially one moving in a circle; in extreme cases a whirlpool.

❑ edge wave

An ocean wave parallel to the COAST, with crests normal to the shoreline. An EDGE WAVE may be standing or progressive. Its height diminishes rapidly seaward and is negligible at a distance of one wave length offshore.

❑ eelgrass

A submerged marine plant with very long narrow leaves.

❑ ekaman transport

Resultant flow at right angles to and to the right of the wind direction (in the northern hemisphere) referred to as UPWELLING and DOWNWELLING.

❑ elevation

The distance of a point above a specified surface of constant potential; the distance is measured along the direction of gravity between the point and the surface.

❑ elutriation

The process by which a granular material can be sorted into its constituent particle sizes by means of a moving stream of fluid (usually air or water). Elutriators are extensively used in studies of sediments for determining PARTICLE SIZE distribution. Under certain circumstances wind, rivers and streams may act as natural elutriating agents.

❐ embankment

An artificial BANK, mound, DIKE, or the like, built to hold back water or to carry a roadway.

❐ embayed

Formed into a BAY or bays; as an embayed shore.

❐ embayment

(1) An indentation in a shoreline forming an open BAY.

(2) The formation of a BAY.

❐ emergent coast

A COAST in which land formerly under water has recently been placed above sea level, either by uplift of the land or by a drop in sea level.

❐ endemic

Native to a specific geographic area.

❐ entrance

The entrance to a navigable BAY, HARBOR or CHANNEL, INLET or mouth separating the ocean from an inland water body.

❐ equatorial currents

(1) Ocean currents flowing westerly near the equator. There are two such currents in both the Atlantic and Pacific Oceans. The one to the north of the equator is called the North Equatorial Current and the one to the south is called the South Equatorial Current. Between these two currents there is an easterly flowing stream known as the Equatorial Countercurrent.

(2) Tidal currents occurring semimonthly as a result of the moon being over the equator. At these times the tendency of the moon to produce DIURNAL INEQUALITY in the current is at a minimum.

❐ equatorial tides

Tides occurring semimonthly as the result of the moon being over the equator. At these times the tendency of the moon to produce a DIURNAL INEQUALITY in the tide is at a minimum.

❐ erosion

Wearing away of the land by natural forces. On a BEACH, the carrying away of beach material by wave action, tidal currents or by DEFLATION. (2) (SMP) The wearing away of land by the action of natural forces.

❐ escarpment

A more or less continuous line of CLIFFS or steep slopes facing in one general direction which are caused by EROSION or faulting, also called SCARP.

❑ estuary

(1) A semi-enclosed coastal body of water which has a free connection with the open sea. The seawater is usually measurably diluted with freshwater. (2) The part of the river that is affected by tides. (3) (SMP) The zone or area of water in which freshwater and saltwater mingle and water is usually brackish due to daily mixing and layering of fresh and salt water.

❑ eulittoral

That part of the LITTORAL ZONE less than 50 m in DEPTH.

❑ event

An occurrence meeting specified conditions, e.g. damage, a threshold wave height or a threshold water level.

F

❑ facies

The sum total of features such as sedimentary rock type, MINERAL content, SEDIMENTARY STRUCTURES, BEDDING characteristics, fossil content, etc. which characterise a sediment as having been deposited in a given environment.

❑ fairway

The parts of a waterway kept open, and unobstructed, for navigation.

❑ fathom

A measure of water DEPTH equal to 1.83 m (6 feet).

❑ fault

A fracture in rock along which there has been an observable amount of displacement. Faults are rarely single planar units; normally they occur as parallel to sub-parallel sets of planes along which movement has taken place to a greater or lesser extent. Such sets are called fault or fracture-zones.

❑ fauna

The entire group of animals found in an area.

❑ feeder beach

An artificially widened beach serving to nourish DOWNDRIFT beaches by natural LITTORAL CURRENTS or other forces.

❑ feeder current

The currents which flow parallel to shore before converging and forming the NECK of a RIP CURRENT.

❑ fetch

The length of unobstructed open sea surface across which the wind can generate waves (GENERATING AREA).

❑ fetch length

(1) The horizontal distance (in the direction of the wind) over which a wind generates SEAS or creates WIND SETUP. (2) The horizontal distance along open water over which the wind blows and generates waves.

❑ fjord

A long, narrow arm of the sea, usually formed by entrance of the sea into a deep glacial trough.

❑ floating bog

A grass or moss growth floating on a pool of water with high content of decayed vegetation, grass and moss roots.

❑ flocculation

The change which takes place when the dispersed phase of a colloid forms a series of discrete particles which are capable of settling out from the DISPERSION medium. In geological processes, flocculation is almost inevitably a result of a colloidal solution mixing with a solution containing electrolytes, e.g., sea water.

❑ flood

(1) Period when tide level is rising; often taken to mean the flood current which occurs during this period. (2) A flow above the CARRYING CAPACITY of a CHANNEL.

❑ flood current

The movement of a tidal current toward the shore or up a tidal stream. In the semidiurnal type of reversing current, the terms greater flood and lesser flood are applied respectively to the flood currents of greater and lesser velocity each day. The terms maximum flood and minimum flood are applied to the maximum and minimum velocities of a flood current the velocity of which alternately increases and decreases without coming to slack or reversing. The expression maximum flood is also applicable to any flood current at the time of greatest velocity.

❑ flood interval

The interval between the transit of the moon over the meridian of a place and the time of the following flood.

❑ flood mark

Proof of any kind on the shoreline used to determine the highest level attained by the water surface during the flood (note: the height of the flood mark usually includes the WAVE RUN-UP).

❑ flood routing

The attenuating effect of storage on a river-flood passing through a valley by reason of a feature acting as control (e.g. a reservoir with a spillway capacity less than the flood inflow, or the widening or narrowing of a valley).

❐ flood strength

The FLOOD CURRENT at time of maximum velocity.

❐ flood tidal delta

The bulge of sand formed at the LANDWARD mouth of TIDAL INLETS as a result of flow expansion.

❐ flood tide

A nontechnical term used for rising tide or flood current. In technical language flood refers to current. The portion of the tidal cycle between LOW WATER and the following high water.

❐ floodplain

(1) A flat tract of land bordering a river, mainly in its lower reaches, and consisting of ALLUVIUM deposited by the river. It is formed by the sweeping of the meander belts downstream, thus widening the valley, the sides of which may become some kilometers apart. In time of flood, when the river overflows its banks, sediment is deposited along the valley banks and plains. (2) (SMP) Synonymous with 100-year floodplain.

❐ flora

The entire group of plants found in an area.

❐ flushing time

The time required to replace all the water in an ESTUARY, HARBOR, etc., by action of current and tide.

❐ foam

The light-colored substance which is made up of an aggregation of bubbles, formed on the surface of liquids by violent agitation.

❐ foam line

(1) The front of a wave as it advances shoreward, after it has broken. (2) Lines of foam such as those which move around the HEAD of a RIP.

❐ fog

Vapor condensed to fine particles of water and obscuring vision near the ground.

❐ following wind

A wind blowing in the same direction as the waves are travelling.

❐ foreshore

(1) The part of the shore, lying between the BERM CREST and the ordinary LOW WATER mark, which is ordinarily traversed by the UPRUSH and BACKRUSH of the waves as the tides rise and fall. (2) The same as the BEACH FACE where unconsolidated material is present. (3) (SMP) In general terms, the BEACH between MEAN HIGHER HIGH WATER and MEAN LOWER LOW WATER.

❒ freeboard

The additional height of a structure above design high water level to prevent overflow. Also, at a given time, the vertical distance between the water level and the top of the structure.

❒ frontager

Person or persons owning, and often living in, property immediately landward of the beach.

❒ fully-developed sea

The waves that form when wind blows for a sufficient period of time across the open ocean. The waves of a fully developed sea have the maximum height possible for a given wind speed, FETCH and DURATION of wind.

G

❒ gabion

(1) Steel wire-mesh basket to hold stones or crushed rock to protect a BANK or bottom from EROSION.

(2) (SMP) Structures composed of masses of ROCKS, rubble or masonry held tightly together usually by wire mesh so as to form blocks or walls. Sometimes used on heavy erosion areas to retard wave action or as a foundation for BREAKWATERS or JETTIES.

❒ gale

A wind between a strong breeze and a storm. A continuous wind blowing in degrees of moderate, fresh, strong, or whole gale and varying in velocity from 28 to 30 nautical miles per hour.

❒ gauge

A device for measuring the water level relative to a DATUM.

❒ generating area

In wave forecasting, a continuous area of the water surface over which the wind blows in essentially the same direction. Sometimes used synonymously with FETCH LENGTH.

❒ generation of waves

(1) The creation of waves by natural or mechanical means.

(2) The creation of and growth of waves by a wind blowing over a water surface for a certain length of time. The area involved is called the GENERATING AREA or FETCH.

❒ geodesy (or geodetics)

The science of dealing with the investigation of scientific questions connected with the shape and dimensions of the Earth.

❑ **geographic information system (gis)**

A system of spatially referenced information, including computer programs that acquire, store, manipulate, analyze, and display spatial data.

❑ **geoid**

The equipotential surface of the Earth's gravity field which best fits, in the least squares sense, MEAN SEA LEVEL.

❑ **geology**

The science which treats of the origin, history and structure of the Earth, as recorded in rocks; together with the forces and processes now operating to modify rocks.

❑ **geomorphology**

(1) That branch of physical geography which deals with the form of the Earth, the general configuration of its surface, the distribution of the land, water, etc. (2) The investigation of the history of geologic changes through the interpretation of topographic forms.

❑ **geophysics**

The study of the physical characteristics and properties of the Earth.

❑ **global positioning system (gps)**

A navigational and positioning system developed by the U.S. Department of Defense, by which the location of a position on or above the Earth can be determined by a special receiver at that point interpreting signals received simultaneously from several of a constellation of special satellites.

❑ **graded bedding**

An arrangement of particle sizes within a single bed, with coarse grains at the bottom of the bed and progressively finer grains toward the top of the bed.

❑ **gradient**

(1) A measure of slope (soil- or water-surface) in meters of rise or fall per meter of horizontal distance.

(2) More general, a change of a value per unit of distance, e.g. the GRADIENT in long shore transport causes EROSION or ACCRETION. (3) With reference to winds or currents, the rate of increase or decrease in speed, usually in the vertical; or the curve that represents this rate.

❑ **gravel**

(1) Loose, rounded fragments of rock, larger than sand, but smaller than cobbles. (2) Small stones and pebbles, or a mixture of these with sand.

❑ **gravity wave**

A wave whose velocity of propagation is controlled primarily by gravity. Water waves more than 5 cm long are considered gravity waves. Waves longer than 2.5

cm and shorter than 5 cm are in an indeterminate zone between CAPILLARY and GRAVITY WAVES.

❑ great diurnal range

The difference in height between MEAN HIGHER HIGH WATER and MEAN LOWER LOW WATER. The expression may also be used in the contracted form diurnal range.

❑ groin

(1) A shore-protection structure (built usually to trap LITTORAL DRIFT or retard EROSION of the shore). It is narrow in width (measured parallel to the shore) and its length may vary from tens to hundreds of meters (extending from a point landward of the shoreline out into the water). GROINS may be classified as permeable (with openings through them) or impermeable (a solid or nearly solid structure).

(2) (SMP) A barrier-type structure extending from the BACKSHORE or stream bank into a water body for the purpose of the protection of a SHORELINE and adjacent upland by influencing the movement of water and/or deposition of materials.

❑ groin bay

The BEACH compartment between two groins.

❑ groin system

A series of GROINS acting together to protect a section of BEACH. Commonly called a GROIN field.

❑ ground penetrating radar (gpr)

A geophysical technology that uses radio waves to detect buried objects and the internal structure of landforms.

❑ ground swell

(1) Long high SWELL in DEEP WATER. (2) Also, this swell as if rises to prominent height in shoal water.

❑ ground water

Subsurface water occupying the zone of saturation. In a strict sense the term applied only to water below the WATER TABLE.

❑ group velocity

The velocity at which a wave group travels. In DEEP WATER, it is equal to one-half the individual wave velocity.

❑ gulf

A relatively large portion of sea, partly enclosed by land.

❑ gut

(1) A narrow passage such as a strait or INLET. (2) A CHANNEL in otherwise shallow water, generally formed by water in motion.

H

❑ habitat

The place where an organism lives.

❑ halcocline

A zone in which salinity changes rapidly.

❑ half-tide level

A plane midway between MEAN HIGH WATER and MEAN LOW WATER, also called MEAN TIDE LEVEL.

❑ harbor

A water area nearly surrounded by land, sea walls, BREAKWATERS or artificial dikes, forming a safe anchorage for ships.

❑ hard defenses

A general term applied to impermeable COASTAL DEFENSE structures of concrete, timber, steel, masonry, etc, which reflect a high proportion of INCIDENT WAVE energy.

❑ head

(1) A comparatively high promontory with either a CLIFF or steep face. It extends into a large body of water, such as a sea or lake. An unnamed HEAD is usually called a headland. (2) The section of RIP CURRENT which has widened out seaward of the BREAKERS, also called head of rip.

❑ headland

A land mass having a considerable ELEVATION.

❑ heave

(1) The vertical rise or fall of the waves or the sea. (2) The translational movement of a craft parallel to its vertical axis. (3) The net transport of a floating body resulting from wave action.

❑ heavy sea

A sea in which the waves run high.

❑high seas

This term, in municipal and international law, denotes all that continuous body of salt water in the world that is navigable in its character and that lies outside territorial waters and maritime belts of the various countries, also called OPEN SEA.

❒ high water (hw)

Maximum height reached by a rising tide. The height may be solely due to the periodic tidal forces or it may have superimposed upon it the effects of prevailing meteorological conditions. Nontechnically, also called the HIGH TIDE.

❒ high water line

The intersection of MEAN HIGH WATER with the shore. The shoreline delineated on the nautical charts of the U.S. Coast and Geodetic Survey is an approximation of the mean high water line.

❒ high water mark

A reference mark on a structure or natural object, indicating the maximum stage of tide or flood.

❒ higher high water (hhw)

The higher of the two high waters of any tidal day. The single high water occurring daily during periods when the tide is DIURNAL is considered to be HIGHER HIGH WATER.

❒ higher low water (hlw)

The higher of the two low waters of any tidal day.

❒ hindcasting

In wave prediction, the retrospective forecasting of waves using measured wind information. See also WAVE HINDCASTING.

❒ hinterland

The region lying inland from the COAST.

❒ historic event analysis

Extreme analysis based on hind casting typically ten events over a period of 100 years.

❒ hook

A spit or narrow CAPE, turned landward at the outer end, resembling a hook in form.

❒ horizon

(1) The line or circle which forms the apparent boundary between Earth and sky. (2) (Geological) A plane in rock strata characterized by particular features, as occurrence of distinctive fossil species. One of the series of distinctive layers found in a vertical cross-section of any well-developed soil.

❒ hurricane

A cyclonic storm, usually of tropic origin, covering an extensive area, and containing winds in excess of 75 miles per hour.

❒ hydrography

(1) The description and study of seas, lakes, rivers and other waters. (2) The science of locating aids and dangers to navigation. (3) The description of physical properties of the waters of a region.

I

❒ incident wave

Wave moving landward.

❒ infragravity waves

Waves with periods above about 30 seconds generated by wave groups breaking in the surf zone.

❒ inlet

(1) A narrow strip of water running into the land or between islands. (2) An arm of the sea (or other body of water) that is long compared to its width, and that may extend a considerable distance inland.

❒ inlet gorge

Generally, the deepest region of an INLET.

❒ inshore

(1) The region where waves are transformed by interaction with the sea bed. (2) In beach terminology, the zone of variable width extending from the LOW WATER LINE through the BREAKER ZONE.

❒ inshore current

Any current inside the SURF ZONE.

❒ intertidal

The zone between the high and LOW WATER marks.

❒ irregular waves

Waves with random wave periods (and in practice, also heights), which are typical for natural wind-induced waves.

❒ isobath

Line connecting points of equal water DEPTH on a chart; a seabed contour.

❒ isostasy

The tendency of the Earth's crust to maintain a state of near equilibrium, i.e., if anything occurs to modify the existing state, a compensation change will occur to maintain a balance.

❒ isotope

An atom with a specified number of protons and a specified number of neutrons.

J

❐ jetty

(1) On open seacoasts, a structure extending into a body of water to direct and confine the stream or tidal flow to a selected CHANNEL, or to prevent shoaling. Jetties are built at the mouth of a river or ENTRANCE to a BAY to help deepen and stabilize a CHANNEL and facilitate navigation. (2) (SMP) A structure usually projecting out into the SEA at the mouth of a river for the purpose of protecting a navigational channel, a harbor or to influence water currents.

K

❐ key

A low, insular BANK of sand, coral, etc., as one of the islets off the southern coast of Florida.

❐ knot

The unit of speed used in navigation. It is equal to one nautical mile (6076.115 feet or 1852 meters) per hour.

L

❐ lagging of tide

The periodic retardation in the time of occurrence of high and LOW WATER due to changes in the relative positions of the moon and sun.

❐ lagoon

A shallow body of water, as a pond or lake, which usually has a shallow restricted INLET from the sea.

❐ laminar flow

Slow, smooth flow, with each drop of water traveling a smooth path parallel to its neighboring drops. Laminar flow is characteristic of low velocities, and particles of sediment in the flow zones are moved by rolling or SALTATION.

❐ landlocked

Enclosed by land, or nearly enclosed, as a HARBOR.

❐ landmark

A conspicuous object, natural or man-made, located near or on land, which aids in fixing the position of an observer.

❐ leadline

A line, wire or cord used in sounding. It is weighted at one end with a plummet (sounding lead).

❐ ledge

A rocky formation continuous with and fringing the shore.

❐ leeward

The direction toward which the prevailing wind is blowing; the direction toward which waves are travelling.

❐ levee

(1) An EMBANKMENT to prevent inundation. (2) (SMP) A large DIKE or EMBANKMENT, often having an access road along the top, which is designed as part of a system to protect land from floods.

❐ light breeze

A wind with velocity from 4 to 6 nautical miles per hour.

❐ limit of backwash

The seaward limit of the BACKWASH at any given tide stage.

❐ limit of uprush

The landward limit of UPRUSH at any given tide stage.

❐ littoral

(1) Of, or pertaining to, a shore, especially a seashore. (2) (SMP) Living on, or occurring on, the SHORE.

❐ littoral currents

A current running parallel to the BEACH and generally caused by waves striking the shore at an angle.

❐ littoral drift

(1) The sedimentary material moved in the LITTORAL ZONE under the influence of waves and currents.

(2) (SMP) The mud, SAND, or GRAVEL material moved parallel to the SHORELINE in the NEARSHORE ZONE by waves and CURRENTS.

❐ littoral transport

The movement of LITTORAL DRIFT in the LITTORAL ZONE by waves and currents. Includes movement parallel (long shore drift) and sometimes also perpendicular (CROSS-SHORE transport) to the shore.

❐ littoral transport rate

The rate of transport of sedimentary material parallel to or perpendicular to the shore in the LITTORAL ZONE. Usually expressed in cubic meters (yards) per year. Commonly used as synonymous with LONGSHORE TRANSPORT RATE.

❐ littoral zone

An indefinite zone extending seaward from the shoreline to just beyond the BREAKER ZONE.

❑ load

The quantity of sediment transported by a current. It includes the suspended load of small particles in the water, and the bedload of large particles that move along the bottom.

❑ locally generated waves

Waves generated within the immediate vicinity, within approximately 50 km, of the point of interest.

❑ long waves

Waves with periods above about 30 seconds; can be generated by wave groups breaking in the surf zone.

❑ longcrested waves

A wave, the crest length of which is long compared to the wave length.

❑ longshore

Parallel and close to the COASTLINE.

❑ longshore bar

A sand ridge or ridges, extending along the shore outside the trough, that may be exposed at LOW TIDE or may occur below the water level in the offshore.

❑ longshore current

A current located in the surf zone, moving generally parallel to the shoreline, generated by waves breaking at an angle with the shoreline, also called the ALONGSHORE CURRENT.

❑ longshore drift

Movement of sediments approximately parallel to the COASTLINE.

❑ longshore transport rate

Rate of transport of sedimentary material parallel to the shore. Usually expressed in cubic meters (yards) per year.

❑ longshore trough

An elongate DEPRESSION or series of depressions extending along the lower BEACH or in the offshore zone inside the BREAKERS.

❑ loop

That part of a STANDING WAVE where the vertical motion is greatest and the horizontal velocities are least.

❑ low tide terrace

A flat zone of the BEACH near the LOW WATER level.

❑ low water (lw)

The minimum height reached by each falling tide. Nontechnically, also called LOW TIDE.

❒ **low water line**

The line where the established LOW WATER DATUM intersects the shore. The plane of reference that constitutes the LOW WATER DATUM differs in different regions.

❒ **lower high water (lhw)**

The lower of the two high waters of any tidal day.

❒ **lower low water (llw)**

The lower of the two low waters of any tidal day. The single LOW WATER occurring daily during periods when the tide is DIURNAL is considered to be LLW.

❒ **lower low water datum**

An approximation to the plane of MEAN LOWER LOW WATER that has been adopted as a standard reference plane for a limited area and is retained for an indefinite period regardless of the fact that it may differ slightly from a better determination of MEAN LOWER LOW WATER from a subsequent series of observations.

❒ **lunar day**

The time of rotation of the Earth with respect to the moon, or the interval between two successive upper transits of the moon over the meridian of a place. The mean lunar day is approximately 24.84 solar hours in length, or 1.035 times as great as the mean solar day. Also called TIDAL DAY.

❒ **lunar tide**

The portion of the tide that can be attributed directly to attraction to the Moon.

M

❒ **macro-tidal**

Tidal range greater than 4 m.

❒ **managed retreat**

The deliberate setting back of the existing line of defense in order to obtain engineering and/or environmental advantages.

❒ **margin, continental**

A zone separating a continent from the deep-sea bottom.

❒ **marigram**

A graphic record of the rise and fall of the tide. The record is in the form of a curve in which time is represented by abscissas and the height of the tide by ordinates.

❒ **marker, reference**

A mark of permanent character close to a survey station, to which it is related by an accurately measured distance and azimuth (or bearing).

❑ marker, survey

An object placed at the site of a station to identify the surveyed location of that station.

❑ marsh

(1) A tract of soft, wet land, usually vegetated by reeds, grasses and occasionally small shrubs.

(2) (SMP) Soft, wet area periodically or continuously flooded to a shallow depth, usually characterized by a particular subclass of grasses, cattails and other low plants.

❑ marsh, diked

A former salt marsh which has been protected by a DIKE.

❑ marsh, salt

A MARSH periodically flooded by salt water.

❑ mass transport, shoreward

The movement of water due to wave motion, which carries water through the BREAKER ZONE in the direction of wave propagation.

❑ mean depth

The average DEPTH of the water area between the still water level and the SHOREFACE profile from the waterline to any chosen distance seaward.

❑ mean high water (mhw)

The average ELEVATION of all high waters recorded at a particular point or station over a considerable period of time, usually 19 years. For shorter periods of observation, corrections are applied to eliminate known variations and reduce the result to the equivalent of a mean 19-year value. All high water heights are included in the average where the type of tide is either semidiurnal or mixed. Only the higher high water heights are included in the average where the type of tide is DIURNAL. So determined, MEAN HIGH WATER in the latter case is the same as MEAN HIGHER HIGH WATER.

❑ mean high water springs (mhws)

The average height of the high water occurring at the time of spring tides.

❑ mean higher high water (mhhw)

The arithmetic average of the elevations of the higher high waters of a mixed tide over a specific 19-year period. For shorter periods of observation, corrections are applied to eliminate known variations and reduce the result to the equivalent of a mean 19-year interval.

❑ mean low water (mlw)

The average height of the low waters over a 19-year period. For shorter periods of observation, corrections are applied to eliminate known variations and reduce the result to the equivalent of a mean 19-year value.

▢ mean low water springs (mlws)

The average height of the low waters occurring at the time of the spring tides.

▢ mean lower low water (mllw)

The average height of the lower low waters over a 19-year period. For shorter periods of observation, corrections are applied to eliminate known variations and reduce the result to the equivalent of a mean 19-year value.

▢ mean range of tide

The difference in height between MEAN HIGH WATER and MEAN LOW WATER.

▢ mean rise of the tide

The height of MEAN HIGH WATER above the plane of reference or DATUM of chart.

▢ mean sea level

The average height of the surface of the sea for all stages of the tide over a 19-year period, usually determined from hourly height readings (see sea level datums).

▢ mean steepness

The ratio of the MEAN DEPTH to the horizontal distance over which the MEAN DEPTH was determined.

▢ mean tide level

Same as HALF-TIDE LEVEL.

▢ mean water level

The mean surface level as determined by averaging the heights of the water at equal intervals of time, usually at hourly intervals.

▢ mean wave period

The mean of all individual waves in an observation interval of approximately half an hour.

▢ meso-tidal

Tidal range between 2 m and 4 m.

▢ meteorological tides

Tidal constituents having their origin in the daily or seasonal variation in weather conditions which may occur with some degree of periodicity.

▢ micro-tidal

Tidal range less than 2 m.

▢ middleground shoal

A shoal formed by EBB and flood tides in the middle of the CHANNEL of the LAGOON or ESTUARY end of an INLET.

❐ mid-extreme tide

A plane midway between the extreme high water and the extreme LOW WATER occurring in any locality.

❐ mineral

A naturally occurring, inorganic, crystalline solid that has a definite chemical composition and possesses characteristic physical properties.

❐ minimum fetch

The least distance in which steady state wave conditions will develop for a wind of given speed blowing a given duration of time.

❐ mist

Water vapor suspended in the air in very small drops finer than rain, larger than FOG.

❐ mixed current

Type of tidal current characterized by a conspicious velocity difference between the two floods or two ebbs usually occurring each tidal day. See also MIXED TIDE.

❐ mixed tide

Type of tide which the presence of a DIURNAL wave is conspicious by a large inequality in either the high or LOW WATER heights with two high waters and two low waters usually occurring each tidal day. In strictness, all tides are mixed, but the name is usually applied without definite limits to the tide intermediate to those predominantly semidiurnal and those predominantly DIURNAL.

❐ mole

In coastal terminology, a massive solid-filled structure (generally revetted) of earth, masonry or large stone.

❐ monochromatic waves

A series of waves generated in a laboratory, each of which has the same length and period.

❐ moraine

An accumulation of earth, stones, etc., deposited by a glacier, usually in the form of a mound, ridge or other prominence on the terrain.

❐ morphodynamics

(1) The mutual interaction and adjustment of the seafloor topography and fluid dynamics involving the motion of sediment.

(2) The coupled suite of mutually interdependent hydrodynamic processes, seafloor morph-ologies and sequences of change.

❐ morphology

River/ESTUARY/lake/seabed form and its change with time.

❐ mud flat

A muddy, low-lying strip of ground by the shore, or an island, usually submerged more or less completely by the rise of the tide.

N

❐ national tidal datum epoch (ntde)

A period of 19 years adopted by the National Ocean Service as the period over which observations of tides are to be taken and reduced to average values for tidal datums.

❐ nautical mile

Also known as GEOGRAPHICAL MILE, its length is 1852 meters (6076.115 feet), approximately 1.15 times as long as the statute mile of 5280 feet.

❐ neap tidal current

Tidal current of decreased velocity occurring semimonthly as the result of the moon being in quadrature.

❐ neap tide

Tide of decreased range occurring semimonthly as the result of the moon being in quadrature. The NEAP RANGE of the tide is the average semidiurnal range occurring at the time of neap tides and is most conveniently computed from the harmonic constants. The NEAP RANGE is typically 10 to 30 percent smaller than the mean range where the type of tide is either semidiurnal or mixed and is of no practical significance where the type of tide is DIURNAL. The average height of the high waters of the neap tide is called NEAP HIGH WATER or HIGH WATER NEAPS (MHWN), and the average height of the corresponding LOW WATER is called NEAP LOW WATER or LOW WATER NEAPS (MLWN).

❐ nearshore

(1) In beach terminology an indefinite zone extending seaward from the SHORELINE well beyond the BREAKER ZONE.

(2) The zone which extends from the swash zone to the position marking the start of the offshore zone, typically at water DEPTHS of the order of 20 m.

❐ nearshore circulation

The ocean circulation pattern composed of the NEARSHORE CURRENTS and the COASTAL CURRENTS.

❐ nearshore current

The current system caused by wave action in and near the BREAKER ZONE, and which consists of four parts: the shoreward mass transport of water; longshore currents; rip currents; and the LONGSHORE movement of the expanding heads of rip currents.

❐ neck

(1) The narrow strip of land which connects a peninsula with the mainland, or connects two ridges.

(2) The narrow band (rip) of water flowing seaward through the surf.

❐ ness

Roughly triangular promontory of land jutting into the sea, often consisting of mobile material, i.e. a beach form.

❐ network

A set consisting of: (a) stations for which geometric relationships have been determined and which are so related that removal of one station from the set will affect the relationships (distances, directions, coordinates, etc.) between the other stations; and (b) lines connecting the stations to show this interdependence.

❐ nip

The cut made by waves in a shoreline of emergence.

❐ nodal zone

An area in which the predominant direction of the LONGSHORE TRANSPORT changes.

❐ nourishment

The process of replenishing a BEACH. It may be brought about naturally, by LONGSHORE TRANSPORT, or artificially by the deposition of dredged materials.

O

❐ ocean

The great body of salt water which occupies two-thirds of the surface of the Earth, or one of its major subdivisions.

❐ ocean current

A nontidal current constituting a part of the general oceanic circulation.

❐ oceanography

That science treating of the oceans, their forms, physical features and phenomena.

❐ offshore

(1) In beach terminology, the comparatively flat zone of variable width, extending from the SHOREFACE to the edge of the CONTINENTAL SHELF. It is continually submerged.

(2) The direction seaward from the shore.

(3) The zone beyond the nearshore zone where sediment motion induced by waves alone effectively ceases and where the influence of the sea bed on wave action is small in comparison with the effect of wind.

(4) The BREAKER ZONE directly seaward of the LOW TIDE line.

❐ offshore breakwater

A BREAKWATER built towards the seaward limit of the LITTORAL ZONE, parallel (or nearly parallel) to the shore.

❐ offshore currents

(1) Currents outside the surf zone. (2) Any current flowing away from the shore.

❐ offshore wind

A wind blowing seaward from the land in the coastal area.

❐ onshore

A direction landward from the sea.

❐ onshore current

Any current flowing towards the shore.

❐ onshore wind

A wind blowing landward from the sea.

❐ open sea

Same as HIGH SEAS.

❐ opposing winds

A wind blowing in the opposite direction to that in which the waves are travelling.

❐ ordinary high water mark (ohwm)

(SMP) That mark that will be found by examining the bed and banks and ascertaining where the presence and action of waters are so common and usual, and so long continued in all ordinary years, as to mark upon the soil a character distinct from that of the abutting upland, in respect to vegitation as that condition exists on June 1, 1971, as it may naturally change thereafter, or as it may change thereafter in accordance with permits issued by a local government or the department.

❐ ordinary tide

This expression is not used in a technical sense by the U.S. Coast and Geodetic Survey, but the word "ordinary" when applied to tides, may be taken as the equivalent of the word "mean". Thus "ordinary HIGH WATER LINE" may be assumed to be the same as "mean high water line".

❐ orthophotograph

A photograph prepared from a perspective photograph by removing distortions and displacements of points caused by tilt, relief, and perspective.

❐ orthophotomap

A map made by assembling a number of orthophotographs into a single, composite picture.

❐ oscillation

A periodic motion backward and forward. To vibrate or vary above and below a mean value.

❐ outcrop

A surface exposure of bare rock, not covered by soil or vegetation.

❐ outfall

(1) The vent of a river, drain, etc. (2) A structure extending into a body of water for the purpose of discharging sewage, storm runoff or cooling water.

❐ outflanking

EROSION behind or around the inner end of a GROIN or BULKHEAD, usually causing failure of the structure.

❐ overfalls

Breaking waves caused by a conflict of currents, or by the wind moving against the current.

❐ oversplash

The water that splashes over the top of a BREAKWATER, seawall, etc.

❐ overtopping

Water carried over the top of a COASTAL DEFENSE due to wave run-up or surge action exceeding the crest height.

❐ overwash

(1) The part of the UPRUSH that runs over the crest of a BERM or structure and does not flow directly back to the ocean or lake.
(2) The effect of waves overtopping a COASTAL DEFENSE, often carrying sediment landwards which is then lost to the beach system.

P

❐ partially mixed estuary

An ESTUARY that shows a small to moderate salinity change with DEPTH.

❐ particle size

in dealing with sediments and sedimentary rocks it is necessary that precise dimensions should be applied to such terms as CLAY, sand, pebble, etc. Numerous scales have been suggested, but in this work, the Wentworth-Udden scale is used, as it is widely accepted as an international standard. In the table which follows, particle size limits are shown, but within most groups further subdivision is possible;

for example, sand may be described as very fine, medium, coarse, very coarse. particle size is normally determined by hand measurement of pebbles, cobbles, and BOULDERS, sieving of GRAVEL, sand and silt and ELUTRIATION of silt and CLAY (7).

Size Range	Particle
>256 mm.	BOULDER
64 - 256 mm.	Cobble
4 - 64 mm.	Pebble
2 - 4 mm.	'Granule', gravel
1/16 - 2 mm.	Sand
1/256 - 1/16 mm.	Silt
<1/256 mm.	CLAY

❒ peak period

The wave period determined by the inverse of the frequency at which the wave energy spectrum reaches its maximum.

❒ pebbles

Beach material usually well-rounded and between about 4 mm to 64 mm diameter.

❒ peninsula

An elongated portion of land nearly surrounded by water and connected to a larger body of land, usually by a NECK or an ISTHMUS.

❒ perigean range

The average semidiurnal range occurring at the time of the PERIGEAN TIDES and most conveniently computed from the harmonic constants. It is larger than the mean range where the type of tide is either semidiurnal or mixed and is of no practical significance where the type of tide is DIURNAL.

❒ perigean tidal currents

Tidal currents of increased velocity occurring monthly as the result of the moon being in perigree (i.e., at the point in its orbit nearest the Earth).

❒ perigean tides

tides of increased range occurring monthly as the result of the moon being in perigree.

❒ periodic current

A current caused by the tide-producing forces of the Moon and the Sun, a part of the same general movement of the sea that is manifested in the vertical rise and fall of the tides.

❒ permanent current

A current that runs continuously independent of the tides and temporary cause. Permanent currents include the fresh water discharge of a river and the currents that form the general circulatory systems of the ocean.

- **permeability**

The property of bulk material (sand, crushed rock, soft rock in situ) which permit movement of water through its pores.

- **permeable groin**

A GROIN with openings large enough to permit passage of appreciable quantities of LITTORAL DRIFT.

- **petrology**

That branch of GEOLOGY which treats of the scientific study of rocks.

- **photic zone**

The zone extending downward from the ocean surface within which the light is sufficient to sustain photosynthesis. The DEPTH of this layer varies with water clarity, time of year and cloud cover, but is about 100 m in the open ocean. It may be considered the DEPTH to which all light is filtered out except for about one percent and may be calculated as about two and one-half times the DEPTH of a SECCHI DISK reading.

- **photogrammetry**

The science of deducing the physical dimensions of objects from measurements on images (usually photographs) of the objects.

- **photomosaic**

An assemblage of photographs, each of which shows part of a region, put together in such a way that each point in the region appears once and only once in the assemblage, and scale variation is minimized.

- **physical geology**

A large division of GEOLOGY concerned with earth materials, changes of the surface and interior of the earth, and the forces that cause those changes.

- **pier**

A structure, usually of open construction, extending out into the water from the shore, to serve as a landing place, recreational facility, etc., rather than to afford coastal protection.

- **pile**

A long substantial pole of wood, concrete or metal, driven into the earth or sea bed to serve as a support or protection.

- **piling, sheet**

Interlocking member of wood, steel, concrete, etc., subject to lateral pressure, driven individually to form an obstruction to percolation, to prevent movement of material for seawalls, stabilization of foundations, etc.

❐ piping

EROSION of closed flow CHANNELS (tunnels) by the passage of water through soil; flow underneath structures, carrying subsoil particles, may endanger the stability of the structure.

❐ pixel

An element of surface resulting from subdividing an image into the smallest identically shaped figures that give information about the location, intensity and perhaps color of the source, but such that no smaller subdivision will provide more information.

❐ placer deposits

MINERAL deposits consisting of dense, resistant and often economically valuable minerals which have been weathered from TERRIGENOUS ROCKS, transported to the sea and concentrated in marine sediments by wave or current action.

❐ placer mine

Surface mines in which valuable MINERAL grains are extracted from stream bar or beach deposits.

❐ planform

The outline or shape of a body of water as determined by the stillwater line.

❐ plateau

(1) (Geographical) an elevated plain, tableland or flat-topped region of considerable extent.

(2) (Oceanographical) an ELEVATION from the bottom of the ocean with a more or less flat top and steep sides.

❐ pleistocene

An epoch of the Quaternary Period characterized by several glacial ages.

❐ plunge point

(1) For a PLUNGING WAVE, the point at which the wave curls over and breaks.

(2) The final breaking point of the waves just before the water rushes up the BEACH.

❐ pocket beach

A BEACH, usually small, between two headlands.

❐ point

(1) The extreme end of a CAPE, or the outer end of any land area protruding into the water, usually less prominent than a CAPE.

(2) (SMP) A low profile SHORELINE PROMONTORY of more or less triangular shape, the top of which extends SEAWARD.

▫ point bar

A stream bar deposited on the inside of a curve in the stream, where the water velocity is low.

▫ port

A place where vessels may discharge or receive cargo.

▫ preservation

Static protection of an area or element, attempting to perpetuate the existence of a given 'state'.

▫ progressive wave

A wave which is manifested by the progressive movement of the wave form.

▫ promontory

A high point of land extending into a body of water; a HEADLAND.

Q

▫ quaternary

(1) The youngest geologic period; includes the present time.

(2) The latest period of time in the stratigraphic column, 0 - 2 million years, represented by local accumulations of glacial (Pleistocene) and post-glacial (Holocene) deposits which continue, without change of fauna, from the top of the Pliocene (Tertiary). The QUATERNARY appears to be an artificial division of time to separate pre-human from post-human sedimentation. As thus defined, the QUATERNARY is increasing in duration as man's ancestry becomes longer.

R

▫ radar

An instrument for determining the distance and direction to an object by measuring the time needed for radio signals to travel from the instrument to the object and back, and by measuring the angle through which the instrument's antenna has traveled.

▫ radioactive dating (radiometric dating)

the most reliable method of obtaining a 'date' for a rock depends upon the observation that the rate of decay of a radioactive element is a constant. The earliest methods, using uranium and thorium minerals as the starting material, yielded evidence that the extent of geological time was at least 2 million years. The development of knowledge concerning radioactive processes since 1939 has made available a number of refined techniques for radioactive dating which are nowadays routine processes.

❐ raised beach

A wave-cut platform, with or without a covering of beach materials, which is now raised above the present sea-level.

❐ random waves

The laboratory simulation of irregular sea states that occur in nature.

❐ range of tide

The difference in height between consecutive high and low waters. The MEAN RANGE is the difference between MEAN HIGH WATER and MEAN LOW WATER. The GREAT DIURNAL RANGE or DIURNAL RANGE is the difference in height between MEAN HIGHER HIGH WATER (MHHW) and MEAN LOWER LOW WATER (MLLW). Where the type of tide is DIURNAL, the mean range is the same as the DIURNAL range.

❐ ravinement

UNCONFORMITIES in a transgressional sequence of deposits which take on a variety of forms based on wave energy and sediment supply.

❐ reach

(1) An arm of the ocean extending into the land.

(2) A straight section of restricted waterway of considerable extent; may be similar to a NARROWS, except much longer in extent.

❐ recent

(Geological) A synonym of Holocene.

❐ recession

(1) A continuing landward movement of the shoreline. (2) A net landward movement of the shoreline over a specified time.

❐ recharge

The addition of new water to an AQUIFER or to the zone of saturation.

❐ rectification

The process of producing, from a tilted or oblique photograph, a photograph from which displacement caused by tilt has been removed.

❐ red tide

Discoloration of surface waters, most frequently in COASTAL ZONES, caused by large concentrations of microorganisms.

❐ reef

A ridge of rock or other material lying just below the surface of the sea.

❐ reef breakwater

Rubble mound of single-sized stones with a crest at or below sea level which is allowed to be (re)shaped by the waves.

❐ reference plane

The plane to which sounding and tidal data are referred.

❐ reference point

(1) A specified location (in plan elevation) to which measurements are referred. (2) In beach material studies, a specified point within the REFERENCE ZONE.

❐ reference station

A tide or current station for which tidal or tidal current constants have previously been determined and which is used as a standard for the comparison of simultaneous observations at a second station; also a station for which independent daily predictions are given in the tide or current tables from which corresponding predictions are obtained for other stations by means of differences or factors.

❐ reference zone

In regard to beach measuring procedure, the part of the FORESHORE subject to wave action (between the LIMIT OF UPRUSH and the LIMIT OF BACKWASH) at mid-tide stage. In areas of great tidal range a more complex definition is needed.

❐ reflected wave

That part of an INCIDENT WAVE that is returned (reflected) seaward when a wave impinges on a BEACH, seawall or other reflecting surface.

❐ reflection

The process by which the energy of the wave is returned SEAWARD.

❐ reflector

A surface, usually a rock or sediment layer, that strongly reflects seismic (sound) waves.

❐ refraction

The process by which the direction of a wave moving in shallow water at an angle to the bottom contours is changed. The part of the wave moving shoreward in shallower water travels more slowly than that portion in deeper water, causing the wave to turn or bend to become parallel to the contours.

❐ regular waves

Waves with a single height, period and direction.

❐ residual (water level)

The components of water level not attributable to astronomical effects.

❐ resolution

(1) In general, a measure of the finest detail distinguishable in an object or phenomenon.
(2) In particular, a measure of the finest detail distinguishable in an image.

❐ retardation

The amount of time by which corresponding tidal phases grow later day by day (about 50 minutes).

❐ return period

Average period of time between occurrences of a given EVENT.

❐ reversing tidal current

A tidal current that flows alternately in approximately opposite directions with a SLACK WATER at each reversal of direction. Currents of this type usually occur in rivers and straits where the direction of flow is more or less restricted to certain CHANNELS. When the movement is towards the shore, the current is said to be flooding, and when in the opposite direction it is said to be ebbing.

❐ revetment

(1) A facing of stone, concrete, etc., to protect an EMBANKMENT, or shore structure, against EROSION by wave action or currents.

(2) A retaining wall. (3) (SMP) Facing of stone, concrete, etc., built to protect a SCARP, EMBANKMENT or shore structure against erosion by waves of CURRENTS.

❐ rill marks

Small drainage CHANNELS forming in the lower portion of a BEACH at LOW TIDE.

❐ rip channel

A CHANNEL cut by seaward flow of RIP CURRENT, usually crosses a LONGSHORE BAR.

❐ rip current

A strong surface current of short duration flowing seaward from the shore. It usually appears as a visible band of agitated water and is the return movement of water piled up on the shore by incoming waves and wind. A rip current consists of three parts: the FEEDER CURRENT flowing parallel to the shore inside the BREAKERS; the NECK, where the FEEDER CURRENTS converge and flow through the breakers in a narrow band or "rip"; and the HEAD, where the current widens and slackens outside the breaker line.

❐ riparian

(1) Pertaining to the banks of a body of water. (2) (SMP) Of, on or pertaining to the banks of a river.

❐ ripple

(1) The light fretting or ruffling on the surface of the water caused by a breeze.

(2) The smallest class of waves and one in which the force of restoration is, to a significant degree, both surface tension and gravity.

❐ ripple marks

Undulations produced by fluid movement over sediments. Oscillatory currents produce symmetric ripples whereas a well-defined current direction produces asymmetrical ripples. The crest line of ripples may be straight or sinuous. The characteristic features of ripples depend upon current velocity, particle size, persistence of current direction and whether the fluid is air or water. Sand DUNES may be regarded as a special kind of 'super'-ripple.

❐ riprap

(1) Broken stones used for revetment, toe protection for BLUFFS, or structures exposed to wave action, foundations, etc.

(2) Foundation of wall or stones placed together irregularly.

(3) (SMP) A layer, facing or protective mound of stones placed to prevent EROSION, scour or sloughing of a structure or EMBANKMENT; also the stone so used.

❐ rips

Agitation of water caused by the meeting of currents or by rapid current setting over an irregular bottom.

❐ risk analysis

Assessment of the total risk due to all possible environmental inputs and all possible mechanisms.

❐ river

A natural stream of water larger than a brook or CREEK.

❐ rocks

An aggregate of one or more minerals rather large in area. The three classes of rocks are the following:

(1) Igneous rock - crystalline rocks formed from molten material. Examples are granite and basalt.

(2) Sedimentary rock - A rock resulting from the consolidatrion of loose sediment that has accumulated in layers. Examples are sandstone, shale and limestone.

(3) Metamorphic rock - Rock that has formed from preexisting rock as a result of heat or pressure.

❐ roller

(1) An indefinite term, sometimes considered to denote one of a series of long-crested waves which roll in upon a COAST, as after a storm.

(2) Long, high SWELL, also called a GROUND SWELL.

❐ rotary current, tidal

A tidal current that flows continually with the direction of flow changing through all points of the compass during the tidal period. Rotary currents are usually found offshore where the direction of flow is not restricted by any barriers. The tendency for the rotation in direction has its origin in the deflecting force of the earth's rotation and, unless modified by local conditions, the change is clockwise in the Northern Hemisphere and counterclockwise in the Southern Hemisphere. The velocity of the current usually varies throughout the tidal cycle, passing through two maxima in approximately opposite directions and two minima with the direction of the current at approximately ninety degrees from the direction at the time of maximum velocity.

❐ **runnel**

A corrugation or trough formed in the foreshore or in the bottom just offshore by waves or tidal currents.

❐ **run-up**

The rush of water up a structure or BEACH on the breaking of a wave. The amount of run-up is the vertical height above still water level that the rush of water reaches.

S

❐ **salient**

Coastal formation of beach material developed by wave refraction and DIFFRACTION and longshore drift comprising of a bulge in the COASTLINE towards an offshore island or BREAKWATER, but not connected to it as in the case of a tombolo. See also ness, CUSP.

❐ **salinity**

Number of grams of salt per thousand grams of sea water, usually expressed in parts per thousand.

❐ **salinity gradient**

Change in salinity with DEPTH, expressed in parts per thousand per foot.

❐ **saltation**

A term used to describe the movement of a particle being transported by wind or water which is too heavy to remain in suspension. The particle is rolled forward by the current, generates lift and rises, loses the forward momentum supplying the lift and settles to the floor, where the process is repeated. The size of the particles which can be saltated depends upon the velocity of the current and its density, e.g., water will saltate larger particles than air at the same velocity.

❐ **salt-wedge estuary**

In this circulation type, the density-driven component dominates and two well-mixed layers are separated by a sharp HALOCLINE. The seawater entering the CHANNEL appears as a tongue or wedge.

❐ **sand**

An unconsolidated (geologically) mixture of inorganic soil (that may include disintegrated shells and coral) consisting of small but easily distinguishable grains ranging in size from about .062 mm to 2.0 mm.

❐ sand bar

(1) See BAR.

(2) In a river, a ridge of sand built to or near the surface by river currents.

❐ sand spit

A narrow sand EMBANKMENT, created by an excess of deposition at its seaward terminus, with its distal end (the end away from the point of origin) terminating in open water.

❐ sand waves

(1) Longshore sand waves are large-scale features that maintain form while migrating along the shore with speeds on the order of kilometers per year. (2) Large-scale asymmetrical bedforms in sandy river beds having high length to height ratios and continuous crestlines.

❐ scour protection

Protection against EROSION of the seabed in front of the toe.

❐ sea

(1) See OCEAN.

(2) A large body of salt water, second in rank to an ocean, more or less LANDLOCKED and generally part of, or connected with, an ocean or a larger sea.

(3) Waves caused by wind at the place and time of observation. (4) State of the ocean or lake surface, in regard to waves.

❐ sea breeze

A breeze that blows from the sea toward the land caused by unequal heating of land and water masses.

❐ sea cliff

A CLIFF situated at the seaward edge of the coast.

❐ sea defenses

Works to prevent or alleviate flooding by the sea.

❐ sea grass

Members of marine seed plants that grow chiefly on sand or sand-mud bottom. They are most abundant in water less than 9 m deep. The common types are: Eel grass (Zostera), Turtle grass (Thallasia) and Manatee grass (Syringodium).

❐ sea level rise

The long-term trend in MEAN SEA LEVEL.

❐ sea puss

A dangerous LONGSHORE CURRENT; a rip current caused by return flow; loosely, the submerged CHANNEL or INLET through a BAR caused by those currents.

❒ seacoast

The coast adjacent to the sea or ocean.

❒ seamount

Conical mountain rising 1000 m or more above the sea floor.

❒ seashore

(1) (Law) All ground between the ordinary high-water and low-water mark. (2) The SHORE of the SEA or OCEAN.

❒ seawall

(1) A structure built along a portion of a coast primarily to prevent EROSION and other damage by wave action. It retains earth against its shoreward face.

(2) (SMP) A structure separating land and water areas primarily to prevent EROSION and other damage by wave action. Generally more massive and capable of resisting greater wave forces than a BULKHEAD.

❒ sechhi disk

Visibility disk used to measure the transparency of the water column.

❒ sediment

(1) Loose, fragments of rocks, minerals or organic material which are transported from their source for varying distances and deposited by air, wind, ice and water. Other sediments are precipitated from the overlying water or form chemically, in place. Sediment includes all the unconsolidated materials on the sea floor.

(2) (SMP) The fine grained material deposited by water or wind.

❒ sediment cell

In the context of a strategic approach to COASTAL MANAGEMENT, a length of COASTLINE in which interruptions to the movement of sand or shingle along the beaches or nearshore sea bed do not significantly affect beaches in the adjacent lengths of COASTLINE.

❒ sediment sink

A point or area at which beach material is irretrievably lost from a coastal cell, such as an ESTUARY, or a deep CHANNEL in the seabed.

❒ sediment source

A point or area on a COAST from which beach material arises, such as an eroding CLIFF, or river mouth.

❒ sediment transport

The main agencies by which sedimentary materials are moved are: gravity (gravity transport); running water (rivers and streams); ice (glaciers); wind; the sea (currents and LONGSHORE DRIFT). Running water and wind are the most widespread transporting agents. In both cases, three mechanisms operate, although the particle size of the transported material involved is very different, owing to the differences in density and viscosity of air and water. The three processes are: rolling or TRACTION, in which the particle moves along the BED but is too heavy to be lifted from it; SALTATION; and SUSPENSION, in which particles remain permanently above the BED, sustained there by the TURBULENT FLOW of the air or water.

❐ sediment transport paths

The routes along which net sediment movement occurs.

❐ seismic reflection

The return of part of the energy of seismic waves to the earth's surface after the waves bounce off a rock boundary.

❐ seismic refraction

The bending of seismic waves as they pass from one material to another.

❐ seismic waves

A long-period wave caused by an underwater seismic disturbance or volcanic eruption.

❐ semidiurnal

Having a period or cycle of approximately one-half of a tidal day (12.4 hours). The predominating type of tide throughout the world is semidiurnal, with two high water and two low waters each tidal day. The tidal current is said to be semidiurnal when there are two flood and two EBB periods each day.

❐ semidiurnal tide

Tides occurring twice daily. There are two high and two lows per tidal day.

❐ sensing, remote

The response of an instrument or organism to stimuli from a distant source.

❐ set (of current)

The direction towards which a current flows.

❐ setback

(SMP) A required open space, specified in shoreline master programs, measured horizontally upland from an perpendicular to the ordinary high water mark.

❐ shallow water

Water of such DEPTH that surface waves are noticeably affected by bottom topography. Typically this implies a water DEPTH equivalent to less than half the wave length.

❐ shallow water wave

A PROGRESSIVE GRAVITY WAVE which is in water less than 1/25 the wave length in DEPTH.

❒ sheet erosion

The removal of a thin layer of surface material, usually topsoil, by a flowing sheet of water.

❒ sheet flow

Sediment grains under high sheer stress moving as a layer that extends from the bed surface to some distance below (on the order of a few cm). Grains are transported in the direction of fluid flow.

❒ sheet, smooth

A sheet on which field control and hydrographic data such as soundings, depth curves, and regions surveyed with a wire drag are finally plotted before being used in making a final chart.

❒ shingle

A loose term for coarse beach material, a mixture of GRAVEL, pebbles and larger material, often well-rounded and of hard rock, e.g. chert, flint etc.

❒ shoal

(1) (noun) A detached area of any material except rock or coral. The DEPTHS over it are a danger to surface navigation. Similar continental or insular shelf features of greater DEPTHS are usually termed BANKS.

(2) (verb) To become shallow gradually. (3) To cause to become shallow. (4) To proceed from a greater to a lesser DEPTH of water.

❒ shore

That strip of ground bordering any body of water which is alternately exposed, or covered by tides and/or waves. A SHORE of unconsolidated material is usually called a BEACH.

❒ shore terrace

A terrace made along a COAST by the action of waves and shore currents; it may become land by the uplifting of the shore or the lowering of the water.

❒ shoreface

The narrow zone seaward from the LOW TIDE SHORELINE permanently covered by water, over which the beach sands and GRAVELS actively oscillate with changing wave conditions.

❒ shoreline

(1) The intersection of a specified plane of water with the shore.

(2) (SMP) All of the water areas of the state, including reservoirs and their associated uplands, together with the lands underlying them, except those areas excluded under RCW 90.58.030(2)(d).

❒ shoreline management

The development of strategic, long-term and sustainable COASTAL DEFENSE and land-use policy within a sediment cell.

❐ short-crested wave

A WAVE, the CREST LENGTH of which is of the same order of magnitude as the WAVE LENGTH. A system of SHORT-CRESTED WAVES has the appearance of hills being separated by troughs.

❐ significant wave

A statistical term relating to the one-third highest waves of a given wave group and defined by the average of their heights and periods.

❐ significant wave height

Average height of the highest one-third of the waves for a stated interval of time.

❐ significant wave period

Average period of the highest one-third of the waves for a stated interval of time.

❐ silt

sediment particles with a grain size between 0.004 mm and 0.062 mm, i.e. coarser than CLAY particles but finer than sand.

❐ slack water (slack tide)

The state of a tidal current when its velocity is near zero, especially the moment when a reversing current changes its direction and its velocity is zero. The term is also applied to the entire period of low velocity near the time of turning of the current when it is too weak to be of any practical importance in navigation. The relation of the time of slack water to the tidal phases varies in different localities. In some places slack water occurs near the times of high and low water, while in other localities the slack water may occur midway between high and LOW WATER.

❐ slide

In mass wasting, movement of a descending mass along a plane approximately parallel to the slope of the surface.

❐ slip face

The steep, downwind slope of a dune; formed from loose, cascading sand that generally keeps the slope at the ANGLE OF REPOSE (about 34 degrees).

❐ slope

The degree of inclination to the horizontal. Usually expressed as a ratio, such as 1:25, indicating one unit rise in 25 units of horizontal distance; or in a decimal fraction (0.04).

❐ slough

A small muddy marshland or tidal waterway which usually connects other tidal areas.

❐ slump

In mass wasting, movement along a curved surface in which the upper part moves vertically downward while the lower part moves outward.

❐ small diurnal range

Difference in height between MEAN LOWER LOW WATER (MLLW) and MEAN HIGHER HIGH WATER (MHHW). Applicable only when the type of tide is either SEMIDIURNAL or MIXED.

❐ soft defenses

Usually refers to beaches (natural or designed) but may also relate to energy-absorbing beach-control structures, including those constructed of rock, where these are used to control or redirect COASTAL PROCESSES rather than opposing or preventing them.

❐ soil

A layer of weathered, unconsolidated material on top of bed rock; often also defined as containing organic matter and being capable of supporting plant growth.

❐ soil horizons

Layers of soil that are distinguishable by characteristic physical or chemical properties.

❐ solitary wave

A wave consisting of a single ELEVATION (above the water surface) of height not necessarily small compared to the DEPTH, and neither followed or preceded by another ELEVATION or DEPRESSION of the water surface.

❐ sorting

Process of selection and separation of sediment grains according to their grain size (or grain shape or specific gravity).

❐ sound

(1) (noun) a relatively long arm of the sea or ocean forming a CHANNEL between an island and a mainland or connecting two larger bodies, as a sea and the ocean, or two parts of the same body; usually wider and more extensive than a STRAIT.

(2) (verb) To measure the DEPTH of the water.

❐ sounding

A measured DEPTH of water. On hydrographic charts the soundings are adjusted to a specific plane of reference (SOUNDING DATUM).

❐ sounding datum

The plane to which SOUNDINGS are referred.

❐ sounding line

A line, wire or cord used in SOUNDING. It is weighted at one end with a plummet.

❐ spit

(1) A long narrow accumulation of sand or shingle, lying generally in line with the COAST, with one end attached to the land the other projecting into the sea or across the mouth of an ESTUARY. See also ness. (2) (SMP) An accretion shoreform which extends seaward from and parallel to the SHORELINE.

❐ spring range

The average SEMIDIURNAL RANGE occurring at the time of SPRING TIDES and most conveniently computed from the harmonic constants. It is larger than the MEAN RANGE where the type of tide is either SEMIDIURNAL or MIXED, and is of no practical significance where the type of tide is DIURNAL.

❐ spring tidal currents

Tidal currents of increased velocity occurring semi-monthly as the result of the moon being new or full.

❐ spring tide

A tide that occurs at or near the time of new or full moon, and which rises highest and falls lowest from the MEAN SEA LEVEL (MSL).

❐ stand of tide

An interval at high or LOW WATER when there is no discernable change in the height of the tide. The water level is stationary at high and LOW WATER for only an instant, but the change in level near these times is so slow that it is not usually perceptible.

❐ standing wave

(1) A type of wave in which the surface of the water oscillates vertically between fixed nodes without progressing. (2) A wave of essentially stable form which does not move with respect to a selected reference point.

❐ station, control

A point on the ground whose horizontal or vertical location is used as a basis for obtaining locations of other points.

❐ step

The nearly horizontal section which more or less divides the BEACH from the SHOREFACE.

❐ stillwater level (swl)

The surface of the water if all wave and wind action were to cease. In DEEP WATER this level approximates the midpoint of the wave height. In shallow water it is nearer to the trough than the crest. Also called the UNDISTURBED WATER LEVEL.

- **stone**

Quarried or artificially broken rock for use in construction.

- **storm surge**

A rise or piling-up of water against shore, produced by strong winds blowing ONSHORE. A storm surge is most severe when it occurs in conjunction with a high tide.

- **strait**

A relatively narrow waterway between two larger bodies of water. See SOUND.

- **strand**

The SHORE or BEACH of the ocean or a large lake. The land bordering any large body of water, especially a sea or an arm of the ocean.

- **strand line**

An accumulation of debris (e.g. seaweed, driftwood and litter) cast up onto a BEACH, and lying along the limit of wave up rush.

- **stratigraphy**

(1) The study of stratified rocks (sediments and volcanics) especially their sequence in time. (2) The character of the rocks and the correlation of beds in different localities.

- **stream**

(1) Any flow of water; a current. (2) A course of water flowing along a bed in the earth.

- **stream current**

A narrow, deep and swift ocean current, such as the Gulf Stream. Opposite of DRIFT CURRENT.

- **structural geology**

The branch of GEOLOGY concerned with the internal structure of bed rock and the shapes, arrangement, and interrelationships of rock units.

- **sub-aerial beach**

That part of the BEACH which is uncovered by water (e.g. at LOW TIDE sometimes referred to as DRYING BEACH).

- **subduction zone**

Elongate region in which the sea floor slides beneath a continent or island arc.

- **submarine canyon**

V-shaped valleys that run across the CONTINENTAL SHELF and down the CONTINENTAL SLOPE.

❐ submergent coast

A COAST in which formerly dry land has been recently drowned, either by land subsidence or a rise in seal level.

❐ subordinate station

A tide or current station at which a short series of observations has been obtained, which is to be reduced by comparison with simultaneous observations at another station having well-determined tidal or current constants.

❐ subsidence

Sinking or down warping of a part of the earth's surface.

❐ sub-tidal beach

The part or the BEACH (where it exists) which extends from LOW WATER out to the approximate limit of storm EROSION. The latter is typically located at a maximum water DEPTH of 8 to 10 m for moderate wave environments and is often identifiable on surveys by a break in the slope of the BED.

❐ surf

(1) Collective term for BREAKERS. (2) The wave activity in the area between the SHORELINE and the outermost limit of breakers. (3) The term surf in literature usually refers to the breaking waves on shore and on reefs when accompanied by a roaring noise caused by the larger waves breaking.

❐ surf beat

irregular oscillations of water level within the SURF ZONE with periods in the order of several minutes.

❐ surf zone

The nearshore zone along which the waves become BREAKERS as they approach the shore.

❐ surf zone

The zone of wave action extending from the water line (which varies with tide, surge, set-up, etc.) out to the most seaward point of the zone (BREAKER ZONE) at which waves approaching the COASTLINE commence breaking, typically in water DEPTHS of between 5 m and 10 m.

❐ surface gravity wave (progressive)

(1) This is the term which applies to the WIND WAVES and SWELL of lakes and oceans, also called SURFACE WATER WAVE, SURFACE WAVE or DEEP WATER WAVE,

(2) A progressive GRAVITY WAVE in which the disturbance is confined to the upper limits of a body of water. Strictly speaking this term applies to those progressive GRAVITY WAVES whose celerity depends only upon the wave length.

❐ surge

(1) Long-interval variations in velocity and pressure in fluid flow, not necessarily periodic, perhaps even transient in nature.

(2) The name applied to wave motion with a period intermediate between that of an ordinary wind wave and that of the tide.

(3) Changes in water level as a result of meteorological forcing (wind, high or low barometric pressure) causing a difference between the recorded water level and that predicted using harmonic analysis, may be positive or negative.

❐ survey, control

A survey that provides coordinates (horizontal or vertical) of points to which supplementary surveys are adjusted.

❐ survey, hydrographic

A survey that has as its principal purpose the determination of geometric and dynamic characteristics of bodies of water.

❐ survey, photogrammetric

A survey in which monuments are placed at points that have been determined photogrammetrically.

❐ survey, topographic

A survey which has, for its major purpose, the determination of the configuration (relief) of the surface of the land and the location of natural and artificial objects thereon.

❐ suspended load

The finest of the beach sediments, light enough in weight to remain lifted indefinitely above the bottom by water turbulence.

❐ swash

(1) Same as UPRUSH. (2) A body of dashing, splashing water. (3) A BAR over which the ocean washes.

❐ swash bars

Low broad sandy BARS formed by sediment in the surf and swash zones, separated by linear depressions, or RUNNELS, running parallel to the SHORE.

❐ swash channel

A narrow sound or CHANNEL of water lying within a sandbank, or between a sandbank and a shore.

❐ swash mark

The thin wavy line of fine sand left by the UPRUSH when it recedes from its upward limit of movement on the BEACH FACE.

❐ swash zone

The zone of wave action on the beach, which moves as water levels vary, extending from the limit of run-down to the limit of run-up.

❐ swell

Waves that have traveled a long distance from their GENERATING AREA and have been sorted out by travel into LONG WAVES of the same approximate period.

T

❐ tectonic forces

Forces generated from within the earth that result in uplift, movement, or deformation of part of the earth's crust.

❐ tectonics

The study of the major structural features of the Earth's crust or the broad structure of a region.

❐ terrace

A horizontal or nearly horizontal natural or artificial topographic feature interrupting a steeper slop, sometimes occurring in a series.

❐ terrigenous sediments

Literally 'land-formed' sediment that has found its way to the sea floor. The term is applied (a) to sediments formed and deposited on land (e.g., soils, sand DUNES) and (b) to material derived from the land when mixed in with purely marine material (e.g., sand or CLAY in a shelly limestone).

❐ thalweg

The line down the center of the main CHANNEL of a stream.

❐ threshold velocity

The maximum orbital velocity at which the sediment on the BED begins to move as waves approach shallow water.

❐ tidal current

The alternating horizontal movement of water associated with the rise and fall of the tide caused by ASTRONOMICAL TIDE-producing forces.

❐ tidal flats

(1) Marshy or muddy areas covered and uncovered by the rise and fall of the tide. A TIDAL MARSH. (2) (SMP) Marshy or muddy areas of the seabed which are covered and uncovered by the rise and fall of tidal water.

❐ tidal period

The interval of time between two consecutive like phases of the tide or tidal current.

❒ tidal pool

A pool of water remaining on a BEACH or REEF after recession of the tide.

❒ tidal prism

(1) The total amount of water that flows into a HARBOR or out again with movement of the tide, excluding any fresh water flow.

(2) (SMP) The volume of water present between MEAN LOW and MEAN HIGH TIDE.

❒ tidal rise

The height of tide as referred to the DATUM of a chart.

❒ tidal stand

An interval at high or LOW WATER when there is no observable change in the height of the tide. The water level is stationary at high and LOW WATER for only an instant, but the change in level near these times is so slow that it is not usually perceptible.

❒ tidal wave

(1) A wave, in the oceans and seas, produced by tides and tidal currents.

(2) Non-technical term in popular usage for an unusually high and destructive water level along a shore. It usually refers to STORM SURGE or TSUNAMI.

❒ tidally driven circulation

The movement of fresh water and seawater that are mixed by the sloshing back and forth of the ESTUARY in response to ocean tides.

❒ tide

The periodic rising and falling of the water that results from gravitational attraction of the moon and sun acting upon the rotating earth. Although the accompanying horizontal movement of the water resulting from the same cause is also sometimes called the tide, it is preferable to designate the latter as TIDAL CURRENT, reserving the name TIDE for the vertical movement.

❒ tide gage

A device for measuring the rise and fall, and the current height of the tide.

❒ tide level

The height of the tide above a specified level.

❒ tide staff

A tide gage consisting of a vertical graduated staff from which the height of the tide can be read directly. It is called a fixed staff when it is secured in place so that it cannot be easily removed. A portable staff is one that is designed for removal from the water when not in use.

❐ tide station

The geographic location at which tidal observations are made. It is called a primary tide station when continuous observations are to be taken over a number of years to obtain basic tidal data for the locality. A secondary tide station is one operated over a short period of time to obtain data for a specific purpose.

❐ tide tables

Tables which give daily predictions of the times and heights of the tide. These predictions are usually supplemented by tidal differences and constants by means of which additional predictions can be obtained for numerous other places.

❐ tide, astronomic

The periodic change in magnitude and direction of gravity as caused by attraction of the Sun, Moon, and other members of the Solar system.

❐ tides, types of

The characteristic form of the tide with special reference to the relation of the DIURNAL and semidiurnal waves. Tides are sometimes classified as DIURNAL, SEMIDIURNAL and MIXED, but there are no sharply defined limits separating the groups. The tide is said to be DIURNAL when the diurnal wave predominated and only a single high and single LOW WATER occur each day during the greater part of the month. The tide is SEMIDIURNAL when the semidiurnal wave predominates and two high and two low waters occur each tidal day with a relatively small inequality in the high and LOW WATER heights. In the MIXED type of tide the DIURNAL and semidiurnal waves are both important factors and the tide is characterized by large inequality in the high or LOW WATER heights or in both. There will usually be two high and two low waters each day, but occasionally the tide will become DIURNAL.

❐ toe

(1) Lowest part of sea- and portside breakwater slope, generally forming the transition to the seabed.

(2) The point of break in slope between a dune and a beach face.

❐ tombolo

(1) Coastal formation of beach material developed by refraction, DIFFRACTION and LONGSHORE DRIFT to form a 'neck' connecting a COAST to an offshore island or BREAKWATER (see also salient).

(2) (SMP) A causeway-like ACCRETION SPIT that connects an offshore rock or island to the main shore, or to another island.

❐ tongue

A long narrow strip of land, projecting into a body of water.

❐ topographic map

A map on which elevations are shown by means of CONTOUR LINES.

❏ topography

The form of the features of the actual surface of the Earth in a particular region considered collectively.

❏ training wall

A wall or JETTY to direct current flow.

❏ transgression, marine

The invasion of a large area of land by the sea in a relatively short space of time (geologically speaking). Although the observable result of a marine transgression may suggest an almost 'instantaneous' process, it is probable that the time taken is in reality to be measured in millions of years. The plane of marine transgression is a plane of UNCONFORMITY. The reverse of a transgression is a REGRESSION.

❏ transitional water (zone)

In regard to PROGRESSIVE GRAVITY WAVES, water whose DEPTH is less than one-half, but more than 1/25, the wave length, also called a SHALLOW WATER WAVE.

❏ transverse bar

A BAR which extends approximately right angles to SHORELINES.

❏ travel time

The time necessary for waves to travel a given distance from the GENERATING AREA.

❏ trench

A long narrow submarine DEPRESSION with relatively steep sides.

❏ trough

A long and broad submarine DEPRESSION with gently sloping sides.

❏ truncated landform

A landform cut off, especially by EROSION, and forming a steep side or CLIFF.

❏ tsunami

A large, high-velocity wave generated by displacement of the sea floor (such as sudden faulting, land sliding, or volcanic activity); also called seismic sea wave. Commonly misnamed TIDAL WAVE.

❏ turbidity

(1) A condition of a liquid due to fine visible material in suspension, which may not be of sufficient size to be seen as individual particles by the naked eye but which prevents the passage of light through the liquid.

(2) A measure of fine suspended matter in liquids.

❐ turbidity current

A flowing mass of sediment-laden water that is heavier than clear water and therefore flows down slope along the bottom of the sea or a lake.

❐ turbulent flow

Any flow which is not LAMINAR, i.e., the stream lines of the fluid, instead of remaining parallel, become confused and intermingled.

U

❐ unconformity

A surface that represents a break in the geologic record, with the rock unit immediately above it being considerably younger than the rock beneath. There are three major aspects to consider:

(1) **Time.** An unconformity develops during a period of time in which no sediment is deposited. This concept equates deposition and time, and an unconformity represents unrecorded time.

(2) **Deposition.** Any interruption of deposition, whether large or small in extent, is an unconformity. This aspect of unconformity pre-supposes a standard 'scale' of deposition which is complete. Major breaks in sedimentation can usually be demonstrated easily, but minor breaks may go unrecorded until highly detailed investigations are made.

(3) **Structure.** Structurally, unconformity may be regarded as planar structures separating older rocks below from younger rocks above, representing the 'break' as defined in (1) and (2) above. A plane of unconformity may be a surface of weathering, EROSION or denudation, or a surface of non-deposition, or possibly some combination of these factors. It may be parallel to the upper strata, make an angle with the upper strata, or be irregular. Subsequent Earth movements may have folded or faulted it.

❐ unconsolidated

In referring to sediment grains, loose, separate, or unattached to one another.

❐ undercutting

EROSION of material at the foot of a CLIFF or BANK, e.g., a sea cliff, or river bank on the outside of a meander. Ultimately, the overhang collapses, and the process is repeated.

❐ undertow

(1) A current below water surface flowing seaward; the receding water below the surface from waves breaking on a shelving beach.

(2) Actually undertow is largely mythical. As the BACKWASH of each wave flows down the BEACH, a current is formed which flows seaward. However, it is a periodic phenomenon. The most common phenomena expressed as undertow are actually RIP CURRENTS.

❐ underwater gradient

The slope of the sea bottom.

❐ undisturbed water level

Same as STILL WATER LEVEL.

❐ updrift

The direction to which the predominant longshore movement of beach material approaches.

❐ upland

(SMP) Generally described as the dry land area above and landward of the ORDINARY HIGH WATER MARK (OHWM).

❐ uprush

The rush of water up the FORESHORE following the breaking of a wave, also called SWASH or RUNUP.

❐ upwelling

The process by which water rises from a deeper to a shallower DEPTH, usually as a result of offshore surface water flow. It is most prominent where persistent wind blows parallel to a COASTLINE so that the resultant Ekman transport moves surface water away from the COAST.

V

❐ valley

An elongated DEPRESSION, usually with an outlet, between BLUFFS or between ranges of hills or mountains.

❐ variability of waves

(1) The variation of heights and periods between individual waves within a wave train.

(2) The variation in direction of propagation of waves leaving the GENERATING AREA.

(3) The variation in height along the crest, usually called "variation along the wave".

❐ velocity profile

The velocity gradient within the BOTTOM BOUNDARY LAYER, displayed as a graph of height above the BED against the velocity of the flow.

❐ viscosity

Resistance to flow.

W

❐ water depth

Distance between the seabed and the still water level.

❐ water level

The ELEVATION of a particular point or small patch on the surface of a body of water above a specific point or surface, averaged over a period of time sufficiently long to remove the effects of short period disturbances.

❐ water line

(1) The juncture of land and sea. This line migrates, changing with the tide or other variation of the water level. Where waves are present on the BEACH, this line is also known as the limit of BACKRUSH.

(2) The common boundary between the water surface and any immersed structure.

❐ water mark

A line or mark left on the shores of a body of water by the water as an indication of the water's former ELEVATION.

❐ water table

The upper surface of a zone of saturation, where the body of groundwater is not confined by an overlying impermeable formation. Where an overlying confining formation exists, the AQUIFER in question has no water table.

❐ water type

A water having well-defined temperature, salinity and nutrient characteristics.

❐ water, navigable

The waters which are or can be used as water highways for commerce.

❐ wave

(1) An oscillatory movement in a body of water manifested by an alternate rise and fall of the surface.

(2) A disturbance of the surface of a liquid body, as the ocean, in the form of a ridge, swell or hump.

(3) The term wave by itself usually refers to the term SURFACE GRAVITY WAVE (PROGRESSIVE).

❐ wave age

The ratio of wave velocity to wind velocity.

❐ wave base

The plane or DEPTH to which waves may erode the bottom in shallow water.

❐ wave climate

Average condition of the WAVES at a place, over a period of years, as shown by height, period, direction, etc.

❐ wave climate atlas

Series of maps showing the variability of wave conditions over a long COASTLINE.

❐ wave crest

(1) The highest part of the wave.

(2) That part of the wave above still water level.

❐ wave direction

The direction from which the waves are coming.

❐ wave drift

The small net forward displacement of water in the direction of the wave travel, particularly in waves of large AMPLITUDE, so that the orbits are not quite closed, and the water, while in the crests, moves slightly further forward than it moves backward while in the troughs.

❐ wave generation

Growth of wave energy by wind.

❐ wave group

A series of waves in which the distance between crests, and the AMPLITUDE, vary only slightly.

❐ wave height

The vertical distance between the crest (the high point of a wave) and the trough (the low point).

❐ wave hindcast

The calculation from historic synoptic weather charts of the wave characteristics that probably occurred at some past time.

❐ wave length

The distance, in meters, between equivalent points (CRESTS or TROUGHS) on waves.

❐ wave period

(1) The time required for two successive wave crests to pass a fixed point.

(2) The time, in seconds, required for a wave crest to traverse a distance equal to one wave length.

❐ wave propagation

The transmission of waves through water.

❐ wave recorder

A meter which records either the surface time history of GRAVITY WAVES, or the subsurface pressure time history due to these waves.

❐ wave rose

Diagram showing the long-term distribution of wave height and direction.

❐ wave set-up

ELEVATION of the still-water level due to breaking waves.

❐ wave staff

An instrument consisting of a graduated vertical pole for measuring wave heights, and, by introducing a timing device, wave periods. The staff may support a strip or series of electrical contacts for activating a recorder.

❐ wave steepness

The ratio of wave height to its length. Not the same thing as the slope between a wave crest and its adjacent trough.

❐ wave train

A series of waves from the same direction.

❐ wave transformation

Change in wave energy due to the action of physical processes.

❐ wave trough

The lowest part of the wave form between crests. Also that part of a wave below STILL WATER LEVEL.

❐ wave variability

(1) The variation of heights and periods between individual waves within a wave train. Wave trains are not composed of waves of equal heights and periods, but rather of heights and periods which vary in a statistical manner.

(2) The variability in direction of wave travel when leaving the GENERATING AREA.

(3) The variation in height along the crest.

❐ wave velocity

Speed at which the individual wave form advances, defined as the wave length divided by the wave period (in meters per second).

❐ wave wash

The erosive action on shores or embankments caused by the lapping or breaking of waves.

❐ wave-cut platform

A horizontal BENCH of rock formed beneath the surf zone as a COAST retreats because of wave EROSION.

❑ weather shore

The land lying in the direction from which the wind is coming. The WINDWARD side.

❑ weir

A low DAM or wall across a stream to raise the upstream water level. Termed fixed crest weir when uncontrolled.

❑ well

A hole, generally cylindrical and usually walled or lined with pipe, that is dug or drilled into the ground to penetrate an AQUIFER below the zone of saturation.

❑ well mixed estuary

In this circulation type, tidal fluctuations dominate, and the water column is mixed vertically.

❑ wetlands

Lands whose saturation with water is the dominant factor determining the nature of soil development and the types of plant and animal communities that live in the soil and on its surface (e.g. Mangrove forests).

❑ whitecap

The white froth on crests of waves in a wind (caused by the wind blowing the crest forward and over).

❑ wind current

A current created by the action of the wind. From theoretical considerations, currents produced by winds in the open sea will set to the right of the direction towards which the wind is blowing if in the Northern Hemisphere and to the left of this direction if in the Southern Hemisphere.

❑ wind ripple

Small, low ridge of sand produced by the saltation of windblown sand.

❑ wind rose

Diagram showing the long-term distribution of wind speed and direction.

❑ wind sea

wave conditions directly attributable to recent winds, as opposed to swell.

❑ wind setup

(1) The vertical rise in the still water level on the LEEWARD side of a body of water caused by wind stresses on the surface of the water.

(2) The difference in still water levels on the windward and the LEEWARD sides of a body of water caused by wind stresses on the surface of the water.

(3) synonymous with WIND TIDE and STORM SURGE. STORM SURGE is usually reserved for use on the OCEAN and large bodies of water. Wind

setup is usually reserved for use on reservoirs and smaller bodies of water.

❒ windward

The direction from which the wind is blowing.

❒ witness mark

A material mark placed at a known distance and direction from a property corner, an instrument station or a survey station, as an aid in its recovery and identification.

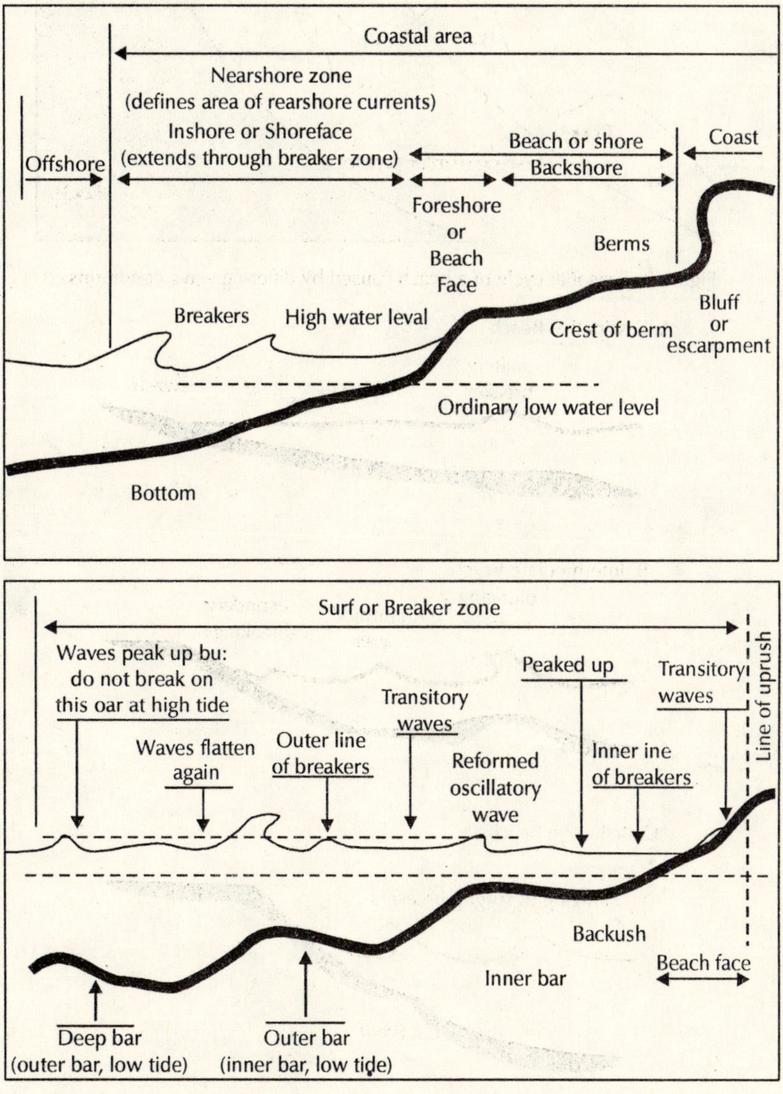

Figure 1:-(a) The coastal zone and (b) the surf or breaker zone.

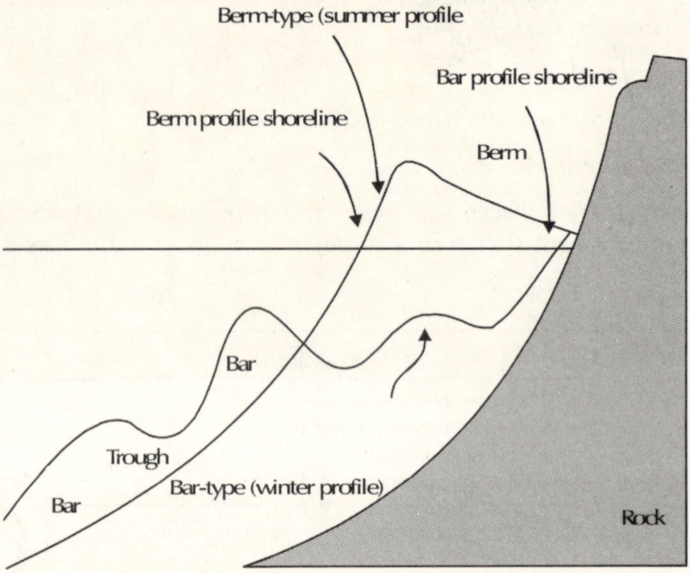

Figure 2: Seasonal cycle of a beach caused by differing wave conditions.

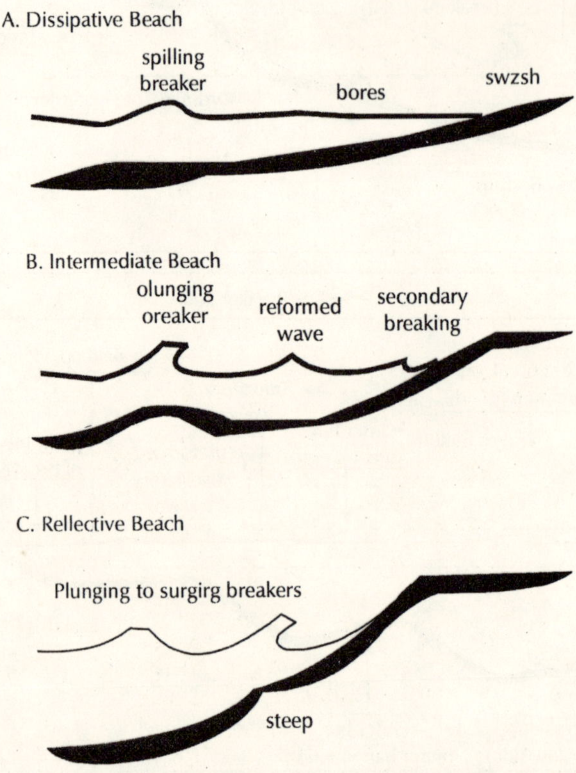

Figure 3:(a) Dissipative mild sloping beach with spilling breakers, (b) intermediate beach with plunging and secondary breakers, and (c) steep sloping reflective beaches.

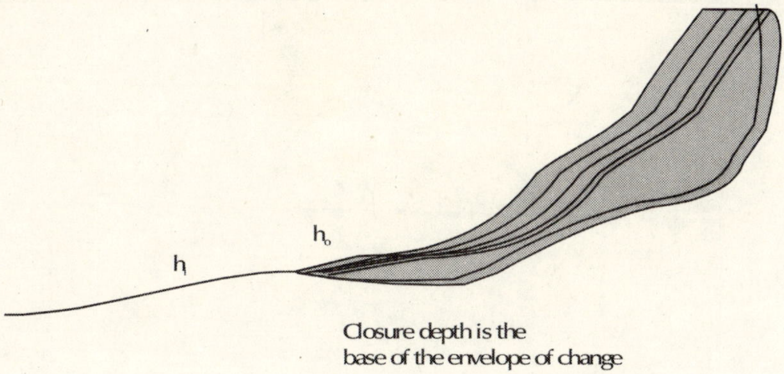

Figure 4: The depth of closure of an upper shoreface cross-shore profile. The shaded area indicates the total envelope of change for a time-series of profiles.

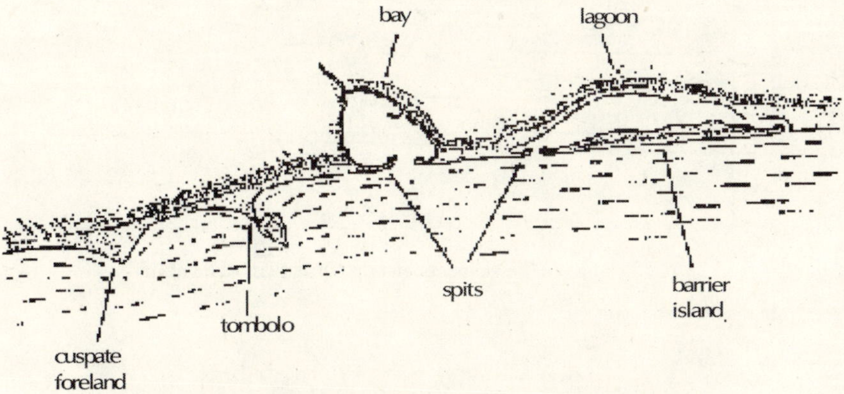

Figure 5: Features associated with accreting coasts.

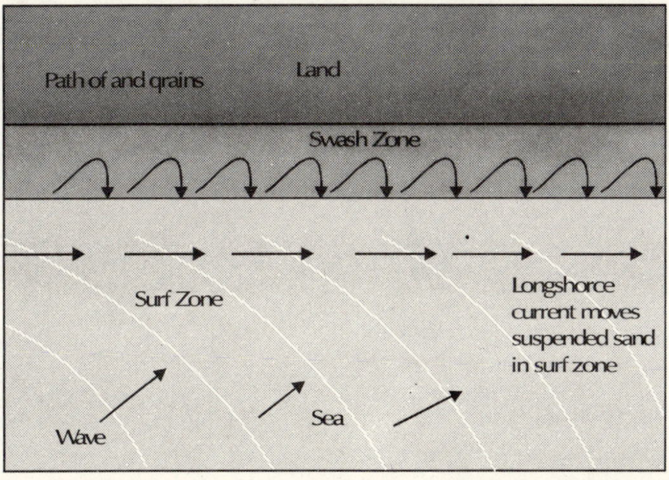

Figure 6: Longshore drift of sand on the beach face and by a longshore current within the surf zone.

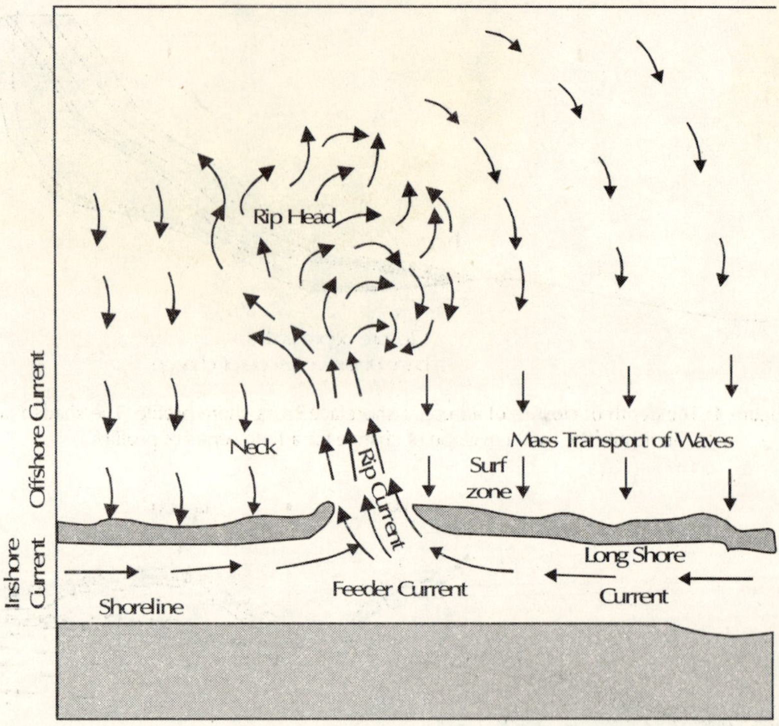

Figure 7: The elements of a rip current embayment.

Budget of Sediments

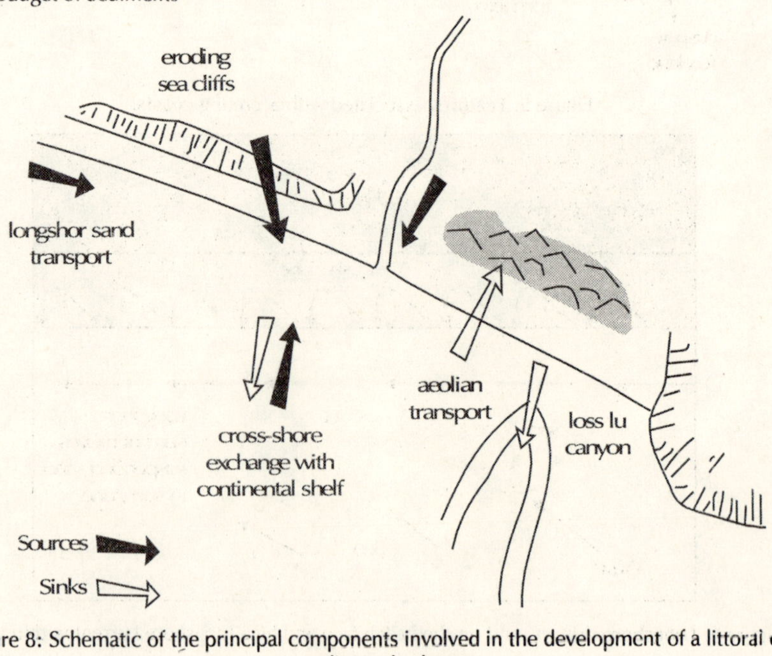

Figure 8: Schematic of the principal components involved in the development of a littoral cell sediment budget.

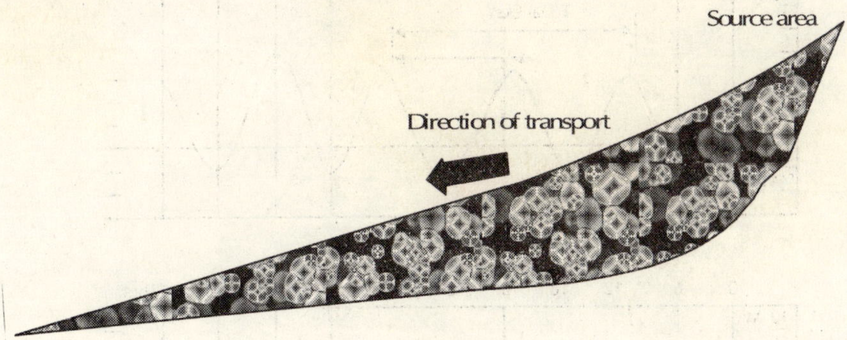

Figure 9: Sediment deposits usually become thinner away from the source area, and sediment grains become finer and more rounded.

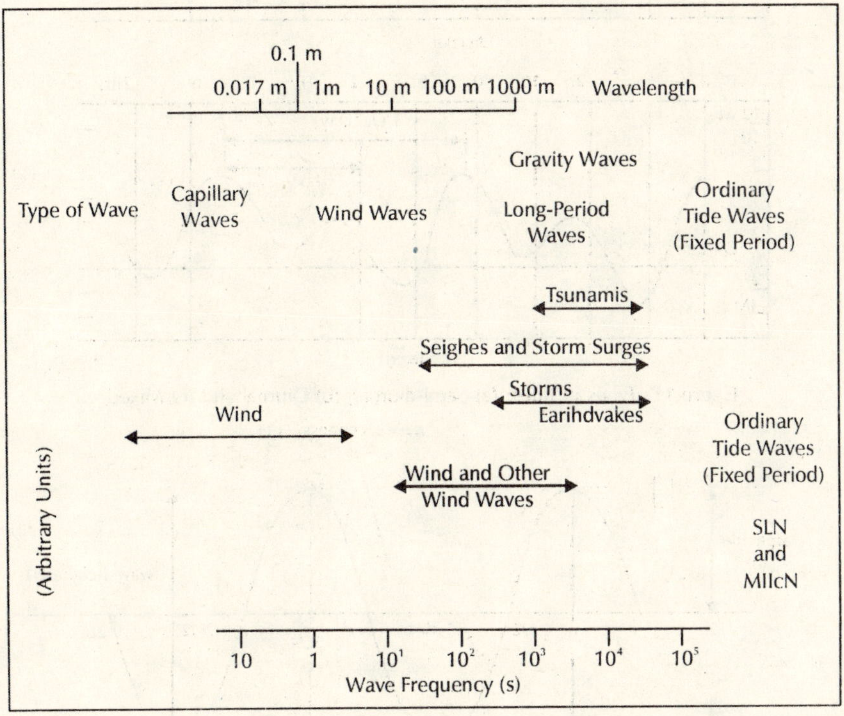

Figure 10: Types of surface waves, showing the relationships between wave length, wave frequency, the nature of the displacing forces and the relative amounts of energy in each type of wave.

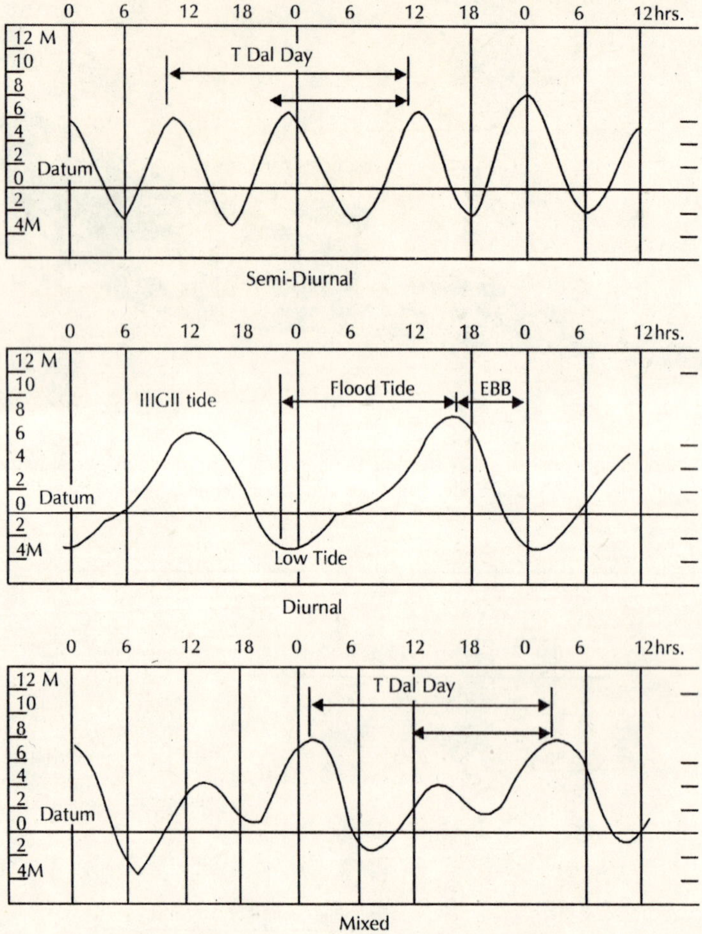

Figure 11: Types of tides: (a) Semi-diurnal, (b) Diurnal and (c) Mixed.

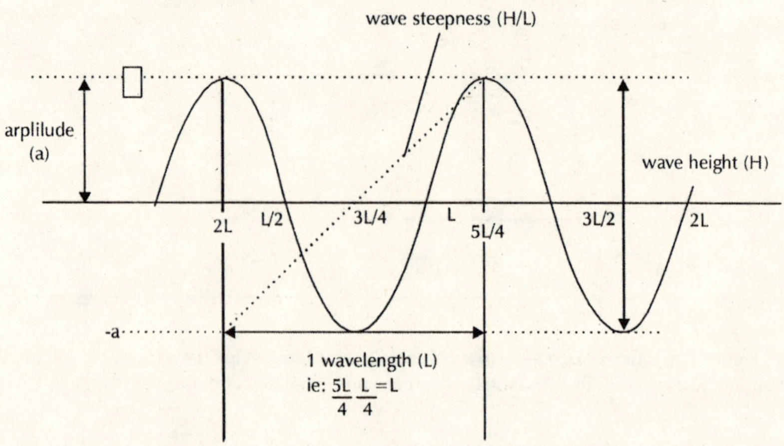

Figure 12: A vertical profile, showing the linear dimensions, of two successive idealized ocean waves.

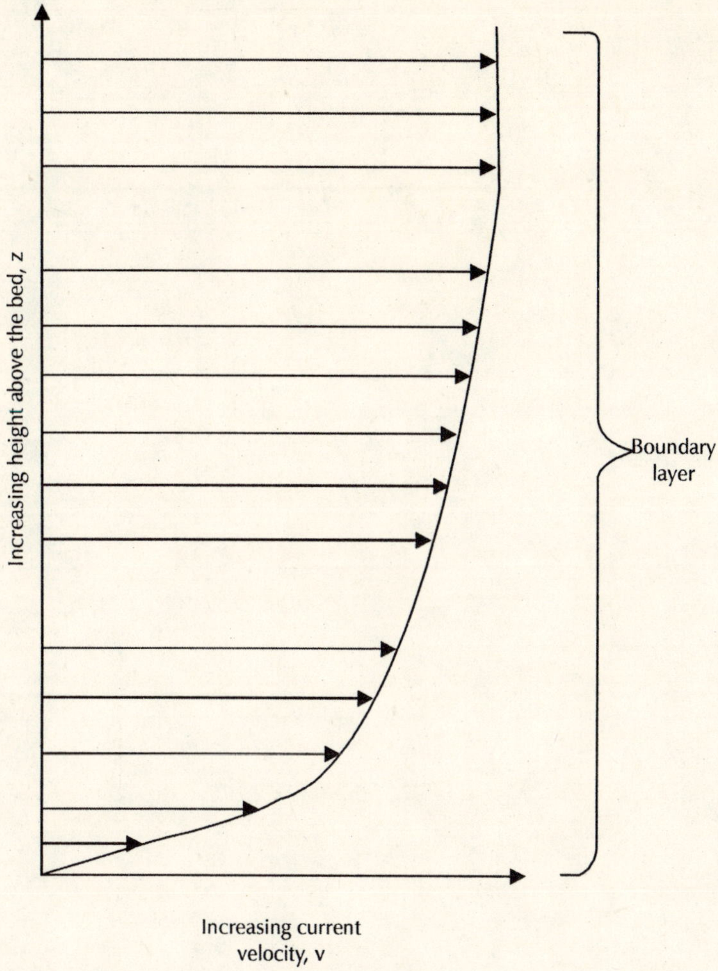

Figure 13: The velocity profile for a steady current flow over a bed. The arrows indicate the direction of flow and the length of each arrow is proportional to the current velocity at the height above the bed.

Ecosystem

A

❐ abiotic

Non-living. Climate is an abiotic component of ecosystems.

❐ acute toxicity

The ability of a substance to cause poisonous effects that result in severe biological harm or death soon after a single exposure or dose. (See chronic toxicity.)

❐ adaptive management

A type of natural resource management that implies making decisions as part of an on-going process. Monitoring the results of actions will provide a flow of information that may indicate the need to change a course of action. Scientific findings and the needs of society may also indicate the need to adapt resource management to new information.

Adaptive management is based upon the premise that managed natural systems are complex and unpredictable. While there are numerous definitions of adaptive management, most include adaptive management is the process of adjusting management actions and/or directions as new and better information emerges about the ecosystem.

❐ adsorption

The adhesion of molecules of gas, liquid, or dissolved solids to a surface.

❐ advisory

A nonregulatory document that communicates risk information.

❐ aerial logging

Removing logs from a timber harvest area by helicopter. Fewer roads are required, so the impact to an area is minimized.

❐ affected environment

The natural environment that exists at the present time in an area being analyzed.

❐ age class

An age grouping of trees according to an interval of years, usually 20 years. A single age class would have trees that are within 20 years of the same age, such as 1-20 years or 21-40 years.

❐ air pollutant

Any substance in air that could, if in high enough concentration, harm living things.

❐ airshed

A geographic area that shares the same air.

❐ algae

Simple rootless plants that grow in sunlit waters in relative proportion to the amounts of light and nutrients available. They are food for fish and small aquatic animals.

❐ allotment (range allotment)

The area designated for use by a prescribed number of livestock for a prescribed period of time. Though an entire Ranger District may be divided into allotments, all land will not be grazed, because other uses, such as recreation or tree plantings, may be more important at a given time.

❐ allowable sale quantity

The amount of timber that may be sold within a certain time period from an area of suitable land. The suitability of the land and the time period are specified in the Forest Plan.

❐ anadromous fish

Species of fish that mature in the sea and migrate into streams to spawn.

❐ aquifer

A body of rock that is saturated with water or transmits water. When people drill wells, they tap water contained within an aquifer.

❐ aspect

The direction a slope faces. A hillside facing east has an eastern aspect.

❐ atmospheric deposition

Pollution from the atmosphere associated with dry deposition in the form of dust, wet deposition in the form of rain and snow, or as a result of vapor exchanges.

❐ aum (animal unit month)

The quantity of forage required by one mature cow and her calf (or the equivalent, in sheep or horses, for instance) for one month.

B

❐ bacteria

A group of universally distributed, rigid, essentially unicellular microscopic organisms lacking chlorophyll. Some bacteria can aid in pollution control by consuming or breaking down organic matter in sewage or by similarly acting on oil spills or other water pollutants. Bacteria in soil, water, or air can also cause human, animal, and plant health problems.

❑ bark beetle

An insect that bores through the bark of forest trees to eat the inner bark and lay its eggs. Bark beetles are important killers of forest trees.

❑ basal area

The area of the cross section of a tree trunk near its base, usually 4 and 1/2 feet above the ground. Basal area is a way to measure how much of a site is occupied by trees. The term basal area is often used to describe the collective basal area of trees per acre.

❑ benthic organism (benthos)

A form of aquatic plant or animal life that is found near the bottom of a stream, lake, or ocean. Benthic populations are often indicative of sediment quality. The benthos comprise:

1. Sessile animals, such as sponges, some worms and many attached algae.
2. Creeping forms, such as snails and flatworms.
3. Burrowing forms, which include most clams, worms, mayflies and midges.

❑ benthic region

The bottom layer of a body of water.

❑ big game

Large mammals, such as deer, elk, and antelope that are hunted for sport.

❑ bioaccumulative substances

Substances that increase in concentration in living organisms (that are very slowly metabolized or excreted) as they breathe contaminated air or water, drink contaminated water, or eat contaminated food.

❑ bioassay

An evaluation using organisms to measure the effect of a substance, factor, or condition by comparing before and after data.

❑ biological control

The use of natural means to control unwanted pests. Examples include introduced or naturally occurring predators such as wasps, or hormones that inhibit the reproduction of pests. Biological controls can sometimes be alternatives to mechanical or chemical means.

❑ biological diversity

The number and abundance of species found within a common environment. This includes the variety of genes, species, ecosystems, and the ecological processes that connect everything in a common environment.

❑ biological integrity

The biotic composition, structure, and functioning at genetic, organism, and community levels consistent with natural conditions, including the natural biological processes that shape genomes, organisms, and communities.

❑ biological magnification

Refers to the process whereby certain substances become more concentrated in tissues at each successive stage up the food web.

❑ biomass

All the living material in a given area often refers to vegetation. Algal biomass is often indicative of the trophic status of a water body.

❑ biome

The complex of living communities maintained by the climate of a region and characterized by a distinctive type of vegetation. Example of biomes in North America include the tundra, desert, prairie, and the western coniferous forests.

❑ biomes

Biome is a term that describes areas on the earth with similar climate, plants, and animals at a global scale. Biomes are classified according to the predominant vegetation and characterized by adaptations of organisms to that particular environment. In the Great Lakes-Big Rivers Region, there are three distinct biomes - aquatic, forest, and grassland biomes. Biomes are composed of many smaller ecosystems - communities of plants and animals and their habitats (the physical parts of their environment that affect them). Whereas the boundaries of a biome are determined by climate, the boundaries of ecosystems are physical features, such as ridges and or riverbanks that separate one community from another. The ecosystems of a particular biome tend to have plants with similar growth forms and animals with similar feeding habits.

❑ biosphere

The term "Biosphere" was coined by the Russian scientist Vladimir Vernadsky in1929. It refers to the relatively thin vertical zone of air, soil, and water that extends from the deepest ocean floor up roughly 6 miles in to the atmosphere. The biosphere is often divided into distinct biomes that represent the interactions between groups of organisms forming a trophic pyramid and the environment or habitat in which they live.

❑ biota

The plant and animal life of a particular region.

❑ biotic

Living. Green plants and soil microorganisms are biotic components of ecosystems.

❒ bmp

Practices designed to prevent or reduce water pollution.

❒ board foot

A measurement term for lumber or timber. It is the amount of wood contained in an unfinished board 1 inch thick, 12 inches long, and 12 inches wide.

❒ broadcast burn

A prescribed fire that burns a designated area. These controlled fires can reduce wildfire hazards, improve forage for wildlife and livestock, or encourage successful regeneration of trees.

❒ browse

Twigs, leaves, and young shoots of trees and shrubs that animals eat. Browse is often used to refer to the shrubs eaten by big game, such as elk and deer.

❒ buffer

A land area that is designated to block or absorb unwanted impacts to the area beyond the buffer. Buffer strips along a trail could block views that may be undesirable. Buffers may be set aside next to wildlife habitat to reduce abrupt change to the habitat.

❒ byproduct

Material, other than the principal product, that is generated as a consequence of an industrial process.

C

❒ cable logging

Logging that involves the transport of logs from stump to collection points by means of suspended steel cables. Cable logging reduces the need for the construction of logging roads.

❒ canopy

The part of any stand of trees represented by the tree crowns. It usually refers to the uppermost layer of foliage, but it can be use to describe lower layers in a multi-storied forest.

❒ carcinogen

Any substance that can cause or contribute to the production of cancer.

❒ causes of erosion

Erosion of landscapes as a result of human activity can take many forms. Often, human activity begins to erode a landscape and natural processes accelerate the process.

Farming
Farming activities, whether manual or mechanized, can result in general wear and tear of land. Soils are exposed to the elements and altered in ways which allow erosion to affect the landscape.

Transportation routes
The construction of roads and railways can often allow the elements to begin erosion of the land because construction tampers with the natural structure of the terrain. The vehicles themselves can also erode the surface of a transport route.

Deforestation
The removal of forest from the landscape can have a dramatic effect as trees give cover to soils as well as providing stability to the land with their roots. The absence of forest allows rainwater and gravity to combine forces, resulting in marked erosion.

Flooding
When water levels rise soils become increasingly waterlogged, resulting in instability and the likelihood of landslides and similar types of erosion. Damming valleys, land irrigation and higher rainfall from global warming all contribute to these effects.

Denudation
This is the total removal of natural land cover, whether it be grass, scrub or forest. Many human activities result in denudation of land, and the result is that the elements are free to begin erosion of the exposed and unprotected surface.

Acid Rain
Rain polluted by sulfur and nitrogen oxide gases becomes acid rain. When it falls on calcium-based rock it erodes the surface by a chemical action known as corrosion. These medieval stone sculptures have been eaten away by acid rain.

Quarrying
Open-cast mining and quarrying for minerals directly erodes the landscape, resulting in gaping holes where material has been removed. These activities affect the stability of the surrounding landscape, making it more vulnerable to erosion.

Walking
People tend to follow the same routes across landscapes, creating footpaths which are the result of feet gradually eroding the surface, as well as exposing it to the elements. Modern 'wilderness trekking' tourism can also erode the land.

❑ cavity

A hole in a tree often used by wildlife species, usually birds, for nesting, roosting, and reproduction.

❐ chemical control

The use of pesticides and herbicides to control pests and undesirable plant species.

❐ chlorophyll-a

The photosynthetic pigment found in most algae. Chlorophyll-a is used to measure the rate of photosynthesis in a body of water.

❐ chronic toxicity

The capacity of a substance to cause poisonous effects in an organism after long-term exposure. (See acute toxicity).

❐ clear cut

A harvest in which all or almost all of the trees are removed in one cutting.

❐ climax

The culminating stage in plant succession for a given site. Climax vegetation is stable, self-maintaining, and self-reproducing.

❐ coarse filter management

Land management that addresses the needs of all associated species, communities, environments, and ecological processes in a land area.

❐ collector roads

These roads serve small land areas and are usually connected to a Forest System Road, a county road, or a state highway.

❐ combined sewers

A sewer system that carries both sewage and storm water runoff. Normally, its entire flow goes to a waste treatment plant, but during a heavy storm, the stormwater volume may be so great as to cause overflows (combined sewer overflow). When this happens, untreated mixtures of stormwater and sewage may flow into receiving waters. Stormwater runoff may also carry toxic chemicals from industrial areas or streets into the sewer system.

❐ community

All the groups of organisms living together in the same area, usually interacting or depending on each other for existence.

❐ composition

What an ecosystem is composed of. Composition could include water, minerals, trees, snags, wildlife, soil, microorganisms, and certain plant species,

❐ conifer

A tree that produces cones, such as a pine, spruce, or fir tree.

❑ connectivity (of habitats)

The linkage of similar but separated vegetation stands by patches, corridors, or "stepping stones" of like vegetation. This term can also refer to the degree to which similar habitats are linked.

❑ consequences of erosion

Primary consequences

- Water and wind erosion often result in the thinning and eventual loss of soils.

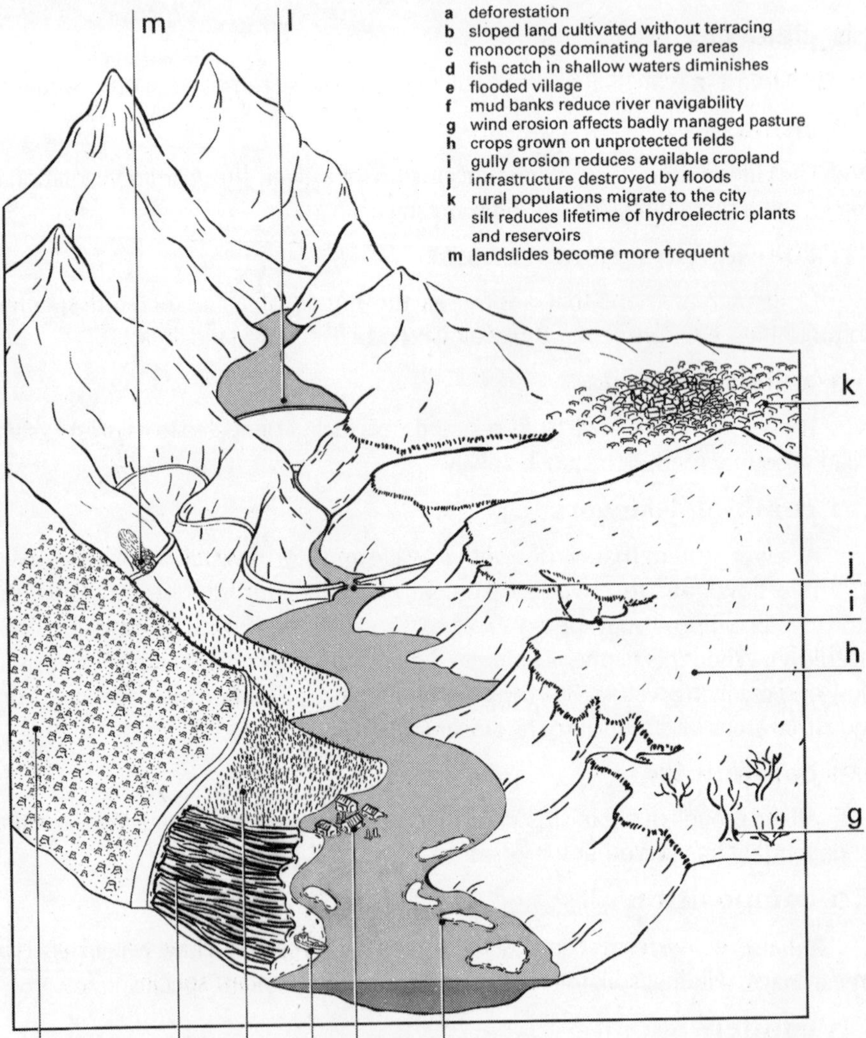

- Soils may also become unstable because of a lack of vegetation, resulting in landslides.
- Soils transported by water and wind can build up as deposits in new locations.

Secondary consequences

- Loss of soil makes land infertile and renders it unsuitable for farming.
- Unstable land is no longer suitable or safe on which to build houses.
- Deposited soils can clog up irrigation systems, hampering farming productivity.

Tertiary consequences

- Land that is no longer suitable for farming or settlement is often abandoned by people.
- The actions of people can therefore eventually result in their own or other's loss of living space.
- The end result can be famine and people made refugees in search of new places to settle.

❐ consumptive use

Use of resources that reduces the supply, such as logging and mining.

❐ contour

A line drawn on a map connecting points of the same elevation.

❐ conventional pollutants

Such contaminants as organic waste, sediment, acid, bacteria and viruses, nutrients, oil and grease, or heat.

❐ corridor

Elements of the landscape that connect similar areas. Streamside vegetation may create a corridor of willows and hardwoods between meadows where wildlife feed.

❐ cover

Any feature that conceals wildlife or fish. Cover may be dead or live vegetation, boulders, or undercut stream banks. Animals use cover to escape from predators, rest, or feed.

❐ cover forage ratio

The ratio of hiding cover to foraging areas for wildlife species.

❐ cover type (forest cover type)

Stands of a particular vegetation type that are composed of similar species. The aspen cover type contains plants distinct from the pinyon-juniper cover type.

❐ created opening

An opening in the forest cover created by the application of even-aged silvicultural practices.

❒ critical habitat

Areas designated for the survival and recovery of federally listed threatened or endangered species.

❒ crown height

The distance from the ground to the base of the crown of a tree.

❒ cultural resource

The remains of sites, structures, or objects used by people in the past; this can be historical or pre-historic.

❒ cumulative effects

Effects on the environment that result from separate, individual actions that, collectively, become significant over time.

D

❒ dbh (diameter at breast height)

The diameter of a tree 4 and 1/2 feet above the ground on the uphill side of the tree.

❒ decay

The breakdown of organic matter by bacteria and fungi.

❒ decision criteria

The rules and standards used to evaluate alternatives to a proposed action on National Forest land. Decision criteria are designed to help a decision maker identify a preferred choice from the array of alternatives.

❒ decking area

A site where logs are collected after they are cut and before they are taken to the landing area where they are loaded for transport.

❒ deforestation: global data-107 (la)

Loss of forest

- An estimated 50 percent of the Earth's surface was once forested.
- About 30 percent of the Earth's surface is covered by forest today.
- About 7.5 percent of the total forest cover is tropical forest, about 50 percent of the area of tropical forest that existed 50 years ago.
- The average rate of tropical deforestation worldwide during the 1980s was around 0.8 of one percent each year; this percentage has grown each year since the area lost is roughly constant but the area left becomes smaller.

Collecting data

- Calculations about deforestation rates are difficult to make; absolute areas are made up from thousands of pieces of forest lost all over the world.

- Fragmentation of forests can go unnoticed.
- Some methods for collecting data have proved impractical and inaccurate; deforestation rates are therefore based on estimates rather than established facts.

Satellites

- Satellite mapping is the accepted technique for making calculations in the 21st century.
- It is a relatively accurate means of gathering deforestation data because scans of the landscape are analyzed by computer.

Annual change in the natural forest area by region 1990–2000

In just ten years Africa lost over 12 million acres, and Asia nearly 10 million acres of their natural forests. Only Europe managed to increase its forest cover.

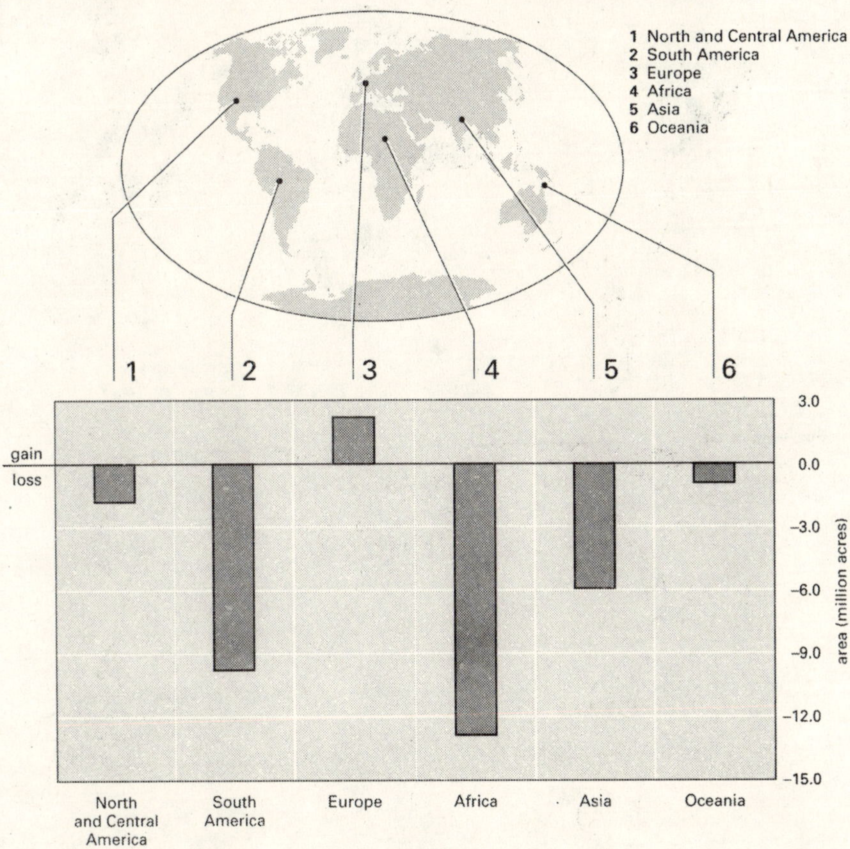

☐ deforestation: world crisis areas

Deforestation

- The removal of forests has become a worldwide cause for concern, especially the removal of tropical forests which comprise complex and unique ecosystems.

- Fragmentation of tropical forests also accounts for much of the damage because many species require uninterrupted cover to survive.

Motives

- Rising populations in developing countries require more space to settle and live.
- New populations require wood for building and cooking.
- Tropical timber can provide a valuable source of income through international trade.

Areas most at risk from deforestation

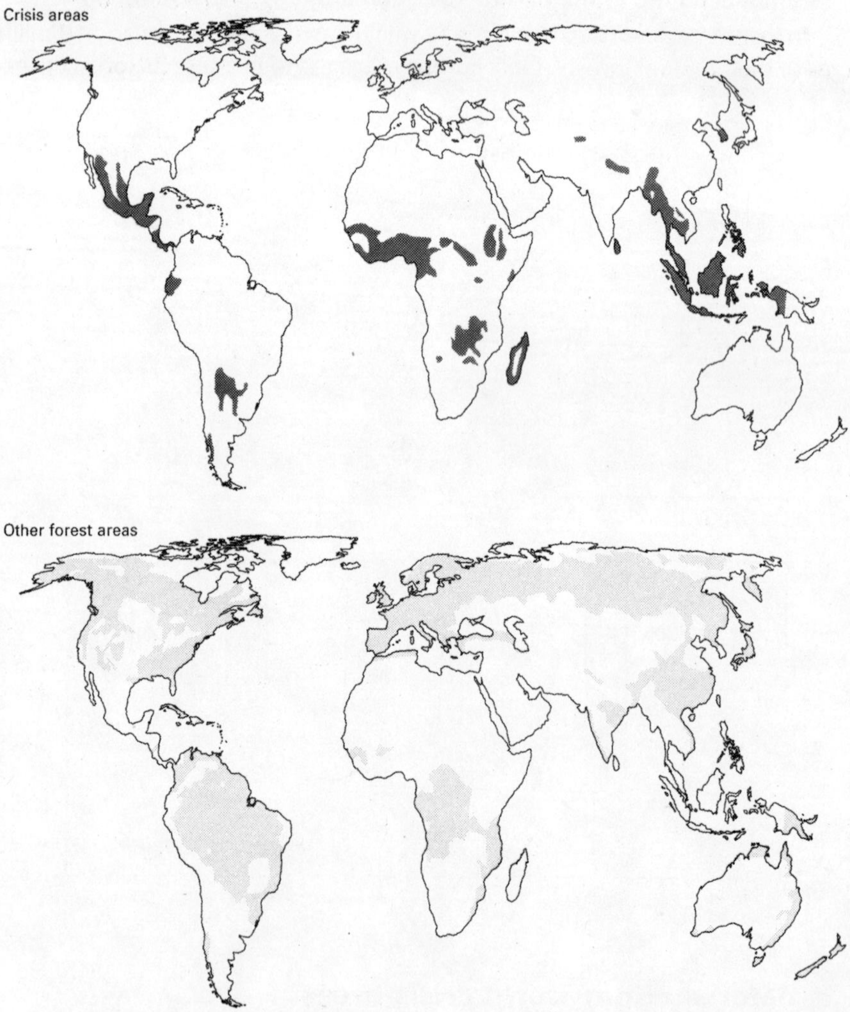

- Land clearance for farms, factories, and open cast mines is often required.
- Transport routes are often cut through forests to improve communications networks.

- Forest fires, started deliberately or accidentally, destroy large areas of forest every year.

❐ deis (draft environmental impact statement)

The draft version of the Environmental Impact Statement that is released to the public and other agencies for review and comment

❐ desertification in africa

Threats

- Desertification has its greatest impact in Africa.

Vulnerability to desertification in Africa
Human-induced desertification is especially likely to occur to the north and south of the Sahara Desert.

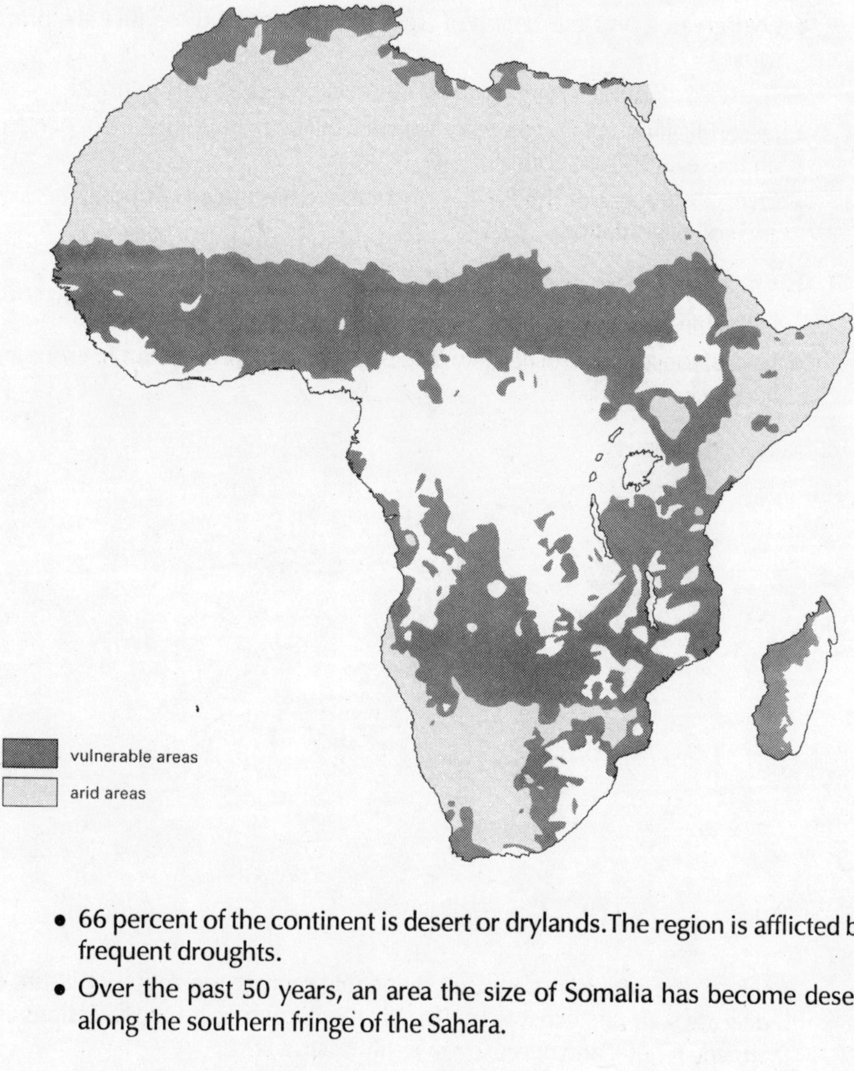

- 66 percent of the continent is desert or drylands.The region is afflicted by frequent droughts.
- Over the past 50 years, an area the size of Somalia has become desert along the southern fringe of the Sahara.

- Desertification now threatens more than 33 percent of the African continent.
- The Sahelian countries at the southern fringe of the Sahara are particularly vulnerable.
- About 22 million people live in the area of very high vulnerability.
- A further 330 million people (over 40 percent of the African population) live in areas of high or moderate vulnerability.
- The Sahara Desert extends south by an average of 3 miles every year.
- About 180 tons of soil per acre may be blown awayfrom denuded land around the Sahara every year.

Causes

- The main cause of desertification is not drought but mismanagement of land.
- Overgrazing and the felling of trees and brushwood for fuel are prime causes.
- These susceptible drylands cover 40 percent of the Earth's surface.
- Desertification puts at risk more than one billion people who are dependent on these lands for survival.
- Over 25 percent of the Earth's land surface has already suffered erosion and soil degradation.

▢ desertification worldwide

- Desertification is the loss and destruction of healthy fertile soils.
- Loss of topsoil and soil fertility results in declining production of crops and livestock.

Current extent of desertification

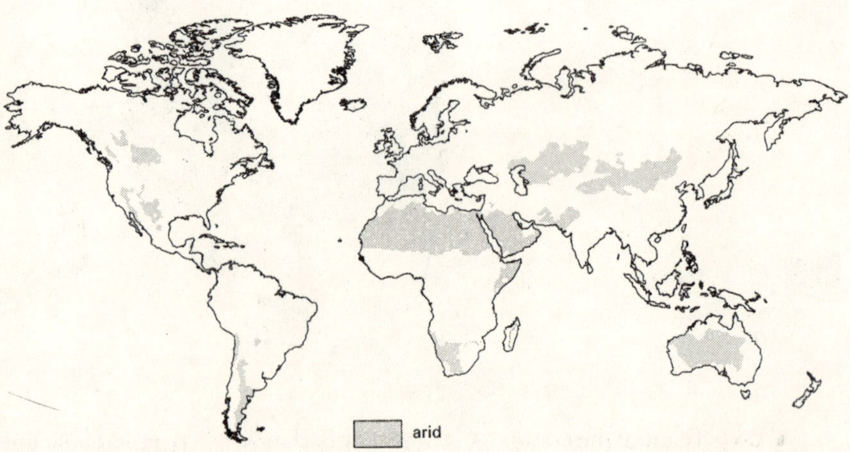

- Often, the problem is caused by people putting too much pressure on delicate soils and ecosystems. In places where there is little rain, the soil is already fragile, and overuse can easily destroy it.

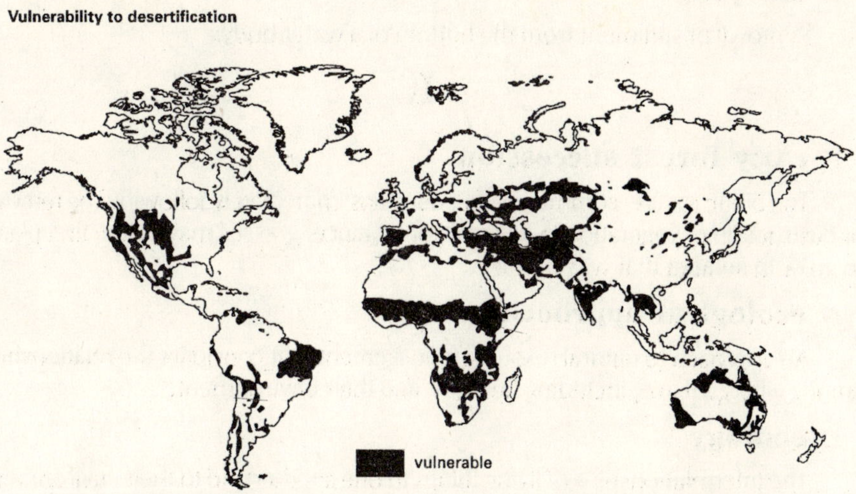

- The United Nations Environment Program (UNEP) calls desertification 'one of the most serious global environment problems.
- Desertification currently threatens the health and livelihoods of over 1 billion people – mostly poor rural families in developing countries.

❒ desired future condition

Land or resource conditions that are expected to result if goals and objectives are fully achieved.

❒ developed recreation

Recreation that requires facilities that, in turn, result in concentrated use of the area.

❒ dispersed recreation

Recreation that does not occur in a developed recreation site, such as hunting, backpacking, and scenic driving.

❒ dissolved oxygen (do)

The oxygen freely available in water. Dissolved oxygen is vital to fish and other aquatic life. Traditionally, the level of dissolved oxygen has been accepted as the single most important indicator of a water body's ability to support desirable aquatic life.

❒ disturbance

Any event, such as forest fire or insect infestations that alter the structure, composition, or functions of an ecosystem.

❒ drainage basin

A water body and the land area drained by it.

❒ dredging

Removal of sediment from the bottom of a water body.

E

❒ early forest succession

The biotic (or life) community that develops immediately following the removal or destruction of vegetation in an area. For instance, grasses may be the first plants to grow in an area that was burned.

❒ ecological approach

An approach to natural resource management that considers the relationships among all organisms, including humans, and their environment.

❒ ecology

The interrelationships of living things to one another and to their environment, or the study of these interrelationships.

❒ ecoregion

An area over which the climate is sufficiently uniform to permit development of similar ecosystems on sites that have similar properties. Ecoregions contain many landscapes with different spatial patterns of ecosystems.

❒ ecosystem

An arrangement of living and non-living things and the forces that move among them. Living things include plants and animals. Non-living parts of ecosystems may be rocks and minerals. Weather and wildfire are two of the forces that act within ecosystems.

❒ ecosystem approach

A method for sustaining or restoring ecological systems and their functions and values. It is goal driven and it is based on a collaboratively developed visions applied within geographic frameworks defined primarily by ecological boundaries.

❒ ecosystem degradation

Ecosystems are degraded when recovery to original conditions is unlikely under normal circumstances (without management). Degraded ecosystems can be distinguished by those that can be restored, rehabilitated, or reclaimed. All ecosystems found today within the Region have experienced some form of anthropogenic disturbance resulting in degradation in different degrees. This degradation may have resulted from fragmentation, fire suppression, invasive species invasions, and/or over-grazing by deer, among other things.

❒ ecosystem integrity

Ecological processes that are essential for ecosystems to function in a defined and predictable manner.

❒ ecosystem management

An ecological approach to natural resource management to assure productive, healthy ecosystems by blending social, economic, physical, and biological needs and values

❒ ecosystem management

The integration of ecological, social, and economic objectives for natural resource planning and management.

❒ ecotone

The transition zone between two biotic communities, such as between the Ponderosa pine forest type and the mixed conifer forest, which is found at higher elevations than the pine.

❒ ecotype

A population of a species in a given ecosystem that is adapted to a particular set of environmental conditions.

❒ edge

The margin where two or more vegetation patches meet, such as a meadow opening next to a mature forest stand, or a ponderosa pine stand next to an aspen stand.

❒ effluent

Wastewater--treated or untreated--that flows from a treatment plant, sewer, or industrial outfall. Generally refers to discharges into surface waters.

❒ element (of ecosystems)

An identifiable component, process, or condition of an ecosystem.

❒ emission

Discharges into the atmosphere from such sources as smokestacks, residential chimneys, motor vehicles, locomotives, and aircraft.

❒ endangered species

A plant or animal that is in danger of extinction throughout all or a significant portion of its range.

❒ endemic plant/organism

A plant or animal that occurs naturally in a certain region and whose distribution is relatively limited geographically.

❒ environmental analysis

An analysis of alternative actions and their predictable long and short-term environmental effects. Environmental analyses include physical, biological, social, and economic factors.

❑ **environmental assessment**

A brief version of an Environmental Impact Statement.

❑ **environmental impact statement**

A statement of environmental effects of a proposed action and alternatives to it. The EIS is released to other agencies and the public for comment and review.

❑ **ephemeral streams**

Streams that flow only as the direct result of rainfall or snowmelt. They have no permanent flow.

❑ **erosion**

The wearing away of land surface by wind or water. Erosion occurs naturally but can be caused by farming, residential or industrial development, mining, or timber-cutting.

❑ **escape cover**

Vegetation of sufficient size and density to hide an animal, or an area used by animals to escape from predators.

❑ **eutrophication**

The process of fertilization that causes high productivity and biomass in an aquatic ecosystem. Eutrophication can be a natural process or it can be a cultural process accelerated by an increase of nutrient loading to a lake by human activity.

❑ **even aged management**

Timber management actions that result in the creation of stands of trees in which the trees are essentially the same age.

❑ **exotic species**

Species that are not native to the Great Lakes and that have been intentionally introduced to or have inadvertently infiltrated the system. Exotics prey upon native species and compete with them for food or habitat.

F

❑ **fauna**

The animal life of an area.

❑ **felling**

Cutting down trees.

❑ **fertilizer**

Materials, including nitrogen and phosphorus, that provide nutrients for plants.

❐ final cut

The removal of the last seed bearers or shelter trees after regeneration of new trees has been established in a stand being managed under the shelter wood system of silviculture.

❐ fine filter management

Management that focuses on the welfare of a single or only a few species rather than the broader habitat or ecosystem. (See coarse filter management.)

❐ fire cycle

The average time between fires in a given area.

❐ fire regime

The characteristics of fire in a given ecosystem, such as the frequency, predictability, intensity, and seasonality of fire.

❐ fire supression

Fire suppression is an issue facing many ecosystems today. Fire is important in determining natural characteristics, such as habitat structure and composition. Absence of fire in an area can result in profound changes to community characteristics and often leads to reduced species diversity. For instance, fire absence in prairies and oak savannas can lead to woody vegetation encroachment and severe invasion of non-native grasses which can eliminate many prairie plants and associated fauna.

❐ fisheries habitat

Streams, lakes, and reservoirs that support fish, or have the potential to support fish.

❐ flood plain

A lowland adjoining a watercourse. At a minimum, the area is subject to a 1% or greater chance of flooding in a given year.

❐ flora

The plant life of an area.

❐ food chain

A sequence of organisms, each of which uses the next, lower member of the sequence as a food source. Members of a chain are interdependent so that a disturbance to one species can disrupt the entire hierarchy.

❐ food web

The complex feeding network occurring within and between food chains in an ecosystem, whereby members of one food chain may belong to one or more other food chains.

❐ forage

All browse and non-woody plants that are eaten by wildlife and livestock.

❐ forb

A broadleaf plant that has little or no woody material in it.

❐ foreground

The part of a scene or landscape that is nearest to the viewer.

❐ forest health

A measure of the robustness of forest ecosystems. Aspects of forest health include biological diversity; soil, air, and water productivity; natural disturbances; and the capacity of the forest to provide a sustaining flow of goods and services for people.

❐ forest loss in costa rica

Room for people

- The countries of Central America, including Costa Rica, are all relatively small in area.
- Expanding human populations have resulted in forests being cleared for living and farming space.
- About 65 percent of Central America's forests have been lost in this way since the arrival of Europeans.
- Most of the cleared land is used as pasture for rearing cattle.
- The greater part of Costa Rica's forest was destroyed by 1983.

Deforestation of Costa Rica

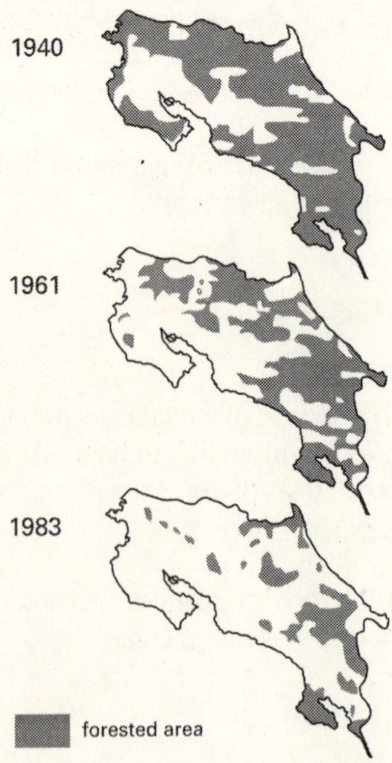

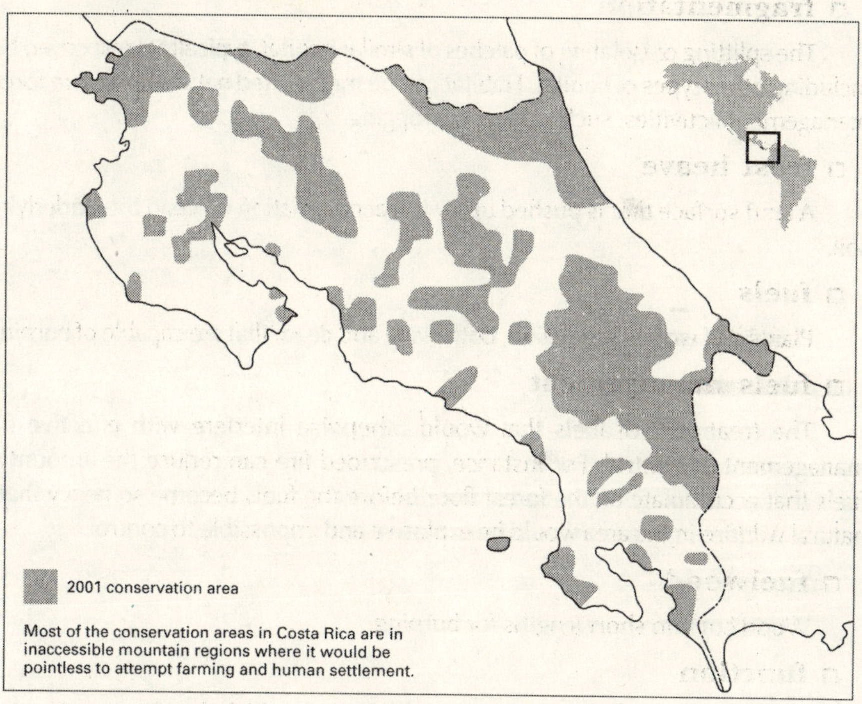

Most of the conservation areas in Costa Rica are in inaccessible mountain regions where it would be pointless to attempt farming and human settlement.

What is being lost

- Costa Rica contains one of highest concentrations of plant and animal species per square mile in the world.
- It has over 11,000 species of plant.
- It has more than 100 kinds of mammal.
- It has in excess of 400 species of bird.
- It has countless types of insects and other invertebrates.

Conservation

- Conservation projects in Costa Rica aim to replace as much of the lost tropical forest as possible, and to conserve what remains.
- Due to the terrain some 50 percent of the country is regarded as being only suitable for trees. This fact has helped conservation teams to protect many areas.

❐ forest roads and trails

Roads and trails under the jurisdiction of the Forest Service.

❐ forest supervisor

The official responsible for administering National Forest lands on an administrative unit, usually one or more National Forests. The Forest Supervisor reports to the Regional Forester.

❑ **fragmentation**

The splitting or isolating of patches of similar habitat, typically forest cover, but including other types of habitat. Habitat can be fragmented naturally or from forest management activities, such as clear cut logging.

❑ **frost heave**

A land surface that is pushed up by the accumulation of ice in the underlying soil.

❑ **fuels**

Plants and woody vegetation, both living and dead, that are capable of burning.

❑ **fuels management**

The treatment of fuels that would otherwise interfere with effective fire management or control. For instance, prescribed fire can reduce the amount of fuels that accumulate on the forest floor before the fuels become so heavy that a natural wildfire in the area would be explosive and impossible to control.

❑ **fuelwood**

Wood cut into short lengths for burning.

❑ **function**

All the processes within an ecosystem through which the elements interact, such as succession, the food chain, fire, weather, and the hydrologic cycle.

G

❑ **game fish**

Fish species caught for sport, such as trout, salmon, or bass.

❑ **game species**

Any species of wildlife or fish that is harvested according to prescribed limits and seasons.

❑ **geomorphic processes**

Processes that change the form of the earth, such as volcanic activity, running water, and glacial action.

❑ **geomorphology**

The science that deals with the relief features of the earth's surface.

❑ **gis (geographic information systems)**

GIS is both a database designed to handle geographic data as well as a set of computer operations that can be used to analyze the data. In a sense, GIS can be thought of as a higher order map.

❑ goals

Broad statements of direction; end results or positions to be achieved.

❑ ground fire

A fire that burns along the forest floor and does nor affect trees with thick bark or high crowns.

❑ ground water

The supply of fresh or saline water found beneath the Earth's surface, usually in aquifers, often supplying wells and springs.

❑ group selection

A method of tree harvest in which trees are removed periodically in small groups. This silvicultural treatment results in small openings that form mosaics of age class groups in the forest.

❑ habitat

The place where a population (e.g., human, animal, plant, micro-organism) lives and its surroundings.

❑ habitat capability

The ability of a land area or plant community to support a given species of wildlife.

❑ habitat diversity

A number of different types of wildlife habitat within a given area.

❑ habitat diversity index

A measure of improvement in habitat diversity.

❑ habitat fragmentation

Habitat fragmentation is a serious problem with potentially devastating effects for biological diversity. The issue of fragmentation has been identified recently as one of the most pressing issues in wildlife management and the conservation of biodiversity. Habitat fragmentation refers to a phenomenon where habitats are broken up into small, isolated pieces which results in new landscapes being developed that differ substantially from the previous landscape. The shape, size, proximity, and contrast of each new landscape piece determines how the fragmented habitat affects wildlife. For instance, the splintering of wetlands, prairies, and forests into small isolated "pieces" makes it difficult for isolated populations of plants and animals to move from one "piece" to another, thus impacting their ability to breed with one another. Problems associated with isolated populations of plants and animals have been described by researchers in conservation biology and island biogeography. Isolated groups often have trouble maintaining the genetic integrity and variability needed for their continued evolutionary viability and prospects for long-term survival (if they cannot move to or be reached by other populations of their species). Such situations occur if there are no corridors of appropriate habitat for the species to move through. This is the situation for many forest interior species in fragmented forest landscapes throughout the Midwest.

❑ **habitat type**

A way to classify land area . A habitat type can support certain climax vegetation, both tree and undergrowth species. Habitat typing can indicate the biological potential of a site.

❑ **heavy metals**

Metallic elements with high atomic weights (e.g., mercury, chromium, cadmium, arsenic, and lead) that tend to be toxic and bioaccumulate.

❑ **herbicide**

A chemical pesticide designed to control or destroy plants, weeds, or grasses.

❑ **hiding area/cover**

Vegetation capable of hiding 90% of an adult elk or deer from human's view at a distance of 200 feet or less.

❑ **horizontal diversity**

The distribution and abundance of different plant and animal communities or different stages of plant succession across an area of land; the greater the numbers of communities in a given area, the higher the degree of horizontal diversity.

❑ **hydrologic cycle**

Also called the water cycle, this is the process of water evaporating, condensing, falling to the ground as precipitation, and returning to the ocean as run-off.

❑ **hydrology**

The science dealing with the study of water on the surface of the land, in the soil and underlying rocks, and in the atmosphere.

I

❑ **igneous rock**

Rocks formed when high temperature, molten mineral matter cooled and solidified.

❑ **indicator**

An organism, species, or community whose characteristics show the presence of specific environmental conditions.

❑ **indicator species**

A plant or animal species related to a particular kind of environment. Its presence indicates that specific habitat conditions are also present.

❑ **indigenous (species)**

Any species of wildlife native to a given land or water area by natural occurrence.

▫ individual tree selection

The removal of individual trees from certain size and age classes over an entire stand area. Regeneration is mainly natural, and an uneven aged stand is maintained.

▫ insecticide

A chemical specifically used to kill or control the growth of insects.

▫ instream flow

The quantity of water necessary to meet seasonal stream flow requirements to accomplish the purposes of the National Forests, including, but not limited to fisheries, visual quality, and recreational opportunities.

▫ integrated pest management▫

IPM evaluates alternatives for managing forest pest populations, based on consideration of pest-host relationships.

▫ interdisciplinary team

A team of individuals with skills from different disciplines that focuses on the same task or project.

▫ interdisciplinary, coordinated approach

Successful coordination and collaboration by the Service means focusing multiple stakeholder resources on priority natural resource issues that benefits listed species, migratory birds, and biological integrity, diversity, and environmental health within Ecosystems. In doing so, the Service is able to leverage conservation efforts, make more efficient use of resources, improve communication, and develop consistency in natural resource conservation. The Service uses a variety of expertise at the local community level.

▫ intermediate cut

The removal of trees from a stand sometime between the beginning or formation of the stand and the regeneration cut. Types of intermediate cuts include thinning, release, and improvement cuttings.

▫ intermittent stream

A stream that flows only at certain times of the year when it receives water from streams or from some surface source, such as melting snow.

▫ intermountain region

The portion of the USDA Forest Service, also referred to as Region Four, that includes National Forests in Utah, Nevada, southern Idaho, and southwestern Wyoming.

▫ introduction land degradation

- Land degradation is a human-induced or natural process which negatively affects the land's ability to accept, store, and recycle water, energy, and nutrients.

- Desertification is land degradation occurring in the arid, semi-arid, and sub-humid areas of the world.
- These susceptible drylands cover 40 percent of the Earth's surface.
- Desertification puts at risk more than 1 billion people who are dependent on these lands.
- Over 25 percent of the Earth's land surface has already suffered erosion and soil degradation.

Land degradation high-risk areas at the end of the 20th century

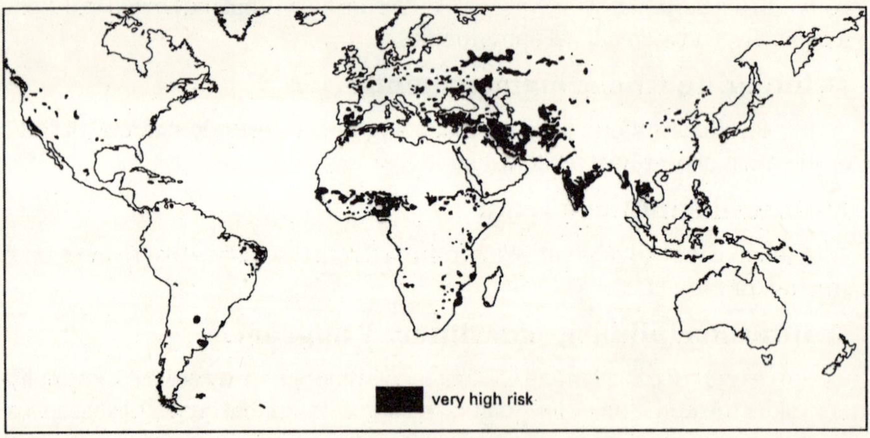

Land degradation high–moderate-risk areas at the end of the 20th century

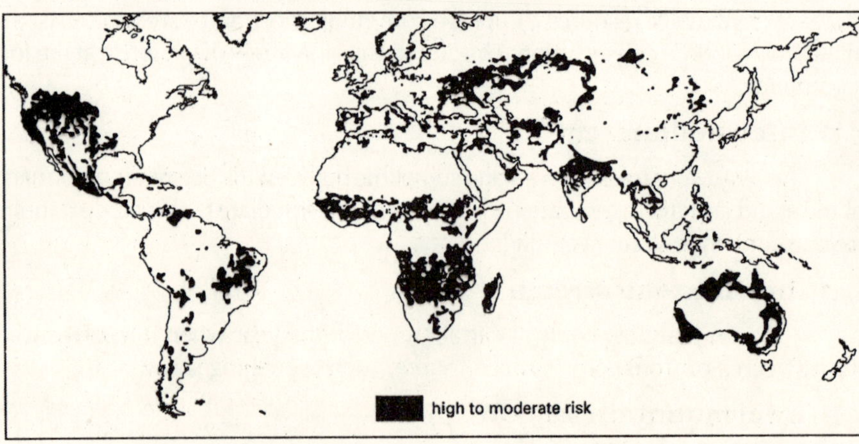

Causes of land degradation

- land clearing and deforestation
- agricultural mining of soil nutrients
- urban conversion
- irrigation
- pollution

Major stresses that degrade land

- loss of organic matter
- destruction of soil structure
- alkalinization
- salinization
- acidity increase
- removal of nutrients
- accelerated erosion by wind and water

❐ irretrievable

One of the categories of impacts mentioned in the National Environmental Policy Act to be included in statements of environmental impacts. An irretrievable effect applies to losses of production or commitment of renewable natural resources. For example, while an area is used as a ski area, some or all of the timber production there is irretrievably lost. If the ski area closes, timber production could resume; the loss of timber production during the time that the area was devoted to winter sports is irretrievable. However, the loss of timber production during that time is not irreversible, because it is possible for timber production to resume if the area is no longer used as a ski area.

❐ irreversible

A category of impacts mentioned in statements of environmental impacts that applies to non-renewable resources, such as minerals and archaeological sites. Irreversible effects can also refer to effects of actions that can be renewed only after a very long period of time, such as the loss of soil productivity.

❐ irrigation: types and problems

Scale of irrigation

- Irrigation of agricultural land is practiced in many parts of the world.
- Only about 17 percent of all croplands are under irrigation, but these lands produce over 33 percent of the world's food.
- About 70 percent of all fresh water used every year throughout the world is used for irrigated agriculture.
- Only about half of all water withdrawn for irrigation reaches the crops.
- The rest soaks into unlined irrigation canals, leaks out of pipes, or evaporates on its way to the fields.

Types of irrigation: There are four general categories of irrigation: surface, sprinkler, direct, and localized.

a Surface irrigation

- Surface irrigation is the oldest and most traditional method used.
- It typically involves the use of ditches and channels to drain and direct water from the water source to the crops.

- Modern variations involve the use of pipes and tanks to prevent the water from being wasted by soaking into the ground, or evaporating into the air before it reaches its destination.

Water lost in irrigation: 55 percent water loss

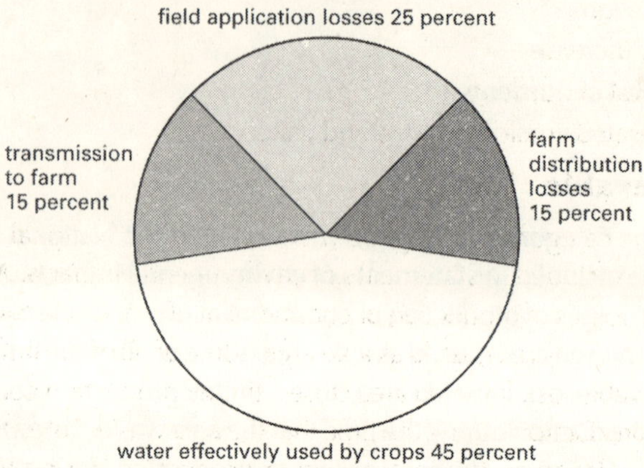

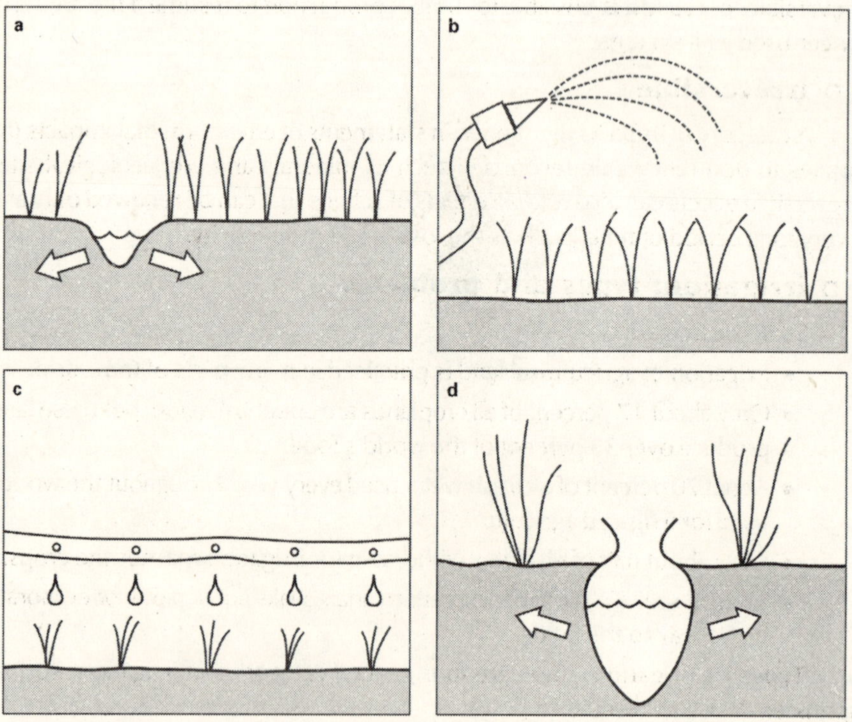

b Sprinkler irrigation

- This is where the water is carried by pipe to a device that sprinkles or sprays the water over the crop.

- It is a more efficient way of irrigating crops, though much water is still lost to evaporation.

c Direct irrigation

- In the interests of conserving as much water as possible direct irrigation aims to supply water straight to the crop plants at a steady rate.
- Small pipes convey the water from a larger conduit pipe and then drip or trickle it from tap holes right by the roots of the plants.

d Localized irrigation

- Attempts have been made to improve upon direct irrigation techniques.
- Pitcher irrigation involves the slow release of water from earthenware jars buried near the plant roots.
- Condensation irrigation uses piped coolant to trap water vapor from the air.

K

❐ key summer range

The portion of a wildlife species' summer range that is essential for the animal's pre, post, and reproduction cycles. Deer require "fawning areas" where does give birth and hide their fawns for an essential period of time in the spring.

❐ key winter range

That portion of big game's range where the animals find food and cover during severe winter weather.

L

❐ ladder fuels

Vegetation located below the crown level of forest trees which can carry fire from the forest floor to tree crowns. Ladder fuels may be low-growing tree branches, shrubs. or smaller trees.

❐ lampricide

A chemical used to kill sea lamprey.

❐ land class

The topographic relief of a unit of land. Land classes are separated by slope; this coincides with the timber inventory process. The three land classes used in the Forest Plan are defined by the following slope ranges: 0 to 35 percent; 36 to 55 percent; and greater than 55 percent.

❐ land degradation in dryland regions

Low rainfall

- Shifts in climate have caused parts of the world to become much drier.
- Prolonged periods of drought result in the death of vegetation.

- The soil then dries up and becomes vulnerable to erosion.
- As the soil turns to dust it can be blown away, leaving the land degraded.
- Sudden rainfall then washes away any remaining nutrients.
- Farming using irrigation methods further degrade soils by salinization.
- The combined effects of natural elements and human activity eventually exhaust the land.
- The degraded land that results is ultimately left to the process of

Land degradation in dryland regions

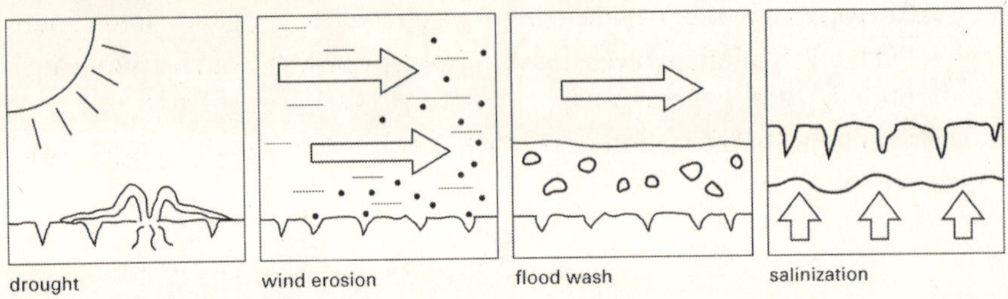

Main causes of dryland soil degradation by region at end of the 20th century

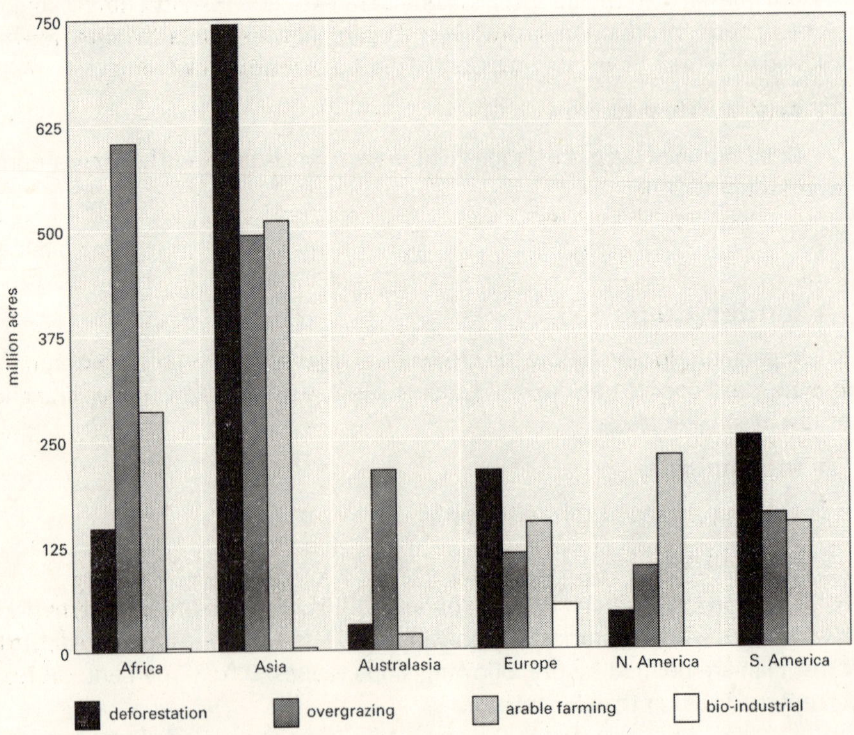

❐ land use planning

The process of organizing the use of lands and their resources to best meet people's needs over time, according to the land's capabilities.

❏ landfills

1. Land disposal sites for nonhazardous solid wastes at which the waste is spread in layers, compacted to the smallest practical volume, and covered with material applied at the end of each operating day.
2. Land disposal sites for hazardous waste designed to minimize the chance of release of hazardous substances into the environment.

❏ landing

Any place where cut timber is assembled for further transport from the timber sale area.

❏ landline

The boundary lines for National Forest land.

❏ landscape

A large land area composed of interacting ecosystems that are repeated due to factors such as geology, soils, climate, and human impacts. Landscapes are often used for coarse grain analysis.

❏ late forest succession

The stage of forest succession in which most of the trees are mature or overmature.

❏ life zone

Areas or "belts" of land that have distinct plant and animal characteristics determined by elevation, latitude, and climate. When ascending a high mountain, you will pass through these life zones. Examples of life zones include the Upper Sonoran, where Cedar City is located and gramma grasses, sagebrush, and scattered pinyon juniper predominate, and the Transition zone, where Ponderosa pine is predominant.

❏ litter (forest litter)

The freshly fallen or only slightly decomposed plant material on the forest floor. This layer includes foliage, bark fragments, twigs, flowers, and fruit.

❏ loading

The addition of a substance to a water body.

❏ logging residue (slash)

The residue left on the ground after timber cutting. It includes unutilized logs, uprooted stumps, broken branches, bark, and leaves. Certain amounts of slash provide important ecosystem roles, such as soil protection, nutrient cycling, and wildlife habitat.

M

❏ m

Thousand. Five thousand board feet of timber can be expressed as 5M board feet.

❑ macro climate

The general, large scale climate of a large area, as distinguished from the smaller scale micro climates within it.

❑ management action

Any activity undertaken as part of the administration of the National Forest.

❑ marsh

A type of wetland that does not accumulate appreciable peat deposits and is dominated by herbaceous vegetation. Marshes may be either freshwater or saltwater and tidal or nontidal.

❑ mass balance approach

An analytic method, based on conservation of mass, used to assess the quantity and cycling of contaminants throughout a water system.

❑ mass movement/wasting

The down-slope movement of large masses of earth material by the force of gravity. Also called a landslide.

❑ matrix

The least fragmented, most continuous pattern element of a landscape; the vegetation type that is most continuous over a landscape.

❑ mature timber

Trees that have attained full development, especially height, and are in full seed production.

❑ mean annual increment of growth

The total increase in size or volume of individual trees. Or, it can refer to the increase in size and volume of a stand of trees at a particular age, divided by that age in years.

❑ metabolite

A substance that is the product of biological changes to a chemical.

❑ micro climate

The climate of a small site. It may differ from the climate at large of the area due to aspect, tree cover (or the absence of tree cover), or exposure to winds.

❑ middleground

A term used the management of visual resources, or scenery. It refers to the visible terrain beyond the foreground where individual trees are still visible but do not stand out distinctly from the stand.

❑ mineral soil

Soil that consists mainly of inorganic material, such as weathered rock, rather than organic matter.

❑ mis (management indicator species)

A wildlife species whose population will indicate the health of the ecosystem in which it lives and, consequently, the effects of forest management activities to that ecosystem. MIS species are selected by land management agencies. (See "indicator species".)

❑ mission (of the usda forest service)

"To Care for the Land and Serve the People". As set forth in law, the mission is to achieve quality land management under the sustainable multiple-use management concept to meet the diverse needs of people.

❑ mitigation

Actions taken to avoid, minimize, or rectify the impact of a land management practice.

❑ mixed stand

A stand consisting of two or more tree species.

❑ monitoring

A scientifically designed system of continuing standardized measurements and observations and the evaluation thereof.

❑ monitoring and evaluation

The periodic evaluation of forest management activities to determine how well objectives were met and how management practices should be adjusted.

❑ mortality

Trees that were merchantable and have died within a specified period of time. The term mortality can also refer to the rate of death of a species in a given population or community.

❑ mosaic

Areas with a variety of plant communities over a landscape, such as areas with trees and areas without trees occurring over a landscape.

❑ mountain pine beetle

A tiny black insect, ranging from 1/8 to 3/4 inch in size, that bores through a pine tree's bark. It stops the tree's intake and transport of the food and nutrients it must have to stay alive, thus killing the tree.

❑ multiple use management

The management of all the various renewable surface resources of National Forest lands for a variety of purposes such as recreation, range, timber, wildlife and fish habitat, and watershed.

N

❐ national park service

The agency of the US Department of the Interior responsible for the administration of National Parks, Monuments, and Historic Sites. It is distinct from the USDA Forest Service both administratively and by mission.

❐ natural barrier

A natural feature, such as a dense stand of trees or downfall, that will restrict animal travel.

❐ natural conditions

Conditions thought to exist from the end of the Medieval Warm Period to the advent of the industrial era (app. 950AD to 1800AD), based upon scientific study and sound professional judgment.

❐ natural range of variability

See range of variability

❐ natural resource

A feature of the natural environment that is of value in serving human needs.

❐ nepa (national environmental policy act)

Congress passed NEPA in 1969 to encourage productive and enjoyable harmony between people and their environment. One of the major tenets of NEPA is its emphasis on public disclosure of possible environmental effects of any major action on public lands. Section 102 of NEPA requires a statement of possible environmental effects to be released to the public and other agencies for review and comment.

❐ nest survey

A way to estimate the size of a bird population by counting the number of nests in a given area.

❐ nflrmp (national forest land and resource management plan)

Also called the Forest Plan or just the Plan, this document guides the management of a particular National Forest and establishes management standards and guidelines for all lands of that National Forest.

❐ nfma (national forest management act)

This law was passed in 1976 and requires the preparation of Regional Guides and Forest Plans.

❐ nfrs

National Forest recreation sites that have been inventoried.

❐ nitrate

A compound containing nitrogen and oxygen that can exist in the atmosphere or in water and that can have harmful effects on humans and animals at high concentrations.

❐ no action alternative

The most likely condition expected to exist in the future if management practices continue unchanged.

❐ noncommercial vegetative treatment

The removal of trees for reasons other than timber production.

❐ nonconsumptive use

The use of a resource that does not reduce the supply. For instance, bird watching is a non-consumptive use of wildlife. Boating and fishing are non-consumptive uses of water.

❐ nongame

Wildlife species that are not hunted for sport.

❐ nonpoint source

Pollution sources that are diffuse and do not have a single point of origin or are not introduced into a receiving stream from a specific outlet. The pollutants are generally carried off land by stormwater runoff. Commonly used categories for nonpoint sources are agriculture, forestry, urban, mining, construction, dams and channels, and land disposal.

❐ nonpoint source pollution

Pollution whose source is not specific in location. The sources of the discharge are dispersed, not well defined, or constant. Rain storms and snowmelt often make this type of pollution worse. Examples include sediments from logging activities and runoff from agricultural chemicals.

❐ non-renewable resource

A resource whose total quantity does not increase measurably over time, so that each use of the resource diminishes the supply.

❐ notice of intent

A notice in the federal register of intent to prepare an environmental impact statement on a proposed action.

❐ nutrient

Any substance assimilated by living organisms that promotes growth. The term is generally applied to nitrogen and phosphorous, but is also applied to other essential trace elements.

❑ nutrient cycle

The circulation of chemical elements and compounds, such as carbon and nitrogen, in specific pathways from the non-living parts of ecosystems into the organic substances of the living parts of ecosystems, and then back again to the non-living parts of the ecosystem. For instance, nitrogen in wood is returned to the soil as the dead tree decays; the nitrogen again becomes available to living organisms in the soil, and upon their death, the nitrogen is available to plants growing in that soil.

O

❑ objectives

Intermediate-term targets necessary for the satisfaction of goals; quantifiable measures that serve as indicators against which attainment, or progress toward attainment, of goals can be measured.

❑ old growth

Old forests often containing several canopy layers, variety in tree sizes and species, decadent old trees, and standing and dead woody material.

❑ organic soil

Soil at least partly derived from living matter, such as decayed plant material.

❑ orv

Off-road vehicles, such as motor cycles, 4-wheel drive vehicles, and 4-wheelers.

❑ over-grazing by deer

Over grazing by deer produces significant changes to forest ecosystem structure and composition. Many grazing-sensitive species have probably been eliminated from many forest remnants, while those more tolerant to grazing (e.g., thorn-bearing taxa such as red haw, honey locust, gooseberry, blackberry) have probably become more abundant. Non-native species also tend to increase from over-grazing, such as garlic mustard, buckbrush, and poison ivy.

❑ overmature timber

Trees that have attained full development, particularly in height, and are declining in vigor, health, and soundness.

❑ overstory

The upper canopy layer; the plants below comprise the understory.

P

❑ parent material

The mineral or organic matter from which the upper layers of soil are formed.

❐ park-like structure

Stands with large scattered trees and open growing conditions, usually maintained by ground fires.

❐ partial retention

A visual quality objective which, in general, means man's activities may be evident but must remain subordinate to the characteristic landscape.

❐ patch

An area of homogeneous vegetation, in structure and composition.

❐ patch cut

A clearcut that creates small openings in a stand of trees, usually between 15 and 40 acres in size. On the Dixie National Forest and elsewhere, patchcuts are used to provide the disturbance needed to regenerate aspen.

❐ percolation

Downward flow or infiltration of water through the pores or spaces of rock or soil.

❐ perennial stream

A stream that flows throughout the year and from source to mouth.

❐ permitted grazing

Grazing on a National Forest range allotment under the terms of a grazing permit.

❐ personal use

The use of a forest product, such as firewood, for home use and not for commercial use.

❐ pesticide

A substance intended for preventing, destroying, repelling, or mitigating any pest. Also, any substance or mixture of substances intended for use as a plant regulator, defoliant, or desiccant.

❐ phosphorus

An essential chemical food element that can contribute to the eutrophication of lakes and other water bodies.

❐ photosynthesis

A process occurring in the cells of green plants and some micro-organisms in which solar energy is transformed into stored chemical energy.

❐ phytoplankton

That portion of the plankton community comprising tiny plants (e.g., algae, diatoms).

❑ plankton

Microscopic plants and animals that live in water.

❑ planning area

The area of National Forest land covered by a Regional Guide or Forest Plan.

❑ planning period

The 50 year time frame for which goods, services, and effects were projected in the development of the Forest Plan.

❑ point source

A stationary facility from which pollutants are discharged or emitted. Also, any single identifiable source of pollution (e.g., a pipe, ditch, ship, ore pit, factory smokestack).

❑ pole timber

Trees at least 5 inches in diameter, but smaller than the minimum size for sawtimber.

❑ pole/sapling

The stage of forest succession in which trees are between 3 and 7 inches in diameter and are the dominant vegetation.

❑ pollutant

Any substance introduced into the environment that adversely affects the usefulness of a resource.

❑ pollution prevention

Measures taken to reduce the generation of a substance that could be harmful to living organisms if released to the environment. Pollution prevention can be achieved in many ways.

❑ precommercial thinning

Removing some of the trees from a stand that are too small to be sold for lumber or house logs, so the remaining trees will grow faster.

❑ predator

An animal that lives by preying on other animals. Predators are at or near the tops of food chains.

❑ predator

Any organism that lives by capturing and feeding on another animal.

❑ pre-existing use

Land use that may not conform to a zoning ordinance but existed prior to the enactment of the ordinance.

❐ preparatory cut

The removal of trees near the end of a rotation to open the canopy so the crowns of seed bearing trees can enlarge. This improves seed production and encourages natural regeneration.

❐ prescribed fire

Fire set intentionally in wild land fuels under prescribed conditions and circumstances. Prescribed fire can rejuvenate forage for livestock and wildlife or prepare sites for natural regeneration of trees.

❐ prescription

Management practices selected to accomplish specific land and resource management objectives.

❐ present net value (pnv), also called present net worth

The measure of the economic value of a project when costs and revenues occur in different time periods. Future revenues and costs are "discounted " to the present by an interest rate that reflects the changing value of a dollar over time. The assumption is that dollars today are more valuable than dollars in the future. PNV is used to compare project alternatives that have different cost and revenue flows.

❐ presuppression

Activities carried out in advance of fire occurrence to ensure effective suppression when the need arises.

❐ pretreatment

Processes used to reduce, eliminate, or alter pollutants from nonresidential sources before they are discharged into publicly owned sewage treatment systems.

❐ prevention over mitigation or restoration

The importance of protecting existing, relatively intact biological systems cannot be overstated, for it is these systems that will ultimately provide the biota and other natural materials for future restorations.

❐ primary waste treatment

This treatment consists of the first steps in wastewater treatment during which screens and sedimentation tanks are used to remove most materials that float or will settle. Primary treatment results in the removal of about 30 percent of carbonaceous biochemical oxygen demand from domestic sewage.

❐ primitive ros (recreation opportunity spectrum)

A classification of wilderness and recreation opportunity. It is characterized by an essentially unmodified environment, where trails may be present but structures are rare, and where it is highly probable to be isolated from the sights and sounds of people.

❏ productive

The ability of an area to provide goods and services and to sustain ecological values.

❏ prognosis

A computer model for timber growth and yield. It projects per-acre growth and volume yield for commercial timber stands.

❏ public domain

The territory ceded to the Federal government by the original thirteen states, plus additions by treaty, cession, and purchase.

❏ public involvement

The use of appropriate procedures to inform the public, obtain early and continuing public participation, and consider the views of interested parties in planning and decision making.

❏ public land

Land for which title and control rests with a government---Federal, state, regional, county, or municipal.

R

❏ range

Land on which the principle natural plant cover is composed of native grasses, forbs, and shrubs that are valuable as forage for livestock and big game.

❏ range management

The art and science of planning and directing range use intended to yield the sustained maximum animal production and perpetuation of the natural resources.

❏ range of variability (also called the historic range of variability or natural range of variation.)

The components of healthy ecosystems fluctuate over time. The range of sustainable conditions in an ecosystem is determined by time, processes (such as fire), native species, and the land itself. For instance, ecosystems that have a 10 year fire cycle have a narrower range of variation than ecosystems with 200-300 year fire cycle. Past management has placed some ecosystems outside their range of variability. Future management should move such ecosystems back toward their natural, sustainable range of variation.

❏ ranger district

The administrative sub-unit of a National Forest that is supervised by a District Ranger who reports directly to the Forest Supervisor.

▫ raptor

A bird of prey, such as a eagle or hawk.

▫ rare ii

Roadless Area Review and Evaluation. The national inventory of roadless and undeveloped areas within the National Forests and Grasslands.

▫ recharge

The addition of water to ground water by natural or artificial processes.

▫ reforestation

The restocking of an area with forest trees, by either natural or artificial means, such as planting.

▫ regeneration

The renewal of a tree crop by either natural or artificial means. The term is also used to refer to the young crop itself.

▫ regional forester

The official of the USDA Forest Service responsible for administering an entire region of the Forest Service.

▫ release cutting

Removal of competing vegetation to allow desired tree species to grow.

▫ removal cut

The removal of the last seed bearers or shelter trees after regeneration is established.

▫ repairing damaged lands

Is repair possible?

- Although erosion is difficult to remedy once it has occurred, it is still possible to halt its progress by repairing much of the damage originally done to the land by people.
- The extent of possible repair depends on the area of land to be dealt with, and the level of resources made available, such as time, money, manpower, and materials.

Methods of repair

Each type of damage has its own causes, so the appropriate method of repair must be chosen.

Soil stability

- Deep soil on sloping land which is susceptible to landslides needs to be stabilized to prevent gravity and water from moving it.
- Replanting the soil with trees and shrubs is the best solution.
- Stapling the soil with stakes can hold the soil while the roots of the plants grow.

- On level areas, where wind erosion affects the soil, planting grasses is a good method of prevention.
- Succession is the term applied to allowing plants to naturally colonize an area of land.

Physical prevention

- Drainage channels can be dug to divert ground water from causing erosion.
- Wind breaks in the form of walls, hedges, or stands of trees can all help to prevent wind erosion.
- Covering soil temporarily with weatherproof sheeting can give plants time to take a hold.

1

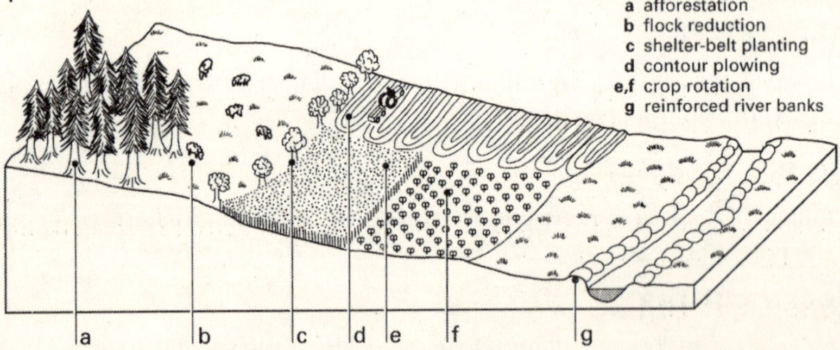

1 Soil reclamation on eroded lands
a afforestation
b flock reduction
c shelter-belt planting
d contour plowing
e,f crop rotation
g reinforced river banks

2

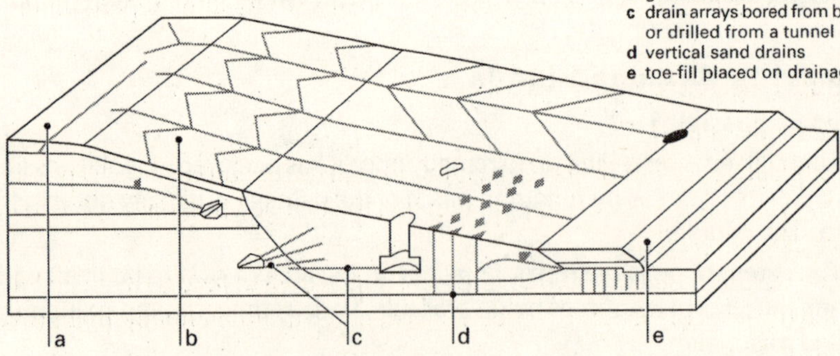

2 Drains can be installed in a variety of ways to stabilize slopes
a cut-off drain at slope crest
b ground surfaces seeded or planted
c drain arrays bored from base of shaft, or drilled from a tunnel
d vertical sand drains
e toe-fill placed on drainage blanket

Soil thickness

- Eroded soils can be replaced by importing soil from other places and spreading it over the land.
- This soil must be stabilized quickly though to prevent erosion from recurring.

Management

- Any repairs will be short-lived if land management techniques are not changed.
- Rotation systems for arable farming, grazing, and timber felling are often effective at preventing a recurrence of erosion.

- Practices such as limited use of resources can avoid the problem of land becoming vulnerable to the effects of erosion because the land's own defences are maintained.

❑ residual stand

The trees remaining standing after an event such as selection cutting.

❑ resilience

The ability of an ecosystem to maintain diversity, integrity, and ecological processes following a disturbance.

❑ responsible official

The Forest Service employee who has been delegated the authority to carry out a specific planning action.

❑ restoration (of ecosystems)

Actions taken to modify an ecosystem to achieve a desired, healthy, and functioning condition.

❑ resuspension (of sediment)

The remixing of sediment particles and pollutants back into the water by storms, currents, organisms, and human activities, such as dredging.

❑ retention time

The time it takes for the volume of water in a lake to exit through its outlet (i.e., total volume/outlet flow = retention time).

❑ revegetation

The re-establishment and development of a plant cover by either natural or artificial means, such as re-seeding.

❑ riparian area

The area along a watercourse or around a lake or pond.

❑ riparian ecosystem

The ecosystems around or next to water areas that support unique vegetation and animal communities as a result of the influence of water.

❑ risk assessment

qualitative and quantitative evaluation to define the hazards posed to human health and/or the environment.

❑ rod

Record of Decision. A official document in which a deciding official states the, alternative that will be implemented from a prepared EIS.

❑ ros

Recreation Opportunity Spectrum. The land classification system that categorizes land by its setting and the probable recreation experiences and activities it affords.

❒ rotation

The number of years required to establish and grow timber crops to a specified condition of maturity.

❒ roundwood

Timber and fuelwood prepared in the round state, such as house logs and telephone poles.

❒ run-off

The portion of precipitation that flows over the land surface or in open channels; That part of precipitation, snow melt, or irrigation water that drains off land into surface water. It can carry sediments and pollutants into the receiving waters.

S

❒ sacrifice area/site

In range management, a site allowed to be overgrazed to obtain efficient overall use of the management area. In cultural resource management, it may refer to a site intentionally sacrificed to extensive public use in order to preserve the larger cultural area.

❒ sanitation salvage

The removal of dead, damaged or susceptible trees primarily to prevent the spread of pests or disease and promote forest health.

❒ sapling

A loose term for a young tree more than a few feet tall and an inch or so in diameter that is typically growing vigorously.

❒ sawtimber

Trees that are 9 inches in diameter at breast height or larger that can be made into lumber.

❒ scale

In ecosystem management, it refers to the degree of resolution at which ecosystems are observed and measured.

❒ scale

A unit of measure.

❒ scoping

The ongoing process to determine public opinion, receive comments and suggestions, and determine issues during the environmental analysis process. It may involve public meetings, telephone conversations, or letters.

❒ second growth

Forest growth that was established after some kind of interference with the previous forest crop, such as cutting, fire, or insect attack.

❐ secondary waste treatment

The second step in most waste treatment systems in which bacteria consume the organic parts of the waste. It is accomplished by bringing together waste, bacteria, and oxygen in trickling filters or in the activated sludge process. This removes floating and settleable solids and about 90 percent of the oxygen-demanding substances and suspended solids. Disinfection is the final stage of secondary treatment.

❐ sediments

Soil, sand, and minerals eroded from land by water or air. Sediments settle to the bottom of surface water.

❐ seed tree harvest

Removal of the mature timber crop from an area in one cut, except for a certain number of seed bearers.

❐ sensitive species

Plant or animal species which are susceptible to habitat changes or impacts from activities. The official designation is made by the USDA Forest Service at the Regional level and is not part of the designation of Threatened or Endangered Species made by the US Fish and Wildlife Service.

❐ seral

The stage of succession of a plant or animal community that is transitional. If left alone, the seral stage will give way to another plant or animal community that represents a further stage of succession.

❐ sewage

The waste and wastewater discharged into sewers from homes and industry.

❐ sewer

A channel or conduit that carries wastewater and stormwater runoff from its source to a treatment plant or receiving stream. Sanitary sewers carry household, industrial, and commercial waste; storm sewers carry runoff from rain or snow; and combined sewers carry both.

❐ shelterwood

A cutting method used in a more or less mature stand, designed to establish a new crop under the protection of the old.

❐ silvicultural system

The cultivation of forests; the result is a forest of a distinct form. Silvicultural systems are classified according to harvest and regeneration methods and the type of forest that results.

❐ silviculture

The art and science that promotes the growth of single trees and the forest as a biological unit.

❑ site preparation

The general term for removing unwanted vegetation, slash, roots, and stones from a site before reforestation. Naturally occurring wildfire, as well as prescribed fire can prepare a site for natural regeneration.

❑ size class

One of the three intervals of tree stem diameters used to classify timber in the Forest Plan data base. The size classes are: Seedling/Sapling (less than 5 inches in diameter); Pole Timber (5 to 7 inches in diameter); Sawtimber (greater than 7 inches in diameter)

❑ skidding

Hauling logs by sliding, not on wheels, from stump to a collection point.

❑ skyline logging

A logging system used to remove timber from steep slopes. Logs are brought up-slope on a suspended cable, or skyline. Since the weight of the log is completely or partially supported by the cable, there is little disturbance to soil or other vegetation.

❑ slash

The residue left on the ground after timber cutting or left after a storm, fire, or other event. Slash includes unused logs, uprooted stumps, broken or uprooted stems, branches, bark, etc.

❑ slump

A landslide where the underlying rock masses tilt back as they slide from a cliff or escarpment.

❑ small game

Birds and small animals normally hunted or trapped.

❑ snag

A standing dead tree. Snags are important as habitat for a variety of wildlife species and their prey.

❑ soil compaction

The reduction of soil volume. For instance, the weight of heavy equipment on soils can compact the soil and thereby change it in some ways, such as in its ability to absorb water.

❑ soil degradation by continent

Around the world

- Soil degradation is a common feature of landscapes in developing and poor countries.
- The combination of poverty and climate result in soils being exhausted of nutrients and minerals.

- They also become contaminated with pollutants.
- Degradation of soils does occur in other places often as a result of intensive farming methods or where population densities are very high.
- The soils become exhausted of nutrients because they are used too often for growing crops and never have the chance to recover.
- Pollutants come in the form of high concentrations of chemicals that are not used by growing crops, so that the soil becomes infertile.
- Additionally, watering by irrigation tends to bring salts to the surface of the soil, compounding the problem by making the soil sterile.

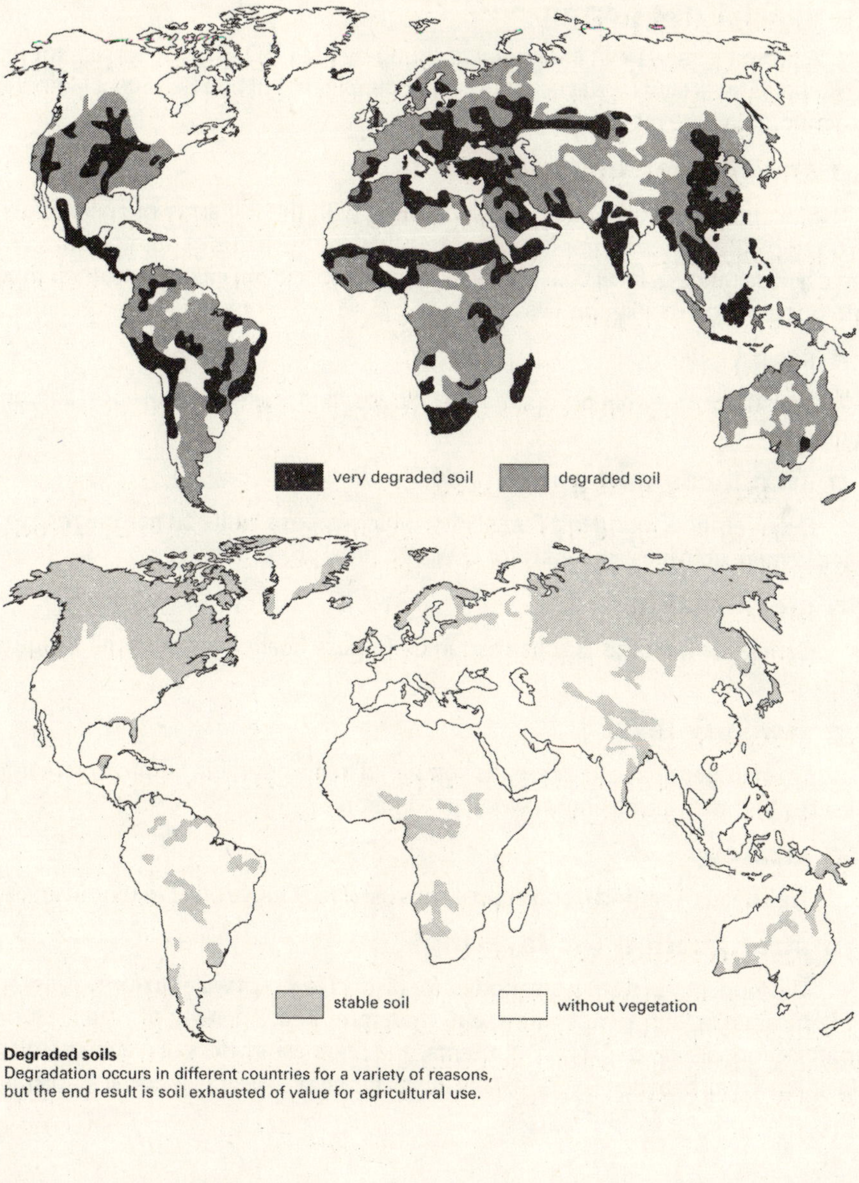

Degraded soils
Degradation occurs in different countries for a variety of reasons, but the end result is soil exhausted of value for agricultural use.

❒ soil productivity

The capacity of a soil to produce a specific crop. Productivity depends on adequate moisture and soil nutrients, as well as favorable climate.

❒ sound wood

Timber that is in solid, whole, good condition. Sound wood is free from damage, decay, or defects.

❒ spatial scales

Consideration of appropriate planning unit (i.e., legal, political, ecological, watershed, ownership boundaries, etc).

❒ special use permit

A permit issued to an individual or group by the USDA Forest Service for use of National Forest land for a special purpose. Examples might be a Boy Scout Jamboree or a mountain bike race.

❒ stakeholder involvement

Stakeholder involvement is collaborating with the full array of stakeholders affected by the issues and/or opportunities that are the focus of our activities. We serve a wide range of local constituencies and will solicit and incorporate their input into our decision making process.

❒ stand

A group of trees that occupies a specific area and is similar in species, age, and condition.

❒ standards and guidelines

Requirements found in a Forest Plan which impose limits on natural resource management activities, generally for environmental protection.

❒ stewardship

Caring for the land and its resources to pass healthy ecosystems to future generations.

❒ stocking level

The number of tree in an area as compared to the desirable number of trees for best results, such as maximum wood production.

❒ strategy

The means, methods, and approaches used to achieve objectives and goals.

❒ stratification (or layering)

The tendency in deep water bodies for distinct layers of water to form as a result of vertical change in temperature and, therefore, in the density of water. During stratification, dissolved oxygen, nutrients, and other parameters of water chemistry do not mix well between layers, establishing chemical as well as thermal gradients.

❑ **stringer**

A strip of vegetation different from surrounding vegetation, such as a stringer of aspen in a area of spruce.

❑ **structure**

How the parts of ecosystems are arranged, both horizontally and vertically. Structure might reveal a pattern, or mosaic, or total randomness of vegetation.

❑ **succession**

The natural replacement, in time, of one plant community with another. Conditions of the prior plant community (or successional stage) create conditions that are favorable for the establishment of the next stage.

❑ **successional stage**

A stage of development of a plant community as it moves from bare ground to climax. The grass-forb stage of succession precedes the woody shrub stage.

❑ **suitability**

The appropriateness of certain resource management to an area of land. Suitability can be determined by environmental and economic analysis of management practices.

❑ **surface resources**

Renewable resources that are on the surface of the earth, such as timber and forage, in contrast to ground water and minerals which are located beneath the surface.

❑ **surface water**

All water open to the atmosphere (e.g., rivers, lakes, reservoirs, streams, impoundments, seas, estuaries) and all springs, wells, or other collectors that are directly influenced by surface water.

❑ **sustainability**

The ability of an ecosystem to maintain ecological processes and functions, biological diversity, and productivity over time.

❑ **sustainable**

The yield of a natural resource that can be produced continually at a given intensity of management is said to be sustainable.

❑ **sustained yield**

The yield that a renewable resource can produce continuously at a given intensity of management.

❑ **swamp**

A type of wetland that is dominated by woody vegetation and that does not accumulate appreciable peat deposits. Swamps may be freshwater or saltwater and tidal or nontidal.

T

❐ target

A National Forest's annual goals for accomplishment for natural resource programs. Targets represent the commitment the Forest Service has with Congress to accomplish the work Congress has funded, and are often used as a measure of the agency's performance.

❐ temporal scales

Consideration of future and historical perspectives of time. Typically, future perspectives might involve time horizons from 20-50 years, while historical perspectives might focus on time intervals of 100-400 years.

❐ the burning of amazônia

Fire and logging

- When trees are felled in the Amazon, the unwanted vegetation and offcut is burned.
- Fires lit for this purpose often get out of hand and burn large areas of virgin forest.
- The result is significant, unnecessary damage to the tropical forest.
- This damage is measured by forest lost and the fragmentation it causes.
- Estimates for the area of forest damaged each year vary between 6,500–23,000 square miles.

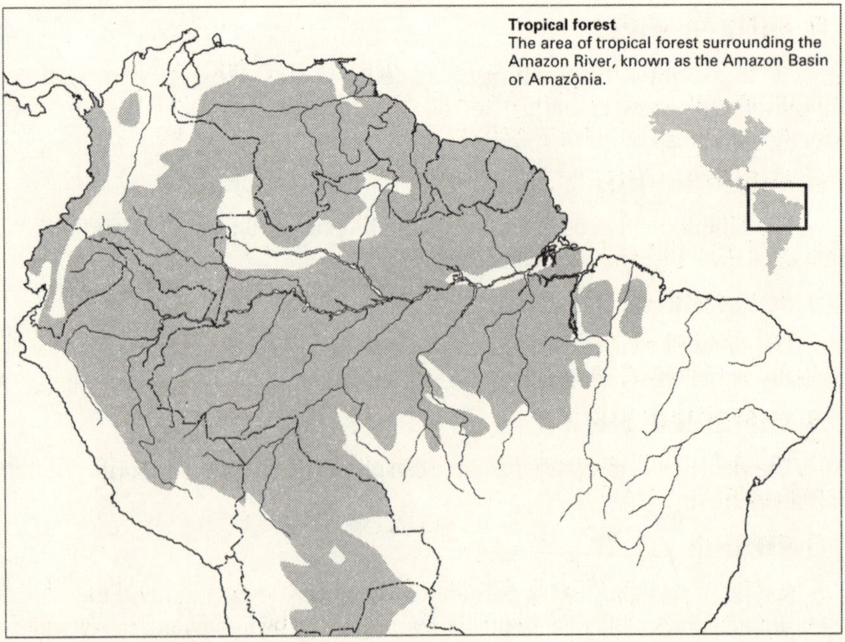

Protecting Amazônia
The Amazon Basin lies over five countries, which all share the responsibility for protecting the tropical forest.

Country and area of forest

1	Brazil	1.7 million sq.miles
2	Peru	310,000 sq.miles
3	Bolivia	270,000 sq.miles
4	Venezuela	232,000 sq.miles
5	Colombia	193,000 sq.miles

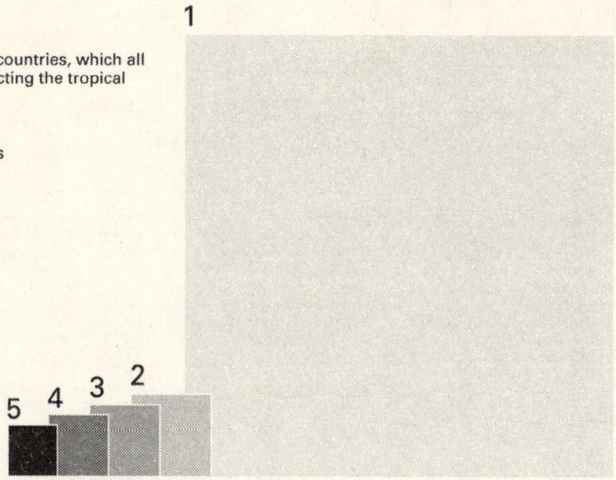

❐ the causes of wetland loss

Land use

- Wetlands are usually found in lowland regions where more land is needed by people.
- People drain the water away so that the land becomes dry.
- The land is then suitable for settlement and farming.
- This is called land reclamation, because the land is reclaimed from nature.

Global distribution of wetlands at end of 20th century

Causes of wetland loss

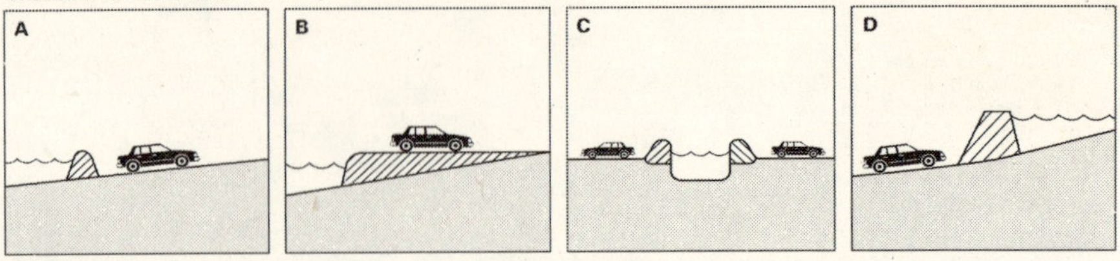

A Polders
- When coastal wetlands, marshes, bogs, lakes, or river beds are reclaimed the areas are called polders.
- Dams are used to keep the water out and pumps are used to drain the wetland.

B Landfill
- Sometimes fill is used to raise the level of wetland so that it naturally drains.
- Networks of ditches are also used to improve the drainage of this land.

C Levees
- Flood plains of rivers are prevented from becoming waterlogged with levees.
- Levees are ridges built up on either side of the river to contain any rise in water level.

D Dams
- Building reservoirs and hydroelectric dams can also result in the loss of wetland.
- The flow of the river beneath is decreased so any wetland areas dry up.
- Also some wetland areas simply become engulfed by the body of water held behind the dam.

❑ the consequences of forest loss

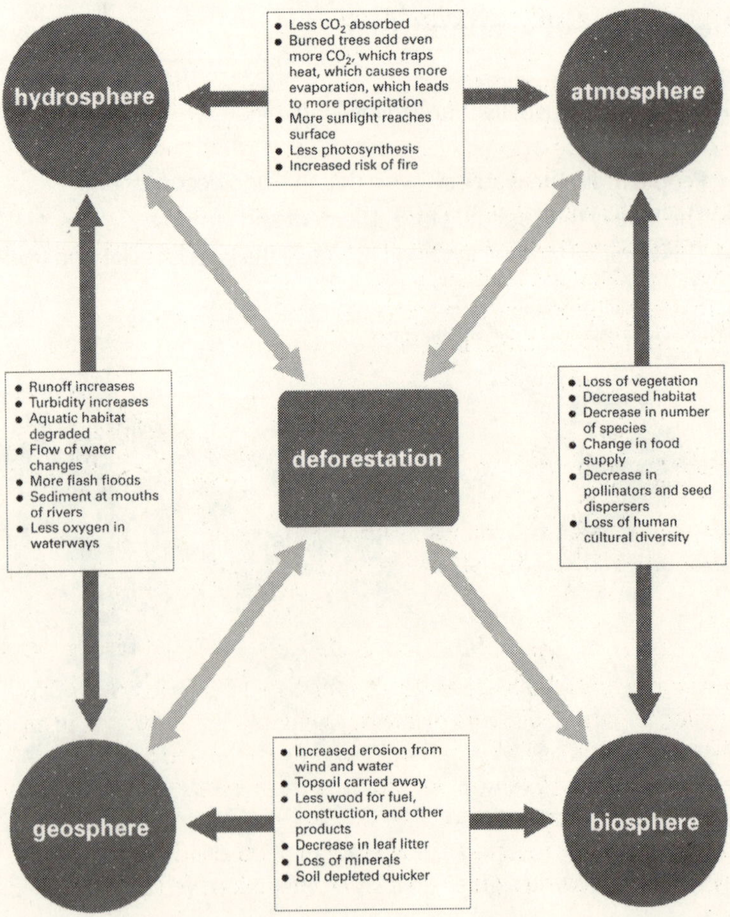

- When trees are removed, topsoil is lost to wind and water erosion.
- Surface evaporation increases, reducing the moisture content of the soil.
- The nutrient content of the soil is also decreased.
- Forests act as a sponge, and their removal increases rainwater runoff.
- This in turn leads to increased probability of floods and droughts.
- Deforestation is the primary cause of loss of biodiversity, as habitats for many species are lost.
- Depletion of tree stocks reduces the replenishment of atmospheric oxygen.
- The burning of forests contributes to greater atmospheric levels of greenhouse gases (up to 30 percent of the global release of CO_2).

□ the disappearing mangroves

- Mangrove forests are distributed throughout the world's tropical and subtropical ocean coastlines.
- They form the basis of unique and rich ecosystems, but are threatened by many human activities.
- Dredging, land reclamation, clearance for agriculture, shrimp ponds, and salt farming all take their toll.

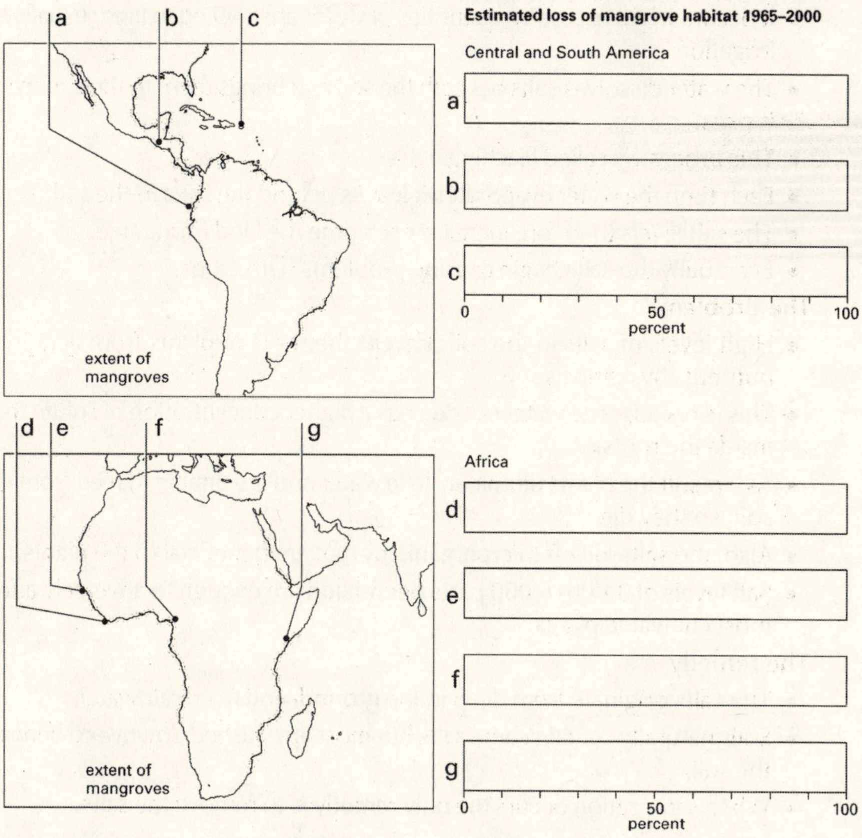

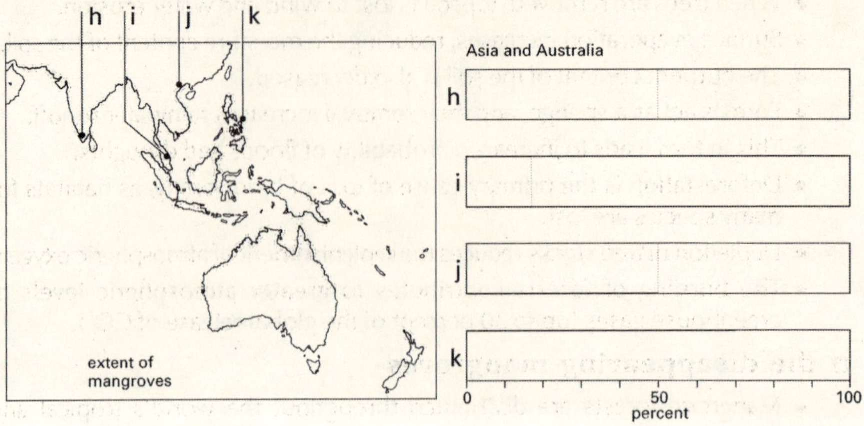

▫ the salinization problem

Salinization

- The word 'saline' indicates high levels of salts.
- Salinization is the buildup of salts in soil.
- It occurs when excessive quantities of water are applied to land, usually by irrigation.
- The water dissolves salts beneath the soil and brings them to the surface as it rises.
- This process is called leaching.
- Each time the water evaporates it leaves behind the salts in the soil.
- The salt levels therefore increase each time the land is irrigated.
- Eventually the salts begin causing problems with crops.

The problems

- High levels of salts in the soil prevent the roots of plants from acquiring nutrients by osmosis.
- This is because the water outside has a higher concentration of solute than inside the roots.
- As a result the plants ultimately lose water and are unable to feed from the soil, so they die.
- Also, the salts kill off microorganisms that are beneficial to the plants.
- Salt levels of 3,000–6,000 parts per million are enough to adversely affect most cultivated plants.

The remedy

- The salts originate from deep in the ground, and from rainwater.
- Soils naturally contain some salts but most are washed downward beneath the soil.
- When salinization occurs the only remedy is to remove the salts.

- The method is to flush the soil with even more water so that the salts are transported away.
- This requires very large amounts of fresh water, and the salts can then cause problems in rivers and lakes where the water drains.
- Desalinization factories are therefore required in some places to remove the salts from this water.
- Globally about 20 percent of irrigated farmland is affected by salinization to some degree.
- Crop yields from irrigated land in the United States may be lowered by 25–30 percent due to salinization.

Regional distribution of salt-affected soils
(total area of saline soils = 980.5 million acres worldwide)

The salinization problem

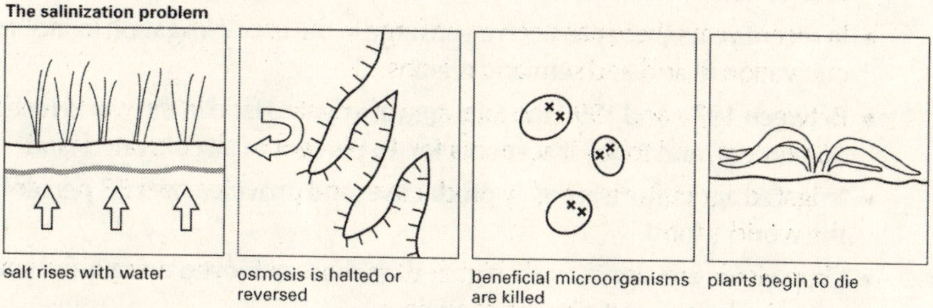

❒ thermal cover

Cover used by animals against weather. For elk, thermal cover can be found in a stand of coniferous trees at least 40 feet tall with a crown closure of at least 70%.

❒ thinning

A cutting made in an immature stand of trees to accelerate growth of the remaining trees or to improve the form of the remaining trees.

❒ threatened species

Those plant or animal species likely to become endangered throughout all or a specific portion of their range within the foreseeable future as designated by the U.S. Fish and Wildlife Service under the Endangered Species Act of 1973.

❒ timber classification

The classification of forested lands into land management alternatives according to how the land relates to management of the timber resource there.

❒ toxic substance (or toxicant)

A substance that can cause death, disease, behavioral abnormalities, cancer, genetic mutations, physiological or reproductive malfunctions, or physical deformities in any organism or its offspring. The quantities and length of exposure necessary to cause these effects can vary widely.

❒ tractor logging

A logging method that uses tractors to carry or drag logs from the stump to a collection point.

❒ treatment area

The site specific location of a resource improvement activity.

❒ tree opening

An opening in the forest created by even-aged silvicultural practices.

❒ trends in irrigation

Irrigation

- Irrigation involves the conveyance of water from a source of supply to an area of land where it is needed for the cultivation of crops.
- In recent years there has been an upsurge in the use of irrigation to facilitate cultivation in arid and semiarid regions.
- Between 1970 and 1990 the total area of irrigated land in the world rose by 36 percent, and today it accounts for 16 percent of all cultivated land.
- Irrigated agriculture is highly productive, and provides over 33 percent of the world's food.
- Crop yields are significantly higher than those achieved on rain-fed lands with similar soil and climatic conditions.
- It is unlikely that there will be a significant expansion of irrigated land in the future, mainly due to a shortage of suitable water.
- The current emphasis is on increasing the efficiency of irrigation methods, which due to leakage often lose up to 60 percent of the water.

Sources of water used in irrigation

- In semi-arid and arid regions irrigation systems rely on the import of surface water, often over large distances. Water is drawn from rivers and reservoirs in surrounding areas where rainfall is high, and transported to the irrigated area usually by pipeline.

- Groundwater may be a suitable source of supply depending upon availability, and the cost of pumping it to the surface. However, groundwater abstraction requires careful management as overexploitation can lead to serious problems of depletion. This is particularly serious in coastal regions where a drop in the level of the water table below that of the sea can result in the incursion of salty water into groundwater aquifers.
- About 70 percent of all fresh water withdrawn from rivers, lakes, and groundwater supplies is used for irrigation.

Irrigated land as a percentage of agricultural land 1965–2001

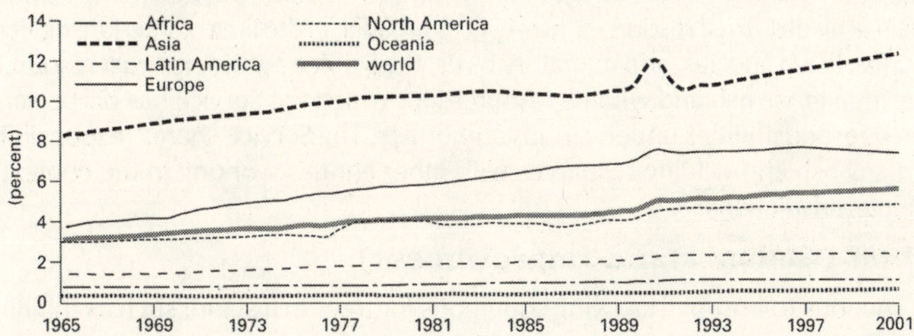

Irrigated land as a percentage of cropland at the end of the 20th century

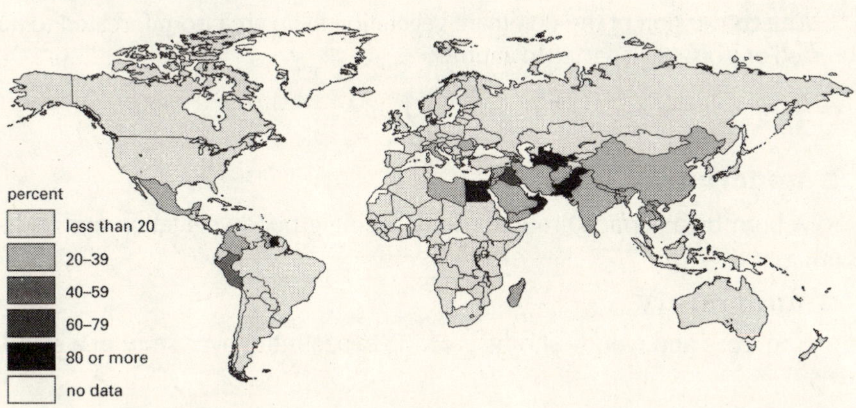

☐ trust resources

When the Service uses the term "trust resources," we are talking about the Service's responsibilities as defined by legislation, treaty, or similar authority. The responsibility for the conservation of fish and wildlife is a shared responsibility, and the degree to which the Service is involved will varies with species and situations. A number of Congressional enactments and court decisions have addressed State and Federal responsibilities for fish and wildlife with the general effect of expanding

Federal jurisdiction over certain species and uses of fish and wildlife traditionally managed by the States. Congress has charged the Secretary of the Interior with responsibilities for the management of certain fish and wildlife resources, e.g., endangered and threatened species, migratory birds, certain marine mammals, and certain aspects of the management of some fish. Even in these specific instances, with the limited exception of marine mammals, State jurisdiction remains concurrent with Federal authority. In general, the States possess broad trustee and police powers over fish and wildlife within their borders. The Service has not officially defined the term in agency guidance. The Service's use of the term "trust resources" includes but goes beyond certain explicit trust responsibilities, such as the federal government's fiduciary responsibility for Indian trust resources and the Secretary of the Interior's responsibility to act as trustee for certain resources in litigation under environmental restoration statutes. These "trust responsibilities" require the Service to conduct certain activities. Trust resources, however, is broader. In the area of interjurisdictional fisheries, listed species, and migratory birds, the Service considers "trust resources" to include those fish and wildlife resources for which the Service has particular or shared responsibilities under various authorities. The Service shares responsibility for many fish and wildlife resources with other entities working in the context of multiple jurisdictions.

□ tsi (timber stand improvement)

Actions to improve growing conditions for trees in a stand, such as thinning, pruning, prescribed fire, or release cutting.

□ type conversion

The conversion of the dominant vegetation in an area from forested to non-forested or from one species to another.

U

□ underburn

A burn by a surface fire that can consume ground vegetation and "ladder" fuels.

□ understory

The trees and woody shrubs growing beneath the over story in a stand of trees.

□ uneven-aged management

Actions that maintain a forest or stand of trees composed of intermingling trees that differ markedly in age. Cutting methods that develop and maintain uneven-aged stands are single-tree selection and group selection.

□ unregulated harvest

Tree harvest that is not part of the allowable sale quantity (ASQ). It can include the removal of cull or dead material or non-commercial species. It also includes volume removed from non-suitable areas for research, to meet objectives other than timber production (such as wildlife habitat improvement), or to improve administrative sites (such as campgrounds.)

❐ unsuitable lands

Forest land that is not managed for timber production. Reasons may be matters of policy, ecology, technology, silviculture, or economics

❐ urban runoff

Stormwater from city streets and adjacent domestic or commercial properties that may pickup terrestrial contamination and carry pollutants of various kinds into sewer systems and/or receiving waters.

❐ use, allowable

An estimate of proper range use. Forty to fifty percent of the annual growth is often used as a rule of thumb on ranges in good to excellent condition. It can also mean the amount of forage planned to be used to accelerate range rehabilitation.

V

❐ vaporization

The change of a substance from a liquid to a gas.

❐ variety class

A way to classify landscapes according to their visual features. This system is based on the premise that landscapes with the greatest variety or diversity have the greatest potential for scenic value.

❐ vegetation management

Activities designed primarily to promote the health of forest vegetation for multiple-use purposes.

❐ vegetation type

A plant community with distinguishable characteristics.

❐ vertical diversity

The diversity in a stand that results from the different layers or tiers of vegetation.

❐ viable population

The number of individuals of a species sufficient to ensure the long-term existence of the species in natural, self-sustaining populations that are adequately distributed throughout their range.

❐ viable population

A viable population is one which has such numbers and distribution of reproductive individuals as to provide a high likelihood that a species will continue to exist and be well-distributed throughout its natural range.

❐ virgin forest

A natural forest virtually uninfluenced by human activity.

❐ visual quality objective

A set of measurable goals for the management of forest visual resources.

❐ visual resource

A part of the landscape important for its scenic quality. It may include a composite of terrain, geologic features, or vegetation

❐ volatile substance

A substance that evaporates readily.

W

❐ waste treatment plant

A facility containing a series of tanks, screens, filters, and other processes by which pollutants are removed from water.

❐ wastewater

The spent or used water from individual homes, a community, a farm, or an industry that often contains dissolved or suspended matter.

❐ water and wind erosion

- Water and wind are the principal agents of erosion.
- Their effect on a landscape can be initiated or accelerated by human actions that make the land vulnerable to erosion.

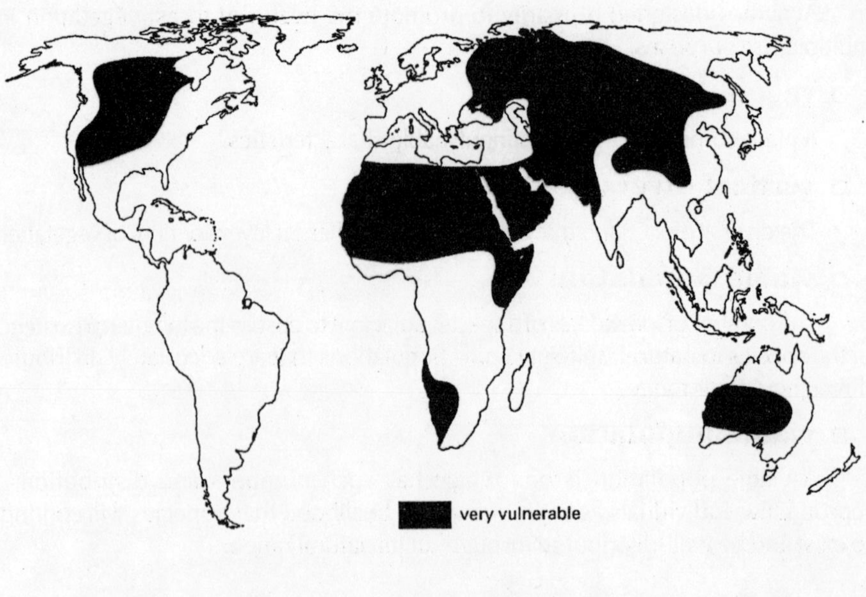

Regions at risk of wind erosion

The Sahara, for example, is the largest desert on Earth, and is continually expanding because of wind erosion.

- Deserts result when the soil covering an area of land is stripped of its vegetation.
- Without the roots of plants to hold it in place, soil turns to dust if it dries out and then the wind blows it away, leaving only sand.
- Overgrazing and poor farming methods cause deserts to grow larger.

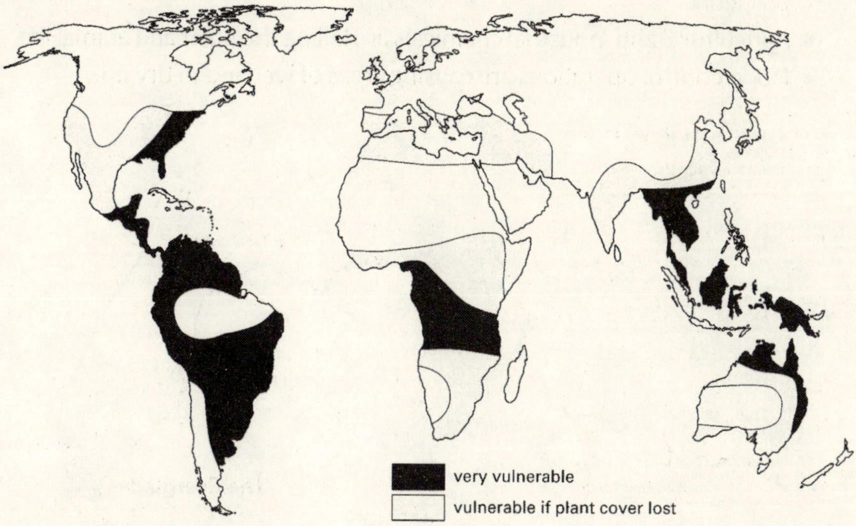

❒ water table

The upper surface of groundwater. Below it, the soil is saturated with water; The level of groundwater.

❒ water yield

The runoff from a watershed, including groundwater outflow.

❒ watershed

The entire region drained by a waterway (or into a lake or reservoir. More specifically, a watershed is an area of land above a given point on a stream that contributes water to the stream flow at that point; The drainage basin contributing water, organic matter, dissolved nutrients, and sediments to a water body; The land area that drains into a river, stream, reservoir or lake.

❒ wetland

An area that is regularly saturated by surface water or groundwater and is characterized by a prevalence of vegetation that is adapted for life in saturated soil conditions (e.g., swamps, bogs, fens, marshes, and estuaries).

□ wetland losses in selected us states

Everglades

- The state of Florida is dominated by an area of wetland known as the Everglades.
- Much of it is being drained and reclaimed so the land can be put to other uses.
- Farming and urban development are the main uses for the reclaimed land.
- A significant proportion of the Everglades wetland is also being lost to pollution.
- Agricultural and industrial chemicals are killing the plant and animal life.
- Flood control operations are causing areas of wetland to dry up.

Areas of wetland in the United State
a The Everglades
b Yellowstone National Park

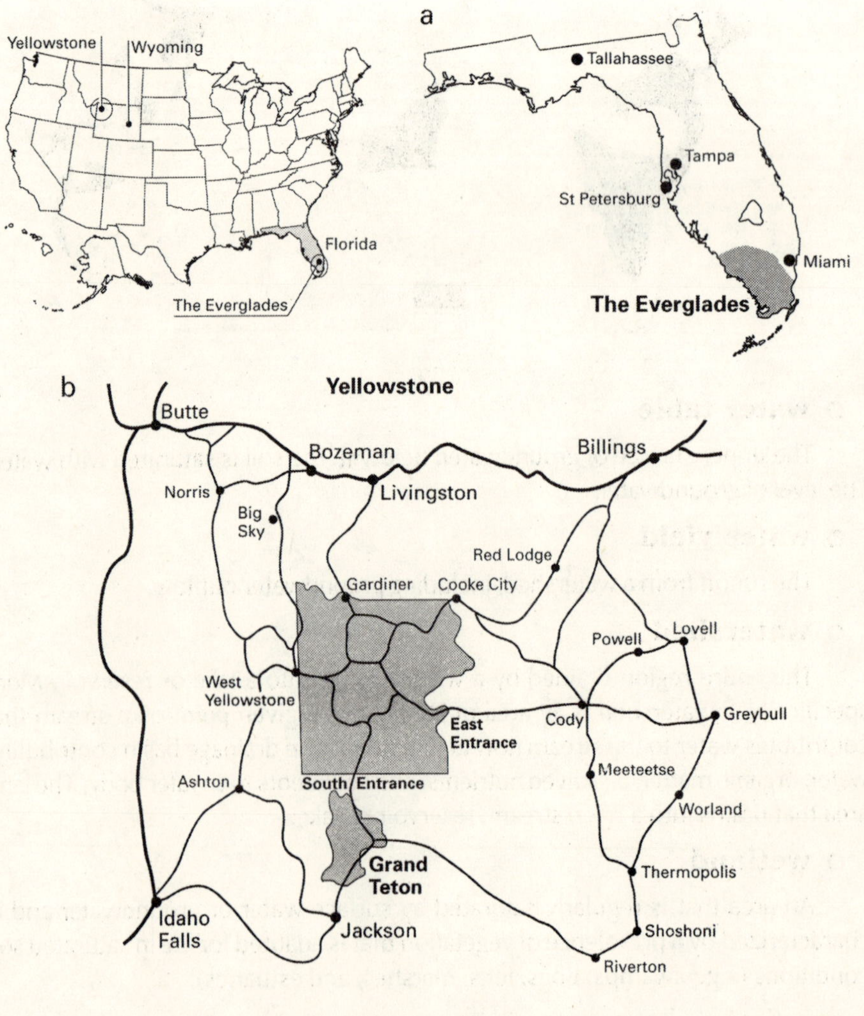

Yellowstone

- The states of Wyoming, Montana, and Idaho all contain a part of Yellowstone National Park.
- Wetland areas in Yellowstone include rivers, lake fringes, and thermal springs.
- Water pollution is a threat to these habitats in some areas.
- The pollution comes from contamination by sewage, waste, and mining operations.

❑ wetlands

Areas that are permanently wet or are intermittently covered with water.

❑ why mangroves matter

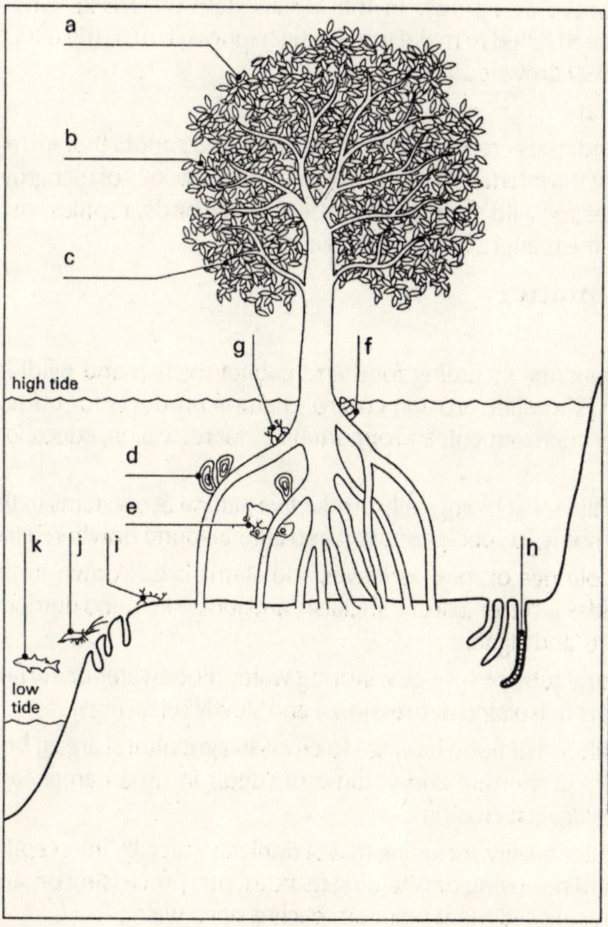

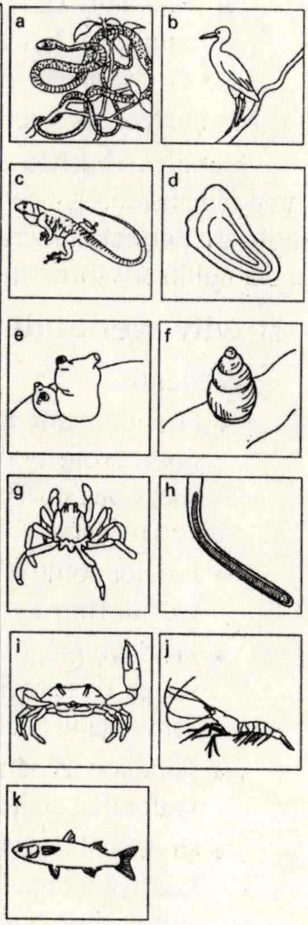

a tree snake
b egret
c iguana
d oyster
e tunicate
f periwinkle
g mangrove crab
h polychaete worm
i fiddler crab
j prawns
k mullet

- Over the past century 62 million acres of mangrove forest are estimated to have been destroyed or grossly degraded.
- Most tropical countries have lost more than half of their mangrove forests to logging, urban development, or conversion to rice fields, cropland, or fish-farming ponds. In the Philippines, for example, the area of mangrove forest fell by 90 percent from two and a half million acres in 1960 to around 250,000 acres in 2000.
- Mangrove wetlands provide a habitat for over 2,000 species of fish, shellfish, invertebrates, and plants.
- Mangroves protect shorelines from devastation by storms. They shield the land from wind and trap sediment in their roots, maintaining a shallow slope on the seabed that absorbs tidal surges.
- Their loss can prove disastrous. In the Indian state of Orissa, where mangroves had been felled to make way for tiger-prawn farms, the effects of a cyclone in 1999 drowned 10,000 people.

Mangroves as habitats

Mangrove bushes and trees create a network of roots and shoots that form a three-dimensional lattice at the interface between land and sea. Thickets of mangrove roots are perfect nurseries for wild shrimp and fishes. Other birds, reptiles, fish, and amphibians thrive in the mangrove's branches and roots.

❑ why wetlands matter

Wetlands

- Provide many benefits, including food and habitat for fish and wildlife; flood protection; shoreline erosion control; natural products for human use; water quality improvement; and opportunities for recreation, education, and research.
- Provide some of the most biologically productive natural ecosystems in the world. They are home to species of plant and animal found nowhere else.
- Produce great volumes of food as leaves and stems break down in the water. This provides food for insects, shellfish, and forage fish, and nutrients for wetland plants and algae.
- Function as natural tubs or sponges, storing water (floodwater or surface water that collects in isolated depressions) and slowly releasing it.
- Reduce the likelihood of flood damage to crops in agricultural areas, help control increases in the rate and volume of runoff in urban areas, and buffer shorelines against erosion.
- Help improve water quality, including that of drinking water, by intercepting surface runoff and removing or retaining its nutrients, processing organic wastes, and reducing sediment before it reaches open water.

Biodiversity

- Wetlands, especially tropical, contain extremely rich biodiversities.
- They are also highly productive, and may support the ecosystem of a much wider area.

- Wetland loss can therefore have a profound effect on the ecological wellbeing of a region.

Value of wetlands

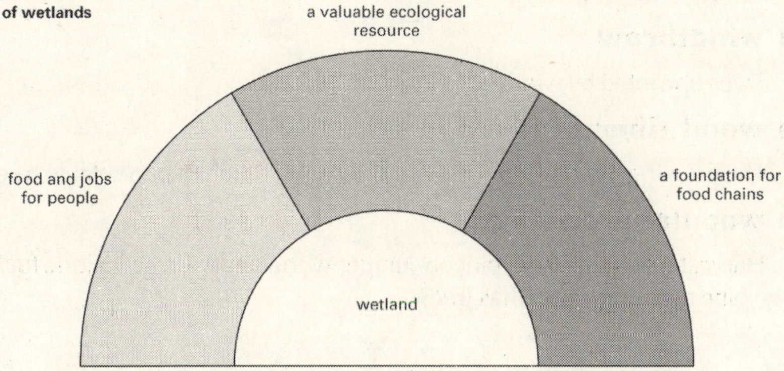

Wetlands as food provider
Wetlands usually support a rich food web. This web ranges from microscopic algae and dragonfly larvae on the smaller end of the scale to alligators and black bears on the other end.

❐ wilderness (wilderness area)

Undeveloped federal land retaining its primeval character, without permanent human habitation or improvements. It is protected and managed to preserve its natural condition. Wilderness Areas are designated by Congress.

❐ wildfire

Any wild land fire that is not a prescribed fire.

❐ wildlife habitat diversity

The distribution and abundance of different plant and animal communities and species within a specific area.

❐ wildlife refuge

An area designated for the protection of wild animals, within which hunting and fishing are either prohibited or strictly controlled.

❐ windthrow

Trees uprooted by wind.

❐ wood fiber production

The growing, tending, harvesting, and regeneration of harvestable trees.

❐ woodland products

Harvestable items from pinyon-juniper woodlands. These include fuelwood, posts, pine nuts and Christmas trees.

Y

❐ yarding

Moving the cut trees from where they fell to a centralized place (landing) for hauling away from the stand.

Z

❐ zoi (zone of influence)

The area influenced by Forest Service management activities.

❐ zooplankton

Microscopic aquatic animals.

Ecosystem Management

A

❐ abiotic

Non-living. Climate is an abiotic component of ecosystems.

❐ adaptive management

A type of natural resource management that implies making decisions as part of an on-going process. Monitoring the results of actions will provide a flow of information that may indicate the need to change a course of action. Scientific findings and the needs of society may also indicate the need to adapt resource management to new information.

❐ aerial logging

Removing logs from a timber harvest area by helicopter. Fewer roads are required, so the impact to an area is minimized.

❐ affected environment

The natural environment that exists at the present time in an area being analyzed.

❐ age class

An age grouping of trees according to an interval of years, usually 20 years. A single age class would have trees that are within 20 years of the same age, such as 1-20 years or 21-40 years.

❐ airshed

A geographic area that shares the same air.

❐ allotment (range allotment)

The area designated for use by a prescribed number of livestock for a prescribed period of time. Though an entire Ranger District may be divided into allotments, all land will not be grazed, because other uses, such as recreation or tree plantings, may be more important at a given time.

❐ allowable sale quantity

The amount of timber that may be sold within a certain time period from an area of suitable land. The suitability of the land and the time period are specified in the Forest Plan.

❐ anadromous fish

Species of fish that mature in the sea and migrate into streams to spawn.

❑ aquifer

A body of rock that is saturated with water or transmits water. When people drill wells, they tap water contained within an aquifer.

❑ aspect

The direction a slope faces. A hillside facing east has an eastern aspect.

❑ aum (animal unit month)

The quantity of forage required by one mature cow and her calf (or the equivalent, in sheep or horses, for instance) for one month.

B

❑ bark beetle

An insect that bores through the bark of forest trees to eat the inner bark and lay its eggs. Bark beetles are important killers of forest trees.

❑ basal area

The area of the cross section of a tree trunk near its base, usually 4 and 1/2 feet above the ground. Basal area is a way to measure how much of a site is occupied by trees. The term basal area is often used to describe the collective basal area of trees per acre.

❑ big game

Large mammals, such as deer, elk, and antelope that are hunted for sport.

❑ biodiversity hotspots

- Biodiversity hotspots are areas supporting a rich assortment of endemic species of plants and animals faced with exceptionally high levels of habitat loss, particularly tropical rain forest.
- They are the 25 richest and most threatened reservoirs of plant and animal life on Earth.
- These 25 hotspots, covering just 1.4 percent of earth's land area, support 44 percent of all known vascular plants and 35 percent of all mammals, birds, reptiles, and amphibians.
- British ecologist Norman Myers conceived the 'hotspots' concept in 1988 as a way to pinpoint key areas of global conservation concern requiring high levels of funding.
- Oceans, which cover 71 percent of the earth's surface, are areas of particularly high biodiversity.
- 9,000 species of fish are exploited for food and provide the primary source of animal protein for large numbers of coastal communities throughout the world.

Pressures

- In the Asia-Pacific region severe population pressures and industrialization create a high demand for land and natural resources.

- Logging, land clearance and associated forest fires have devastated areas of high biodiversity in Indonesia (Sundaland hotspot).
- Logging, road building, mining and oil exploration threaten some of world's most biologically and culturally diverse areas in the tropical Andes of South America.
- As an example, 43 species of ant were found living in a single tree in Peru.

World's 25 biodiversity hotspots

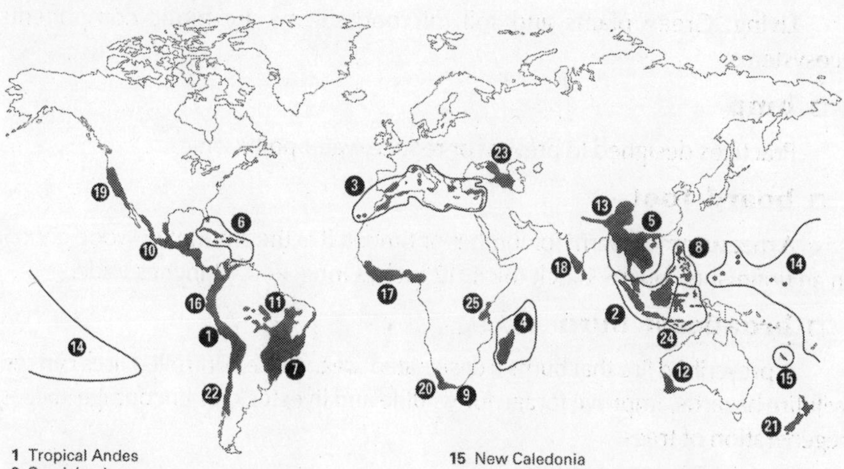

1 Tropical Andes
2 Sundaland
3 Mediterranean Basin
4 Madagascar & Indian Ocean Islands
5 Indo-Burma
6 Caribbean
7 Atlantic Forest Region
8 Philippines
9 Cape Floristic Province
10 Mesoamerica
11 Brazilian Cerrado
12 Southwest Australia
13 Mountains of south-central China
14 Polynesia/Micronesia
15 New Caledonia
16 Chocó-Darién-Western Ecuador
17 Guinean Forests of West Africa
18 Western Ghats & Sri Lanka
19 California Floristic Province
20 Succulent Karoo
21 New Zealand
22 Central Chile
23 Caucasus
24 Wallacea
25 Eastern Arc Mountains & Costal Forests of Tanzania & Kenya

▫ biological control

The use of natural means to control unwanted pests. Examples include introduced or naturally occurring predators such as wasps, or hormones that inhibit the reproduction of pests. Biological controls can sometimes be alternatives to mechanical or chemical means.

▫ biological diversity

The number and abundance of species found within a common environment. This includes the variety of genes, species, ecosystems, and the ecological processes that connect everything in a common environment.

▫ biomass

The total weight of all living organisms in a biological community.

❑ biome

The complex of living communities maintained by the climate of a region and characterized by a distinctive type of vegetation. Example of biomes in North America include the tundra, desert, prairie, and the western coniferous forests.

❑ biota

The plant and animal life of a particular region.

❑ biotic

Living. Green plants and soil microorganisms are biotic components of ecosystems.

❑ bmp

Practices designed to prevent or reduce water pollution.

❑ board foot

A measurement term for lumber or timber. It is the amount of wood contained in an unfinished board 1 inch thick, 12 inches long, and 12 inches wide.

❑ broadcast burn

A prescribed fire that burns a designated area. These controlled fires can reduce wildfire hazards, improve forage for wildlife and livestock, or encourage successful regeneration of trees.

❑ browse

Twigs, leaves, and young shoots of trees and shrubs that animals eat. Browse is often used to refer to the shrubs eaten by big game, such as elk and deer.

❑ buffer

A land area that is designated to block or absorb unwanted impacts to the area beyond the buffer. Buffer strips along a trail could block views that may be undesirable. Buffers may be set aside next to wildlife habitat to reduce abrupt change to the habitat.

C

❑ cable logging

Logging that involves the transport of logs from stump to collection points by means of suspended steel cables. Cable logging reduces the need for the construction of logging roads.

❑ canopy

The part of any stand of trees represented by the tree crowns. It usually refers to the uppermost layer of foliage, but it can be use to describe lower layers in a multi-storied forest.

❑ **cavity**

A hole in a tree often used by wildlife species, usually birds, for nesting, roosting, and reproduction.

❑ **chemical control**

The use of pesticides and herbicides to control pests and undesirable plant species.

❑ **clear cut**

A harvest in which all or almost all of the trees are removed in one cutting.

❑ **climax**

The culminating stage in plant succession for a given site. Climax vegetation is stable, self-maintaining, and self-reproducing.

❑ **coarse filter management**

Land management that addresses the needs of all associated species, communities, environments, and ecological processes in a land area.

❑ **collector roads**

These roads serve small land areas and are usually connected to a Forest System Road, a county road, or a state highway.

❑ **composition**

What an ecosystem is composed of. Composition could include water, minerals, trees, snags, wildlife, soil, microorganisms, and certain plant species,

❑ **conifer**

A tree that produces cones, such as a pine, spruce, or fir tree.

❑ **connectivity (of habitats)**

The linkage of similar but separated vegetation stands by patches, corridors, or "stepping stones" of like vegetation. This term can also refer to the degree to which similar habitats are linked.

❑ **consumptive use**

Use of resources that reduces the supply, such as logging and mining.

❑ **contour**

A line drawn on a map connecting points of the same elevation.

❑ **corridor**

Elements of the landscape that connect similar areas. Streamside vegetation may create a corridor of willows and hardwoods between meadows where wildlife feed.

❑ **cover**

Any feature that conceals wildlife or fish. Cover may be dead or live vegetation, boulders, or undercut stream banks. Animals use cover to escape from predators, rest, or feed.

❒ cover forage ratio

The ratio of hiding cover to foraging areas for wildlife species.

❒ cover type (forest cover type)

Stands of a particular vegetation type that are composed of similar species. The aspen cover type contains plants distinct from the pinyon-juniper cover type.

❒ created opening

An opening in the forest cover created by the application of even-aged silvicultural practices.

❒ critical habitat

Areas designated for the survival and recovery of federally listed threatened or endangered species.

❒ crown height

The distance from the ground to the base of the crown of a tree.

❒ cultural resource

The remains of sites, structures, or objects used by people in the past; this can be historical or pre-historic.

❒ cumulative effects

Effects on the environment that result from separate, individual actions that, collectively, become significant over time.

D

❒ dbh (diameter at breast height)

The diameter of a tree 4 and 1/2 feet above the ground on the uphill side of the tree.

❒ decision criteria

The rules and standards used to evaluate alternatives to a proposed action on National Forest land. Decision criteria are designed to help a decision maker identify a preferred choice from the array of alternatives.

❒ decking area

A site where logs are collected after they are cut and before they are taken to the landing area where they are loaded for transport.

❒ deis (draft environmental impact statement)

The draft version of the Environmental Impact Statement that is released to the public and other agencies for review and comment

❒ desired future condition

Land or resource conditions that are expected to result if goals and objectives are fully achieved.

❐ developed recreation

Recreation that requires facilities that, in turn, result in concentrated use of the area.

❐ dispersed recreation

Recreation that does not occur in a developed recreation site, such as hunting, backpacking, and scenic driving.

❐ disturbance

Any event, such as forest fire or insect infestations that alter the structure, composition, or functions of an ecosystem.

E

❐ early forest succession

The biotic (or life) community that develops immediately following the removal or destruction of vegetation in an area. For instance, grasses may be the first plants to grow in an area that was burned.

❐ ecological approach

An approach to natural resource management that considers the relationships among all organisms, including humans, and their environment.

❐ ecology

The interrelationships of living things to one another and to their environment, or the study of these interrelationships.

❐ ecoregion

An area over which the climate is sufficiently uniform to permit development of similar ecosystems on sites that have similar properties. Ecoregions contain many landscapes with different spatial patterns of ecosystems.

❐ ecosystem

An arrangement of living and non-living things and the forces that move among them. Living things include plants and animals. Non-living parts of ecosystems may be rocks and minerals. Weather and wildfire are two of the forces that act within ecosystems.

❐ ecosystem management

An ecological approach to natural resource management to assure productive, healthy ecosystems by blending social, economic, physical, and biological needs and values

❐ ecotone

The transition zone between two biotic communities, such as between the Ponderosa pine forest type and the mixed conifer forest, which is found at higher elevations than the pine.

❑ endangered species 1: north america and caribbean

	mammals	birds	reptiles	amphibia	fishes	mollusks	other invertebrates	plants	total
Caribbean Islands									
Anguilla	0	0	4	0	0	0	0	3	7
Antigua and Barbuda	0	1	5	0	0	0	0	4	10
Aruba	1	0	2	0	0	0	1	0	4
Bahamas	5	4	5	0	1	0	1	4	20
Barbados	0	1	2	0	0	0	0	2	5
Bermuda	2	2	2	0	1	0	25	4	36
Cayman Islands	0	1	2	0	0	1	0	2	6
Cuba	11	18	7	0	7	0	3	160	206
Dominica	1	3	4	0	0	0	0	11	19
Dominican Republic	5	15	10	1	0	0	2	29	62
Grenada	0	1	4	0	1	0	0	3	9
Guadeloupe	5	1	5	0	0	1	0	7	19
Haiti	4	14	7	1	0	0	2	27	55
Jamaica	5	12	8	4	0	0	5	206	240
Martinique	0	2	5	0	0	1	0	8	16
Montserrat	1	2	4	0	0	0	0	3	10
Netherlands Antilles	3	1	6	0	0	0	0	2	12
Puerto Rico	2	8	8	3	0	0	1	48	70
Saint Kitts and Nevis	0	1	3	0	0	0	0	2	6
Saint Lucia	1	5	6	0	0	0	0	6	18
Saint Vincent and the Grenadines	2	2	4	0	0	0	0	4	12
Trinidad and Tobago	1	1	5	0	0	0	0	1	8
Turks and Caicos Islands	0	3	4	0	0	0	0	2	9
Virgin Islands, British	0	2	6	1	0	0	0	4	13
Virgin Islands, United States	1	2	5	0	0	0	0	7	15
North America									
Canada	14	8	2	1	16	0	10	1	52
United States	37	54	27	25	131	255	300	168	997

St Vincent parrot

ocelot

Bermuda blue-eye grass

whooping crane

□ endangered species 2: south and central america

	mammals	birds	reptiles	amphibia	fishes	mollusks	other invertebrates	plants	total
SOUTH AMERICA									
Argentina	32	38	6	5	2	0	11	41	135
Bolivia	23	27	2	1	0	0	1	70	124
Brazil	79	113	22	6	16	21	13	338	608
Chile	21	15	0	3	4	0	0	40	83
Colombia	36	77	13	0	6	0	0	213	345
Ecuador	31	60	10	0	3	23	0	197	324
Falkland Islands (Malvinas)	4	3	0	0	0	0	0	0	7
French Guiana	9	0	7	0	0	0	0	16	32
Guyana	9	2	6	0	0	0	1	23	41
Paraguay	9	26	2	0	0	0	0	10	47
Peru	47	71	6	1	0	0	2	269	396
Suriname	11	1	6	0	0	0	0	27	45
Uruguay	6	11	3	0	3	0	1	1	25
Venezuela	25	24	12	0	5	0	1	67	134
CENTRAL AMERICA									
Belize	4	2	4	0	4	0	1	28	43
Costa Rica	14	13	7	1	0	0	9	109	153
El Salvador	2	0	4	0	1	0	1	23	31
Guatemala	6	6	8	0	1	2	6	77	106
Honduras	9	5	6	0	1	0	2	108	131
Mexico	69	38	18	4	88	4	36	161	418
Nicaragua	6	5	7	0	1	2	0	39	60
Panama	20	16	7	0	3	0	2	193	241

Lear's macaw

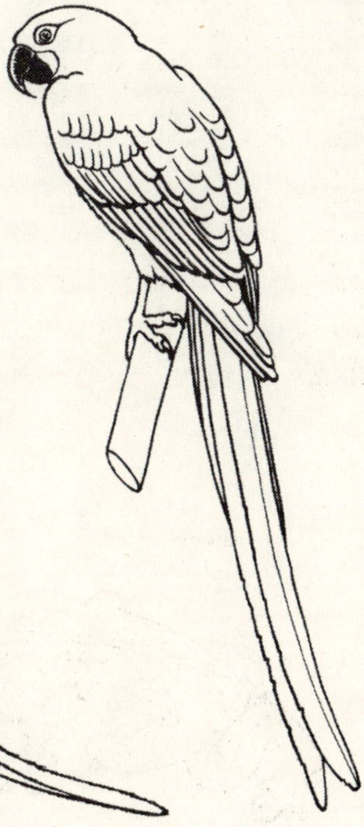

giant otter

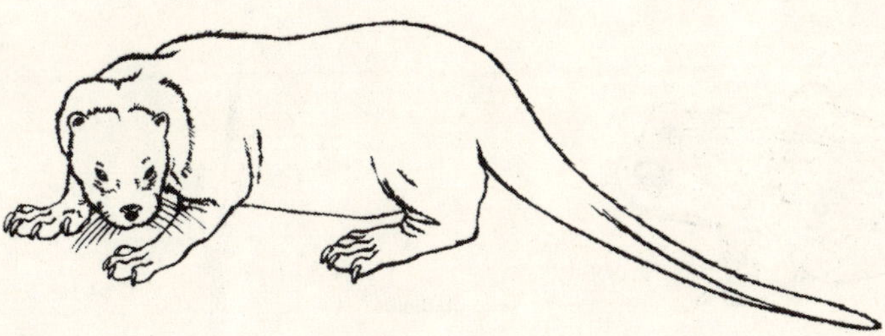

endangered species 3: europe

	mammals	birds	reptiles	amphibia	fishes	mollusks	other invertebrates	plants	total
Albania	3	3	4	0	7	0	4	0	21
Andorra	1	0	0	0	0	1	3	0	5
Austria	9	3	0	0	7	22	22	3	66
Belarus	5	3	0	0	0	0	6	6	20
Belgium	11	2	0	0	0	4	8	0	25
Bosnia/Herzegovina	10	3	1	1	6	0	8	1	30
Bulgaria	15	10	2	0	10	0	9	0	46
Croatia	9	4	1	1	21	0	8	0	44
Czech Republic	8	2	0	0	7	2	18	3	40
Denmark	5	1	0	0	0	1	10	3	20
Estonia	5	3	0	0	0	0	3	0	11
Faroe Islands	3	0	0	0	0	0	0	0	3
Finland	6	3	0	0	0	1	8	1	19
France	18	5	3	2	5	34	30	2	99
Germany	12	5	0	0	6	9	23	12	67
Gibraltar	0	1	0	0	0	2	0	0	3
Greece	14	7	6	1	19	1	9	2	59
Greenland	7	0	0	0	0	0	0	1	8
Hungary	9	8	1	0	8	1	25	1	53
Iceland	6	0	0	0	0	0	0	0	6
Ireland	5	1	0	0	0	1	1	1	9
Italy	14	5	4	4	8	16	41	3	95
Latvia	5	3	0	0	1	0	8	0	17
Liechtenstein	3	1	0	0	0	0	5	0	9
Lithuania	5	4	0	0	1	0	5	0	15
Luxembourg	6	1	0	0	0	2	2	0	11
Macedonia,	11	3	2	0	4	0	3	0	23
Malta	3	1	0	0	0	3	0	0	7
Moldova	3	5	1	0	9	0	5	0	23
Netherlands	11	4	0	0	0	1	7	0	23
Norway	10	2	0	0	0	1	8	2	23
Poland	15	4	0	0	1	1	14	4	39
Portugal	17	7	0	1	9	67	15	15	131
Romania	17	8	2	0	10	0	21	1	59
Russian Fed.	42	38	6	0	14	1	28	0	129
Slovakia	9	4	1	0	8	6	15	1	44
Slovenia	9	1	0	1	8	27	39	0	85
Spain	24	7	6	3	11	0	35	14	100
Sweden	8	2	0	0	0	1	12	3	26
Switzerland	6	2	0	0	4	0	29	2	43
Ukraine	17	8	2	0	12	0	15	1	55
United Kingdom	12	2	0	0	1	2	9	13	39
Yugoslavia	11	5	1	0	10	0	21	1	49

natterjack toad

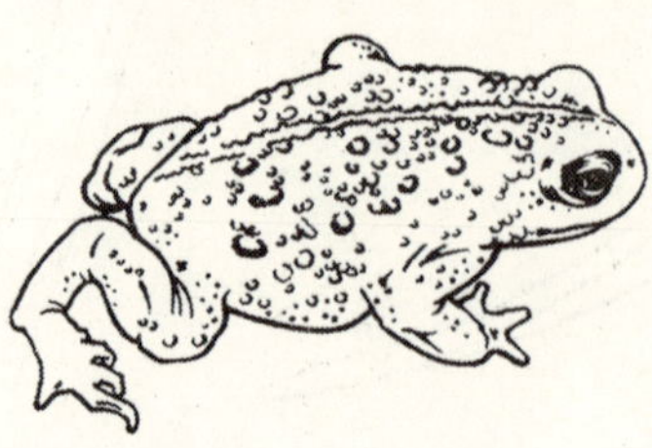

wild gladiolus

❐ endangered species 4: asia

	mammals	birds	reptiles	amphibia	fishes	mollusks	other invertebrates	plants	total
East Asia									
China	76	73	31	1	33	1	3	167	385
Hong Kong	1	11	1	0	0	1	0	4	18
Japan	37	32	11	10	13	25	20	11	159
Korea, Democratic People's Republic of	13	19	0	0	0	0	1	3	36
Korea, Republic of	13	25	0	0	0	0	1	0	39
Macao	0	1	0	0	0	0	0	0	1
Mongolia	12	16	0	0	0	0	3	0	31
Taiwan, Province of China	13	20	8	0	6	0	0	58	105
South and Southeast Asia									
Bangladesh	21	23	21	0	0	0	0	12	77
Bhutan	20	12	0	0	0	0	1	7	40
British Indian Ocean Territory	0	0	2	0	0	0	0	1	3
Brunei Darussalam	9	15	3	0	2	0	0	99	128
Cambodia	21	19	10	0	7	0	0	29	86
East Timor	0	6	0	0	0	0	0	0	6
India	86	70	25	3	8	2	21	244	459
Indonesia	140	113	28	0	67	3	28	384	763
Lao People's Democratic Republic	27	19	12	0	7	0	0	18	83
Malaysia	47	37	21	0	16	1	2	681	805
Maldives	0	1	2	0	0	0	0	0	3
Myanmar	36	35	20	0	1	1	1	37	131
Nepal	27	26	5	0	0	0	1	6	65
Philippines	50	67	8	22	28	3	16	193	387
Singapore	3	7	3	0	2	0	1	54	70
Sri Lanka	20	14	8	0	9	0	2	280	333
Thailand	34	37	18	0	19	1	0	78	187
Vietnam	37	35	24	1	6	0	0	126	229
West and Central Asia									
Afghanistan	13	11	1	1	0	0	1	1	28
Armenia	7	4	5	0	1	0	7	0	24
Azerbaijan	13	8	5	0	5	0	6	0	37
Bahrain	1	6	0	0	1	0	0	0	8
Cyprus	3	3	3	0	0	0	0	1	10
Georgia	14	3	7	1	6	0	10	0	41
Iran	23	13	8	2	7	0	3	1	57
Iraq	10	11	2	0	2	0	2	0	27
Israel	14	12	4	0	0	5	5	0	40
Jordan	8	8	1	0	0	0	3	0	20
Kazakhstan	18	15	2	1	7	0	4	0	47
Kuwait	1	7	1	0	0	0	0	0	9
Kyrgyzstan	7	4	2	0	0	0	3	0	16
Lebanon	6	7	1	0	0	0	1	0	15
Oman	9	10	4	0	4	0	1	6	34
Pakistan	18	17	9	0	1	0	0	2	47
Qatar	0	6	1	0	0	0	0	0	7
Saudi Arabia	7	15	2	0	0	0	1	3	28
Syrian Arab Republic	4	8	3	0	0	0	3	0	18
Tajikistan	9	7	1	0	3	0	2	1	23
Turkey	17	11	12	3	22	0	13	3	81
Turkmenistan	13	6	2	0	7	0	4	0	32
United Arab Emirates	3	8	1	0	1	0	0	0	13
Uzbekistan	9	9	2	0	4	0	1	0	25
Yemen	4	12	2	0	0	2	0	52	72

❐ endangered species 5: africa

	mammals	birds	reptiles	amphibia	fishes	mollusks	other invertebrates	plants	total
North Africa									
Algeria	13	6	2	0	0	0	11	2	34
Egypt	12	7	6	0	0	0	1	2	28
Libya	9	1	3	0	0	0	0	1	14
Morocco	16	9	2	0	0	0	7	2	36
Tunisia	11	5	3	0	0	0	5	0	24
Western Sahara	3	0	0	0	0	0	0	0	3
sub-Saharan Africa									
Angola	18	15	4	0	0	5	1	19	62
Benin	7	2	1	0	0	0	0	11	21
Botswana	5	7	0	0	0	0	0	0	12
Burkina Faso	7	2	1	0	0	0	0	2	12
Burundi	5	7	0	0	0	0	3	2	17
Cameroon	37	15	1	1	27	1	3	155	240
Cape Verde	3	2	0	0	1	0	0	2	8
Central African Republic	12	3	1	0	0	0	0	10	26
Chad	17	5	1	0	0	1	0	2	26
Comoros	2	9	2	0	1	0	4	5	23
Congo	12	3	1	0	1	1	0	33	51
Congo, The Democratic Republic of the	40	28	2	0	1	41	4	55	171
Côte d'Ivoire	17	12	2	1	0	1	0	101	134
Djibouti	4	5	0	0	0	0	0	2	11
Equatorial Guinea	15	5	2	1	0	0	2	23	48
Eritrea	12	7	6	0	0	0	0	3	28
Ethiopia	34	16	1	0	0	3	1	22	77
Gabon	15	5	1	0	1	0	1	71	94
Gambia	3	2	1	0	1	0	0	3	10
Ghana	13	8	2	0	0	0	0	115	138
Guinea	11	10	1	1	0	0	3	21	47
Guinea-Bissau	2	0	1	0	1	0	1	4	9
Kenya	51	24	5	0	18	12	3	98	211
Lesotho	3	7	0	0	1	0	1	0	12
Liberia	16	11	2	0	0	1	1	46	77
Madagascar	50	27	18	2	13	22	8	162	302
Malawi	8	11	0	0	0	8	0	14	41
Mali	13	4	1	0	1	0	0	6	25
Mauritania	10	2	2	0	0	0	0	0	14
Mauritius	3	9	4	0	0	27	5	87	135
Mayotte	0	3	2	0	0	0	1	0	6
Mozambique	15	16	5	0	3	6	1	36	82
Namibia	14	9	3	1	3	1	0	5	36
Niger	11	3	0	0	0	0	1	2	17
Nigeria	25	9	2	0	2	0	1	119	158
Réunion	3	5	2	0	1	14	2	14	41
Rwanda	8	9	0	0	0	0	2	3	22
Saint Helena	1	9	1	0	8	0	2	9	30
São Tome and Principe	3	9	1	0	0	1	1	27	42
Senegal	11	4	6	0	1	0	0	7	29
Seychelles	4	10	4	4	0	2	2	43	69
Sierra Leone	11	10	3	0	0	0	4	43	71
Somalia	19	10	2	0	3	1	0	17	52
South Africa	41	20	19	9	30	10	101	45	275
Sudan	24	6	2	0	0	0	1	17	50
Swaziland	4	5	0	0	0	0	0	3	12
Tanzania, United Republic of	43	33	5	0	15	41	6	236	379
Togo	9	0	2	0	0	0	0	9	20
Uganda	19	13	0	0	27	7	3	33	102
Zambia	12	11	0	0	0	4	2	8	37
Zimbabwe	12	10	0	0	0	0	2	14	38

□ endangered species 6: australia, new zealand, pacific

	mammals	birds	reptiles	amphibia	fishes	mollusks	other invertebrates	plants	total
American Samoa	3	2	2	0	0	5	0	1	13
Australia	63	32	39	25	44	175	106	38	522
Christmas Island	0	3	2	0	0	0	0	1	6
Cook Islands	1	7	2	0	0	0	0	1	11
Fiji	5	12	6	1	0	2	0	65	91
French Polynesia	3	22	1	0	3	29	0	47	105
Guam	2	2	2	0	1	5	0	3	15
Kiribati	0	3	1	0	0	1	0	0	5
Marshall Islands	1	1	2	0	0	1	0	0	5
Micronesia, Fed. States of	6	5	2	0	0	4	0	4	21
Nauru	0	2	0	0	0	0	0	0	2
New Caledonia	6	9	2	0	0	10	1	214	242
New Zealand	8	49	11	1	8	5	8	21	111
Niue	0	1	1	0	0	0	0	0	2
Norfolk Island	0	7	2	0	0	12	0	1	22
Northern Mariana Islands	2	8	2	0	0	2	0	4	18
Palau	3	2	2	0	0	5	0	3	15
Papua New Guinea	58	32	9	0	12	2	10	142	265
Pitcairn	0	8	0	0	0	5	0	7	20
Samoa	3	7	1	0	0	1	0	2	14
Solomon Islands	21	23	4	0	0	2	4	16	70
Tokelau	0	1	2	0	0	0	0	0	3
Tonga	1	3	2	0	0	2	0	2	10
Tuvalu	0	1	1	0	0	1	0	0	3
Vanuatu	4	7	2	0	0	0	0	9	22
Wallis and Futuna Islands	0	1	0	0	0	0	0	1	2

pygmy possum

Kagu

wombat

❐ endemic plant/organism

A plant or animal that occurs naturally in a certain region and whose distribution is relatively limited geographically.

❐ environmental analysis

An analysis of alternative actions and their predictable long and short-term environmental effects. Environmental analyses include physical, biological, social, and economic factors.

❐ environmental assessment

A brief version of an Environmental Impact Statement.

❐ environmental impact statement

A statement of environmental effects of a proposed action and alternatives to it. The EIS is released to other agencies and the public for comment and review.

❐ ephemeral streams

Streams that flow only as the direct result of rainfall or snowmelt. They have no permanent flow.

❐ erosion

The wearing away of the land surface by wind or water.

❐ escape cover

Vegetation of sufficient size and density to hide an animal, or an area used by animals to escape from predators.

❐ even aged management

Timber management actions that result in the creation of stands of trees in which the trees are essentially the same age.

F

❐ fauna

The animal life of an area.

❐ felling

Cutting down trees.

❐ final cut

The removal of the last seed bearers or shelter trees after regeneration of new trees has been established in a stand being managed under the shelter wood system of silviculture.

❐ fine filter management

Management that focuses on the welfare of a single or only a few species rather than the broader habitat or ecosystem. (See coarse filter management.)

❑ fire cycle

The average time between fires in a given area.

❑ fire regime

The characteristics of fire in a given ecosystem, such as the frequency, predictability, intensity, and seasonality of fire.

❑ fisheries habitat

Streams, lakes, and reservoirs that support fish, or have the potential to support fish.

❑ flood plain

A lowland adjoining a watercourse. At a minimum, the area is subject to a 1% or greater chance of flooding in a given year.

❑ flora

The plant life of an area.

❑ forage

All browse and non-woody plants that are eaten by wildlife and livestock.

❑ forb

A broadleaf plant that has little or no woody material in it.

❑ foreground

The part of a scene or landscape that is nearest to the viewer.

❑ forest health

A measure of the robustness of forest ecosystems. Aspects of forest health include biological diversity; soil, air, and water productivity; natural disturbances; and the capacity of the forest to provide a sustaining flow of goods and services for people.

❑ forest roads and trails

Roads and trails under the jurisdiction of the Forest Service.

❑ forest supervisor

The official responsible for administering National Forest lands on an administrative unit, usually one or more National Forests. The Forest Supervisor reports to the Regional Forester.

❑ fragmentation

The splitting or isolating of patches of similar habitat, typically forest cover, but including other types of habitat. Habitat can be fragmented naturally or from forest management activities, such as clear cut logging.

❑ frost heave

A land surface that is pushed up by the accumulation of ice in the underlying soil.

❒ fuels

Plants and woody vegetation, both living and dead, that are capable of burning.

❒ fuels management

The treatment of fuels that would otherwise interfere with effective fire management or control. For instance, prescribed fire can reduce the amount of fuels that accumulate on the forest floor before the fuels become so heavy that a natural wildfire in the area would be explosive and impossible to control.

❒ fuelwood

Wood cut into short lengths for burning.

❒ function

All the processes within an ecosystem through which the elements interact, such as succession, the food chain, fire, weather, and the hydrologic cycle.

G

❒ game species

Any species of wildlife or fish that is harvested according to prescribed limits and seasons.

❒ geomorphic processes

Processes that change the form of the earth, such as volcanic activity, running water, and glacial action.

❒ geomorphology

The science that deals with the relief features of the earth's surface.

❒ gis (geographic information systems)

GIS is both a database designed to handle geographic data as well as a set of computer operations that can be used to analyze the data. In a sense, GIS can be thought of as a higher order map.

❒ ground fire

A fire that burns along the forest floor and does nor affect trees with thick bark or high crowns.

❒ ground water

The supply of fresh water under the earth's surface in an aquifer or in the soil.

❒ group selection

A method of tree harvest in which trees are removed periodically in small groups. This silvicultural treatment results in small openings that form mosaics of age class groups in the forest.

H

❑ habitat

The area where a plant or animal lives and grows under natural conditions.

❑ habitat capability

The ability of a land area or plant community to support a given species of wildlife.

❑ habitat diversity

A number of different types of wildlife habitat within a given area.

❑ habitat diversity index

A measure of improvement in habitat diversity.

❑ habitat type

A way to classify land area . A habitat type can support certain climax vegetation, both tree and undergrowth species. Habitat typing can indicate the biological potential of a site.

❑ hiding area/cover

Vegetation capable of hiding 90% of an adult elk or deer from human's view at a distance of 200 feet or less.

❑ horizontal diversity

The distribution and abundance of different plant and animal communities or different stages of plant succession across an area of land; the greater the numbers of communities in a given area, the higher the degree of horizontal diversity.

❑ hydrologic cycle

Also called the water cycle, this is the process of water evaporating, condensing, falling to the ground as precipitation, and returning to the ocean as run-off.

❑ hydrology

The science dealing with the study of water on the surface of the land, in the soil and underlying rocks, and in the atmosphere.

I

❑ igneous rock

Rocks formed when high temperature, molten mineral matter cooled and solidified.

❑ indicator species

A plant or animal species related to a particular kind of environment. Its presence indicates that specific habitat conditions are also present.

❑ indigenous (species)

Any species of wildlife native to a given land or water area by natural occurrence.

❑ individual tree selection

The removal of individual trees from certain size and age classes over an entire stand area. Regeneration is mainly natural, and an uneven aged stand is maintained.

❑ instream flow

The quantity of water necessary to meet seasonal stream flow requirements to accomplish the purposes of the National Forests, including, but not limited to fisheries, visual quality, and recreational opportunities.

❑ integrated pest management

IPM evaluates alternatives for managing forest pest populations, based on consideration of pest-host relationships.

❑ interdisciplinary team

A team of individuals with skills from different disciplines that focuses on the same task or project.

❑ intermediate cut

The removal of trees from a stand sometime between the beginning or formation of the stand and the regeneration cut. Types of intermediate cuts include thinning, release, and improvement cuttings.

❑ intermittent stream

A stream that flows only at certain times of the year when it receives water from streams or from some surface source, such as melting snow.

❑ intermountain region

The portion of the USDA Forest Service, also referred to as Region Four, that includes National Forests in Utah, Nevada, southern Idaho, and southwestern Wyoming.

❑ international biosphere reserves

- Biosphere reserves are areas of terrestrial and coastal/marine ecosystems which are internationally recognized under UNESCO's Man and the Biosphere (MAB) Program.
- Each reserve contains a diverse, natural ecosystem large enough to be an effective conservation unit.
- It must have a core area for research and conservation. In outer buffer zones, traditional land uses or ecosystèm rehabilitation may be permitted.
- Biosphere reserves are designed to promote and demonstrate a balanced relationship between people and nature.

States participating in the UNESCO Man and the Biosphere Program

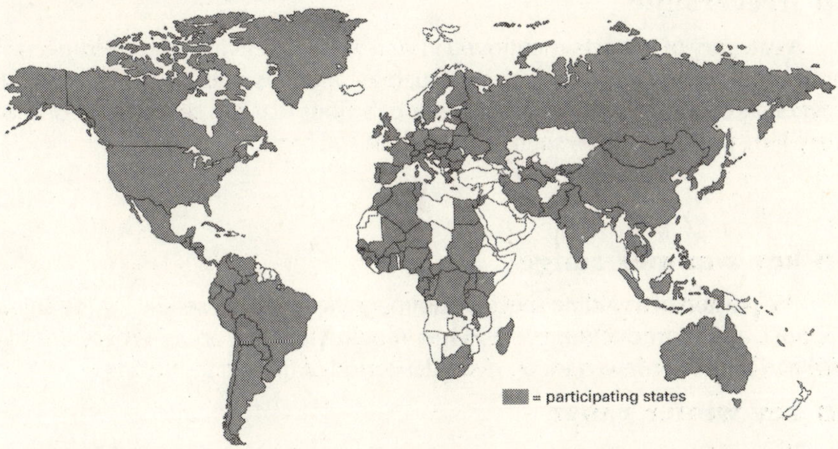

Location of reserves

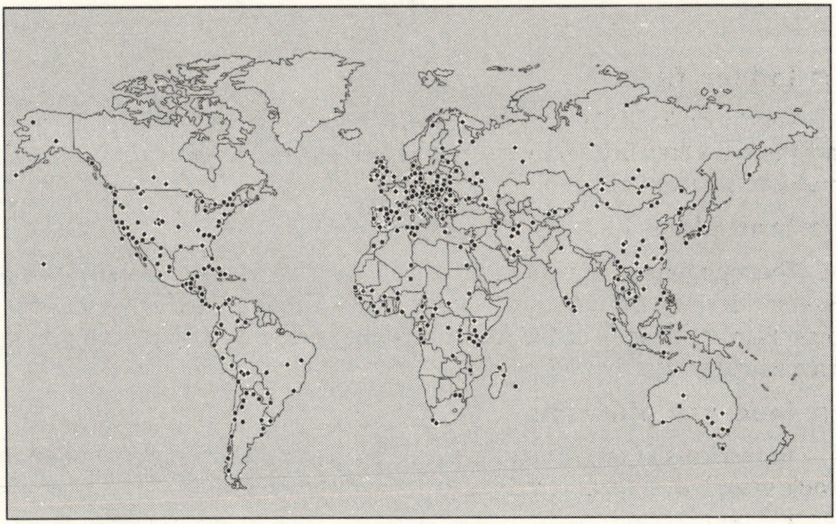

❐ irretrievable

One of the categories of impacts mentioned in the National Environmental Policy Act to be included in statements of environmental impacts. An irretrievable effect applies to losses of production or commitment of renewable natural resources. For example, while an area is used as a ski area, some or all of the timber production there is irretrievably lost. If the ski area closes, timber production could resume; the loss of timber production during the time that the area was devoted to winter sports is irretrievable. However, the loss of timber production during that time is not irreversible, because it is possible for timber production to resume if the area is no longer used as a ski area.

❐ irreversible

A category of impacts mentioned in statements of environmental impacts that applies to non-renewable resources, such as minerals and archaeological sites. Irreversible effects can also refer to effects of actions that can be renewed only after a very long period of time, such as the loss of soil productivity.

K

❐ key summer range

The portion of a wildlife species' summer range that is essential for the animal's pre, post, and reproduction cycles. Deer require "fawning areas" where does give birth and hide their fawns for an essential period of time in the spring.

❐ key winter range

That portion of big game's range where the animals find food and cover during severe winter weather.

L

❐ ladder fuels

Vegetation located below the crown level of forest trees which can carry fire from the forest floor to tree crowns. Ladder fuels may be low-growing tree branches, shrubs. or smaller trees.

❐ land class

The topographic relief of a unit of land. Land classes are separated by slope; this coincides with the timber inventory process. The three land classes used in the Forest Plan are defined by the following slope ranges: 0 to 35 percent; 36 to 55 percent; and greater than 55 percent.

❐ land use planning

The process of organizing the use of lands and their resources to best meet people's needs over time, according to the land's capabilities.

❐ landing

Any place where cut timber is assembled for further transport from the timber sale area.

❐ landline

The boundary lines for National Forest land.

❐ landscape

A large land area composed of interacting ecosystems that are repeated due to factors such as geology, soils, climate, and human impacts. Landscapes are often used for coarse grain analysis.

❒ late forest succession

The stage of forest succession in which most of the trees are mature or overmature.

❒ life zone

Areas or "belts" of land that have distinct plant and animal characteristics determined by elevation, latitude, and climate. When ascending a high mountain, you will pass through these life zones. Examples of life zones include the Upper Sonoran, where Cedar City is located and gramma grasses, sagebrush, and scattered pinyon juniper predominate, and the Transition zone, where Ponderosa pine is predominant.

❒ litter (forest litter)

The freshly fallen or only slightly decomposed plant material on the forest floor. This layer includes foliage, bark fragments, twigs, flowers, and fruit.

❒ logging residue (slash)

The residue left on the ground after timber cutting. It includes unutilized logs, uprooted stumps, broken branches, bark, and leaves. Certain amounts of slash provide important ecosystem roles, such as soil protection, nutrient cycling, and wildlife habitat.

M

❒ m

Thousand. Five thousand board feet of timber can be expressed as 5M board feet.

❒ macro climate

The general, large scale climate of a large area, as distinguished from the smaller scale micro climates within it.

❒ management action

Any activity undertaken as part of the administration of the National Forest.

❒ mass movement/wasting

The down-slope movement of large masses of earth material by the force of gravity. Also called a landslide.

❒ matrix

The least fragmented, most continuous pattern element of a landscape; the vegetation type that is most continuous over a landscape.

❒ mature timber

Trees that have attained full development, especially height, and are in full seed production.

❒ mean annual increment of growth

The total increase in size or volume of individual trees. Or, it can refer to the increase in size and volume of a stand of trees at a particular age, divided by that age in years.

❒ micro climate

The climate of a small site. It may differ from the climate at large of the area due to aspect, tree cover (or the absence of tree cover), or exposure to winds.

❒ middleground

A term used the management of visual resources, or scenery. It refers to the visible terrain beyond the foreground where individual trees are still visible but do not stand out distinctly from the stand.

❒ mineral soil

Soil that consists mainly of inorganic material, such as weathered rock, rather than organic matter.

❒ mis (management indicator species)

A wildlife species whose population will indicate the health of the ecosystem in which it lives and, consequently, the effects of forest management activities to that ecosystem. MIS species are selected by land management agencies. (See "indicator species".)

❒ mission (of the usda forest service)

"To Care for the Land and Serve the People". As set forth in law, the mission is to achieve quality land management under the sustainable multiple-use management concept to meet the diverse needs of people.

❒ mitigation

Actions taken to avoid, minimize, or rectify the impact of a land management practice.

❒ mixed stand

A stand consisting of two or more tree species.

❒ monitoring and evaluation

The periodic evaluation of forest management activities to determine how well objectives were met and how management practices should be adjusted.

❒ mortality

Trees that were merchantable and have died within a specified period of time. The term mortality can also refer to the rate of death of a species in a given population or community.

❒ mosaic

Areas with a variety of plant communities over a landscape, such as areas with trees and areas without trees occurring over a landscape.

❑ most endangered species: rivers and oceans

- More than 1,200 marine and aquatic species (over 600 in the United States alone) are listed as threatened or endangered.
- Fishing operations kill many whales and dolphins.
- In the early 1970s more than 600,000 dolphins were killed by tuna fishing operations.

18 endangered marine and aquatic species

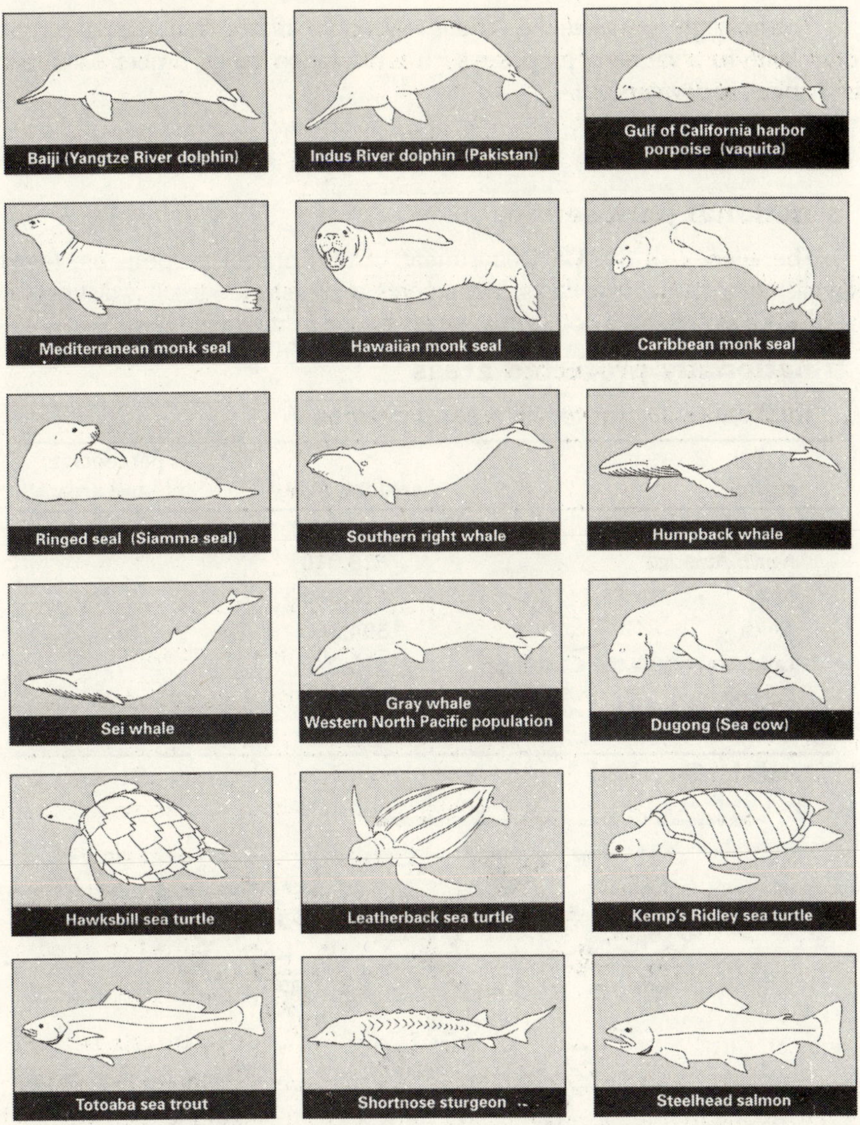

- Since then, fewer have died, because fishing has been more strictly controlled by the introduction of the United States Marine Mammal Protection Act.

- Turtles are hunted for their shells and meat, and many species are critically endangered.

❐ mountain pine beetle

A tiny black insect, ranging from 1/8 to 3/4 inch in size, that bores through a pine tree's bark. It stops the tree's intake and transport of the food and nutrients it must have to stay alive, thus killing the tree.

❐ multiple use management

The management of all the various renewable surface resources of National Forest lands for a variety of purposes such as recreation, range, timber, wildlife and fish habitat, and watershed.

N

❐ national park service

The agency of the US Department of the Interior responsible for the administration of National Parks, Monuments, and Historic Sites. It is distinct from the USDA Forest Service both administratively and by mission.

❐ nationally protected areas

Total nationally protected areas, by region

region	area (sq miles)	as percentage of land area
world*	3,247,260	6.6
North America	829,010	11.7
Asia	628,868	5.3
Africa	594,760	5.2
Latin America & the Caribbean	555,240	7.2
Europe	406,211	4.7
Oceania	299,135	7.1

*excludes Greenland

Protected areas and proportion of country protected (1997 UN list)

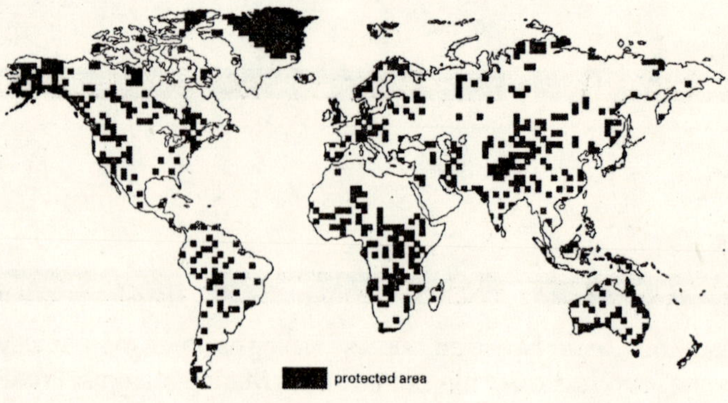

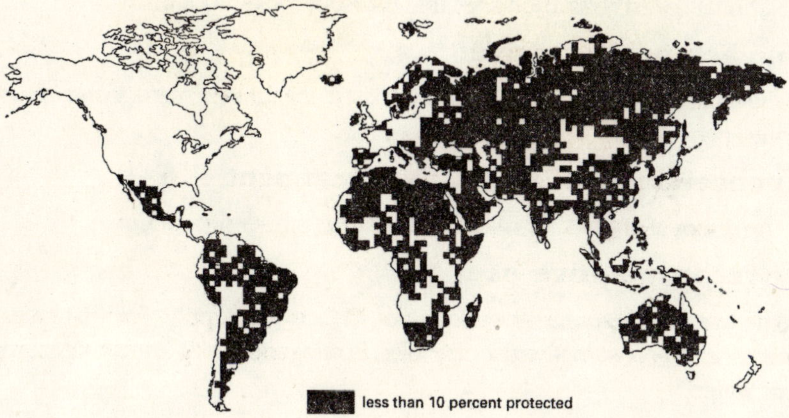

❐ natural barrier

A natural feature, such as a dense stand of trees or downfall, that will restrict animal travel.

❐ natural range of variability

See range of variability

❐ natural resource

A feature of the natural environment that is of value in serving human needs.

❐ nepa (national environmental policy act)

Congress passed NEPA in 1969 to encourage productive and enjoyable harmony between people and their environment. One of the major tenets of NEPA is its emphasis on public disclosure of possible environmental effects of any major action on public lands. Section 102 of NEPA requires a statement of possible environmental effects to be released to the public and other agencies for review and comment.

❐ nest survey

A way to estimate the size of a bird population by counting the number of nests in a given area.

❐ nflrmp (national forest land and resource management plan)

Also called the Forest Plan or just the Plan, this document guides the management of a particular National Forest and establishes management standards and guidelines for all lands of that National Forest.

❐ nfma (national forest management act)

This law was passed in 1976 and requires the preparation of Regional Guides and Forest Plans.

❑ nfrs

National Forest recreation sites that have been inventoried.

❑ no action alternative

The most likely condition expected to exist in the future if management practices continue unchanged.

❑ noncommercial vegetative treatment

The removal of trees for reasons other than timber production.

❑ nonconsumptive use

The use of a resource that does not reduce the supply. For instance, bird watching is a non-consumptive use of wildlife. Boating and fishing are non-consumptive uses of water.

❑ nongame

Wildlife species that are not hunted for sport.

❑ nonpoint source pollution

Pollution whose source is not specific in location. The sources of the discharge are dispersed, not well defined, or constant. Rain storms and snowmelt often make this type of pollution worse. Examples include sediments from logging activities and runoff from agricultural chemicals.

❑ non-renewable resource

A resource whose total quantity does not increase measurably over time, so that each use of the resource diminishes the supply.

❑ notice of intent

A notice in the federal register of intent to prepare an environmental impact statement on a proposed action.

❑ numbers of extinctions by group

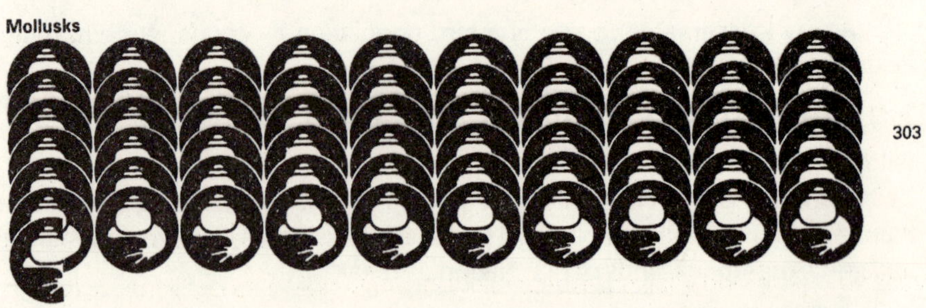

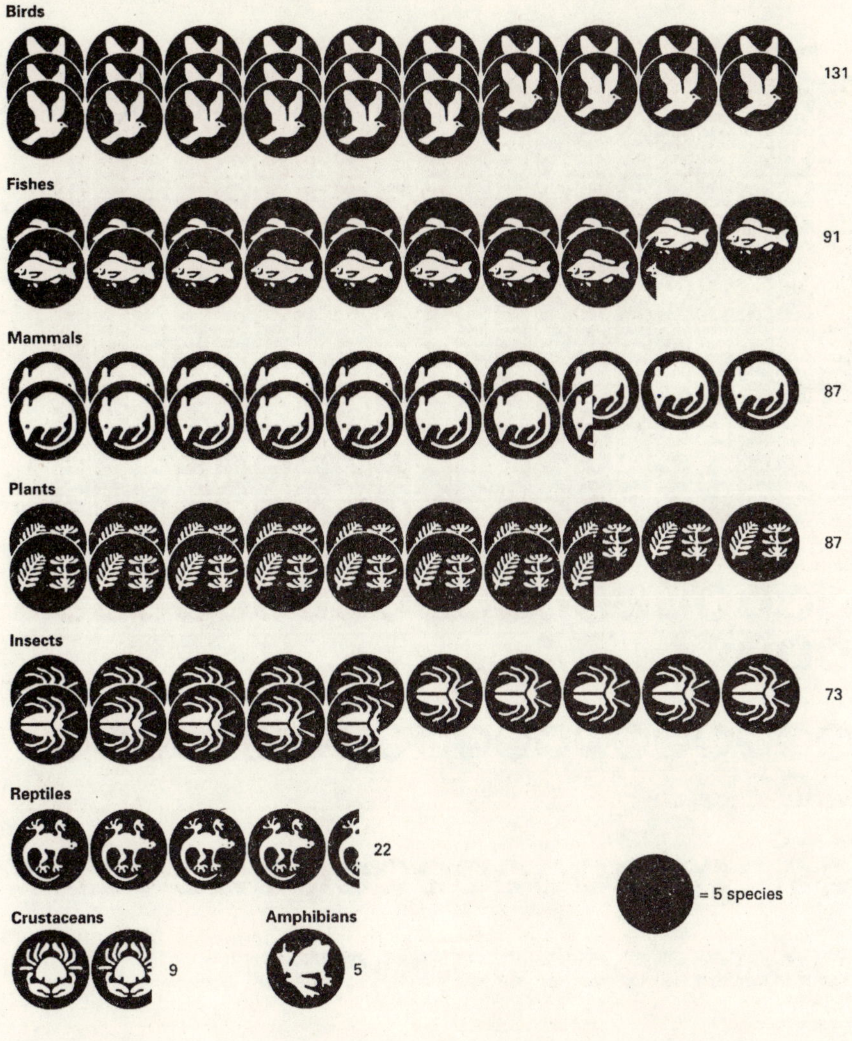

▫ numbers of species (and percentage) threatened, by group

An endangered species is an organism in danger of disappearing from the face of the earth if its situation is not improved. When its race has not been seen in the wild for over fifty years, we say that it is extinct. Those species that may soon become endangered are called threatened species. Rare animals are species with small populations that may also be at risk. The main causes of species extinction or endangerment are habitat destruction, commercial exploitation (such as plant collecting, hunting, and trade in animal parts), damage caused by non-native plants and animals introduced into an area, and pollution. Of these causes, direct habitat destruction endangers the most species. Many organisms have already been wiped out. The only certain thing is that we must act quickly.

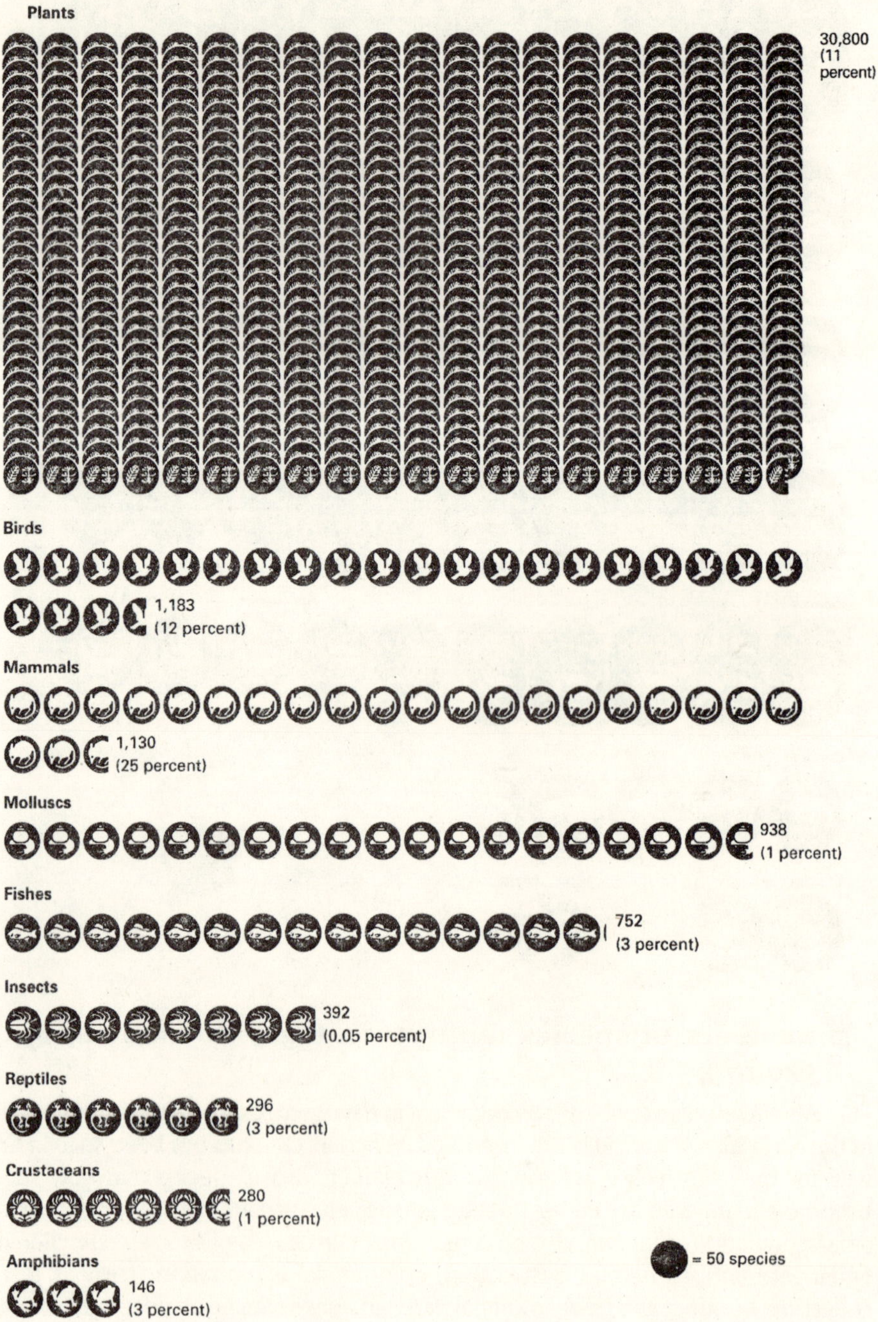
Plants
30,800 (11 percent)
Birds
1,183 (12 percent)
Mammals
1,130 (25 percent)
Molluscs
938 (1 percent)
Fishes
752 (3 percent)
Insects
392 (0.05 percent)
Reptiles
296 (3 percent)
Crustaceans
280 (1 percent)
Amphibians
146 (3 percent)
= 50 species

▫ nutrient cycle

The circulation of chemical elements and compounds, such as carbon and nitrogen, in specific pathways from the non-living parts of ecosystems into the organic substances of the living parts of ecosystems, and then back again to the non-living parts of the ecosystem. For instance, nitrogen in wood is returned to the soil as the dead tree decays; the nitrogen again becomes available to living organisms in the soil, and upon their death, the nitrogen is available to plants growing in that soil.

O

▫ old growth

Old forests often containing several canopy layers, variety in tree sizes and species, decadent old trees, and standing and dead woody material.

▫ organic soil

Soil at least partly derived from living matter, such as decayed plant material.

▫ orv

Off-road vehicles, such as motor cycles, 4-wheel drive vehicles, and 4-wheelers.

▫ overfishing 1: peruvian anchoveta

- One of the most productive fishing areas in the world is a 300 by 30 mile area off the coast of Peru.
- The coastal upwelling in this region forces nutrientrich cool water to the surface.
- The phytoplankton that thrives there is fed upon by a variety of creatures including the anchoveta.
- The huge biomass feeds many creatures including predatory fish, guano birds, and mammals.
- This enormous natural resource began to be exploited after World War II, and factories were opened to process the anchoveta into fishmeal.
- A similar upwelling off the coast of California was overfished. The fisheries there closed, and much of the equipment used to overfish California was sold to Peru.
- More efficient and larger fishing vessels led to larger and larger Peruvian catches.
- The Peruvian Government is trying to ensure that the anchoveta is fished sustainably, but the industry's capacity is much bigger than the sustainable harvest.
- The currents change direction in El Niño years, causing the disappearance of the nutrient-rich upwelling, and a collapse in the anchoveta population.

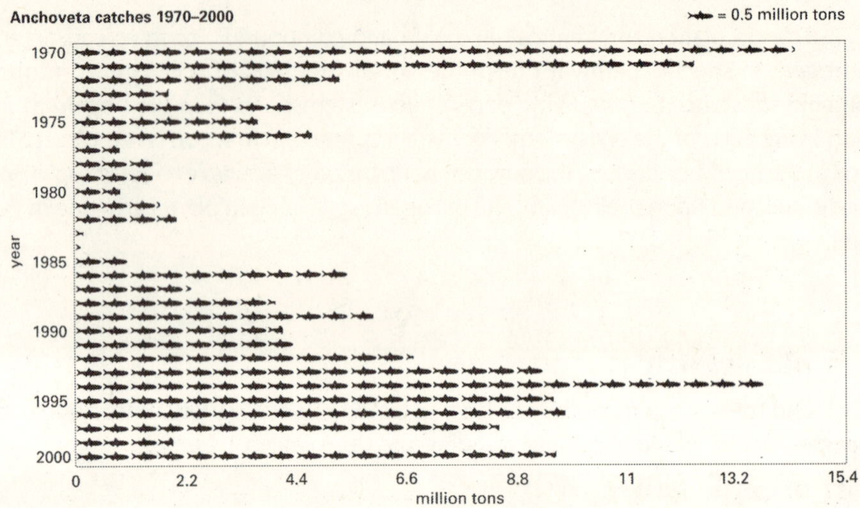

1. In normal years, cool, nutrient-rich water (**a**) rising off the coast of South America is warmed, and flows westwards towards Australia, providing food for anchoveta off Peru.

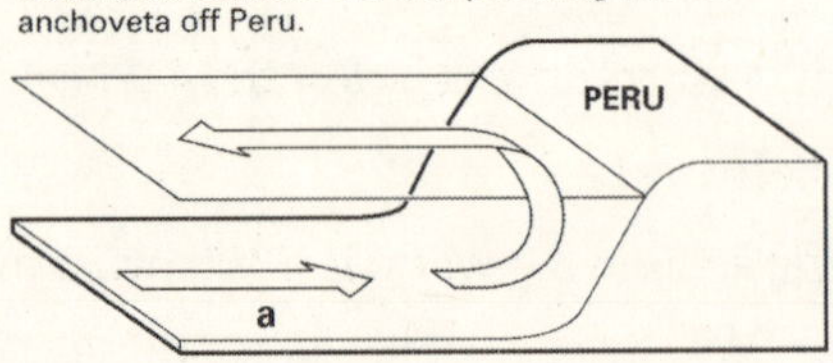

2. In El Niño years (which occur about 14 times every century) warmer water (**b**) flows towards South America stopping the rise of the cooler (**c**
the Peruvian coast and reducing the anchoveta population.

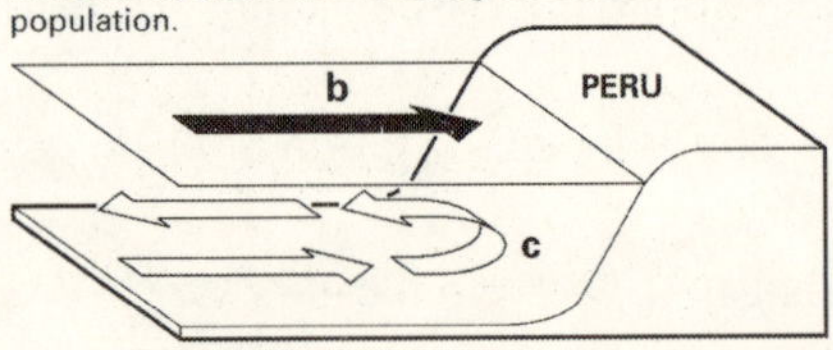

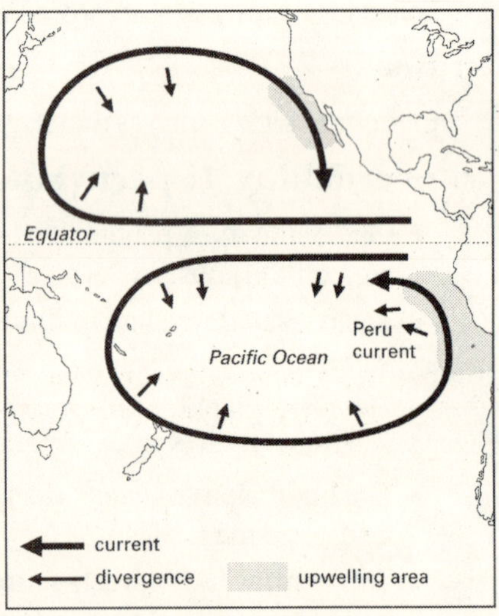

❑ overfishing 2: north sea herring and atlantic cod

Herring

- The North Sea is a shallow sea, kept free of ice by warm ocean currents.
- Overfishing in the North Sea now threatens the entire marine ecosystem and since the 1990s has placed virtually all commercial fish species, including cod, hake, herring, haddock, mackerel, plaice, saithe, and whiting at dangerously low levels and at risk of stock collapse.
- Overfishing in the 1970s reduced the total catch from over 4.4 million tons in 1966 to under 992,000 tons in 1979. However, catches have recovered somewhat in recent years.

Cod

- For 450 years, cod were fished sustainably, until new fishing techniques and refrigeration led to overfishing.
- Total catches fell off sharply from record levels set in the late 1960s, as deep-sea factory trawlers decimated the stocks.
- Between 1962 and 1977, North Atlantic cod declined 82 percent due to relentless overfishing.
- Northern cod were on the verge of extinction as a commercial species when Canada extended its jurisdiction to the 200-mile limit in 1977.
- This reduced the catches for a while, but Canada expanded its fleet and overfishing began again.

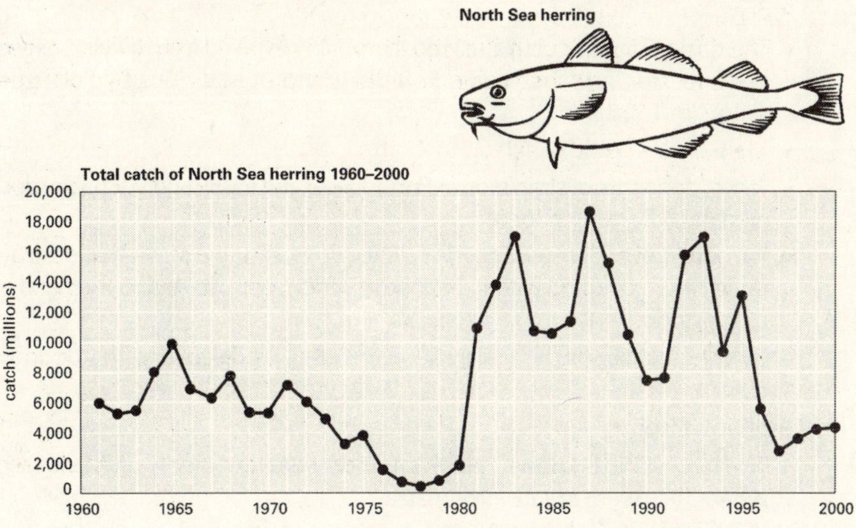

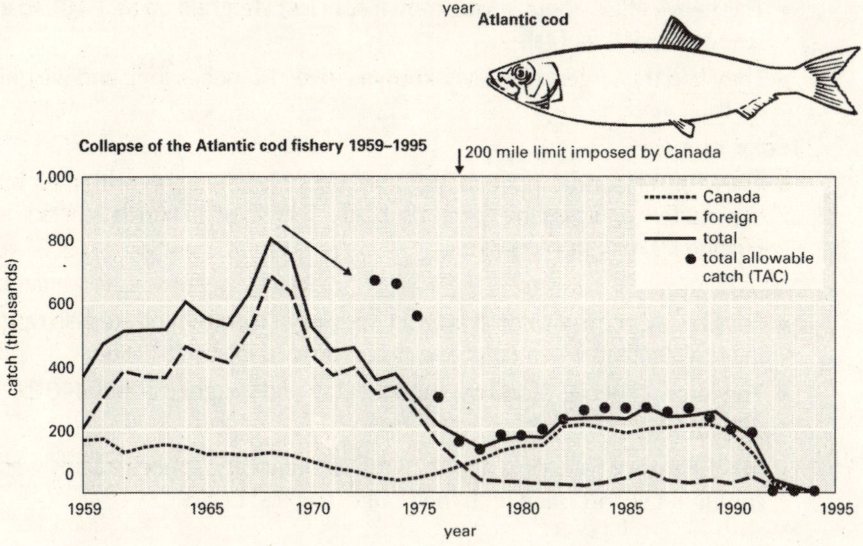

☐ overhunting 1: the dodo and other flightless birds

Vulnerability

- Flightless birds appeared early in the Cenozoic Era, around 65 million years ago, when dinosaurs vanished and new creatures evolved.
- An absence of naturally occurring predators on oceanic islands favored the evolution of flightlessness in birds.
- With no anatomical constraints of forelimb development for flight and lack of competition from other species, flightless birds often evolved greater body size.
- Flightlessness, large size and lack of awareness of the dangers of humans made island birds easy prey for hunters and domestic animals.

Dodos

- The dodo (Raphus cucullatus) most probably evolved from a flying, pigeon-like ancestor, was first sighted on the island of Mauritius by Portuguese sailors in 1598.
- Male dodos could weigh up to 46 lb.
- Tame (dodo meant simpleton in Portuguese) and easily approachable, dodos were heavily hunted for food.
- Introduction of domestic animals – cats, dogs, rats and pigs – that destroyed dodo nests, coupled with habitat destruction, contributed to the rapid end of the dodo.
- By the mid-1660s, less than 80 years after its discovery, the dodo was extinct.

Elephant birds

- The largest birds known, elephant birds were found only on Madagascar, where they were called vorompatra.
- The heaviest (of about seven known species) weighed up to 1,100 lb and stood over 9.8 feet tall.
- They laid the largest eggs ever known – over 12 inches long and weighing 20 lb.

Terror cranes

- These birds stood 6 foot 6 in high and probably killed prey with their huge clawed feet and massive 'parrot's' beak. They lived in North America and Europe 50 million years ago.

Moas

- Fossil evidence indicates at least 11 species of moa (which are also ratites) in New Zealand when Polynesians first arrived in about 1250.
- The largest species stood over 6.6 feet tall and weighed about 440 lb.

Phorusracids

- Inhabitants of Patagonia about 20 million years ago, phorusracids were 6 foot 6 in tall and killed with their huge 'eagle's' beaks.

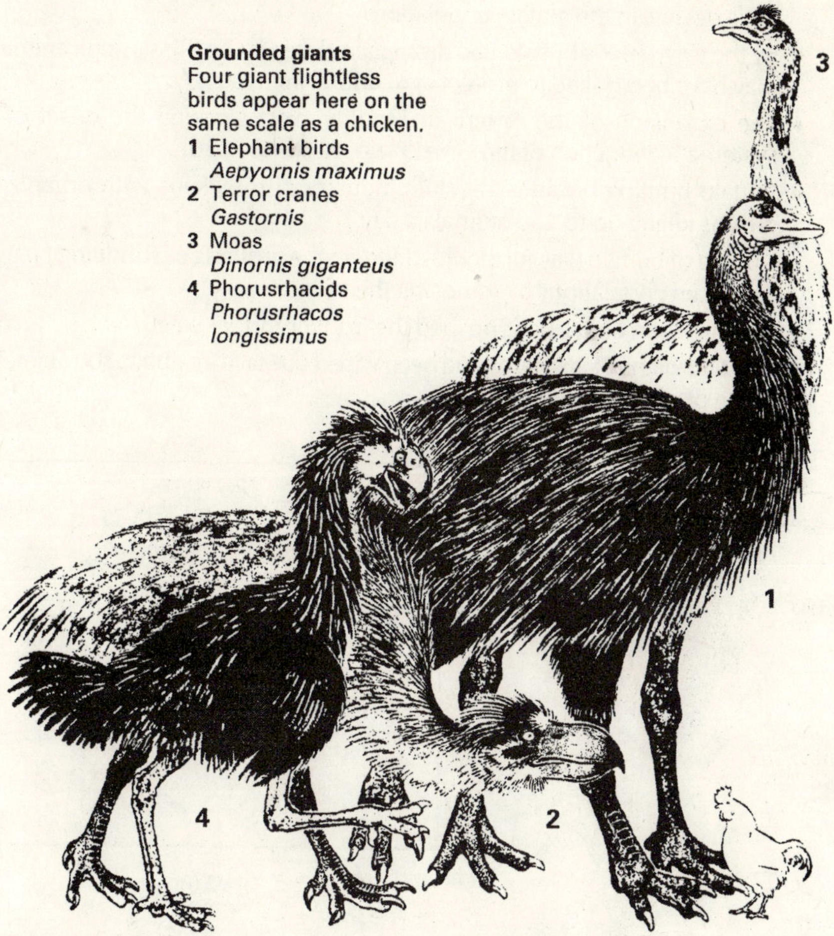

Grounded giants
Four giant flightless birds appear here on the same scale as a chicken.
1 Elephant birds *Aepyornis maximus*
2 Terror cranes *Gastornis*
3 Moas *Dinornis giganteus*
4 Phorusrhacids *Phorusrhacos longissimus*

▢ overhunting 2: north american buffalo

- The buffalo (Bison bison) is believed to have crossed over from Asia into America via the Bering Sea land bridge.
- Its range extended as far south as Mexico and Florida, and east to the Atlantic.
- The largest herds were found on the Great Plains.
- The herds were nearly destroyed before organized surveys of their numbers were attempted.
- Estimates of the original bison population range from 35 to 70 million.

Slaughter

- Great Plains Native Americans relied on bison for meat and hides, killing them in small numbers.

- Commercialized and subsistence hunting by arriving Europeans caused a huge decline in the buffalo population.
- By the early 1800s buffalo had disappeared east of the Mississippi; animals may have been killed to protect agricultural interests.
- The expansion of the American Frontier west signaled the onset of a systematic reduction of the Great Plains herds in 1830.
- Buffalo hunting became the chief industry of the Plains with organized groups killing up to 250 animals a day.
- Culling of buffalo was further justified on the grounds of subduing Native American populations by removing their food supply.
- Construction of railroads pushed the frontier further west.
- By 1893 all the Plains herds had been wiped out save for about 300 animals, some of which were in Canada.

Range of the buffalo 1600–1875

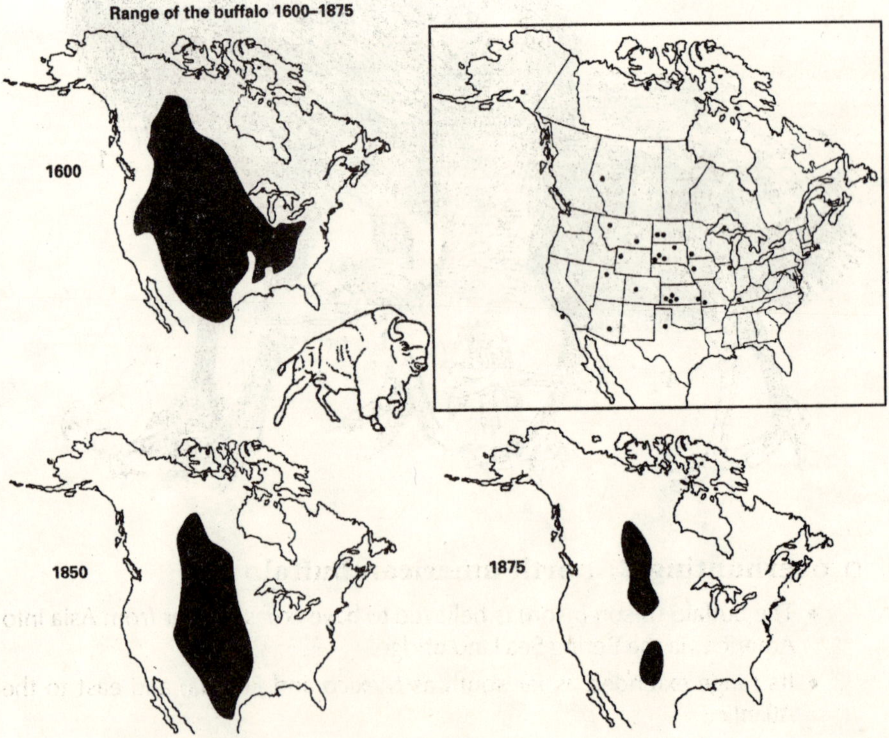

Preservation work

- Private and governmental efforts at buffalo conservation were slow to start.
- In 1894 Congress enacted a law to protect buffalo.
- In 1902 a captive herd of 41 animals from a private source was placed under government protection in Yellowstone National Park.
- Today there are around 4,000 animals in the park.

- Elsewhere, in Nebraska, Iowa, Montana, Oklahoma and North Dakota, buffalo herds have been re-established with total numbers estimated at over 250,000.
- But the herds have low genetic diversity because of the massive depopulation in the 19th century.

▫ overhunting 3: baleen and toothed whales

- Baleen whales include the largest whales.
- They feed on plankton and krill, tiny sea creatures that they strain in millions through sieve-like plates in their mouths (the baleen).

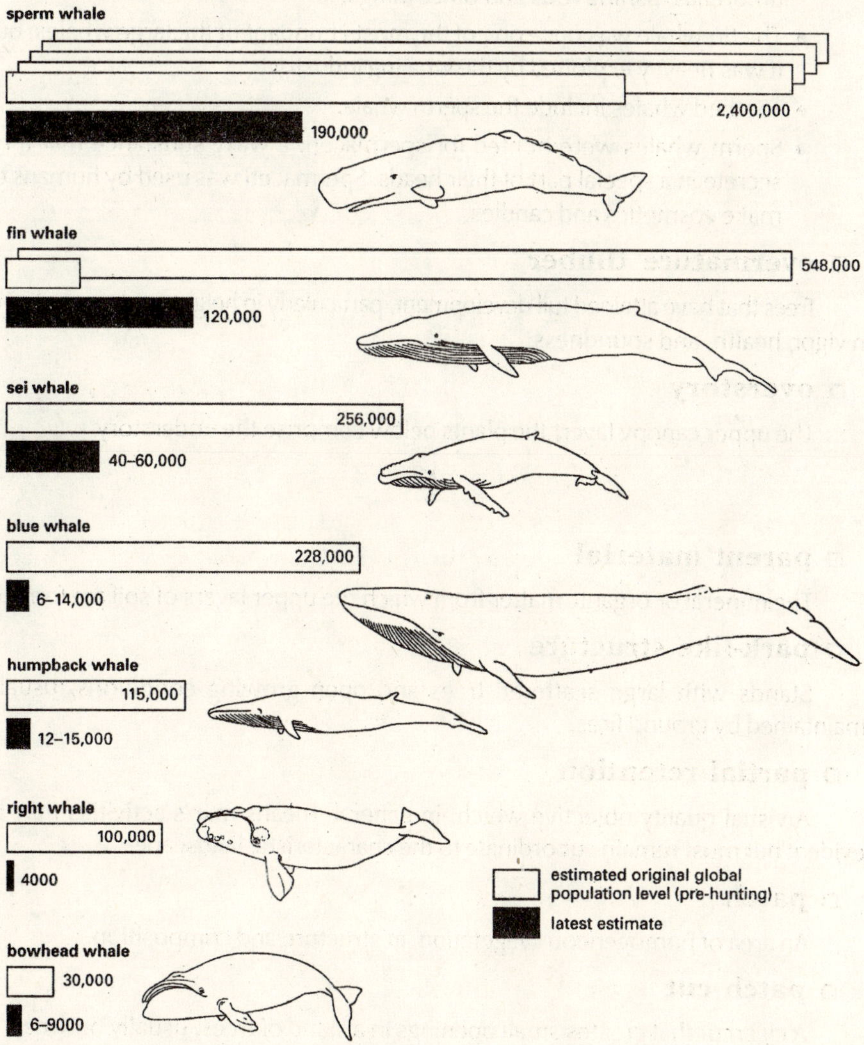

- These whales were originally the most hunted because of their valuable baleen and large size.

- Six species of baleen whales are listed for protection under the United States Endangered Species Act of 1973: the blue, bowhead, fin, humpback, right, and sei whales.
- The largest of the baleen whales, the blue whale, may reach more than 85 feet long and weigh over 90 tons.
- There may be as few as 500 blue whales left in the southern hemisphere.
- They were hunted for their blubber (fat) for oil in lamps and to make candles and soap.
- The plates from baleen whales were used for making women's clothing, umbrellas, fishing rods and other things.
- The fin whale was once one of the most abundant of the large whales, but it was heavily exploited by the whaling industry.
- Toothed whales include the sperm whale.
- Sperm whales were hunted for spermaceti, a waxy substance that they secrete in a special part of their heads. Spermaceti was used by humans to make cosmetics and candles.

❑ overmature timber

Trees that have attained full development, particularly in height, and are declining in vigor, health, and soundness.

❑ overstory

The upper canopy layer; the plants below comprise the understory.

P

❑ parent material

The mineral or organic matter from which the upper layers of soil are formed.

❑ park-like structure

Stands with large scattered trees and open growing conditions, usually maintained by ground fires.

❑ partial retention

A visual quality objective which, in general, means man's activities may be evident but must remain subordinate to the characteristic landscape.

❑ patch

An area of homogeneous vegetation, in structure and composition.

❑ patch cut

A clearcut that creates small openings in a stand of trees, usually between 15 and 40 acres in size. On the Dixie National Forest and elsewhere, patchcuts are used to provide the disturbance needed to regenerate aspen.

❑ percolation

Downward flow or infiltration of water through the pores or spaces of rock or soil.

❑ perennial stream

A stream that flows throughout the year and from source to mouth.

❑ permitted grazing

Grazing on a National Forest range allotment under the terms of a grazing permit.

❑ personal use

The use of a forest product, such as firewood, for home use and not for commercial use.

❑ planning area

The area of National Forest land covered by a Regional Guide or Forest Plan.

❑ planning period

The 50 year time frame for which goods, services, and effects were projected in the development of the Forest Plan.

❑ poaching case study 1: africa's elephants

Ivory trade

- For centuries elephant ivory has been an important trading commodity, especially in the Far East.
- Legal and illegal movement of ivory totaled 770 tons (75,000 pairs of tusks) during the 1970s.

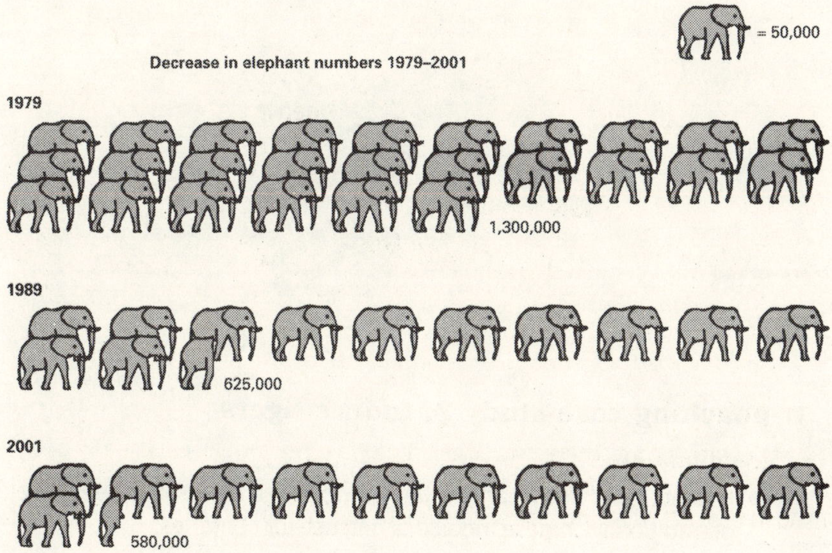

- A combination of habitat loss and high levels of illegal poaching threatened the African elephant with extinction.
- The African elephant population declined from around 1.3 million in 1979 to 625,000 in ten years.
- Global concern for elephant conservation resulted in CITES declaring a total ban on ivory trade in 1989.
- World demand for ivory remained high and ivory prices escalated from $ 50 per lb in 1975 to as much as $ 500 per lb during the ban period.
- Poaching intensity increased with the use of sophisticated automatic weapons; poachers were killing an average of 200 elephants a day.
- The CITES ban was only partially effective andpoaching continued uncontrolled in some countries.

African elephant ranges 2001

- By 2001, Botswana, Namibia, South Africa and Zimbabwe all showed increases in their elephant populations, some to such an extent that elephants were beginning to cause serious habitat damage. A conflict became apparent between the interests of conservationists and the needs of those living and farming in the elephant-populated areas of those countries.
- The excess elephants resulting from the ban on poaching had to be culled, increasing the tonnage of ivory stockpiles in each country.
- CITES later allowed these countries to enter into an experimental, limited trade in ivory with Japan.
- Revenue generated from ivory sales supports community development and improved elephant conservation and management.

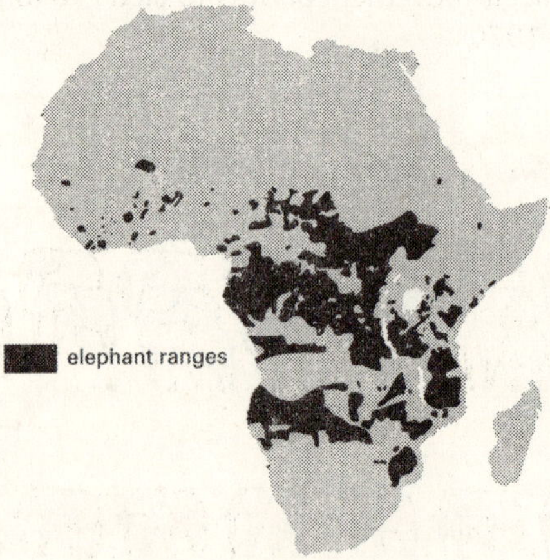

❑ poaching case study 2: india's tigers

Traditional uses

- Use of tigers' body parts, especially forelimb bones, has been an element of traditional Chinese medicinal practice for centuries.

- Taiwanese pharmacies use skulls and bones.
- Tiger penises are used in aphrodisiac potions.
- Skins and heads find buyers in the Middle East for robes used by royalty.
- In other areas of Asia tiger parts are used for soup.
- In the past, levels of use of tiger parts were probably rational and sustainable.
- As tiger populations have declined demand for their parts in traditional Chinese medicine has remained high.

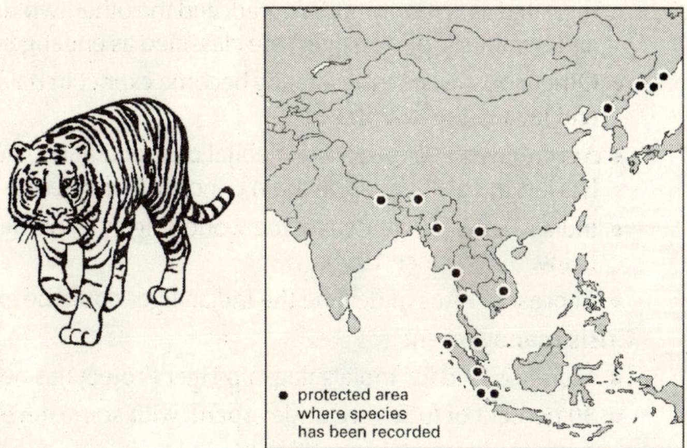

Range of all surviving Asian tigers

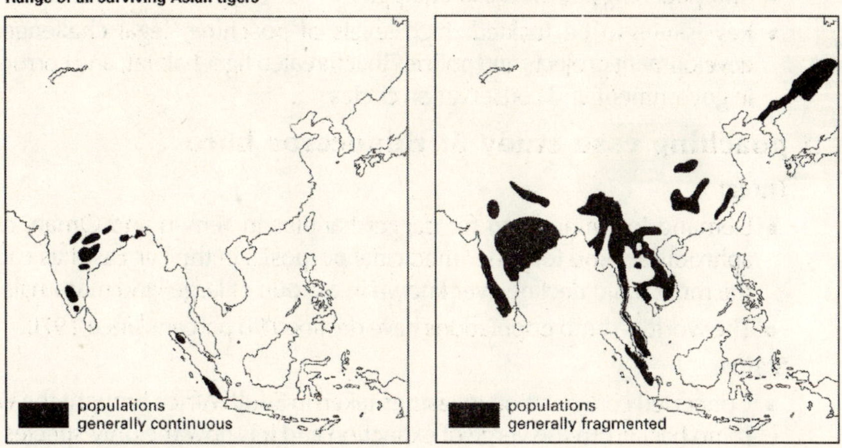

Growth of poaching

- Market value of tiger parts has dramatically increased.
- Tiger bones may sell for $ 500 per lb.
- Illegal poaching is now the only way tiger parts can be supplied to consumer markets.
- Far Eastern traditional medicine manufacturers arenow targeting India's tiger populations as sources of tiger parts dry up elsewhere in Asia.

- Poaching of tigers for traditional Chinese medicine industry began in northern India in the mid-1980s.

Population crash

- Throughout its Asian range, the tiger's numbers have declined by 95 percent in last 100 years.
- Shooting and hunting of tigers for sport has also played a role in their historical decline.
- Three subspecies – the Amur, Sumatran and South China tigers – are regarded as critically endangered, and the other two (Indian or Bengal tiger and the Indo-Chinese tiger) are classified as endangered.
- Other subspecies have already become extinct in Bali, the Caspian region, and Java in the last 50 years.
- A combination of poaching, habitat destruction and human encroachment has left India's tiger population critically endangered.
- India's tigers (60 percent of the world's tigers) have declined from 4,500 to below 3,000 since 1998.
- At present rates of decline the Indian tiger could be extinct by 2010.

Crisis management

- Money raised for India's flagship Tiger Project has been mismanaged.
- 30 percent of funds are under-spent, with some never reaching field staff.
- Anti-poaching patrols are ill-equipped.
- Key issues to be tackled: high levels of poaching, legal challenges to development projects and policies that threaten tiger habitat, and corruption in government and conservation circles.

❑ poaching case study 3: rhinoceros horn

Threat

- Demand for rhino horn for dagger handles in Yemen and Oman, as an aphrodisiac, and for other medicinal purposes in the Far East has caused the most rapid decline ever known in a group of large land mammals.
- The world's rhino populations have declined 90 percent since 1970.

Status

- Concerted conservation measures taken in South Africa brought the white rhino back from the verge of extinction and it is now the only species that is not endangered.
- The two-horned African black rhino suffered the most dramatic decline, from 100,000 animals in 1960 to fewer than 3,000 in 2001.
- The Javan rhino is found in only one park in Indonesia, with an additional 15 animals in Vietnam.
- Fewer than 300 Sumatran rhino survive in small forested areas of Malaysia and Indonesia.
- Efforts to save the Indian rhino have resulted in a small increase in numbers, but heavy poaching continues.

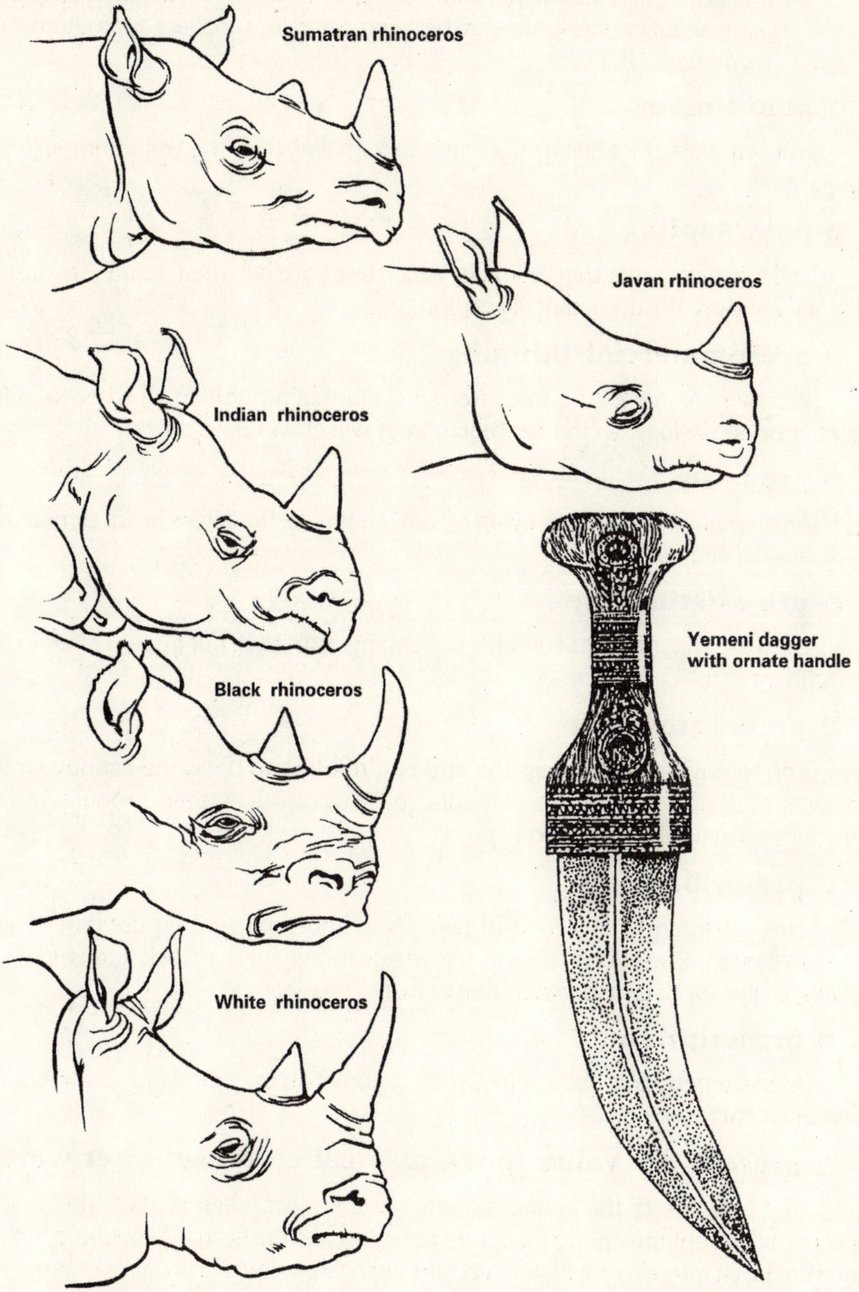

Innovations

- The plight of rhinos has produced innovative conservation measures. Black rhinos in Zimbabwe and Namibia have been systematically caught and dehorned to discourage poachers.

- The principal market for rhino horn is traditional eastern medicine. Pharmaceutical substitutes are being prepared for suppliers as an alternative to powdered horn.

❐ pole timber

Trees at least 5 inches in diameter, but smaller than the minimum size for sawtimber.

❐ pole/sapling

The stage of forest succession in which trees are between 3 and 7 inches in diameter and are the dominant vegetation.

❐ precommercial thinning

Removing some of the trees from a stand that are too small to be sold for lumber or house logs, so the remaining trees will grow faster.

❐ predator

An animal that lives by preying on other animals. Predators are at or near the tops of food chains.

❐ pre-existing use

Land use that may not conform to a zoning ordinance but existed prior to the enactment of the ordinance.

❐ preparatory cut

The removal of trees near the end of a rotation to open the canopy so the crowns of seed bearing trees can enlarge. This improves seed production and encourages natural regeneration.

❐ prescribed fire

Fire set intentionally in wild land fuels under prescribed conditions and circumstances. Prescribed fire can rejuvenate forage for livestock and wildlife or prepare sites for natural regeneration of trees.

❐ prescription

Management practices selected to accomplish specific land and resource management objectives.

❐ present net value (pnv), also called present net worth

The measure of the economic value of a project when costs and revenues occur in different time periods. Future revenues and costs are "discounted " to the present by an interest rate that reflects the changing value of a dollar over time. The assumption is that dollars today are more valuable than dollars in the future. PNV is used to compare project alternatives that have different cost and revenue flows.

❐ presuppression

Activities carried out in advance of fire occurrence to ensure effective suppression when the need arises.

❑ primitive ros (recreation opportunity spectrum)

A classification of wilderness and recreation opportunity. It is characterized by an essentially unmodified environment, where trails may be present but structures are rare, and where it is highly probable to be isolated from the sights and sounds of people.

❑ productive

The ability of an area to provide goods and services and to sustain ecological values.

❑ prognosis

A computer model for timber growth and yield. It projects per-acre growth and volume yield for commercial timber stands.

❑ public domain

The territory ceded to the Federal government by the original thirteen states, plus additions by treaty, cession, and purchase.

❑ public involvement

The use of appropriate procedures to inform the public, obtain early and continuing public participation, and consider the views of interested parties in planning and decision making.

❑ public land

Land for which title and control rests with a government---Federal, state, regional, county, or municipal.

R

❑ range

Land on which the principle natural plant cover is composed of native grasses, forbs, and shrubs that are valuable as forage for livestock and big game.

❑ range management

The art and science of planning and directing range use intended to yield the sustained maximum animal production and perpetuation of the natural resources.

❑ range of variability (also called the historic range of variability or natural range of variation.)

The components of healthy ecosystems fluctuate over time. The range of sustainable conditions in an ecosystem is determined by time, processes (such as fire), native species, and the land itself. For instance, ecosystems that have a 10 year fire cycle have a narrower range of variation than ecosystems with 200-300 year fire cycle. Past management has placed some ecosystems outside their range of variability. Future management should move such ecosystems back toward their natural, sustainable range of variation.

❑ ranger district

The administrative sub-unit of a National Forest that is supervised by a District Ranger who reports directly to the Forest Supervisor.

❑ raptor

A bird of prey, such as a eagle or hawk.

❑ rare ii

Roadless Area Review and Evaluation. The national inventory of roadless and undeveloped areas within the National Forests and Grasslands.

❑ recharge

The addition of water to ground water by natural or artificial processes.

❑ reforestation

The restocking of an area with forest trees, by either natural or artificial means, such as planting.

❑ regeneration

The renewal of a tree crop by either natural or artificial means. The term is also used to refer to the young crop itself.

❑ regional forester

The official of the USDA Forest Service responsible for administering an entire region of the Forest Service.

❑ release cutting

Removal of competing vegetation to allow desired tree species to grow.

❑ removal cut

The removal of the last seed bearers or shelter trees after regeneration is established.

❑ residual stand

The trees remaining standing after an event such as selection cutting.

❑ resilience

The ability of an ecosystem to maintain diversity, integrity, and ecological processes following a disturbance.

❑ responsible official

The Forest Service employee who has been delegated the authority to carry out a specific planning action.

❑ restoration (of ecosystems)

Actions taken to modify an ecosystem to achieve a desired, healthy, and functioning condition.

❐ **revegetation**

The re-establishment and development of a plant cover by either natural or artificial means, such as re-seeding.

❐ **riparian area**

The area along a watercourse or around a lake or pond.

❐ **riparian ecosystem**

The ecosystems around or next to water areas that support unique vegetation and animal communities as a result of the influence of water.

❐ **rod—record of decision**

A official document in which a deciding official states the alternative that will be implemented from a prepared EIS.

❐ **ros—recreation opportunity spectrum**

The land classification system that categorizes land by its setting and the probable recreation experiences and activities it affords.

❐ **rotation**

The number of years required to establish and grow timber crops to a specified condition of maturity.

❐ **roundwood**

Timber and fuelwood prepared in the round state, such as house logs and telephone poles.

❐ **run-off**

The portion of precipitation that flows over the land surface or in open channels.

S

❐ **sacrifice area/site**

In range management, a site allowed to be overgrazed to obtain efficient overall use of the management area. In cultural resource management, it may refer to a site intentionally sacrificed to extensive public use in order to preserve the larger cultural area.

❐ **sanitation salvage**

The removal of dead, damaged or susceptible trees primarily to prevent the spread of pests or disease and promote forest health.

❐ **sapling**

A loose term for a young tree more than a few feet tall and an inch or so in diameter that is typically growing vigorously.

❑ sawtimber

Trees that are 9 inches in diameter at breast height or larger that can be made into lumber.

❑ scale

In ecosystem management, it refers to the degree of resolution at which ecosystems are observed and measured.

❑ scoping

The ongoing process to determine public opinion, receive comments and suggestions, and determine issues during the environmental analysis process. It may involve public meetings, telephone conversations, or letters.

❑ second growth

Forest growth that was established after some kind of interference with the previous forest crop, such as cutting, fire, or insect attack.

❑ seed tree harvest

Removal of the mature timber crop from an area in one cut, except for a certain number of seed bearers.

❑ sensitive species

Plant or animal species which are susceptible to habitat changes or impacts from activities. The official designation is made by the USDA Forest Service at the Regional level and is not part of the designation of Threatened or Endangered Species made by the US Fish and Wildlife Service.

❑ seral

The stage of succession of a plant or animal community that is transitional. If left alone, the seral stage will give way to another plant or animal community that represents a further stage of succession.

❑ shelterwood

A cutting method used in a more or less mature stand, designed to establish a new crop under the protection of the old.

❑ silvicultural system

The cultivation of forests; the result is a forest of a distinct form. Silvicultural systems are classified according to harvest and regeneration methods and the type of forest that results.

❑ silviculture

The art and science that promotes the growth of single trees and the forest as a biological unit.

❑ site preparation

The general term for removing unwanted vegetation, slash, roots, and stones from a site before reforestation. Naturally occurring wildfire, as well as prescribed fire can prepare a site for natural regeneration.

❒ size class

One of the three intervals of tree stem diameters used to classify timber in the Forest Plan data base. The size classes are: Seedling/Sapling (less than 5 inches in diameter); Pole Timber (5 to 7 inches in diameter); Sawtimber (greater than 7 inches in diameter)

❒ skidding

Hauling logs by sliding, not on wheels, from stump to a collection point.

❒ skyline logging

A logging system used to remove timber from steep slopes. Logs are brought up-slope on a suspended cable, or skyline. Since the weight of the log is completely or partially supported by the cable, there is little disturbance to soil or other vegetation.

❒ slash

The residue left on the ground after timber cutting or left after a storm, fire, or other event. Slash includes unused logs, uprooted stumps, broken or uprooted stems, branches, bark, etc.

❒ slump

A landslide where the underlying rock masses tilt back as they slide from a cliff or escarpment.

❒ small game

Birds and small animals normally hunted or trapped.

❒ snag

A standing dead tree. Snags are important as habitat for a variety of wildlife species and their prey.

❒ soil compaction

The reduction of soil volume. For instance, the weight of heavy equipment on soils can compact the soil and thereby change it in some ways, such as in its ability to absorb water.

❒ soil productivity

The capacity of a soil to produce a specific crop. Productivity depends on adequate moisture and soil nutrients, as well as favorable climate.

❒ sound wood

Timber that is in solid, whole, good condition. Sound wood is free from damage, decay, or defects.

❒ special use permit

A permit issued to an individual or group by the USDA Forest Service for use of National Forest land for a special purpose. Examples might be a Boy Scout Jamboree or a mountain bike race.

❐ stand

A group of trees that occupies a specific area and is similar in species, age, and condition.

❐ standards and guidelines

Requirements found in a Forest Plan which impose limits on natural resource management activities, generally for environmental protection.

❐ stewardship

Caring for the land and its resources to pass healthy ecosystems to future generations.

❐ stocking level

The number of tree in an area as compared to the desirable number of trees for best results, such as maximum wood production.

❐ stringer

A strip of vegetation different from surrounding vegetation, such as a stringer of aspen in a area of spruce.

❐ structure

How the parts of ecosystems are arranged, both horizontally and vertically. Structure might reveal a pattern, or mosaic, or total randomness of vegetation.

❐ succession

The natural replacement, in time, of one plant community with another. Conditions of the prior plant community (or successional stage) create conditions that are favorable for the establishment of the next stage.

❐ successional stage

A stage of development of a plant community as it moves from bare ground to climax. The grass-forb stage of succession precedes the woody shrub stage.

❐ suitability

The appropriateness of certain resource management to an area of land. Suitability can be determined by environmental and economic analysis of management practices.

❐ surface resources

Renewable resources that are on the surface of the earth, such as timber and forage, in contrast to ground water and minerals which are located beneath the surface.

❐ sustainability

The ability of an ecosystem to maintain ecological processes and functions, biological diversity, and productivity over time.

T

☐ the world's rarest land mammals

World's rarest land mammals and their present habitat

● = 10 animals

- **Vancouver Island Marmot** (Canada) at least 24
- **Seychelles Sheath-tailed Bat** fewer than 50 individuals
- **Northern Hairy-nosed Wombat** (Australia) 65
- **Javan Rhino** (Indonesia) fewer than 70
- **Hispid Hare** (Assam Rabbit, India, Nepal) 110
- **Tamaraw** (Dwarf Water Buffalo, Philippines) thought to be about 30–200
- **Dwarf Blue Sheep** (Sichuan, China) approximately 200
- **Tonkin Snub-nosed Monkey** (North Vietnam) about 200
- **Yellow-tailed Woolly Monkey** (Northern Peru) fewer than 250
- **Kouprey** (Cambodian Forest Ox – Laos, Cambodia, Thailand, Vietnam) generally assumed to be less than 250
- **Malabar Large Spotted Civet** (SW India) fewer than 250 mature individuals are thought to survive
- **Philippine Spotted Deer,** Panay and Negros islands a few hundred
- **Saola** (Vu Quang Ox, Vietnam, Laos) estimated at several hundred
- **Sumatran Rhino** (Indonesia) estimated population: 250–300
- **Hirola** (Hunter's Hartebeest, Kenya, Somalia) by 1995 numbers were down to 300
- **African Wild Ass** (Eritrea, Ethiopia, Somalia) possibly only a few hundred animals
- **Addax** (Niger, Chad, Mauritania) may not exceed a few hundred individuals
- **Black-faced Lion Tamarin** (Brazil) as few as 400
- **Ethiopian Wolf** only 500 individuals left
- **Arabian Oryx** (Oman, Saudi Arabia) approximately 500 animals in the wild (reintroduced)
- **Iberian Lynx** (Portugal, Spain) about 600
- **Hairy-eared Dwarf Lemur** (Madagascar) under 1,000 survive
- **Riverine Rabbit** (Karoo, South Africa) fewer than 1,000 animals
- **Muriqui** (Woolly Spider Monkey, Brazil) 700–1,000
- **Giant Panda** (W China) about 1,000
- **Golden Lion Tamarin** (Brazil) about 1,000
- **Golden-rumped Lion Tamarin** (Brazil) about 1,000
- **Greater Bamboo Lemur** (Madagascar) about 1,000

❑ the world's rarest plants

- There are approximately 25,000 plant species at risk of extinction in the wild today.
- Some 8,700 of these are trees – 10 percent of all those known to science.
- Malaysia has 681 threatened plant species, much the greatest of any country. A large proportion of these are tropical timber trees.
- Indonesia, Brazil and Sri Lanka follow with 384, 338, and 280 threatened species respectively.
- Plants designated as 'endangered' by the World Conservation Union (IUCN) are listed below.

Arizona agave (Agave arizonica)	United States
Ash meadows milk-vetch (Astragalus phoenix)	United States
Bankouale palm (Wissmania carinensis)	Somalia, Yemen, Djibouti
Bois de fer (Vateria seychellarum)	Seychelles
Bois de prune blanc (Drypetes caustica)	Réunion, Indian Ocean
Dall's pittosporum (Pittosporum dallii)	New Zealand
Dotalu palm (Loxococcus rupicola)	Sri Lanka
Egyptian papyrus (Cyperus papyrus hadidii)	Egypt
Fibrosa she oak (Casuarina fibrosa)	Australia
Fragrant prickly-apple cactus (Cereus eriophorus)	United States
Furbish lousewort (Pedicularis furbishiae)	United States
Gigasiphon (Gigasiphon macrosiphon) see below	East Africa
Ginkgo (Ginkgo biloba) see below	China
Green pitcher plant (Sarracenia oreophila)	United States
Horseshoe fern (Marattia salicina)	Australia
Javan phalaenopsis orchid (Phalaenopsis javanica)	Indonesia
Large-flowered fiddleneck (Amsinckia grandiflora)	United States
Leafless acacia (Acacia aphyela) see below	Australia
Lord Howe Island passionfruit (Passiflora herbertiana insulae-howei)	Pacific
Louisiana quillwort (Isoetes louisianensis)	United States
Madeira net-leaf orchid (Goodyera macrophylla)	Madeira
Mauritius ebony (Diospyros angulata)	Mauritius
Mellblom's spider orchid (Calenia hastata)	Australia
Okeechobee gourd (Cucurbita okeechobeensis) see below	United States
Pondberry (Lindera melissifolia) see below	United States
San Clemente Island Indian paintbrush (Castilleja grisea)	United States
Schimper's toxocarpus (Toxocarpus schimperianus)	Seychelles

Silver gum (Eucalyptus crenulata)	Australia
Star cactus (Astrophytum asterias) see below	United States, Mexico
Sunshine diuris orchid (Diuris fragrantissima)	Australia
Tennessee purple coneflower (Echinacea tennesseensis)	United States
Tree cactus (Cereus robinii)	United States
Virginia round-leaf birch (Betula uber)	United States
West Himalayan elm (Ulmus wallichiana)	India
Western lily (Lilium occidentale)	United States
White gum (Eucalyptus argophloia)	Australia
Yellow sobralia orchid (Sobralia xantholeuca)	Mexico, Guatemala
Zhoushan hornbeam (Carpinus putoensis)	China

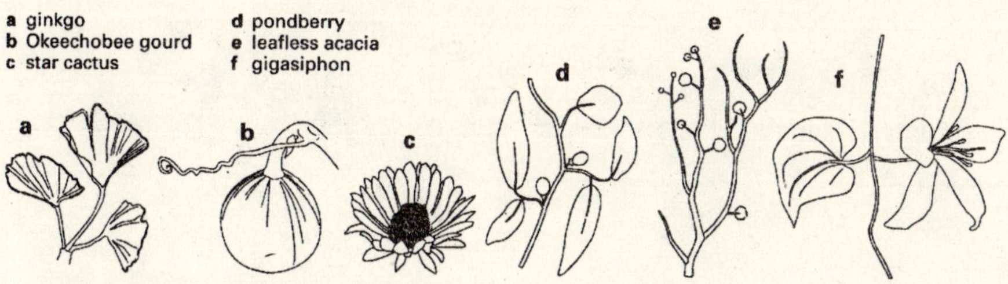

❒ thermal cover

Cover used by animals against weather. For elk, thermal cover can be found in a stand of coniferous trees at least 40 feet tall with a crown closure of at least 70%.

❒ thinning

A cutting made in an immature stand of trees to accelerate growth of the remaining trees or to improve the form of the remaining trees.

❒ threatened species

Those plant or animal species likely to become endangered throughout all or a specific portion of their range within the foreseeable future as designated by the U.S. Fish and Wildlife Service under the Endangered Species Act of 1973.

❒ timber classification

The classification of forested lands into land management alternatives according to how the land relates to management of the timber resource there.

❒ top 10 coral reef hotspots

- The 10 coral reef hotspots, ranked according to the degree of threat, are:
  1. Philippines
  2. Gulf of Guinea
  3. Sunda Islands

4. Southern Mascarene Islands
5. Eastern South Africa
6. Northern Indian Ocean
7. Southern Japan, Taiwan and southern China
8. Cape Verde Islands
9. Western Caribbean
10. Red Sea and Gulf of Aden

10 coral reef hotspots

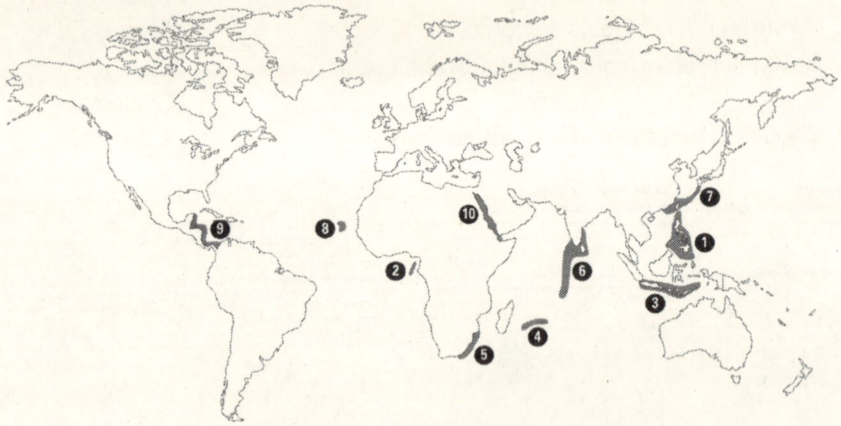

Areas with high concentrations of species

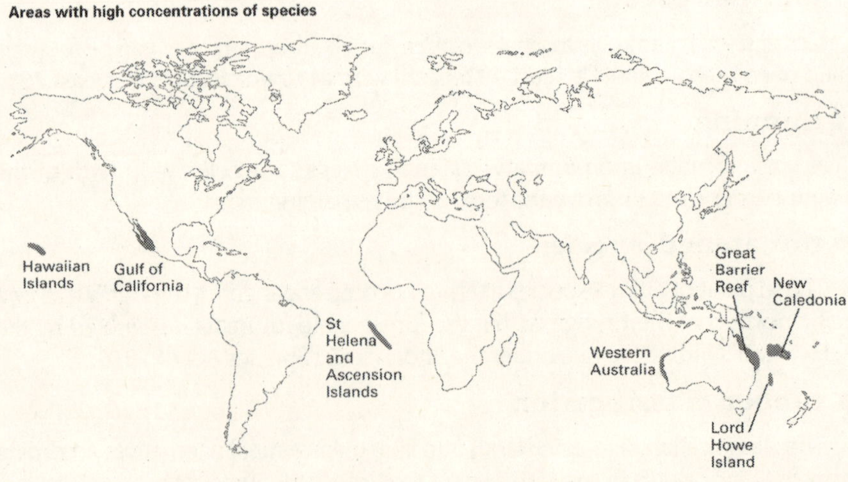

- These 10 hotspots account for 34 percent of restricted-range coral reef species.
- Eight of them are close to a biodiversity hotspot.
- Activities destroying habitat in the biodiversity hotspots are also contributing to coral reef destruction.

- Nearly 60 percent of the world's reefs are threatened by human activities.
- The leading causes of reef destruction, along with intense fishing and climate change, are sediment, nutrients and other pollutants from farming, logging, and development.
- Global warming is believed to be responsible for destroying or severely degrading 25 percent of the world's coral reefs.

❑ tractor logging

A logging method that uses tractors to carry or drag logs from the stump to a collection point.

❑ treatment area

The site specific location of a resource improvement activity.

❑ tree opening

An opening in the forest created by even-aged silvicultural practices.

❑ tsi (timber stand improvement)

Actions to improve growing conditions for trees in a stand, such as thinning, pruning, prescribed fire, or release cutting.

❑ type conversion

The conversion of the dominant vegetation in an area from forested to non-forested or from one species to another.

U

❑ underburn

A burn by a surface fire that can consume ground vegetation and "ladder" fuels.

❑ understory

The trees and woody shrubs growing beneath the over story in a stand of trees.

❑ uneven-aged management

Actions that maintain a forest or stand of trees composed of intermingling trees that differ markedly in age. Cutting methods that develop and maintain uneven-aged stands are single-tree selection and group selection.

❑ unregulated harvest

Tree harvest that is not part of the allowable sale quantity (ASQ). It can include the removal of cull or dead material or non-commercial species. It also includes volume removed from non-suitable areas for research, to meet objectives other than timber production (such as wildlife habitat improvement), or to improve administrative sites (such as campgrounds.)

❐ unsuitable lands

Forest land that is not managed for timber production. Reasons may be matters of policy, ecology, technology, silviculture, or economics

❐ use, allowable

An estimate of proper range use. Forty to fifty percent of the annual growth is often used as a rule of thumb on ranges in good to excellent condition. It can also mean the amount of forage planned to be used to accelerate range rehabilitation.

V

❐ variety class

A way to classify landscapes according to their visual features. This system is based on the premise that landscapes with the greatest variety or diversity have the greatest potential for scenic value.

❐ vegetation management

Activities designed primarily to promote the health of forest vegetation for multiple-use purposes.

❐ vegetation type

A plant community with distinguishable characteristics.

❐ vertical diversity

The diversity in a stand that results from the different layers or tiers of vegetation.

❐ viable population

The number of individuals of a species sufficient to ensure the long-term existence of the species in natural, self-sustaining populations that are adequately distributed throughout their range.

❐ virgin forest

A natural forest virtually uninfluenced by human activity.

❐ visual quality objective

A set of measurable goals for the management of forest visual resources.

❐ visual resource

A part of the landscape important for its scenic quality. It may include a composite of terrain, geologic features, or vegetation

W

❐ water table

The upper surface of groundwater. Below it, the soil is saturated with water.

❐ **water yield**

The runoff from a watershed, including groundwater outflow.

❐ **watershed**

The entire region drained by a waterway (or into a lake or reservoir. More specifically, a watershed is an area of land above a given point on a stream that contributes water to the stream flow at that point.

❐ **wetlands**

Areas that are permanently wet or are intermittently covered with water.

❐ **wetlands of international importance**

States participating in the Ramsar Convention

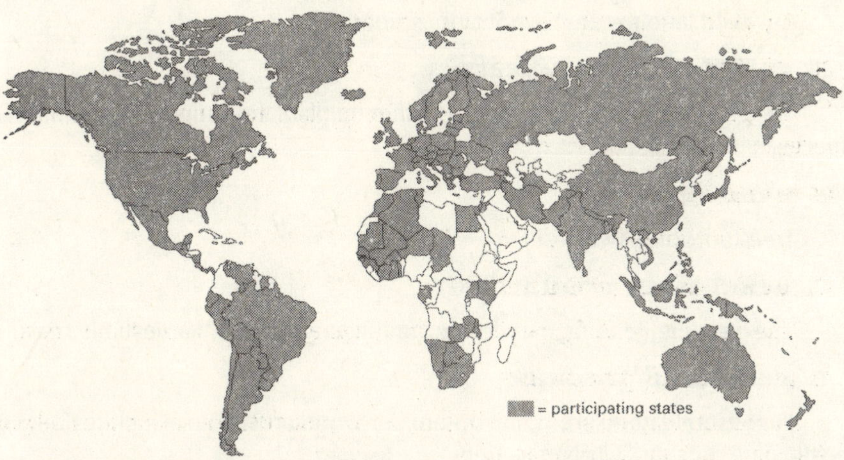

Location of wetlands of international importance

- The Convention on Wetlands of International Importance was signed in Ramsar (Iran) in 1971, and came into force in December 1975.
- Known as the Ramsar Convention, this provides a framework for international cooperation for the conservation of wetland habitats.
- Wetlands are defined as areas of marsh, fen, peatland or water, with water that is static or flowing, fresh, brackish or salt, including areas of marine waters, the depth of which at low tide does not exceed 19 feet.

❐ wilderness (wilderness area)

Undeveloped federal land retaining its primeval character, without permanent human habitation or improvements. It is protected and managed to preserve its natural condition. Wilderness Areas are designated by Congress.

❐ wildfire

Any wild land fire that is not a prescribed fire.

❐ wildlife habitat diversity

The distribution and abundance of different plant and animal communities and species within a specific area.

❐ windthrow

Trees uprooted by wind.

❐ wood fiber production

The growing, tending, harvesting, and regeneration of harvestable trees.

❐ woodland products

Harvestable items from pinyon-juniper woodlands. These include fuelwood, posts, pine nuts and Christmas trees.

❐ world heritage natural sites

States participating in the World Heritage Convention

- World Heritage sites, as defined under the World Heritage Convention which came into force in 1975, include natural sites of outstanding universal value as well as cultural sites.
- Some constitute the habitat of threatened species of animals and plants.

Listed natural sites

Y

❑ yarding

Moving the cut trees from where they fell to a centralized place (landing) for hauling away from the stand.

Z

❑ zoi (zone of influence)

The area influenced by Forest Service management activities.

Engine Emissions

A

❐ additives

Chemicals added to fuel in very small quantities to improve and maintain fuel quality and/or to lower emissions.

❐ aftercooling/intercooling

Cooling the engine intake air after the turbocharger and prior to introduction into the cylinder. Aftercooling increases engine power and lowers NOx emissions.

❐ aftertreatment devices

Devices which remove pollutants from exhaust gases after the gas leaves combustion chamber (e.g., catalytic converters or diesel particulate filters). The term "exhaust gas aftertreatment" is considered derogatory by some in the emission control industry, but there is no consensus on the use of such alternatives as "post-combustion treatment" or "exhaust emission control".

❐ air quality management district (aqmd)

Administrative districts organized in California to control air pollution. Nationwide in the U.S., AQMDs are parallel to the areas designated for classification against the National Ambient Air Quality Standards (NAAQS). Generally, AQMDs and their national parallel encompass multiple jurisdictions and closely follow the definition of Consolidated Metropolitan Statistical Areas and Metropolitan Statistical Areas.

❐ air toxics

Toxic air pollutants, as classified by pertinent regulations. Examples of substances classified as air toxics by the US Clean Air Act include acetaldehyde, benzene, 1,3-butadiene, formaldehyde, and polycyclic organic matter (POM). California air toxics regulations also classify diesel exhaust particulates as a toxic air contaminant.

❐ alternative fuel

Fuel other than petroleum diesel or gasoline.

❐ american society for testing and materials (astm)

A non-profit organization that establishes specifications and standard test methods for a broad range of materials and products. ASTM standards are recognized as definitive guidelines for quality of motor fuels.

❐ articulated pistons

Two-piece pistons incorporating an entirely separate piston crown or dome with a separate skirt, and linking the two together with the piston pin. Many 1994 and later engines incorporate steel crown/aluminum skirt articulated pistons.

B

❐ bi-fueled vehicle

A vehicle with two separated fuel systems designed to run on either conventional fuel or an alternative fuel using only one fuel at a time.

❐ biodiesel

The mono alkyl esters of long chain fatty acids derived from renewable lipid feedstocks, such as vegetable oils and animal fats, for use in compression ignition (diesel) engines. Manufactured by transestrification of the organic feedstock by methanol.

❐ brake mean effective pressure (bmep)

The work accomplished during one engine cycle divided by the engine swept volume. It is essentially the engine torque normalized by the engine displacement. The word "brake" denotes the actual torque/power available at the engine flywheel as measured on a dynamometer. Thus, BMEP is a measure of the useful power output of the engine.

❐ brake specific fuel consumption (bsfc)

BSFC is the ratio of the engine fuel consumption to the engine power output (as measured at the flywheel). BSFC has units of grams of fuel per kilowatt-hour (g/kWh) or pounds mass of fuel per brake horsepower-hour (lb/bhp-hr). BSFC is a measure of engine efficiency.

C

❐ california air resources board (carb)

A state regulatory agency charged with regulating the air quality in California.

❐ carbon dioxide (CO_2)

A colorless, odorless, non-toxic gas. It is one of main products of fossil-fuel combustion. Carbon dioxide is a greenhouse gas that contributes to the potential for global warming.

❐ carbon monoxide (CO)

A colorless, odorless and toxic gas. It blocks the lungs' ability to obtain oxygen. CO is produced by incomplete combustion of fossil fuels and is a major part of air pollution. Compression ignition (diesel) engines generate significantly lower CO emissions than spark ignited engines.

❐ carcinogens

Substances known to cause cancer.

❐ catalyst

A substance which influences the rate of a chemical reaction but is not one of the original reactants or final products, i.e. it is not consumed or altered in the reaction. Catalysts are used in many processes in the chemical and petroleum industries. Emission control catalysts are used to promote reactions that change exhaust pollutants from internal combustion engines into harmless substances.

❐ cetane index

A calculated value, derived from fuel density and volatility, giving a reasonably close approximation to cetane number.

❐ cetane number

A measure of ignition quality of diesel fuel. The higher the cetane number the easier the fuel ignites when injected into an engine. Cetane number is determined by an engine test using two reference fuel blends of known cetane numbers. The reference fuels are prepared by blending normal cetane (n-hexadecane), having a value of 100, with heptamethyl nonane, having a value of 15.

❐ cfr (cooperative fuel research) engine

A single cylinder, overhead valve, variable compression ratio engine used for measuring octane or cetane quality.

❐ clean air act (caa)

In the U.S., the fundamental legislation to control air pollution. The original Clean Air Act was signed in 1963. The law set emissions standards for stationary sources, such as factories and power plants. Criteria pollutants included lead, ozone, CO, SO_2, NOx and PM, as well as air toxics. The CAA was amended several times, most recently in 1990. The Amendments of 1970 introduced motor vehicle emission standards for automobiles and trucks.

❐ clean-fuel vehicle (cfv)

A vehicle that has been certified to meet clean-fuel standards of the Clean Air Act Amendments of 1990.

❐ cloud point (cp)

A measure of the ability of a diesel fuel to operate under cold weather conditions. Defined as the temperature at which wax first becomes visible when diesel fuel is cooled under standardized test conditions.

❐ cold filter plugging point (cfpp)

A measure of the ability of a diesel fuel to operate under cold weather conditions. Defined as the lowest temperature at which diesel fuel will pass through a fine wire mesh screen of the test apparatus.

❒ common rail injection

A diesel fuel injection system employing a common pressure accumulator, called the rail, which is mounted along the engine block. The rail is fed by a high pressure fuel pump. The injectors, which are fed from the common rail, are activated by solenoid valves. The solenoid valves and the fuel pump are electronically controlled. In the common rail injection system the injection pressure is independent from engine speed and load. Therefore, the injection parameters can be freely controlled. Usually a pilot injection is introduced, which allows for reductions in engine noise and NOx emissions.

❒ compressed natural gas (cng)

Natural gas compressed to a volume and density that is practical as a portable fuel supply.

❒ compression ignition (ci)

The form of ignition that initiates combustion in a diesel engine. The rapid compression of air within the cylinders generates the heat required to ignite the fuel as it is injected.

❒ converted vehicle

A vehicle, originally designed to operated on gasoline or diesel, that has been modified or altered to operate on an alternative fuel.

❒ cordierite

A ceramic material of the formula $2MgO\text{-}2Al_2O_3\text{-}5SiO_2$ which is used for automotive flow-through catalyst substrates and ceramic wall-flow diesel filters.

D

❒ dedicated vehicle

A vehicle designed to operate solely on one alternative fuel, as opposed to a converted vehicle which was later altered to use an alternative fuel.

❒ diesel oxidation catalyst (doc)

Catalyst promoting oxidation processes in diesel exhaust. Usually designed to reduce emissions of the organic fraction of diesel particulates, gas-phase hydrocarbons, and carbon monoxide.

❒ diesel particulate filter (dpf)

A device which physically captures diesel particulates preventing their discharge from the tailpipe. Collected particulates need to be removed from the filter, usually by continuous or periodic oxidation in a process called "regeneration".

❒ diesel particulate matter (dpm)

Sub-micron size particles found in diesel exhaust. Most emission regulations specify DPM measurement methods in which particulates are sampled on filters from cooled exhaust gas. The cooling causes condensation of vapors in the gas sampling train. Thus, the DPM is composed of both solid and liquid particles and is generally classified into three fractions: (1) inorganic carbon (soot), (2) organic fraction (often referred to as SOF or VOF), and (3) sulfate fraction (hydrated sulfuric acid).

- **dimethyl ether (dme)**

The simplest ether CH_3-O-CH_3. Can be manufactured from natural gas or from a renewable organic feedstock. DME is a prospective alternative diesel fuel.

- **direct injection (di)**

In diesel engines with direct injection the combustion chamber is not divided and fuel is injected directly to the cylinder.

- **dod**

The U.S. Department of Defense.

- **doe**

The U.S. Department of Energy.

- **dot**

The U.S. Department of Transportation.

- **dual-fuel vehicle**

A vehicle designed to operate on a combination of alternative fuel, such as compressed natural gas (CNG) or liquefied petroleum gas (LPG), and conventional fuel, such as diesel or gasoline. These vehicles have two separate fuel systems, which inject both fuels simultaneously into the engine combustion chamber.

E

- **electronic control module (ecm)**

A microprocessor that determines the beginning and end of each injection cycle on every cylinder. The ECM determines both fuel metering and injection timing in response to such parameters as engine crankshaft position and rpm, engine coolant and intake air temperature, and absolute intake air boost pressure.

- **elemental carbon (ec)**

Inorganic carbon, as opposed to carbon in organic compounds, sometimes used as a surrogate measure for diesel particulate matter, especially in occupational health environments. Elemental carbon usually accounts for 40-60% of the total DPM mass.

- **emission credit trading**

A program administered by the Environmental Protection Agency under which low polluters are awarded credits which may be traded on a regulated market and purchased by polluters who are in noncompliance for emissions until compliance can be achieved.

❐ **epact**

The U.S. Energy Policy Act of 1992.

❐ **ev**

Electric vehicle.

❐ **evaporative emissions**

Hydrocarbon vapors that escape from a fuel storage tank or a vehicle fuel tank or vehicle fuel system.

F

❐ **federal test procedure (ftp)**

Test cycle(s) used in the U.S. for emission testing and certification of engines and vehicles. The chassis dynamometer cycle for light duty vehicle testing is commonly referred to as FTP-75. The engine dynamometer cycle for testing of heavy-duty (HD) engines is known as HD FTP, or FTP Transient cycle.

❐ **flash point**

The temperature at which a combustible liquid gives off just enough vapor to produce a vapor/air mixture that will ignite when a flame is applied. The flash point is measured in a standardized apparatus using standard test methods, such as ASTM D93 or ISO 2719.

❐ **flexible-fueled vehicle**

A vehicle with the ability to operate on alternative fuels, 100 percent petroleum-based fuels, or a mixture of alternative fuel and petroleum-based fuels.

❐ **fossil fuel**

Any naturally occurring organic fuel formed in the Earth's crust, such as petroleum, coal and natural gas.

❐ **fuel cycle**

The processes involved in extracting a fuel in its native form, converting it to a useful product, transporting it to market, and consuming it at its final destination.

G

❐ **gas turbine**

A turbine powered by a gas, such as the hot gas produced in combustion processes.

❐ **gas turbine engine**

A combustion engine utilizing a turbine, such as a jet engine.

❑ genset

Generator set, a generating system comprising a combustion engine driving an electrical generator.

❑ geometric surface area (gsa)

In monolith catalyst substrates, the total channel surface area per unit of substrate volume.

❑ gross vehicle weight (gvw)

The maximum loaded weight (including the vehicle itself, passengers, and cargo) for which a vehicle is designed, as specified by the manufacturer. Often used as a criterion of vehicle size for the purpose of legislation; the exact definition may vary depending on the jurisdiction.

H

❑ hybrid electric vehicle (hev)

Hybrid-electric vehicle. Various types of electric vehicles that use another power source to propel the vehicle or generate power for an electric drive train, or a combination of the two types.

❑ hydraulic/electronic unit injector (heui)

A type of unit injector actuated by engine oil pressure rather than the camshaft. A very high oil pressure (up to 3,000 psi) is created by a separate oil pump. This high pressure is routed to every injector through a gallery. The engine's Electronic Control Module varies the pressure in response to engine speed and other parameters.

❑ hydromechanical injection

An injection system in which mechanical parts work through hydraulic pressure to meter and time the injection of fuel. No electronics are incorporated into hydromechanical injection systems.

I

❑ ignition delay

The length of time or number of degrees of crankshaft rotation between the beginning of injection and ignition of the fuel.

❑ in-direct injection (idi)

In diesel engines with in-direct injection the fuel is injected to an auxiliary pre-chamber. Combustion starts in the prechamber and propagates to the cylinder.

❑ inherently low emission vehicle (ilev)

Any vehicle that is certified to meet transitional low-emission vehicle standards established by the California Air Resources Board (CARB) and does not emit any evaporative emissions.

❐ injection period

The time, measured in degrees of crankshaft rotation, between the beginning and end of injection. On engines with hydromechanical injection systems, it is controlled by the opening and closing of ports in the injector body or by the action of a plunger forcing fuel out of a cup. On electronic injection systems, it is determined directly or indirectly by the action of a solenoid valve.

❐ in-line injection pump

An injection pump with a separate cylinder and plunger for each engine cylinder. Each plunger is rotated by a rack to determine metering via ports in the body of the pump and helical cuts on the pump plungers. The plungers are driven off a camshaft, which usually incorporates a centrifugal or electronically controlled timing advance mechanism.

❐ inspection and maintenance programs (im)

In-use emission testing programs, primarily for passenger cars, used by some U.S. states to identify vehicles that cause excessive emissions which contribute to air pollution.

❐ intumescent mat

Ceramic fiber mat which irreversibly expands after exposed to high temperature. Usually contains vermiculite. Intumescent mats are used in the canning of catalytic converters and diesel filters to hold the ceramic substrate inside the steel canister.

❐ in-use deterioration (iud)

The effect of time and use on vehicle performance and emissions.

L

❐ lean nox catalyst (lnc)

Catalyst designed to reduce nitrogen oxides from diesel or spark-ignited engine exhaust gases under net oxidizing conditions, i.e., in the presence of excessive amount of oxygen.

❐ light duty vehicle (ldv)

A passenger car or small truck.

❐ liquefied natural gas (lng)

Natural gas that has been refrigerated to cryonic temperatures where the gas condenses into a liquid.

❐ liquefied petroleum gas (lpg)

Liquefied Petroleum Gas (LPG) is a mixture of low-boiling hydrocarbons that exists in a liquid state at ambient temperatures when under moderate pressures (less than 1.5 MPa or 200 psi). LPG is a by-product from the processing of natural gas and from petroleum refining. Major components of LPG are propane (min. 85% content in the U.S.), butane and propylene.

❐ low emission vehicle (lev)

A vehicle that is certified to meet the LEV emission standards set by the California Air Resources Board (CARB).

N

❐ national ambient air quality standards (naaqs)

Ambient standards for six pollutants including ozone, carbon monoxide, nitrogen dioxide, lead, particulate matter, and oxides of sulfur specifically regulated under the U.S. Clean Air Act of 1990. Urban areas are required to achieve attainment in regards to ambient concentrations of these criteria pollutants.

❐ natural gas (ng)

A mixture of hydrocarbon compounds and small quantities of various non hydrocarbon components existing in the gas phase or in solution with crude oil in natural underground reservoirs. The main component of natural gas is methane.

❐ nitrogen oxides (nox)

Several air-polluting gases composed of nitrogen and oxygen which play an important role in the formation of photochemical smog. Nitrogen oxides are collectively referred to as "NOx", where "x" represents a changing proportion of oxygen to nitrogen. Internal combustion engines are significant contributors to the worldwide nitrogen oxide emissions. For the purpose of emission regulations, NOx is composed of colorless nitric oxide (NO), and the reddish-brown, very toxic and reactive nitrogen dioxide (NO_2). Other nitrogen oxides, such as nitrous oxide N_2O (the anesthetic "laughing gas"), are not regulated emissions.

❐ nmhc

Non-Methane Hydrocarbons.

❐ nmog

Non-Methane Organic Gases.

❐ nonattainment area

A region that exceeds the U.S. National Ambient Air Quality Standards (NAAQS) for one or more criteria pollutants. Such regions, or areas, are required to seek modifications to their State Implementation Plans (SIPs), setting forth a reasonable timetable using means that are approved by the Environmental Protection Agency (EPA) to achieve attainment of NAAQS by a certain date. Under the Clean Air Act, if a nonattainment area fails to attain NAAQS, the EPA may superimpose a Federal Implementation Plan (FIP) with stricter requirements. Also, the EPA may impose fines, construction bans, or cut-offs in Federal grant revenues until the area achieves applicable NAAQS.

O

❑ on-board diagnostics (obd)

A system on board of the vehicle that monitors emission control components and alerts the driver (e.g., by a dashboard light) if malfunctions or emission deterioration occurs. The OBD system involves a number of sensors and a data processor, which is typically integrated with the vehicle's electronic management system.

❑ open frontal area (ofa)

In monolith catalyst substrates, the part of the total substrate cross-section area which is available for the flow of gas. The OFA is frequently expressed as a percentage of the total substrate cross-section and sometimes also called the substrate void fraction.

❑ original equipment manufacturer (oem)

Manufacturers of equipment (such as engines, vehicles, etc.) that provide the original product design and materials for its assembly and manufacture. OEMs are directly responsible for manufacturing and modifying the products, making them commercially available, and providing the warranty.

❑ overhead cam

A camshaft used for operating both valves and unit injectors, located on top of or within the cylinder head. Such camshafts are driven by a multi-gear gear train off the crankshaft. They simplify the design of the cylinder head and eliminate pushrods, allowing for much larger, open intake and exhaust ports and better breathing.

❑ oxygen storage capacity (osc)

A capacity of the catalyst washcoat to store oxygen at lean and to release it at rich condition. Typically provided by cerium oxide (ceria), which oscillates between an oxidized and reduced state, depending on the exhaust gas chemistry. The OSC is an important component of three-way catalysts, used to extend the catalyst window.

❑ oxygenated fuel

Any fuel substance containing oxygen, such as ethanol, methanol, or biodiesel. Oxygenated fuel tends to give a more complete combustion of its carbon into carbon dioxide (CO_2), thereby reducing emissions of hydrocarbons and carbon monoxide. Oxygenated fuels may result in increased nitrogen oxides emissions.

❑ ozone (O_3)

An oxygen molecule with three oxygen atoms. The stratosphere ozone layer, which is a concentration of ozone molecules located at 10 to 50 kilometers above sea level, is in a state of dynamic equilibrium. Oxygen molecules absorb ultraviolet (UV) light to form ozone which, in turn, decomposes back to oxygen. These processes absorb most of the ultraviolet light from the sun, shielding life from the harmful effects of UV radiation. Ozone is normally present at ground level in low concentrations. In cities where high level of air pollutants is present, the action of the sun's ultraviolet light can, through a complex series of reactions, produces harmful concentrations of the ground level ozone. The resulting air pollution is known as photochemical smog.

P

❑ particulate matter (pm)

Particles formed by incomplete combustion of fuel. Compression ignition (diesel) engines generate significantly higher PM emissions than spark ignited engines. The particles are composed of elemental carbon, heavy hydrocarbons (SOF), and hydrated sulfuric acid ("sulfate particulates").

❑ petroleum

A generic term applied to oil and oil products in all forms. Examples include crude oil, lease condensate, unfinished oil, refined petroleum products, and natural gas plant liquids.

❑ polycyclic organic matter (pom)

A class of air toxics defined in the US Clean Air Act as compounds with more than one benzene ring and a boiling point of 100°C and higher. Includes practically all of diesel PAH material.

❑ polynuclear aromatic hydrocarbons (pah)

Aromatic hydrocarbons with two or more (up to five or six) benzene rings joined in various, more or less clustered forms.

❑ pour point

A measure of the ability of a diesel fuel to operate under cold weather conditions. Defined as the temperature at which the amount of wax out of solution is sufficient to gel the fuel when tested under standard conditions.

❑ precombustion chamber

A small, auxiliary combustion chamber connected by a narrow orifice with the main chamber. Fuel is injected into the prechamber and ignites there, causing hot gases to expand into the main chamber (cylinder).

❑ propane (C_3H_8)

A normally gaseous straight-chain hydrocarbon. Propane is a colorless paraffinic gas that boils at a temperature of -42°C. It is extracted from natural gas or refinery gas streams.

❑ pump-line-nozzle fuel system

A fuel system using a single injection pump driven off the geartrain on the front of the engine that also drives the camshaft. The central injection pump feeds a separate injection nozzle located in the cylinder head above each cylinder. Lines which must be of exactly equal length link each pump plunger with the associated nozzle. Each nozzle incorporates a needle valve and the orifices which actually handle atomization.

❒ purge test

A test used to determine if fuel vapors are properly drawn from the evaporative canister and the fuel tank into the engine for combustion. If the purge system is not working properly, the evaporative canister can become saturated and vent hydrocarbons into the atmosphere.

R

❒ reformulated gasoline (rfg)

Gasoline whose composition has been changed, from that of gasolines sold in 1990, to (1) include oxygenates, (2) reduce the content of olefins, aromatics and volatile components, and (3) reduce the content of heavy hydrocarbons to meet performance specifications for ozone-forming tendency and for release of toxic substances (benzene, formaldehyde, acetaldehyde, 1,3-butadiene, and polycyclic aromatic hydrocarbons) into the air from both evaporation and tailpipe emissions.

❒ renewable energy

Energy obtained from sources that are essentially inexhaustible, unlike fossil fuels. It includes conventional hydro-electric, wood, bio-feedstocks, waste, geothermal, wind, photovoltaic, and solar thermal energy.

❒ respirable combustible dust (rcd)

A method of measuring ambient DPM exposures using a combustion process. Used in underground mines in Canada.

❒ rotary injection pump

A lower-cost injection pump used with pump-line-nozzle systems. The pump has a central plunger system (usually consisting of two opposing plungers) that provides fuel to every cylinder during the required injection period. A plate located near the top of the pump rotates, opening an appropriate orifice at the right time for distribution to each cylinder's injection nozzle through a separate line. It is usually used with automotive or agricultural engines that have lower performance and durability requirements than the heavy-duty truck diesels.

S

❒ scaqmd

South Coast Air Quality Management District.

❒ selective catalytic reduction (scr)

Term frequently used as a synonym for catalytic reduction of NOx in diesel exhaust or flue gases by nitrogen containing compounds, such as ammonia or urea. Such SCR systems are commercially available for stationary applications and are being developed for mobile diesel engines. Since "selective catalytic reduction" is a generic term which also applies to other reactions, its use may lead to confusion in some situations.

❐ size selective sampling (sss)

A method of measuring ambient DPM exposures through a mechanical separation (inertial impaction) of particles into two fractions: above and below (typically) 0.8 μm, followed by gravimetric analysis. Based on air quality research in underground mining, it is believed that the particle fraction below the 0.8 μm cut point can be attributed to diesel exhaust.

❐ soluble organic fraction (sof)

The organic fraction of diesel particulates. SOF includes heavy hydrocarbons derived from the fuel and from the engine lubricating oil. The term "soluble" originates from the analytical method used to measure SOF which is based on extraction of particulate matter samples using organic solvents.

❐ super ultra low emission vehicle (sulev)

A vehicle that meets the SULEV emission standards set by the California Air Resources Board. The SULEV standards are more stringent than the ULEV standards.

❐ supercharger

A compressor device to compress the combustion air or the air/fuel mixture before it enters the engine cylinder. Superchargers are typically driven by the engine itself, through a system of gears, a belt drive, or by an electrical motor.

❐ swirl combustion

A combustion chamber configuration which uses curved mixing ridges in the intake ports and/or a re-entrant piston bowl (a bowl whose top edges curve inward). Some swirl combustion chambers have a larger rim around the outside of the piston and a more compact combustion chamber or bowl. The swirl is used to reduce particulate emissions.

T

❐ threshold limit value (tlv)

Time-weighted average concentration of an air pollutant at the workplace for a conventional 8-hour workday and a 40-hour workweek, to which nearly all workers may be repeatedly exposed without adverse health effects.

❐ total carbon (tc)

The sum of the elemental carbon and organic carbon associated with diesel particulates. Typically amounts to 80-85% of the total DPM mass.

❐ total particulate matter (tpm)

The total particulate matter emissions including all fractions of diesel particulates, i.e. the carbonaceous, organic (SOF), and sulfate particulates.

❑ **transitional low emission vehicle (tlev)**

A vehicle that is certified to meet the TLEV emission standards established by the California Air Resources Board (CARB). The TLEV standards are less stringent than the LEV standards.

❑ **turbine**

A rotary machine which extracts mechanical shaft power from the working fluid (gas or liquid) using rotor vanes.

❑ **turbo generator**

Gas turbine combined with an electrical generator.

❑ **turbo lag**

The time delay between injecting fuel to accelerate and delivering air to the intake manifold by the turbocharger. This phenomenon may cause black smoke emissions in some turbocharged diesel engines during acceleration.

❑ **turbocharging**

A process of compressing the engine intake air charge in order to allow more air and fuel into the cylinder and, thus, to increase the engine power output. The compressor, called the turbocharger, is driven by an exhaust gas propelled turbine.

U

❑ **ultra-low emission vehicle (ulev)**

A vehicle that meets the ULEV emission standards set by the California Air Resources Board. The ULEV standards are more stringent than the LEV standards.

❑ **unit injector**

An injector which is camshaft-driven and incorporates a plunger. The plunger works in conjunction with orifices in the injector body to determine the beginning and end of injection. The plungers has a helix and is rotated by a rack so the beginning and end of injection can occur closer together or farther apart, thus shortening or lengthening the injection period and changing metering. A special pump which supplies fuel through an orifice to the injector operates at a pressure which is precisely controlled depending on the changes in engine speed.

V

❑ **vmt**

Vehicle miles traveled.

❑ **volatile organic compounds (voc)**

Hydrocarbon-based emissions released through evaporation or combustion. The term VOC is usually used in regard to stationary emission sources.

❑ volatile organic fraction (vof)

The organic fraction of diesel particulate matter as determined by vacuum evaporation. It may or may not be equivalent to the SOF fraction. Depending on the exact analytical procedure, the VOF may include the organic material (SOF) as well as some of the sulfate particulates which, being composed primarily of hydrated sulfuric acid, are also volatile.

W

❑ white smoke

The smoke emitted during a cold start from a diesel engine, composed mainly of unburnt fuel and particulate matter.

Z

❑ zero emission vehicle (zev)

A vehicle that is certified to meet the most stringent emission standards established by the California Air Resources Board (CARB). These standards require zero regulated emissions of nonmethane organic gases (NMOGs), carbon monoxide (CO) and nitrogen oxides (NOx). A ZEV is most likely powered by electricity, fuel cells or hydrogen.

Environment Health

A

❐ absorption

The process of taking in, as when a sponge takes up water. Chemicals can be absorbed into the bloodstream after breathing or swallowing. Chemicals can also be absorbed through the skin into the bloodstream and then transported to other organs. Not all of the chemical breathed, swallowed, or touched is always absorbed.

❐ access to sanitation

- Throughout the world, access to sanitation is more limited than access to safe water.
- About one quarter of the population of the developing countries lack access to safe water, but more than half have no access to sanitation.
- Some 2.5 billion people in the developing countries lack access to sanitation, as against 1.9 billion who have access.
- In over half of the African countries and more than one quarter of the Asian and Oceanic countries, the majority of the population lacks access to sanitation.

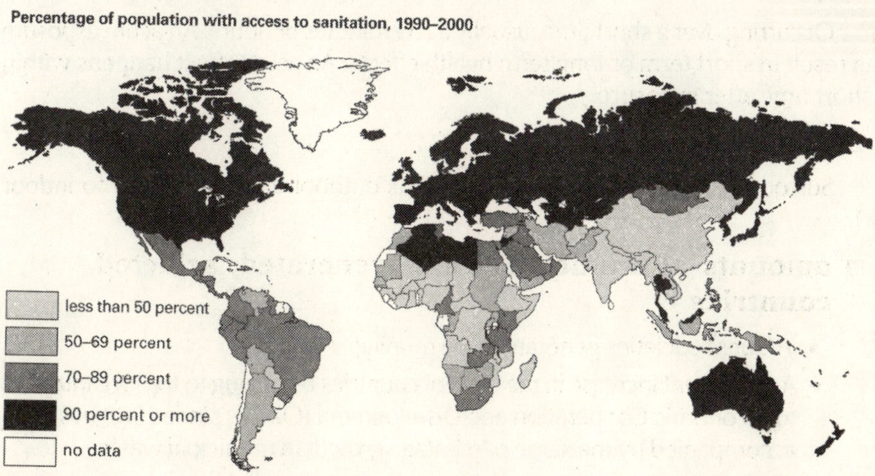

- This has severe implications for health, especially children's health.
- Poor sanitation encourages the spread of cholera, typhoid, diarrhea and intestinal parasites.

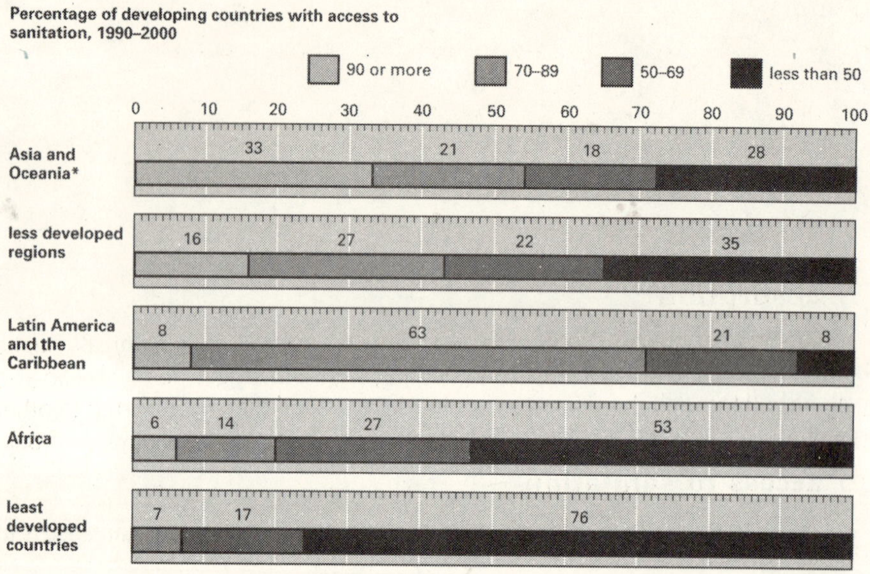

- Diarrheal diseases are the biggest killer of children under five, accounting for 2 million deaths every year.
- Substantial investment in sanitation facilities is needed in order to achieve the goal of universal access to sanitation, especially in the poorest countries.

❐ acute

Occurring over a short time, usually a few minutes or hours. An acute exposure can result in short term or long term health effects. An acute effect happens within a short time after exposure.

❐ ambient

Surrounding. Ambient air usually means outdoor air (as opposed to indoor air).

❐ amounts of municipal waste generated, selected countries

- Affluent societies generate more municipal waste.
- A 40 percent increase in the GDP of countries belonging to the Organization for Economic Cooperation and Development (OECD) since 1980 has been accompanied by the same percentage growth in municipal waste.
- The OECD predicts that there will be a further 70–100 percent increase in GDP in its region by 2020.
- Unless the link between waste generation and GDP is severed, there could be a commensurate increase in waste.

- Certain social trends, such as the increase in singleperson households due to higher divorce rates and the ageing population, will lead to higher per capita waste generation.
- As the developing world industrializes and grows more affluent, it too can be expected to increase waste generation.
- The level of recycling in most industrialized countries is very low, and is not keeping pace with waste increases in most countries.
- The highest levels are in Switzerland and Austria (40 percent and 38 percent).
- Switzerland recycles over 80 percent of its glass , the USA and Canada only around 20 percent.

Municipal waste in OECD countries: quantities and destinations

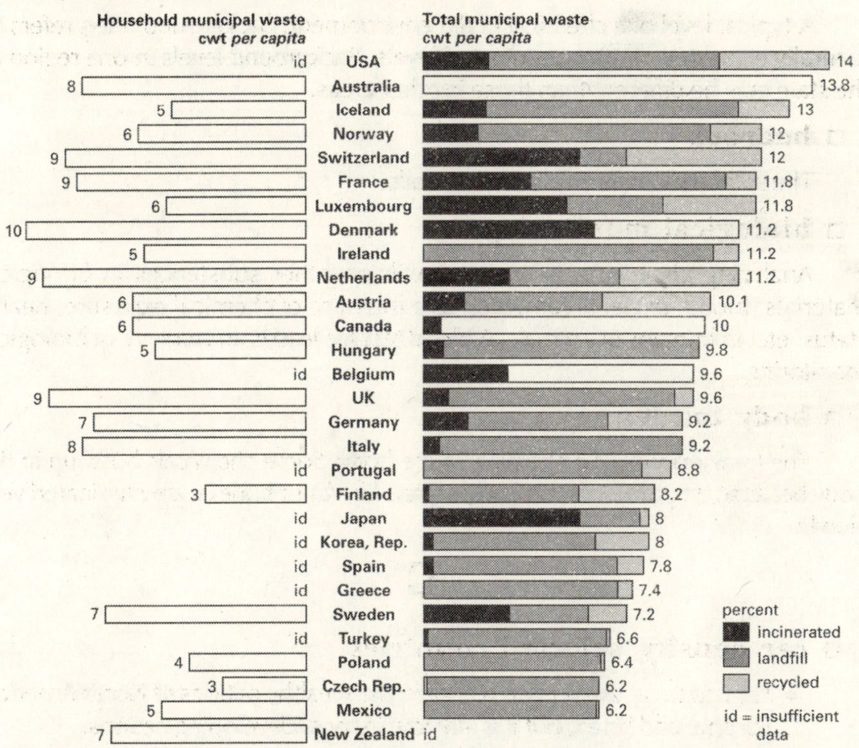

Glass waste recycling as a proportion of total consumption in selected countries

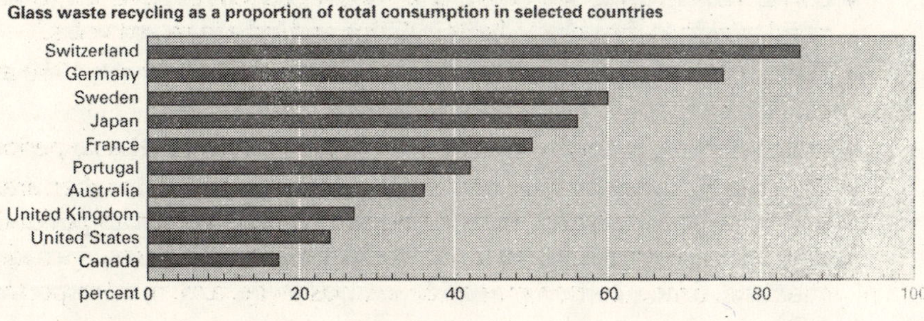

▫ analyte

A chemical for which a sample (such as water, air, blood, urine or other substance) is tested. For example, if the analyte is mercury, the laboratory test will determine the amount of mercury in the sample.

▫ aquifer

An underground source of water. This water may be contained in a layer of rock, sand or gravel.

B

▫ background level

A typical level of a chemical in the environment. Background often refers to naturally occurring or uncontaminated levels. Background levels in one region of the state may be different than those in other areas.

▫ bedrock

The solid rock underneath surface soils.

▫ biological monitoring

Analyzing chemicals, hormone levels or other substances in biological materials (blood, urine, breath, etc.) as a measure of chemical exposure, health status, etc. in humans or animals. A blood test for lead is an example of biological monitoring.

▫ body burden

The total amount of a chemical in the body. Some chemicals build up in the body because they are stored in body organs like fat or bone or are eliminated very slowly.

C

▫ car density, selected countries

- Per capita car ownership is high in the wealthy nations of North America, Europe, and Japan, but it is still low in most developing nations.
- Lower vehicle ownership costs and robust economies have led to very rapid growth in the vehicle fleets in China and India in recent years.
- The number of vehicles in China grew by 300 percent between 1980 and 1998.
- India's fleet expanded by nearly 9 percent per year over the same period.
- The growing use of internal combustion vehicles, especially in urban areas, will increase congestion, raise the demand for oil, worsen air pollution, and increase emissions of a variety of greenhouse gases, including methane, ozone, carbon monoxide, nitrous oxide, and, most important, CO_2.

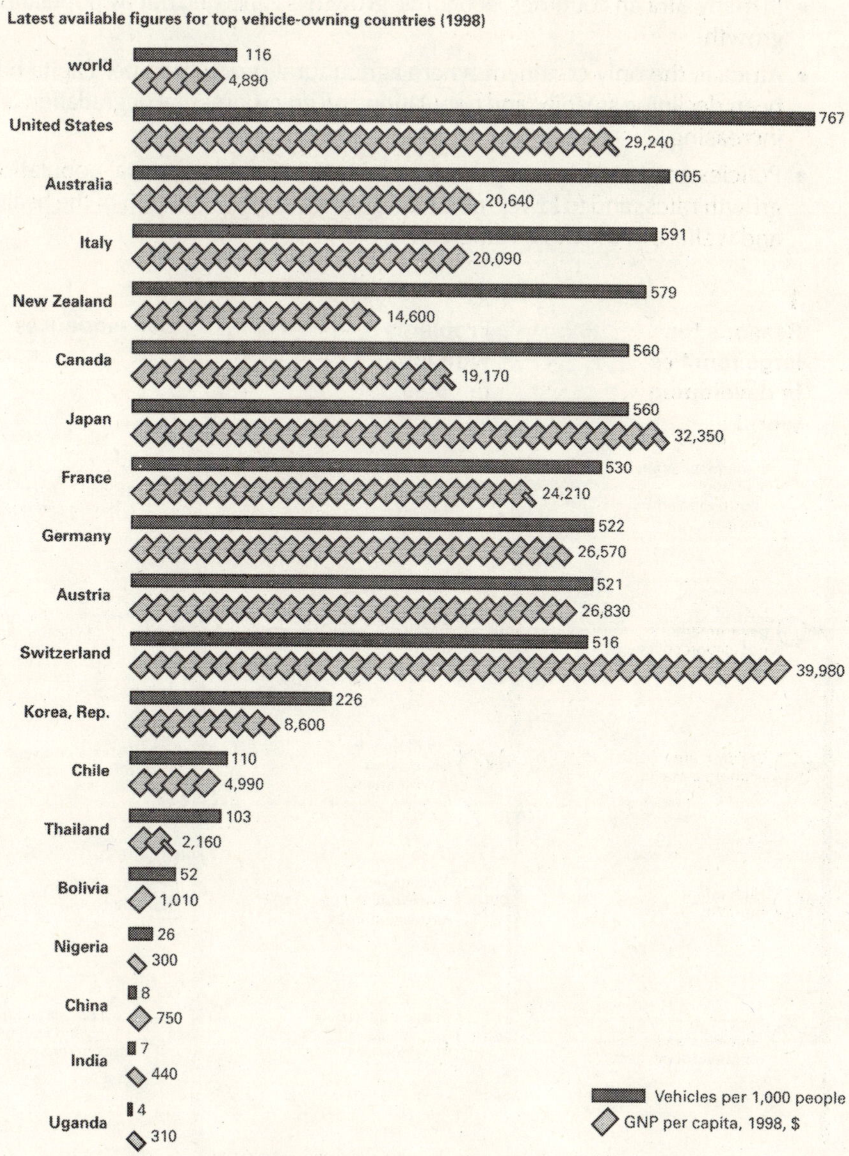

❑ case control study

A study in which people with a disease (cases) are compared to people without the disease (controls) to see if their past exposures to chemicals or other risk factors were different.

❑ causes and consequences of high population growth

- If the unmet need for family planning was satisfied, population growth rates would fall.

- In many African countries, economic growth is cancelled out by population growth.
- Africa is the only continent where agricultural production per capita has been declining steadily, and population and environmental degradation are increasing.
- Policies are needed to empower women; to lower high annual population growth rates; and to lower mortality rates, through promotion of the health and welfare of the most vulnerable.

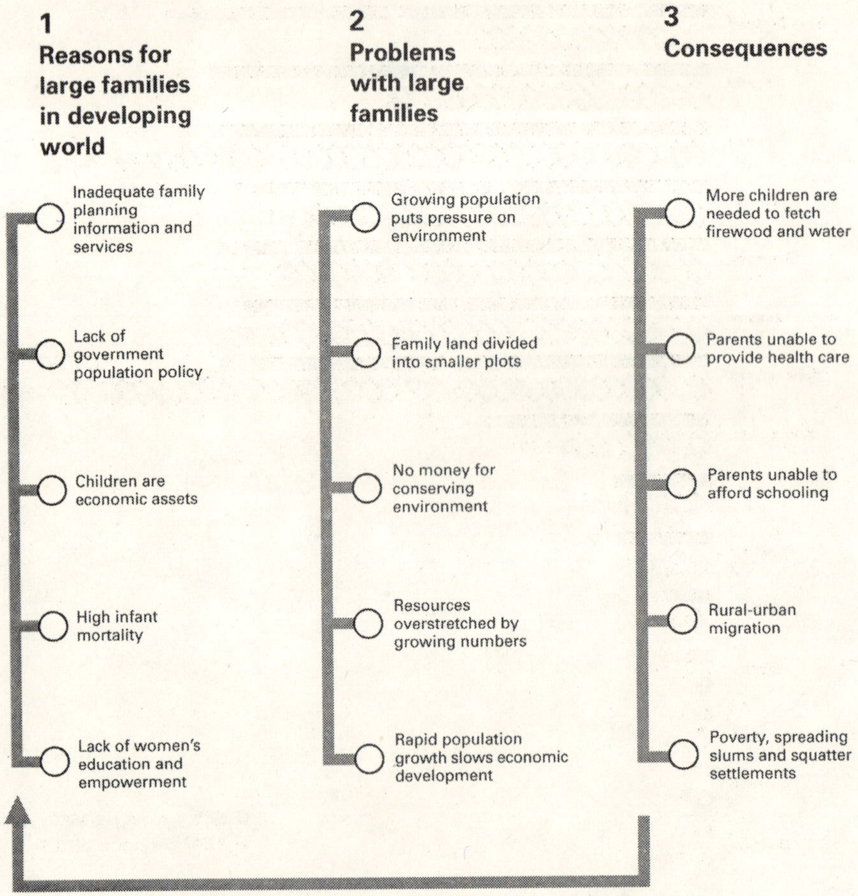

❑ causes of death and disability in developed and developing countries

- In the developing regions where 80 percent of the planet's people live, noncommunicable diseases such as heart disease and conditions such as depression are fast replacing infectious diseases and malnutrition as the leading causes of disability and premature death.

- Noncommunicable diseases are expected to account for seven out of every ten deaths in the developing regions by the year 2020, compared with less than 50 percent today.
- In 1990 the three leading causes of healthcare dependence were: pneumonia; diarrheal diseases, and perinatal (birth-related) conditions.
- The three causes projected to take their place by 2020 are heart disease, depression and injuries sustained in road traffic collisions.
- By 2020 it is expected that the healthcare costs attributable to illness caused by tobacco use will outweigh those caused by any single other disease.

Changes in healthcare dependence worldwide 1990–2020

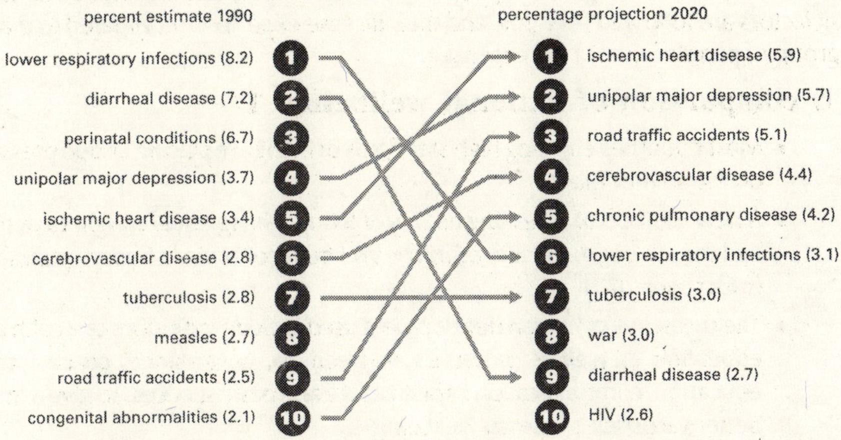

Latest available figures for death by main causes in developed and developing countries, 1985 and 1997

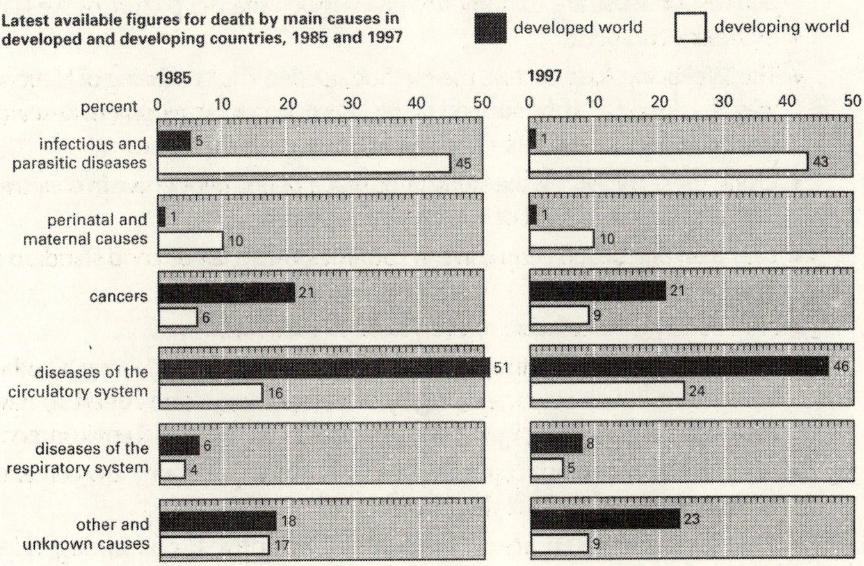

❑ central nervous system (cns)

The part of the nervous system that includes the brain and the spinal cord.

❑ cercla

Comprehensive Environmental Response, Compensation and Liability Act. See "Superfund."

❑ chronic

Occurring over a long period of time, several weeks, months or years.

❑ cohort study

A study in which a group of people with a past exposure to chemicals or other risk factors are followed over time and their disease experience compared to that of a group of people without the exposure.

❑ comparison of national wellbeing: 1

- Most countries enjoying high standards of living are placing undue pressure on the environment.
- A new method of assessing quality of living, giving equal weight to people and the environment, is used in The Wellbeing of Nations Report, published in October 2001.
- The measures of human development used include indicators of wealth and education as well as measures on freedom, governance, peace, order, education, communication infrastructure and basic services, to give a much better yardstick of human wellbeing.
- Similarly, measures on the broad themes of land, water, air, biodiversity and resource use are combined to give a much sharper picture of the state of the environment.
- The Wellbeing Assessment, the method used in The Wellbeing of Nations, was developed with the support of the International Development Research Center (IDRC) and IUCN – the World Conservation Union.
- Using these measures, the study finds that 4 billion people live in countries with a poor or bad level of human development.
- Less than one billion people live in countries with a fair or good standard of living.
- No country is sustainable or even close to sustainability.
- The leaders (Sweden, Finland, Norway, Iceland, and Austria) and 32 other ecosystem deficit countries (largely in Europe and North America) have high standards of living but excessive impacts on the global environment.
- Some 27 human deficit countries (chiefly in Africa) have fairly low demands on the ecosystem but are desperately poor.
- The remaining 116 are double deficit countries, combining weak environmental performance and inadequate development.

- In more than 140 nations, ecosystem stress is higher than human wellbeing–evidence that most people's efforts to improve their lot are inefficient and overexploit the environment.

The Wellbeing Index

- The Wellbeing Index (WI) is the point on the Barometer of Sustainability where the Human Wellbeing Index (HWI) and Ecosystem Wellbeing Index (EWI) intersect. It is expressed here as the average of the HWI and the EWI.

The table shows countries ranked in order of the Wellbeing Index

Small island nations and others with insufficient data are omitted.

A ranking (= joint)

B country

C wellbeing index

D human wellbeing index

E ecosystem wellbeing index

A	B	C	D	E
1	Sweden	64.0	79	49
2	Finland	62.5	81	44
3	Norway	62.5	82	43
4	Iceland	61.5	80	43
5	Austria	61.0	80	42
7=	Canada	60.5	78	43
7=	Switzerland	60.5	78	43
9	Belize	57.0	50	64
10	Guyana	57.0	51	63
11	Uruguay	56.5	61	52
12	Germany	56.5	77	36
13	Denmark	56.0	81	31
14	New Zealand	55.5	73	38
15	Suriname	55.0	52	58
16	Latvia	54.0	62	46
17	Ireland	54.0	76	32
18	Australia	53.5	79	28
19	Peru	53.0	44	62
20	Slovenia	53.0	71	35
22	Lithuania	52.5	61	44
23	Cyprus	52.5	67	38
24	Japan	52.5	80	25
27	United States	52.0	73	31
28	Italy	52.0	74	30
29	France	52.0	75	29

(Contd.)

A	B	C	D	E
30=	Czech R	51.5	70	33
30=	Greece	51.5	70	33
32	Portugal	51.5	72	31
33	United Kingdom	51.5	73	30
34	Belgium	51.5	80	23
35	Botswana	51.0	34	68
36	Slovakia	50.5	61	40
37	Luxembourg	50.5	77	24
38=	Armenia	50.0	45	55
38=	Netherlands	50.0	78	22
41	Ecuador	49.5	43	56
42	Mongolia	49.5	39	60
43	Singapore	49.0	66	32
44	Hungary	49.0	65	33
45	Mauritius	49.0	54	44
47	Benin	49.0	27	71
48	Costa Rica	48.5	56	41
49	Sri Lanka	48.5	40	57
50	Bolivia	48.5	34	63
51	Estoniaz	48.0	62	34
52	Fiji	48.0	50	46
53	Belarus	48.0	46	50
54	Poland	47.5	65	30
55	Argentina	47.5	55	40
56	Dominican R	47.5	49	46
58	Korea, R	47.0	67	27
59	Barbados	47.0	62	32
60	Cape Verde	47.0	47	47
61	Spain	46.5	73	20

❑ composite sample

A sample which is made by combining samples from two or more locations. The sample can be of water, soil or another medium.

❑ concentration

The amount of one substance dissolved or contained in a given amount of another substance or medium. For example, sea water has a higher concentration of salt than fresh water does.

❑ contaminant

Any substance that enters a system (the environment, human body, food, etc.) where it is not normally found. Contaminants are usually referred to in a "negative" sense and include substances that spoil food, pollute the environment or cause other adverse effects.

❒ contraceptive use

- The availability and quality of family planning services, social traditions of the acceptability of contraceptive use, and the costs of creating the organization needed have all often hindered developing countries in achieving a comprehensive contraception program.

World map of contraceptive use

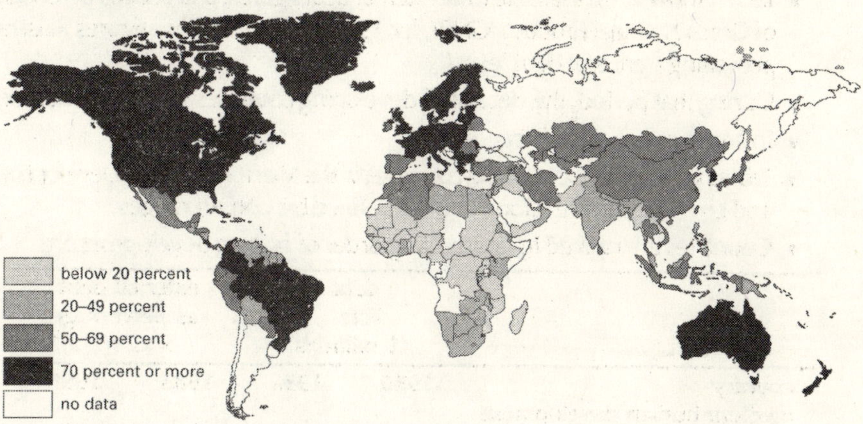

Types of contraception and proportions of people using them

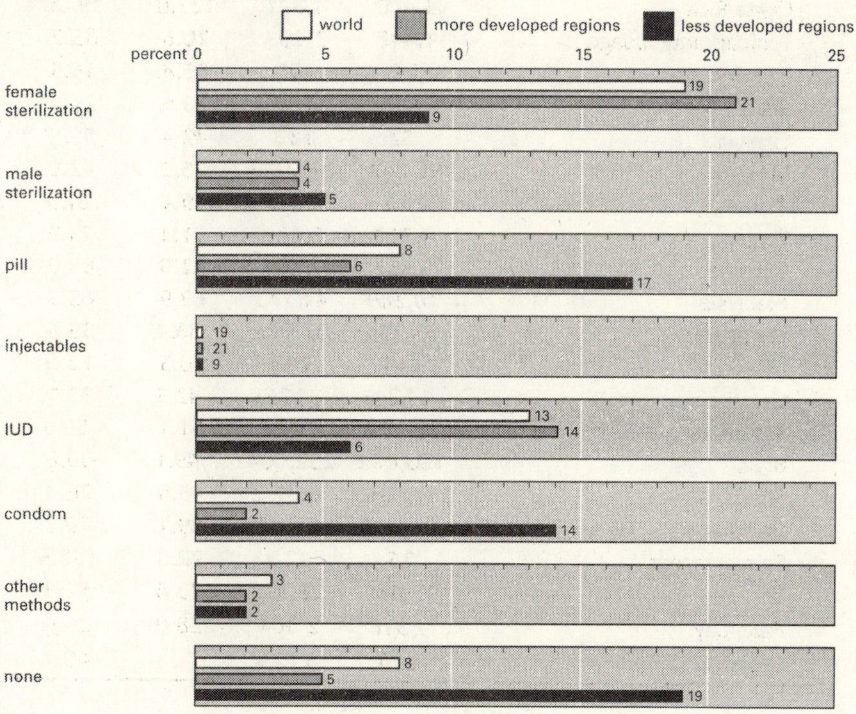

- Female sterilization is the most popular contraceptive method worldwide, followed by the intrauterine device (IUD).
- However, access to modern methods is still very limited in Africa and in parts of other regions.

D

❑ debt by country: 1

- Tables show latest available total external debt figures, and debt as percentage of Gross National Product (GNP), for 1985 and 1998. These figures illustrate prevailing trends in debt levels.
- During that period, the debt of all developing countries more than doubled.
- In some cases, debt exceeds GNP.
- Tables show most developing countries in the Medium Human Development and Low Human Development (as defined by UNDP) ranges.
- Countries are ranked in descending order of human development.

	debt Total ($ millions)		external debt as percentage of GNP ($)	
country medium human development	1985	1998	1985	1998
Saint Kitts and Nevis	13	1,151	6.7	43.2
Costa Rica	4,400	3,971	121.0	39.06
Trinidad and Tobago	1,448	2,193	20.6	35.7
Dominica	54	109	55.8	46.5
Seychelles	97	187	59.6	36.3
Grenada	52	183	42.4	55.9
Mexico	96,862	159,959	55.2	42.0
Belize	118	338	59.4	51.9
Panama	4,758	6,689	91.4	78.0
Bulgaria	3,852	9,907	22.0	83.0
Malaysia	20,269	44,773	69.9	65.3
Venezuela	35,334	37,003	58.4	39.6
Fiji	444	193	40.5	12.6
Colombia	14,245	33,263	42.5	33.1
Mauritius	629	2,482	61.1	59.6
Brazil	103,602	232,004	49.1	30.6
Thailand	17,546	86,172	45.9	76.4
Philippines	26,637	47,817	89.1	70.1
Saint Vincent	25	420	22.3	138.9
Peru	12,884	32,397	73.0	52.9
Paraguay	1,817	2,304	58.0	26.6
Lebanon	870	6,725	..	40.7

	debt Total ($ millions)		external debt as percentage of GNP ($)	
country medium human development	**1985**	**1998**	**1985**	**1998**
Jamaica	4,103	3,995	225.6	63.1
Sri Lanka	3,540	8,526	59.5	54.9
Turkey	26,013	102,074	38.4	50.0
Oman	2,329	3,629	26.3	..
Dominican Rep.	3,502	4,451	74.1	29.8
Saint Lucia	23	184	12.4	31.9
Maldives	83	180	116.3	58.1
Ecuador	8,703	15,140	58.9	82.5
Jordan	4,022	8,484	78.7	146.9
Samoa (Western)	76	180	88.8	102.1
Guyana	1,496	1,653	388.8	248.6
Iran	6,057	14,391	3.4	12.7
China	16,696	154,599	5.5	16.4
Tunisia	4,884	11,078	60.6	58.0
South Africa	..	24,711	..	18.9
El Salvador	1,851	3,633	50.2	30.8
Cape Verde	97	244	..	49.8
Algeria	18,260	30,665	32.4	67.5
Vietnam	61	22,359	..	82.3
Indonesia	36,715	150,875	44.4	176.5
Syria	10,843	22,435	66.5	137.9
Swaziland	243	251	60.8	18.7
Honduras	2,730	5,002	78.5	96.9
Bolivia	4,805	6,0771	67.2	72.8
Nicaragua	5,758	5,968	229.0	335.9
Mongolia	..	739	..	74.7
Vanuatu	16	63	13.0	28.3
Egypt	36,102	31,964	115.0	37.3
Guatemala	2,677	4,565	28.0	24.3
Solomon Islands	66	152	42.6	51.6

❑ debt by country: 2

	debt Total ($ millions)		external debt as percentage of GNP ($)	
country medium human development (contd.)	**1985**	**1998**	**1985**	**1998**
Botswana	351	548	33.1	11.8
Gabon	1,206	4,425	39.0	90.7

(Contd.)

	debt Total ($ millions)		external debt as percentage of GNP ($)	
country medium human development (contd.)	**1985**	**1998**	**1985**	**1998**
Morocco	15,779	20,687	129.2	60.3
Myanmar (Burma)	3,098	5,680	..	..
Lesotho	175	692	36.7	64.7
India	40,951	98,232	17.7	23.0
Ghana	2,256	6,884	51.0	91.8
Zimbabwe	2,415	4,716	43.9	79.8
Equatorial Guinea	132	306	175.7	75.7
São Tomé & Principe	63	246	..	684.0
Papua New Guinea	2,112	2,692	90.4	76.9
Cameroon	3,174	9,829	40.2	119.4
Pakistan	13,465	32,229	43.9	52.8
Cambodia	7	2,210	..	77.7
Comoros	134	203	118.4	103.3
Kenya	4,181	7,010	70.8	61.5
Congo	3,050	5,119	150.7	306.9
low human development				
Laos	619	2,437	26.1	199.1
Madagascar	2,529	4,394	92.7	119.5
Bhutan	9	120	5.6	32.1
Sudan	8,955	16,843	75.1	182.7
Nepal	590	2,646	22.2	54.2
Togo	935	1,448	128.9	97.4
Bangladesh	6,870	16,376	32.1	37.1
Mauritania	1,454	2,589	198.7	272.5
Yemen	3,339	4,138	..	104.8
Djibouti	144	288	..	..
Haiti	717	1,048	36.1	27.1
Nigeria	18,643	30,315	68.1	78.8
Congo, Dem. Rep.	6,171	12,929	93.0	208.2
Zambia	4,499	6,865	226.5	217.4
Côte d'Ivoire	9,659	14,852	153.4	145.4
Senegal	2,566	3,861	104.7	83.1
Tanzania	9,107	7,603	..	94.3
Benin	854	1,647	83.3	72.2
Uganda	1,232	3,935	35.5	58.2

	debt Total ($ millions)		external debt as percentage of GNP ($)	
country medium human development (contd.)	**1985**	**1998**	**1985**	**1998**
Eritrea	..	149	..	19.4
Angola	2,993	12,173	47.6	297.1
Gambia	245	477	113.7	116.7
Guinea	1,466	3,546	..	102.0
Malawi	1,021	2,444	94.6	137.5
Rwanda	366	1,226	21.4	60.8
Mali	1,456	3,201	113.1	120.4
Central African Republic	344	921	40.1	88.8
Chad	217	1,091	20.9	65.5
Mozambique	2,871	8,208	65.9	223.0
Guinea-Bissau	318	964	199.6	503.7
Burundi	455	1,119	40.2	128.3
Ethiopia	5,206	10,352	78.0	160.4
Burkina Faso	511	1,399	35.9	54.5
Niger	1,195	1,659	85.5	82.1
Sierra Leone	709	1,243	60.4	197.7
all developing countries	**857,562**	**2,051,435**	**41.3**	**42.8**
least developed countries	**71,341**	**145,635**	**66.9**	**99.5**

❑ demands on the environment

- Many human demands on the environment are incompatible with the damage being inflicted on ecosystems by human population pressure.
- The health of the marine, freshwater and forest ecosystems has suffered a considerable decline since 1970.

Environmental health as used by the WHO Regional Office for Europe, includes both the direct pathological effects of chemicals, radiation and some biological agents, and the effects (often indirect) on health and well being of the broad physical, psychological, social and aesthetic environment which includes housing, urban development, land use and transport.

Nutrition, pollution, waste control and public health are related concerns. When well-being of a whole population is measured, these become economic and political concerns. Increasingly wellness concerns are affecting fiscal policy and prompting some advocates to call for monetary reform (to end systematic pollution credit, governments actually paying to create human health harms).

clean air
fish from the sea, rivers and lakes
stable soils for crop growing and animal husbandry
clean water for drinking, washing and irrigation
stable climatic conditions for agriculture
minerals for industry
good roads and telecommunications
fossil fuels and water power for energy
timber for housing and wood for fuel
amenity, contemplation and recreation

❐ demographic profiles 1: africa

1. country/region
2. population mid-2001 (millions)
3. births per 1,000 population
4. deaths per 1,000 population
5. rate of natural increase (percent)
6. total fertility rate (number of children per woman)
7. life expectancy at birth
8. projected population change 2001–2050 (percent)

1	2	3	4	5	6	7	8
NORTHERN AFRICA	**177**	**28**	**7**	**2.1**	**3.6**	**66**	**72**
Algeria	31	25	6	1.9	3.1	69	66
Egypt	69.8	28	7	2.1	3.5	66	64
Libya	5.2	28	4	2.4	3.9	75	106
Morocco	29.2	26	6	2	3.4	69	66
Sudan	31.8	34	11	2.4	4.9	56	100
Tunisia	9.7	19	6	1.3	2.3	72	46
Western Sahara	0.3	46	17	2.9	6.8	-	128
WESTERN AFRICA	**240**	**42**	**15**	**2.7**	**5.8**	**51**	**142**
Benin	6.6	45	15	3	6.3	50	174
Burkina Faso	12.3	47	17	3	6.8	47	180
Cape Verde	0.4	37	7	3	4	68	-6
Côte d'Ivoire	16.4	36	16	2	5.2	46	118
Gambia	1.4	43	14	3	5.9	52	195
Ghana	19.9	32	10	2.2	4.3	58	61
Guinea	7.6	41	19	2.3	5.5	45	138
Guinea-Bissau	1.2	42	20	2.2	5.8	45	167
Liberia	3.2	49	17	3.1	6.6	50	210
Mali	11	50	20	3	7	46	230
Mauritania	2.7	43	15	2.8	6	51	208
Niger	10.4	53	24	2.9	7.5	41	175
Nigeria	126.6	41	14	2.8	5.8	52	140
Senegal	9.7	41	13	2.8	5.7	52	135
Sierra Leone	5.4	47	20	2.6	6.3	45	189
Togo	5.2	40	11	2.9	5.8	55	89
EASTERN AFRICA	**252**	**42**	**16**	**2.5**	**5.7**	**50**	**128**
Burundi	6.2	42	17	2.5	6.5	47	158
Comoros	0.6	47	12	3.5	6.8	56	208
Djibouti	0.6	43	16	2.7	6.1	46	67
Eritrea	4.3	43	13	3	6	55	209
Ethiopia	65.4	44	15	2.9	5.9	52	164

(Contd.)

1	2	3	4	5	6	7	8
Kenya	29.8	34	14	2	4.4	48	26
Madagascar	16.4	43	13	3	5.8	54	186
Malawi	10.5	46	23	2.3	6.4	39	110
Mauritius	1.2	17	7	1	2	71	24
Mayotte	0.2	41	10	3.1	5.6	59	259
Mozambique	19.4	43	22	2.1	5.6	72	18
Reunion	0.7	20	5	1.5	2.3	75	39
Rwanda	7.3	39	21	1.8	5.8	39	22
Seychelles	0.1	18	7	1.1	2	70	15
Somalia	7.5	48	19	3	7.3	46	240
Tanzania	36.2	41	13	2.8	5.6	53	144
Uganda	24	48	19	2.9	6.9	42	251
Zambia	9.8	45	22	2.3	6.1	37	108
Zimbabwe	11.4	29	20	0.9	4	40	-18
CENTRAL AFRICA	**99**	**46**	**16**	**2.9**	**6.6**	**48**	**203**
Angola	12.3	50	25	2.4	6.9	38	141
Cameroon	15.8	39	12	2.7	5.2	55	119
Central African Rep.	3.6	38	18	2	5.1	45	78
Chad	8.7	49	16	3.3	6.6	50	282
Congo	3.1	46	16	3	6.3	50	245
Congo, Dem. Rep. Of	53.6	47	16	3.1	7	48	239
Equatorial Guinea	0.5	45	14	3.1	5.9	50	193
Gabon	1.2	32	16	1.6	4.3	52	49
Sao Tome & Principe	0.2	43	8	3.5	6.2	65	204
SOUTHERN AFRICA	**50**	**27**	**14**	**1.3**	**3.1**	**52**	**-18**
Botswana	1.6	31	20	1	3.9	41	-26
Lesotho	2.2	33	13	2	4.3	53	31
Namibia	1.8	36	17	1.9	5	46	37
South Africa	43.6	25	14	1.2	2.9	53	-25
Swaziland	1.1	41	20	2	5.9	40	84

▫ demographic profiles 2: americas

Demographic data for the subregions and countries in the Americas and the Caribbean

1. country/region
2. population mid-2001 (millions)
3. births per 1,000 population
4. deaths per 1,000 population
5. rate of natural increase (percent)
6. total fertility rate (number of children per woman
7. life expectancy at birth
8. projected population change 2001–2050 (percent)

1	2	3	4	5	6	7	8
NORTH AMERICA	**316**	**14**	**9**	**0.5**	**2**	**77**	**43**
Canada	31	11	8	0.3	1.4	79	18
United States	284.5	15	9	0.6	2.1	77	45
LATIN AMERICA & THECARIBBEAN	**525**	**24**	**6**	**1.7**	**2.8**	**71**	**55**
CENTRALAMERICA	**138**	**26**	**5**	**2.1**	**.1**	**3.1**	**73 65**
Belize	0.3	25	6	1.9	3.2	72	132
Costa Rica	3.7	22	4	1.8	2.6	77-51	
El Salvador	6.4	30	7	2.3	3.5	70	93
Guatemala	13	36	7	2.9	4.8	66	143
Honduras	6.7	33	6	2.8	4.4	66	81
Mexico	99.6	24	5	1.9	2.8	75	50
Nicaragua	5.2	35	6	3	4.3	68	122
Panama	2.9	25	5	2.1	2.6	74	48
CARIBBEAN	**37**	**21**	**8**	**1.3**	**2.6**	**68**	**36**
Antigua and Barbuda	0.1	22	6	1.6	2.4	70	3
Bahamas	0.3	21	5	1.5	2.4	72	46
Barbados	0.3	14	9	0.5	1.6	73	3
Cuba	11.3	14	7	0.6	1.6	75	-3
Dominica	0.1	16	8	0.8	1.8	73	14
Dominican Rep.	8.6	26	5	2.1	3.1	69	74
Grenada	0.1	21	8	1.3	2.4	65	-2
Guadeloupe	0.5	17	6	1.2	1.9	77	23
Haiti	7	33	15	1.7	4.7	49	70
Jamaica	2.6	20	5	1.5	2.4	71	48
Martinique	0.4	14	6	0.8	1.8	79	14
Netherlands Antilles	0.2	17	6	1.1	2.1	74	13
Puerto Rico	3.9	15	7	0.8	1.9	75	8
St. Kitts-Nevis	0.04	20	11	0.9	2.5	69	33
Saint Lucia	0.2	19	6	1.3	2.1	71	50
St. Vincent and the Grenadines	0.1	19	7	1.2	2.2	72	19
Trinidad and Tobago	1.3	14	8	0.7	1.7	71	6
SOUTH AMERICA	**350**	**23**	**7**	**1.6**	**2.6**	**70**	**53**
Argentina	37.5	19	8	1.1	2.6	73	45
Bolivia	8.5	32	9	2.4	4.2	62	100
Brazil	171.8	22	7	1.5	2.4	68	44
Chile	15.4	18	5	1.3	2.3	75	26
Colombia	43.1	24	6	1.8	2.6	71	66
Ecuador	12.9	28	6	2.2	3.3	71	92
French Guiana	0.2	27	4	2.3	3.4	76	78
Guyana	0.7	21	8	1.3	2.5	65	-34
Paraguay	5.7	32	5	2.7	4.3	73	155
Peru	26.1	24	6	1.8	2.9	69	62
Suriname	0.4	26	7	1.9	3	71	-11
Uruguay	3.4	16	10	0.7	2.3	74	34
Venezuela	24.6	25	5	2	2.9	73	63

▫ demographic profiles 3: asia

Demographic data for the subregions and countries in Asia

1. country/region
2. population mid-2001 (millions)
3. births per 1,000 population
4. deaths per 1,000 population
5. rate of natural increase (percent)
6. total fertility rate (number of children per woman)
7. life expectancy at birth
8. projected population change 2001–2050 (percent)

1	2	3	4	5	6	7	8
ASIA	**3,720**	**22**	**8**	**1.4**	**2.7**	**67**	**41**
ASIA (Excl. China)	**2,447**	**25**	**8**	**1.7**	**3.2**	**64**	**59**
WESTERN ASIA	**193**	**28**	**7**	**2.1**	**3.9**	**68**	**107**
Armenia	3.8	9	6	0.3	1.1	73	0
Azerbaijan	8.1	15	6	0.9	2	72	42
Bahrain	0.7	21	3	1.9	2.8	72	300
Cyprus	0.9	13	8	0.6	1.8	77	10
Georgia	5.5	9	9	0	1.2	73	-23
Iraq	23.6	37	10	2.7	5.3	59	127
Israel	6.4	22	6	1.6	3	78	64
Jordan	5.2	27	5	2.2	3.6	70	128
Kuwait	2.3	20	2	1.8	4.2	73	181
Lebanon	4.3	23	7	1.7	2.5	71	35
Oman	2.4	39	4	3.5	6.1	71	218
Palestinian Territory	3.3	42	5	3.7	5.9	72	239
Qatar	0.6	31	4	2.7	3.9	72	45
Saudi Arabia	21.1	35	6	2.9	5.7	67	185
Syria	17.1	31	6	2.6	4.1	70	106
Turkey	66.3	22	7	1.5	2.5	69	47
United Arab Emirates	3.3	18	4	1.4	3.5	74	54
Yemen	18	44	11	3.3	7.2	59	295
SOUTH CENTRAL ASIA	**1,505**	**27**	**9**	**1.8**	**3.4**	**61**	**66**
Afghanistan	26.8	43	19	2.4	6	45	151
Bangladesh	133.5	28	8	2	3.3	59	56
Bhutan	0.9	40	9	3.1	5.6	66	127
India	1,033	26	9	1.7	3.2	61	58
Iran	66.1	18	6	1.2	2.6	70	52
Kazakhstan	14.8	15	10	0.5	1.8	66	-5
Kyrgyzstan	5	20	7	1.3	2.4	69	52
Maldives	0.3	41	9	3.2	5.8	61	162
Nepal	23.5	35	11	2.4	4.8	57	111

1	2	3	4	5	6	7	8
Pakistan	145	39	11	2.8	5.6	60	138
Sri Lanka	19.5	18	6	1.2	2.1	72	19
Tajikistan	6.2	19	4	1.4	2.4	68	40
Turkmenistan	5.5	19	5	1.3	2.2	67	29
Uzbekistan	25.1	22	5	1.7	2.7	70	61
SOUTH EAST ASIA	**519**	**23**	**7**	**1.6**	**2.8**	**66**	**49**
Brunei	0.3	22	3	2	2.7	74	73
Cambodia	13.1	28	11	1.7	4	56	38
East Timor	0.8	33	16	1.6	4.4	48	84
Indonesia	206.1	23	6	1.7	2.7	67	48
Laos	5.4	39	14	2.5	5.4	52	72
Malaysia	22.7	25	4	2	3.2	73	94
Myanmar	47.8	28	12	1.6	3.3	56	43
Philippines	77.2	29	6	2.2	3.5	67	67
Singapore	4.1	14	5	0.9	1.6	78	151
Thailand	62.4	14	6	0.8	1.8	72	15
Vietnam	78.7	20	6	1.4	2.3	66	49
EAST ASIA	**1,503**	**15**	**7**	**0.8**	**1.8**	**72**	**5**
China	1,273	15	6	0.9	1.8	71	8
China, Hong Kong	6.9	8	5	0.3	1	80	8
China, Macao	0.4	9	3	0.6	1.2	77	82
Japan	127.1	9	8	0.2	1.3	81	-21
Korea, North	22	21	7	1.5	2.3	70	20
Korea, South	48.8	14	5	0.9	1.5	74	5
Mongolia	2.4	20	7	1.4	2.2	63	61
Taiwan	22.5	14	6	0.8	1.7	75	12

❑ demographic profiles 4: europe and oceania-611 (th)

Demographic data for the subregions and countries in Europe and Oceania

1. country/region
2. population mid-2001 (millions)
3. births per 1,000 population
4. deaths per 1,000 population
5. rate of natural increase (percent)
6. total fertility rate (number of children per woman)
7. life expectancy at birth
8. projected population change 2001–2050 (percent)

1	2	3	4	5	6	7	8
EUROPE	**727**	**10**	**11**	**-0.1**	**1.4**	**74**	**-9**
NORTHERN EUROPE	**96**	**11**	**10**	**0.1**	**1.6**	**77**	**5**

(Contd.)

1	2	3	4	5	6	7	8
Denmark	5.4	13	11	0.2	1.7	76	16
Estonia	1.4	9	13	-0.4	1.3	71	-36
Finland	5.2	11	10	0.2	1.7	77	-8
Iceland	0.3	15	7	0.8	2	79	18
Ireland	3.8	14	9	0.6	1.9	77	18
Latvia	2.4	8	14	-0.6	1.2	71	-25
Lithuania	3.7	9	11	-0.1	1.3	73	-16
Norway	4.5	13	10	0.3	1.8	78	15
Sweden	8.9	10	11	0	1.5	80	7
United Kingdom	60	12	11	0.1	1.7	77	7
WESTERN EUROPE	**184**	**11**	**10**	**0.1**	**1.6**	**78**	**-2**
Austria	8.1	10	9	0	1.3	78	1
Belgium	10.3	11	10	0.1	1.6	78	-3
France	59.2	13	9	0.4	1.9	79	10
Germany	82.2	9	10	-0.1	1.3	78	-14
Liechtenstein	0.03	12	7	0.6	1.4	-	18
Luxembourg	0.4	13	9	0.4	1.7	78	33
Monaco	0.03	20	17	0.3	-	-	15
Netherlands	16	13	9	0.4	1.7	78	12
Switzerland	7.2	11	9	0.2	1.5	80	2
EASTERN EUROPE	**303**	**9**	**14**	**-0.5**	**1.2**	**68**	**-14**
Belarus	10	9	14	-0.4	1.3	68	-15
Bulgaria	8.1	9	14	-0.5	1.2	72	-35
Czech Republic	10.3	9	11	-0.2	1.1	75	-9
Hungary	10	10	14	-0.4	1.3	71	-19
Moldova	4.3	11	11	-0.1	1.4	68	0
Poland	38.6	10	10	0	1.4	73	-12
Romania	22.4	10	12	-0.1	1.3	71	-14
Russia	144.4	9	15	-0.7	1.2	66	-12
Slovakia	5.4	10	10	0	1.3	73	-13
Ukraine	49.1	8	15	-0.7	1.1	68	-22
SOUTHERN EUROPE	**145**	**10**	**10**	**0**	**1.3**	**77**	**-16**
Albania	3.4	17	5	1.2	2.8	72	51
Andorra	0.1	13	4	0.9	1.2	-	118
Bosnia-Herzegovina	3.4	12	8	0.4	1.6	68	-1
Croatia	4.7	10	11	-0.2	1.4	74	-16
Greece	10.9	10	10	0	1.3	78	-11
Italy	57.8	9	10	0	1.3	79	-20
Macedonia	2	14	8	0.5	1.9	73	3
Malta	0.4	11	8	0.3	1.7	77	11
Portugal	10	12	11	0.1	1.5	76	-18
San Marino	0.03	12	8	0.4	1.3	80	11

1	2	3	4	5	6	7	8
Slovenia	2	9	10	-0.1	1.2	76	-15
Spain	39.8	10	9	0	1.2	78	-23
Yugoslavia	10.7	12	11	0.1	1.6	72	-4
OCEANIA	**31**	**18**	**7**	**1.1**	**2.5**	**74**	**49**
Australia	19.4	13	7	0.6	1.7	79	29
Fed. States of Micronesia	0.1	31	6	2.5	4.6	66	108
Fiji	0.8	25	6	1.9	3.3	67	9
French Polynesia	0.2	21	5	1.6	2.6	72	58
Guam	0.2	29	4	2.5	4.2	74	71
Kiribati	0.1	32	8	2.4	4.5	62	148
Marshall Islands	0.1	26	4	2.2	6.6	65	343
Nauru	0.01	19	5	1.4	3.7	61	100
New Caledonia	0.2	22	5	1.7	2.6	73	54
New Zealand	3.9	15	7	0.8	2	77	29
Palau	0.02	18	8	1	2.5	67	38
Papua-New Guinea	5	34	11	2.3	4.8	56	124
Solomon Islands	0.5	41	7	3.4	5.7	67	217
Tonga	0.1	27	6	2.1	4.2	71	86
Tuvalu	0.01	30	9	2.1	3.1	67	82
Vanuatu	0.2	36	6	3	4.6	65	51
Western Samoa	0.2	30	6	2.4	4.5	68	0

❑ dermal

Referring to the skin. For example, dermal absorption means absorption through the skin.

❑ detection limit

The smallest amount of substance that a laboratory test can reliably measure in a sample of air, water, soil or other medium.

❑ dose

The amount of substance to which a person is exposed.

E

❑ environmental effects of transportation

The environmental effects of transportation include more than simply habitat loss from road construction and air pollution from motor vehicle use.

- Other environmental consequences are associated with transportation, such as the effects of vehicle manufacture and supporting facilities.
- All four modes of transportation – road, rail, aviation and maritime – have environmental impacts.

Main environmental effects

- Criteria air pollutants: pollutants that adversely affect human health and welfare: carbon monoxide (CO), ground level ozone (O_3) and its precursors, lead (Pb), nitrogen dioxide (NO_2), particulate matter (PM), and sulfur dioxide (SO_2);
- Toxics: pollutants that cause or may cause cancer or other serious problems, such as reproductive effects or birth defects, or adverse ecological effects;
- Greenhouse gases: compounds that contribute to global climate change by trapping heat within the Earth's atmosphere: carbon dioxide, methane, nitrous oxide, and several other compounds;
- Chlorofluorocarbons and stratospheric ozone depletion: human-made chemicals that are known to deplete the stratospheric ozone;
- Habitat and land use: adverse effects on wildlife, ecosystems, and endangered species such as habitat destruction or fragmentation;
- Water quality and aquatic resources: adverse effects to the water quality in lakes, streams, rivers, or groundwater and the ecosystems that depend on them;

Effects of transportation and related activities on the environment

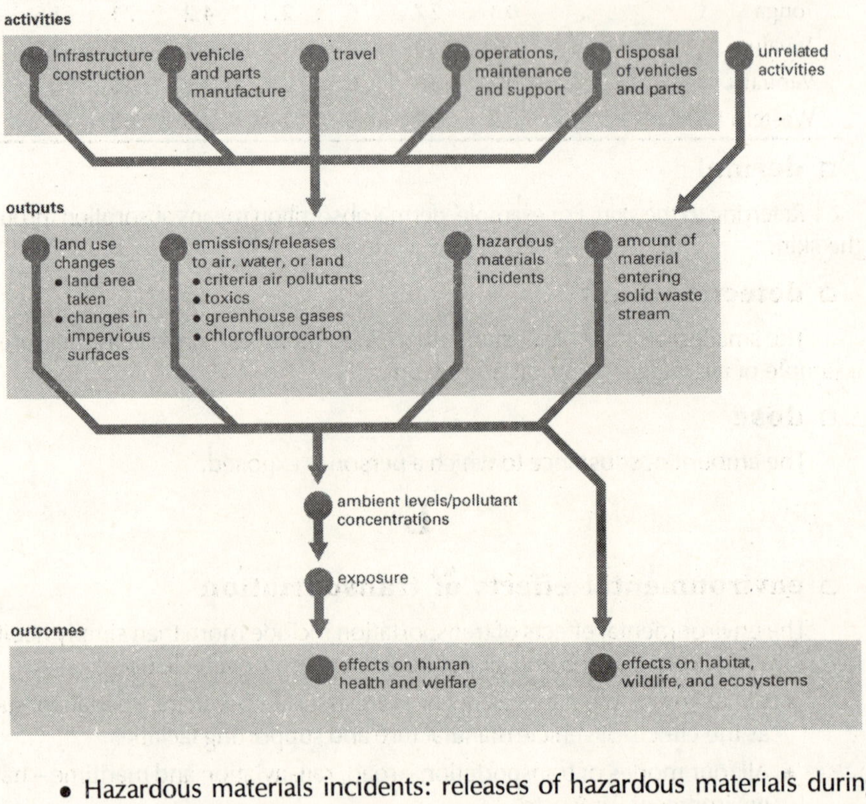

- Hazardous materials incidents: releases of hazardous materials during transportation;

- Noise: unwanted sounds that interfere with communication or cause adverse health effects; and
- Solid waste: disposal of materials and parts through recycling, reuse, or shipment to landfills.

❑ environmental refugees 1: narmada valley dam, india-

- The Narmada Valley Development Project is the single largest river development scheme in India.
- It is one of the largest hydroelectric projects in the world and is likely to displace a total of approximately 1.5 million people from their land in three Indian states (Gujarat, Maharashtra, and Madhya Pradesh).
- The environmental costs of the project, which involves the construction of more than 3,000 large and small dams, are immense.
- The largest of the dams under construction is the Sardar Sarovar, with a proposed height of 455 feet, which, if completed, will flood more than 91,000 acres of forest and agricultural land, displacing more than half a million people and destroying some of India's most fertile land.
- The Indian Government claims that the multipurpose Sardar Sarovar Project (SSP) would irrigate more than 4.4 million acres (mostly in Gujarat, some in Rajasthan) and quench the thirst of the drought prone areas of Kutch and Saurashtra in Gujarat.

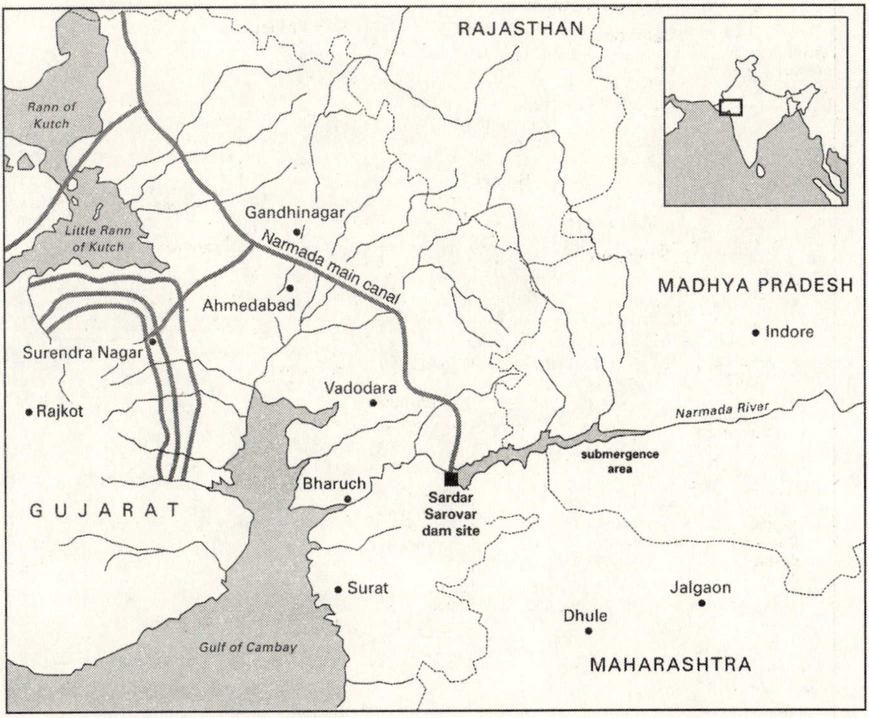

- The proposed dams will affect millions of people but only a certain percentage of them will be privy to the government's resettlement and rehabilitation (R & R) programs.
- Hundreds of thousands whose lands or livelihoods are affected by either project-related developments or downstream impacts are not entitled to any compensation at all.
- The Narmada Dam has generated intense opposition, both within India and abroad.

❑ environmental refugees 2: forced resettlement, indonesia

- Since 1999, serious conflict and population displacement have occurred in the Moluccas, Kalimantan, Sulawesi, Aceh, Timor, and Irian Jaya, exacerbated by Indonesia's political and economic crises.

Total estimated displaced persons throughout Indonesia mid-2001
This total of 1.2 million includes many displaced by conflict, as well as those forcibly resettled.

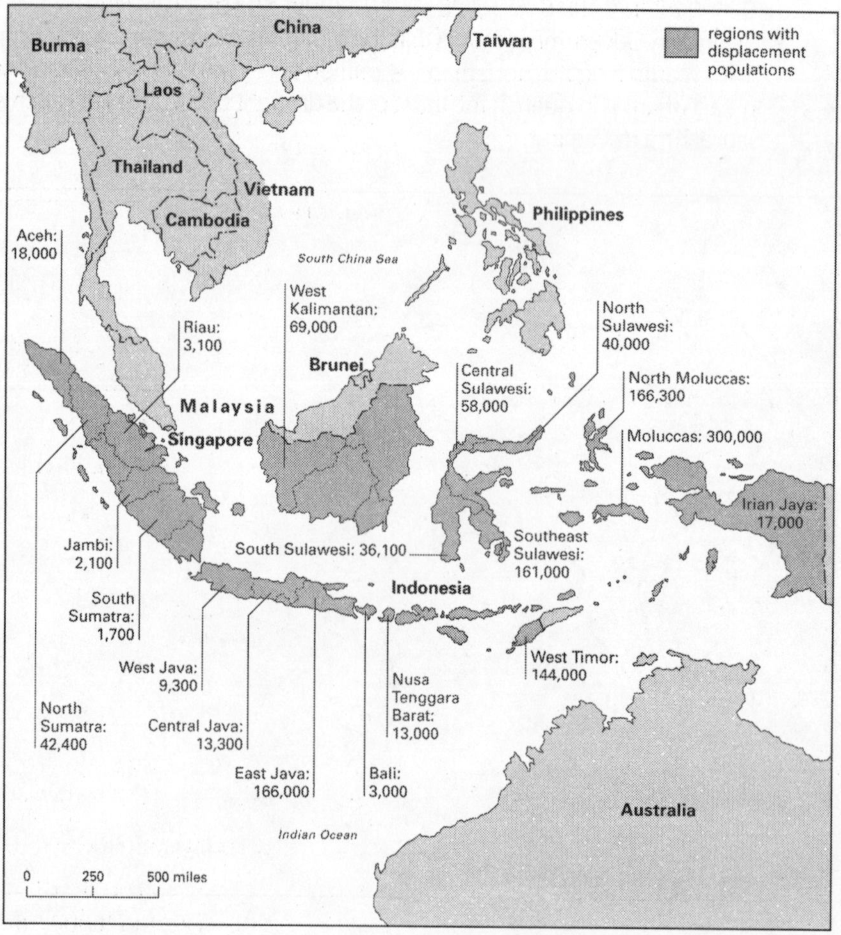

- Conflict in each of these regions has been fueled by resentments stemming from Indonesia's nowabandoned policy of transmigration, through which residents of Java and other heavily crowded islands were relocated to less populated areas.
- Under this controversial program of transmigration, millions of people were forcibly moved from the crowded islands of Java and Bali to less populated provinces.
- The program also involved forcing indigenous people on the islands that received the transmigrants to cede social and economic rights, and traditional rights to land, water, and other natural resources in favor of the transmigrants.
- In 2001, there were more than 1.2 million InternallyDisplaced Persons (IDPs) throughout Indonesia.
- These included at least 460,000 IDPs in the Moluccas; more than 179,000 IDPs in Central and East Java; approximately 295,000 IDPs in Sulawesi; up to 18,000 IDPs in Aceh and more than 42,000 Acehnese IDPs in other areas of North Sumatra; over 16,000 IDPs in Irian Jaya; and an estimated 144,000 East Timorese refugees in West Timor.

▫ environmental refugees 3: flight from famine, horn of africa

- Several years of intermittent rains resulted in poor harvests and have produced a severe drought in the Horn of Africa that has affected millions of people throughout the region.
- The worst affected populations are pastoralists and in southern and eastern Ethiopia, southern Somalia, and northern Kenya.

Drought-induced migration movements in Somalia and neighboring countries

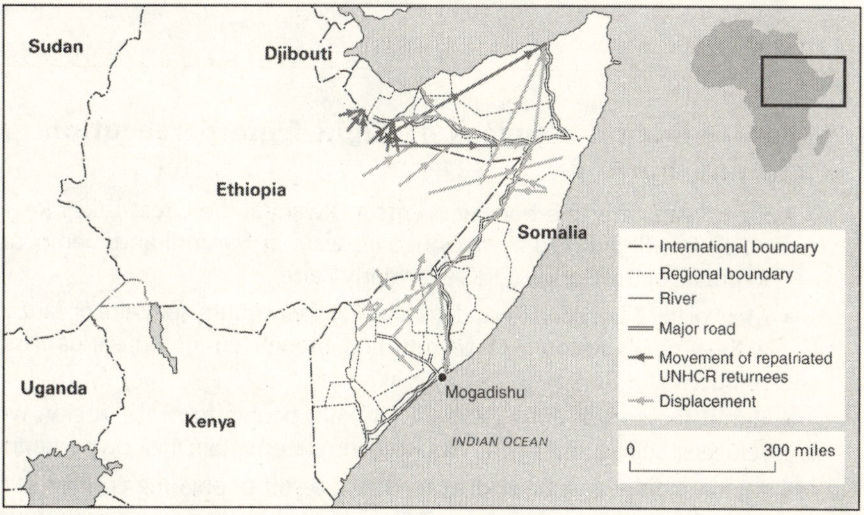

- Other countries in the region affected by the drought include Eritrea, Djibouti, and Sudan.
- The current drought conditions have resulted in a serious deterioration of food security, the loss or sale of assets such as livestock and agricultural inputs, high food prices, and decreased livestock prices.
- Stress migration and cross border movements have further strained resources as people search for additional pasture, improved conditions, and external assistance.
- Increased competition for decreased resources has raised tensions in many drought-affected areas.
- The drought-affected populations by country in 2001 were: 10 million in Ethiopia, 2.2 million in Kenya, 750,000 in Somalia, 335,000 in Eritrea, and 150,000 in Djibouti.

Areas of food insecurity, Horn of Africa

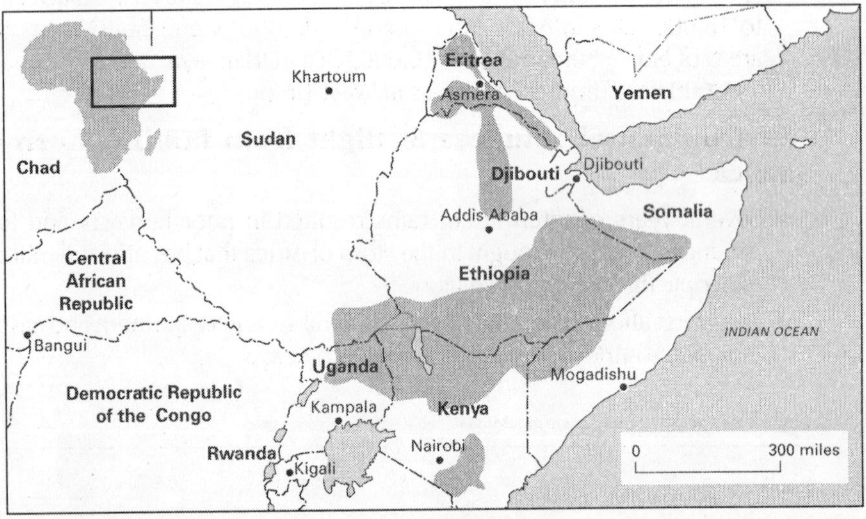

▢ environmental refugees 4: flight from persecution, rwanda/burundi

- Eight years after the horrific events in Rwanda, the Great Lakes Region (GLR) is still afflicted by the ongoing wars in Burundi and Democratic Republic of the Congo (DRC – formerly Zaire).
- Most of the 1.2 million Rwandans who left their country for Zaire or Tanzania at the time of the conflict have returned, though tens of thousands are still refugees.
- But at the end of 2001, some 1,130,400 people from the region were refugees, and a further 3,012,900 were displaced within their own countries.
- Many people have been displaced as a result of ongoing conflict in the Democratic Congo (formerly Zaire), as well as those who are refugees from ethnic persecution in Rwanda and Burundi.

Main groups of refugees and internally displaced persons in the African Great Lakes Region (late 2001)

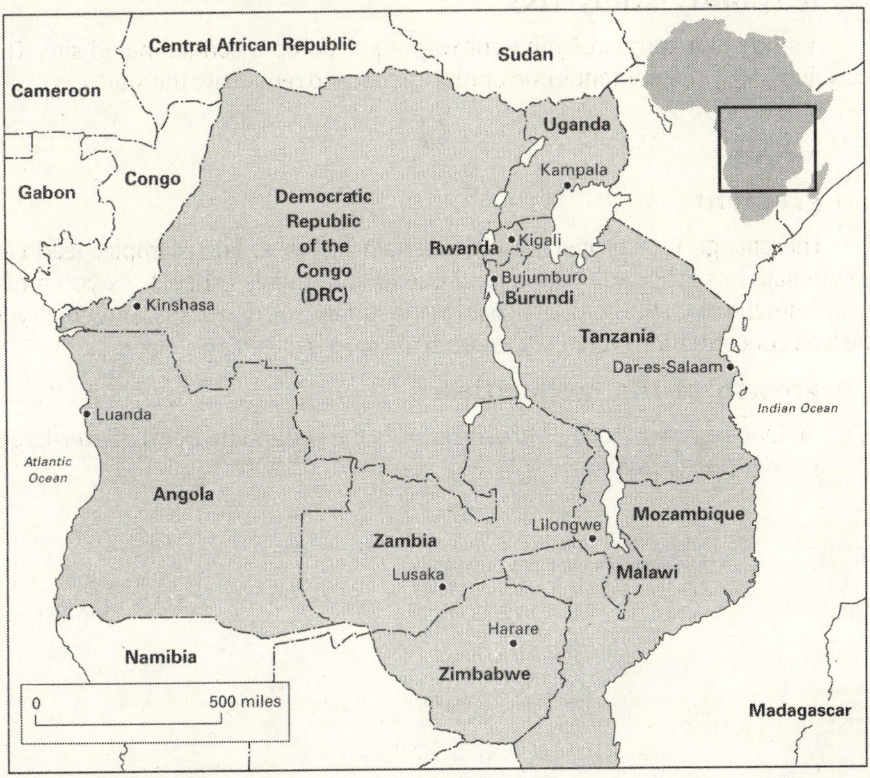

❑ epidemiology

The study of the occurrence and causes of health effects in human populations. An epidemiological study often compares two groups of people who are alike except for one factor such as exposure to a chemical or the presence of a health effect. The investigators try to determine if the factor is associated with the health effect.

❑ exposure

Contact with a chemical by swallowing, by breathing or by direct contact (such as through the skin or eyes). Exposure may be either short term (acute) or long term (chronic).

❑ exposure assessment

A process that estimates the amount of a chemical that enters or comes into contact with people or animals. An exposure assessment also describes how often and for how long an exposure occurred, and the nature and size of a population exposed to a chemical.

F

❒ feasibility study (fs)

A study that compares different ways to clean up a contaminated site. The feasibility study recommends one or more actions to remediate the site.

G

❒ gradient

The change in a property over a certain distance. For example, lead can accumulate in surface soil near a road due to automobile exhaust. As you move away from the road, the amount of lead in the surface soil decreases. This change in the lead concentration with distance from the road is called a gradient.

❒ growth of the megacities

- One feature of today's urbanization is the continuing trend of ever-larger metropolitan areas.

Population of cities with 10 million inhabitants or more, 1950, 1975, 2000 and 2015 (in millions)

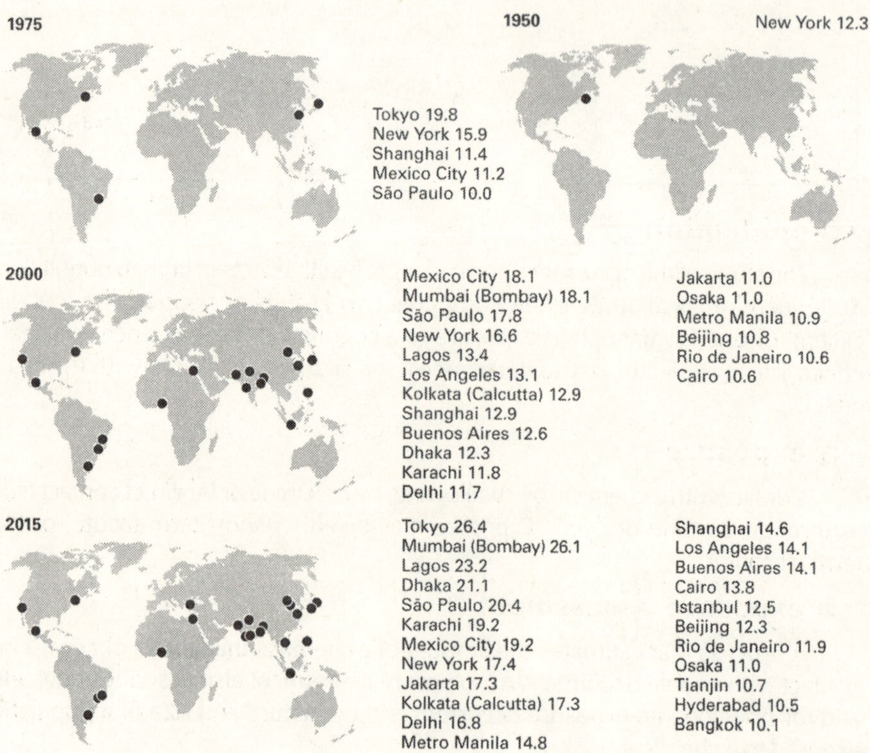

- The term 'megacity' is given to an urban agglomeration with 10 million inhabitants or more.
- The number of megacities rose from just 1 in 1950 (New York) to 19 in 2000, with 15 of them in the developing world.
- In 2000, 4.3 percent of the world's population lived in megacities.
- The number of megacities will increase from 19 in 2000 to 23 in 2015.
- Most of the large cities are located in developing countries.
- In 2000 just 10 of the 41 large cities belonged to developed countries and only one more, of a total of 11, will be located in developed countries by 2015.
- The growth rates of the megacities are expected to be much higher in the developing than the developed world.

Problems

- The current pace and scale of change often strains the capacity of local and national governments to provide even the most basic services to urban residents.
- Many countries are developing policies to address the new demands created by the increasing concentration of people in cities while capitalizing on the benefits of urbanization, such as economic growth and efficient delivery of services.

H

❐ health assessment for contaminated sites

Determination of actual or possible health effects due to environmental contamination or exposure. It includes a health-based interpretation of all the information known about the situation. The information may come from site investigations (environmental sampling and studies), exposure assessments, risk assessments, biological monitoring or health effects studies. The health assessment is used to advise people how to prevent or reduce their exposures, to determine remedial actions or the need for additional studies.

❐ health effects studies related to contaminants

Studies of the health of people who may have been exposed to contaminants. They include, but are not limited to, epidemiological studies, reviews of health status of people in exposure or disease registries, and doing medical tests.

❐ health registry

A record of people exposed to a specific substance (such as a heavy metal), or having a specific health condition (such as cancer or a communicable disease). New York State maintains several health registries.

I

▫ influences on the environment

- Humans have been altering their environment for thousands of years.
- The dominant technological influences in the pastV10,000 years have been the use of timber for building and the spread of crop cultivation.
- In the past 150 years these practices have accelerated, during which time the rising population has doubled the area of arable land in use on the Earth's surface.

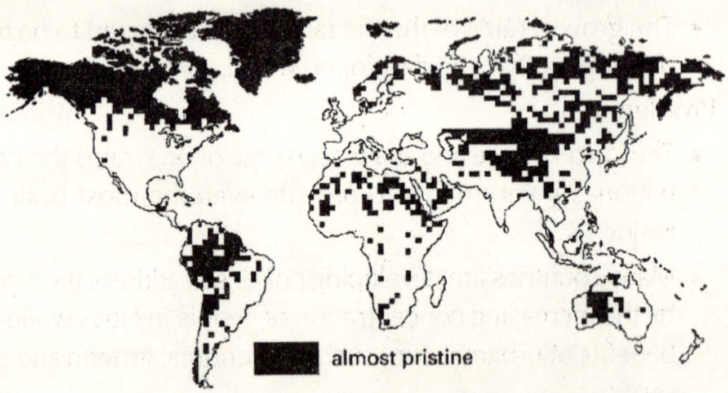

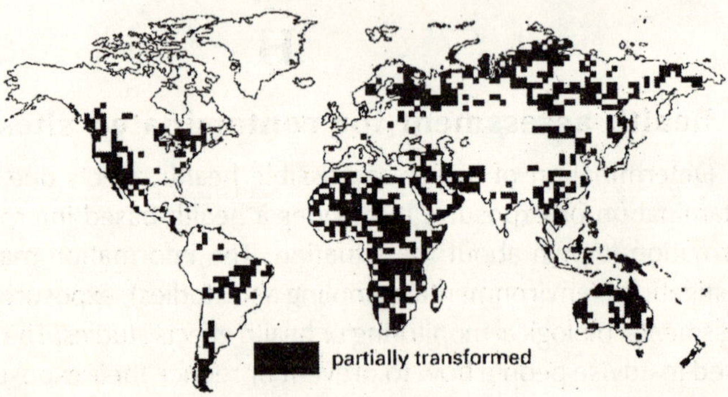

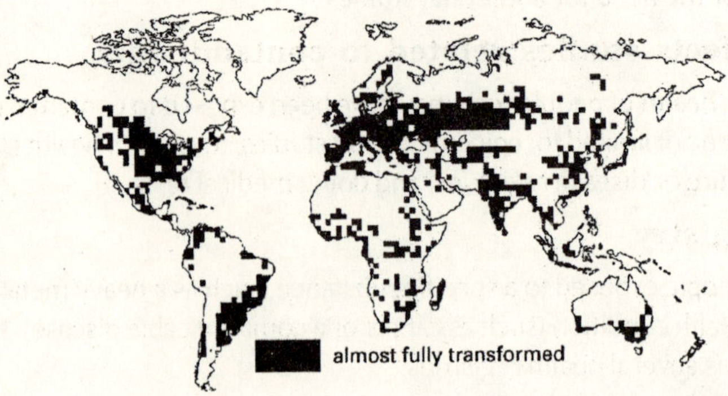

- The burning of fossil fuels during this same period has for the first time had a major effect on ecosystems through pollution and, most recently, climate change.
- The widespread saturation of ecosystems in the past 30 years with nitrogen compounds such as ammonia and nitrogen oxides from agricultural fertilizers and air pollution, has emerged as a new global-scale threat.
- The extent of ecosystem loss and alteration is closely related to population density, which is very uneven across the planet.
- Today, 50 percent of the human population lives on less than 10 percent of the Earth's land, and 75 percent of people on only 20 percent.
- The most heavily populated regions of the planet now and for much of human history, and the most ecologically disturbed, have been Europe and South and East Asia.
- The population densities of the Americas and Africa have only now risen to those achieved in Europe and India by 1750.
- In India today, population density is more than 780 people per square mile, seven times the global average, and almost 80 percent of the original forest cover there has been lost.

Human transformation of the land around 2000

- While the map shows a considerable proportion of the world's land mass as 'almost pristine', inreality very little land has completely escaped human activity.
- Human-induced climate change and the effects of pesticides and other chemicals are expected to change profoundly those parts of the planet that have escaped deliberate transformation.

❐ ingestion

Swallowing (such as eating or drinking). Chemicals in or on food, drink, utensils, cigarettes, hands, etc. can be ingested. After ingestion, chemicals may be absorbed into the blood and distributed throughout the body.

❐ inhalation

Breathing. People can take in chemicals by breathing contaminated air.

❐ interim remedial measure (irm)

An action taken at a contaminated site to reduce the chances of human or environmental exposure to site contaminants. Interim remedial measures are planned and carried out before comprehensive remedial studies. They can prevent additional damage during the study phase, but don't interfere in any way with the need to develop a complete remedial program. An example of an interim remedial measure is removing drums of chemicals to a storage facility from a site that has drums sitting in an empty field.

❑ international aid data

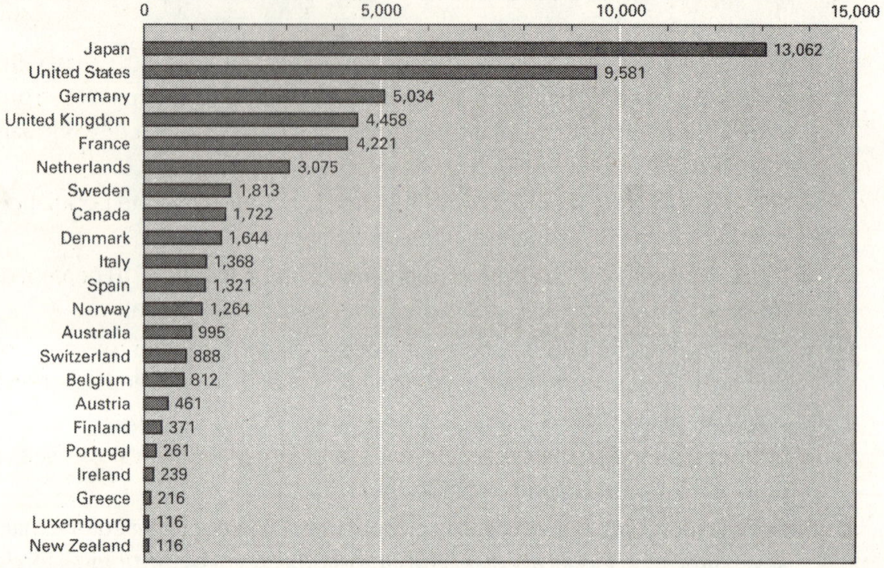

Total OECD Development Assistance Committee members (DAC) 53,058

Net ODA in 2000 as a percentage of Gross National Income

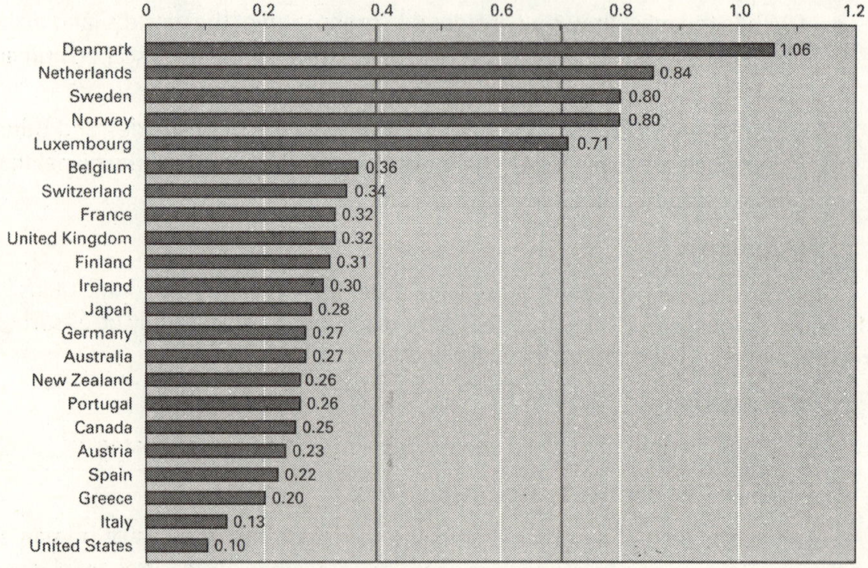

total DAC 0.22 average country effort 0.39 UN Target 0.7

❑ international aid flow

- economic aid from wealthier countries to the developing world can help poorer countries invest in their environmental health.

- The more the developing countries stay in debt, the more they will feel that they need to milk the Earth's resources for the hard cash they can bring in.
- Poverty also results in cutbacks in social, educational, health, environmental conservation, employment and other important programs.

Major donor countries (ODA = over $1 billion in 2000):

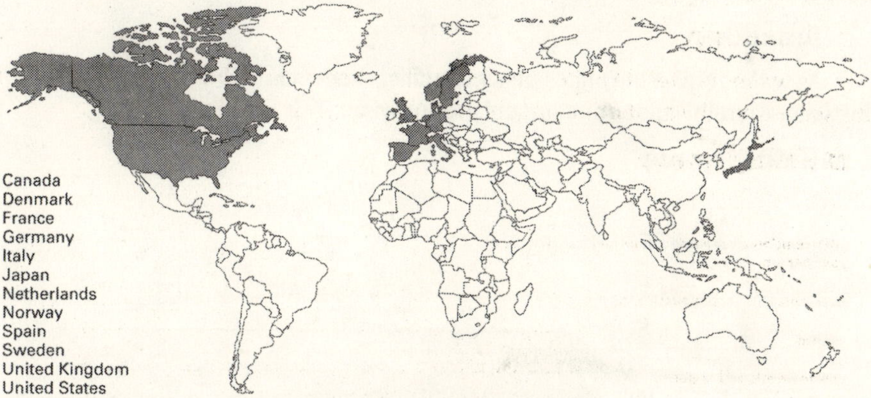

Canada
Denmark
France
Germany
Italy
Japan
Netherlands
Norway
Spain
Sweden
United Kingdom
United States

Major recipient countries (net ODA = over $100 million in 2000):

Albania
Algeria
Angola
Armenia
Azerbaijan
Bangladesh
Barbados
Belize
Benin
Bolivia
Bosnia/Herzegovina
Brazil
Bulgaria
Burkina Faso
Cambodia
Cameroon
Chad
China
Colombia
Congo, Dem.Rep.
Côte d'Ivoire
Czech Republic
Dominican Republic
Ecuador
Egypt
El Salvador
Eritrea
Ethiopia
Georgia
Ghana
Guatemala
Guinea
Guyana
Haiti
Honduras
India
Indonesia
Iran
Iraq
Israel
Jordan
Kazakhstan
Kenya
Kyrgyz Rep.
Laos
Lebanon
Lithuania
Macedonia
Madagascar
Malawi
Mali
Mauritania
Moldova
Mongolia
Morocco
Mozambique
Myanmar
Namibia
Nepal
Nicaragua
Niger
Nigeria
Pakistan
Papua New Guinea
Peru
Philippines
Poland
Romania
Russia
Rwanda
Senegal
Sierra Leone
Slovak Republic
Somalia
South Africa
Sri Lanka
Sudan
Suriname
Swaziland
Syria
Tajikistan
Tanzania
Thailand
Tunisia
Turkey
Uganda
Ukraine
Uzbekistan
Vietnam
West Bank & Gaza
Yemen
Yugoslavia
Zambia
Zimbabwe

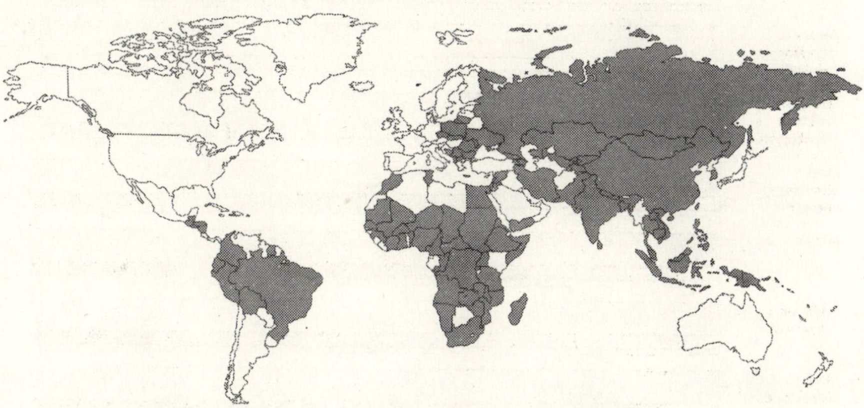

L

❑ latency period

The period of time between exposure to something that causes a disease and the onset of the health effect. Cancer caused by chemical exposure may have a latency period of 5 to 40 years.

❑ leaching

As water moves through soils or landfills, chemicals in the soil may dissolve in the water thereby contaminating the groundwater. This is called leaching.

❑ living space

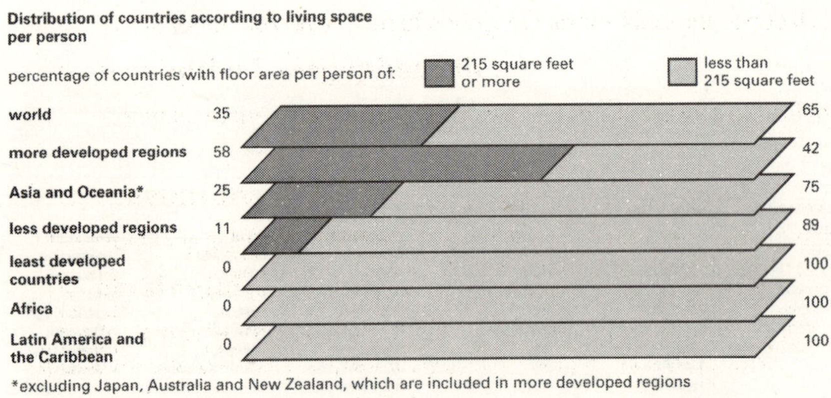

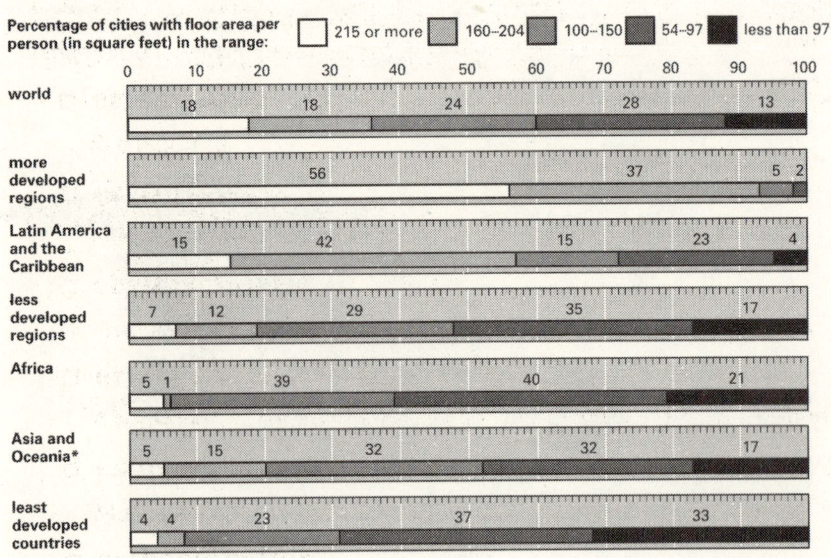

- Human settlement conditions in many parts of the world are deteriorating, mainly as a result of low levels of investment.
- Housing policies, particularly in urban areas, greatly affect the living conditions of people.
- In low-income settlements, reduced space per person can be associated with certain health risks.
- About 60 percent of industrialized countries report floor area per person of 215 square feet or larger.
- In the case of the less developed regions, only 10 percent of the countries do so.
- Floor area per person is less than 215 square feet for all of African countries, and for 75 percent of Asian and Pacific countries.
- Overcrowding is much more common in cities.
- In the less developed countries, only 7 percent of the cities have an average floor area of 215 square feet or more, while over half have average floor area under 100 square feet per person.
- In nearly 60 percent of Latin American and Caribbean cities, the average floor area is at least 160 square feet per person, while this is the case in only 6 percent of African cities and 20 percent of the cities in Asia and Oceania.

M

❑ malaria and its links to environmental change

- Malaria is by far the world's most lethal tropical parasitic disease, and kills around 1 million people a year – more than any other communicable disease except tuberculosis.
- Worldwide prevalence of the disease is estimated to be in the order of 300–500 million clinical cases each year – 90 percent in Africa.
- The Plasmodium parasite which causes the disease is spread by the Anopheles mosquito.
- The geographical area affected by malaria has shrunk considerably over the past 50 years, but control is becoming more difficult and gains are being eroded.
- Increased risk of the disease is linked with changes in land use associated with activities such as road building, mining, logging and agricultural and irrigation projects.
- Other causes of its spread include global climatic change, disintegration of health services, armed conflicts and mass movements of refugees.
- El Niño events have an impact on malaria because the associated weather disturbances influence mosquito breeding sites, and hence transmission of the disease.

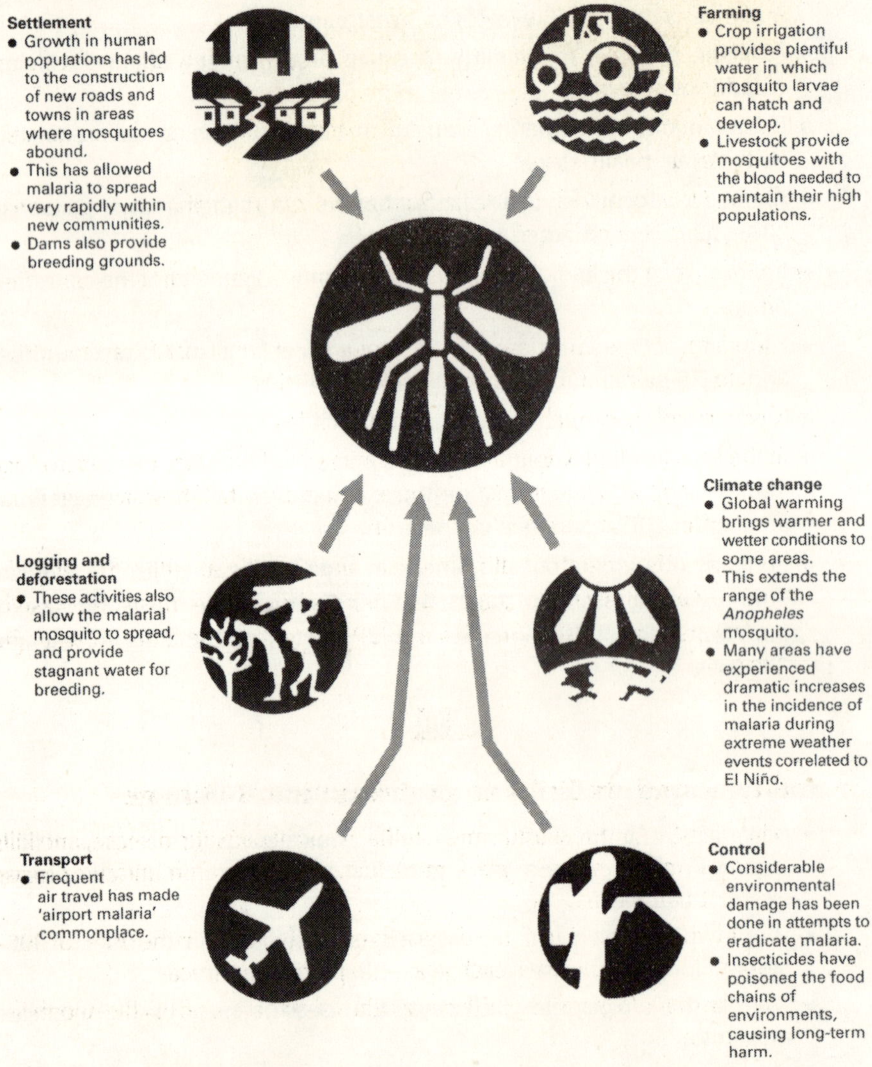

❑ maximum contaminant level (mcl)

The highest (maximum) level of a contaminant allowed to go uncorrected by a public water system under federal or state regulations. Depending on the contaminant, allowable levels might be calculated as an average over time, or might be based on individual test results. Corrective steps are implemented if the MCL is exceeded.

❑ media

Elements of a surrounding environment that can be sampled for contamination; usually soil, water, or air. Plants, as well as humans (when sampling blood, urine, etc) and animals (such as sampling fish to update fish consumption advisories) can also be considered media. The singular of "media" is "medium".

❒ metabolism

All the chemical reactions that enable the body to work. For example, food is metabolized (chemically changed) to supply the body with energy. Chemicals can be metabolized by the body and made either more or less harmful.

❒ morbidity

Illness or disease. A morbidity rate for a certain illness is the number of people with that illness divided by the number of people in the population from which the illnesses were counted.

N

❒ national priority list (npl)

A list maintained by the U.S. Environmental Protection Agency (EPA) of certain inactive hazardous waste sites. The list is produced and updated periodically by the EPA. See "Superfund".

❒ odor threshold

The lowest concentration of a chemical that can be smelled. Different chemicals have different odor thresholds. Also, some people can smell a chemical at lower concentrations than others can.

O

❒ organic

Generally considered as originating from plants or animals, and made primarily of carbon and hydrogen. Scientists use the term organic to mean those chemical compounds which are based on carbon.

P

❒ patient-doctor ratios

- The world distribution of physicians (doctors) is very unequal.
- It ranges for one to every 200 patients in Italy, to around one for every 50,000 in Liberia.

Number of people per physician worldwide

- The world distribution of physicians (doctors) is very unequal.

- Even within countries, patient-doctor ratios vary widely between rural and urban areas.
- In poor countries, trained nurses may be almost as valuable in treating patients and saving lives as doctors.

- It ranges for one to every 200 patients in Italy, to around one for every 50,000 in Liberia.

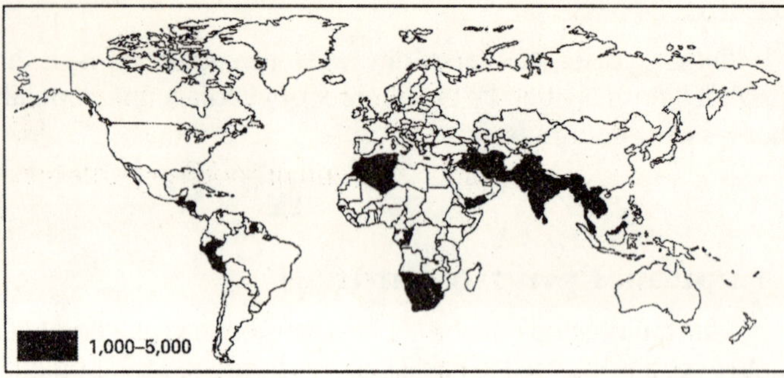

- Even within countries, patient-doctor ratios vary widely between rural and urban areas.

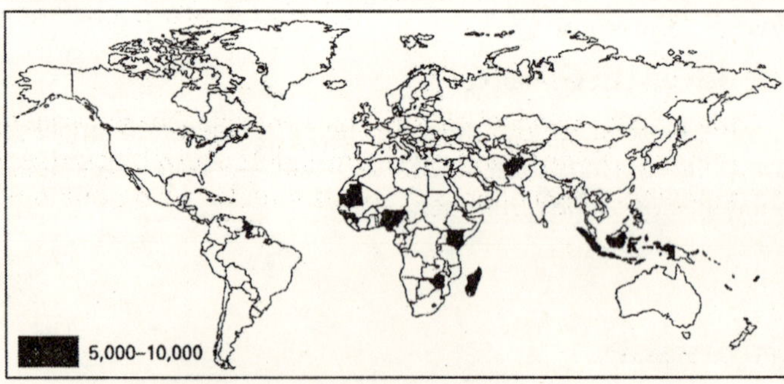

- In poor countries, trained nurses may be almost as valuable in treating patients and saving lives as doctors.

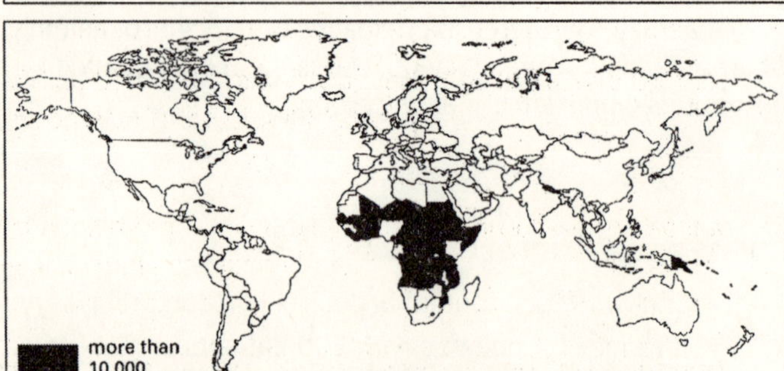

▫ people and resources

- The connections between human population growth, human society, and maintaining the Earth's resources are complex.
- Population is not the only factor to be considered.
- Government policies, technological developments and external events such as war and trade agreements interact in many different ways.
- Sustainable development–taking no more from the Earth than it can replace in the timescales human societies evolve in– decreases environmental damage and could help avert ecological disaster.

- Slowing population growth further would help improve living standards and would buy time to protect natural resources.

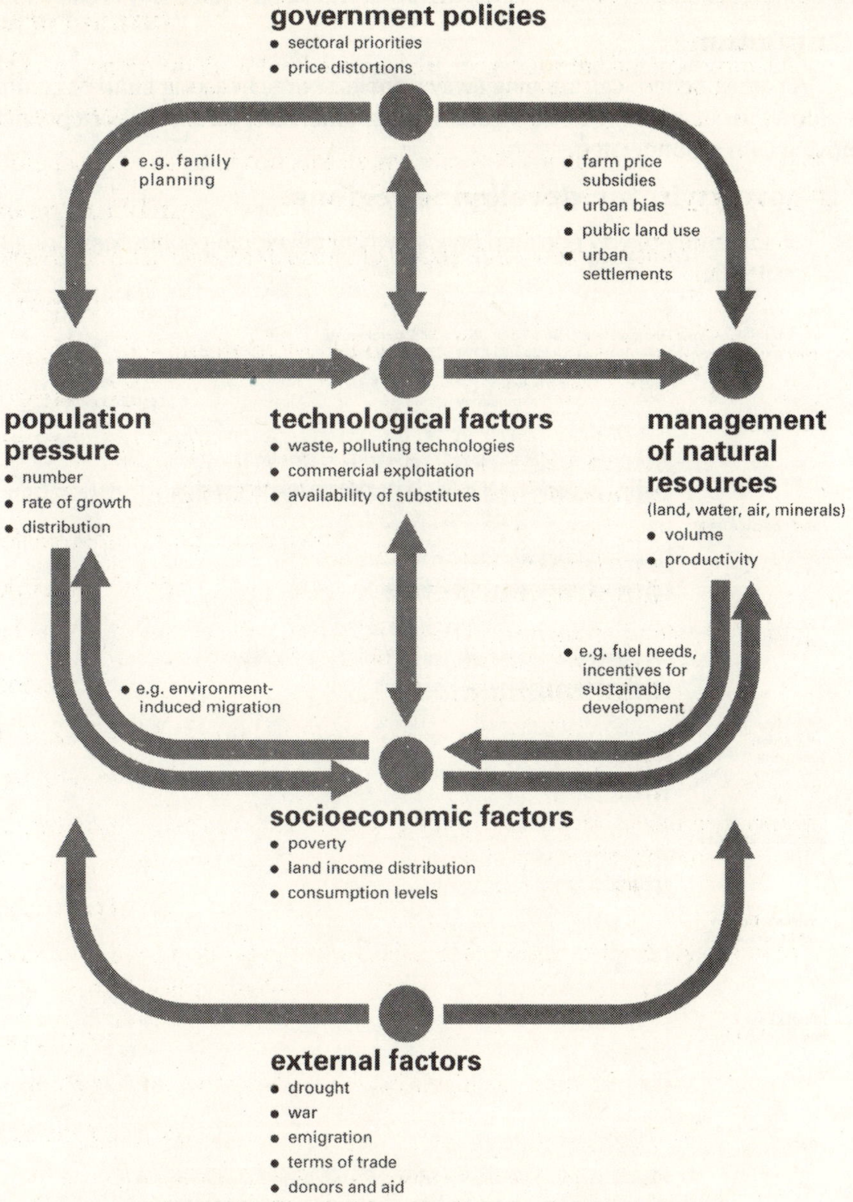

❑ permeability

The property of permitting liquids or gases to pass through. A highly permeable soil, such as sand, allows a liquid to pass through quickly. Clay has a low permeability.

❐ persistence

The quality of remaining for a long period of time (such as in the environment or the body). Persistent chemicals (such as DDT and PCBs) are not easily broken down.

❐ plume

An area of chemicals moving away from its source in a long band or column. A plume, for example, can be a column of smoke from a chimney or chemicals moving with groundwater.

❐ poverty in the developing regions

- Extreme poverty declined only slowly in developing countries during the 1990s.

Population by regions living on less than $1 per day (in 1993 purchasing power parity terms), and regional proportions, in 1987, 1990 and 1998

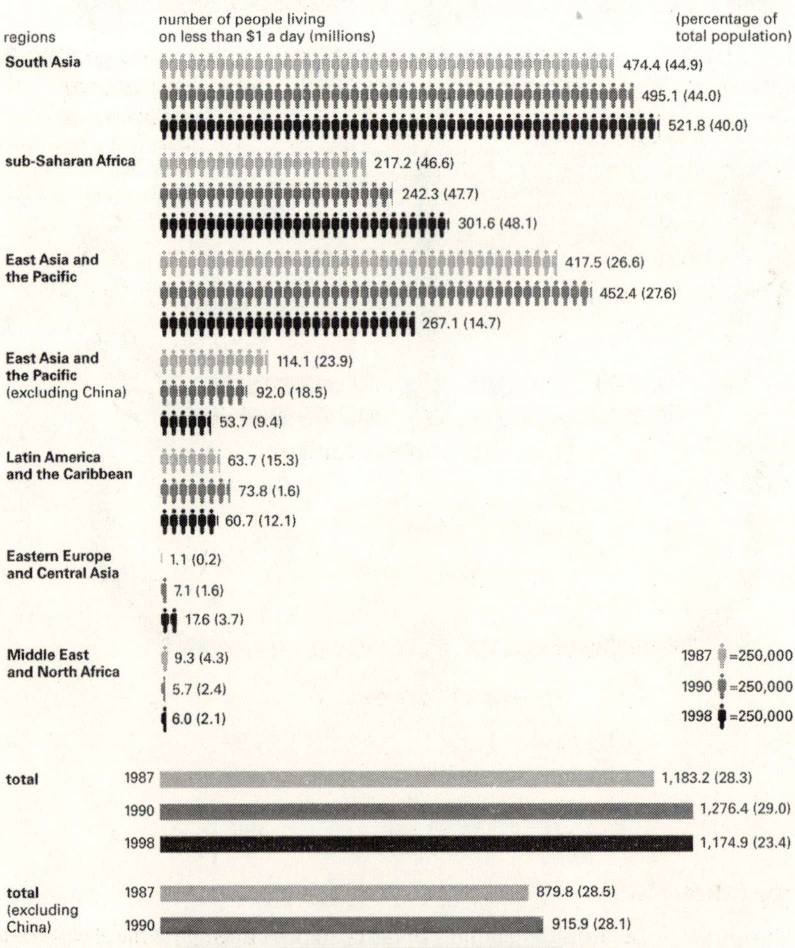

- The share of the developing world population living on less than $1 a day fell from 28 percent in 1987 to 23 percent in 1998.
- The number of poor people remained roughly constant, at nearly 1.2 billion, as the world population increased.
- Trends in social indicators show that, while there has been steady progress in average indicators of health and education, in all aspects of life the income poor are systematically worse off than the non-poor.
- The share (56 percent) and number (2.8 billion) of developing world people living on less than $2 per day – a more relevant threshold for middle-income economies such as those of East Asia and Latin America – showed roughly similar trends.

❐ problems of the urban environment

- Urban populations generally have greater access toeducation, health, clean water and sanitation than their rural counterparts.

Urban environment problems

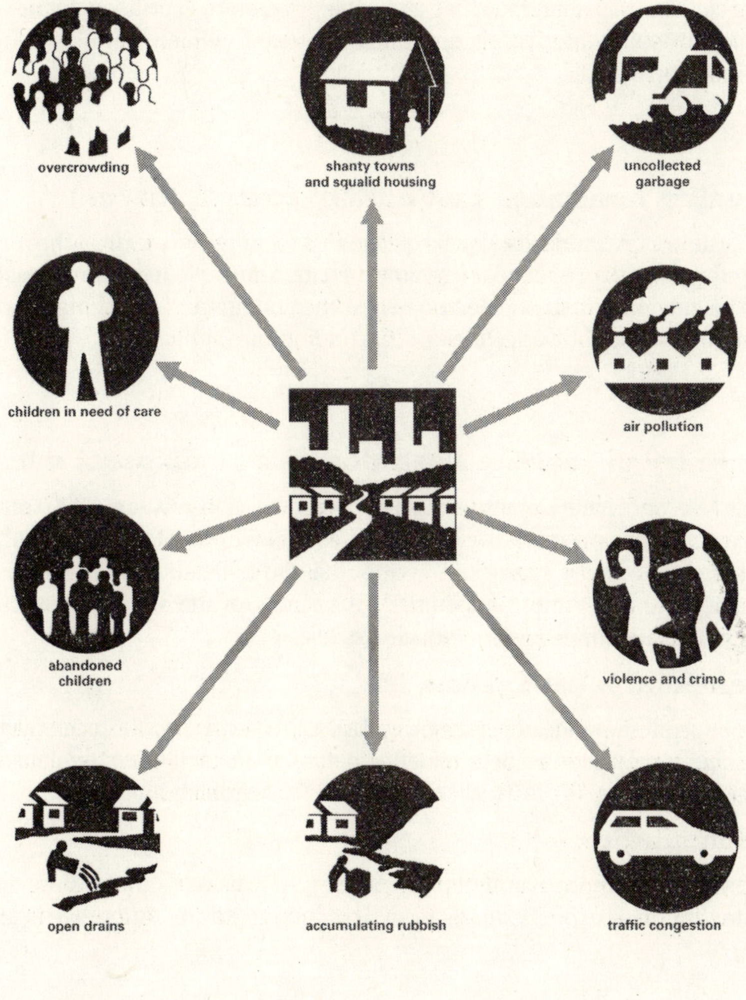

- An estimated 25–50 percent of urban inhabitants in developing countries live in impoverished slums and squatter settlements, with little or no access to adequate water, sanitation, or refuse collection.
- Overcrowding is common because of land shortage and high rents.
- This encourages epidemics of tuberculosis, diarrhea and other communicable diseases.
- Traditional family structures break down, increasing the numbers of children in need of care.
- Worldwide, more than 1 billion people live in urban areas where air pollution exceeds acceptable levels.
- Traffic congestion, violence and crime are other problems that affect cities more than rural areas.
- These problems occur particularly frequently insquatter settlements where new urban immigrants have made their homes.

❑ protocol

The detailed plan for conducting a scientific procedure. A protocol for measuring a chemical in soil, water or air describes the way in which samples should be collected and analyzed.

Q

❑ quality assurance and quality control (qa/qc)

A system of procedures, checks and audits to judge and control the quality of measurements and reduce the uncertainty of data. Some quality control procedures include having more than one person review the findings and analyzing a sample at different times or laboratories to see if the findings are similar.

R

❑ registry of inactive hazardous waste disposal sites

The New York State Department of Environmental Conservation (DEC) maintains a list of inactive hazardous waste disposal sites in New York State. When DEC finds that a site may contain hazardous waste, the site is listed in the registry and a preliminary site assessment is planned. The status of the site is updated in the registry as investigations and remediation occur.

❑ remedial investigation (ri)

An in-depth study (including sampling of air, soil, water and waste) of a contaminated site needing remediation to determine the nature and extent of contamination. The remedial investigation (RI) is usually combined with a feasibility study (FS).

❑ remediation

Correction or improvement of a problem, such as work that is done to clean up or stop the release of chemicals from a contaminated site. After investigation of a site, remedial work may include removing soil and/or drums, capping the site or collecting and treating the contaminated fluids.

❒ risk

Risk is the possibility of injury, disease or death. For example, for a person who has measles, the risk of death is one in one million.

❒ risk assessment

A process which estimates the likelihood that exposed people may have health effects. The four steps of a risk assessment are: hazard identification (Can this substance damage health?); dose-response assessment (What dose causes what effect?); exposure assessment (How and how much do people contact it?); and risk characterization (combining the other three steps to characterize risk and describe the limitations and uncertainties).

❒ risk management

The process of deciding how and to what extent to reduce or eliminate risk factors by considering the risk assessment, engineering factors (Can procedures or equipment do the job, for how long and how well?), social, economic and political concerns.

❒ route of exposure

The way in which a person may contact a chemical substance. For example, drinking (ingestion) and bathing (skin contact) are two different routes of exposure to contaminants that may be found in water. See "Exposure".

S

❒ safe

Strictly, free from harm or risk. Exposure to a chemical usually has some risk associated with it, although the risk may be very small. However, many people use the word safe to mean something that has a very low risk or one that is acceptable to them.

❒ schistosomiasis - links to environmental change

- Schistosomiasis (also called bilharzia) is the disease that ranks second behind malaria in socioeconomic and public health importance in tropical and subtropical areas.
- It is caused by a water-borne parasitic worm that lives in fresh water in the tropics.
- To infect humans, the worms must first infect and mature in freshwater snails, which are their 'intermediate hosts.'
- The disease is endemic in 74 developing countries, infecting more than 200 million people in rural agricultural and peri-urban areas.
- In many areas, schistosomiasis infects a large proportion of under-14 children, causing anemia, retarding growth and affecting school performance.

- An estimated 500–600 million people worldwide are at risk from the disease, which is treatable with safe and effective drugs.
- The prevalence of schistosomiasis has increased largely because of the expansion of irrigation systems and dams.
- These systems support more water snails, and bring more people into closer contact with worms.
- Between 1950–1990, the world's dams increased from about 5,000 to over 36,000, with a consequent increase in schistosomiasis, particularly in sub-Saharan Africa.
- In the second half of the 20th century, the area under irrigation worldwide doubled to more than 640 million acres.

Current worldwide distribution of three main types of schistosomiasis parasite

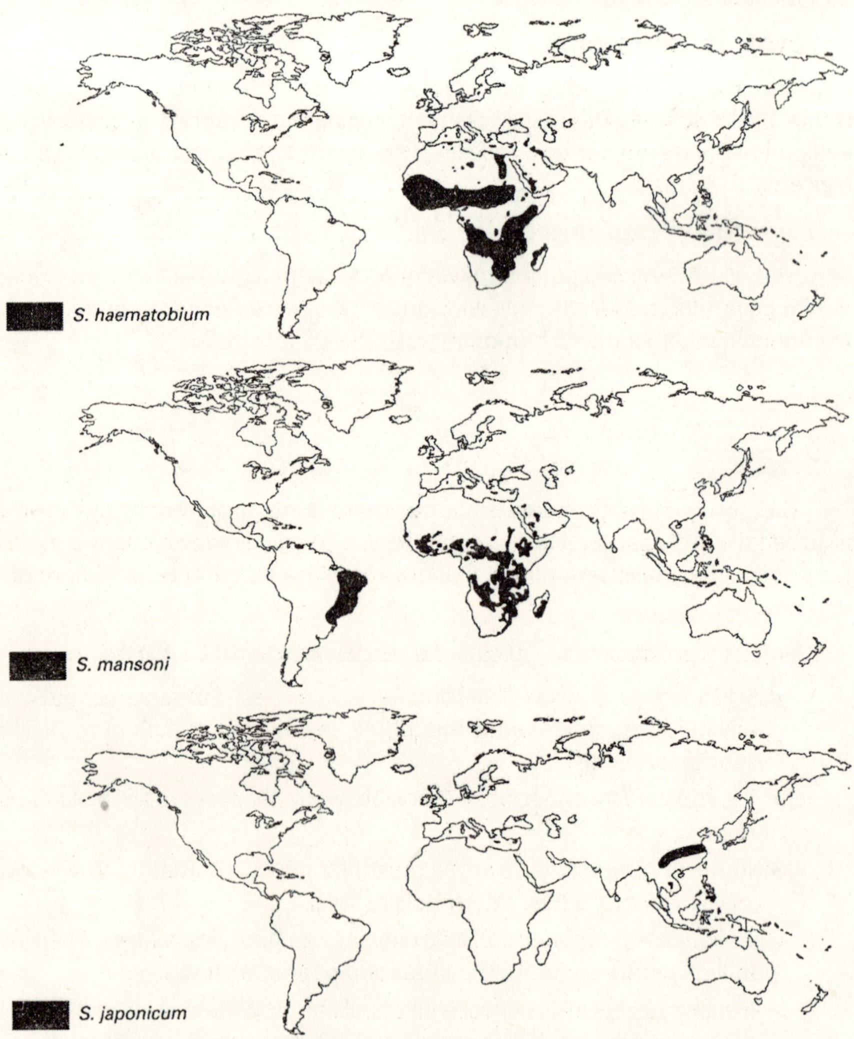

❐ site inspection

A Department of Health visit to a site to evaluate the likelihood of human exposure to toxic chemicals, and to do an exposure assessment. See "Exposure assessment."

❐ solubility

The largest amount of a substance that can be dissolved in a given amount of a liquid, usually water. For a highly water-soluble compound, such as table salt, a lot can dissolve in water. Motor oil is only slightly soluble in water.

❐ superfund (federal and state)

The federal and state programs to investigate and clean up inactive hazardous waste sites.

T

❐ target organ

An organ (such as the liver or kidney) that is specifically affected by a toxic chemical.

❐ the fresh water crisis: 1

- The supply of fresh water is finite, but demand is rising rapidly as population grows and as water use per capita increases.
- Globally, between 16 and 18 billion cubic yards of water are available for human use on an annual basis.
- In 1989, this amount equaled about 12,000 cubic yards per person per year and by 2000 had dropped to around 10,000 cubic yards per person.
- Over the past 70 years, the world's population has tripled to 6.1 billion people, but water-use has increased sixfold.
- Today, some 508 million people live in 31 countries that lack water.
- By 2025, when global population will have risen to 8 billion, that number is expected to reach 2.8 billion in 48 countries.

Projected water scarcity 2025

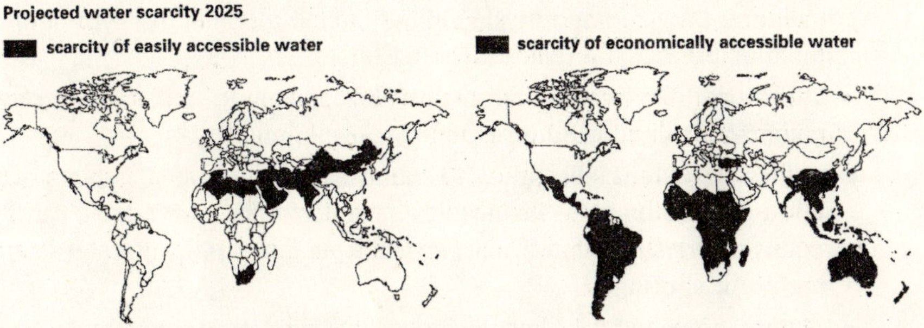

- By 2050, the number of water-short countries will be 54, affecting 4 billion people, or 40 percent of the projected global population.

- The worst affected areas will be the Middle East, North Africa and sub-Saharan Africa.
- Over 200 million sub-Saharan Africans already live in drought-affected countries.
- This figure will rise to 700 million by 2025.

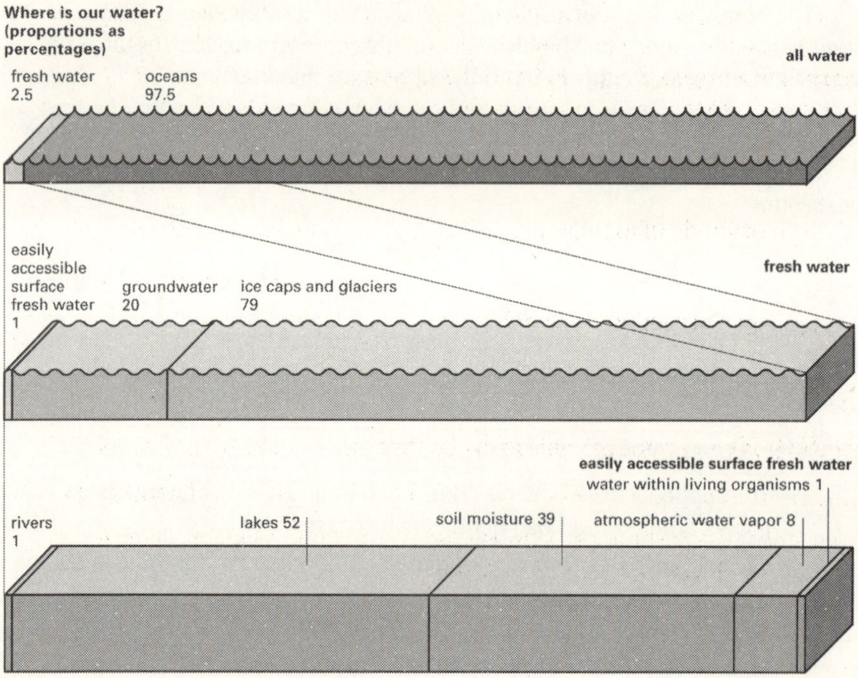

▫ the fresh water crisis: 2

- Water use from rivers and underground reserves has grown by 2.5–3 percent annually since 1940, significantly greater than population growth.
- Little of the water in some of the world's major rivers now reaches the sea, including the Colorado in the United States, the Nile in Egypt and, for much of the year, the Yellow River in China.
- The World Bank warns that fresh water is likely to become one of the major factors limiting future economic development.
- At the start of the 21st century, 49 countries with around 35 percent of the world population were believed to have less than 2,600 cubic yards of renewable fresh water available per capita per year, implying water scarcity or chronic shortage.
- Major nations in the list include India, Ethiopia, Nigeria and Kenya.
- More than 60 percent of the water used in the world each year is diverted for irrigating crops.

- Egypt, which must irrigate all its crops, uses more than five times as much water per capita as Switzerland.
- In Asia, which has 66 percent of the world's irrigated land, 85 percent of water goes for irrigation.
- A worldwide doubling in the area under irrigation to more than 1 million acres played a major role in the 'green revolution' that increased world food production in the late 20th century.

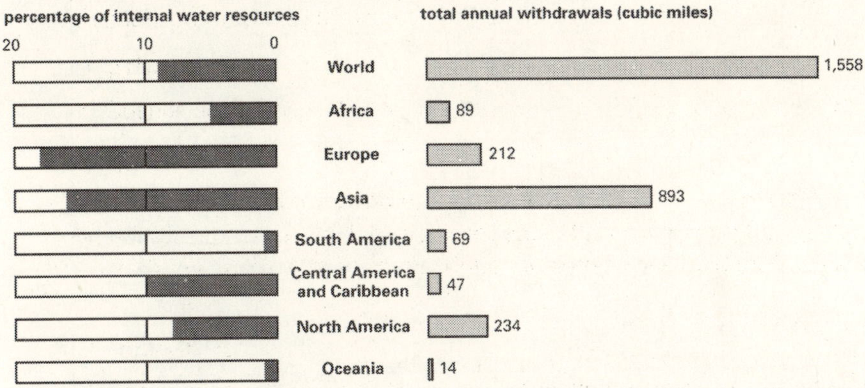

❐ the fuelwood crisis

- Wood provides nearly all of the energy needs of sub-Saharan African nations.
- 50–70 percent of all wood consumed on Earth is used for daily cooking purposes.
- 92 percent of all wood harvested in South Asia and 73 percent in Southeast Asia is used for fuel.
- The UN Food and Agriculture Organization (FAO) estimated that by the year 2000, some 2,400 million people were either short of fuelwood or were using it faster than it was being regenerated.
- By the year 2000 the world fuelwood deficit hadreached 1,255 million cubic yards a year – the energy equivalent of 240 million tons of oil a year.
- Spectacular economies can be achievedimmediately with relatively simple and inexpensive equipment.
- About 5 million fuel-saving stoves could be built for the price of one dam project in Africa.
- Use of these stoves combined would save 5 times the energy the dam could produce.
- Traditional wood-burning stoves (left) operate at an efficiency as low as 6–8 percent.
- Improved fuelwood stoves are at least three times as efficient.
- Other benefits:
  - reduction in indoor pollution

- reduced drudgery for women and children
- saving of firewood leading to conservation of forests.

Fuelwood consumption gathered by region in 1988 and 1998, shown as percentages

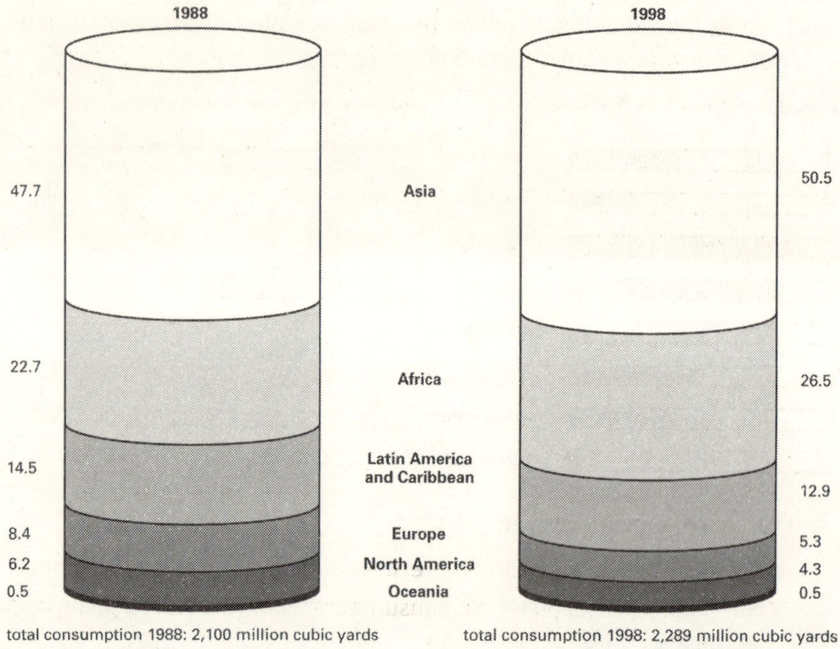

❐ the spread of aids: 1

- Twenty years after the first clinical evidence of acquired immunodeficiency syndrome was reported, AIDS has become one of the most devastating diseases to affect humans.
- Since the epidemic began, more than 60 million people have been infected with the human immunodeficiency virus (HIV).
- HIV/AIDS is now the leading cause of death in sub- Saharan Africa.
- Worldwide, it is the fourth biggest killer of people.
- At the end of 2001, an estimated 40 million people globally were living with HIV.
- In many parts of the developing world, the majority of new infections occurred in young adults, young women emerging as especially vulnerable.

- About 33 percent of those currently living with HIV/AIDS are aged 15–24.
- Most of them do not know they carry the virus. Many millions more know nothing or too little about HIV to protect themselves against it.

Adults and children estimated to be living with HIV/AIDS 2001

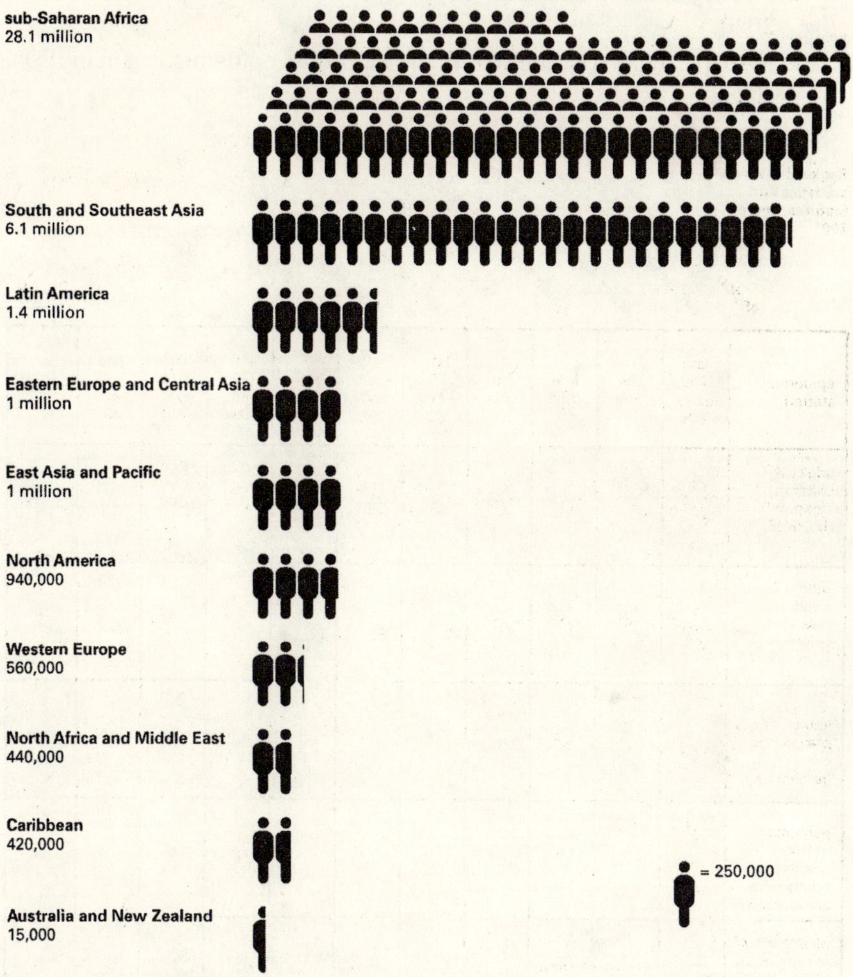

❐ the spread of aids: 2

- The impact of the AIDS epidemic is being increasingly felt in many countries across the world.
- Southern Africa continues to be the worst affected area, with adult prevalence rates still rising in several countries.
- Over half those HIV-positive in Africa are women.
- The AIDS epidemic will have a profound impact on growth, income, and poverty worldwide.

- It is estimated that the annual per capita growth in half the countries of sub-Saharan Africa is falling by 0.5–1.2 percent as a direct result of AIDS.
- By 2010, per capita GDP in some of the hardest hit countries may drop by 8 percent and per capita consumption may fall even farther.
- Heavily affected countries could lose more than 20 percent of GDP by 2020.
- In many regions of the world, the HIV/AIDS epidemic is still in its early stages.

Regional HIV/AIDS statistics and features, end of 2001

Region	sub-Saharan Africa	North Africa and Middle East	South and South East Asia	East Asia and Pacific	Latin America	Caribbean	Eastern Europe and Central Asia	Western Europe	North America	Australia and New Zealand	total
epidemic started	late '70s, early '80s	late '80s	late '80s	late '80s	late '70s early '80s	late '70s early '80s	early '90s	late '70s early '80s	late '70s early '80s	late '70s early '80s	
adults & children living with HIV/AIDS	28.1 million	440,000	6.1 million	1 million	1.4 million	420,000	1 million	560,000	940,000	15,000	40 million
adults & children newly infected with HIV	3.4 million	80,000	800,000	270,000	130,000	60,000	250,000	30,000	45,000	500	5 million
adult prevalence rate percent(*)	8.4	0.2	0.6	0.1	0.5	2.2	0.5	0.3	0.6	0.1	1.2
percentage of HIV-positive adults who are women	55	40	35	20	30	50	20	25	20	10	48
Main mode of transmission (#) for those living with HIV/AIDS	hetero	hetero, IDU	hetero, IDU	IDU, hetero, MSM	MSM, IDU, hetero	hetero, MSM	IDU	MSM, IDU	MSM, IDU, hetero	MSM	

* The proportion of adults (15–49 years of age) living with HIV/AIDS in 2001, using 2001 population numbers.
# Hetero (heterosexual transmission), IDU (transmission through injecting drug use), MSM (sexual transmission among men who have sex with men).

- Low national prevalence rates can disguise serious epidemics that are initially concentrated in certain localities or among specific population groups.
- In combating AIDS, young people must be a priority.
- Twenty years into the epidemic, millions of young people know little about HIV/AIDS.

- According to UNICEF, over 50 percent of young people aged 15–24 in more than a dozen countries, including Bolivia, Botswana, Côte d'Ivoire, the Dominican Republic, Ukraine, Uzbekistan and Vietnam, have never heard of AIDS or have very inaccurate information about how HIV is transmitted.
- Providing young people with candid information and practical preventative advice is vital for success in any attempt to combat the spread of AIDS.

▢ the urbanization of world population

Urbanization trends

- For the first time in history more people live in urban areas than rural areas.
- Cities occupy less than 2 percent of the Earth's land surface, but house almost half of the human population and use 75 percent of the resources we take from the Earth.

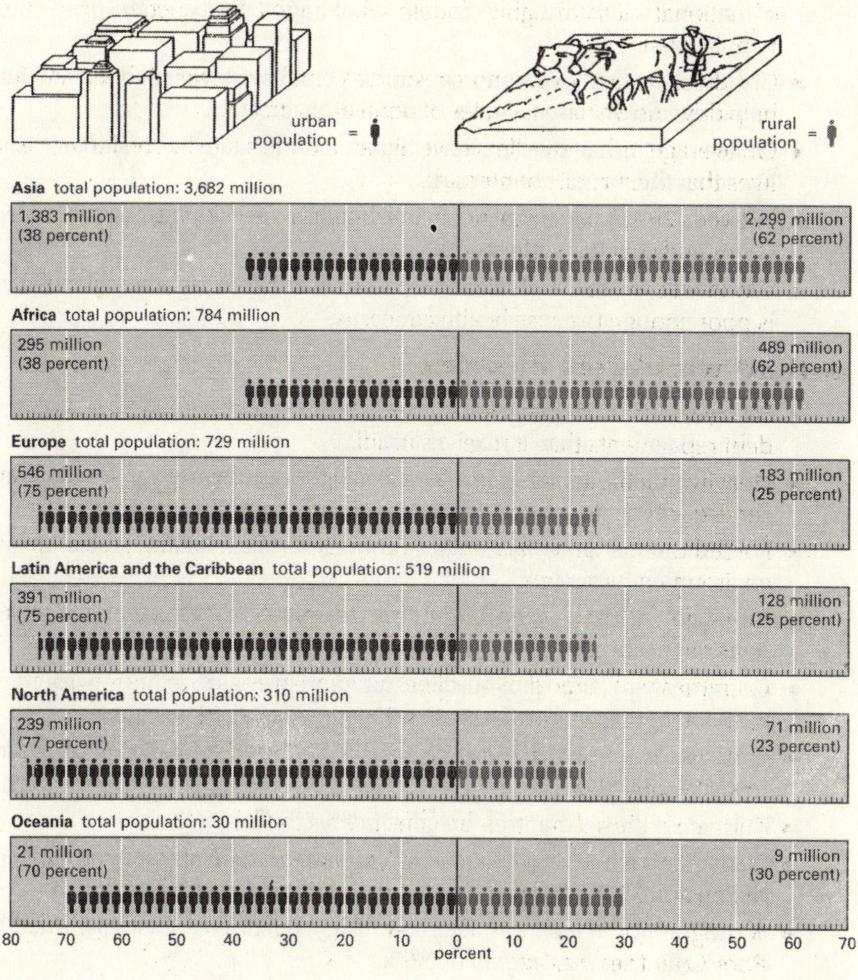

- Virtually all the population growth expected during 2000–2030 will be concentrated in the urban areas of the world.
- During 2000–2030 the world's urban population will grow at an annual rate of 1.8 percent, nearly double the rate expected for the total world population.
- Growth will be particularly rapid in the urban areas of less developed regions, averaging 2.3 percent per year during 2000–2030.
- The rural population is expected to grow less quickly, at 0.1 percent per year for the same period.
- The urban population of more developed regions is expected to increase very slowly.

Urbanization factors

- Cities are economically very productive: the World Bank estimates that urban areas in the developing world account for between 65 and 80 percent of national GDP (roughly double what might be expected from their populations).
- City dwellers tend to use more resources, but have fewer children and thus help drive down national rates of population growth.
- On average, urban dwellers have higher incomes and live healthier, easier lives than their rural counterparts.
- Between 26–50 percent of urban inhabitants in developing countries live in slums and squatter settlements.
- Worldwide, more than 1 billion people live in urban areas where air pollution is poor enough to cause health concerns.

❑ third world debt overview

- Each year the developing world gives Western countries nine times more in debt repayments than it receives in aid.
- Heavily indebted poor countries are caught in a downward spiral of debt service, which diverts resources away from economic development.
- Forced to meet debt payments, countries sacrifice health, education and environment programs.
- Africa, for example, spends four times as much on debt payments than it does on healthcare.
- Countries with large debt burdens put more pressure on the environment as the need to generate foreign exchange to pay debts increases.
- Forty-two countries have been defined by the World Bank and IMF as having unsustainable debts.
- Thirteen of these countries are affected by conflict.
- Debt relief for the 20 worst affected countries would cost between US$5.5 billion and $7.7 billion, less than the cost of one Stealth bomber.
- A program of 'deeper, broader and faster' relief for the Highly Indebted Poor Countries was agreed in 1999.
- By the end of 2001, over 20 countries, mostly in Africa, had benefited.

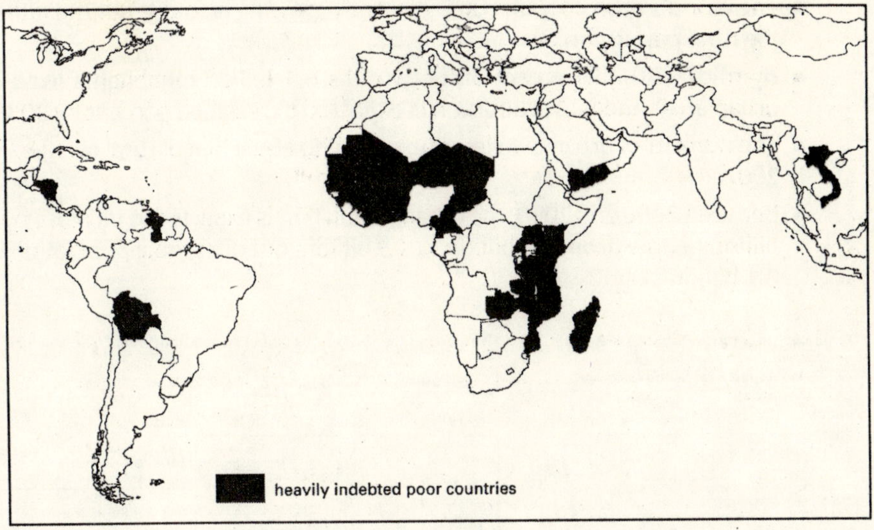

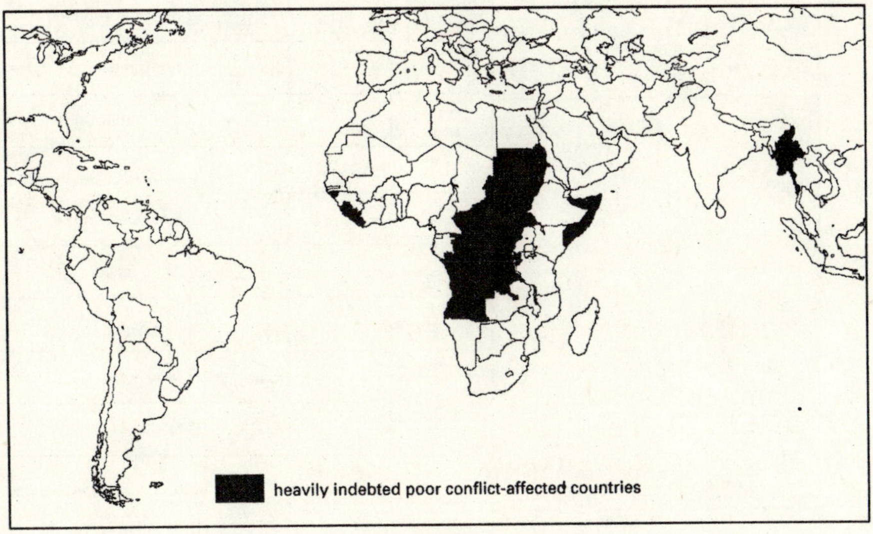

U

☐ urban and rural population growth

- The world is steadily becoming more urban, as people move to cities and towns in search of employment, educational opportunities and higher standards of living.
- Some leave agricultural communities and land that, for whatever reason, can no longer support them.
- The share of the global population living in urban areas has increased from 33 percent in 1960 to 47 percent (2.8 billion people) in 1999.

- The world's urban population is currently growing at four times the rate of the rural population.
- By mid-2000, 47 percent of the world's 6.1 billion inhabitants lived in urban areas and that proportion is expected to reach 60 percent by 2030.
- The number of urban dwellers is expected to equal that of rural dwellers in 2007.
- Between 2000 and 2030, the urban population is expected to increase by 2 billion people (from 2.9 billion to 4.9 billion, out of a global population of 8.1 billion).

Projected urban growth rate 1990–2030

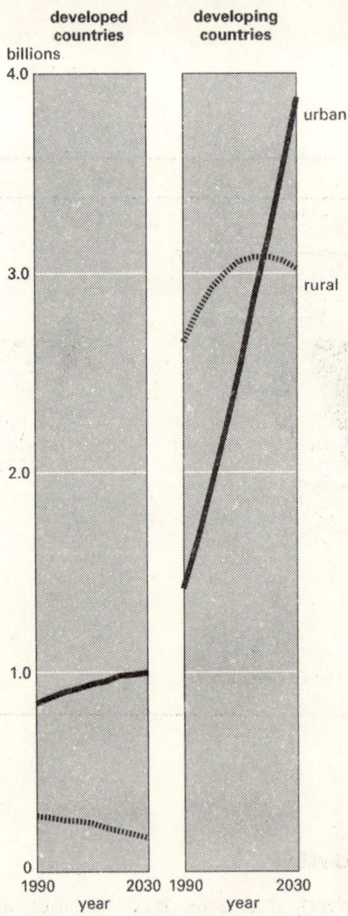

Urban/rural distribution of global population 1975, 2000, and 2015

(population (millions) / percentage)

	1975	2000	2015
world			
5 million or more — population (millions)	195	418	623
5 million or more — percentage	4.8	6.9	8.7
1 to 5 million — population (millions)	327	704	1,006
1 to 5 million — percentage	8.0	11.6	14.1
less than 1 million — population (millions)	1,022	1,723	2,189
less than 1 million — percentage	25.1	28.5	30.6
rural areas — population (millions)	2,531	3,210	3,337
rural areas — percentage	62.1	53.0	46.6
total population — population (millions)	4,075	6,055	7,154
total population — percentage	100	100	100
more developed regions			
5 million or more — population (millions)	98	112	120
5 million or more — percentage	9.3	9.5	9.9
1 to 5 million — population (millions)	145	219	250
1 to 5 million — percentage	13.9	18.5	20.6
less than 1 million — population (millions)	491	571	598
less than 1 million — percentage	46.8	48.1	49.3
rural areas — population (millions)	315	285	246
rural areas — percentage	30.0	24.0	20.3
total population — population (millions)	1,048	1,188	1,214
total population — percentage	100	100	100
less developed regions			
5 million or more — population (millions)	97	305	503
5 million or more — percentage	3.2	6.3	8.5
1 to 5 million — population (millions)	182	485	756
1 to 5 million — percentage	6.0	10.0	12.7
Less than 1 million — population (millions)	531	1 152	1 591
Less than 1 million — percentage	17.6	23.7	26.8
rural areas — population (millions)	2 217	2,925	3,091
rural areas — percentage	73.2	60.1	52.0
total population — population (millions)	3,026	4,867	5,940
total population — percentage	100	100	100

- The growth of the world rural population has been slowing markedly.
- In 1950, 7 of every 10 persons on Earth lived in rural areas, and they numbered 1.8 billion.

- Over the course of the next 50 years, rural population growth averaged 1.19 percent per year and the rural population nearly doubled, reaching an estimated 3.2 billion by 2000.
- But by 2030, rural population growth is expected to be minimal so that by 2030 the number of rural inhabitants will scarcely have changed, remaining at 3.2 billion.

V

❐ volatile

Evaporating readily at normal temperatures and pressures. The air concentration of a highly volatile chemical can increase quickly in a closed room.

❐ volatile organic compound (voc)

An organic chemical that evaporates readily. Petroleum products such as kerosene, gasoline and mineral spirits contain VOCs. Chlorinated solvents such as those used by dry cleaners or contained in paint strippers are also VOCs. See "organic" and "volatile".

W

❐ world distribution of wealth

- The last century witnessed major improvements in the healthcare and education available to many, reflected in declining infant mortality rates, increasing life expectancy, and higher literacy rates.
- 1.2 billion people are estimated to still live on less than $1 per day, and almost 3 billion on less than $2 per day.
- 110 million primary school age children are out of school, 60 percent of them girls.
- 31 million people are infected with HIV/AIDS.
- Many more live without adequate food, shelter, clean water, and sanitation.
- One of the central challenges in today's global economy and society is poverty in the midst of plenty.
- Fighting poverty is both a moral imperative and a necessity for a stable world.
- Two important ways of reducing poverty are development assistance and debt relief.
- The international community adopted International Development Goals during the 1990s.
- Progress in meeting these goals has been slow and uneven across the world.
- Poverty has increased in countries of the former Soviet Union, and mortality rates are deteriorating in Africa because of HIV/AIDS.

World Bank estimates of 2000 GNP per capita

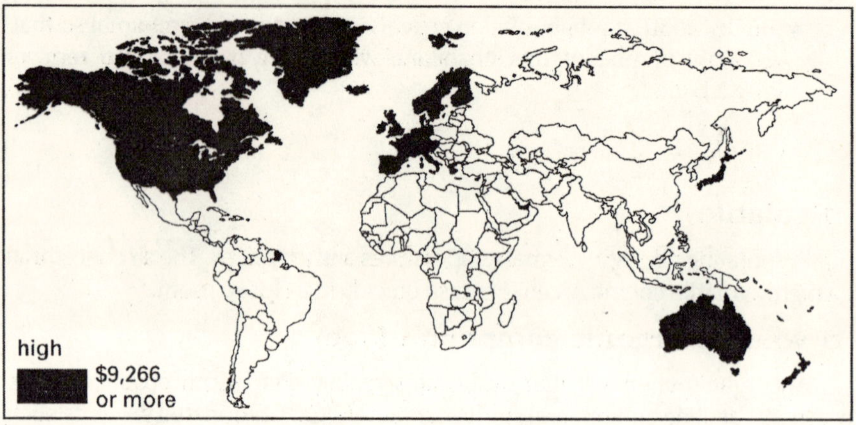

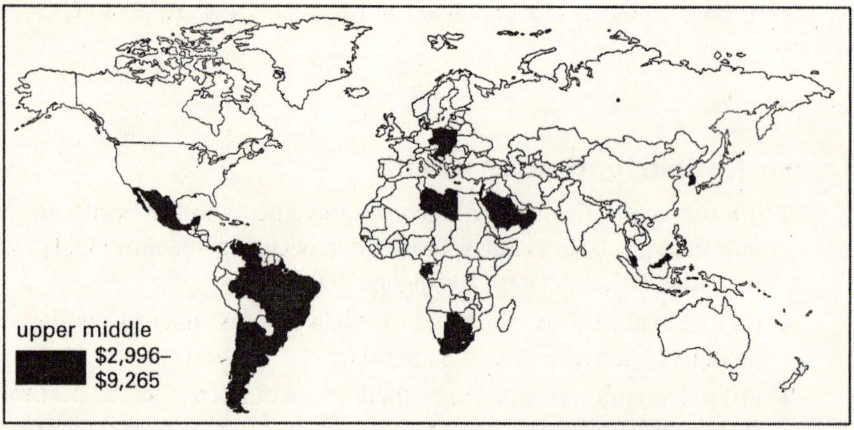

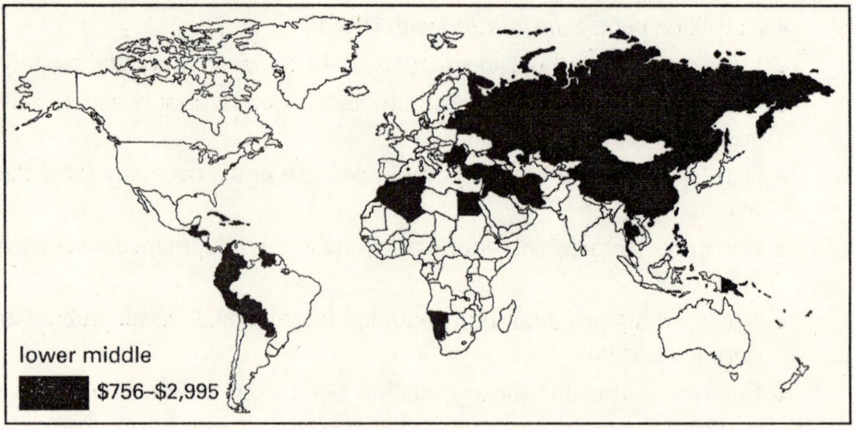

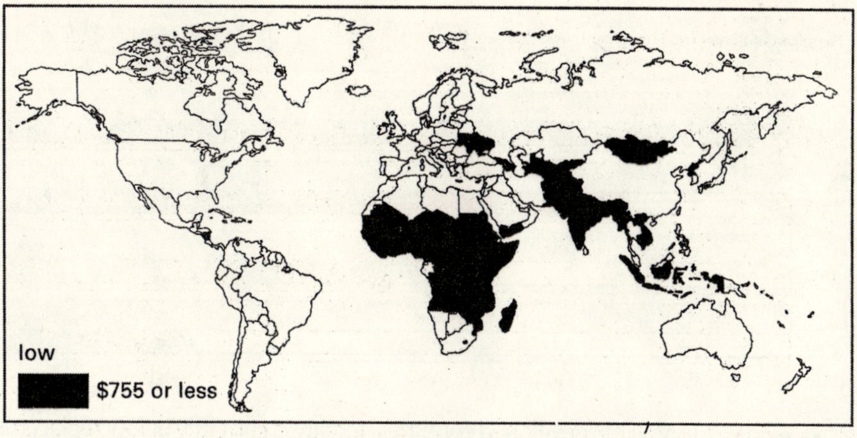

- Debt relief under the enhanced Heavily Indebted Poor Countries Initiative, under which more countries will be able to obtain deeper debt relief faster, is an important dimension of assistance to developing countries.

❑ world fertility rates and trends

- 'Total fertility' describes the number of children a woman has on average during her lifetime.
- The level of fertility at which the generations are replaced ('replacement level') is normally 2.1 children per woman.

Infant deaths per 1,000 live births 2000

100 or more

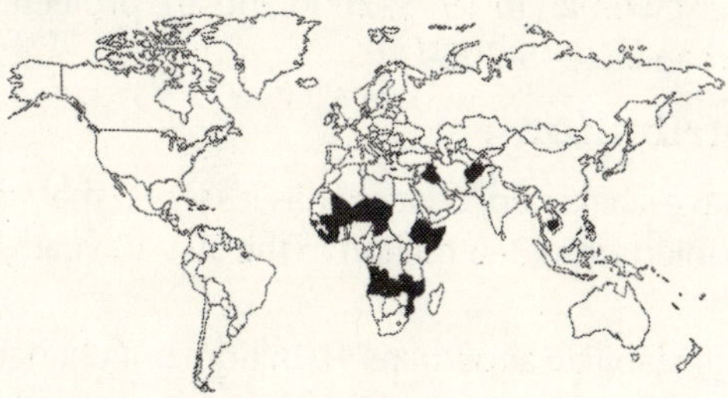

Fertility rate (births per woman) 2000

4 or more

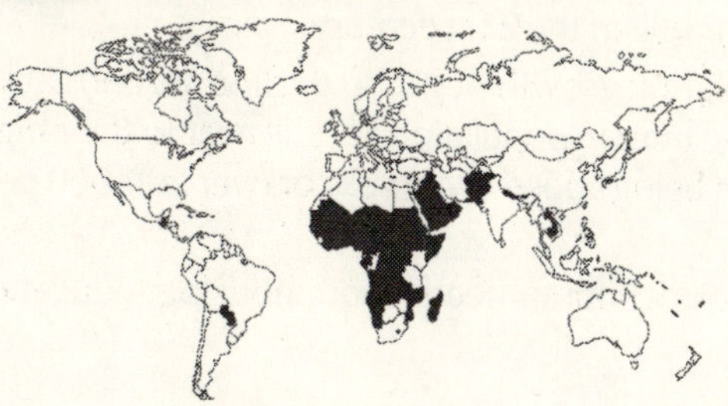

Fertility trends
actual and projected levels of childbearing

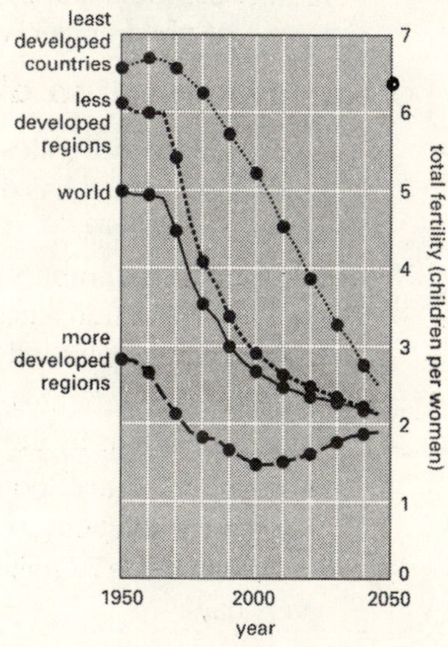

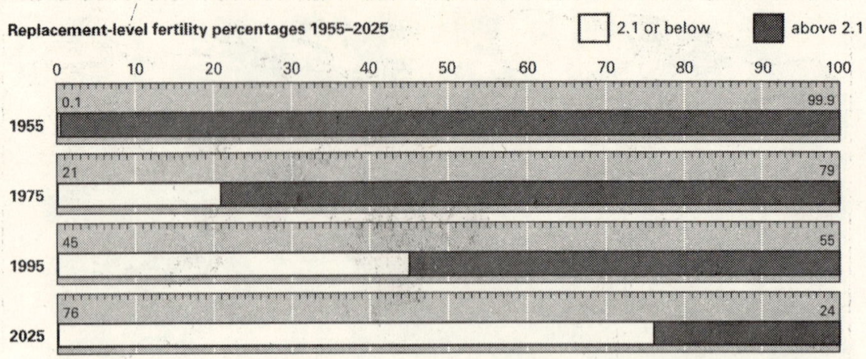

- In virtually all the developed countries, fertility is currently below replacement level and is expected to remain below that level.
- In the less developed regions as a whole, fertility is still above replacement level (at 3.1 children per woman in 1995–2000) and is projected to remain above replacement level until 2045–2050, when it reaches 2.17 children per woman.
- Most high-fertility countries belong to the group of least developed countries.
- In 1995–2000, the 48 least developed countries had a total fertility of 5.74 children per woman, which is projected to decline to 2.51 children per woman in 2045–2050, a level still well above the replacement level of 2.1 children per woman.
- Remaining countries in the less developed regions had an average total fertility of 3.06 children per cwoman in 1995–2000 and are projected to have 2.06 children per woman in 2045-2050.

❐ world population distribution

- As worldwide birth rates have slackened, concern has lessened that world population, which has doubled from 3–6 million in the last 40 years, will outstrip resources.
- It is estimated numbers will stabilize at perhaps 10 billion, but changes in the world situation, such as war, or technological innovation, could alter the outcome significantly.
- At current growth rates, countries such as Pakistan and Nigeria will triple their populations by the middle of the 21st century.
- Although the rate of population growth began to decline around 30 years ago, annual additions to the human population are still near to their highest level, with some 77 million being added every year, or over 200,000 people every day.
- Every one of these newborn children needs food, housing, education, a job, and healthcare.

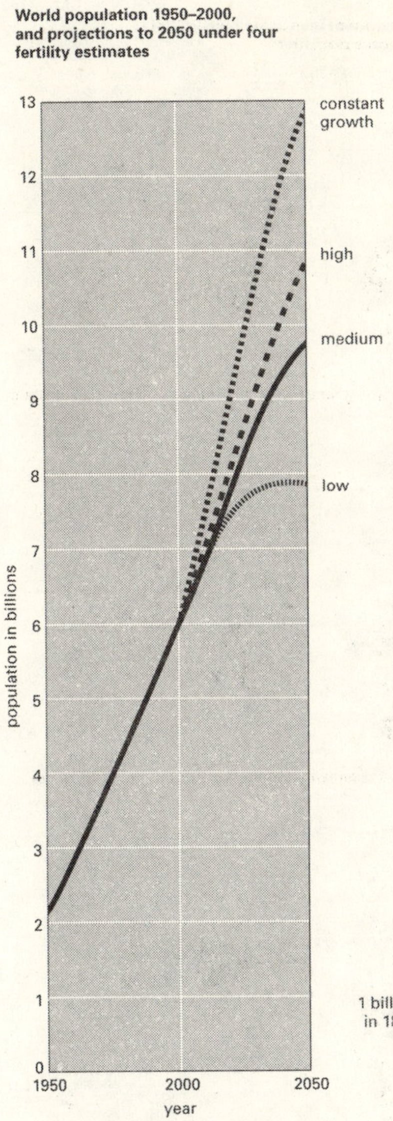

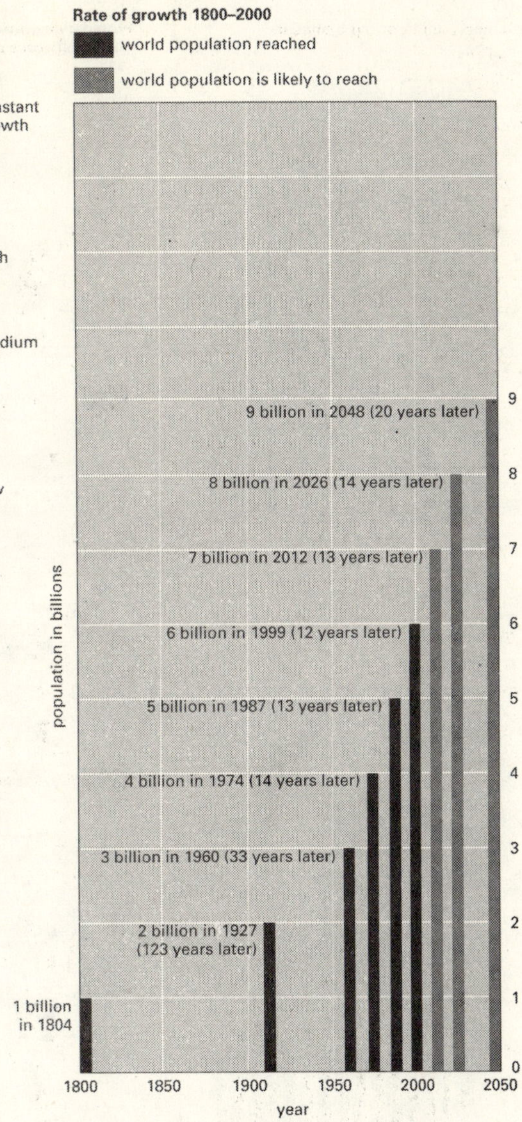

❑ world population growth by percentage

- The rapid growth of the world population is a recent phenomenon.
- The population of the world 2,000 years ago was about 300 million.
- It took more than 1,600 years for the world population to double to 600 million.
- Today, although the rate of global population increase peaked over 30 years ago, the latest data we have available indicates that the world population is still growing by over 200,000 people every day.

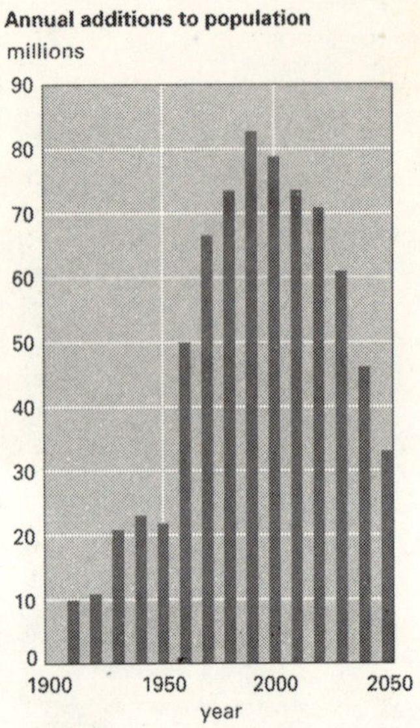

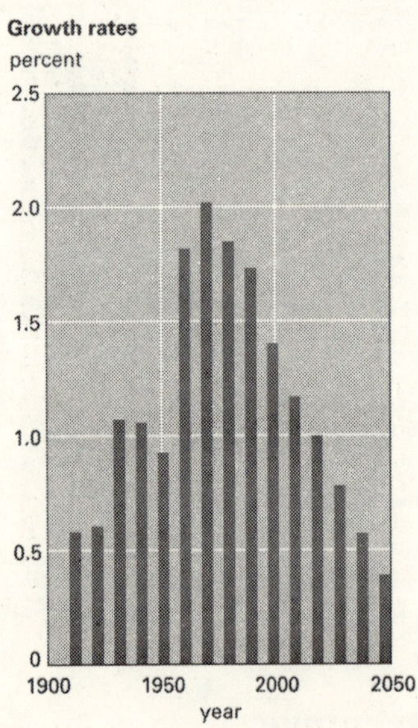

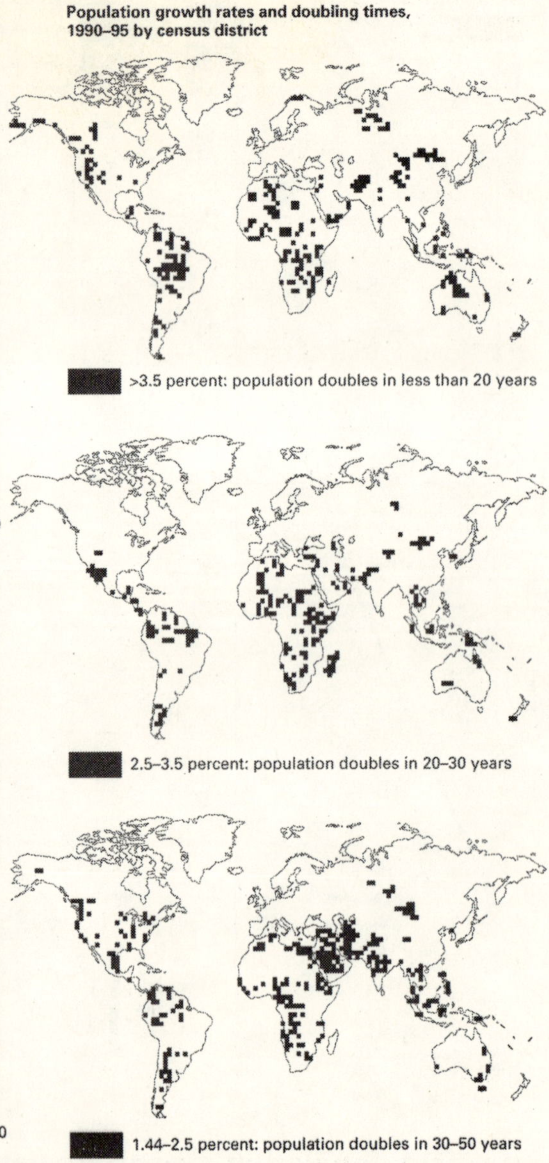

- The momentum of this growth derives from the huge numbers of people entering their reproductive age group who are now having children.

❑ world population growth by region

- Different demographic growth rates are changing the distribution of the world's people.

Population of the major regions of the world 1950, 2000 and 2050, in millions

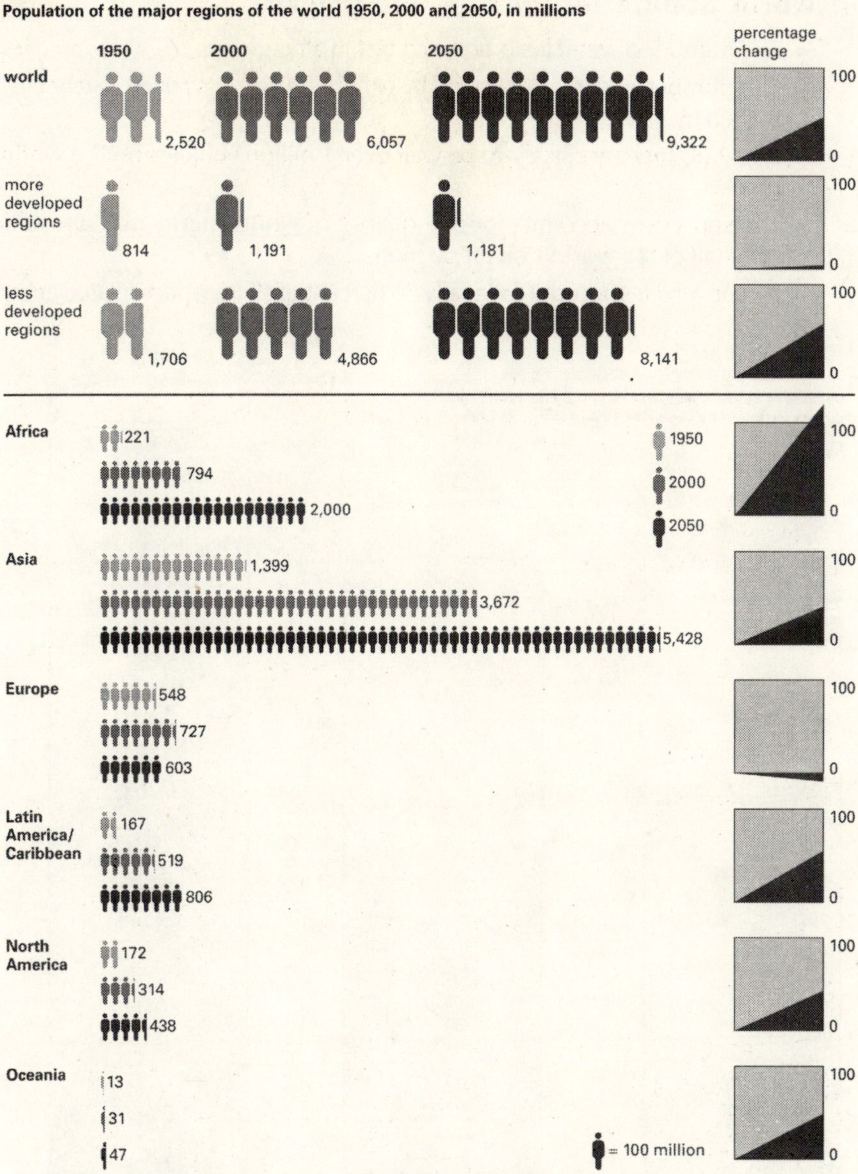

- While in 1950, Europe and North America accounted for 28.5 percent of the world population, that share fell to 17.2 percent in 2000, and it is expected to further decline to 11.2 percent in 2050.
- Africa's percentage of world population rose from 8.8 in 1950 to 13.1 in 2000 and is projected as 21.5 in 2050, although the current AIDS epidemic may not be fully integrated into this projection.
- Asia and Latin America in 2050 are relatively more stable, at approximately 58 percent and 9 percent respectively.

❑ world population of cars and trucks

- The number of vehicles is growing fast in all regions.
- The number of cars, trucks and buses on the world's roads reached 676 million by 1996.
- By 2025, there are likely to be well over 1 billion vehicles on the world's roads.
- Transport now accounts for one-quarter of world energy use, and about one-half of the world's oil production.
- Motor vehicles account for nearly 80 percent of all transport-related energy.

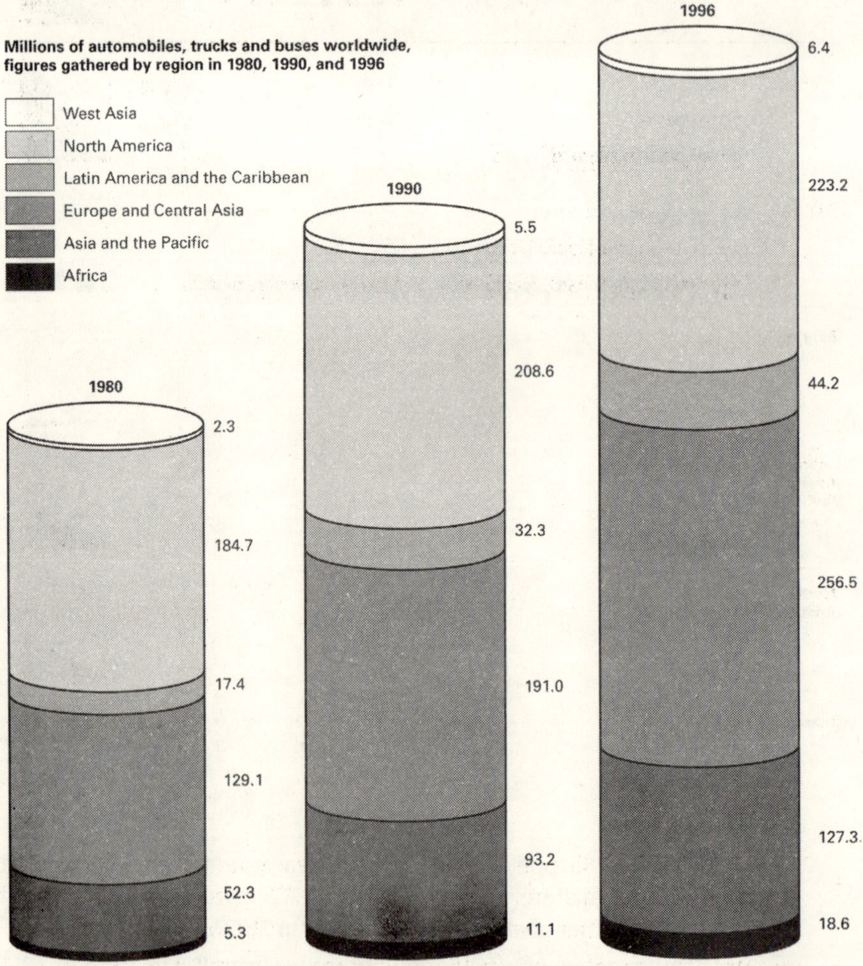

- Transport is thus a major contributor to greenhouse gas emissions and urban air pollution.
- With urbanization, motorized transport becomes essential for commuting, shopping and many other activities.

- As public transport is often poor, most people aspire to own an automobile.
- In 1998 the world automobile fleet exceeded 500 million for the first time – one for every 12 of the world's people and 10 times the figure of half a century ago.
- Two-thirds are in Western Europe and North America.
- One countervailing trend is that some rich, developed nations can afford such good urban mass transit systems that people prefer them to driving cars in congested city streets, while poorer services in more thinly populated rural areas lead to greater car ownership and heavier use than in cities.

Forest Genetics

A

❐ additive genes

A form of allelic interaction in which there is no dominance. The heterozygote is intermediate in phenotype between homozygotes for the alternative alleles. For multiple-gene traits, approximately equal contributions made by many loci.

❐ allele

One of several alternative forms of a gene occupying the same locus on a particular chromosome.

❐ analysis of variance

A statistical analysis by which F values (variance ratios) are composed in such a manner as to determine the probability that differences among populations or treatments are too large to be due to chance.

B

❐ backcross

Crossing a hybrid back to one of the original parents, called the recurrent parent. In agronomy, the recurrent parent is usually a single genotype, often a named variety. In forestry, the recurrent parent is usually a different genotype from the same species or population. (This is done to prevent inbreeding depression.) Example: collecting pollen from a pitch pine x loblolly pine F1 hybrid and pollinating a pure loblolly pine. The seed produced would be called first backcross and the loblolly pine the recurrent parent.

❐ backward selection

Selection of parent trees based on results from a progeny test.

❐ best linear prediction (blp)

A statistical method which utilities matrix algebra to predict breeding values for any trait or selection index; in BLP fixed effects are assumed to be known. BLP is especially suited for analyses of messy or unbalanced data.

❐ best linear unbiased prediction (blup)

A statistical method which predicts breeding values for any trait or selection index; unlike BLP, in BLUP the fixed effects are estimated. Like BLP, BLUP is well suited for analyses of messy or unbalanced data.

❐ biodiversity

The variety and variability (both in numbers and frequency) of the organisms and the genetic variability within each species. The term can be used to describe a particular site, a general habitat type, a small or large geographic region, or (less correctly), the genetic diversity of a particular species or population.

❐ block

In a field study arranged in a randomized complete-block design, a block is a unit of land upon which at least one plot of all of the entries (families, clones provenances) are planted. (In this case a replication and a block are the same.)

❐ breeding orchard

(breeding seed orchard, breeding arboretum) A planting of selected trees, usually clonally propagated, designed to ease breeding work.

❐ breeding population

A group of individuals selected from a wild population for use in a breeding program. Usually phenotypically selected for desirable traits. In species with large natural ranges, there are usually several to many, more-or-less separate breeding populations, each designed to provide progeny suited to a particular geographic region.

❐ breeding value

The genetic value of an individual determined by the mean value of its progeny. May be on the basis of individual traits or a selection index.

❐ breeding zone

An area within which a single population of improved trees can be planted without fear of mis-adaptation.

C

❐ candidate tree

A tree that has been tentatively selected for inclusion in a breeding program, but has not yet been measured or compared with surrounding trees.

❐ character (trait)

A distinctive but not necessarily invariable feature exhibited by all individuals of a group and capable of being described or measured; e.g. color, size, performance. A character of a given individual will have a certain phenotype as determined by the individual's genotype and environment.

❐ chromosome number

The number of chromosomes found in a typical individual of a particular species. Typically the sex cells of a species have one-half the number of chromosomes found in the vegetative portion and are said to be haploid.

▫ cline

An environmental gradient (temperature, rainfall, soil ph) and a corresponding phenotypic gradient in a population of plants or animals. Where these have been evaluated by provenance tests, clines are often found to have a genetic basis.

▫ clonal test

A field planting of several to many vegetatively propagated plants. Such tests furnish estimates of the relative performance of different genotypes, but do not necessarily provide information on breeding behavior.

1. Clone-To propagate a plant asexually usually by grafting, rooting cuttings, tissue culture, or apomictic seed. Except for an extremely low level of mutation, all plants from a clone are genetically identical. (This is the usage most common in agronomy, horticulture and forestry.)
2. To regenerate a whole plant from a single cell.(This usage is common in tissue culture research.)
3. To identify and isolate a gene controlling a specific trait from an organism. (This usage is common in molecular biology.)
4. A group of plants produced from cuttings, stump or root sprouts, tissue culture, or some other method that produces offspring genetically identical to the original plant. Most commonly used in forestry to establish grafted seed orchards, but becoming more common for commercial plantations.

▫ combining ability

The ability of an individual, when crossed, to produce progeny with strong expression of a particular trait.

▫ combining ability, general

Good general combining ability (GCA) is the ability of an individual to produce progeny with high genetic quality, when crossed to many other individuals in the population.

▫ combining ability, specific

Good specific combining ability (SCA) is when the progeny from a particular full-sib cross perform better than what would be predicted from the general combining ability of either parent.

▫ conelet

An immature cone (strobilus) in the conifers.

▫ control pollination

To purposely pollinate the female flowers of a tree with pollen from a known source, usually one specific tree. Usually the flowers are Protected from undesirable pollen by covering them with a pollen-tight cloth or paper bag before they are receptive and adding known-source pollen at receptivity. Full-sib families are produced.

❒ cross

(1) To collect pollen from one tree and pollinate a second tree. (2) The progeny of a control pollination.

❒ crossing groups

Any group of individuals which comprises a unique set of parents. A diallel CG is one in which controlled crosses are made between each pair of parents in the group but crosses with parents outside the group are excluded. A factorial CG is one in which a limited number of parents are used as male testers in controlled crosses with an unlimited number of female parents. An open-pollinated CG is one in which all parents in a breeding population are included in a progeny test or series of tests and which may be used for reforestation with genetically improved material.

❒ cross-pollination

Pollination by genetically different plants

D

❒ diallel, complete or full

A mating design and subsequent progeny test resulting from the crossing of 'n' parents in all possible 'n2' combinations including selfs and reciprocals. (Because of severe inbreeding depression in the selfs, these are often skipped and the test still called a "full" diallel.)

❒ diallel, incomplete or partial

A partial sampling - any individual family or type of family may be omitted. In either the complete or incomplete diallel, identities of both seed and pollen parents are maintained for each family.

❒ dominance

In classical Mendelian genetics, the masking of the action of one allele by another. If an individual with red flowers is crossed with an individual with white flowers and all progeny have red flowers, then the allele for red pigment is fully dominant over the allele for white flowers.

❒ dominance, partial

The incomplete masking of the action of one allele by another. If an individual with red flowers is crossed with an individual with white flowers and all progeny have pink flowers, then the allele for red pigment is partially dominant over the allele for white flowers.

❒ dysgenic

An action or process that is detrimental to the genetic qualities of a population. Usually applied to human actions, such as high grading or land clearing, which would reduce the local gene pool.

E

❐ ecotype

A sub-population of a species that occurs in a particular well-defined environment, usually showing better adaptation to that environment than the species as a whole.

❐ elite tree

A tree that has been shown by progeny testing to produce superior offspring.

❐ endemic

A plant or animal species or subspecies native to a small region.

❐ ex situ management

The management of planted stands of trees to protect them and insure survival, growth and identity; the conservation or preservation of trees as seed, pollen, tissue culture, or excised plant parts.

❐ exotic

Broad definition: A non-native population introduced into a new area. Narrow definition: a species introduced from another country.

F

❐ f1

The first (filial) generation of a cross. Usually the progeny from an F1 cross will be phenotypically fairly uniform.

❐ f2

The second (filial) generation of a cross, produced by intercrossing individuals from the F1 generation. The progeny of an F2 cross will be more variable phenotypically than the F1.

❐ family

A group of closely related genotypes.

❐ family selection

The selection of progeny families on their mean performance. In addition, the best individuals are usually selected in the best families.

❐ fertilization

Union of the nucleus and other cellular constituents of a male gamete (sperm) with those of an egg to form a zygote. In some species, fertilization may occur months after pollination.

❑ fitness

In classical genetics, the ability of an individual or population to both survive and reproduce in a particular environment.

❑ fitness, adaptive

The ability of a population to survive, reproduce, and adapt genetically to a changing environment for at least a modest number (usually fewer than 50) generations.

❑ fitness, evolutionary

The ability of a population (usually a whole species or ecotype) to survive, reproduce, and adapt genetically to a changing environment for many (usually more than 50) generations.

❑ flower

Angiosperm reproductive structure bearing pistils, stamens, or both, and usually also sepals and petals. So-called flowers of conifers are the male and female strobili before and during pollination.

❑ forest genetics

The study of heredity in forest trees.

❑ forest tree improvement

The application of genetic principles to the improvement and management of forest trees.

❑ forward selection

Choosing the best individuals out of a progeny test for use in seed orchards and/or subsequent generations of breeding.

G

❑ gene

The basic unit of inheritance.

❑ gene bank

A collection of germplasm (seeds, pollen, whole plants, extracted DNA) collected and maintained so as to sample as much as possible of the genetic variability in a population.

❑ gene flow

The movement of specific alleles among different populations of a species or among related species.

❑ gene pool

The sum total of all the genetic variation in the breeding population of a species and closely related species capable of crossing with the species.

❐ gene pool, primary

The sum total of all the genetic variation in the breeding population of a species and closely related species that commonly interbreed with, or can be routinely crossed with, the species.

❐ gene pool, secondary

The sum total of all the genetic variation in the breeding populations of related species that can be crossed with the species using mentor pollen, embryo rescue or other unusual measures.

❐ gene pool, tertiary

The sum total of all the genetic variation in other organisms that cannot be crossed with the species. (With the development of genetic engineering, it is theoretically possible to transfer genes isolated from any organism (plant, animals, virus, or bacterium) into a plant. This makes the line between the secondary and tertiary gene pools somewhat fuzzy.)

❐ genetic diversity

General concept: the amount of genotypic variability in a population. Quantitative definition: the number of different alleles per loci and the proportion of loci with more than one allele in a species or population.

❐ genetic drift

Changes in gene frequency in small populations due to random chance. Usually a loss of rare alleles.

❐ genetic gain

The change achieved by artificial selection in a specific trait. Gain is usually expressed as the change per generation or the change per year. Gain is influenced by selection intensity, parental variation, and heritability.

❐ genetics

Genetics is the basic science dealing with causes of resemblances and differences among organisms related by descent. It takes into account the effects of genes and the environment.

❐ genotype

The specific set of genes possessed by an individual, both expressed and recessive.

❐ genotype-environment interaction

Changes in rank or levels of performance among individuals when tested in different environments.

❐ geographic variation

The phenotypic differences among native trees growing in different portions of a species' range. If the differences are largely genetic rather than environmental, the variation is usually specified as racial, ecotypic, or clinal.

❐ germplasm conservation

Maintenance of the genetic variability of a population. (The term is used in preference to oegermplasm preservation" to reflect the ever-changing nature of living populations.)

❐ germplasm conservation, ex situ

Maintaining the genetic variability of a population in a different environment or geographic location than where it evolved, i.e., plantations of exotic species, breeding orchards, cold storage of seed or pollen.

❐ germplasm conservation, in situ

Maintaining the genetic variability of a population in approximately the same geographic and ecological conditions under which it evolved. (This should always be the first choice.)

H

❐ heritability

General concept: The degree to which progeny resemble their parents. Quantitative definition: A ratio of genetic relative to environmental factors that influence the expression of a trait.

❐ heritability, broad sense

The proportion of the total phenotypic variability for which genetic differences are responsible.

❐ heritability, narrow sense

The proportion of the total phenotypic variability that is due to additive genetic variability.

❐ heterozygous

Having two different alleles at a locus. When used to refer to the whole genotype, indicates that the individual has different alleles at most loci. When used to refer to a species as having low or high heterozygosity relative to other species, indicates that the species has a relatively high number of variable loci.

❐ homozygous

Having two identical alleles at a locus, at most loci, or in the entire species.

❐ hybrid

Progeny from a cross among dissimilar genotypes. In forestry, the term is usually used for crosses between species.

I

❐ ideotype

The ideal type or perfect specimen. A description or illustration of what the final goal of genetic improvement for a species would look like. The ideotype is developed without regard to heritabilities for the individual traits involved. It is not intended to be a practical field guide for selecting "+" trees, but as a starting point for developing a field-selection guide.

❐ inbreeding

Matings between related individuals. In open pollinated tree species, inbreeding usually leads to mild to poor seed set, low germination, and severe growth reduction.

❐ inbreeding depression

The reduction in vigor often observed in progeny from matings between close relatives. Inbreeding depression is due to the expression of recessive deleterious alleles and is usually severe in open-pollinated out crossing species (most forest trees) that occur in fairly high population densities.

❐ introgression

The movement of genes from one population into another through hybridization followed by backcrossing. Usually refers to movement of genes from one species to another or among sub-species that have been geographically isolated then brought back together by changes in the species ranges or planting of exotic populations.

L

❐ land race

A population of trees of a non-native species that has undergone one or more generations of natural selection in a new environment.

❐ locus

The position of a gene on a chromosome.

M

❐ mutation

A sudden change in the genotype, usually caused by a small change in the DNA sequences in the chromosomes. It may also be caused by changes in chromosome number or breakage of individual chromosomes.

N

❐ neighborhood

The largest population in which mating occurs at random. Generally, a small population within which individuals can be expected to be closely related.

Neighborhoods may be created by barriers to pollen and seed movement, or other isolating factors.

O

- **open pollinated**

Pollination occurring due to wind or insects.

- **ortet**

The original plant from which a clone is started through rooted cuttings, grafting, or tissue culture, or other means of vegetative propagation. The original plus tree used to start a grafted clone for inclusion in a seed orchard is the ortet.

- **outcrossing**

Matings (controlled or natural) among unrelated individuals. May also refer to a species that has specific barriers to selfing, or exhibits such inbreeding depression that inbred individuals never reach maturity.

P

- **pedigree**

Record of ancestry.

- **pedigree breeding**

A breeding system where all the ancestors of any individual in the population are known back to the original "+" trees selected in wild stands. The system depends on control pollination for all matings.

- **pedigree, full**

All parents, grandparents of a particular genotype are known all the way back to some natural population.

- **pedigree, partial**

Some, but not all, ancestors of a particular genotype are known, usually the female parents. Partial pedigrees are most common where open-pollinated or polycross seed is used for progeny tests.

- **phenotype**

The visible characteristics of a tree. The phenotype is determined by the genotype interacting with the environment in which it is grown.

- **plot**

In a field study, a group of trees, all from the same entry (family, clone, provenance . . .) planted together (usually).

- **plot, contiguous**

In a field study, a group of trees, all from the same entry (family, clone, provenance . . .) planted together. A five-tree row-plot is probably the most common design for forest genetics experiments.

❒ plot, non-contiguous

In a field study, a group of trees, all from the same entry (family, clone, provenance . . .) planted in the same block, but not planted together. The non-contiguous interlocked-plot design is a recent innovation that allows tests to be thinned systematically and still maintaining uniform spacing.

❒ plus tree

A phenotypically superior but untested tree.

❒ pollination

Deposition of pollen on the receptive part of the female flower.

❒ polycross test

A progeny test to assess general combining ability from crosses among selected parents. Identities can be maintained only for the seed parents. A mixture of pollen is artificially applied to each female parent, or the selections naturally and randomly open-pollinate themselves.

❒ polyploid

A cell, tissue, individual, population, or whole species having more than twice the basic number oex" of chromosomes of the ancestral species. Polyploidy may lead to increased growth rates (Populus) or severe dwarfing (Pinus). Three sets of chromosomes is called triploid, four sets-tetraploid, six sets-hexaploid, etc.

❒ population

A group of individual trees having some characteristics in common, either location, family ancestry, or intended use.

❒ progeny test

A test to compare the offspring of different parents.

❒ propagule

Any type of plant to be used for reproduction. It might be a seedling, a rooted or unrooted cutting, a graft, or a tissue-cultured plantlet.

❒ provenance test

A test comparing trees grown from seed or cuttings collected in many parts of a species range.

R

❒ race

A sub-population, usually fairly large, of a species exhibiting some degree of phenotypic (and presumably genotypic) uniformity among individuals within the population and distinct from the species as a whole.

❑ ramet

A vegetatively reproduced copy of a plant. Each ramet will have almost precisely the same genotype as the original parent tree, known as the ortet.

❑ randomized complete-block design

The most common experimental design used in progeny, provenance and clonal field tests. Each entry in the test is replicated once in each block. All the entries are arranged randomly within a block and a new randomization pattern is used for each block.

❑ reciprocal cross

The repetition of a cross where the sexual function of the parents is reversed, i.e., female B x male A is the reciprocal of female A x male B.

❑ recurrent selection

Selection of individuals or families and intermating them or allowing them to interpollinate to produce the next generation. Usually the process is repeated several times (for long-lived perennials such as trees) or many times (for annuals.) The technique is used to improve the overall performance of the population rather than to select outstanding varieties for immediate use.

❑ replication

In a genetics test a replication contains one plot of each entry in the test.

❑ resilience

The ability of a population to persist in a given environment despite disturbance or reduced population size. The resilience of a population is based upon the ability of individuals within the population to survive (fitness) and reproduce (fecundity) in a changed environment and the genetic variability of the population which allows the production of new genotypes.

❑ rogue

To remove trees that have an undesirable phenotype, or that have been shown through progeny tests to have a less desirable genotype from a seed orchard, seed production area, or nursery bed.

❑ rootstock

A rooted plant, usually a seedling, on which a scion is grafted.

S

❑ scion

A twig, bud, or other vegetative cutting to be grafted onto the root-system of another plant.

❑ screening

Selecting for a particular trait, often used when referring to the application of a specific treatment such as white pine blister rust spores or a herbicide, and then looking for resistant individuals. Usually implies checking large number of individuals at one time.

❐ seed orchard

A plantation established for the production of tree seed.

❐ seed orchard, clonal

A seed orchard established from vegetatively propagated trees, usually grafts. Usually clonal seed orchards are established with single-tree plots with different ramets of each clone located as far apart as possible to reduce self pollination.

❐ seed orchard, seedling

A seed orchard established from seedlings (as opposed to grafts). Usually seedling seed orchards are established with multiple-tree plots so that selection can take place first among families and then among the individuals within each family plot, reducing each plot to a single tree.

❐ seed production area

A stand or plantation not originally planted to produce seed, but rogued of inferior trees and treated in such a manner as to produce large quantities of seed. Usually the stand or plantation is from a known origin or selected seedlot.

❐ seed source

The location where a seedlot was collected. If different seedlots of an exotic species are collected and tested, the test is called a seed-source test to distinguish it from a provenance test.

❐ seed transfer guide

A set of rules for collecting seed and planting seedlings so that genotypes are not moved to microclimates or soils within which they will be mis-adapted. Usually seed transfer guides describe the maximum movement from the point of collection in miles east and west, north and south as well as feet in elevation.

❐ seed zone

An area within which seed can be collected from any natural stand and planted in any new site without fear of mis-adaptation.

❐ seedlot

A designation of a group of seeds with some factor in common, i.e., year of collection, stand or seed orchard, individual "+" trees, point of origin in a provenance test, a half- or full-sib family.

❐ selection

Choosing individual trees or populations with desirable characters to obtain genetic improvement.

❐ selection differential

The difference between a selected tree, family, or clone and the average of the population from which it is taken.

❒ selection, clonal

Choosing the best clones from a clonal test.

❒ selection, index

Choosing parents on the basis of a weighted score which combines economic value and heritability of several desirable traits.

❒ selection, mass

(phenotypic selection) Choosing trees on the basis of their phenotypes (appearance) alone.

❒ selection, recurrent

Choosing individuals out of families, making crosses, then progeny testing. The process is usually repeated several to many times. Most tree improvement programs are based on recurrent selection.

❒ selection, tandem

Selecting for two or more traits sequentially rather than simultaneously. The technique is useful when one trait can be evaluated early and a second trait only after a number of years, or when one trait is much more expensive to measure.

❒ self pollination

The natural or artificial pollination of a female flower with pollen from the same genotype.

❒ sib

(sibling) A term meaning brother or sister. Half sibs have one parent in common. Full sibs have both parents in common.

❒ sublining

Dividing a breeding population into several to many smaller populations. All controlled crosses for forward selection are made within a subline, leading to inbreeding within sublines. Production seed orchards are planted with clones or seedlings from several to many sublines to minimize inbreeding in the production orchard.

T

❒ tissue culture

A technique for cultivation cells, tissues, or organs of plants in a sterile, synthetic medium; includes the tissues excised from a plant and the culture of pollen or seeds.

U

❒ unbalanced

In experimental design, refers to an experiment or set of data in which all treatments or treatment combinations are not equally represented. The most common cause of unbalanced experiments is unequal mortality among entries in a test. Also used to describe any plant breeder specializing in an organism with nearly as long a life cycle as his or her own.

V

❐ variability

Differing from the average value. Absence of uniformity, usually referring to lack of genetic uniformity in a population.

❐ variance

A statistical measure of variability.

❐ variety

A distinctive seedling population or clone, usually one that possesses enough desirable characteristics to be commonly cultivated. In agriculture and horticulture all the plants within a variety would usually be quite uniform genetically. In forestry the term is more loosely used and the variability within a seedling variety is usually much greater.

❐ vegetative propagation

Propagation of a plant by asexual means, as in budding, dividing, grafting, rooting, and air layering.

❐ viability

A populations ability to live, grow and develop. It is affected by physical habitat factors (climate, geology, topography, and aquatic features) and by biotic habitat factors (plant and animal populations and communities).

W

❐ wind pollination

Pollination by wind-borne pollen.

Hurricane

A

❑ anatomy of a volcano

- A volcano is most commonly a conical hill or mountain that has grown as a result of volcanic deposition around a vent that connects with reservoirs of molten rock below the surface of the Earth.

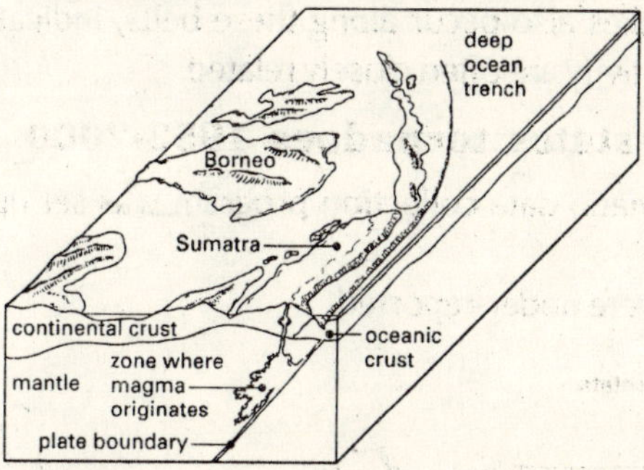

Island-arc volcanic environment
- In a typical island-arc environment, volcanoes lie along the crest of an arc-shaped, crustal ridge bounded on its convex side by a deep oceanic trench.

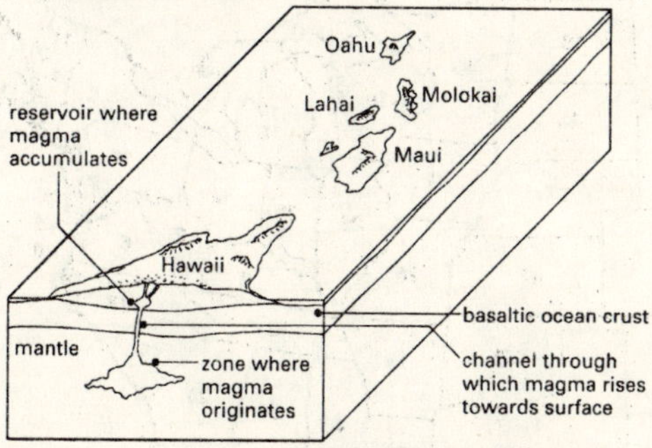

Oceanic volcanic environment
- Volcanoes are aligned along the crest of a broad ridge that marks an active fracture system in the oceanic crust in a typical oceanic environment.
- Generated in the upper mantle beneath the ridge, basaltic magmas rise along fractures through the basaltic layer.

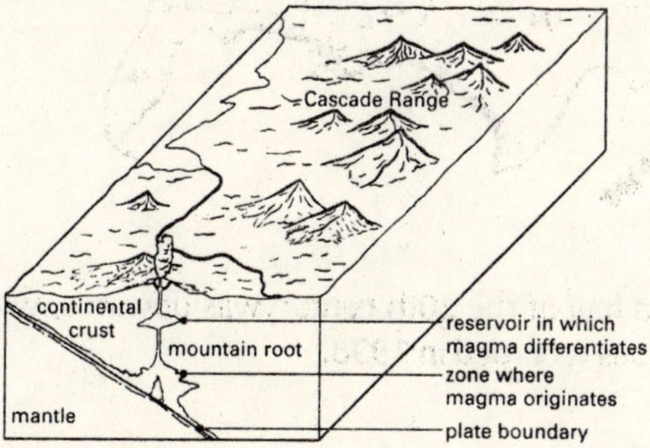

Continental volcanic environment
- In the typical continental environment, volcanoes are located in unstable, mountainous belts that have thick roots of granite or granite-like rock.
- Generated near the base of the mountain root, magmas rise slowly or intermittently along fractures in the crust.

- The molten rock, lighter than the surrounding solid rock, is driven by buoyancy and gas pressure and forces its way upward and breaks through areas of weakness in the Earth's crust.
- During an eruption, the molten rock may pour from the vent as nonexplosive lava flows, or it may shoot violently into the air as dense clouds of lava fragments.
- Larger lava fragments fall back around the vent, and accumulations of fallback fragments may move downslope as ash flows under the force of gravity.
- Some volcanoes form chains of islands in the deep ocean basins, while others crown island areas lying near the continents.
- Volcanoes tend to cluster along narrow mountainous belts where folding and fracturing of the rocks provide channelways to the surface for the escape of the magma.
- Significantly, major earthquakes also occur along these belts, indicating that volcanism and seismicactivity are often closely related.

❑ average annual united states tornadoes 1953-2000

- A United States national tornado data collection program was set up in 1953.
- Before this date, tornadoes were under-reported.

Annual average number of tornadoes 1950–1995 by state

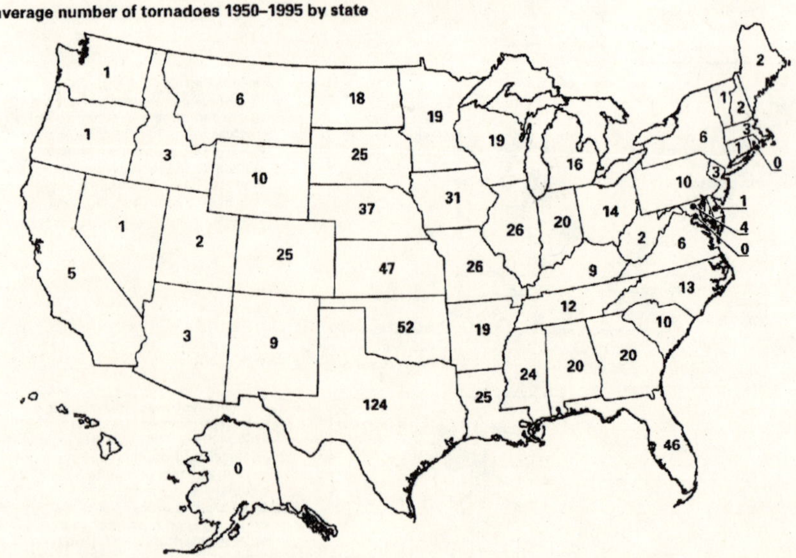

- The trend over the second half of the 20th centurywas upwards, with a peak of over 1,400 tornadoes recorded in 1998.

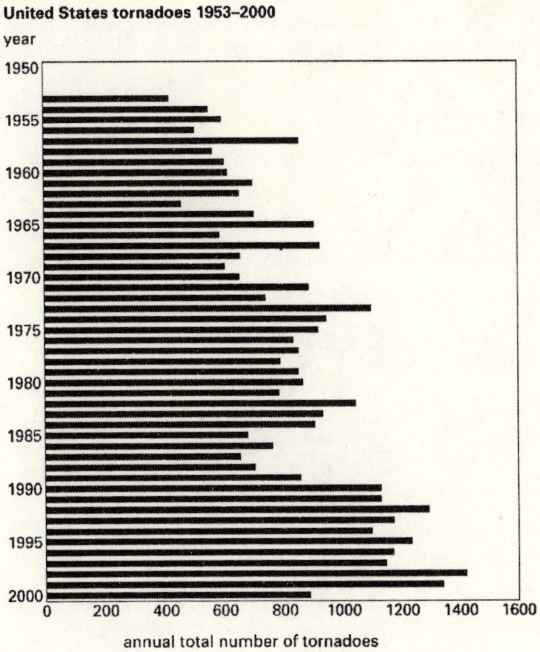

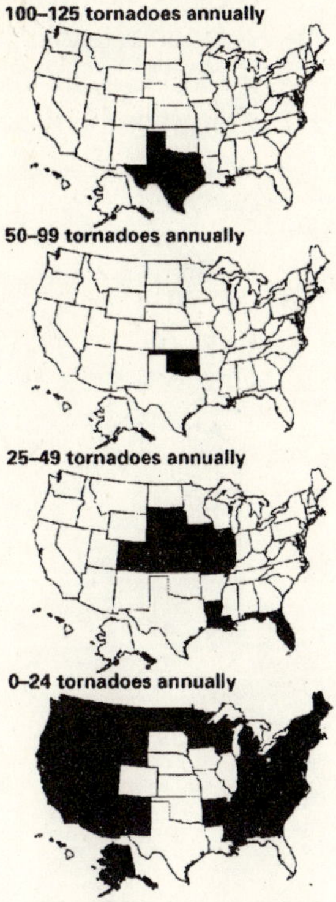

B

❒ billion-dollar United States droughts and heatwaves

- The most severe United States drought of the 20th century was the 1930s Dust Bowl drought which lasted up to seven years in the Great Plains.
- Vividly described in John Steinbeck's novel, The Grapes of Wrath, this drought was so severe that it resulted in a mass migration of millions of people from the Great Plains to the western United States in search of jobs and better living conditions.
- The costs were over $1 billion in 1930s dollars, although it is argued that the 1988–89 drought was economically more devastating.

❒ billion-dollar united states floods

- Flooding was the natural disaster in the United States that killed most people and damaged most property during the 20th century.
- Floods can occur at any time of the year, in any part of the nation, and at any time of the day or night.

- Flood currents also possess tremendous destructive power, as lateral forces can demolish buildings and erosion can undermine bridge foundations and footings leading to the collapse of structures.
- Most lives are lost when people are swept away by flood currents, whereas most property damage results from inundation by sediment-laden water.

20th century floods that caused damage of at least $1 billion (uninflated; B=billion)

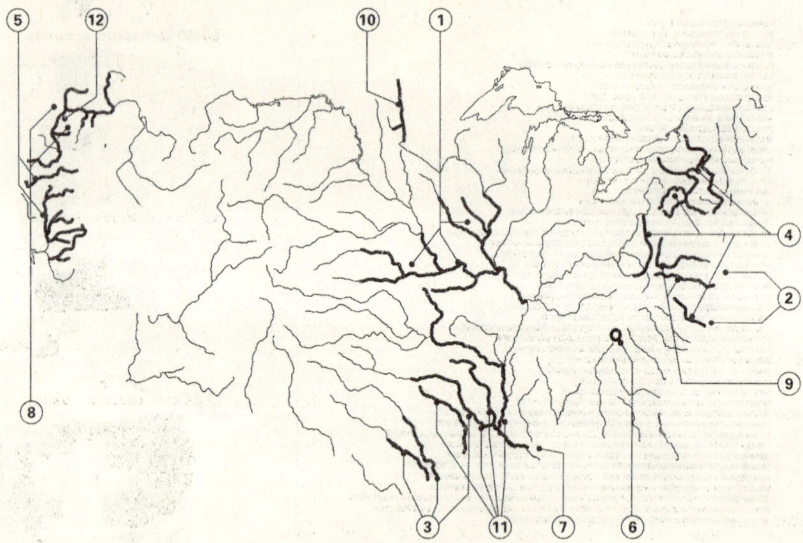

period	location	cost	causes	deaths
		regional floods		
1. 1993 May–Sept	Mississippi Basin	$20B	long period of excessive rainfall	48
2. 1999 Sept.	Eastern N Carolina	$6B	slow-moving Hurricane Floyd	42
3. 1995 May	Southcentral USA	$5-6B	recurring thunderstorms	32
4. 1972 June	NE USA	$3.2B	extratropical remnants of Hurricane Agnes.	117
5. 1995 Jan.–Mar.	California	$3B	frequent winter storms	27
		dam failure flood		
6. 1977 Nov.	Georgia	$2.8B	dam failure after excessive rain	39
		storm-surge flood		
7. 1969 Aug.	Gulf Coast (MI and LA)	$2.4B	Hurricane Camille	259
8. 1996 Dec.–Jan. 1997	Pacific NW and Montana	$2-3B	torrential rains and snowmelt	36
9. 1996 Nov.	Virginia	$1.25B	excessive regional rain	69
10. 1997 Apr.–May	N Dakota, Minnesota	$2B	very rapid snowmelt	8
11. 1990 Apr.	Texas, AR, OK	$1B	recurring intense thunderstorms	17
12. 1996 Feb.	Pacific NW and W Montana	$1B	torrential rains and snowmelt	9

C

❐ causes of tsunamis

- Tsunamis are also referred to as 'tidal waves,' but they have no relation to tides.
- Tsunamis are large, rapidly moving ocean waves caused by a major disturbance of the ocean floor, usually the result of an earthquake but sometimes produced by a submarine landslide or a volcanic eruption.
- Ships at sea cannot detect a passing tsunami, nor can the waves be seen from aircraft.

Development of a tsunami
In these diagrams, the wave heights are greatly exaggerated compared to water depths.

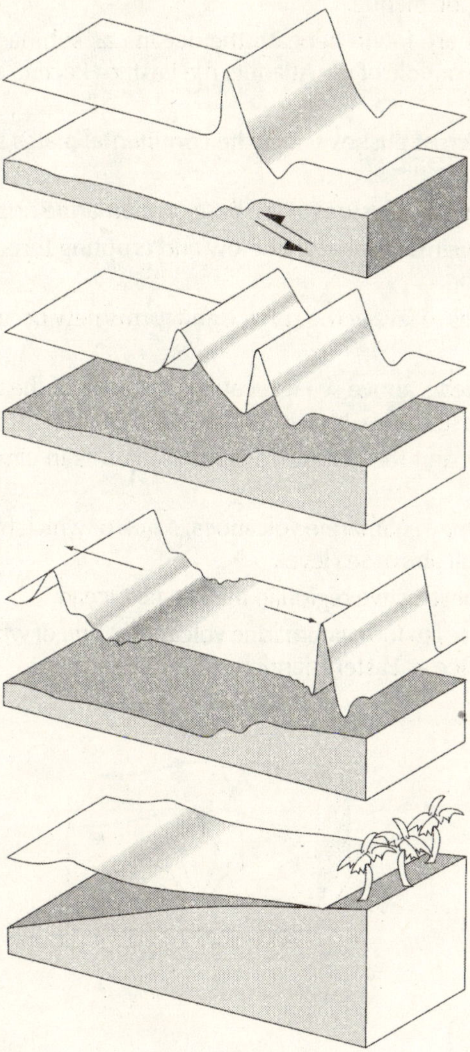

Initiation
- Near the source of submarine earthquakes, the seafloor is lifted up and then dropped down, pushing the entire water column up and down.
- Pushing water above mean sea level releases an enormous amount of potential energy that is then transferred to the horizontal, so propagating wave motion in the form of the tsunami wave (kinetic energy).

Split
- The initial tsunami splits within several minutes of the earthquake into a tsunami that travels out to the deep ocean (distant tsunami) and another tsunami that travels towards the nearby coast (local tsunami).

Amplification
- The amplitude increases and the wavelength decreases as the local tsunami travels over the continental slope. This results in steepening of the leading wave.
- Because of the higher propagation speed the deep ocean tsunami has traveled much farther than the local tsunami.
- Amplification and shortening of the deep ocean tsunami will occur as it approaches a distant shore, just as with the local tsunami shown here.

Runup
- Tsunami runup occurs as the tsunami wave travels from the deep-water, continental slope region to the near-shore region.
- Runup is a measurement of the height of the water onshore observed above a reference sea level.
- Most tsunamis do not result in giant breaking waves but come in much like very strong and very fast tides.
- Because of the complicated behavior of tsunami waves near the coast, the first runup of a tsunami is often not the largest.

- While passing through deep oceans, a tsunami consists of a series of waves that are only a few feet high and a hundred miles or more apart. These waves typically travel at speeds of about 600 mph.
- As they reach shallow water, the waves slow down but greatly increase in height, and the distance between them shrinks.
- When the tsunami finally strikes the coast, the waves may crest to heights of 100 feet and travel inland at speeds of 30 mph.
- A series of waves may reach the coast at intervals of 5 to 40 minutes; the first wave is frequently not the largest.

□ constructive (divergent) plate margins

- The floor of the world's oceans is constantly being renewed in zones where molten rock wells up from the mantle.
- These 'constructive' zones are found beneath the oceans as submarine ridges, such as those in the middle of the Atlantic, the Eastern Pacific, and the Indian Ocean.
- At its far edges, the ocean crust pushes under the continental plates and melts.
- The destroyed crust is replaced in fissures at the base of submarine ridges.
- The fissures are filled by magma rising from below and erupting through the cracks in the ocean floor.
- The magma rapidly solidifies to lava at the surface and forms new oceanic crust.
- The Mid-Atlantic ridge appears above sea level at Iceland, one of the few countries of the world whose surface area is increasing each year.
- Iceland straddles the ridge and the central volcanic plateau is in almost constant eruption.
- Offshore, the ridge has formed submarine volcanoes, some of which have grown large enough to erupt above sea level.
- The East Pacific Rise is a constructive region in the Pacific Ocean.
- The East Pacific Rise also rises to form submarine volcanoes, one of which rises above the ocean surface as Easter Island.

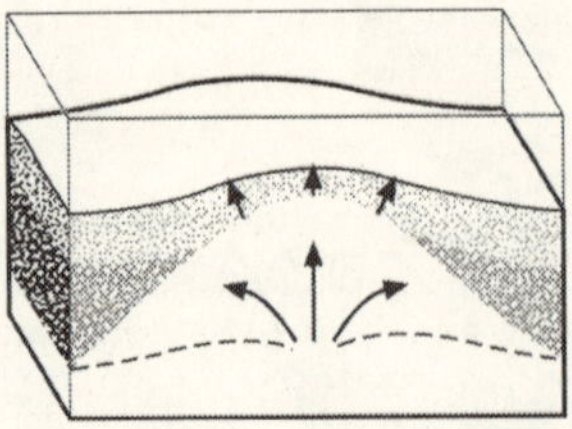

A rising mantle current pushes up and weakens the oceanic crust.

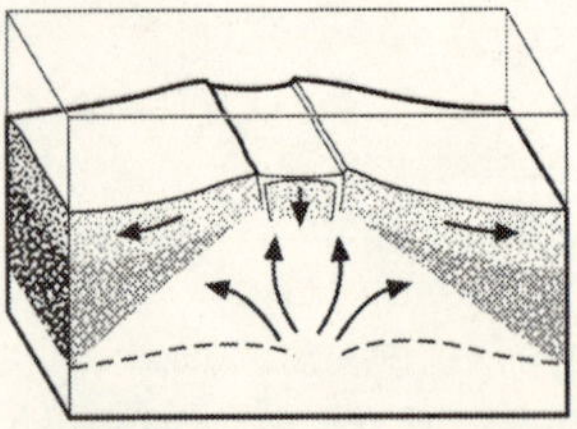

The oceanic crust splits and a central block sinks. Molten rock rises through the cracks around the block.

Iceland straddles the Mid-Atlantic Ridge

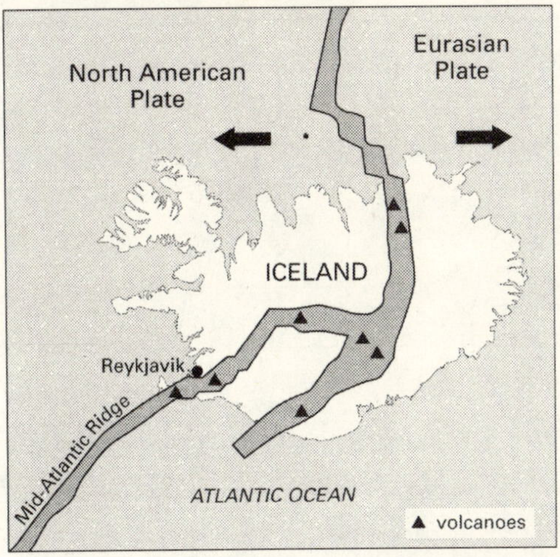

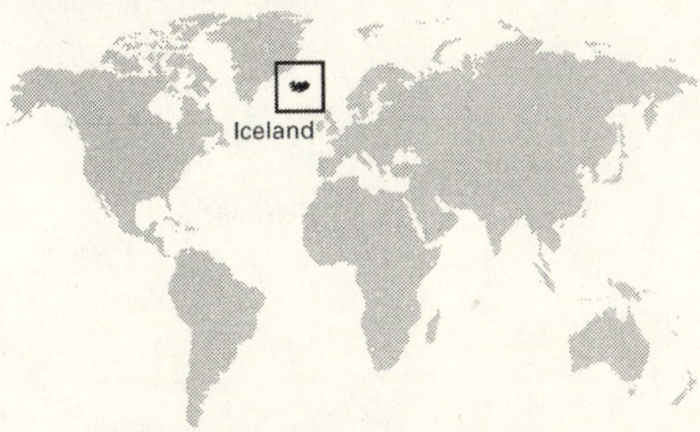

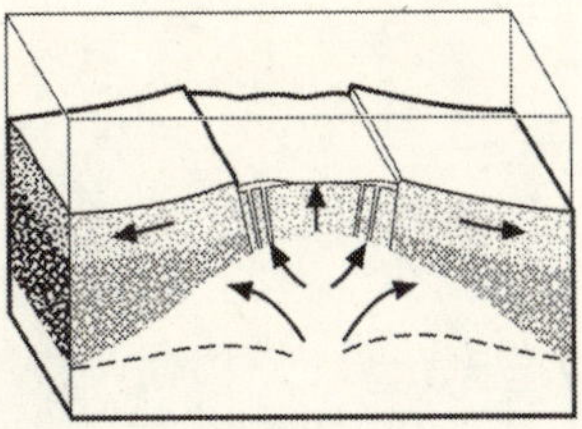

Tension pulls the two parts of the split crust apart and more molten rock wells up into the gaps.

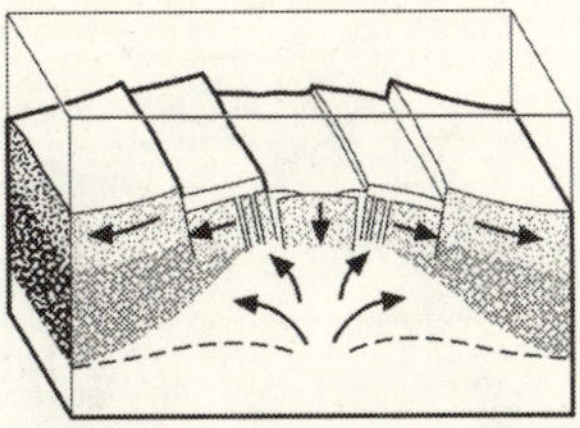

Another central block subsides and the process is repeated.

▢ countries affected by drought in africa

- During the past ten years, 75 percent of the droughts that have occurred in the world have occurred in African countries.
- The countries in Africa reporting the highest frequency of drought include: Ethiopia; Chad; Botswana; Burkina Faso; Kenya; Mozambique, and Mauritania.

Drought is a normal, recurrent feature of climate, although many erroneously consider it a rare and random event. It occurs in virtually all climatic zones, but its characteristics vary significantly from one region to another. Drought is a temporary aberration; it differs from aridity, which is restricted to low rainfall regions and is a permanent feature of climate. Drought is an insidious hazard of nature. Although it has scores of definitions, it originates from a deficiency of precipitation over an extended period of time, usually a season or more. This deficiency results in a water shortage for some activity, group, or environmental sector. Drought should be considered relative to some long-term average condition of balance between precipitation and evapotranspiration (i.e., evaporation + transpiration) in a particular area, a condition often perceived as "normal".

Risk of drought in African countries

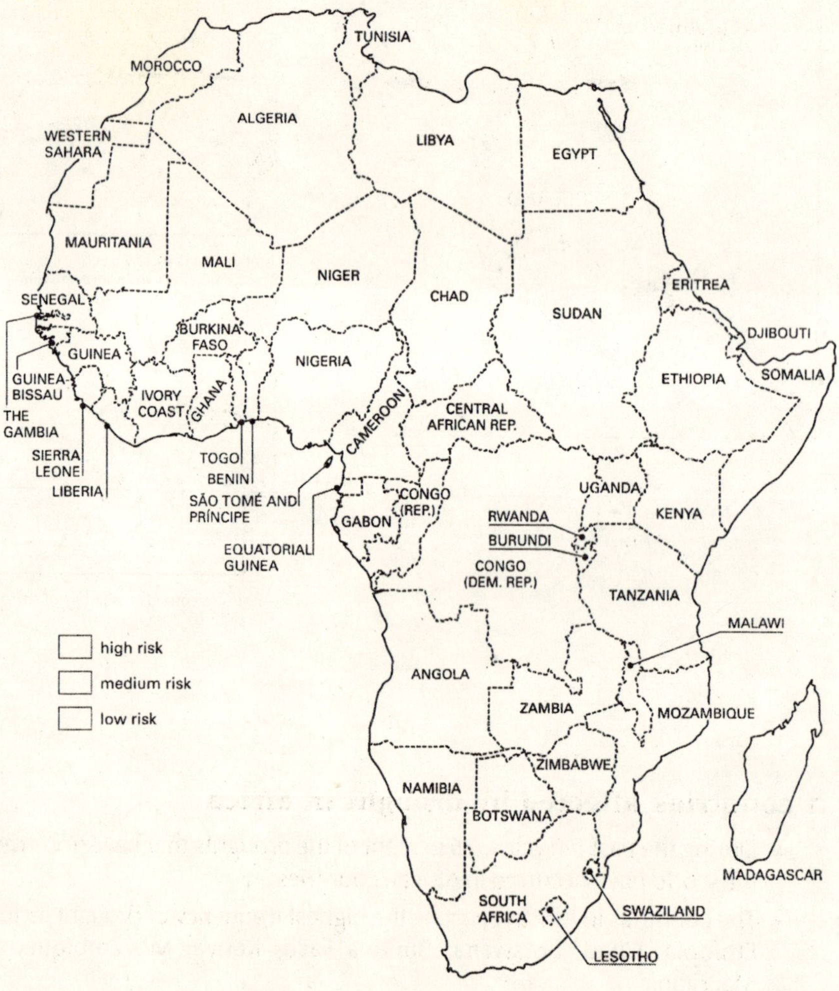

D

❑ destructive (convergent) plate margins

- Destructive plate margins occur where continental and oceanic plates meet or two continental plates collide.
- The thinner oceanic plate sinks beneath the continental plate, forming an ocean trench and subduction zone.
- As the plates converge, friction occurs between them.

- The combined heat from this convergence and from the mantle melts the lower oceanic plate, which then becomes molten magma.
- The newly formed magma rises up through the crust under pressure because it is less dense.
- When it reaches the Earth's surface, volcanoes are formed.
- When pressure is suddenly released, friction and pressure between continental and oceanic plates can also cause earthquakes.

Destructive plate margin at convergence of oceanic and continental plates

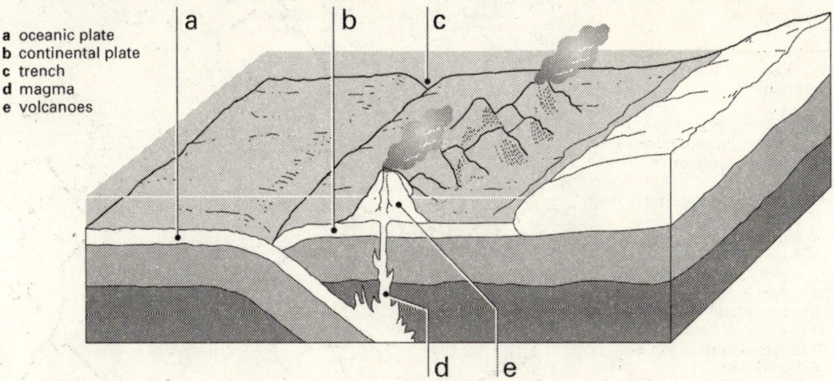

Destructive plate margin at convergence of two continental plates

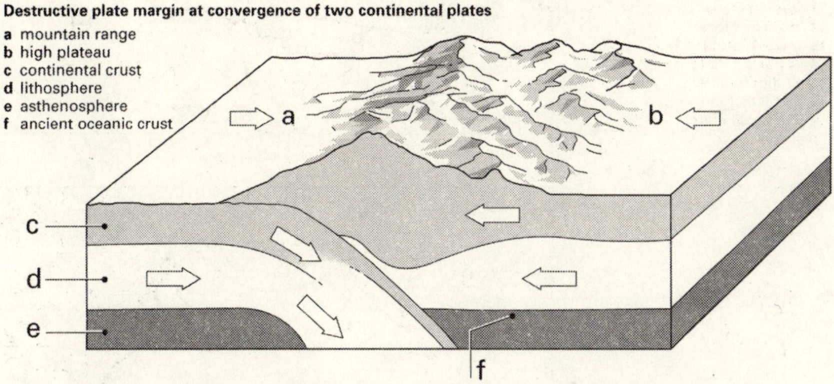

Formation of the Himalayan mountain range by the convergence of two continental plates

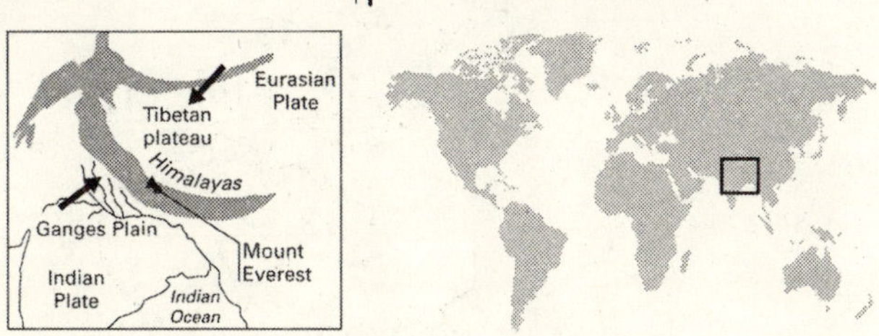

- When two continental plates collide, neither plate sinks because they share the same properties.
- The rock is then compressed and forced upward to form high fold mountains.
- This type of collision can cause earthquakes.

E

❐ earth's active zones

The Earth's lithospheric plates float on the asthenosphere (the semi-fluid upper mantle). Scientists believe the plates' movements are affected by currents of heat in the asthenosphere.

······ convergent plate boundary (where adjacent plates come together)

—— divergent plate boundary (where adjacent plates move apart)

- - - - transform plate boundary (where adjacent plates move past one another)

→ direction of plate movement

1 Pacific Plate
2 Juan de Fuca Plate
3 North American Plate
4 Cocos Plate
5 Nazca Plate
6 Caribbean Plate
7 South American Plate
8 Scotia Plate
9 Eurasian Plate
10 African Plate
11 Iranian Plate
12 Arabian Plate
13 Indo-Australian Plate
14 Antarctic Plate
15 Philippine Plate
16 Caroline Plate
17 Bismarck Plate

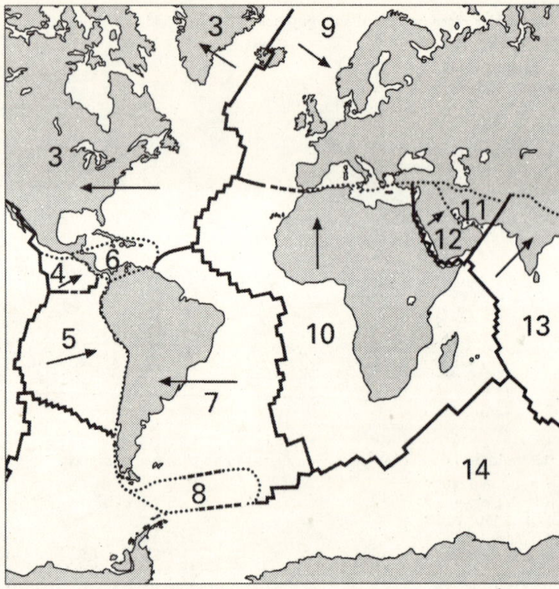

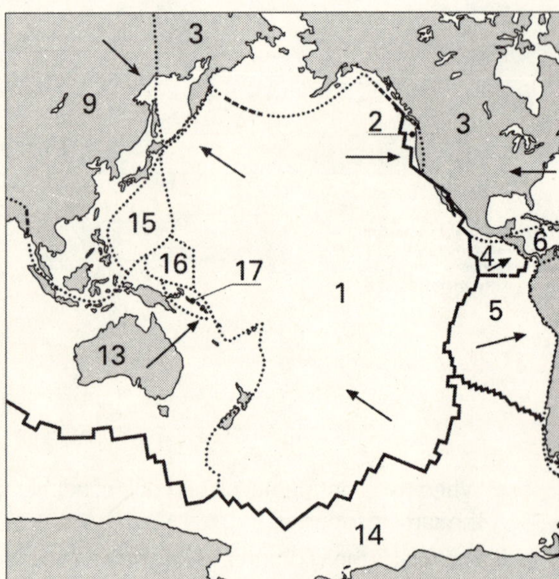

❐ easterly wave

A wavelike disturbance in the tropical easterly winds that usually moves from east to west. Such waves can grow into tropical depressions.

❐ extratropical cyclone

A storm that forms outside the tropics, sometimes as a tropical storm or hurricane changes.

❐ eye

The low pressure center of a tropical cyclone. Winds are normally calm and sometimes the sky clears.

❐ eye wall

The ring of thunderstorms that surrounds a storm's eye. The heaviest rain, strongest winds and worst turbulence are normally in the eye wall.

F

❐ famous earthquakes 1: lisbon 1755

- This huge earthquake occurred on Nov 1, 1755.
- It was of magnitude 8.6 and killed 70,000 people; 10,000–20,000 were killed in Lisbon alone. It leveled the beautiful city of Lisbon, and was felt as far away as southern France and northern Africa.

The disaster was not limited simply to the moments of the earthquake and tsunami: fires raged through the ruins of the city for three days.

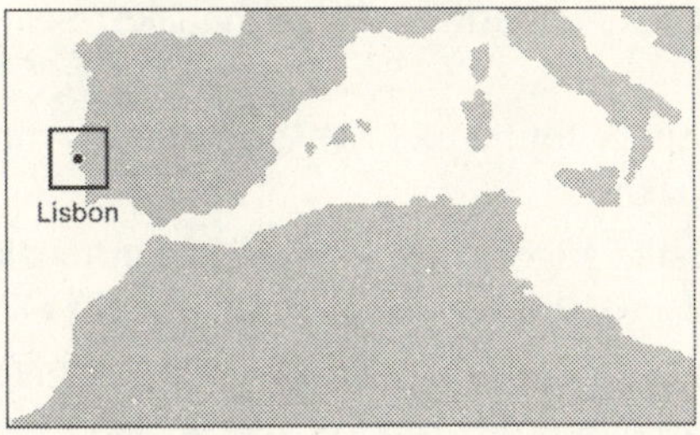

The earthquake, shown here in this contemporary engraving, caused a tsunami (tidal wave), drowning thousands and causing widespread damage.

- Most of Lisbon's population of 250,000 were at church for All Saints' Day when the first shock struck at 9:40 a.m.

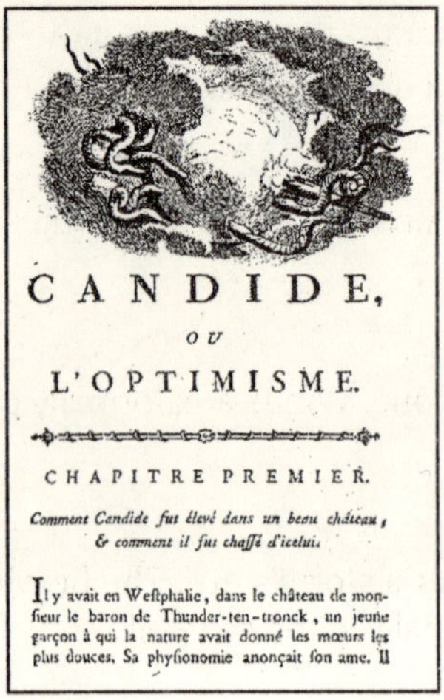

CANDIDE,

OU

L'OPTIMISME.

CHAPITRE PREMIER.

Comment Candide fut élevé dans un beau château, & comment il fut chaſſé d'icelui.

Il y avait en Weſtphalie, dans le château de monſieur le baron de Thunder-ten-tronck, un jeune garçon à qui la nature avait donné les mœurs les plus douces. Sa phyſionomie anonçait ſon ame. Il

The French writer and philosopher Voltaire published a *Poème sur le désastre de Lisbonne* and used the event in his play *Candide* to mock the main character, Dr Pangloss's, philosophy that 'all was for the best in the best of all possible worlds'.

- It was followed by an even more powerful tremor which sent buildings toppling down on the terrified populace.
- The same earthquake and associated tsunamis also caused great destruction and loss of life much farther away – for example, approximately 10,000 people died in Morocco.
- The Lisbon earthquake was of particular importance for earthquake research in Europe.
- It motivated the German philosopher Immanuel Kant to write his three treatises on the causes of earthquakes.
- In Britain John Michell was the first to suggest a link between earthquakes and volcanoes in his explanation of the Lisbon earthquake published in 1760.
- A series of religious publications, sermons and poems were published in the subsequent year in addition to numerous reports in newspapers and more detailed accounts of this event.

☐ famous earthquakes 2: san francisco 1906

- The California earthquake of April 18, 1906 ranks as one of the most notable earthquakes in human history. Today, its importance is based more on the wealth of scientific knowledge derived from it than from its sheer size.
- The earthquake surprised contemporary geologists with its large, horizontal displacements and great rupture length along the northernmost 270 miles of the San Andreas fault from northwest of San Juan Bautista to Cape Mendocino.

- Not until the advent of plate tectonics more than 50 years later would the significance of the fault and recognition of its large cumulative effect be fully appreciated.
- Reid's analysis in 1910 of the 1906 displacements and strain in the surrounding crust led him to formulate his elastic-rebound theory of the earthquake source, which remains today the principal model of the earthquake cycle.
- A foreshock occurred at almost precisely 5:12 a.m., local time, with sufficient force to be felt widely throughout the San Francisco Bay area.
- The great earthquake broke loose some 20–25 seconds later, with an epicenter near San Francisco.
- Violent shocks punctuated the strong shaking which lasted some 45–60 seconds.

Extent of the 1906 rupture seen at the surface

- The earthquake was felt inland as far as central Nevada and on the coast from southern Oregon to south of Los Angeles.
- The highest Modified Mercalli Intensities (MMIs) of VII to IX paralleled the length of the rupture, extending as far as 50 miles inland from the fault trace.

- As well as being arguably the most important study of a single earthquake, the Lawson report of 1908 remains the authoritative work on the San Francisco earthquake.
- The earthquake is remembered most for the fire in San Francisco that followed it, but shaking damage was equally severe in many other places along the fault rupture.
- The often quoted number of 700 deaths caused by the earthquake and fire is now believed to underestimate the total deaths by a factor of 3 or 4.

❒ famous earthquakes 3: mexico 1985

- At 7:17 a.m. September 19, 1985, a powerful earthquake registering magnitude 8.1 rocked central Mexico.
- Thirty six hours later a second earthquake of magnitude of 7.5 occurred.
- The quake caused substantial damage throughout central Mexico, including in Mexico City, some 240 miles east of the epicenter, which was offshore from Acapulco.
- There was widespread light to moderate damage in other parts of the country, but severe damage occurred to about 800 buildings in Mexico City. The total cost was $5 billion.

Mexico City earthquakes 1957–1991

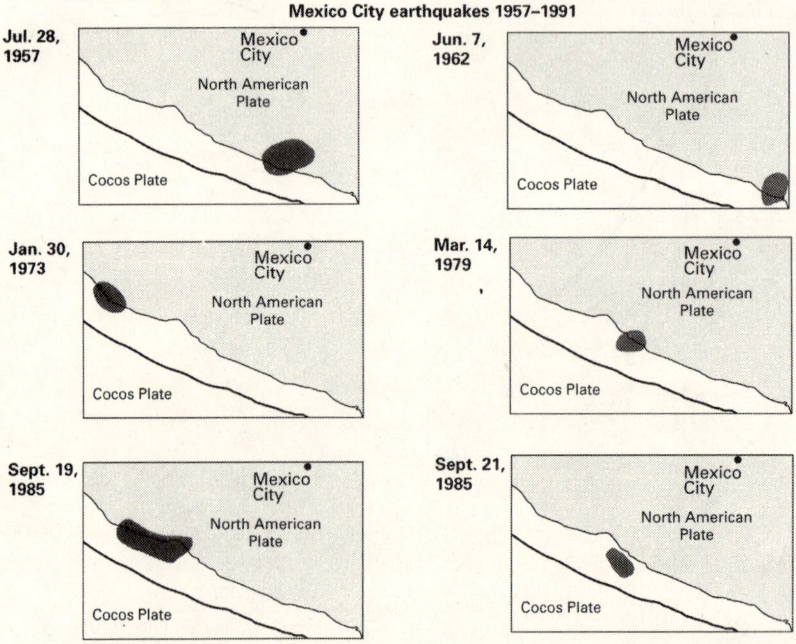

- About 10,000 people were killed, 50,000 injured and 250,000 people were left homeless, but casualties in the city would have been much higher if the quake had occurred just a couple of hours later when the schools and office buildings would have been full with people.

- Mexico City is largely built upon soft, poorly supported lake sediments from the former Lake Texcoco, which was drained by the Spanish following their occupation of the region.
- Soft sediments like these tend to amplify seismic waves and cause the ground to shake much more vigorously than the surrounding bedrock.
- Mercalli Intensity level IX shaking in some areas, lasting up to three minutes, was caused by a resonance between the earthquake waves, the subsoils, and the buildings. The greatest damage happened to buildings with six or more floors.

Proximity to Mexico City of meeting of Cocos and North American Plates

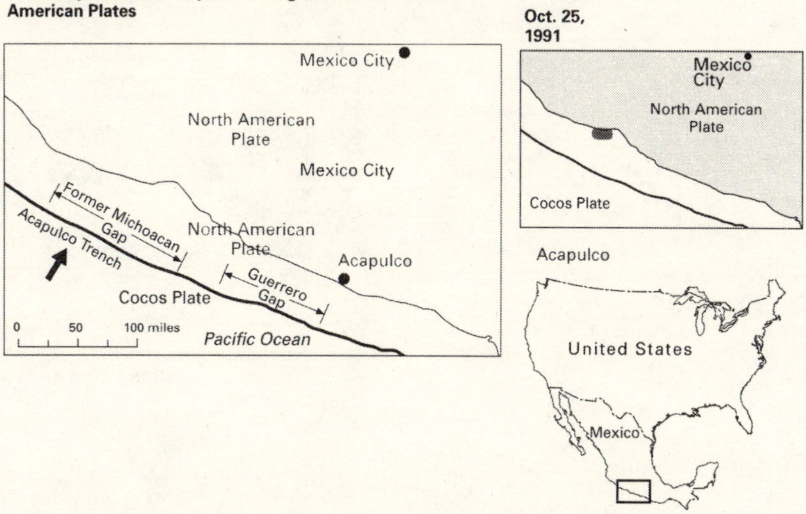

☐ famous earthquakes 4: osaka/kobe 1995

- On Tuesday, January 17, 1995 at 5:46 a.m. local time, an earthquake of magnitude 7.2 struck the region of Kobe and Osaka in south-central Japan.
- After Tokyo, the Osaka region is Japan's secondmost populated and industrialized area, with a total population of about 10 million.
- Some 4,500 people were killed and about 15,000 injured.
- Strong ground shaking caused severe damage over a large area and lasted for about 20 seconds.
- Nearly 180,000 buildings were badly damaged or destroyed. Rescue bodies estimated that more than 300,000 people were homeless on the night following the earthquake.
- The life loss caused by the earthquake was the worst in Japan since the 1923 Great Kanto Earthquake, when about 140,000 people were killed, mostly by the post-earthquake conflagration.
- Leaving aside the indirect economic damage caused by loss of life, business interruption, and loss of production, the direct damage caused by the shaking has been estimated at over 7 trillion yen (about $75 billion).

- Kobe and its immediate region were the areas most severely affected, but damage was recorded over a 60-mile radius from the epicenter, including the cities of Osaka and Kyoto.

Osaka Bay, showing main damage to transport infrastructure and other infrastructure in Kobe

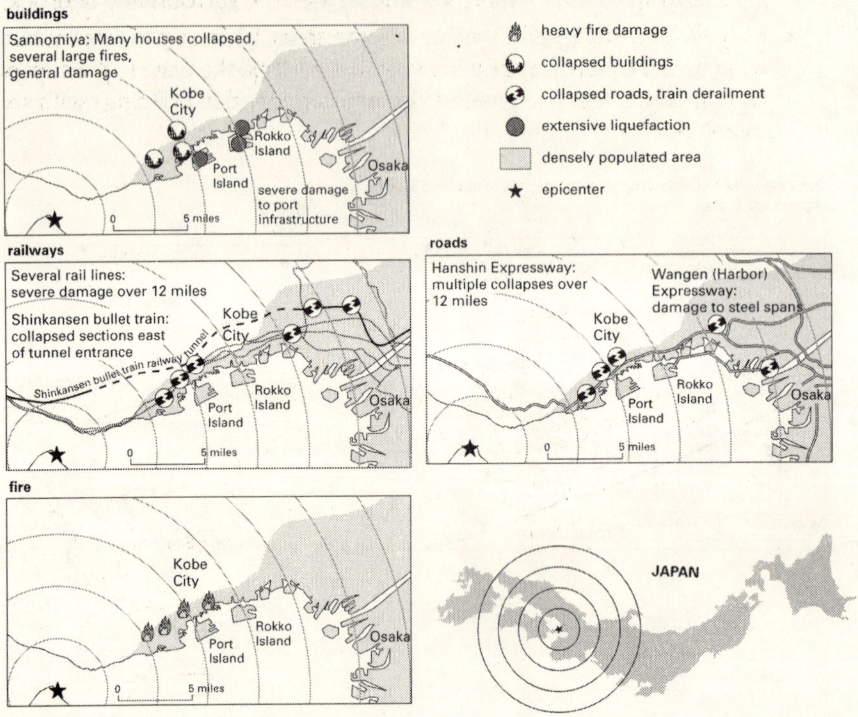

❑ famous earthquakes 5: turkey 1999

- At 3 a.m. on August 17, 1999 a major section of the north Anatolian fault line slipped, sending shockwaves through cities and villages across the northwest of Turkey.
- The quake, measuring 7.4 on the Richter scale, flattened buildings like packs of cards. Its epicenter was Izmit, 65 miles from Istanbul.
- More than 17,000 people were killed in the quake, crushed in their sleep or dying of wounds, dehydration or exhaustion as they lay trapped under the rubble of their homes.
- Some 44,000 people were injured and thousands left homeless. Damages totaled $6.5 billion.
- The earthquake originated at a shallow depth of about 10.5 miles along the northernmost edges of the North Anatolian fault system, which has produced seven earthquakes with magnitudes greater than 7.0 since 1939.
- This was one of the strongest earthquakes ever to occur in western Turkey and is the largest event to have devastated a modern, industrialized area since the 1906 San Francisco and the 1923 Tokyo earthquakes.

- An associated tsunami (tidal wave) flooded Izmit Bay.

Northern Anatolian Fault compared to the San Andreas Fault in California
The two faults have similar slip rate, age, length and straightness

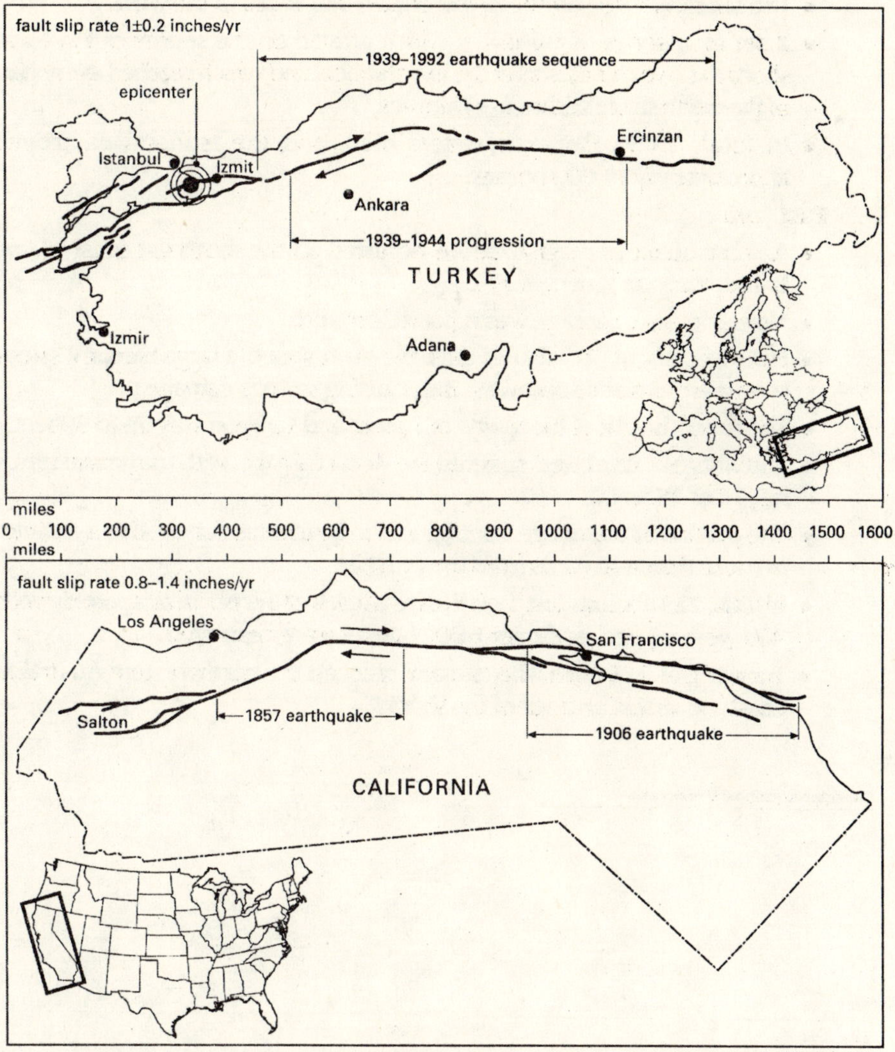

❒ famous tsunamis 1: indonesia

- Indonesia lies on the boundary of the Australian and Eurasian tectonic plates.
- More than 30,000 people died in Java from a tsunami caused by the famous volcanic eruption of Krakatau in 1883 (see Plate 8.27).
- Three major tsunamis occurred in Indonesia in the 1990s.

Geographic position of Indonesia

Flores

- An earthquake of magnitude 7.8 occurred just off the north coast of the eastern part of Flores Island, Indonesia, on December 12, 1992.
- The shock was felt on the island of Bali, 430 miles to the west.
- It set off a series of tsunamis, which arrived on the shores of Flores as shortly as two minutes after the initial shock, and which reached every part of the north shore within five minutes.
- In total, 1,690 Flores Islanders died, and the tsunami destroyed approximately 18,000 houses.

East Java

- An earthquake of magnitude 7.6 occurred off the southeast coast of Java Island, early on June 3, 1994.
- No earthquake damage was reported on land.
- However, about 50 minutes after the main shock, a large tsunami struck southeast Java and southwest Bali, causing serious damage.
- The runup height at the towns of Pancer and Lampon was up to 30 feet.
- The village of Rajekwesi suffered the worst damage, with maximum runup height of 45 feet.
- The portion of the village fronting a wide beach was flattened to a distance of more than 440 yards from the coastline.
- In total, 223 persons lost their lives as a result of the tsunami, approximately 400 were injured, and over 1,000 houses were destroyed.
- Traveling at 215 mph, the tsunami reached the northwestern Australian coastline within an hour of the shock.

Geographic position of Indonesia

Flores

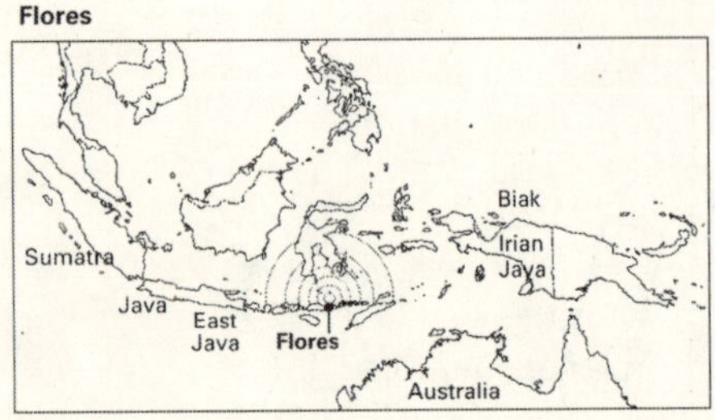

East Java

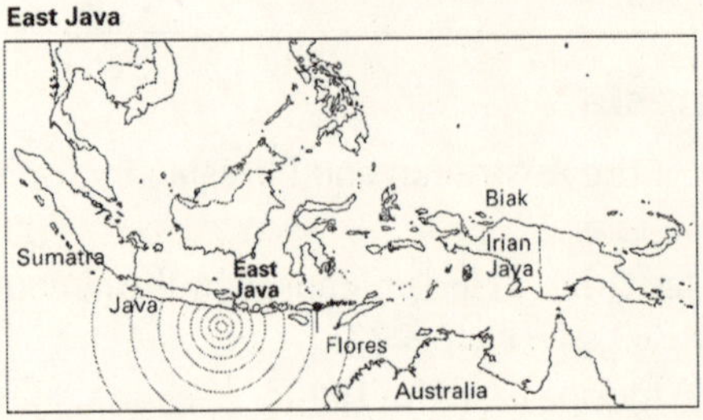

Biak

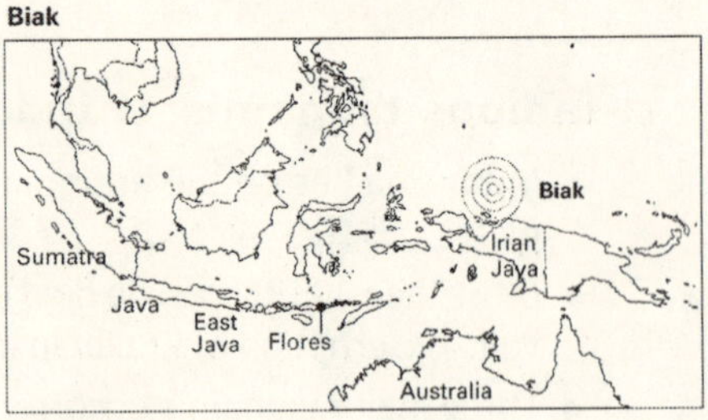

Biak

- On February 17, 1996 a major earthquake of magnitude 8.2 occurred near Biak Island, situated off the north shore of Irian Jaya.
- It generated a tsunami that reached a maximum measured height of 25 feet.
- The most severe damage caused by the tsunami occurred at Korim, where it reached a height of 12–15 feet. The inundation area covered almost 1 square mile.
- About 380 homes were completely destroyed by the inundation.
- Sixty-four of the dead and 18 missing people were reported from the area in and around Korim.
- About 150 people were killed, a further 100 seriously injured, and 10,000 were made homeless.

▫ famous tsunamis 2: okushiri, japan 1993

- Japan is situated near the boundaries of four tectonic plates: the North American, Eurasian, Philippines, and Pacific Plates.
- It has suffered many serious earthquakes and tsunamis.
- In 1896, a tsunami destroyed 175 miles of coastline on the island of Honshu, killing 27,000 people.
- In March 1933, a tsunami generated by an earthquake killed 2,990 at Sanriku.
- The most damaging tsunami in recent times occurred on July 12, 1993, when a large submarine earthquake of magnitude 7.8 occurred in the northeastern part of the Japan Sea near the small island of Okushiri.
- In 2–5 minutes after the shock several large wavesflooded the Okushiri coastline and the central west coast of Hokkaido.
- The average run-up heights along the western coast of the Okushiri Island were between 33 and 66 feet with the 95 feet maximum measured near the small coastal village of Monai, where all 12 houses were destroyed killing 10 persons.
- Tsunami waves were observed over the entire Sea of Japan.
- In about 30 minutes tsunami waves reached Primorie (on the southeast coast of Russia) where their height was from 3.2–13 feet.
- After 90 minutes, the tsunami struck the coast of South Korea.
- On the southwestern coast of Hokkaido, severe damage was caused by the earthquake and accompanying fires, landslides, and the tsunami with run-up heights varying from 13–30 feet.
- The total number of fatalities was 185, with 120 attributed to the tsunamis. Property losses were estimated at US$ 600 billion.

Okushiri submarine earthquake 1993

❑ famous tsunamis 3: alaska

- Alaska and the Aleutian Islands are situated at the boundary of the Pacific Plate and the North American Plate.
- Most earthquakes are produced where these two plates come into contact and slide past each other.
- During the past century, four large and well documented tsunamis were generated in the waters off the Alaskan coast.

1946

- During the early morning of April 1, 1946 an earthquake of magnitude 7.8 occurred in the area of the Aleutian Trench located approximately 90 miles south of Unimak Island, Aleutian Islands.
- During the quake, a large section of seafloor was uplifted along the fault where the quake occurred, producing a large, Pacific-wide tectonic tsunami.
- Approximately 48 minutes after the earthquake, a 100-foot tsunami struck the area of Scotch Cap, Unimak Island.

- The tsunami completely destroyed the newly built US Coast Guard lighthouse, surging over the cliff to a height of 115 feet above sea level and killing all five members of the lighthouse crew.
- Less than five hours after the earthquake in Alaska, the first of several tsunami waves reached the Hawaiian Islands, causing extensive destruction.
- Wave heights across the Islands reached an estimated maximum of 55 feet, 36 feet and 33 feet on Hawaii, O'ahu, and Maui, respectively.
- A total of 149 people were killed as a result of the tsunami.

1957

- On March 9, 1957 a magnitude 8.3 earthquake struck along a section of the Aleutian Trench located south of the Andreanof Islands.
- The vertical seafloor displacement that accompanied this quake generated a major Pacific-wide tectonic tsunami, similar to the 1946 Aleutian event, but fortunately smaller.
- Wave heights reached 40 feet near the Scotch Cap Lighthouse at Unimak Island, and were reportedly up to 75 feet high on the Pacific side of Umnak Island.
- Unlike the situation during 1946, the Hawaiian Islands received advanced warning of the approaching tsunami.
- As a result, there were no fatalities directly associated with the 1957 tsunami.

1958

- Lituya Bay, located within Glacier Bay National Park, along Alaska's southeastern coast, was the site of one of the largest local tsunamis ever recorded.
- On the night of July 7 1958 a magnitude 8.0 earthquake occurred along the Fairweather Fault, with its epicenter only 13 miles from Lituya Bay.
- The earthquake caused a large landslide in the bay, which sent water surging to a height of 1,720 feet across the point opposite the inlet, and produced a local tsunami of over 100 feet high.

1964

- During the evening of March 29, 1964 the area of Prince William Sound, in Southeastern Alaska, was struck by a magnitude 8.4 earthquake, the largest ever recorded in North America.
- The quake was accompanied by large vertical displacements over an approximate 200,000 square mile area along the source fault, much of this beneath Prince William Sound.
- This vertical displacement generated a major tectonic tsunami that struck the southeast coast of Alaska, the Pacific Coast of British Columbia, and the west coast of the United States.
- Many of the coastal communities along Prince William Sound and Kodiak Island were completely wiped out.
- In all, tsunami waves generated by the 1964 quake killed 119 people and caused approximately $300–400 million in damage to Alaska alone.

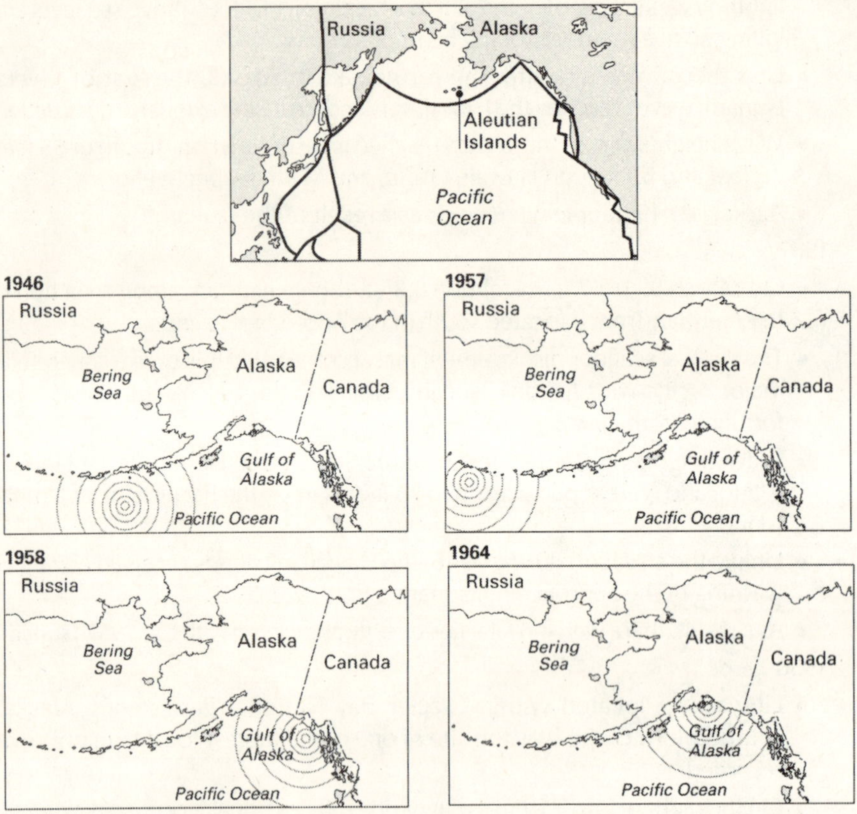

❐ famous volcanic eruptions 1: krakatau 1883

- The renowned Krakatau (Krakatoa) volcano lies in the Sunda Strait between Java and Sumatra.
- Caldera collapse, perhaps in 416 AD, destroyed the ancestral Krakatau edifice, forming a 4-mile-wide caldera.
- Remnants of this volcano formed Verlaten and Lang Islands; subsequently Rakata, Danan and Perbuwatan volcanoes were formed, coalescing to create the pre-1883 Krakatau Island.
- Caldera collapse during the catastrophic 1883 eruption destroyed Danan and Perbuwatan volcanoes, and left only a remnant of Rakata volcano.
- The post-collapse cone of Anak Krakatau (Child of Krakatau), which emerged within the 1883 caldera between the former cones of Danan and Perbuwatan, has been the site of frequent eruptions since 1927.
- The eruption on 26 August, 1883 is often cited as a classic example of caldera formation by collapse following eruption of large volumes of pumice.
- The climactic eruption involved the ejection of hugevolumes of magma, primarily as dense pyroclastic flows.

- These flows were initially erupted from vents in the north of the island of Krakatau itself.
- Some of the pyroclastic flows traveled as far as the coast of Sumatra, where they killed around 3,000 people in areas not affected by the tsunami.
- Later in the eruption, most of the island of Krakatau subsided as the emptying magma chamber collapsed, producing a giant crater or caldera, around 3 miles across and several hundred yards deep.
- A series of tsunamis washed away 165 coastal villages on Java and Sumatra, killing 36,000 people.
- The larger tsunamis were recorded by tide gauges as far away as the southern coast of the Arabian Peninsula – more than 4,000 miles from Krakatau.

Islands of Krakatau and the disappearance of land following the 1883 eruption

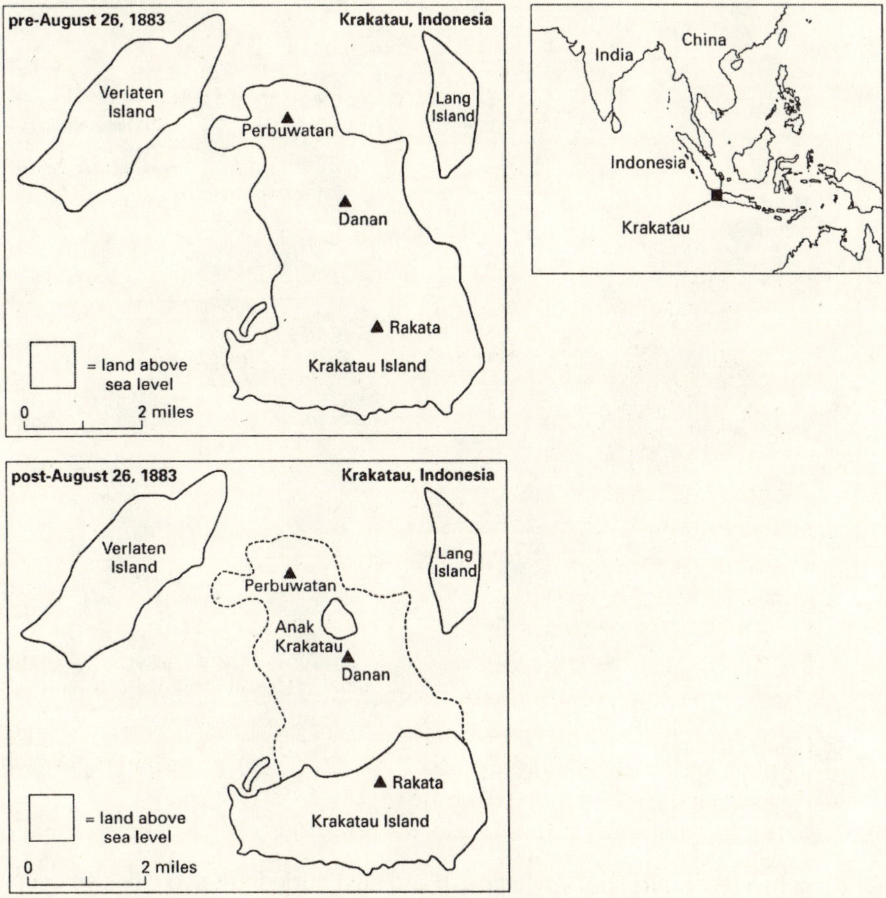

❒ famous volcanic eruptions 2: vesuvius

- On August 24, 79 AD Vesuvius suddenly exploded and destroyed the Roman cities of Pompeii and Herculaneum.
- Although Vesuvius had shown stirrings of life when a succession of earthquakes in 63 AD caused some damage, it had been quiet for hundreds of years and was considered extinct.
- Its surface and crater were covered with vegetation, so the eruption was totally unexpected.

Location of Vesuvius, ash fallout pattern, and the 13-foot layer of deposits at Pompeii

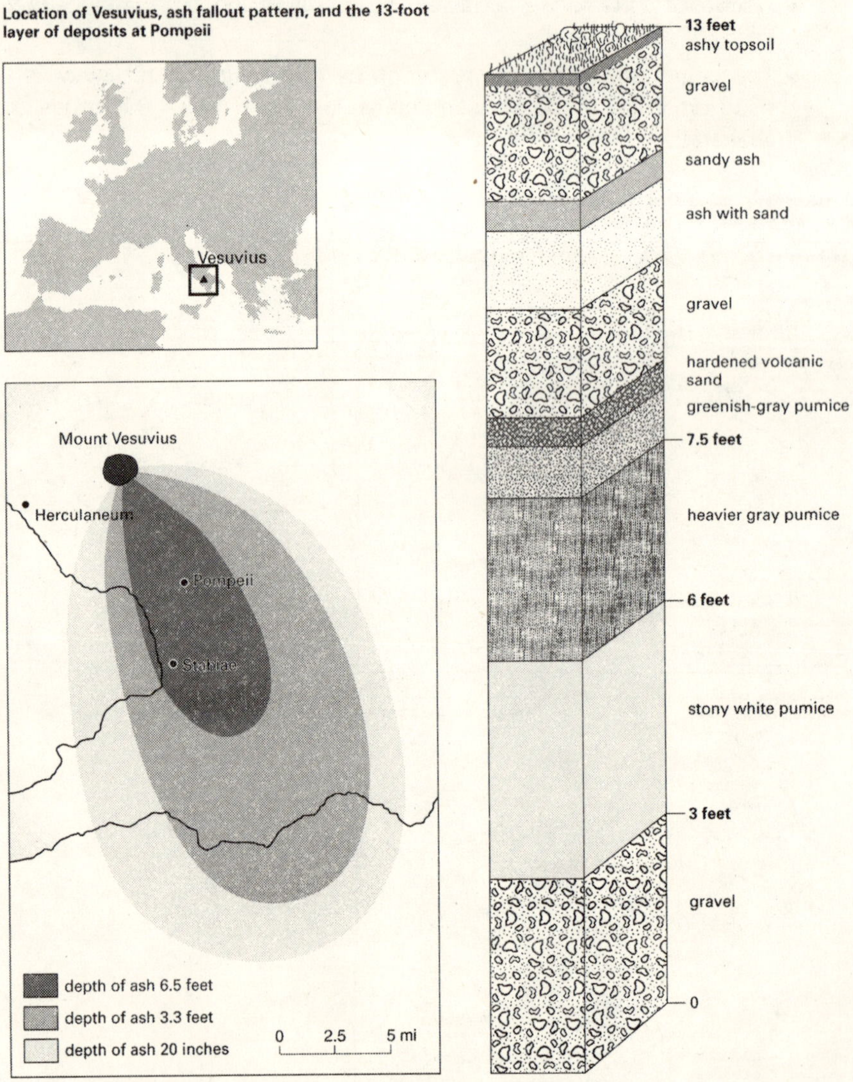

- In a few hours, hot volcanic ash and dust buried the two cities so deeply that their ruins were not uncovered for nearly 1,700 years, when the discovery of an outer wall in 1748 started a period of modern archeology.

- Vesuvius has continued its activity intermittently ever since 79 AD with numerous minor eruptions and several major eruptions occurring in 1631, 1794, 1872, 1906, and in 1944 in the midst of the Italian campaign during World War II.

❑ famous volcanic eruptions 3: mont pelée 1902

- Mont Pelée is the highest mountain in Martinique, West Indies (height approx. 4,200 feet).
- Martinique is part of the Lesser Antilles volcanic arc, formed by the subduction of the North American Plate under the Caribbean Plate.
- When the first European settlers arrived in 1635, they learned that the volcano was called Fire Mountain by the indigenous Caribbean Indians.
- The Europeans named the volcano Montagne Pelée (Peeled Mountain) because a small eruption in 1630 had destroyed the vegetation at the top.

Lesser Antilles volcanic arc and conjunction of North American and Caribbean Plates

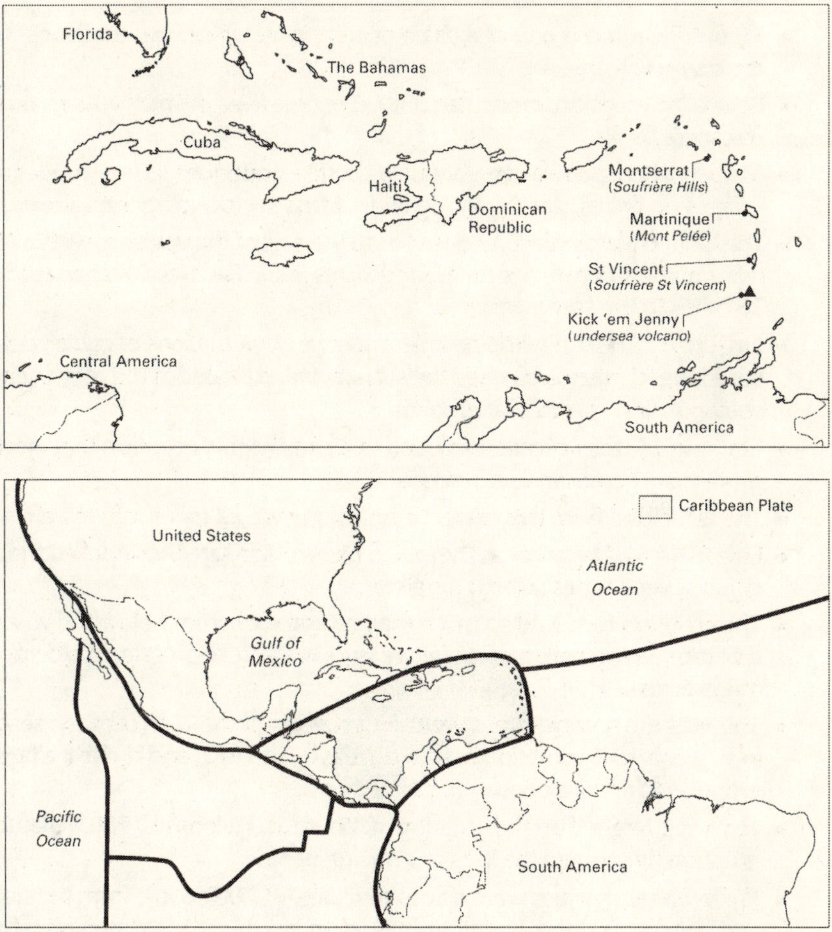

- An extremely destructive eruption followed the growth of a dome on May 8, 1902, which completely destroyed the coastal town of St. Pierre.
- The destruction was caused by a nuée ardente, a type of pyroclastic flow that consists of hot incandescent solid particles.
- The phenomenon of pyroclastic flow was first documented in the 1902 eruption.
- Nearly 30,000 inhabitants were killed within minutes by an incandescent, high-velocity ash flow and associated hot gases and volcanic dust.
- Only two men survived; one because he was in a poorly ventilated, dungeon-like jail cell and a second who somehow made his way unscathed through the burning city.
- The term Peléan is now used to describe the phase of eruptions where material flows down the sides of a volcano as a fast-moving avalanche of gas and dust at speeds of up to 125 mph.

❑ famous volcanic eruptions 4: mount pinatubo 1991

- Mount Pinatubo is one of a chain of composite volcanoes that constitute the Luzon volcanic arc.
- Before the eruption, more than 30,000 people lived in small villages on the volcano's flanks.
- A much larger population – about 500,000 – continues to live in cities and villages on broad, gently sloping alluvial fans surrounding the volcano.
- The United States Clark Air Base lies to the east of the volcano, within 15.5 miles of the summit, and the United States Subic Bay Naval Station is about 25 miles to the southwest.
- On June 12, 1991 (Philippine Independence Day), millions of cubic yards of gas-charged magma reached the surface and exploded in the reawakening volcano's first spectacular eruption.
- On June 15, the volcano exploded in a cataclysmic eruption that ejected more than 1 cubic mile of material.
- The ash cloud from this climactic eruption rose 22 miles into the air.
- Fine ash fell as far away as the Indian Ocean, and satellites tracked the ash cloud several times around the globe.
- The eruption removed so much magma and rock from below the volcano that the summit collapsed to form a large volcanic depression (caldera) 1.6 miles across.
- This was the second largest volcanic eruption of the 20th century, second in size only to an eruption in Katmai, Alaska, in 1912, and by far the largest eruption to affect a densely populated area.
- Ten times larger than the eruption of Mount St. Helens in 1980, Pinatubo's eruption threatened the lives of 1 million people.
- Philippine authorities were able to evacuate 60,000 people from the slopes and valleys, and the American military evacuated 18,000 personnel and their dependents from Clark Air Base below the mountain. These actions

saved many thousands of lives and an estimated billion dollars in property, making this probably the most successful case of volcanic damage limitation in human history.

Location of Mount Pinatubo and effects of eruption

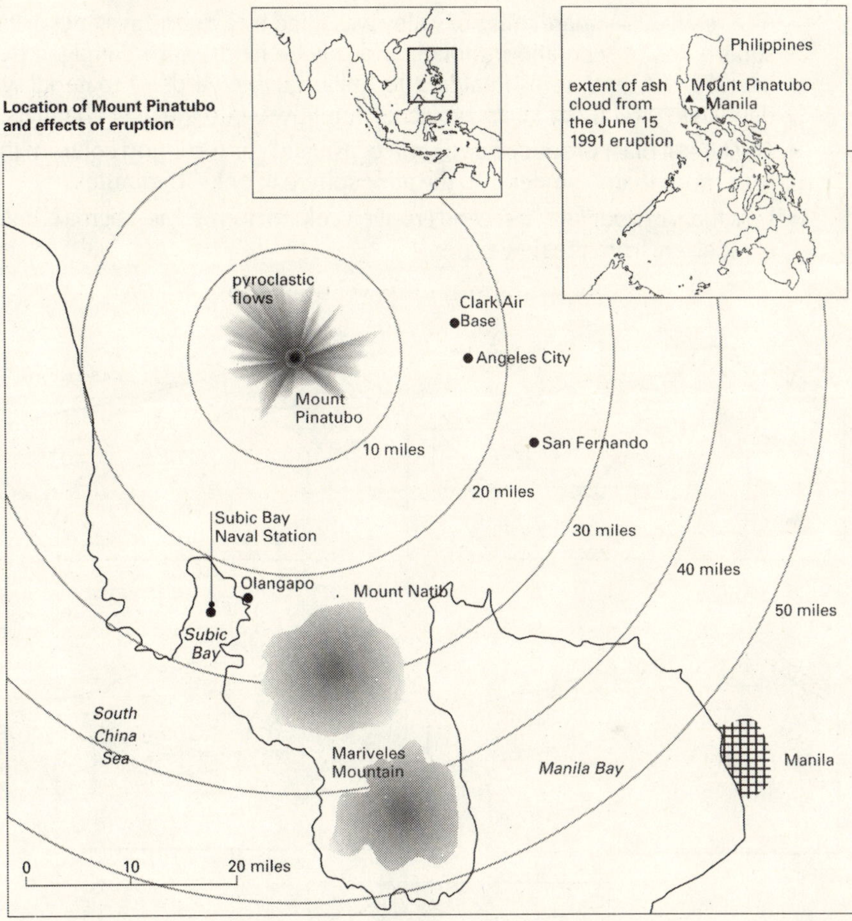

□ famous volcanic eruptions 5: mount st. helens 1980

- On May 18, 1980 at 8:32 a.m. Pacific Daylight Time, a magnitude 5.1 earthquake shook Mount St. Helens, Washington State, the most active volcano in the Cascade Range.
- This resulted in the removal of more than half a cubic mile of material from the mountain, the largest known landslide in human history.
- The landslide depressurized the volcano's magma system, triggering powerful explosions that ripped through the sliding debris.
- Rocks, ash, volcanic gas, and steam were blasted upward and outward to the north.
- This lateral blast of hot material accelerated to at least 300 miles per hour, then slowed as the rocks and ash fell to the ground and spread away from the volcano.

- The blast cloud traveled as far as 17 miles northward from the volcano and the landslide traveled about 14 miles west down the North Fork Toutle River.
- 1,300 feet of the peak collapsed or blew outwards.
- As a result, 24 square miles of valley was filled by a debris avalanche, 250 square iles of recreation, timber, and private lands were damaged by a lateral blast, and an estimated 200 million cubic yards of material was deposited directly by lahars (volcanic mudflows) into the river channels.
- The lateral blast produced a column of ash and gas (eruption column) that rose more than 12 miles into the atmosphere in only 10 minutes.
- Less than an hour later, a second eruption column formed as magma erupted explosively from the new crater.

Massive landslide triggers the Mount St Helens eruption

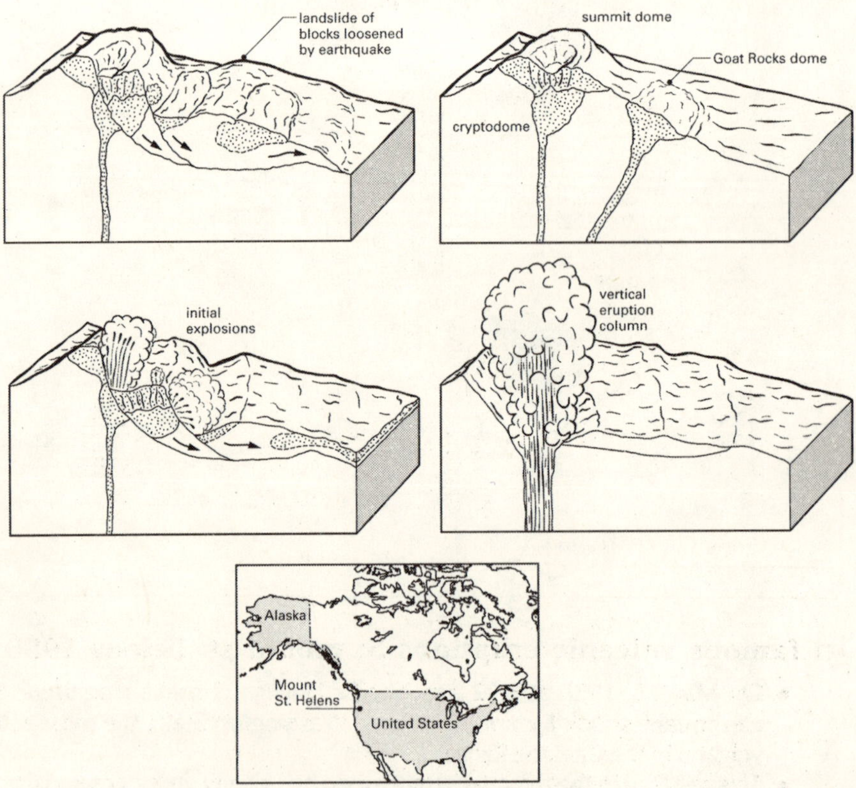

- Then, beginning just after noon, swift avalanches of hot ash, pumice, and gas (pyroclastic flows) poured out of the crater at 50 to 80 miles per hour and spread as far as 5 miles to the north.
- Over the course of the day, prevailing winds blew 520 million tons of ash eastward across the United States and caused complete darkness in Spokane, Washington, 250 miles from the volcano.
- 57 people died as a result of the eruptions.

G

□ great mississippi 1993 flood

- The Mississippi river system experienced several major floods in the 20th century.
- In 1927 flooding made over 700,000 people homeless, 246 people and 165,000 head of livestock drowned, and property damage exceeded $364 million.
- In 1937, heavy rains drenched the lower Mississippi, creating a lake nearly the area of Lake Superior.
- In 1993 flooding was concentrated in the upper Mississippi basin and the middle and lower sections of the Missouri basin.

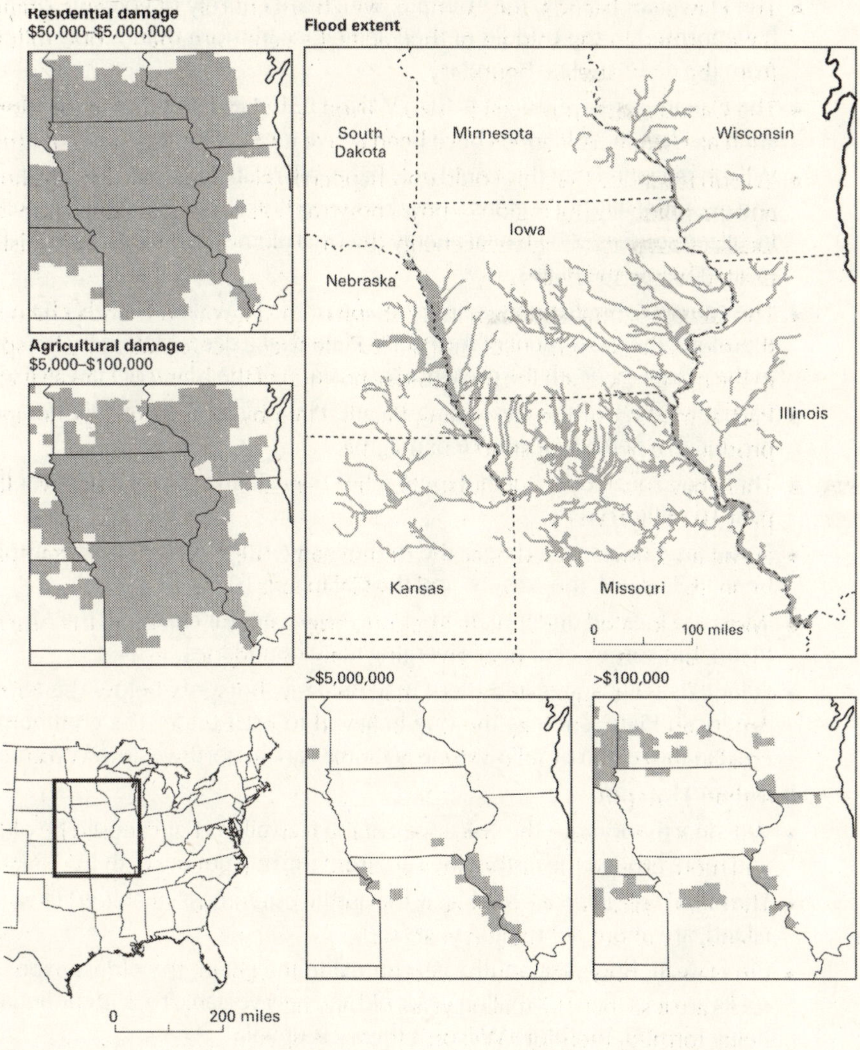

- Over 70,000 people were displaced by the floods.
- Nearly 50,000 homes were damaged or destroyed and 52 people died.
- Over 12,000 square miles of productive farmland were rendered useless.
- Damage to the transportation infrastructure in the region was the most complete ever caused in the United States.
- Total damage was estimated between $15–20 billion.

H

❐ hotspots

- The majority of earthquakes and volcanic eruptions occur near plate boundaries, but there are exceptions.
- The Hawaiian Islands, for example, which are entirely of volcanic origin, have formed in the middle of the Pacific Ocean more than 2,000 milesc from the nearest plate boundary.
- The Canadian geophysicist J. Tuzo Wilson noted in 1963 that in locations such as Hawaii, volcanoes have been active for very long periods of time.
- Wilson reasoned that this could only happen if relatively small, long-lasting, and exceptionally hot regions – now known as hotspots – that would provide localized sources of high heat energy (thermal·plumes) to sustain volcanism, existed below the plates.
- The cause of the distinctive linear shape of the Hawaiian Islands chain is, therefore, the movement of the Pacific Plate over a deep, stationary hotspot in the mantle beneath the present-day position of the island of Hawaii itself.
- Partial melting of the overriding Pacific Plate by heat from this hotspot produced a persistent source of magma.
- There have been over 100 hotspots active beneath the Earth's crust over the past 10 million years.
- Some are concentrated near the mid-oceanic ridge system, for example, beneath Iceland, the Azores, and the Galapagos Islands.
- Most are located under plate interiors, however, (for example, the African Plate), but some occur near diverging plate boundaries.
- Scientists have suggested there may be a few hotspots below the North American Plate, such as the one believed to exist under the continental crust in the region of Yellowstone National Park in northwestern Wyoming.

Hawaiian Hotspot

- Wilson's theory says the volcanoes of the Hawaiian chain should be older and more eroded the farther they are from the hotspot beneath the islands.
- The oldest volcanic rocks on Kauai, the northwesternmost inhabited Hawaiian island, are about 5.5 million years old.
- On Hawaii, however, southeasternmost in the chain, the oldest exposed rocks are less than 0.7 million years old and new volcanic rock is continually being formed, therefore Wilson's theory is upheld.

a. Hawaii: a volcanically active island since 0.8 million years ago (mya)
b. Maui: volcanically inactive island since 0.8 mya
c. Molokai: volcanically inactive island since 1.3 mya
d. Oahu: volcanically inactive island since 2.2 mya
e. Kauai: volcanically inactive island since 3.8 mya

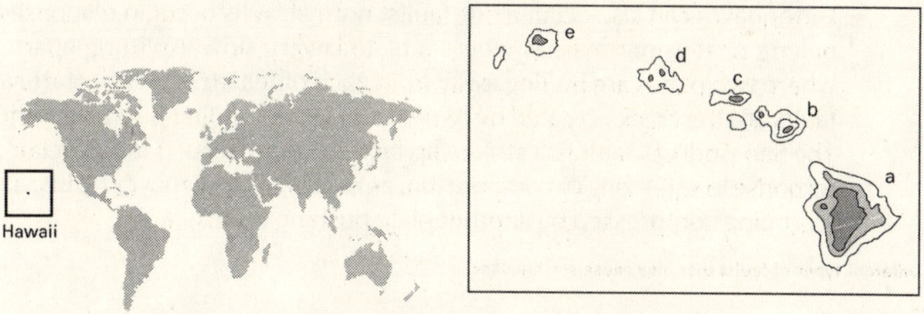

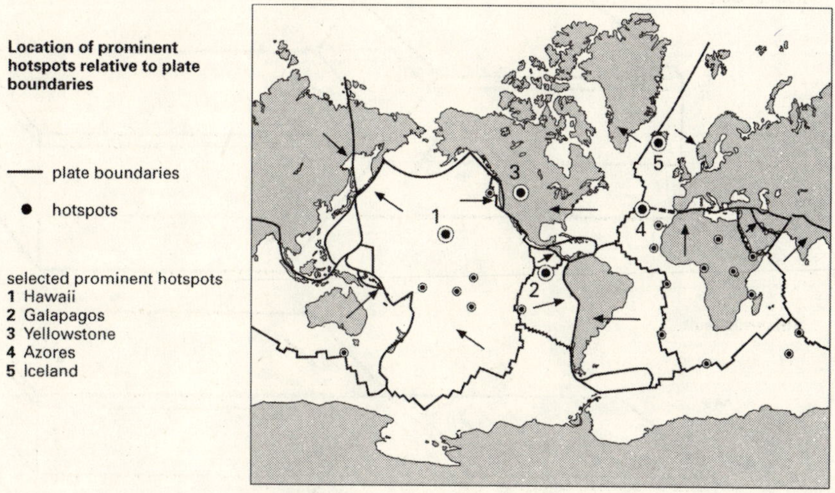

▫ how earthquakes happen

- Earthquakes are very common in areas of volcanic activity, where they either precede or accompany eruptions, and some earthquakes have been linked to specific volcanic eruptions.
- Most earthquakes are caused when underground rock breaks along a fault. The energy released when this happens causes the seismic waves that make the ground shake during an earthquake.
- Some blocks of rock or tectonic plates are continually moving past each other. Sometimes the friction of this motion causes the rock or plates to jam against each other.
- After a while, pressure builds up and forces the blocks to move suddenly, causing an earthquake.

- The underground area where the movement takes place is called the focus of the earthquake.
- The spot on the ground above the focus is called the epicenter of the earthquake.
- Most earthquakes occur on the edges of the oceanic and continental plates (see page 8.05).
- Earthquakes can also occur along faults: normal faults occur in response to pulling or tension in areas where a plate is very slowlysplitting apart or where two plates are pulling away from each other. strike-slip (lateral) faults are the cracks created by two plates that are sliding past each other. The San Andreas fault is a strike-slip fault. reverse (thrust) faults occur in response to squeezing or compression; an overlying block moves up because it is being compressed by another plate pushing against it.

Different types of faults that may cause earthquakes

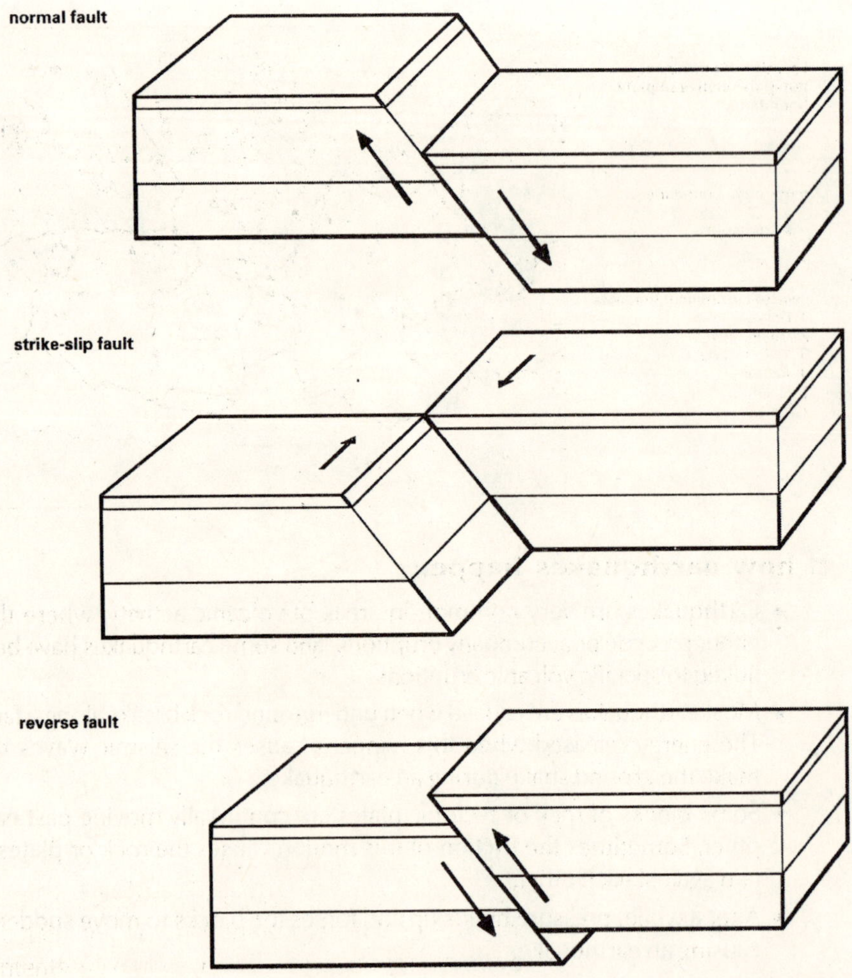

❐ hurricane

A tropical cyclone with winds of 74 mph or more. Normally applied to such storms in the Atlantic Basin and the Pacific Ocean east of the International Date Line.

- A hurricane is a large, rotating tropical weather system with a well defined circulation and sustained winds of up to 74 mph.
- In the western Pacific, hurricanes are called typhoons, and similar storms in the Indian Ocean are called cyclones.

Structure of a hurricane

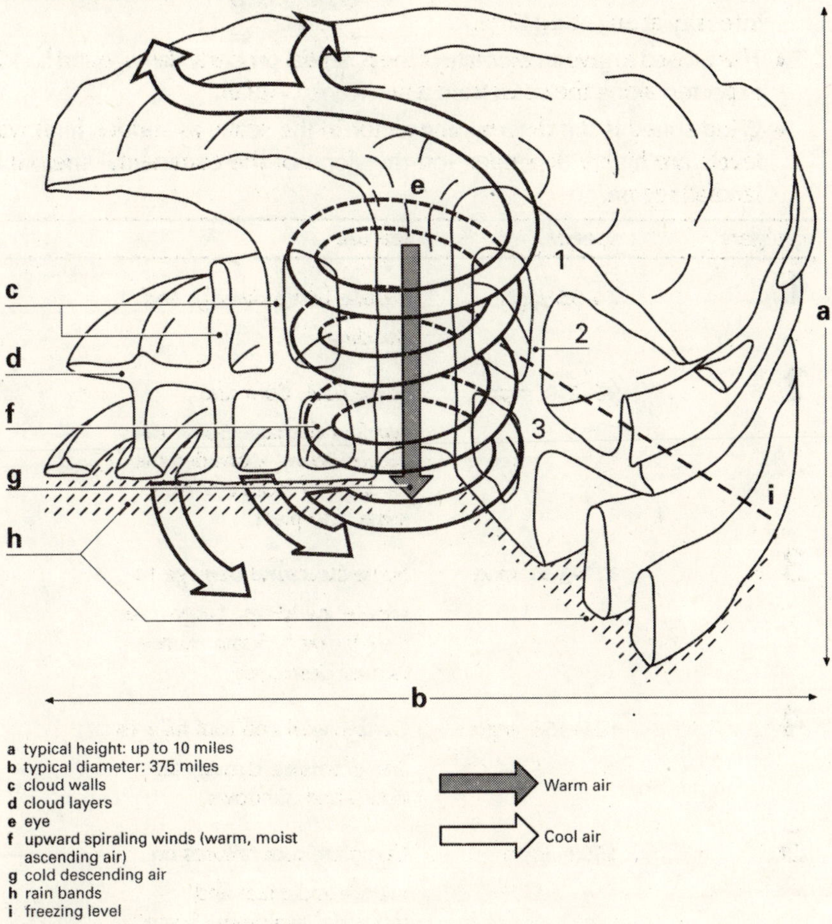

a typical height: up to 10 miles
b typical diameter: 375 miles
c cloud walls
d cloud layers
e eye
f upward spiraling winds (warm, moist ascending air)
g cold descending air
h rain bands
i freezing level

1 ice-crystal clouds
2 mixed water and ice clouds
3 water droplet clouds

- Hurricanes are products of the warm waters of tropical oceans and warm, moist atmospheres.
- The eye (center) of a hurricane is calm, with light winds and few clouds.

- The most powerful winds, with gusts of up to 225 mph and rain, occur in the eyewall, the ring of thunderstorms surrounding the eye.
- Above the eyewall (about 50,000 feet), most of the air is propelled outward, increasing the air's upward motion.
- Some air moves inward, however, and sinks into the eye, creating a cloud-free area.

❑ hurricane categories and united states hurricanes since 1950

Hurricane categories: the Saffir-Simpson Scale

- The Saffir-Simpson Hurricane Scale is a 1–5 rating based on the hurricane's intensity at any given time.
- This is used to give an estimate of the potential property damage and flooding expected along the coast from a hurricane landfall.
- Wind speed is the determining factor in the scale, as sudden high water levels are highly dependent on the slope of the continental shelf in the landfall region.

category	speed	feature
1	74–95 mph	Mobile home damage and flooding.
2	96–110 mph	Some roof, door and window damage. Some trees blown down. Considerable damage to mobile homes, signs, and piers.
3	111–130 mph	Some structural damage to smaller buildings. Large trees blown down. Some mobile homes destroyed
4	131–155 mph	Curtain wall and roof failures and extensive damage to doors and windows
5	155+ mph	Complete roof failures on many residential and industrial buildings. Some complete building failures with small buildings blown over or away. All shrubs, trees, and signs blown down. Complete destruction of mobile homes.

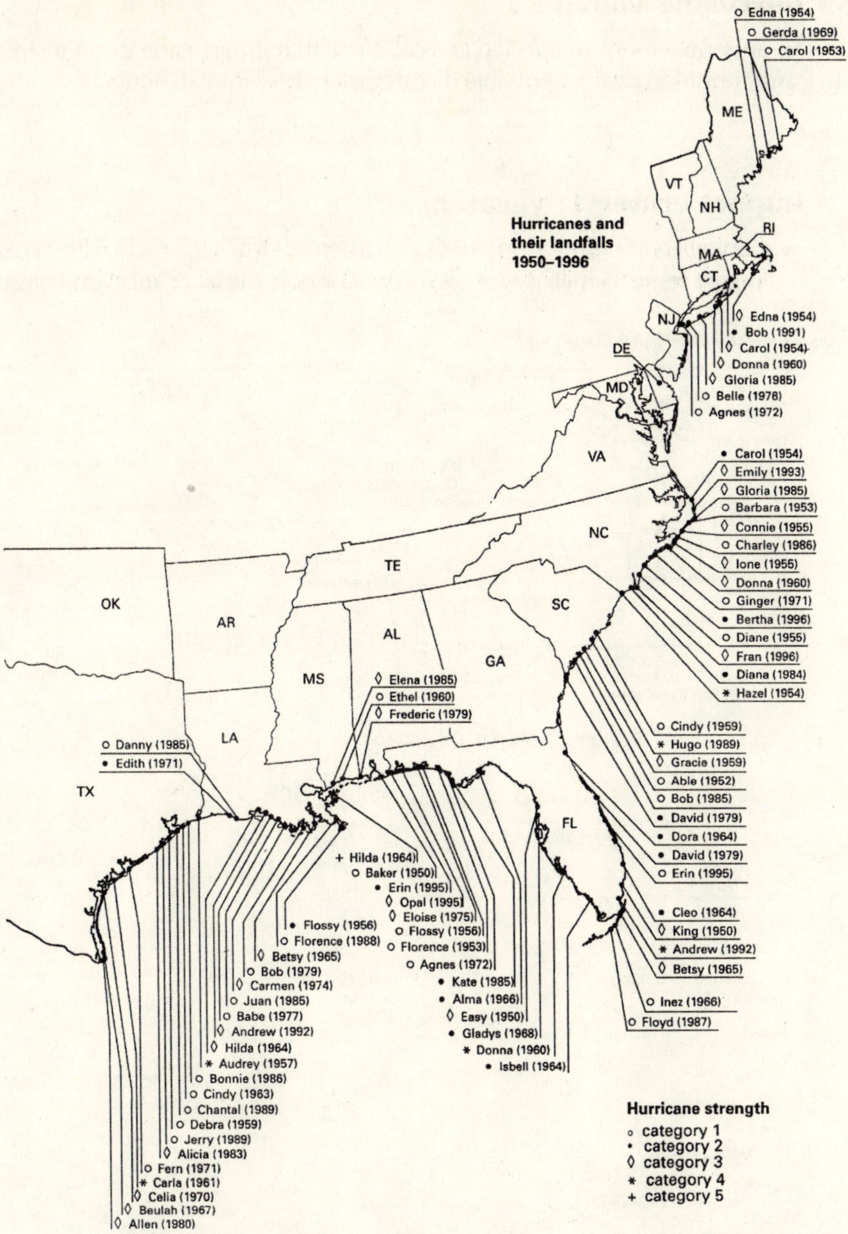

❐ hurricane warning

A warning that sustained winds 64 kt (74 mph or 119 kph) or higher associated with a hurricane are expected in a specified coastal area in 24 hours or less. A hurricane warning can remain in effect when dangerously high water or a combination of dangerously high water and exceptionally high waves continue, even though winds may be less than hurricane force.

❒ hurricane watch

An announcement of specific coastal areas that a hurricane or an incipient hurricane condition poses a possible threat, generally within 36 hours.

I

❒ impact crater 1: yucatan

- Geologists confirmed in 1990 that an asteroid struck the Earth with massive impact some 65 million years ago in what is now Mexico's Yucatán Peninsula.

Location and size of Chicxulub Crater

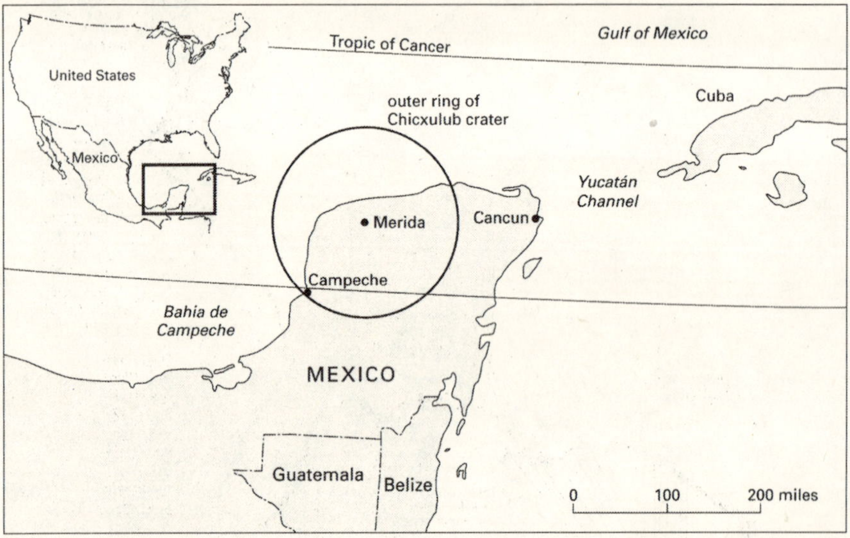

Chicxulub crater cross-section

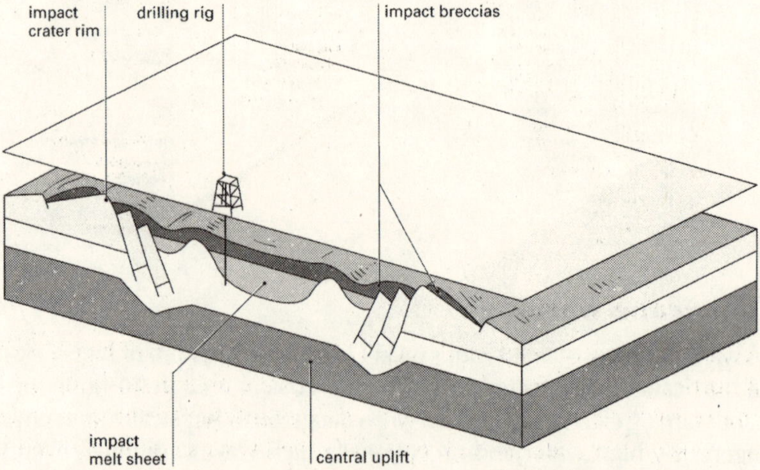

- The geologists named the crater Chicxulub Crater, after a nearby Mayan village.
- The crater, now buried beneath a half-mile-thick sequence of sediments, has been imaged using geophysical techniques that allow geologists to visualize underground structures.
- Radar images from space shuttles can also detect the buried crater.
- It appears to have a diameter of 90–112 miles, which makes it one of the largest confirmed impact structures on Earth.
- Scientists believe the asteroid which caused the crater was 6–12 miles in diameter.
- The asteroid's collision with Earth is believed to have brought total darkness for six months, and may have caused the extinction of the dinosaurs.

❑ impact crater 2: siberia

- On June 30, 1908 a catastrophic explosion occurred in the atmosphere over a remote area in central Siberia, near the Tunguska River.
- Although the cause of this airburst has remained a mystery, nearly 1,000 square miles of forest were instantaneously destroyed.
- Eighty million fallen trees pointed radially outward for a radius of 19–25 miles from the center of the blast.
- The sound of the explosion was heard half-way around the world in London.
- The object was a probably a small comet, traveling at 130,000 mph, consisting of about 30,000 tons of water, methane, and ammonia ice with traces of silicates and iron oxides.
- The energy released by the impact of the Tunguska comet would have been equivalent to 10 to 15 megatons of TNT, putting it in the same range as a magnitude 8 earthquake, or the eruption of Mount St. Helens.

Limits of Tunguska blast damage

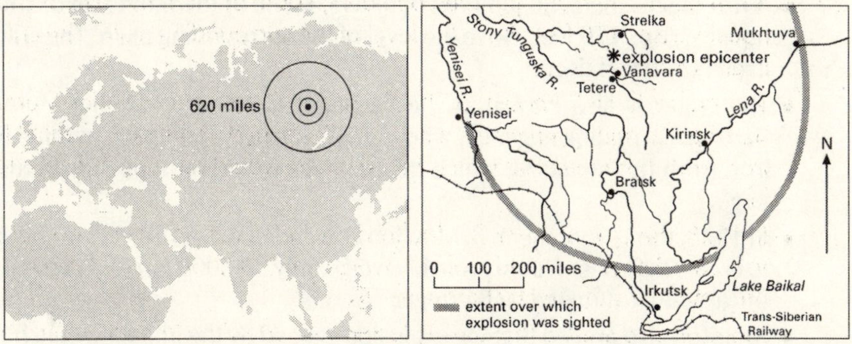

- Had this happened over a densely populated area it would have been one of the greatest disasters in human history.

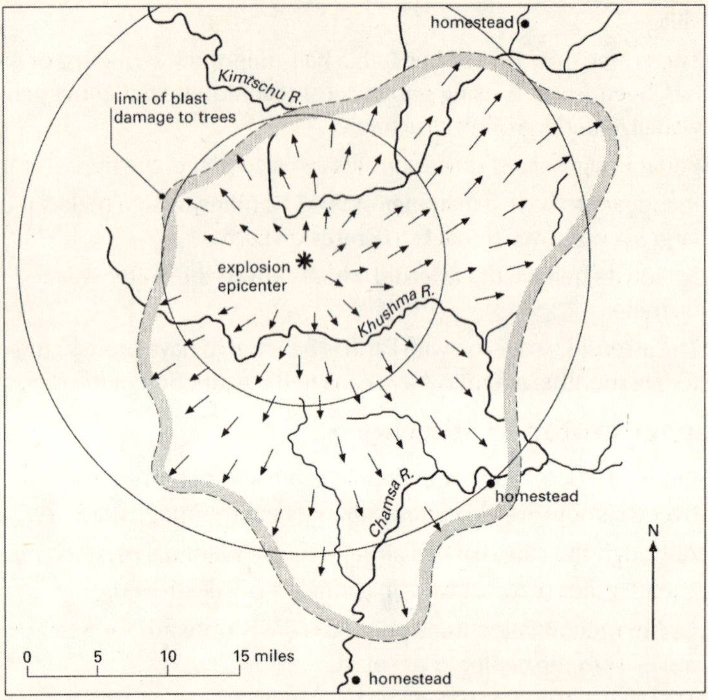

▫ impact crater 3: arizona, united states

- Meteor Crater in the Arizona Desert was the first crater to be identified as an impact crater.
- Between 20,000 to 50,000 years ago, a small asteroid about 80 feet in diameter collided with the Earth and formed the crater.
- The crater is the best preserved crater on Earth and measures nearly a mile in diameter.
- A rim of smashed and jumbled boulders, some of them the size of small houses, rises 150 feet above the level of the surrounding plain. The crater itself is 570 feet deep.
- The crater is also known as the Barringer Crater after Daniel Moreau Barringer, a mining engineer, who in 1902 formed a company to mine the iron from the meteorite which he assumed would have been extremely large.
- In 1928, the astronomer F. R. Moulton concluded that an object big enough to create the crater would probably weigh only 300,000 tons – 3 percent of the amount estimated by Barringer.
- Moulton also argued that the explosion caused by the impact would have vaporized most of the meteorite.
- Barringer was proved correct in his theory that it was an impact crater, but wrong in his assumption of the size of the asteroid.

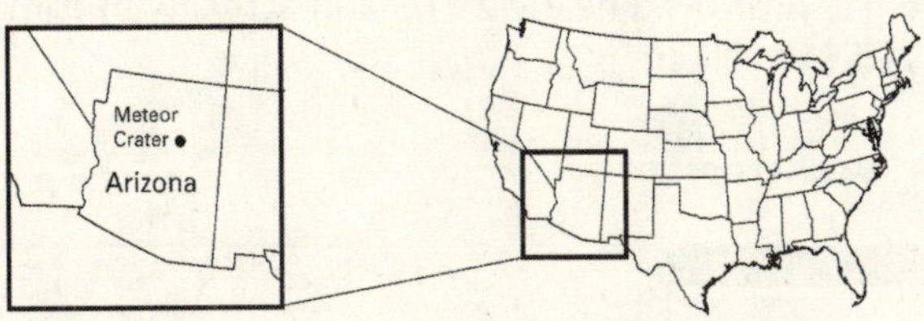

Cross-section of crater, showing Pleistocene-era lake, layer of brecciated rock, and undisturbed strata

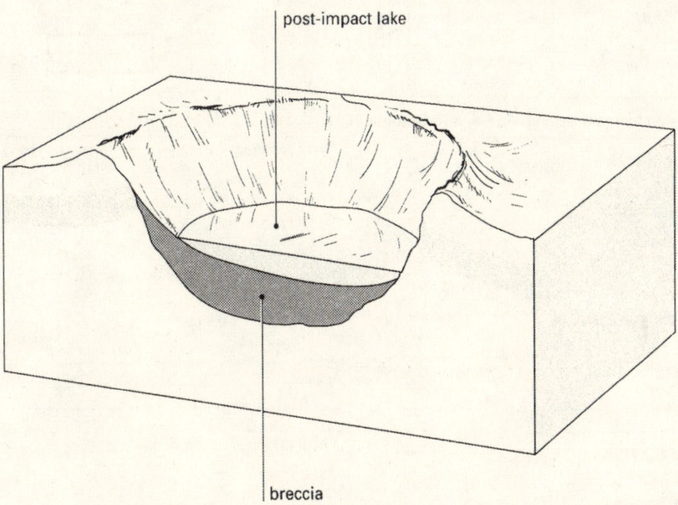

☐ indonesia's catastrophic forest fires

- Massive outbreaks of fire have occurred in Indonesia over the past 20 years.
- These have been caused by the use of fire for land clearance, combined with severe droughts induced by El Niño.

- In 1982–83 the first great fire resulting from the combination of bad forest management and an El Niño-induced drought engulfed 81,000 square miles of East Kalimantan.
- The fires started at the end of 1982 and burned out of control until the rains came in May 1983.
- Nearly 8 million acres were destroyed, including 6.6 million acres of tropical rainforest.
- Major fires occurred several times in the decade following the East Kalimantan fire, burning some 1.23 million acres in 1991 and nearly 12 million acres in 1994.
- The worst fires occurred when an El Niño-induced drought struck Indonesia in 1997–1998, when fires were reported from 23 of Indonesia's 27 provinces.
- By early 1998, fire had destroyed nearly 25 million acres, with damage estimated at nearly US$10 billion.

Health effects from fire-related haze exposure in eight Indonesia provinces Sept.–Nov. 1997

health effects	number of cases
deaths	527
asthma	298,125
bronchitis	58,095
acute respiratory infection	1,446,120
daily activity constraint	4,758,600
increase in outpatient treatments	36,462
increase in hospitalizations	15,822
lost work days	2,446,352

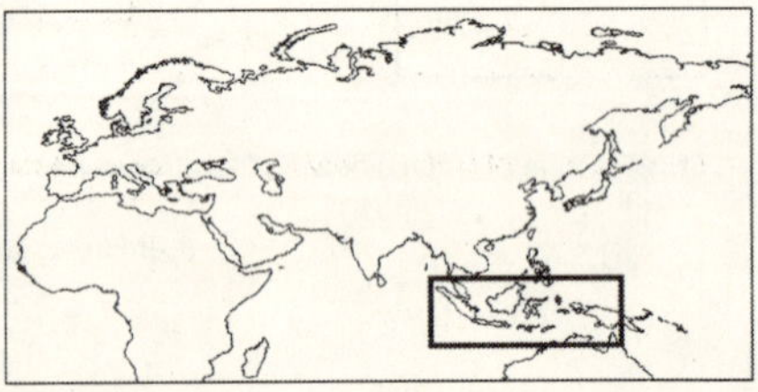

Extent of the Indonesian forest fires and haze in mid-October 1997

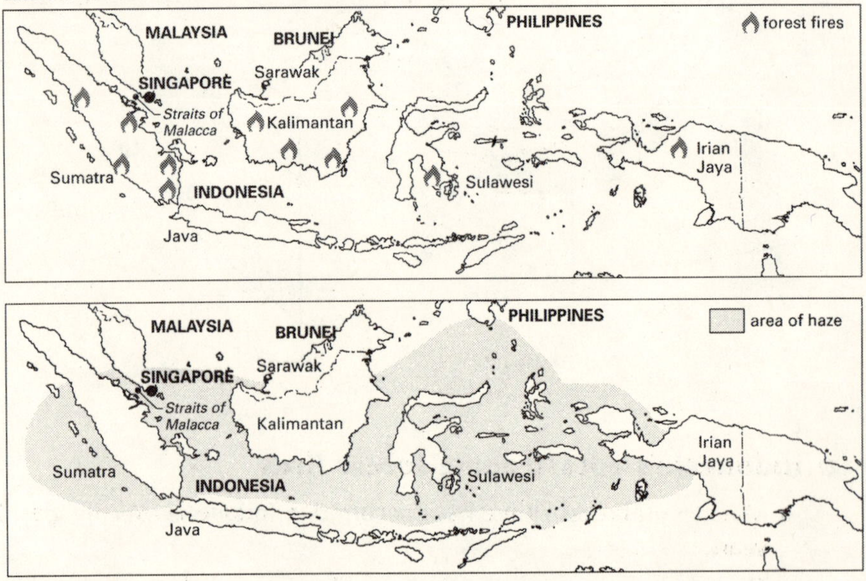

- Smoke from these fires shrouded much of Southeast Asia for months.

- Haze reached Malaysia and Singapore in July, and air quality deteriorated steeply in September.
- The September rains failed, and by the end of that month, approximately 385,000 square miles were haze-covered, affecting about 70 million people.
- There were many transport accidents linked to the poor visibility, including a ship collision in the Straits of Malacca that killed 29 people.
- Hospitals were filled with people seeking treatment for respiratory, eye, and skin ailments.
- Schools, businesses, and airports closed, and tourists stayed away, inflicting economic hardship on the region.
- Total economic losses were estimated at US$ 9.3 billion, including $2.7 billion for lost crops, $1.8 billion for lost timber, $1.6 billion for erosion and siltation, $400 million for flood protection, and $145 million in health costs.
- The excess carbon emissions were high enough to propel Indonesia into the ranks of the largest polluters in the world.

□ introduction

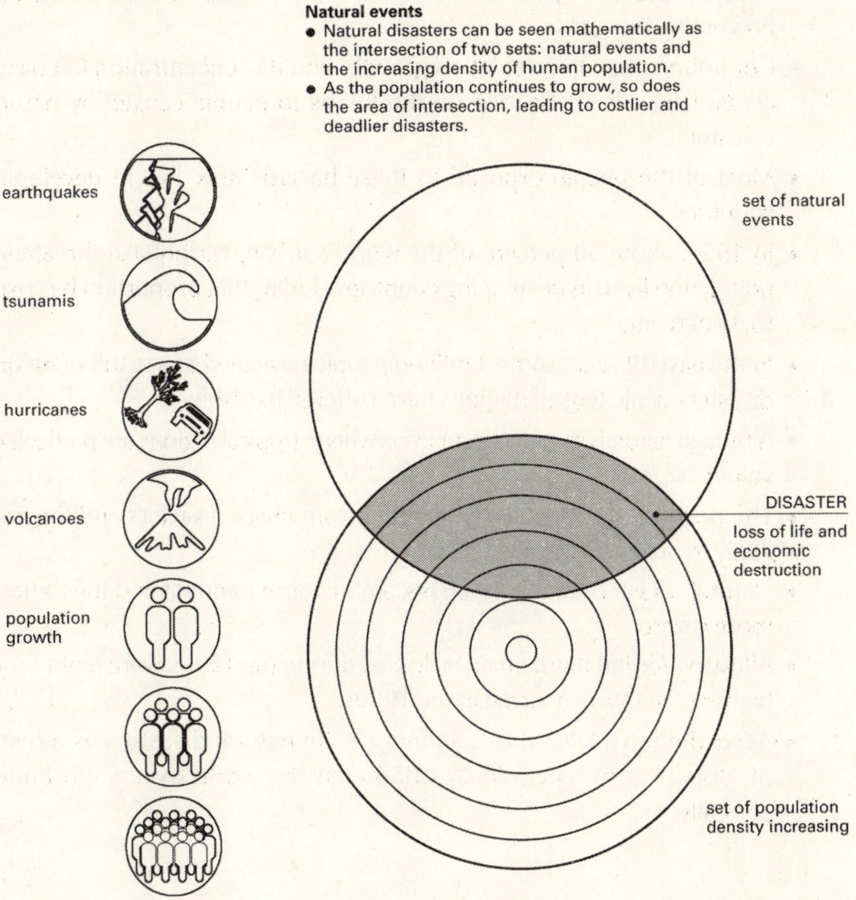

- Geologic disasters, such as earthquakes, volcanic eruptions, tsunamis and landslides, occur in the outer shell of the Earth known as the lithosphere. It consists of plates of rock, known as tectonic plates, that move relative to one another by means of fluid interactions, friction, buckling and fracturing.
- Such mechanisms create strong instabilities in the lithosphere that interact in a complex system.
- Geologic disasters such as earthquakes are critical phenomena in this system.
- Worldwide, the geologic hazards most destructive of human structures are those caused by volcanoes and earthquakes.
- For example, each of over 100 earthquakes during the past 100 or so years have resulted in 1,000 or more fatalities.
- Earthquakes, volcanoes and tsunamis all occur most frequently at points where the Earth's tectonic plates are in friction with each other.
- Study of the mechanisms of these events can help in their prediction, preventing loss of life.
- Continuing human population growth, and its concentration in coastal areas, has a direct bearing on the losses to people caused by natural disasters.
- Most of the people exposed to these hazards now live in developing countries.
- In 1950, about 50 percent of the world's urban, earthquake-threatened population lived in developing countries. Today, that proportion has risen to 85 percent.
- In the past 30 years some 3 million people have died as a result of natural disasters while tens of millions have suffered hardship.
- Although natural disasters occur everywhere, tropical regions are particularly vulnerable.
- The poor are the most likely to suffer from major disasters and the least likely to be insured against loss.
- Natural disasters appear to be becoming more frequent and their effects more severe.
- Allowing for inflation, financial losses during the 1990s were eight times higher than those incurred in the 1960s.
- According to a UN report, an increase in natural disasters as a result of global warming could by 2050 cost the world over $300 billion annually.

World distribution of natural disasters 1975–2001 by region and type of phenomenon

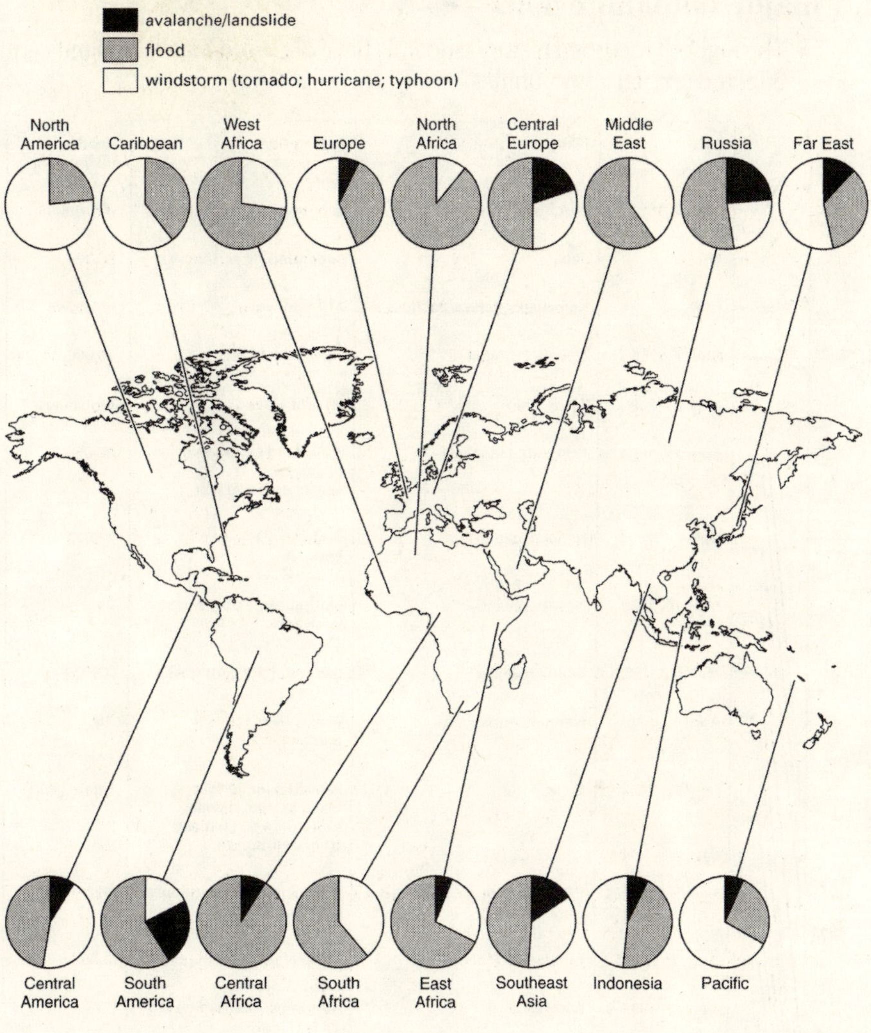

K

▫ knot

A measure of speed. It is one nautical mile per hour. Never refer to "knots per hour" unless you want to describe acceleration. A nautical mile is one minute of one degree of latitude and is slightly longer than the ordinary, or statute, mile used in the United States. To convert nautical miles to miles or knots to miles per hour, multiply by 1.15. To convert miles to nautical miles or miles per hour to knots, divide by 1.15.

M

❐ major tsunami events

- Throughout recorded history tsunamis have occurred that have significantly affected coastal communities.

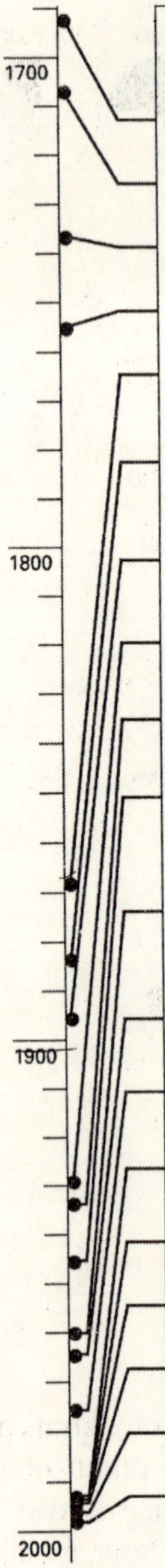

Date	Place	Description	Deaths (estimates)
June 7, 1692	Port Royal, Jamaica	generated by earthquake	thousands
1707	Japan	generated by earthquake	30,000
1737	Kamchatka Peninsula, Russia	210-foot wave	unknown
Nov. 1, 1755	Lisbon, Portugal	waves 20–50 feet	10,000–60,000
Aug. 8, 1868	Arica, Chile	50-foot wave	thousands
Aug. 27, 1883	Krakatau, Indonesia	generated by eruption of volcano waves over 100 feet	36,000
1896	Honshu, Japan	destroyed 175-mile coastline	27,000
Nov. 18, 1929	Newfoundland	earthquake, subsea landslide	29
March 3, 1933	Sanriku, Japan	generated by earthquake	2,990
April 1, 1946	Unimak, Alaska	waves over 100 feet damage in Hawaii	165
May 22, 1960	Chile	caused over 30-foot wave at Hilo, Hawaii, more than 6,200 miles from earthquake	up to 2,000
March 28, 1964	Prince William Sound, Alaska	8.4 magnitude earthquake, 100-foot wave	122
Aug. 17, 1976	Mindanao, Philippines	generated by earthquake	8,000
Sept. 1, 1992	Nicaragua	earthquake caused waves 35 feet high	170
Dec. 12, 1992	Flores, Indonesia	waves over 80 feet	1,000
July 12, 1993	Okushiri, Japan	waves over 95 feet	239
June 3, 1994	East Java, Indonesia	waves over 50 feet	238
Feb. 17, 1996	Irian Jaya	waves over 25 feet	161
July 17, 1998	Papua New Guinea	waves over 50 feet	2,200

- There has been a historical average of 57 tsunamis every decade in recent times throughout the world – although over 80 were recorded in the 1990s.

- The previous table lists some historical major events, as well as the most destructive recent tsunamis.

❒ measuring earthquake activity

- Two types of deformation are caused when an earthquake fault ruptures: static and dynamic.
- Static deformation occurs when the earthquake causes permanent change in the ground.
- While most of the plate-tectonic energy that causes fault ruptures is absorbed by static deformation, up to 10 percent may immediately take the form of seismic waves.
- Seismic waves are modulated into two kinds by the structure of the rocks through which they travel.
- Compressional waves, also known as primary or P waves, travel fastest, at speeds between 1 and 5 miles per second in the Earth's crust.
- Shear waves, also known as secondary or S waves, travel more slowly, usually at 60 percent–70 percent Of the speed of P waves.
- Some earthquake waves, known as surface waves, move only along the surface of the Earth.
- These surface waves are known by geologists as Love and Rayleigh waves, named for the two 20th century British scientists who first described them.
- The instrument that scientists who study earthquakes use most in their work is called a seismograph.

How a seismograph works

- The heart of a seismograph is a simple pendulum attached to a rigid frame and base. During an earthquake, the base and frame respond to the shaking of the Earth and move, but inertia keeps the pendulum still.
- The pendulum has a pen attached to the end. Paper placed immediately below it records a jagged pattern of lines that are a visual record of the motion of the seismograph's frame and of the ground's movement during the earthquake. This visual record is called a seismogram.

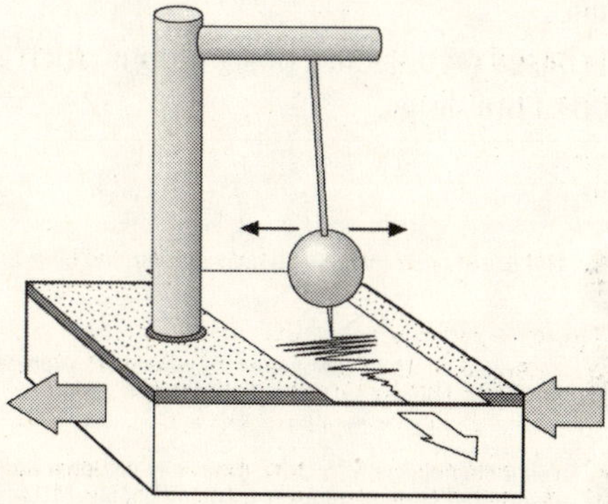

Kinds of seismic wave

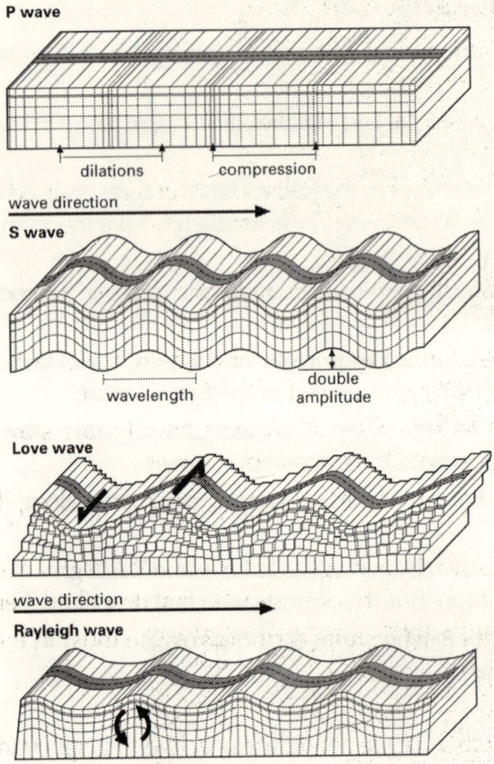

- Seismic P waves, or primary waves, follow a strong back-and-forth pattern of motion that produces areas of compression and expansion in the Earth, spreading outward from the earthquake's epicenter.
- S waves, the secondary or shear waves, are the back-and-forth, up-and-down motion. S waves can be horizontal or vertical.
- Love waves shake the Earth's surface side-to-side.
- Rayleigh waves move the surface of the Earth around in a circle, forward and down then back and up.

❑ measuring earthquake intensity

- While the Richter scale measures the amount of seismic energy released during an earthquake, the measure of an earthquake's intensity is based on its observed effects on its surroundings.
- Earthquake intensity is usually measured using the Modified Mercalli Intensity (MMI) scale.
- This scale is based on objective observations, such as the amount of damage to the fabric of buildings.

scale	effects	
I		Not usually perceived by people; animals and birds may appear uneasy.
II		Perceived only by a few stationary people, particularly on higher floors of buildings. Loosely suspended objects may swing.
III		Felt quite noticeably indoors, especially on upper floors of buildings, but not recognized as an earthquake by many people. Motor vehicles at rest may rock slightly. Vibration like passing truck.

IV		During the day felt indoors by many, outdoors by a few. At night some awakened. Dishes, windows and doors disturbed; walls make creaking sound. Sensations like heavy truck striking building. Standing motorcars rocked noticeably.
V		Felt by nearly everyone; many awakened. Some dishes, windows, etc., broken; a few instances of cracked plaster; unstable objects overturned. Disturbances of trees, poles and other tall objects sometimes noticed. Pendulum clocks may stop.
VI		Felt by all; many frightened and run outdoors. Some heavy furniture moved; a few instances of fallen plaster or damaged chimneys. Damage slight.
VII		Everybody runs outdoors. Damage negligible in buildings of good design and constructions; slight to moderate in well-built ordinary structures; considerable in poorly built or badly designed structures. Some chimneys broken. Noticed by persons driving motor cars.
VIII		Damage slight in specially designed structures; considerable in ordinary substantial buildings, with partial collapse; great in poorly-built structures. Panel walls thrown out of frame structures. Fall of chimneys, factory stacks, columns, monument walls. Heavy furniture overturned. Sand and mud ejected in small amounts. Persons driving motor cars disturbed.
IX		Damage considerable in specially designed structures; well designed frame structures thrown out of plumb; great in ordinary substantial buildings, with partial collapse. Buildings shifted off foundations. Ground cracked conspicuously. Underground pipes broken.
X		Some well built wooden structures remain standing; most masonry and frame structures destroyed with foundations; ground badly cracked. Rails bent. Landslides considerable from river banks.
XI		Few, if any masonry structures remain standing. Bridges destroyed. Broad fissures in the ground. Underground pipelines totally disrupted. Earth slumps and land slips in soft ground.
XII		Damage total. Waves seen on ground surfaces. Lines of sight and level distorted. Objects thrown up and into the air.

❑ measuring earthquake magnitude

- Charles F. Richter of the California Institute of Technology developed the Richter magnitude scale in 1935 as a mathematical tool for comparing the size of earthquakes.
- The magnitude of an earthquake is determined from the logarithm of the amplitude of waves (amplitude x10) recorded by seismographs.
- Each whole number increase in magnitude represents a tenfold increase in measured amplitude because of the logarithmic basis of the scale.
- As an estimate of energy, each whole number step in the magnitude scale corresponds to the release of about 31 times more energy than the amount associated with the preceding whole number value.
- Earthquakes not commonly felt by people and generally recorded only on local seismographs have a magnitude of about 2.0 or less and are usually called microearthquakes.
- There are several thousand events with magnitudes of about 4.5 or greater annually – strong enough to be recorded by sensitive seismographs all over the world.

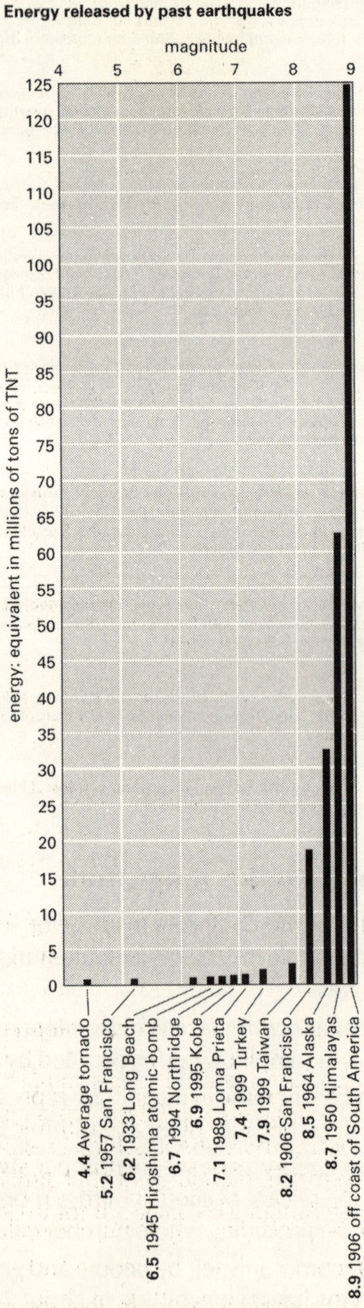

- On average, one earthquake of a magnitude of 8.0 or higher, such as the 1964 Good Friday earthquake in Alaska, occurs somewhere in the world each year.

- The Richter Scale is not used to express damage.
- An earthquake felt as a shock in a remote area that does nothing more than frighten the wildlife may have the same magnitude as one in a densely populated area that results in many deaths and considerable damage.
- Recently, another scale called the Moment-Magnitude Scale has been devised for more precise study of major earthquakes.
- The Moment-Magnitude Scale is applied by multiplying the area of a fault's rupture surface by the distance the Earth's surface moves along the fault.

Richter Scale of Magnitude (M)

The Richter Scale has no upper limit, but for practical purposes does not go above 9.

1 Registered by instruments only.

2 Felt by sensitive people and sensitive animals (10 times #1).

3 Felt by many people – feels like a passing truck (100 times #1).

4 Felt by everyone; pictures fall off the wall (1,000 times #1).

5 Damage – may cause weak walls to crack and fall (10,000 times #1).

6 A destructive earthquake in populated areas; falling smokestacks, etc. (100,000 times #1).

7 A major earthquake causing serious damage (1,000,000 times #1) (recent Taiwan, Turkey, Kobe, Japan, and California earthquakes).

8 A disaster – a great earthquake that produces total destruction to nearby communities (10,000,000 times #1) (1906 San Francisco, 1949 Queen Charlotte Islands)

9 Rare great earthquake, major damage over a 620 mile radius (100,000,000 times #1) (Chile 1960, Alaska 1964, and British Columbia, Washington, Oregon, 1700).

❒ measuring fault movements

- Seismologists use networks of seismograph stations to find where an earthquake begins, and better estimate its other parameters.
- The principal use of seismograph networks is to locate earthquakes.
- Telemetric plots from three or four stations enable seismic events to be accurately located.
- Locating the source of any earthquake is important in assessing the damage that the event may have caused.

Mid-Atlantic Ridge fault movement

This example uses stations in Boston (United States), Edinburgh (United Kingdom), and Manaus (Brazil). All three circles can intersect only at a single point on the Mid-Atlantic Ridge spreading center.

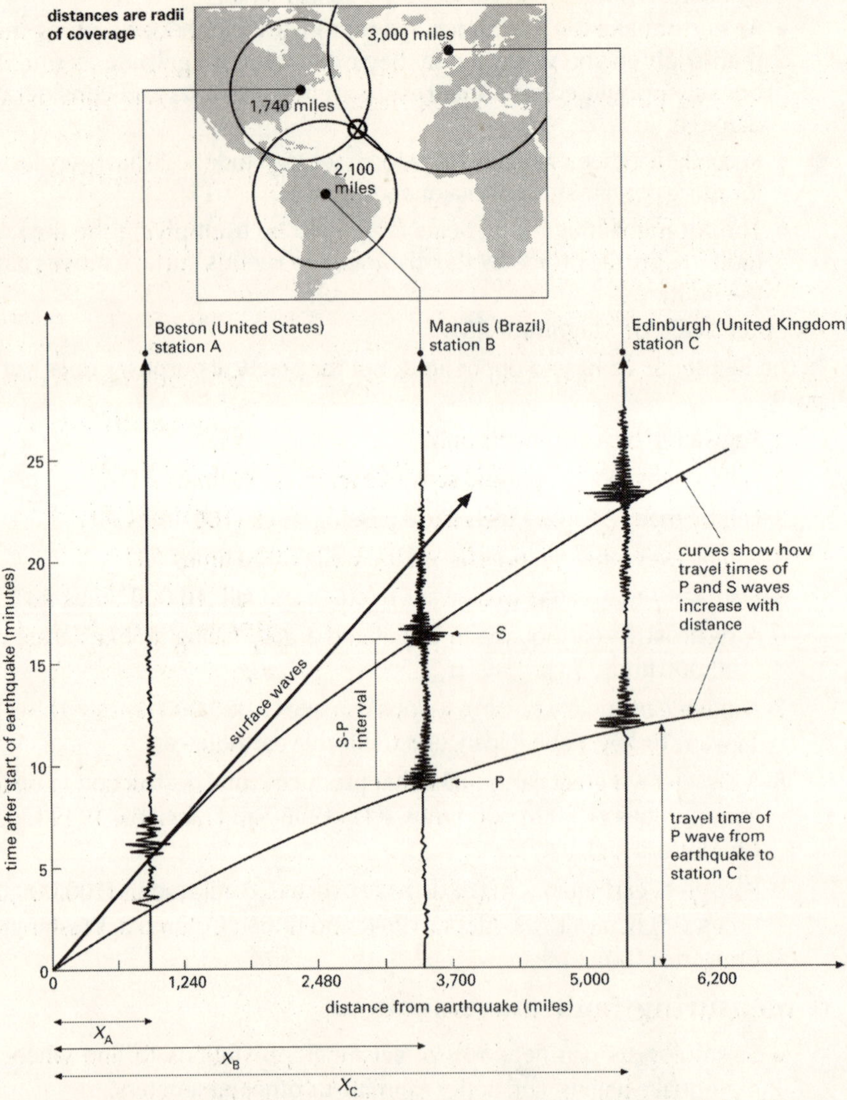

The distance between the station and the event can be calculated from the interval between the P and S waves. The average speed of the wave equals the distance divided by the travel time of the P waves.

❑ millibar

A metric measurement of air pressure.

N

❑ north atlantic basin (sometimes called just the "atlantic basin")

The Atlantic Ocean north of the equator, the Caribbean Sea, and the Gulf of Mexico.

P

❐ plate tectonics

- The Earth's surface is made up of a series of large plates (like pieces of a giant jigsaw puzzle).
- These plates are in constant motion, traveling at a few inches per year.
- The ocean floors are continually moving, spreading from the center and sinking at the edges.

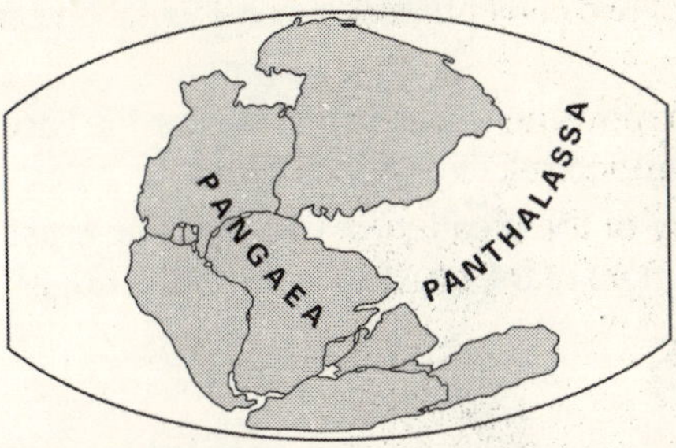

200 million years ago

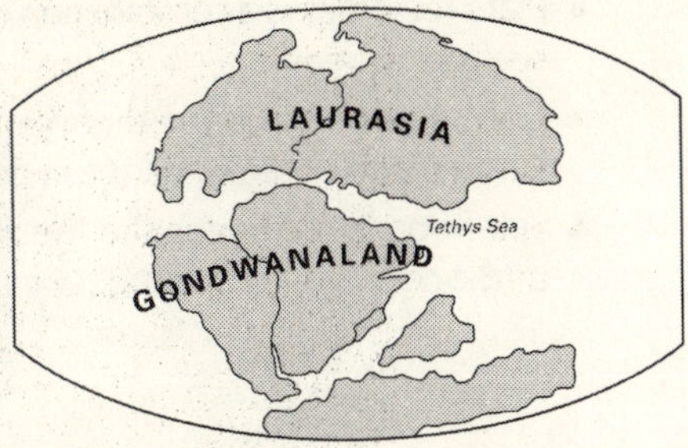

180 million years ago

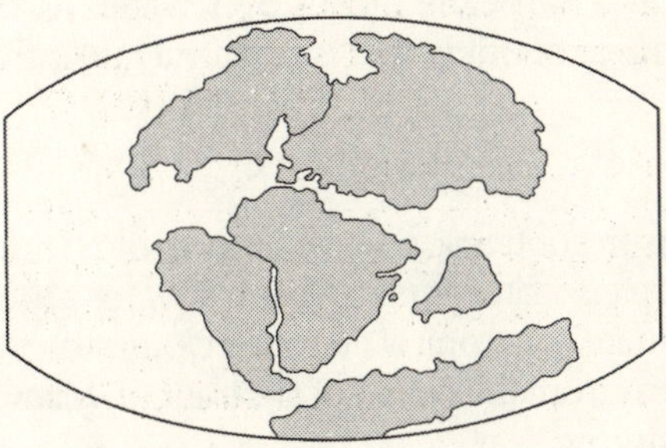

135 million years ago

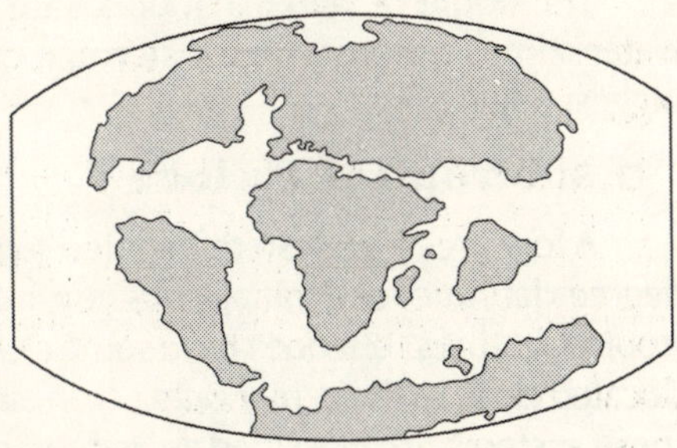

65 million years ago

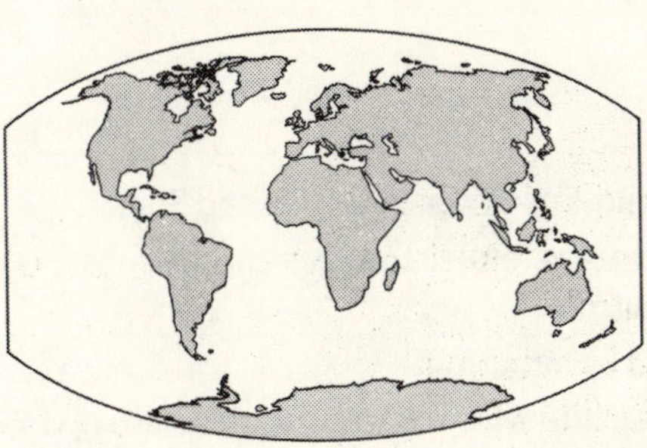

present day

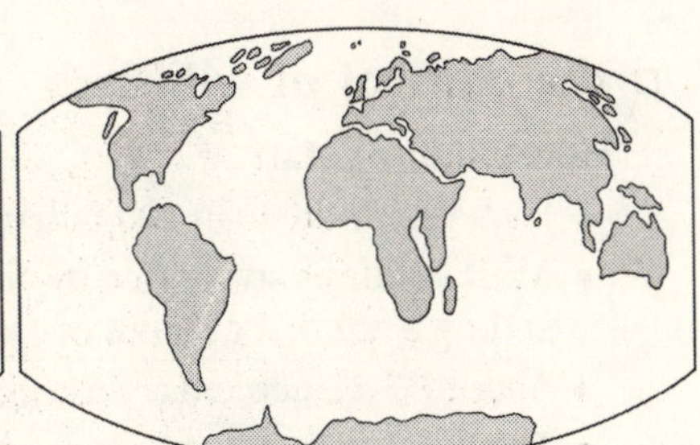

50 million years in the future

- Convection (heat) currents beneath the plates move the plates in different directions.
- The decay of radioactive elements deep within theEarth generates the heat that drives the convection currents.

- In some cases, plates can collide with each other at the plate boundaries, causing them to slip under each other (subduction), leading to earthquakes, volcanic activity, mountain building, and creating oceanic trenches.
- At other plate boundaries, plates may move away from each other because of seafloor spreading, or horizontally move past one another, creating transform faults and earthquakes.
- The theory of plate tectonics was not clearly described until the 1960s when geologists, with the help of ocean surveys, began to understand more of what goes on beneath the Earth's surface.
- Plate tectonics is a combination of two earlier theories: continental drift and seafloor spreading.
- Continental drift is the theory of movement of continents over the Earth's surface and of their change in position relative to each other.
- Sea-floor spreading is the theory of the creation of new oceanic crust at mid-ocean ridges and the movement of the crust away from the ridges.

S

❐ storm surge

The dome of water that builds up as a hurricane moves over water. As this water comes ashore with the storm, it causes flooding that is usually a hurricane's biggest killer.

❐ subtropical cyclone

A low pressure system that develops in subtropical waters (north of 20 north degrees latitude) and initially has non-tropical features (see table below for a list of tropical features) but does have some element of a tropical cyclone's cloud structure (located close to the center rather than away from the center of circulation). Many of these systems are classified as "hybrid" storms.

T

❐ the control of wildfires

How wildfires start

- Each year more than 100,000 wildfires occur in the United States.
- Most wildfires are started by humans, either through carelessness, or by setting a fire that gets out of control.
- About 10–15 percent are started by lightning.
- The 'western wildfire season' starts in May when conditions are driest.
- The United States fire season in the year 2000 was the worst in half a century because of warm and dry conditions following La Niña Pacific weather conditions.
- By late August 2000, 4.9 million acres had burned nationwide, more than double the ten-year average.

The Fire Triangle

- Three components are necessary to start and maintain a fire: oxygen, fuel and heat.
- At least 16 percent oxygen must be in the air for a fire to start (our atmosphere contains 21 percent).
- When there is not enough heat to sustain the process, when the fuel is exhausted or removed, or when the oxygen supply is below a certain level, then a side of the triangle is broken and the fire is suppressed.

Satellite imaging

- Researchers are using the Multi-angle Imaging Spectro-Radiometer (MISR) on NASA's Terra satellite, launched in 1999, to estimate forest canopy structure.

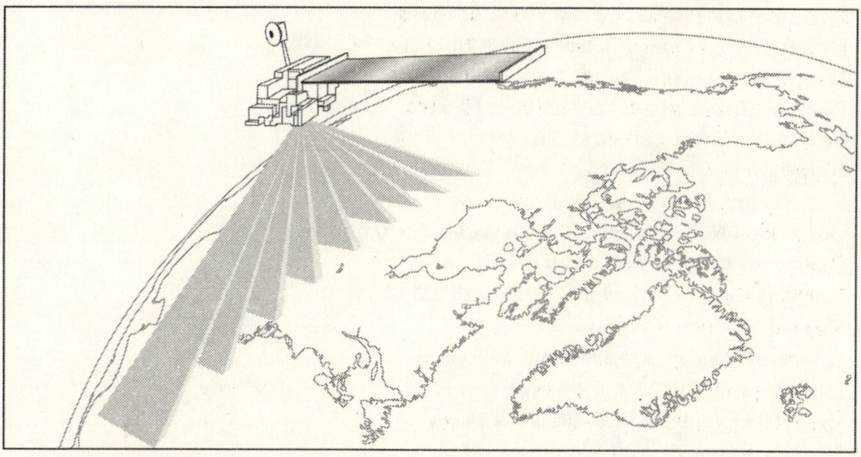

- The MISR instrument can help scientists decipher the structure of the forest canopy. This is important information for estimating if a ground fire will become a more dangerous crown fire (a fire that spreads from tree crown to tree crown).
- In the future, this combination of computer and satellite technologies should become increasingly useful for helping reduce the risk posed by wildfires.

❐ the world's biggest meteorite impacts

- Most terrestrial impact craters have been obliterated by other terrestrial geological processes.

- To date, approximately 150 impact craters have been identified on Earth.
- Almost all known craters have been recognized since 1950 and several new structures are found each year.
- Fifty nine of the biggest craters are listed below.

name	location	age (million years)	diameter (miles)	exposed at surface

Acraman South Australia 450 56 yes
Ames Oklahoma, United States 470 ± 30 10 no
Aorounga Chad, Africa <345 8 yes
Araguainha Dome Brazil 247.0 ± 5.5 25 yes
B.P. Structure Libya <120 2 yes
Barringer Arizona, United States 0.049 ± 0.003 1 yes
Beaverhead Montana, United States ~600 37 yes
Beyenchime-Salaatin Russia <65 5 yes
Bosumtwi Ghana 1.03 ± 0.02 7 yes
Brent Ontario, Canada 450 ± 30 2 yes
Carswell Saskatchewan, Canada 115 ± 10 24 yes
Charlevoix Quebec, Canada 357 ± 15 34 yes
Chesapeake Bay Virginia, United States 35.5 ± 0.6 52 no
Chicxulub Yucatan, Mexico 64.98 ± 0.05 105 no
Clearwater East Quebec, Canada 290 ± 20 16 yes
Clearwater West Quebec, Canada 290 ± 20 21 yes
Decaturville Missouri, United States <300 4 yes
Deep Bay Saskatchewan, Canada 100 ± 50 8 yes
Eagle Butte Alberta, Canada <65 6 no
El'gygytgyn Russia 3.5 ± 0.5 11 yes
Goat Paddock Western Australia <50 3 yes
Gosses Bluff Northern Territory, Australia 142.5 ± 0.8 14 yes
Gweni-Fada Chad, Africa <345 9 yes
Haughton Northwest Territories, Canada 23 ± 1 15 yes
Kara-Kul Tajikistan <5 32 yes
La Moinerie Quebec, Canada 400 ± 50 5 yes
Lappajärvi Finland 77.3 ± 0.4 14 yes
Lawn Hill Queensland, Australia >515 11 yes
Logancha Russia 25 ± 20 12 yes
Lonar India 0.052 ± 0.006 1 yes
Manicouagan Quebec, Canada 214 ± 1 62 yes
Mistastin Newfoundland/Labrador, Canada 36.4 ± 4 17 yes
Mjølnir Norway 143 ± 20 25 no
New Quebec Quebec, Canada 1.4 ± 0.1 2 yes
Nicholson Northwest Territories, Canada <400 8 yes
Oasis Libya <120 11 yes
Pilot Northwest Territories, Canada 445 ± 2 4 yes
Popigai Russia 35 ± 5 62 yes
Pretoria Saltpan South Africa 0.220 ± 0.052 1 yes

name	location	age (million years)	diameter (miles)	exposed at surface
Riachao Ring	Brazil	<200	3	yes
Ries	Germany	15 ± 1	15	yes
Roter Kamm	Namibia	3.7 ± 0.3	1	yes
Saint Martin	Manitoba, Canada	220 ± 32	25	yes
Serpent Mound	Ohio, United States	<320	5	yes
Serra da Cangalha	Brazil	<300	8	yes
Shoemaker (formerly Teague Ring)	Western Australia, Australia	1630 ± 5	19	yes
Shunak	Kazakhstan	12 ± 5	2	yes
Sierra Madera	Texas, United States	<100	8	yes
Siljan	Sweden	368.0 ± 1.1	31	yes
Slate Islands	Ontario, Canada	~450	19	yes
Spider	Western Australia	>570	8	yes
Steinheim	Germany	15 ± 1	2	yes
Sudbury	Ontario, Canada	1850 ± 3	155	yes
Tenoumer	Mauritania	2.5 ± 0.5	1	yes
Upheaval Dome	Utah, United States	<65	6	yes
Vredefort	South Africa	2023 ± 4	186	yes
Wanapitei	Ontario, Canada	35.3 ± 2	5	yes
Wells Creek	Tennessee, United States	200 ± 100	7	yes
Zhamanshin	Kazakhstan	0.9 ± 0.1	8	yes

❑ the world's deadliest volcanic eruptions

- The table below lists all eruptions with more than 500 known human fatalities.
- Other eruptions have been as big or bigger than these, but no-one lived nearby to be threatened (for example the Valley of 10,000 Smokes eruption in Alaska in 1912).

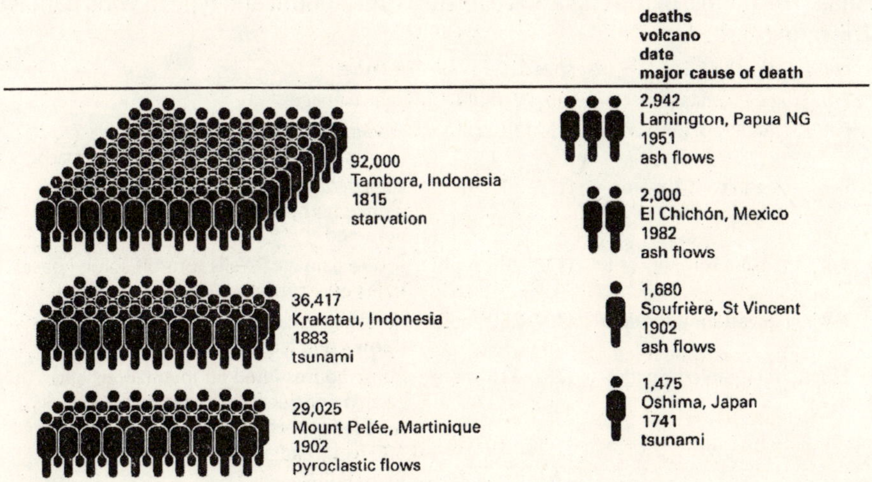

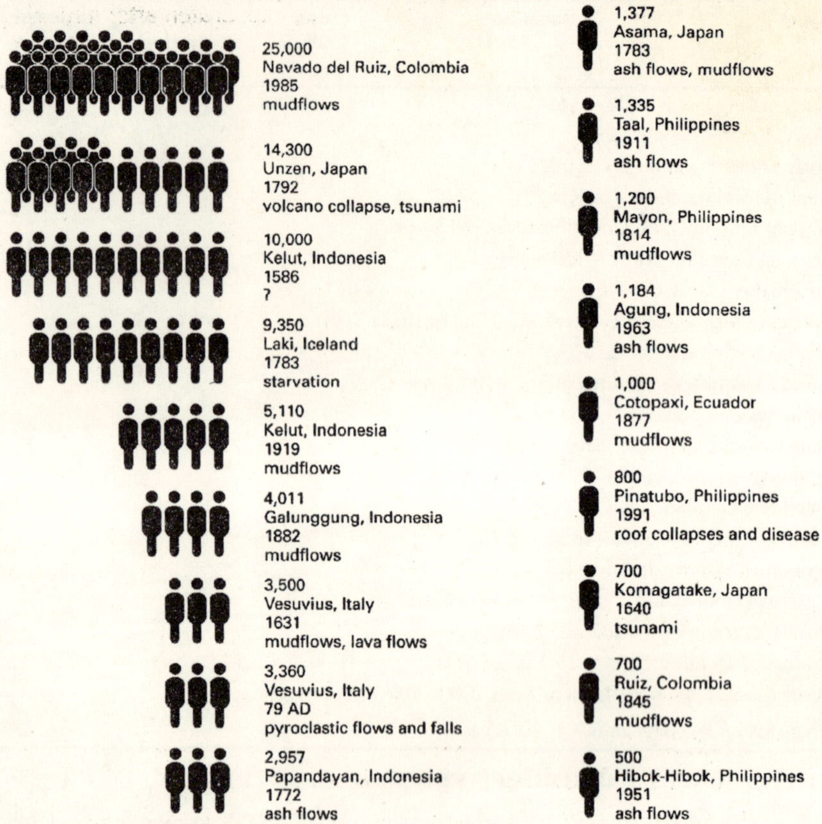

❐ tornadoes

Tornado classification

Tornadoes are classified according to a wind damage scale created by T. Theodore Fujita. The ratings of the Fujita Scale relate to the amount and type of wind damage. The ratings are:

class	type	speed	feature
F-0	gale tornado	(40–72 mph)	light damage.
F-1	moderate tornado	(73–112 mph)	moderate damage. Mobile homes overturned; moving cars blown off roads.
F-2	significant tornado	(113–157 mph)	considerable damage. Roofs torn off; mobile homes demolished; large trees uprooted.
F-3	severe tornado	(158–206 mph)	severe damage. Walls torn off some houses; trains overturned; most trees uprooted.
F-4	devastating tornado	(207–260 mph)	devastating damage. Houses leveled; cars thrown and large missiles generated.
F-5	incredible tornado	(261–318 mph)	frame houses lifted off foundations and carried considerable distances; huge missiles propelled through the air for over 330 feet; reinforced concrete structures badly damaged.

1 **Distribution of tornadoes over the United States**
(average for 12 years)

2 **Average number of tornadoes recorded per month in the United States**
(average over 45 years)

1

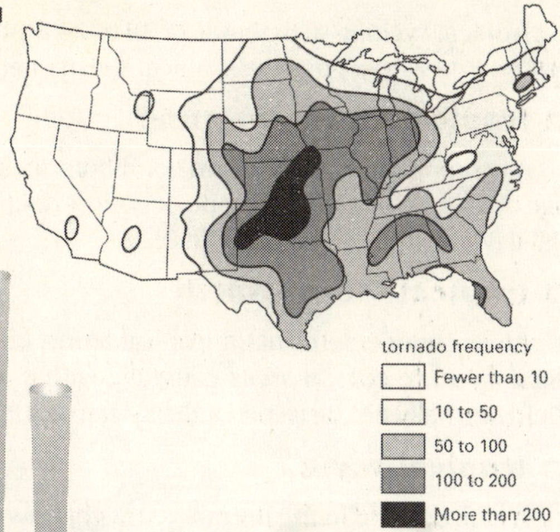

2

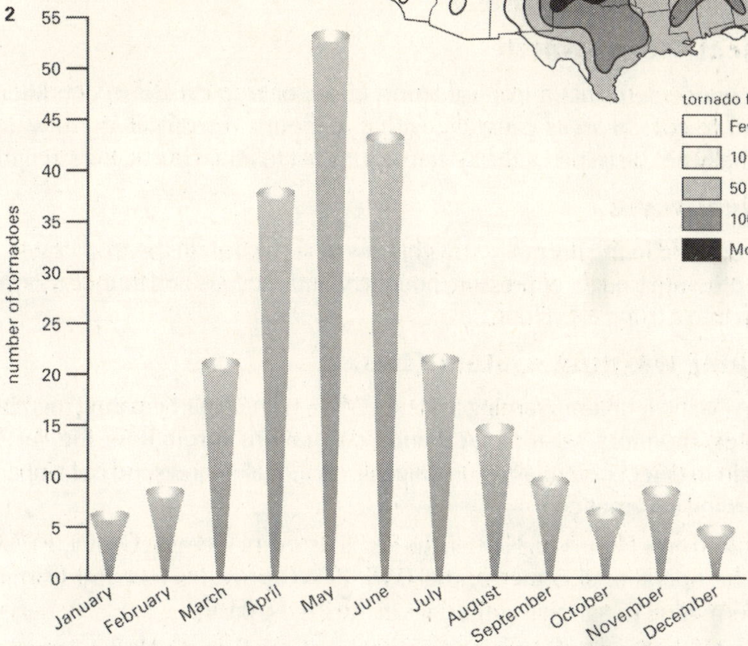

▫ tropical cyclone

A low-pressure weather system in which the central core is warmer than the surrounding atmosphere. See the table below for differences between tropical and extratropical cyclones. The term "tropical cyclone" is also used in the Indian Ocean and around the Coral Sea off northeastern Australia to describe storms called "hurricanes" and "typhoons" in other areas.

▫ tropical depression (td)

A tropical cyclone with maximum sustained winds near the surface of less than 39 mph. Tropical Depressions are listed only with a number, not a name.

▫ tropical disturbance

Often the earliest stages of a tropical cyclone. Normally an organized area of thunderstorms that forms in the tropics and persists for more than 24 hours. Low pressure might form at the surface, but winds around remain below 30 mph.

❐ tropical storm

Tropical cyclone with winds of 39 to 74 mph. In most of the world, a storm is given a name when it reaches tropical storm intensity.

❐ tropical storm warning

A warning for tropical storm conditions including sustained winds within the range of 34 to 63 kt (39 to 73 mph or 63 to 118 kph) that are expected in a specified coastal area within 24 hours or less.

❐ tropical storm watch

An announcement that a tropical storm poses or tropical storm conditions pose a threat to coastal areas generally within 36 hours. A tropical storm watch should normally not be issued if the system is forecast to attain hurricane strength.

❐ tropical wave

A kink or bend in the normally straight flow of surface air in the tropics which forms a low pressure trough, or pressure boundary, and showers and thunderstorms. Can develop into a tropical cyclone.

❐ tsunami warning system (tws)

- The Pacific Tsunami Warning System (TWS), with 26 participating member states, monitors seismological and tidal stations throughout the Pacific Basin to detect potentially tsunamigenic earthquakes and send out tsunami warning information.
- Located near Honolulu, Hawaii, the Pacific Tsunami Warning Center (PTWC) is the operational center of the TWS. PTWC provides tsunami warning information to national authorities in the Pacific Basin.
- The Alaska Tsunami Warning Center (ATWC) in Palmer, Alaska, serves as the regional Tsunami Warning Center for Alaska, British Columbia, Washington, Oregon, and California.
- The Pacific Tsunami Warning Center receive a constant flow of information from seismometers and tide-gauge stations all around the Pacific Ocean.
- When an earthquake with the potential to generate a tsunami occurs, the Warning Center puts out a 'tsunami watch' within 30–40 minutes, alerting Civil Defense and other authorities of a possible tsunami.
- Linked to the TWS, DART operates a number of tsunami detection buoys in the Pacific.
- A DART system comprises a seafloorbottom pressure-recording system capable of detecting tsunamis as small as 0.4 inches, and a moored surface buoy for realtime communications.
- An acoustic link transmits data from the seafloor to the surface buoy. The data are then relayed via a Geostationary Operational Environmental Satellite (GOES) link to ground stations.

DART system detects minute variations in ocean-bottom pressure caused by tsunamis

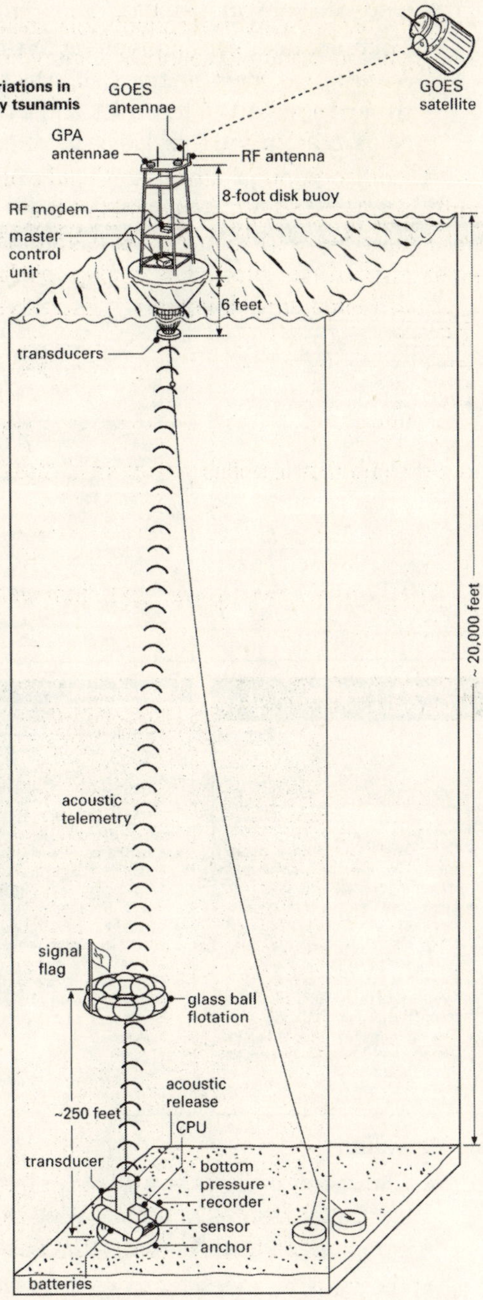

Earthquake detectors, Deep-ocean Assessment and Reporting of Tsunamis (DART) detectors, and related earthquake zones

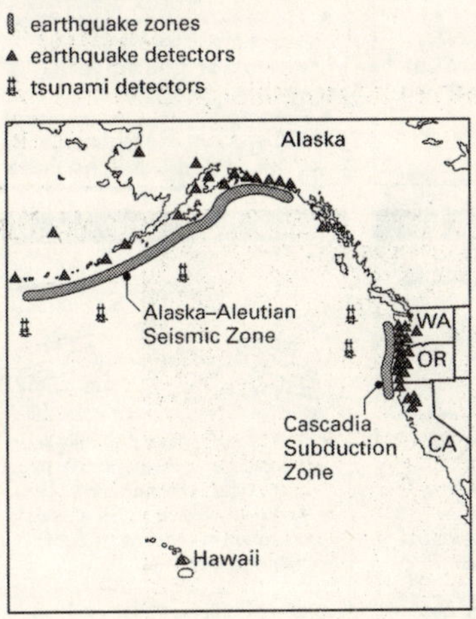

◻ types of volcanoes and eruptions

Types of volcanic eruption

Gas

- Unusual: only gas is emitted during eruption, usually carbon dioxide (CO_2).
- Example: Lake Nyos (Cameroon, West Africa) is a volcanic crater lake. On August 21, 1986 it emitted CO2 which is heavier than air that moved like a river of smoke 164 feet thick and 110 miles long to engulf the towns of Nyos, Subum and Cha, killing more than 1,700 people and 3,000 cattle.

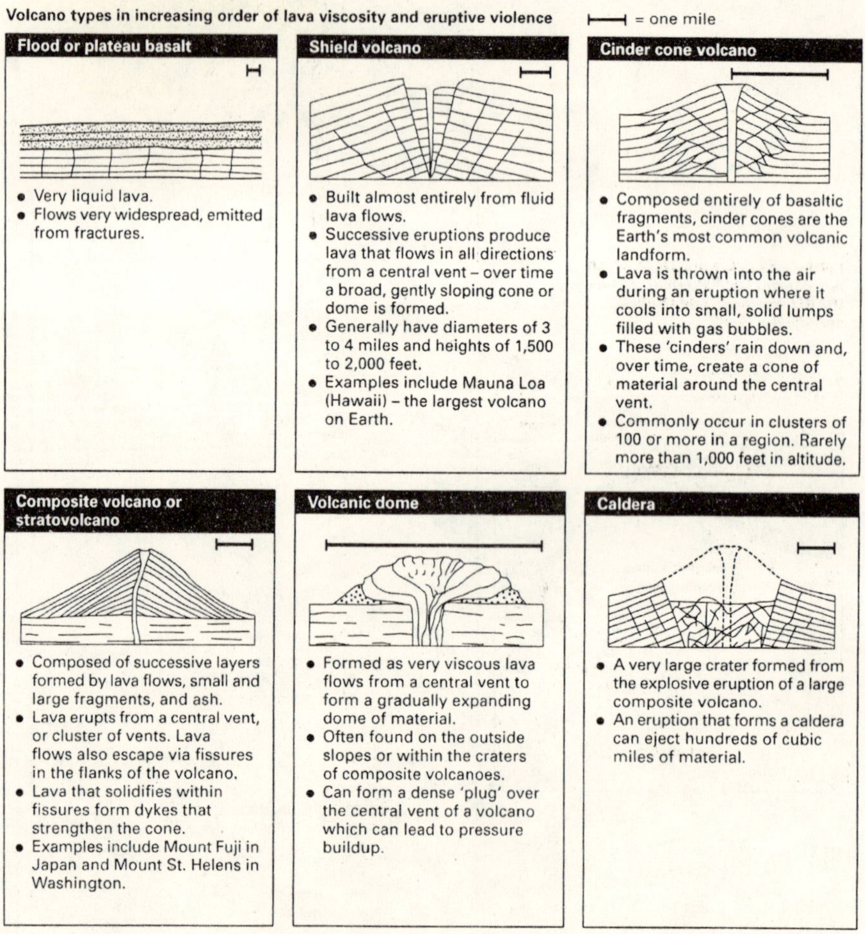

Hawaiian

- Commonly occur as lava fountains that generate lava rivers emerging from elongated fissures or tubular vents as roaring jets of gas, carrying liquid granules and blobs to form fountains reaching high in the sky.
- The jetlike incandescent gases stream rapidly upwards through the magma, tearing it apart into liquid blobs that partially or completely congeal while flying through the air.

Strombolian

- High, arcing, incandescent plume.
- Cinders and lava blobs rain down around a vent to construct cinder cones.

W

◻ world's most devastating earthquakes

Deadliest earthquakes on record (50,000 deaths or more) (skull) = 10,000 deaths

Date	Location	Deaths	Richter magnitude
Jan. 23 1556	Shansi, China	830,000	n.a.
July 27 1976	Tangshan, China	255,000[1]	8.0
Aug. 9 1138	Aleppo, Syria	230,000	n.a.
May 22 1927	near Xining, China	200,000	8.3
Dec. 22 856[2]	Damghan, Iran	200,000	n.a.
Dec. 16 1920	Gansu, China	200,000	8.6
March 23 893[2]	Ardabil, Iran	150,000	n.a.
Sept. 1 1923	Kwanto, Japan	143,000	8.3
Oct. 5 1948	Ashgabat, Turkmenistan	110,000	7.3
Dec. 28 1908	Messina, Italy	70,000–100,000[3]	7.5
Sept. 1290	Chihli, China	100,000	n.a.
Nov. 1667	Shemakha, Caucasia	80,000	n.a.
Nov. 18 1727	Tabriz, Iran	77,000	n.a.
Nov. 1 1755	Lisbon, Portugal	70,000	8.7
Dec. 25 1932	Gansu, China	70,000	7.6
May 31 1970	Peru	66,000	7.8
1268[4]	Silicia, Asia Minor	60,000	n.a.
Jan. 11 1693	Sicily, Italy	60,000	n.a.
May 30 1935	Quetta, Pakistan	30,000–60,000	7.5
Feb. 4 1783	Calabria, Italy	50,000	n.a.
June 20 1990	Iran	50,000	7.7

1 Official. Estimated death toll as high as 655,000.
2 Note that these dates are prior to 1000 AD. No digit is missing.
3 Estimated.
4 No month and day available.

Meterology

A

❐ air discharge

A form of lightning discharge probably similar to a cloud discharge in which the lightning channel propagates away from a cloud charge center into apparently clear air where it terminates. Thus, cloud charge is moved away from its original location and space charge of opposite sign outside the cloud may be neutralized.

❐ aircraft electrification

The accumulation of a net electrical charge on the surface of aircraft; or the separation of charge into two concentrations of opposite sign on distinct portions of the aircraft surface. Net charges appears as a result of autogenous electrification when aircraft fly through clouds of ice crystals or dust. Charge separation occurs by induction when aircraft fly through regions of strong atmospheric electrical field, as in thunderstorms. Charging may also occur by the engine exhaust carrying away a net charge leaving the aircraft charged. Development of large local charges on aircraft parts may lead to appearance of corona discharge or St. Elmo's fire and is almost always accompanied by poor radio communication due to so-called precipitation static.

❐ air--earth conduction current (also called fair-weather current)

That part of the air-earth current contributed by the electrical conduction of the atmosphere itself. It is represented as a downward current in storm-free regions all over the world. The conduction current is the largest portion of the air-earth current, far outweighing the contributions made by the precipitation current and convection current which are zero in storm-free regions. Its magnitude is approximately 3 x 10^-12+ amp/m^2+, or about 1800 amperes for the entire earth. Such observations of the vertical variation of the conduction current as have been made indicate that it is approximately uniform throughout the troposphere, a condition that is consistent with the generally accepted view that the conduction current flows from a positively charged conducting region in the lower ionosphere downward to the negatively charged earth. Only in areas of temporarily disturbed weather does the conduction current become replaced by reverse flow. Accumulating evidence points to the conclusion that the conduction current continues to exist only because of the action of thunderstorms scattered at all times over the earth, which supply the positive charge to the upper atmosphere and negative charge to the earth.

❐ air--earth current

The transfer of electric charge from the positively-charged atmosphere to the negatively-charged earth. This current is made up of the air-earth conduction current ,a point-discharge current, a precipitation current , a convection current , and miscellaneous smaller contributions. Of these, the first is by far the largest. The existence of this quasi-steady current in fair weather and the observed maintenance of the earth's net negative charge are both better established than the nature of the supply current which must replenish the positive charge in the upper atmosphere and the negative charge on the earth.

❐ arc discharge

A luminous electrical discharge in a gas where the current flows more of less continuously along a narrow channel of high ionization. An arc discharge requires a continuous source of electric potential difference across the terminals of the arc. This steady-state high current discharge is to be distinguished from the low current and visually diffuse corona discharge and point discharge, as well as from the transient, high luminosity, high-current spark discharge.

❐ atmospheric electric field

A quantitative term, denoting the electric field strength of the atmosphere at any specified point in space and time. In areas of fair weather, the atmospheric electric field near the earth's surface typically is about 100 volts per meter and is directed vertically in such a sense as to drive positive charges downward to the earth. In areas of fair weather this field decreases in magnitude with increasing altitude, falling, for example, to only about 5 volts per meter at an altitude of about 10 km. Near thunderstorms, and under clouds of vertical development, the surface electric field varies widely in magnitude and direction, usually reversing its direction immediately beneath active thunderstorms. In areas of minimum local disturbance, a characteristic diurnal variation of electric field strength is observed. This variation is characterized by a maximum which occurs at about 19 hr UTC for all points on the earth, and is now believed to be produced by thunderstorms which, for geographic regions, are more numerous for the world as a whole at that universal time than at any other. It is now believed that thunderstorms, by replenishing the negative charge to the earth's surface, provide the supply current to maintain the fair-weather electric field in spite of the continued flow of the air-earth current that tends to neutralize that field.

❐ atmospheric electricity - 1

Electrical phenomena, regarded collectively, which occur in the earth's atmosphere. These phenomena include not only such striking manifestations as lightning and St. Elmo's fire, but also less noticeable but more ubiquitous effects such as atmospheric ionization, the air-earth current, and other quiescent electrical processes. The existence of separated electric charges in the atmosphere is a consequence of many minor processes (spray electrification, dust electrification, etc.) and a few major processes (cosmic ray ionization, radiactive-particle ionization, and thunderstorm electrification). The details of thunderstorm charge separation are poorly understood at the present time, which is unfortunate in that the maintenance of the prevailing atmospheric electric field is now widely believed to be due to thunderstorm effects.

❐ atmospheric electricity - 2

The study of electrical processes occurring within the atmosphere.

❐ atmospherics

The radio frequency electromagnetic radiation originating, principally, in the irregular surges of charge in thunderstorm lightning discharges. Atmospherics are heard as a quasi-steady background of crackling noise (static)on certain radio receivers, such as AM radio. Since any acceleration of electric charge leads to emission of electromagnetic radiation, and since the several processes involved in propagation of lightning lead to very large charge accelerations, the lightning channel acts like a huge transmitter, sending out radiations with frequencies of the order of 10 kHz. Atmospherics may occasionally be detected at distances in excess of 2000 miles from their source. Advantage has been taken of this in using radio direction-finding equipment to plot cloud-to-ground lightning locations, to locate active thunderstorm areas in remote regions, and in-between weather reporting stations.

❐ autogenous electrification

The process by which net charge is built up on an object, such as an airplane, moving relative to air containing dust or ice crystals. The electrification is produced by frictional effects (triboelectriciation) accompanying contact between the object and the particulate matter.

B

❐ ball lightning

A relatively rarely seen form of lightning, generally consisting of an orange or reddish ball of the order of a few cm to 30 cm in diameter and of moderate luminosity, which may move up to 1 m/s horizontally with a lifetime of a second or two. Hissing noises sometimes emanate from such balls, and they can explode noisily or disappear noiselessly. The physical nature of ball lighting is not understood. Similar phenomena occasionally occur in high current industrial equipment, but the conditions for such production are not reproducible.

❐ beaded lightning

A particular visual variation of the end of a normal lightning flash where periodic sections of the channel appear to die out slowly because they have greater radius and hence lose heat more slowly, are seen end-on, or for other unknown reasons.

❐ bipolar pattern

A lightning pattern revealed in cloud-to-ground lightning locations in which areas of predominantly negative flash locations are separated horizontally from positive flash locations. This pattern has been identified in midlatitude mesoscale convective systems (MCS). The convective regions of the MCS (those with radar echoes 35 dBZ) typically have negative lightning flashes, whereas the positive flashes are typically located in the stratiform region (<35 dBZ) adjacent to the convective region. The bipolar pattern is hypothesized to arise from

(1) the advection of positively charged particles from the upper portions of the convective region to the stratiform region due to vertical wind shear;
(2) in situ charging of ice particles that acquire positive charge in an environment of super cooled liquid water and mesoscale ascent in the stratiform cloud layer; or
(3) some combination of the two processes.

❐ bishop wave

A striking example of an atmospheric lee wave, formed in the lee of the Sierra Nevada range near Bishop, California. The phenomenon includes a rotor cloud and a series of lenticular clouds parallel to the crest of the range.

❐ blue jet

A narrowly collimated beam of mostly blue, with some green, light that appears to propagate upward from the tops of thunderstorms. The typical blue jet is observed to appear from the apparent top of the anvil and propagate upwards, at speeds of about 100 km/s, in a narrow cone and to flare out as it reaches maximum altitude, approximately 40-50 km, such that it resembles a trumpet. Currently, there is no satisfactory theory of this phenomenon. One possible mechanism may be that jets are discharge following some sort of collimated quasi-straight ion trail emanating from the thunderstorm. Any mechanism producing blue jets will have to continue for at least 200 ms.

❐ blue jets

Weakly luminous upward propagating discharges, blue in color, emanating from the tops of thunderstorms. Following their emergence from the top of the thundercloud, they typically propagate upward in narrow cones of about 15 degrees full width at vertical speeds of roughly 100 km/s (Mach 300), fanning out and disappearing at heights of about 40-50 km. Their intensities are on the order of 800 kR near the base, decreasing to about 10 kR near the upper terminus. These correspond to an estimated optical energy of about 4 kJ, a total energy of about 30 MJ, and an energy density on the order of a few mJ/m ^ 3. Blue jets are not aligned with the local magnetic field.

❐ breakdown

The process by which electrically-stressed air is transformed from an insulator to a conductor. Breakdown involves the acceleration of electrons to ionization potential in the electric field imposed by the thundercloud, and the subsequent creation of new electrons which avalanche and expand the scale of enhanced conductivity. Breakdown precedes the development of lightning.

❐ breakdown field

The electric field necessary to produce breakdown.

❑ breaking-drop theory

A theory of thunderstorm charge separation based upon the suggested occurrence of the Lenard effect in thunderclouds, that is, the separation of electric charge due to the breakup of water drops. This theory, advance by Sir George C. Simpson in 1927, was initially intended to account for a bipolar charge distribution within a thundercloud having the main positive charge center near the base of the cloud and the main negative charge center higher up. Simpson's theory, however, does not explain this phenomenon well because temperatures are below the freezing point in this region of the cloud. Evidence does remain to support the weak positive charge center that lies slightly below the lower main negative charge concentration in many, if not all, thunderclouds where the temperature is above the freezing point. Hence the breaking-drop theory is best ascribed to this localized secondary positive charge center. The mechanism was also proposed as an explanation for waterfall electricity.

C

❑ chain lightning

Same as bead or beaded lightning.

❑ charge separation

The physical process causing cloud electrification. The process can include particle collisions with selective charge transfer and particle capture of small ions at the particle scale. The process can include gravity-driven differential particle motions and convective transport of charged air parcels at the cloud scale.

❑ clayden effect

When a photographic emulsion is given a very brief exposure to light of high intensity, it is desensitized towards a subsequent longer exposure to light of moderate intensity. That is, the second exposure produces less effect than if the pre-exposure had not been given. This phenomenon was observed originally by Clayden when photographing lightning flashes, but it can be produced equally well by any type of light source, provided the intensity is sufficiently high and the duration short enough.

❑ cloud electrification

The process by which clouds become electrified. This process separates positive and negative electric charge and develops potential differences occasionally sufficient to produce lightning.

❑ cloud flash

(also called cloud flash, intracloud flash, cloud-to-cloud flash) a lightning discharge occurring between a positively charged region and a negatively charged region, both of which may lie in the same cloud. The most frequent type of cloud discharge is one between a main positive charged region and a main negative charged region. Cloud flashes tend to outnumber cloud-to-ground flashes. In general, the channel of a cloud flash will be wholly surrounded by cloud. Hence the channel's luminosity typically produces a diffuse glow when seen from outside the cloud and this widespread glow is called sheet lightning.

❐ cloud-to-cloud flash

A type of cloud flash. A lightning discharge that occurs between two electrified clouds.

❐ cloud-to-ground flash

A lightning flash occurring between a charge center (either negative or positive) in the cloud and the ground. Negative discharges are about 10 times more common than positive. The cloud-to-ground discharge is initiated by a downward propagating leader followed by an extremely intense upward-moving return stroke that is the principal source of luminosity. A cloud-to-ground discharge is usually a composite event, called a flash, composed of several distinct leader-return-stroke sequences, the majority using the same channel following each other at intervals of a few hundredths of a second (see return stroke, stepped leader dart leader).

❐ cloud-to-ground flash

A lightning flash occurring between a charge center in the cloud and the ground. On an annual basis, negative charge is lowered to ground in about 95% of the flashes, the remaining flashes lowering positive charge to ground. This type of lightning flash, which can be contrasted with an intracloud flash or cloud-to-cloud flash, consists of one or more return strokes. The first stroke begins with a stepped leader followed by an intense return stroke which is the principal source of luminosity and charge transfer. Subsequent strokes begin with a dart leader followed by another return stroke. Most of the strokes use the same channel to ground. The time interval between strokes is typically 40 milliseconds.

❐ columnar resistance

In atmospheric electricity, the electrical resistance of a column of air one meter square, extending from the earth's surface to some specified altitude. Measurements extending to an altitude of 18 km indicate that the atmospheric columnar resistance to that height amounts to about 10^17 ohm per m^2. Probably, this is only slightly less than the total columnar resistance from earth to ionosphere. In fact, roughly half of the total columnar resistance from earth to 18 km is contributed by the lowest 3 km of the column where, in addition to the greater density of the air, the high concentration of atmospheric particulates leads to a relatively high population of poorly conducting large ions rather than the more mobile small ions. Total columnar resistance does not vary greatly with either time or locality. In contrast to this is the wide variation in columnar resistance of the lowest fraction of a kilometer. It is chiefly because of the latter variations that the sea-level atmospheric electric field fluctuates, especially in industrial areas of highly variable atmospheric pollution.

❐ conduction current

The migration of charged particles in a gaseous medium acted upon by an external electric field.

❒ conductivity

A unit measure of electrical conduction; the facility with which a substance conducts electricity, as represented by the current density per unit electrical potential gradient in the direction of flow. Electrical conductivity is the reciprocal of electrical resistivity and is expressed in units such as mhos (reciprocal ohms) per meter. It is an intrinsic property of a given type of material under given physical conditions (dependent mostly upon temperature.)

❒ conductivity current

same as air-earth conduction current.

❒ continuing current

A sustained current in the lightning stroke that flows to the ground after the return stroke. Continuing currents can have a duration in excess of 100 ms with magnitudes of typically 100 amperes. Continuing currents occur in negative and positive cloud-to-ground flashes.

❒ continuing current

A sustained current in the lightning stroke that flows to ground after the return stroke. Continuing currents can have durations in excess of 100 milliseconds with magnitudes of typically 100 amperes. Continuing currents occur in negative and positive cloud-to-ground lightning flashes.

❒ continuous leader

Same as dart leader.

❒ convection

In atmospheric electricity, a process of vertical charge transfer by transport of air containing a net space charge, or by motion of other media (e.g., rain) carrying net charge. Eddy diffusion of air containing a net charge gradient may also yield a convection current.

❒ convection current

Any net transport of electric charge effected through mass motions of some charged medium; any electric current induced by other than electrical forces. In atmospheric electricity, the convection current is part of the air-earth current of charge transfer vertically between the earth's surface and the upper atmosphere. The term includes not only eddy diffusion currents existing in regions of net space charge but also currents due to fall of charged precipitation particles (precipitation current).

❒ corona current

The electrical current that is equivalent to the rate of charge transferred to the air from a pointed object (or array of objects) experiencing corona discharge. Ordinarily, the corona current form terrestrial objects at times of thunderstorm passage constitutes a transfer of negative charge from air to object.

❐ corona discharge

A luminous, and often audible, electric discharge that is intermediate in nature between a spark discharge (with, usually, its single discharge channel) and a point discharge (with its diffuse, quiescent, and non-luminous character). It occurs from objects, especially pointed ones, when the electric field strength near their surfaces attains a value near 100,000 volts per m. Aircraft flying through active electrical storms often develop corona discharge streamers from antennas and propellers, and even from the entire fuselage and wing structure. So-called precipitation static results. It is seen also, during stormy weather, emanating from the yards and masts of ships at sea.

❐ current

Any movement of electric charge in space, by virtue of which a net transport of charge occurs as, for example (in atmospheric electricity), in a conduction current , convection current , an precipitation current .

D

❐ dark lightning

The present definition for dark lightning is ok.

❐ dart leader

The leader which, after the first stroke, typically initiates each succeeding stroke of a multiple-stroke flash lightning. (The first stroke is initiated by a stepped leader.) The dart leader derives its name from its appearance on photographs taken with streak cameras. The dart leader's brightest luminosity is at its tip which is tens of meters in length, propagating downward at about 10 ^ 7 m/s. In contrast to stepped leaders, dart leaders do not typically exhibit branching because the pre-established channel's low gas density and residual ionization provide a more favorable path for this leader than do any alternative ones.

❐ dielectric strength

A measure of the resistance of a dielectric to electrical breakdown under the influence of strong electric fields; usually expressed in volts per meter. The dielectric strength of dry air at sea-level pressures is about 3,000,000 volts per meter. The exact value for air depends upon geometry of the electrodes between which the electric field is established, upon the humidity, and upon whether or not water drops are present in the air.

❐ direction finder

An instrument consisting of two orthogonal magnetic loop antennas and associated electronics for the purpose of detecting azimuth to a cloud-to-ground lightning flash.

❐ dissipation constant

In atmospheric electricity, a measure of the rate at which a given electrically charged object loses its charge to the surrounding air. If the object bears a charge [s] at time [s], then according to a law established by Coulomb, IFI, where [s] is the object's dissipation constant expressed in reciprocal time units. It is found that [s] depends not only upon the geometry of the charged object but also upon the density of the surrounding air, its humidity, and its motion relative to the object.

▫ double layer

On the surface of a substance, a layer of electric dipoles whose axes have an average orientation normal to the surface. Double layers may appear on interfaces of solid and gas, liquid and gas, liquid and liquid, etc. They arise whenever media with different electron affinities (forces of attraction, or work function) are contiguous, and if dipoles are available. A net potential difference, the electrokinetic potential exists across a double layer.

▫ dust-devil effect

In atmospheric electricity, a rather sudden and short-lived change of the vertical component of the atmospheric electric field that accompanies passage of a dust devil near an instrument sensitive to the vertical gradient. Such changes may be either positive or negative and the charge is probably produced by triboelectrification .

▫ dynamo theory

The hypothesis, first proposed by Balfour Stewart, which explains the regular daily variations in the earth's magnetic field in terms of electrical currents in the lower ionosphere, generated by tidal motions of the ionized air across the earth's magnetic field.

E

▫ earth current

A large-scale surge of electric charge within the conductive earth, associated with a disturbance of the ionosphere. Current patterns of quasi-circular form and extending over areas the size of whole continents have been identified and are known to be closely related to solar-induced variations in the extreme upper atmosphere.

▫ earth--air current

see air-earth conduction current.

▫ earth-current storm

Irregular fluctuations in an earth current, often associated with electric field strengths as large as several volts per kilometer, in the earth's crust, superimposed on the normal diurnal variation of the earth currents. Such storms are closely related to magnetic storms.

▫ earth--ionospheric waveguide

The physical structure formed by the conductive Earth and the conductive lower ionosphere (D region) which together sandwich the highly resistive atmosphere between them. This wave guide sustains global electromagnetic wave propagation from the lower end of the Schumann resonance band (5 Hz) to the wave guide cutoff frequency near 1500 Hz.

❐ electric charge

A fundamental property of matter exhibiting two states, positive and negative, that results in the action of electric forces in the presence of an electric field. These two states were identified and named by Benjamin Franklin. The positive charge on the proton and the negative charge on the electron represent the fundamental charge, 1.6 x 10 ^ -19 Coulombs.

❐ electric discharge

The flow of electricity through a gas, resulting in the emission of radiation which is characteristic of the gas and of the intensity of the current.

❐ electric field strength

The electrical force exerted on a unit positive charge placed at a given point in space. The electric field strength is expressed, in the mks system of electrical units, in terms of volts per m and is a vector quantity. The electric field strength of the atmosphere is commonly referred to as the atmospheric electric field .

❐ electric potential

In electrostatics, the work done in moving unit positive charge from infinity to the point whose potential is being specified.

❐ electrical breakdown

(Term Reference: electrical breakdown or, simply, breakdown). The sudden decrease of resistivity of a substance when the applied electric field strength rises above a certain threshold value (the substance's dielectric strength). For air at normal pressures and temperatures, experiment has shown that the breakdown process occurs at a field strength of about 3,000,000 volts per m. This value decreases approximately linearly with pressure, and is dependent upon humidity and traces of foreign gases. In the region of high field strength just ahead of an actively growing leader in a lightning stroke, breakdown occurs in the form of a rapidly moving wave of sudden ionization (electron avalanche). The dielectric strength in a cloud of water drops is less than that in cloud-free humid air, for the water drops are distorted as a result of the Macky effect.

❐ electrical storm

Name sometimes applied to a relatively rare condition of disturbed atmospheric electric field in the lower atmosphere that arises when strong winds are blowing and much dust is in the air, but involving no thunderstorm activity. Triboelectrification due to the blowing dust may charge fences and other metallic objects to such an extent that slight shocks are felt upon touch.

❐ electrode effect

The accumulation of an excess of ions of positive sign in the neighborhood of a negative electrode, and vice versa, when ions are continuously produced in the space above the electrode and move under the influence of the electrode's field. The electrode effect is observed over water surfaces where radioactivity is not present.

▫ electron avalanche

The process in which a relatively small number of free electrons in a gas that is subjected to a strong electric field accelerate, ionize gas atoms by collision, and thus form new electrons to undergo the same process in cumulative fashion. All streamers in a lightning discharge propagate by formation of electron avalanches in the regions of high electric field strength that move ahead of their advancing tips. Particularly in the case of the intense return streamer , avalanche processes are enhanced by formation of photoelectrons as a result of ultraviolet radiation emitted by the excited molecules in the region just behind the tip. An avalanche cannot possibly begin until the local electric field strength is high enough to accelerate a free electron to the minimum ionizing speed in the space and time interval corresponding to one mean free path of the electron, for upon collision, the electron usually loss its forward motion in the direction of the field. Maintenance of an avalanche requires a large reservoir of charge, such as accumulates more or less periodically in active thunderstorms.

▫ electrosphere

A layer beginning a few tens of kilometers above the surface of the earth and extending to the ionosphere in which the electrical conductivity is so high that the layer is essentially at a constant electric potential. This layer is called the electrosphere.

▫ elve

Transient laterally extensive illumination of the airglow layer, about 90 km, over thunderstorms and associated with the electromagnetic pulse from the return stroke of a lightning flash to ground.

▫ eoso

End-of-Storm-Oscillation; a characteristic in the electric field signature at the ground near the end of a thunderstorm. The electric field changes from upward-directed (foul-weather-polarity) to downward-directed (fair-weather-polarity) and often retains the latter polarity for several tens of minutes before returning momentarily to foul-weather-polarity. Positive ground flashes and spider lightning are occasionally observed during the period of reversed field polarity. The origin of the EOSO is only partly understood.

▫ exogenous electrification

The separation of electric charge in a conductor placed in a preexisting electric field. This is especially applied to the charge separation observed on metal-covered aircraft. It is the result of induction effects, and does not by itself create any net total charge on the conductor. It is to be distinguished, therefore, from autogenous electrification.

F

▫ fair-weather current

same as air-earth conduction current.

▢ fair-weather electricity

The distribution of ions and currents in the atmosphere and at the surface of the earth that occur during "Fair Weather," or in areas where there is no thunderstorm activity. This distribution sets up a downward directed electric field referred to as a fair-weather field.

▢ feawp

A characteristic feature in electric field records directly beneath a thundercloud in which the electric field momentarily reverses polarity during a burst of intense precipitation. The time scale for these events is of the order of minutes and of shorter duration than the EOSO. The physical origins of the FEAWP are still unresolved.

▢ field changes

The rapid variations of the electrical field at the earth's surface, beneath, within and above thunderclouds. Used to determine quantitative estimates of the charge transferred during a lightning discharge, heights of the charge centers, and many other features of thunderclouds.

▢ first-hop wave

The first ionospheric reflection from a lightning stroke which arrives after the original ground-wave.

▢ forked lightning

The common form of cloud-to-ground discharge always visually present to a greater or lesser degree which exhibits downward-directed branches from the main lightning channel. In general, of the many branches of the stepped leader only one is connected to ground defining the primary, bright return stroke path, and the other incomplete channels decay after the ascent of the first return stroke.

G

▢ global circuit

The structure and combination of processes set up by the conductive Earth, the conductive ionosphere, and all agents of electrification within the resistive troposphere. The so-called 'DC' global circuit is characterized by the ionosphere potential and the 'AC' global circuit refers to the Schumann resonances within the Earth-ionosphere wave guide.

▢ ground streamer

An upward advancing column of high ionization (a streamer or arc) which typically ascends from a point on the earth's surface toward a descending stepped leader. The ground streamer usually joins the stepped leader about fifty meters above the ground, after which the upward-propagating light and current of the return stroke begin. Ground streamers occur because of the very high electric field intensities that build up directly below the descending, charged stepped leader.

Often, more than one ground streamer starts up from the general area under a descending leader, but usually only one makes contact with the leader.

❐ ground-to-cloud discharge

A lightning discharge in which the original leader process starts upward from some object on the ground; the opposite of the more common cloud-to-ground discharge. Ground-to-cloud discharges most frequently emanate from very tall structures which, being equipotential with the earth, can exhibit the strong field intensities near their upper extremities necessary to initiate leaders.

H

❐ heat lightning

Non-technically, the luminosity observed from ordinary lightning too far away for its thunder to be heard. Since such observations have often been made with clear skies overhead, and since hot summer evenings particularly favor this type of observation, there has arisen a popular misconception that the presence of diffuse flashes in the apparent absence of thunderclouds implies that lightning is somehow occurring in the atmosphere merely as a result of excessive heat.

❐ hydrometeor charge

The net charge, for example, on a raindrop or a water drop.

I

❐ igneous meteor

In United States weather observing practice, a visible electrical discharge in the atmosphere. Lightning is the most common and important type, but types of corona discharge are also included.

❐ induction charging mechanism

A physical process for particle charging involving the collision of pairs of particles in an ambient electric field. Electric charge induced on the particle surfaces by the ambient electric field is made available for transfer when the two particles come into contact. Subsequent differential particle motions under gravity is postulated to result in large scale charge separation. The specific role of induction charging in the electrification of thunderclouds has not been resolved.

❐ intracloud flash

A lightning discharge occurring between a positive charge center and a negative charge center, both of which lie in the same cloud. Starts most frequently in the region of the high electric field between the upper positive and lower negative space charge regions. In summer thunderstorms, intracloud flashes precede the occurrence of cloud-to-ground flashes; they also outnumber cloud-to-ground flashes. Intracloud lightning develops bidirectionally like a two-ended tree: one end of the tree is a branching negative leader, the other is a branching positive leader. Later in the flash, fast negative leaders similar to dart leaders (also called K-changes) appear in the positive end region and propagate toward the flash origin. = In weather observing, this type of discharge is often mistaken for a cloud-to-cloud flash , but the latter term should be restricted to true intercloud discharges, which are far less common than intracloud discharges. Cloud discharges tend to outnumber cloud-to-ground discharges in semiarid regions where the bases of thunderclouds may be several kilometers above the earth's surface. In general, the channel of a cloud flash will be wholly surrounded by cloud. Hence the channel's luminosity typically produces a diffuse glow when seen from outside the cloud, and this widespread glow is called sheet lightning .

❑ ion

In atmospheric electricity, any of several types of electrically charged submicroscopic particles normally found in the atmosphere. Atmospheric ions are of two principal types, small ions an large ions , although a class of intermediate ions has occasionally been reported. The ionization process which forms small ions depends upon two distinct agencies, cosmic rays and radioactive emanations. Each of these consists of very energetic particles which ionize neutral air molecules by knocking out one or more planetary electrons. The resulting free electron and positively charged molecule (or atom) very quickly attach themselves to one or, at most, a small number of neutral air molecules, thereby forming new small ions. In the presence of Aitken nuclei , some of the small ions will in turn attach themselves to these nuclei, thereby creating new large ions. The two main classes of ions differ widely in mobility. Only the highly mobile small ions contribute significantly to the electrical conductivity of the air under most conditions. The intermediate ions and large ions are important in certain space charge effects, but are too sluggish to contribute much to conductivity. The processes of formation of ions are offset by certain processes of destruction of the ions.

❑ ion mobility

In gaseous electric conduction, the average velocity with which a given ion drifts through a specified gas under the influence of an electric field of unit strength. Mobilities are commonly expressed in units of m/sec per volt/m. In a vacuum, a single gaseous ion subjected to any non-zero potential gradient would accelerate indefinitely; but in the midst of a gas the ion continually experiences collisions with gas molecules. These encounters tend to break up its trajectory into a series of short intervals of acceleration punctuated by deflections. The net result is that the ion's gross motion resembles drift at a uniform velocity. The mobility depends not only upon the nature of the ion and gas but also upon the density of the gas, for the latter controls the mean free path of the ion. In atmospheric electricity, the mobilities of small and large ions weight their relative importance in atmospheric conduction. Small ions have mobilities of about 1.3 x 10[-4] m/sec per volt/m in air at sea level with negative small ions exhibiting slightly greater values than do the positive small ions. High humidities suppress small ion mobilities slightly. Large ions have mobilities of only about 4 x 10[-7] m/sec per volt/m at sea level, their sluggishness being due to their great mass.

❐ ion-capture theory

A theory of thunderstorm charge separation advanced by C. T. R. Wilson. According to this theory, the lower negative charge of a thundercloud is generated by the accumulation there of raindrops which have captured predominantly negative ions in their descent through the cloud. The preferential capture of negative ions by such drops is said to be due to the polarization of the drops in the normal atmospheric electric field existing between the negatively charged earth and positively charged ionosphere. The lower halves of the falling drops therefore would attract and capture negative charges while their upper halves would be unable to draw in positive charges with comparable efficiency; so a net negative charge builds up on the drops. This theory is generally regarded today as incapable of accounting for any important portion of thunderstorm charge separation, for it is quantitatively inadequate in view of typical ion densities.

❐ ionic conduction

Any electrical conduction where the current is sustained by the motion of ions (as opposed to electrons) within the conductor. All electrical conduction in the atmosphere is of this type.

❐ ionization

In atmospheric electricity, the process by which neutral atmospheric molecules (small ions) or other suspended particles (mainly large ions) are rendered electrically charged chiefly by collisions with high-energy particles. Cosmic rays and emanations from radioactive gases are the main sources of atmospheric ionization. In the lower atmosphere, decay electrons of mu-mesons plus alpha particles from radioactive gases, as well as beta particles and gamma rays, serve to ionize air molecules. The rate at which these agencies ionize the air is expressed in units of one ion pair per cubic centimeter per second, symbolized by I. Cosmic rays at sea level yield about 2 I, both over land and at sea. Radioactive gases contribute about 5 I over land areas at sea level, while radioactive materials in the soil and rocks themselves yield about 4 I. At heights above about 5 km, only cosmic rays provide significant ionization, and this contribution finally reaches a maximum at about 13 km, above which the rate decreases, due to decreasing air density and consequent lack of target molecules for the cosmic rays.

❐ ionospheric potential

The potential difference between the conductive Earth and the conductive upper atmosphere, whose value is approximately +250 kilovolts. This voltage difference is maintained by worldwide cloud electrification. Most of the 250 kV is realized well below the formal ionospheric height because of the rapid increase of electrical conductivity with height in the atmosphere.

J

❐ junction streamer

The process by which negative charge centers at successively more distant locations in a thundercloud are "tapped" for discharge by successive strokes of cloud-to-ground lightning.

K

▫ k changes

The K-process is generally viewed as a "recoil streamer" or small return stroke that occurs when a propagating discharge within the cloud encounters a pocket of charge opposite to its own. In this view, the J-process represents a slowly propagating discharge that initiates the K-process. This is the case for K-changes in cloud discharges. It is reasonable to expect that cloud discharge K-changes are similar to the in-cloud portion of ground discharges.

L

▫ leader

The electric discharge which initiates each return stroke in a cloud-to-ground lightning discharge. It is a channel of high ionization which propagates through the air by virtue of the electric breakdown at its front produced by the charge it lowers. The stepped leader initiates the first stroke in a cloud-to-ground flash and establishes the channel for most subsequent strokes of a lightning discharge. The dart leader initiates most subsequent strokes. Dart-stepped leaders begin as dart leaders and end as stepped leaders. The initiating processes in cloud discharges are sometimes also called leaders but their properties are not well measured.

▫ lenard effect

The separation of electric charges accompanying the aerodynamic breakup of water drops, first studied systematically by the German physicist P. {Lenard}. Experiments have shown that the degree of charge separation in spray processes depends upon the drop temperature, presence of dissolved impurities, speed of the impinging air blast, and contact with foreign surfaces. The largest fragments of the broken drops are observed to carry positive changes and the fine spray of drops carried off in the impinging air current carries a net negative charge. Distilled water drops of 4 mm diameter, broken after a five centimeter free fall into an updraft of 1 m/sec, were found by {Chapman} to yield about 10[-10] coulombs of separated charge per drop. The Lenard effect was incorporated by {Simpson} into his breaking-drop theory of thunderstorm charge generation, but many critical details are but poorly understood.

▫ lightning

Lightning is a transient, high-current electric discharge whose path length is measured in kilometers. The most common sources of lightning is the electric charge separated in ordinary thunderstorm clouds (cumulonimbus). Well over half of all lightning discharges occur within the thunderstorm cloud and are called intracloud discharges. The usual cloud-to-ground lightning (sometimes called streaked or forked lightning) has been studied more extensively than other lightning forms because of its practical interest (i.e., as the cause of injuries and death, disturbances in power and communicating systems, and the ignition of forest fires) and because lightning channels below cloud level are more easily photographed and studied with optical instruments. Cloud-to-cloud and cloud-to-air discharges are less common than intracloud or cloud-to-ground lightning. All discharges other than clout-to-ground are often lumped together and called cloud discharges. Lightning is a self-propagating and electrodeless atmospheric discharge that transfers through the induction process the electrical energy of an electrified cloud into electrical charges and current in its ionized and thus conducting channel. Positive and negative leaders are essential components of the lightning. Only when a leader reaches the ground, the ground potential wave (return stroke) affects the lightning process. Natural lightning starts as a bi-directional leader although at different stages of the process uni-directional leader development can occur. Artificially triggered lightning starts on a tall structure or from a rocket with a trailing wire. Most of the lightning energy goes into heat, with smaller amounts transformed into sonic energy (thunder), radiation, and light. (See also cloud-to-ground, intracloud, and air discharges) Lightning, in its various forms, is known by many names such as the common streak lightning , forked lightning , sheet lightning , heat lightning , and the less common air discharge ; also, the rare and mysterious ball lightning and rocket lightning . (For some detailed explanation of lightning processes, see lightning discharge and related terms.) An important effect of world-wide lightning activity is the net transfer of negative charge from the atmosphere to the earth. This fact is of great important in one problem of atmospheric electricity, the question of the source of the supply current . Existing evidence suggests that lightning discharges occurring sporadically at all times in various parts of the earth, perhaps 100 per second, may be the principal source of negative charge that maintains the earth-ionosphere potential difference of several hundred thousand volts in spite of the steady transfer of charge produced by the air-earth current . However, there also is evidence that point discharge currents may contribute to this more significantly than lightning.

❑ lightning channel

The irregular path through the air along which a lightning discharge occurs. The lightning channel is established at the start of a discharge by the growth of a leader, which seeks out a path of least resistance between a charge source and the ground or between two charge centers of opposite sign in the thundercloud or between a cloud charge center and the surrounding air or between charge centers in adjacent clouds.

❑ lightning current

the current flowing in a component of the lightning flash. It is usually considered to be the current in the return stroke.

❑ lightning discharge

The series of electrical processes taking place within one second by which charge is transferred along a discharge channel between electric charge centers of opposite sign within a thundercloud (intracloud discharge) between a cloud charge center and the earth's surface (cloud-to-ground discharge or ground-to-cloud discharge), within two different clouds (intercloud or cloud-to-cloud discharge), or between a cloud charge and the air (air discharge). It is a very large-scale form of the common spark discharge. A single lightning discharge is called a lightning flash.

❒ lightning flash

The total observed lightning discharge generally has a duration less than one second. A single flash is usually composed of many distinct luminous events that often occur in such rapid succession that the human eye cannot resolve them.

❒ lightning mapping system

A network of lightning detection equipment for locating the electromagnetic sources of a lightning flash. The flash, both intracloud and cloud-to-ground, is mapped in three-dimensional space using equipment with a time resolution of less than a microsecond.

❒ lightning stroke

In a cloud-to-ground discharge, a leader plus its subsequent return stroke. In a typical case, a cloud-to-ground discharge is made up of three or four successive lightning strokes, most following he same lightning channel.

❒ lightning suppression

Procedures to prevent the occurrence of lightning. Seeding below cloud base with 10-cm fiber chaff in a Colorado study resulted in corona discharges that caused a discharging current to flow within developing or active thunderstorms. Electric fields below thunderstorms seeded with chaff decayed much faster than electric fields below nonseeded storms, and chaff seeding of existing thunderstorms greatly reduced cloud-to-ground flashes in non-seeded storms. Recent evidence suggests that chaff releases may result in a significant decrease in downwind cloud-to-ground lightning. Another experimental approach is to use lasers to discharge lightning in an overhead cloud in order to divert the flash from striking people or highly sensitive equipment on the ground; more research is needed to make this a realistic method of lightning suppression. In the 1960's, seeding with silveer iodide was considered in order to produce an excess of ice crystals to cause numerous coronal discharges within the thunderstorm and reduce the need for the flash to reach the ground, but the test results were complex and difficult to identify. Finally, electric space charge was released into the atmosphere from a network of high voltage wires on the ground to produce corona discharge, but a field test showed minimal effects on suppressing lightning.

M

❒ m component

M-component is the name given to an increase in channel luminosity accompanied by a rapid electric field variation, itself called a M-electric field changes. M-components occur when the channel is already faintly luminous. Downward-moving leaders have not been observed to precede M-components. M-components may be confused with branch components, the increases in channel luminosity that occur between each branch and ground when the upward-propagating return stroke reaches that branch, since the higher branches are obscured by the cloud. The M-component is a minor surge of current that re-illuminates the channel in a negative cloud-to-ground flash. It may occur within microseconds or up to a few milliseconds of a return stroke. Evidences exists that M-components are a result of the sequence of fast in-cloud negative leader (K-changes) contacting a conducting ground channel and renewed ground potential wave that re-illuminates the channel.

❒ magnetic storm

A worldwide disturbance of the earth's magnetic field. Magnetic storms are frequently characterized by a sudden onset, in which the magnetic field undergoes marked changes in the course of an hour or less, followed by a very gradual return to normalcy, which may take several days. Magnetic storms are caused by solar disturbances, though the exact nature of the link between solar and terrestrial disturbances is not understood. They are more frequent during years of high sunspot number. Sometimes a magnetic storm can be linked to a particular solar disturbance. In these cases, the time between solar flare and onset of the magnetic storm is about one or two days, suggesting that the disturbance is carried to the earth by a cloud of particles thrown out by the sun. When these disturbances are observable only in the auroral zones, they may be termed polar magnetic storms.

❒ mobility

Quantitative measure of the velocity of ions in an electric field of unit strength. The mobility can be expressed as Dmu+3Dv/E where v is the velocity of the ion and E is the electric field strength. The units are m ^ 2 Volts ^ (-1) s ^ (-1). Heavier ions will have lower mobilities than lighter ones and values of mobility vary inversely with density producing slower mobilities near the surface.

N

❒ negative cloud-to-ground lightning

A lightning flash or stroke between a cloud and ground that lowers negative charge to ground.

❒ negative rain

Rain which exhibits a net negative electrical charge.

❒ noninductive charging mechanism

A charging mechanism whereby the electric field increase is independent of the existing electric field.

O

❒ ohmic current

A current that is proportional to the voltage and inversely proportional to the resistance.

P

❑ particle charge

The electric charge, associated with a particle, e.g. a rain drop, ice crystal, hail stone, or aerosol.

❑ peak current

Usually refers to the maximum current in a lightning return stroke.

❑ pilot streamer

A relatively slow-moving, non-luminous lightning streamer, the existence of which has been postulated but not verified, to provide a physical explanation for the observed intermittent mode of advance of a stepped leader as it initiates a cloud-to-ground lightning discharge. Whereas the stepped leader descends at an average speed of the order of 10^5 m/s during its downward motion, it advances only about fifty meters at a time with higher speed and then pauses for fifty to one hundred microseconds before resuming its downward movement. The average downward speed has been associated with an invisible streamer, the pilot streamer, which is postulated to descend at a uniform speed only slightly in excess of ionizing speed of electrons in air and lay down a trail of weak residual ionization along which the stepped leader moves very rapidly in a pulsating manner. The idea of a pilot leader has been supplanted by more modern theory based on laboratory measurements of long spark generation.

❑ point discharge

A silent, non-luminous, gaseous electrical discharge from a pointed conductor maintained at a potential which differs from that of the surrounding gas. In the atmosphere, trees and other grounded objects with points and protuberances may, in disturbed weather, be sources of point discharge current . Close to a pointed and grounded conductor that extends above surrounding objects, the local electric field strength may be many times greater than that existing at the same level far from the elevated conductor. When this local field reaches such a value that a free electron, finding itself acted upon by this field, can be accelerated (in one mean free path) to a sufficiently high velocity to ionize neutral air molecules, point discharge will begin. Different structures will yield point discharge under quite different gross field conditions, for geometry is critically important. Point discharge is recognized as a major process of charge transfer between electrified clouds and Earth, and is a leading item in the charge balance of the global electrical circuit.

❑ point discharge current

The electrical current accompanying any specified source of point discharge . In the electrical budget of the earth-atmosphere system, point discharge currents are of considerable significance as a major component of the supply current . Estimates made by B. F. J. {Schonland} of the point discharge current from trees in arid Southwest Africa suggest that this process accounts for about twenty times as much delivery of negative charge to the earth as do lightning discharges during typical thunderstorms. Although the great height of thundercloud bases in arid regions, such as that referred to in Schonland 's study, tends to favor point discharge over lightning charge transfer, point discharge still seems more significant than lightning even in England, where {Wormell} found for Cambridge a ratio of about five to one in favor of point discharge over lightning charge transfer.

❑ polarization

With respect to particles in an electric field, the displacement of the charge centers within a particle in response to the electric force acting thereon.

❑ positive cloud-to-ground lightning

a positive cloud-to-ground lightning flash lowers positive charge from the cloud to the ground.

❑ positive discharge

a positive discharge lowers positive charge to ground via a lightning flash. The flash may be initiated in the cloud or from the ground.

❑ positive ground flash

a positive ground flash lowers positive charge to ground from the cloud above.

❑ positive rain

Rain having a net positive electrical charge.

❑ potential gradient

In general, the local space rate of change of any potential , as the gravitational potential gradient or the velocity potential gradient. In atmospheric electricity, the electric potential gradient (electric field strength) of the atmosphere is commonly referred to as the atmospheric electric field. The electric field is the negative of the potential gradient.

❑ precipitation current

The downward transport of charge, from cloud region to earth, that occurs in a fall of electrically charged rain or other hydrometeors; a particular case of a convection current. Observations of the charge on individual raindrops during thunderstorms have revealed a complex picture. On the average, more positive than negative charge is brought to earth by precipitation currents, but wide deviations occur both within individual storms and from one storm to another. The reasons for these wide fluctuations are not understood. Precipitation currents in continuous rain generally vary from about 10^[-12] to 10^[-10] amp/(m^2), while thunderstorm currents become as large as 10^[-8] amp/(m^2).

❑ precipitation electricity - 1

That branch of the study of atmospheric electricity concerned with the electrical carried by precipitation particles and with the manner in which these charges are acquired.

❑ precipitation electricity - 2

The electrical charge borne by precipitation particles. A very complex and highly variable picture is obtained when charges are measured on individual raindrops or snow crystals and no present theory approaches a complete explanation of all details. In general, more raindrops are positively than negatively charged. Sometimes the prevailing sign of the charges even shifts in the course of a given storm's lifetime.

Q

❑ q burst

An electromagnetic transient launched by a lightning discharge of large amplitude within the Earth-ionosphere cavity. Originally named by Toshio Ogawa, the 'Q' connotes 'Quiet' and characterizes a transient in which the fundamental 8 Hz mode of the Schumann resonances is the dominant contributor. In general, a mix of Schumann modes makes up a Q-burst.

❑ q noise

Quasi-continuous sequences of VHF (Very High Frequency) radiation emanating from lightning that is associated with high-speed K-change activity along pre-ionized lightning channels.

R

❑ radioactive decay

The primary source of ions in the lowest one or two kilometers of the atmosphere above land (cosmic radiation is dominant over the oceans and at higher altitudes). The ionizing radiation can take the form of (1) emission of [alpha], [beta], or [gamma] radiation by radioactive materials in the soil or (2) from gaseous radioactive daughter products (radon) that emanate from the soil. The [alpha] radiation emitted by materials in the ground is absorbed in the lowest few centimeters of the atmosphere, [beta] radiation penetrates a few meters and [gamma] radiation a few hundred meters into the atmosphere. The ionization produced by radioactive gases in the atmosphere is highly variable and depends on the rate of emission from the soil and also on atmospheric dispersion. [^ 222 Rn], which has a half-life of 3.8 days, is produced by the uranium decay series. [^220 Rn] (thoron) is produced by the thorium decay series and has a half-life of 54 seconds.

❑ radius of protection

The radius of the circle within which a lightning discharge will not strike due to presence of an elevated lightning rod at the center. A rule of thumb is that this radius is equal to the height of the rod, though cases of lightning damage inside this distance are well established.

❑ return stroke

The intense luminosity which propagates upward from earth to cloud base in the last phase of each lightning stroke of a cloud-to-ground discharge. In a typical flash, the first return stroke ascends as soon as the descending stepped leader makes electrical contact with the earth, often aided by short ascending ground streamers . The second and all subsequent return strokes differ only in that they are initiated by a dart leader and not a stepped leader. It is the return stroke which produces almost all of the luminosity and charge transfer in most cloud-to-ground strokes. Its great speed of ascent (about 1 X 10 ^ 8 m/sec) is made possible by residual ionization of the lightning channel remaining from passage of the immediately preceding leader, and this speed is enhanced by the convergent nature of the electrical field in which channel electrons are drawn down toward the ascending tip in the region of the streamer's electron avalanche. Current peaks as high as 300,000 amp have been reported, and values of 30,000 amp are fairly typical. The entire process of the return stroke is completed in a few tens of microseconds, and even most of this is spent in a long decay period following an early rapid rise to full current value in only a few microseconds. Both the current and propagation speed decrease with height. In negative cloud-to-ground flashes the return stroke deposits the positive charge of several coulombs on the preceding negative leader channel, thus charging earth negatively. In positive cloud-to-ground flashes, the return stroke deposits the negative charge of several tens of coulombs on the preceding positive leader channel, thus lowering positive charge to ground. The entire return stroke process is completed in a few tens of microseconds. In negative cloud-to-ground flashes, multiple return strokes are common. Positive cloud-to-ground flashes, in contrast, typically have only one return stroke. The return streamer of cloud-to-ground discharges is so intense because of the high electrical conductivity of the ground, and hence this type of streamer is not to be found in air discharges, cloud discharges or cloud-to-cloud discharges.

❐ reversal temperature

The temperature at which the charge transferred to riming graupel particles during collisions with ice crystals in thunderstorms reverses sign. The reversal temperature is affected by the cloud liquid water content such that graupel charges positively (negatively) at high (low) liquid water contents and temperatures above (below) the reversal temperature. The process can account for the positive charging of ice crystals carried to the top of the clouds, for the negative charge region typically at temperatures between -10C and -20C and for the lower positive charge center.

❐ ribbon lightning

Ordinary cloud-to-ground lightning that appears to be spread horizontally into a ribbon of parallel luminous streaks when a very strong wind is blowing at right angles to the observer's line of sight. Successive strokes of the lightning flash are then displaced by small angular amounts and may appear to the eye or camera as distinct paths. The same effect is readily created artificially by rapid transverse movement of a camera during film exposure.

❐ rocket lightning

A form of cloud discharge, generally horizontal and at cloud base, whose luminous channel appears to advance through the air with visually resolvable speed, often intermittently.

❐ rocket-triggered lightning

A form of artificial lightning discharge initiated with a rocket trailing wire which may or may not be connected to the ground. The first phase of the discharge is a unidirectional leader starting from the tip of the wire. When the low end of the wire is not connected to ground, bi-directional leader development occurs from both ends of the wire, similar to lightning initiation from aircraft. In the case of the negative space charge overhead (usual summer thunderstorm condition), a triggered lightning may only be a positive leader or become a sequence of dart leader/return stroke processes following the initial positive leader. The latter is analogous to the subsequent return stroke process in a negative cloud-to-ground flash with the initial positive leader being analogous to the first return stroke. In case of the positive space charge overehead (usual winter storm condition), the triggered lightning is a single negative leader.

S

❐ schumann resonance

Electromagnetic waves in the ELF range trapped in the spherical cavity formed by the conductive earth and the conductive ionosphere. The fundamental resonance mode represents one wavelength around the Earth with a corresponding frequency of approximately 8 Hz. The resonances are maintained continuously by global lightning activity. Five to seven higher order modes are frequently discernible.

❐ sferics fix

The determination of the bearing to the lightning source usually based on the measurement of the horizontal magnetic field with orthogonal coils or loop antennae.

❐ sferics observation

The detection of electromagnetic radiation from lightning generally in the frequency range 10-30 kHz. The physical measurement can include the electric field, the magnetic field, or both. Sferics are generally attributed to the high current phases of lightning, i.e., to return strokes and K-changes.

❐ sferics source

That portion of a lightning discharge that radiates strongly in the frequency interval 10-30 kHz. The physical source is generally identified with the return stroke in flashes to ground and the K-change in the case of intracloud flashes.

❐ sheet lightning

A diffuse, but sometimes fairly bright, illumination of those parts of a thundercloud that surround the path of a lightning flash, particularly a cloud discharge or cloud-to-cloud discharge. Thus, sheet lightning is no unique form of lightning but only one manifestation of ordinary lightning types in the presence of obscuring clouds.

❐ sky wave

In radio terminology, or for lightning transients, electromagnetic waves received after they have been reflected by the ionosphere.

❐ slow tail

The extremely low frequency (ELF) component of some sferics that lags the initial very low frequency (VLF) arrival because of the lower phase velocity at low frequencies. The cause of the slow tail is still controversial. One explanation attributes it to the continuing current. The lightning radiation arrives after the initiating ground stroke because a) it has components below about 3 k Hz in frequencies (necessary for the propagation in the earth-ionospheere wave guide cavity) and b) because it suffers dispersion as a result of the wave guide propagation which broadens the original vlf impulse making it look like a "tail."

❐ space charge

Any net electrical charge that exists in a given region of space. In electronics, this usually refers to the electrons in the space between the filament and plate of as electron tube. In atmospheric electricity, space charge refers to a preponderance of either negative or positive ions within any given portion of the atmosphere. A net positive space charge is found in fair weather at all altitudes in the atmosphere, and is largest near the earth's surface. The general downward flux of this positive space charge is known as the air-earth conduction current.

❐ spark discharge

That type of gaseous electrical discharge in which the charge transfer occurs transiently along a relatively constricted path of high ion density, resulting in high luminosity. It is of short duration and to be contrasted with the non-luminous point discharge and corona discharge, and also with the continuous arc discharge. The exact meaning to be attached to the term "spark discharge" varies somewhat in the literature. It is frequently applied to just the transient phase of the establishment of any arc discharge. A lightning discharge can be considered a large scale spark discharge.

❐ spider lightning

Lightning with extraordinary lateral extent near cloud base where its dendritic structure is clearly visible. This lightning type is prevalent beneath the stratiform anvil of mesoscale convective systems and is often associated with positive ground flashes. This discharge form is also referred to as 'sheet' lightning.

❐ spray electrification

A process of charge separation associated with the mechanical disruption of liquid drops and believed to be based on the electrical double layer at the air-water interface. Spray electrification is the common explanation for the electric field generation in the vicinity of waterfalls.

❐ sprite

weak luminous emissions that appear directly above an active thunderstorm and are coincident with cloud-to-ground or intracloud lightning flashes. Their spatial structures range from small single or multiple vertically elongated spots, to spots with faint extrusions above and below, to bright groupings which extend from the cloud tops to altitudes up to about 95 km. Sprites are predominantly red. The brightest region lies in the altitude range 65-75 km, above which there is often a faint red glow or wispy structure that extends to about 90 km. Below the bright red region, blue tendril-like filamentary structures often extend downward to as low a s

40 km. High-speed photometer measurements show that the duration of sprites is only a few ms. Current evidence strongly suggests that sprites preferentially occur in decaying portions of thunderstorms and are correlated with large positive cloud-to-ground flashes. The optical intensity of sprite clusters, estimated by comparison with tabulated stellar intensities, is comparable to a moderately bright auroral arc. The optical energy is roughly 10-50 kJ per event, with a corresponding optical power of 5-25 MW. Assuming that optical energy constitutes 1/1000 of the total for the event, the energy and power are on the order of 10-100 MJ and 5-50 GW, respectively. Early research reports for these events referred to them by a variety of names, including "upward lightning," "upward discharges," "cloud-to-stratosphere discharges," and cloud-to-ionosphere discharges." Now they are simply referred to as sprites, a whimsical term that evokes a sense of their fleeting nature, while at the same time remaining nonjudgmental about physical processes that have yet to be determined. (See also blue jets).

❒ stepped leader

The initial leader of a lightning discharge; an intermittently advancing column of high ionization and charge which establishes the channel for a first return stroke. The peculiar characteristic of this type of leader is its step-wise growth at intervals of about fifty to one hundred microseconds. The velocity of growth during the brief intervals of advance, each only about one microsecond in duration, is quite high (about 5 x 10 ^ 7 m/s), but the long stationary phases reduce its effective speed to only about 5 x 10 ^ 5 m/s. To help explain its mode of advance, the concept of a pilot streamer was originally suggested but has been supplanted by analogy to recent work on long laboratory sparks.

❒ streak lightning

Ordinary lightning , of a cloud-to-ground discharge , that appears to be entirely concentrated in a single, relatively straight lightning channel .

❒ streamer

A sinuous channel of very high ion-density which propagates itself though a gas by continual establishment of an electron avalanche just ahead of its advancing tip. In lightning discharge, the stepped leader, dart leader, and return stroke all constitute special types of streamers.

❒ stroke density

The areal density of lightning discharges over a given region during some specified period of time, as number per square mile or per square kilometer per year.

❒ sunspot cycle

A cycle with an average length of 11.1 years, but varying between about 7 and 17 years, in the number and area of sunspots, as given by the relative sunspot number. This number rises from a minimum of 0-10 to a maximum of 50-140 about four years later, and then declines more slowly. Records of sunspot counts have been kept since the early seventeenth century. A period of anomalously low activity was observed between about 1645-1700 and is known as the Maunder minimum. An approximate 11-year cycle has been found or suggested in geomagnetism, frequency of aurora, and other ionospheric characteristics. The n index of geomagnetic intensity variation shows one of the strongest known correlation to solar activity. Eleven-year cycles have been suggested for various tropospheric phenomena, but none of these has been substantiated.

❐ superbolt

A lightning discharge of extraordinary peak luminosity when observed from space.

❐ supply current

The electrical current in the atmosphere which is required to balance the observed air-earth current of fair-weather regions by transporting positive charge upward or negative charge downward The problem of accounting for the supply current has been for many years a key problem of the field of atmospheric electricity and has received much attention. A quasi-steady current of about 1800 amperes for the earth as a whole is estimated to be required to balance the air-earth current. C. T. R. {Wilson} suggested in 1920 that the thunderstorms present in widely scattered regions of the earth at any one time might be responsible for the supply current. Although this suggestion has not been fully confirmed, there is growing conviction that this is correct. When one considers an average over many storms, thunderstorm lightning transports negative charge downward to earth, as does point discharge in the regions below thunderstorms. Also, positive ions flow upward above active thunderstorms.

❐ surface wave

A gravity wave formed on the free surface of a fluid. In classical hydrodynamics, to distinguish surface waves from tidal waves, the condition is imposed that vertical accelerations are not negligible. Dynamically this wave is similar to that on an interface separating two fluids, becoming identical in the case of zero density in the upper fluid.

❐ surge

A short-duration, high-amperage, electric current that may sweep through an electrical network, such as a power transmission network, when some portion of it is strongly influenced by the electrical activity of a thunderstorm. This may take the form of a direct lightning stroke, or may be simply the release of previously induced charge on the line when a thundercloud overhead suddenly discharges itself. Such current surges rapidly through the lines until they find a path to the ground through arc-over at a weak insulator or by entering terminal equipment at the end of the line. Use of grounded guard wires above the power lines of a transmission system reduces the frequency of surge current difficulties, and installation of lightning arresters at sensitive terminal equipment protects the system from damage there.

❐ surge current

A short-duration, high-amperage, electric current wave that may sweep through an electrical network, as a power transmission network, when some portion of it is strongly influenced by the electrical activity of a thunderstorm. This activity may take the form of a direct lightning strike, or it may be simply the release of previously induced charge on the line when a thundercloud overhead suddenly discharges itself. Such current surges flow rapidly through the lines until they find a path to the ground through arc-over at a weak insulator or by entering terminal equipment at the end of the line. Use of grounded guard wires above the power lines of a transmission system reduces the frequency of surge current difficulties, and installation of lightning arresters at sensitive terminal equipment protects the system against damage there.

T

❐ thunderbolt

In mythology, a lightning flash accompanied by a material ``bolt" or dart; this is the legendary cause of the damage done by lightning. It is still used as a popular term for a lightning discharge accompanied by thunder.

❐ thunderstorm charge

The existence of regions of net charge in a thunderstorm. During transient collisions of ice crystals with riming graupel pellets charge is transferred. The separating particles then carry equal and opposite charges, the larger (often negative) particles fall while the smaller ones (often positively charged ice crystals) are carried up in the updraft to produce a vertical electric field that eventually produces lightning. There is no complete understanding of the charge transfer process, but possible processes include charges on the surface layers of the particles, charges on dislocations in the ice lattice, temperature differences along surface features that may be broken off during collisions, and contact potential differences between the surfaces of the interacting particles.

❐ thunderstorm dipole

The simplest representation of the electrostatic structure of an electrified cloud with overall charge neutrality. Ordinary thunderstorms are characterized by upper positive charge and lower negative charge.

❐ thunderstorm tripole

A refinement of the simpler dipole representation for the electrostatic structure of isolated electrified clouds. The tripole structure includes the lower positive charge center that appears in many observations. The tripole picture is generally attributed to C. C. Simpson.

❐ time-of-arrival technique

the time-of-arrival technique refers to locating the source of an emitted signal from a precise recording of the time that a signal is observed. For example, the time interval between an observed lightning flash and the arrival of the thunder can be used to estimate the distance to the lightning flash.

❐ total conductivity

In atmospheric electricity, the sum of the electrical conductivity of the positive and negative ions found in a given portion of the atmosphere.

❐ triboelectrification

A process of charge separation that involves the rubbing together of dissimilar material surfaces. The triboelectric series is a classification scheme for the ordering of the tendency for positive charge acquisition in rubbing. The detailed physical mechanism in triboelectrification is a long unsolved problem.

❐ type-[alpha] leader

A stepped leader that exhibits very little branching and whose individual steps are short and so weakly luminous as to be difficult to discern on high-speed streak photographs.

❐ type-[beta] leader

A stepped leader in which the upper portion of the channel is characterized by longer and brighter steps than those found in the lower portion of the channel. Many researchers report that a preliminary breakdown precedes the stepped leader in electric field records. The beta leader may be this preliminary breakdown.

V

❐ vhf source

typically, a source of very high frequency radiation. Lightning discharges are one example of VHF sources in the atmosphere.

W

❐ workman--reynolds effect

A mechanism for electric charge separation during freezing of slightly impure water, discovered by E. J. Workman and S. E. Reynolds. When a very dilute solution of certain salts freeze rapidly, a strong potential difference is established between the solid and liquid phases. For some salts, the ice attains negative charge, for others, positive. This mechanism was thought to play a role in the charging of thunderstorms but it is now known that it cannot account for the electrification of riming graupel by the shedding of charged surface water because the droplets freeze faster than the time taken for a substantial ice-water freezing potential to be set up. This mechanism has been suggested as one possible mode of thunderstorm charge separation in those portions of a thunderstorm downdraft where snow-pellet or hail particles sweep out super cooled water drops. Partial freezing and partial blow-off of a liquid film could lead to charge separation. The acknowledged predominance of dry growth of graupel in New Mexico thunderclouds led Reynolds to question the viability of this mechanism in the atmosphere. = This should not be confused with the Reynolds effect.

Z

▫ zigzag lightning

Ordinary lightning of a cloud-to-ground discharge that appears to have a single, but very irregular, lightning channel.

Mountain Meterology

A

❐ absolutely stable

An atmospheric stability condition that exists when the environmental lapse rate is less than the moist adiabatic lapse rate.

❐ absolutely unstable

An atmospheric stability condition that exists when the environmental lapse rate is greater than the dry adiabatic lapse rate.

❐ absorption

The process in which incident radiant energy is retained by a substance by conversion to some other form of energy.

❐ acid precipitation

Precipitation, as rain, snow or sleet containing relatively high concentrations of acid-forming chemicals that have been released into the atmosphere and combined with water vapor; harmful to the environment.

❐ adiabat

A line on a thermodynamic chart relating the pressure and temperature of a substance (such as air) that is undergoing a transformation in which no heat is exchanged with its environment.

❐ adiabatic lapse rate

The rate of decrease of temperature experienced by a parcel of air when it is lifted in the atmosphere under the restriction that it cannot exchange heat with its environment. When lifted, unsaturated parcels cool at the dry adiabatic lapse rate, while saturated parcels cool at the moist adiabatic lapse rate.

❐ adiabatic process

A process which occurs with no exchange of heat between a system and its environment.

❐ advection

The horizontal transport of atmospheric properties.

❐ advection fog

A fog that forms when warm air flows over a cold surface and cools from below until saturation is reached.

❑ aeroallergen

Any of a variety of allergens such as pollens, grasses, or dust carried by winds.

❑ aerosol

A system of colloidal particles dispersed in a gas, such as smoke or fog.

❑ air mass

A body of air covering a relatively wide area and exhibiting horizontally uniform properties.

❑ air mass thunderstorm

A thunderstorm produced by local convection within an unstable air mass, as opposede to thunderstorms associated with fronts or instability lines.

❑ air pollutant

Harmful substance or product introduced into the atmosphere.

❑ air pollution potential

The meteorological potential for air pollution problems, considered without regard to the presence or absence of actual pollution sources.

❑ air quality model

Mathematical or conceptual model used to estimate present or future air quality.

❑ air toxic

Toxic air pollutant.

❑ albedo

Reflectivity; the fraction of radiation striking a surface that is reflected by that surface.

❑ along-valley wind system

A closed, thermally driven, diurnal mountain wind circulation whose lower branch blows up or down the axis of a valley. The upper branch blows in the opposite direction, thereby closing the circulation.

❑ altocumulus

A cloud of a class characterized by globular masses or rolls in layers or patches, the individual elements being larger and darker than those of cirrocumulus and smaller than those of stratocumulus. These clouds are of medium altitude, about 8000-20,000 ft (2400-6100 m).

❑ altostratus

A cloud of a class characterized by a generally uniform gray sheet or layer, lighter in color than nimbostratus and darker than cirrostratus. These clouds are of medium altitude, about 8000 to 20,000 ft (2400-6100 m).

❑ ambient

Of the surrounding area or environment.

❑ aneroid barometer

An instrument for measuring atmospheric pressure in which a needle, attached to the top of an evacuated box, is deflected as changes in atmospheric pressure cause the top of the box to bend in or out.

❑ anthropogenic

Caused or produced by humans.

❑ anticyclogenesis

The formation or intensification of an anticyclone or high pressure center.

❑ anticyclone

A large-scale circulation of winds around a central region of high atmospheric pressure, clockwise in the Northern Hemisphere, counterclockwise in the Southern Hemisphere.

❑ anti-wind

The upper or return branch of an along-valley wind system, confined within a valley and blowing in a direction opposite to the winds in the lower altitudes of the valley.

❑ area source

An array of pollutant sources, so widely dispersed and uniform in strength that they can be treated in a dispersion model as an aggregate pollutant release from a defined area at a uniform rate.

❑ aspect (angle)

The cardinal direction or bearing angle toward which a slope faces.

❑ azimuth angle

1. The direction or bearing toward which a sloping surface faces (e.g., a north-facing slope has an azimuth angle of 360°; a northeast-facing slope, an azimuth angle of 45°).
2. The arc of the horizon measured clockwise from north to the point where a vertical circle through a given heavenly body intersects the horizon (e.g., used for solar azimuth angle).

B

❑ backfire

1. A fire started to stop an advancing fire by creating a burned area in its path.
2. A firing pattern or fire ignition pattern for a prescribed fire, in which a fire is ignited across the leeward edge of an area and burns into the wind.

Backfires can be ignited in parallel strips to protect an area from fire escapement.

❒ banner cloud

A cloud plume produced by rising motions in an eddy that forms downwind of steep, isolated mountain peaks; may occur even on otherwise cloud-free days.

❒ barometric pressure

The pressure of the atmosphere as indicated by a barometer.

❒ barrier jet

A jet-like wind current that forms when a stably-stratified low-level airflow approaches a mountain barrier and turns to the left to blow parallel to the longitudinal axis of the barrier.

❒ belt weather kit

Belt-mounted case with pockets fitted for anemometer, compass, sling psychrometer, water bottle, and forms. Used to provide on-site weather observations for the fire weather forecaster or fire behavior analyst.

❒ billow cloud

A cloud consisting of broad parallel bands oriented perpendicular to the wind.

❒ blizzard

A severe weather condition characterized by low temperatures and strong winds (32 mph or higher) bearing a great amount of snow (including snow picked up from the ground) that reduces visibility to less than 500 ft.

❒ blocked (flow)

Flow approaching a mountain barrier at elevations below the barrier height that is too weak or too stable to be carried over the barrier.

❒ blowup fire

A fire that suddenly increases in intensity or rate of spread, usually accompanied by intense convection, and adversely affects fire control activities or fire suppression plans.

❒ bora

A regional down slope wind whose source is so cold that it is experienced as a cold wind, despite compression warming as it descends the lee slope of a mountain range.

❒ boundary layer

The layer of fluid in the immediate vicinity of a bounding surface; in the atmosphere, the air layer near the ground affected by diurnal heat, moisture or momentum transfer to or from the surface.

❒ box model

A computer model used to calculate air pollution concentrations by assuming that pollutants emitted into a box-shaped volume are immediately and uniformly dispersed throughout the volume. The sides and bottom of the box, for example, could be defined by the sidewalls and floor of a valley and the top could be defined by the mixing depth.

❒ brocken specter

An optical effect characterized by concentric rings of color (red outermost and violet innermost) surrounding the shadow of an observer's head when the shadow is cast onto a cloud deck below the observer's elevation.

❒ build

To increase in pressure: a building ridge or high (anticyclone).

C

❒ canopy

The cover formed by the leafy upper branches of the trees in a forest; height stratum containing the crowns of the tallest vegetation present (living or dead).

❒ canyon wind

1. A foehn wind that is channeled through a canyon as it descends the lee side of a mountain barrier. Example: Wasatch wind.
2. Any along-canyon wind.

❒ cap cloud

A stationary lens-shaped cloud forming over a mountain peak, with cloud base below the mountain top.

❒ capping inversion

An elevated inversion layer that caps a convective boundary layer, keeping the convective elements from rising higher into the atmosphere.

❒ ceilometer

A device using a laser or other light source to determine the height of a cloud base. An optical ceilometer uses triangulation to determine the height of a spot of light projected onto the base of the cloud; a laser ceilometer determines the height by measuring the time required for a pulse of light to be scattered back from the cloud base.

❒ centerfire

A firing pattern or fire ignition pattern for prescribed fires, in which a fire is ignited at a chosen spot and then in concentric circles or spirals around the spot to produce a fire with strong central convection, inward spread and high fire intensity. This firing pattern is modified on slopes to account for the faster speed of fire propagation up the slope.

❒ chemistry model

A computer model used in air pollution investigations that simulates chemical and photochemical reactions of pollutants during their transport and diffusion.

❒ chinook

The name given to the foehn in North America, especially used on the eastern side of the Rocky Mountains and Sierras.

❒ chinook arch

A foehn cloud formation appearing as a bank of altostratus clouds east of the Rocky Mountains, heralding the approach of a chinook. It forms in the rising portion of standing waves on the lee side of the mountains. An observer underneath or east of the cloud sees an arch of clear air between the cloud's leading edge and the mountains below. The cloud appears to converge with the mountains to the north and south due to a perspective effect.

❒ cirrocumulus

A cirriform cloud characterized by thin, white patches, each of which is composed of very small granules or ripples. These clouds are of high altitude (20,000-40,000 ft or 6000 -12,000 m).

❒ cirrostratus

A cloud of a class composed of ice crystals and appearing as a whitish and usually somewhat fibrous veil, often covering the whole sky and sometimes so thin as to be hardly discernible. Haloes are often seen in cirrostratus. These clouds are of high altitude (20,000-40,000 ft or 6000 -12,000 m).

❒ cirrus

A cloud of a class characterized by thin white filaments or narrow bands and composed of ice crystals. These clouds are of high altitude (20,000-40,000 ft or 6000 -12,000 m).

❒ class i area

Geographic areas designated by the Clean Air Act where only a small amount or increment of air quality deterioration is permissible.

❒ clear ice

A thin coating of ice on terrestrial objects caused by rain that freezes on impact. The ice is relatively transparent, as opposed to rime ice, because of large drop size, rapid accretion of liquid water, or slow dissipation of latent heat of fusion.

❒ climate

The composite or generally prevailing weather conditions of a region throughout the year, averaged over a series of years.

❒ climatology

The science that deals with the phenomena of climates or climatic conditions.

❒ cloud condensation nucleus

A particle, either liquid or solid, upon which water condenses to form cloud droplets.

❐ cold air avalanche

Down slope flow pulsations that occur at more or less regular intervals as cold air builds up on a peak or plateau, reaches a critical mass, and then cascades down the slopes.

❐ cold air damming

A process in which a shallow cold air mass is carried up the slope of a mountain barrier, but with insufficient strength to surmount the barrier. The cold air, trapped upwind of the barrier alters the effective terrain configuration of the barrier to larger-scale approaching flows.

❐ cold front

A zone separating two air masses, of which the cooler, denser mass is advancing and replacing the warmer.

❐ cold occlusion

A frontal zone formed when a cold front overtakes a warm front and, being colder than the air ahead of the warm front, slides under the warm front, lifting it aloft. Compare warm occlusion.

❐ collection efficiency

The fraction of droplets approaching a surface that actually deposit on that surface.

❐ colorado low

A low pressure storm system that forms in winter in southeastern Colorado or northeastern New Mexico and tracks northeastward across the central plains of the U.S. over a period of several days, producing blizzards and hazardous winter weather.

❐ complex terrain

Mountainous terrain. In general usage, it may also refer to coastal regions and heterogeneous landscapes.

❐ conditionally unstable

An atmospheric stability condition that exists when the environmental lapse rate is less than the dry adiabatic lapse rate but greater than the moist adiabatic lapse rate.

❐ conduction

Flow of heat in response to a temperature gradient within an object or between objects that are in physical contact.

❐ coning

Pattern of plume dispersion in a neutral stability atmosphere, in which the plume attains the form of a cone with its vertex at the top of the stack.

❐ continental divide

The line of summits in the Rocky Mountains that separate streams flowing toward the Gulf of California and Pacific from those flowing toward the Gulf of Mexico, Hudson Bay and the Arctic Ocean.

❐ continentality

The degree to which the climate of a region typifies that of the interior of a large landmass.

❐ contrail

Condensation trail from a jet aircraft.

❐ convection

1. Vertical air circulation in which warm air rises and cool air sinks, resulting in vertical transport and mixing of atmospheric properties.
2. Flow of heat by this circulation.

❐ convection column

A rising column of gases, smoke, fly ash, particulates, ana other debris produced by a fire.

❐ convective boundary layer

The unstable boundary layer that forms at the surface and grows upward through the day as the ground is heated by the sun and convective currents transfer heat upwards into the atmosphere.

❐ convective over development

Convection that covers the sky with clouds, thereby cutting off the sunshine that produces convection.

❐ convergence

A net flow of air into a given region.

❐ coriolis force

A fictitious force used to account for the apparent deflection of a body in motion with respect to the earth, as seen by an observer on the earth. The deflection (to the right in the northern hemisphere) is caused by the rotation of the earth.

❐ corona

A white or colored circle or set of concentric circles of light of small radius seen around a luminous body, esp. Around the sun or moon. The color varies from blue inside to red outside and the phenomenon is attributed to diffraction of light by thin clouds or mist (distinguished from halo).

❐ cross-valley wind system

A thermally driven wind that blows during daytime across the longitudinal axis of a valley toward the heated sidewall.

❐ crown

The top or highest part of the forest canopy.

❐ crown fire

A fire where flames travel from tree to tree at the level of the tree's crown or top.

❐ crowning

1. Movement of a fire from the understory into the crown of a forest canopy.
2. Spectacular fire behavior phenomenon in which a fire spreads rapidly across the overstory of a forest.

❐ cumuliform

Having the appearance or character of cumulus clouds.

❐ cumulonimbus

A cloud of a class indicative of thunderstorm conditions characterized by large, dense towers that often reach altitudes of 30,000 ft (9000 m) or more, cumuliform except for their tops, which appear fibrous because of the presence of ice crystals.

❐ cumulus

A cloud of a class characterized by dense individual elements in the form of puffs, mounds or towers, with flat bases and tops that often resemble cauliflower. They are found at a lower altitude than altocumulus, usually below 8000 ft (2400 m).

❐ cyclogenesis

The formation or intensification of a cyclone or low-pressure storm system.

❐ cyclone

A large-scale circulation of winds around a central region of low atmospheric pressure, counterclockwise in the Northern Hemisphere, clockwise in the Southern Hemisphere.

D

❐ damp haze

Haze in which the individual particles have grown by the absorption of water, often at relative humidities below 100%.

❐ dart leader

A faint, negatively charged channel that travels more or less directly and continuously from cloud to ground.

❐ day length

Duration of the period from sunrise to sunset.

❐ dead fuel moisture

Moisture content of non-living fuels. For dead fuels, moisture content is governed primarily by exposure to precipitation or soil moisture and by the tendency of the dead fuels to approach equilibrium with the relative humidity of the surrounding air.

❐ deepen

To decrease in atmospheric pressure: a deepening trough or low (cyclone).

❒ dew-point temperature

The temperature to which air must be cooled (at constant pressure and constant water vapor content) for saturation to occur.

❒ diagnostic model

A computer model used to calculate air pollution concentrations. A diagnostic model produces a wind field over an area by interpolating in space and time from actual wind observations.

❒ dispersion

Transport and diffusion of pollutants.

❒ diurnal

Daily, especially pertaining to actions which are completed in 24 hours and are repeated every 24 hours.

❒ divergence

The net flow of air from a given region. Compare convergence.

❒ dividing streamline height

In the blocked flow region upwind of a mountain barrier, the height above ground of the streamline that separates the blocked flow region near the ground from the air aloft which flows over the barrier.

❒ domain

In numerical models, the geographical area over which a simulation is performed.

❒ downburst

A strong downdraft of air from a cumulonimbus cloud, of limited duration and often associated with intense thunderstorms

❒ downslope wind

A thermally driven wind directed down a mountain slope and usually occurring at night; part of the slope wind system.

❒ down-valley wind

A thermally driven wind directed down a valley's axis, usually occurring during nighttime; part of the along-valley wind system

❒ downwash

A deflection of air downward behind a flow obstruction.

❒ drainer

A valley or basin from which air drains continuously during nighttime rather than becoming trapped or pooled.

❒ drift

Sprayed or dusted material that does not deposit on the target area.

❐ **drizzle**

Precipitation consisting of numerous minute droplets of water less than 0.5 mm (500 micrometers) in diameter.

❐ **dry adiabatic lapse rate**

The rate (5.4°F per 1000 ft or 9.8°C per km) at which the temperature of a parcel of dry air decreases as the parcel is lifted, under the assumption that no heat is exchanged with its environment.

❐ **dry line**

The roughly north-south boundary between moist air in the Mississippi Valley and dry air on the west side of the Great Plains descending from the Mexican Plateau and Southern Rockies. Thunderstorms often form along this line, which moves eastward during the morning and westward in the evening.

❐ **duff**

Organic matter in various stages of decomposition on the floor of a forest.

E

❐ **eddy**

Swirling currents of air at variance with the main current.

❐ **effective topography**

The topography as seen by an approaching flow, which may include not only the actual terrain but also cold air masses trapped within or adjacent to the actual topography.

❐ **el nino**

Literally, the Christ child, a name given to an extensive ocean warming in the equatorial eastern Pacific along the coast of Peru and Ecuador that often begins around Christmas (hence, the name). The warming brings nutrient-poor tropical water southward along the west coast of South America in major events that recur at intervals of 3-7 years. El Nino is associated with atmospheric circulations that produce wide ranging effects on global weather and climate.

❐ **emissivity**

The ability of a surface to emit radiant energy compared to that of a black body at the same temperature and with the same area.

❐ **entrainment zone**

A shallow region at the top of a convective boundary layer where fluid is entrained into the growing boundary layer from the overlying fluid by the collapse of rising convective plumes or bubbles.

❐ **environmental temperature lapse rate**

The rate of decrease of air temperature with height, usually measured with a radiosonde.

❐ environmental temperature sounding

An instantaneous or near-instantaneous sounding of temperature as a function of height. This sounding or vertical profile is usually obtained by a balloon-borne instrument, but can also be measured using remote sensing equipment.

❐ equinox

The time when the sun crosses the earth's equator, making night and day of approximately equal length all over the earth and occurring about March 21 (the spring or vernal equinox) and September 22 (autumnal equinox).

❐ evaporation-mixing fog

A fog that forms when the evaporation of water raises the dew point of the adjacent air.

❐ extraterrestrial shortwave radiation

The theoretically-calculated radiation flux from the sun at the top of the atmosphere, before losses involved in traversing the atmosphere.

❐ extreme fire behavior

Level of fire behavior that ordinarily precludes direct control of the fire. One or more of the following is usually involved: high rates of spread, prolific crowning or spotting, presence of fire whirls, or an intense convection column.

F

❐ fall line

The line of steepest descent of a slope.

❐ fanning

A pattern of plume dispersion in a stable atmosphere, in which the plume fans out in the horizontal and meanders about at a fixed height.

❐ fill

To increase in atmospheric pressure a filling trough or low (cyclone).

❐ fine fuels

Fast-drying dead fuels, generally less than 1/4 inch in diameter and having a time lag of one hour or less. These fuels (e.g., grass, twigs and needles) ignite readily and are consumed rapidly by fire when dry.

❐ fire behavior

The manner in which a fire reacts to the influences of fuel, weather, and topography. Fire behavior is described in terms of the specific characteristics exhibited by a fire, including its rate of spread, spread direction, flame length, and rate of heat release.

❐ fire behavior model

A computer model that uses a set of mathematical equations to predict certain aspects of fire behavior when provided with data on fuel and environmental conditions unique to a site.

❐ fire danger

The exposure to risk or harm from a fire as determined from factors that affect the start, spread, intensity and difficulty of suppression of wildfires and the damages they cause.

❐ fire intensity

The rate of heat release per unit time and per unit of fire travel distance at the fire front. Numerically, it is the product of the quantity of fuel consumed at the fire front, the heat yield per unit of fuel consumed, and the rate of fire spread.

❐ fire line

A zone along a fire's edge where there is little or no fuel available to the fire.

❐ fire severity

The degree to which a site has been altered or disrupted by fire. Severity is dependent primarily on the product of fire intensity and duration.

❐ fire weather

Weather conditions that influence fire ignition, behavior or suppression.

❐ fire weather watch

A term used to alert land mangers to the potential that weather conditions when coupled with critically dry fuels can lead to dangerous wildfires.

❐ fire whirl

A vortex similar to a dust devil that forms in the fire area, often on the lee edge of the fire. A fire whirl increases fire intensity and the frequency of spot fires.

❐ fire wind

A wind blowing radically inward toward a fire, produced by horizontal temperature differences (and thus pressure differences)between the heated air above the fire and the surrounding cooler free atmosphere.

❐ firebrand

Any source of heat, natural or man made, capable of igniting wild land fuels; flaming or glowing fuel particles that can be carried naturally by wind, convection currents, or gravity into unburned fuels.

❐ firestorm

Raging fire of great intensity that spreads rapidly.

❐ firing pattern

The specific pattern and timing of ignition of a prescribed fire to affect the direction and rate of fire spread and fire intensity.

❑ flankfire

1. A fire started to stop an advancing fire by creating a burned area on its flank.
2. A firing pattern or fire ignition pattern for prescribed fires, in which the lee edge of an area to be burned is first protected by a backfire and then fires are ignited in parallel lines leading to windward from the leeward edge.

❑ flash

A sudden, brief illumination of a conductive channel associated with lightning, which may contain multiple strokes with their associated stepped leaders, dart leaders and return strokes.

❑ flash density

The number of lightning flashes per unit area in a specified time interval (e.g., one year).

❑ flash flood

A sudden and destructive rush of water down a narrow gully or over a sloping surface, caused by heavy rainfall.

❑ flow

1. Wind;
2. Volume of air carried by wind.

❑ flow separation

The process by which a separation eddy forms on the windward or leeward sides of bluff objects or steeply rising hillsides.

❑ flow splitting

The splitting of a stable airflow around a mountain barrier, with branches going around the left and right edges of the barrier, often at accelerated speeds.

❑ flux

The rate of transfer of fluids, particles or energy per unit area across a given surface.

❑ foehn

A warm, dry wind on the lee side of a mountain range, the warmth and dryness of the air being due to adiabatic compression as the air descends the mountain slopes.

❑ foehn pause

A temporary cessation of the foehn at the ground due to the formation or intrusion of a cold air layer which lifts the foehn off the ground.

❑ foehn wall

The steep leeward boundary of an extensive cloud layer that forms as air is lifted over ridges and mountains during foehn conditions.

- **foehn wall cloud**

Same as foehn wall.

- **fog**

A cloud with its base at the earth's surface.

- **föhn**

Alternate spelling of foehn.

- **forced channeling**

Channeling of upper winds along a valley's axis when upper winds are diverted by the underlying topography.

- **fractocumulus**

A cumulus cloud presenting a ragged, shredded appearance, as if torn.

- **fractostratus**

A stratus cloud presenting a ragged, shredded appearance, as if torn. It differs from a fractocumulus cloud in having a smaller vertical extent and darker color.

- **free air winds**

Synoptic-scale wind. In fire and land management, also called general winds.

- **free atmosphere**

The part of the atmosphere that lies above the frictional influence of the earth's surface.

- **freezing drizzle**

A drizzle that falls as a liquid but freezes into glaze or rime upon contact with the cold ground or surface structures.

- **freezing level**

The altitude at which the air temperature first drops below freezing.

- **freezing rain**

Rain that falls as a liquid but freezes into glaze upon contact with the ground.

- **front or frontal zone**

An interface or zone of transition between two dissimilar air masses.

- **frontal inversion**

An elevated temperature inversion that develops above a frontal zone when cold air at the surface is overrun by warmer air aloft.

- **frostbite**

Human tissue damage caused by exposure to intense cold.

- **fuel loading**

The amount of fuel present as expressed in the weight of fuel per unit area.

❐ fugitive dust

Dust that is not emitted from definable point sources such as industrial smokestacks. Sources include open fields, roadways, storage piles, etc.

❐ full-physics numerical model

A computer model used to calculate air pollution concentrations. A full-physics numerical model uses a full set of equations describing the thermodynamic and dynamic state of the atmosphere and can be used to simulate atmospheric phenomena.

❐ fumigation

A pattern of plume dispersion produced when a convective boundary layer grows upward into a plume trapped in a stable layer. The elevated plume is suddenly brought downward to the ground, producing high surface concentrations.

G

❐ gap winds

Strong winds driven through low passes or major breaks in mountain barriers by cross-gap pressure gradients that develop regionally. In north america, the term is used mostly for winds in coastal mountain ranges where major river valleys issue onto seaways.

❐ gaussian plume model

A computer model used to calculate air pollution concentrations. The model assumes that a pollutant plume is carried downwind from its emission source by an average wind. Plume concentrations are obtained by assuming that the highest concentrations occur on the horizontal and vertical midlines of the plume, with the distribution about these mid-lines characterized by gaussian- or bell-shaped concentration profiles.

❐ gaussian plume segment model

A modification of the Gaussian plume model in which wind direction changes during plume transport are handled by breaking the plume into segments.

❐ gaussian puff model

A computer model used to calculate air pollution concentrations. The model assumes that a continuously emitted plume or instantaneous cloud of pollutants can be simulated by the release of a series of puffs that are carried in a time- and space-varying wind field. The puffs are assumed to have gaussian or bell-shaped concentration profiles in their vertical and horizontal planes.

❐ general circulation

The continuous circulation of wind and ocean currents that acts to moderate the temperature differences between the poles and the equator.

❑ general wind

Land management agency term for wind produced by synoptic-scale pressure systems on which a smaller-scale or local convective wind may be superimposed. Also called free air wind.

❑ geostrophic wind

1. A wind blowing parallel to isobars with strength proportional to the pressure gradient (i.e., spacing of the isobars).
2. The wind obtained theoretically by a balance between pressure gradient force and coriolis force,

❑ glaciation

1. The transformation of cloud particles from water drops to ice crystals. Thus, a cumulonimbus cloud is said to have a "glaciated" upper portion.
2. Land surface cover of ice or snow.

❑ glacier wind

A shallow down slope wind above the surface of a glacier, caused by the temperature difference between the air in contact with the glacier and the free air at the same altitude. The glacier wind does not reverse diurnally like slope and along-valley wind systems.

❑ glaze

A smooth, clear coating of ice, but sometimes containing some air pockets.

❑ gradient

A rate of change with respect to distance of a variable quantity, as temperature or pressure, in the direction of maximum change.

❑ graupel

Same as snow pellets or small hail.

❑ gravity waves

A wave created by the action of gravity on density variations in the stratified atmosphere. A generic classification for lee waves, mountains waves, and many other waves that form in the atmosphere.

❑ greenhouse effect

Atmospheric heating caused by solar radiation being readily transmitted inward through the earth's atmosphere but long wave radiation less readily transmitted outward, due to absorption by certain gases in the atmosphere.

❑ greenness

Amount of viable or living material in vegetation. Measurements of greeness are obtained from emissivity data collected from remote sensors on satellites.

❐ ground heat flux

The flux of heat from the ground to the earth's surface; a component of the surface energy budget.

❐ ground stroke

Electrical current propagating along the ground from the point where a direct stroke of lightning hits the ground.

H

❐ hail

Showery precipitation in the form of irregular pellets or balls of ice more than 5 mm in diameter, falling from a cumulonimbus cloud.

❐ halo

Any of a variety of bright circles or arcs centered on the sun or moon, caused by the refraction or reflection of light by ice crystals suspended in the earth's atmosphere and exhibiting prismatic coloration ranging from red inside to blue outside.

❐ haze

An aggregation in the atmosphere of very fine, widely dispersed, solid or liquid particles, or both, giving the air an opalescent appearance that subdues colors. See also regional haze, layered haze and damp haze.

❐ headfire

A firing pattern or fire ignition pattern for prescribed fires, in which a fire is ignited along a line perpendicular to the wind direction and runs to leeward. The headfire is often ignited in strips with the lee edges protected by backfires to guard against fire escapement.

❐ heat exhaustion

A mild form of heat stroke, characterized by faintness, dizziness, and heavy sweating.

❐ heat stroke

A condition resulting from excessive exposure to intense heat, characterized by high fever, collapse, and sometimes convulsions or coma.

❐ hertz

An international unit of frequency equal to one cycle per second, and named after a German physicist.

❐ hydraulic flow

Atmospheric flow that is similar in character to the flow of water over an obstacle.

❑ **hydraulic jump**

A steady disturbance in the lee of a mountain, where the airflow passing over the mountain suddenly changes from a region of low depth and high velocity to a region of high depth and low velocity.

❑ **hydrometeors**

Liquid water or ice in the atmosphere in various forms, including rain, ice crystals, hail, fog or clouds.

❑ **hygrometer**

Any instrument that measures the water vapor content of the atmosphere.

❑ **hygroscopic**

Absorbing or attracting moisture from the air.

❑ **hygrothermograph**

An instrument that records, on one record, the variation with time of both atmospheric humidity and temperature.

❑ **hypothermia**

A rapid, progressive mental and physical collapse that accompanies the lowering of body temperature.

I

❑ **ice crystal**

Precipitation consisting of small, slowly falling crystals of ice.

❑ **ice nucleus**

Any particle that serves as a nucleus in the formation of ice crystals in the atmosphere.

❑ **ice pellets**

Precipitation consisting of particles of ice less than 5 mm in diameter, occurring either as frozen raindrops or as small hailstones.

❑ **icing**

Formation of a coating of ice on a solid object.

❑ **ideal gas laws**

The thermodynamic laws applying to perfect gases.

❑ **inclination angle**

The tilt angle of a surface relative to the horizon.

❑ **insolation**

Solar radiation received at the earth's surface.

- **inter-tropical convergence zone**

The dividing line between the northeast trade winds of the Northern Hemisphere and the southeast trade winds of the Southern Hemisphere where air converges to produce convection and generally rising air.

- **iridescence**

Brilliant colored borders or spots of color in clouds, usually red and green, caused by diffraction of light by small cloud particles. The phenomenon is usually observed in thin cirrus clouds within about 30° of the sun and is characterized by bands of color in the cloud that contour the cloud edges.

- **isobar**

A line of equal or constant pressure; an isopleth of pressure.

- **isotach**

A line on a weather map or chart connecting points where winds of equal speeds have been recorded.

- **isotherm**

A line of equal or constant temperature; an isopleth of temperature.

- **isothermal**

Temperature remaining constant with height or time.

- **isotropic**

Of equal physical properties along all axes.

J

- **jet**

A fast-moving wind current surrounded by slower moving air.

- **jet stream**

Strong, generally westerly winds concentrated in a relatively narrow and shallow stream in the upper troposphere.

- **jet stream cirrus**

A loose term for filamentous cirrus that appears to radiate from a point in the sky, and exhibits characteristics associated with strong vertical wind shear, such as twisted or curved filaments.

- **jet wind speed profile**

A vertical profile of horizontal wind speeds characterized by a relatively narrow current of high winds with slower moving air above and below. Large wind shears occur above and below the jet speed maximum.

K

▫ keetch-byrum drought index

A drought index representing the net effect of evapotranspiration and precipitation in producing cumulative moisture deficiency in deep duff and upper organic soil layers. This index is widely used in the southeastern united states.

▫ kelvin-helmholtz waves

Vertical waves in the air associated with wind shear across statically-stable regions. Can appear as breaking waves and as braided patterns in radar images and cloud photos.

L

▫ lake breeze

A thermally produced wind blowing during the day from the surface of a large lake to the shore, caused by the difference in the rates of heating of water and land.

▫ lake breeze front

The leading edge of a lake breeze, whose passage is often accompanied by showers, a wind shift, or a sudden drop in temperature.

▫ lake-effect storm

A fall or winter storm that produces heavy but localized snowfall over the lee shoreline of a large open-water lake. Air is moistened and warmed as it flows over the lake and snowfall is produced by convergence when the flow encounters increased roughness at the shoreline and is lifted up rising ground.

▫ land breeze

A coastal breeze at night blowing from land to sea, caused by the difference in the rates of cooling of their respective surfaces.

▫ lapse rate

The rate of decrease of air temperature with increase of elevation.

▫ latent heat

Heat absorbed or released during a change of water phase (from gas, liquid or solid phases) at constant temperature and pressure.

▫ latent heat flux

The flux of heat from the earth's surface to the atmosphere that is associated with evaporation or condensation of water vapor at the surface; a component of the surface energy budget.

▫ layered haze

Haze produced when air pollution from multiple line, area or point sources is transported long distances to form distinguishable layers of discoloration in a stable atmosphere.

❐ leaf area index

Projected foliage area per unit area of ground surface.

❐ lee

The side or part that is sheltered or turned away from the wind.

❐ lee wave

The wavelike effect, characterized by a series of updrafts and downdrafts, that occurs in the lee of a mountain range when rapidly flowing air is lifted up the steep front of a mountain range.

❐ leeward

The side away from the wind. Compare windward.

❐ lenticular cloud

A very smooth, round or oval, lens-shaped cloud that is often seen, singly or stacked in groups, near or in the lee of a mountain ridge.

❐ lightning

A visible electrical discharge produced by a thunderstorm. The discharge may occur within or between clouds, between the cloud and air, or between a cloud and the ground.

❐ line source

An array of pollutant sources along a defined path that can be treated in dispersion models as an aggregate uniform release of pollutants along a line. Example: the sum of emissions from individual cars traveling down a highway can be treated as a line source.

❐ live fuel moisture

The ratio of the amount of water to the amount of dry plant material in living plants.

❐ local convective wind

In fire weather terminology, local, diurnal, thermally driven winds arising over a comparatively small area and influenced by local terrain. Examples include sea and land breezes, lake breezes, diurnal mountain wind systems and convective currents.

❐ lofting

A pattern of plume dispersion in a stable boundary layer topped by a neutral stability layer, in which the upper part of the plume disperses upward while the lower part of the plume undergoes little dispersion.

❐ longwave radiation

A term used to describe the infrared energy emitted by the earth and atmosphere at wavelengths between about 5 and 25 micrometers.

❑ long-wave ridge

A ridge in the hemispheric Rossby wave pattern.

❑ long-wave trough

A trough in the hemispheric Rossby wave pattern.

❑ looping

A pattern of plume dispersion in an unstable atmosphere in which the plume undergoes marked vertical oscillations as it is alternately affected by rising convective plumes and the subsiding motions between the plumes.

❑ loran

Long Range Navigation, a system of long range navigation whereby latitude and longitude are determined from the time displacement of radio signals from two or more fixed transmitters.

❑ lower atmosphere stability index

An atmospheric index used to indicate the potential for rapid wildfire growth. This index, also called the haines index, contains a stability term and a dryness term.

❑ low-level jet

A regular, strong, nighttime, northward flow of maritime tropical air over the sloping Great Plains of the central United States, in which the wind increases to a peak in the lowest kilometer and then decreases above.

M

❑ maloja wind

A wind named after the Maloja Pass between the Engadine and the Bergell valleys of Switzerland, which blows down the valley of the Upper Engadine by day and up the valley at night. These wind directions are contrary to mountain wind system theory, and have been explained as the encroachment of winds from the Bergell into the Upper Engadine.

❑ marine air intrusion

Invasion of an air mass with marine characteristics into a continental area.

❑ marine inversion

Temperature inversion produced when cold marine air underlies warmer air.

❑ massif

A compact portion of a mountain range, containing one or more summits.

❑ mercury barometer

An instrument for measuring atmospheric pressure. The instrument contains an evacuated and graduated glass tube in which mercury rises or falls as the pressure of the atmosphere increases or decreases.

▫ mesopause

The top of the mesosphere, corresponding to the level of minimum temperature in the atmosphere found at 70 to 80 km.

▫ mesoscale

Pertaining to meteorological phenomena, such as wind circulations or cloud patterns, that are about 2 to 200 km in horizontal extent.

▫ mesosphere

The atmospheric shell between about 20 km and about 70 to 80 km, extending from the top of the stratosphere (the stratopause) to the upper temperature minimum that defines the mesopause (the base of the thermosphere).

▫ meteorology

The science dealing with the atmosphere and its phenomena.

▫ microburst

An intense, localized downdraft of air that spreads on the ground, causing rapid changes in wind direction and speed; a localized downburst.

▫ microclimate

The climate of a small area such as a cave, house, city or valley that may be different from that in the general region.

▫ microscale

Pertaining to meteorological phenomena, such as wind circulations or cloud patterns, that are less than 2 km in horizontal extent.

▫ mid-flame wind

Wind measured at the midpoint of the flames, considered to be most representative of the wind that is affecting fire behavior.

▫ millibar

A unit of atmospheric pressure equal to 1/1000 bar, or 1000 dynes per square centimeter.

▫ mixed layer

An atmospheric layer, usually the layer immediately above the ground, in which pollutants are well mixed by convective or shear-produced turbulence.

▫ mixing depth

Vertical distance between the ground and the altitude to which pollutants are mixed by turbulence caused by convective currents or vertical shear in the horizontal wind.

▫ moist adiabatic lapse rate

The rate at which the temperature of a parcel of saturated air decreases as the parcel is lifted in the atmosphere. The moist adiabatic lapse rate is not a constant like the dry adiabatic lapse rate but is dependent on parcel temperature and pressure.

❐ monsoon

A thermally driven wind arising from differential heating between a land mass and the adjacent ocean that reverses its direction seasonally.

❐ mountain meteorology

Meteorology of a mountainous or topographically complex area.

❐ mountain wave

1. The wavelike effect, characterized by updrafts and downdrafts, that occurs above and behind a mountain range when rapidly flowing air encounters the mountain range's steep front.
2. Generic term for all gravity waves occurring in the vicinity of or caused by mountains.
3. Specific term for waves that form above, rather than behind, mountains.

❐ mountain wind system

The system of diurnal winds that forms in a complex terrain area, consisting of mountain-plain, along-valley, cross-valley and slope wind systems.

❐ mountainado

A vertical-axis eddy produced in a downslope windstorm by the vertical stretching of horizontal roll vortices produced near the ground by vertical shear of the horizontal wind. Mountainadoes, when carried by the mean wind, can produce strong horizontal wind shears and wind gusts that are much more damaging than the general prevailing wind speeds.

❐ mountain-plain wind system

A closed, large-scale, diurnal, thermally driven circulation between the mountains and the surrounding plain. The mountain-to-plain flow making up the lower branch of the closed circulation usually occurs during nighttime, while the plain-to-mountain flow occurs during daytime.

N

❐ national ambient air quality standards

In the United States, national standards for the ambient concentrations in air of different air pollutants designed to protect human health and welfare.

❐ national fire danger rating system

A uniform fire danger rating system used in the United States that focuses on the environmental factors that impact the moisture content of fuels. Fire danger is rated daily over large administrative areas, such as national forests.

❐ net all-wave radiation

The net or resultant value of the upward and downward longwave and shortwave radiative fluxes through a plane at the earth-atmosphere interface; a component of the surface energy budget.

❐ neutral

An atmospheric stability condition that exists in unsaturated (saturated) air when the environmental lapse rate equals the dry (moist) adiabatic rate.

❐ nieve penitente

A spike or pillar of compacted snow, firn or glacier ice, caused by differential melting and evaporation. The pillars form most frequently on low-latitude mountains where air temperatures are near freezing, dew points are much below freezing and insolation is strong. Penitents are oriented individually toward the noon-day sun, and usually occur in east-west lines.

❐ nimbostratus

A cloud of the class characterized by a formless layer that is almost uniformly dark gray; a rain cloud of the layer type, of low altitude, usually below 8000 ft (2400 m).

❐ nonattainment area

An area that out of compliance with ambient air quality standards.

O

❐ occluded front

A composite of two fronts, formed as a cold front overtakes a warm or quasi-stationary front. Two types of occlusions can form depending on the relative coldness of the air behind the cold front to the air ahead of the warm or stationary front. A cold occlusion results when the coldest air is behind the cold front and a warm occlusion results when the coldest air is ahead of the warm front.

❐ orographic wave

A wavelike airflow produced over and in the lee of a mountain barrier.

❐ ozone

A form of oxygen, O_3. A powerful oxidizing agent that is considered a pollutant in the lower troposphere but an essential chemical in the stratosphere where it protects the earth from high-energy ultraviolet radiation from the sun.

P

❐ palmer drought severity index

An index used to gage the severity of drought conditions by using a water balance equation to track water supply and demand. This index is calculated weekly by the national weather service.

❐ particle trajectory model

A computer sub-model that tracks the trajectories of multiple particles that are released into an atmospheric flow model.

❐ permafrost

A layer of soil at varying depths below the surface in which the temperature has remained below freezing continuously from a few to several thousands of years.

❐ perturbation model

A computer model used to calculate air pollution concentrations. A perturbation model produces a wind field from solutions to a simplified set of equations that describe atmospheric motions.

❐ phenomenological model

A computer model used to calculate air pollution concentrations. A phenomenological model focuses on an individual phenomenon, such as plume impingement or fumigation.

❐ photochemical smog

Air pollution containing ozone and other reactive chemical compounds formed by the reaction of nitrogen oxides and hydrocarbons in the presence of sunlight.

❐ pilot balloon

A small helium-filled meteorological balloon that is tracked as it rises through the atmosphere to determine how wind speed and direction change with altitude.

❐ plume blight

Visibility impairment caused by air pollution plumes aggregated from individual sources.

❐ plume impingement

The collision of an air pollution plume with topography that rises above the plume altitude; often a temporary condition that occurs as the plume sweeps by the face of a hill as the wind shifts.

❐ plume-dominated fire

A fire whose behavior is governed primarily by the local wind circulation produced in response to the strong convection above the fire rather than by the general wind.

❐ point source

A pollutant source that can be treated in a dispersion model as though pollutants were emitted from a single point that is fixed in space. Example: the mouth of a smokestack.

❐ polar front

The variable frontal zone of mid latitudes separating air masses of polar and tropical origin.

❐ **polar front jet**

A strong, generally westerly jet stream wind concentrated in a relatively narrow and shallow current in the upper troposphere above the polar front; a feature of the general circulation.

❐ **powder snow**

Dry, loose, unconsolidated snow.

❐ **prescribed fire**

A management-ignited or natural wildland fire that burns under specified conditions where the fire is confined to a predetermined area and produces the fire behavior and fire characteristics required to attain planned resource management objectives.

❐ **pressure**

The exertion of force upon a surface by a fluid (e.g., the atmosphere) in contact with it.

❐ **pressure gradient force**

The force caused by the change in atmospheric pressure per unit of horizontal distance, and acting in the direction in which pressure changes most rapidly.

❐ **pressure-driven channeling**

Channeling of wind in a valley by synoptic-scale pressure gradients superimposed along the valley's axis.

❐ **prevention of significant deterioration**

A program, specified in the Clean Air Act, whose goal is to prevent air quality from deteriorating significantly in areas of the country that are presently in compliance with ambient air quality standards.

❐ **primary ambient air quality standards**

Air quality standards designed to protect human health.

❐ **primary pollutant**

Substances that are pollutants immediately on entering the atmosphere.

❐ **pyro-cumulonimbus**

Cumulonimbus formed in the convection column of a fire.

❐ **pyro-cumulus**

Cumulus formed in the convection column of a fire.

R

❐ **radar**

Radio detecting and ranging; a device used to detect and determine the range to distant objects (e.g., hydrometeors) or atmospheric discontinuities by measuring the time for the echo of a radio wave to return from it and the direction from which it returns.

❒ radiation

Energy transport through electromagnetic waves.

❒ radiation fog

A fog that forms when outgoing longwave radiation cools the near-surface air below its dew point temperature.

❒ radiosonde

An instrument that is carried aloft by a balloon to send back information on atmospheric temperature, pressure and humidity by means of a small, expendable radio transmitter. Radiosondes can be tracked by radar, radio direction finding, or navigation systems (such as the satellite global positioning system) to obtain wind data. See also rawinsonde.

❒ rain

Precipitation that falls to earth in droplets with diameters greater than 0.5 mm.

❒ rain shadow

An area of reduced precipitation on the lee side of a mountain barrier caused by warming of air and dissipation of cloudiness as air descends the barrier.

❒ red flag warning

A term used by fire weather forecasters to alert land managers to an imminent or ongoing weather event that could cause dangerous fire activity.

❒ regional haze

Haze that is mixed uniformly between the surface and the top of a convective boundary layer.

❒ relative humidity

The ratio of the actual water vapor pressure at a given time to the vapor pressure that would occur if the air were saturated at the same ambient temperature.

❒ residence time

The time, in seconds, required for the flaming front of a fire to pass a stationary point at the surface of the fuel.

❒ residual layer

The elevated portion of a convective boundary layer that remains after a stable boundary layer develops at the ground (usually in late afternoon or early evening) and cuts off convection.

❒ resonance

The state of a system in which an abnormally large vibration is produced in response to an external stimulus, occurring when the frequency of the stimulus is the same, or nearly the same, as the natural vibration frequency of the system.

- **retrograding**

A westward shifting of a Rossby wave pattern that normally propagates to the east.

- **return stroke**

An electrical discharge that propagates upward along a lightning channel from the ground to the cloud.

- **ridge**

On a weather chart, a narrow elongated area of relatively high pressure.

- **ridging**

The building or intensification of a ridge or high pressure center.

- **rime ice**

An opaque coating of tiny, white, granular ice particles caused by the rapid freezing of supercooled water droplets on impact with an object.

- **ringfire**

A firing pattern or fire ignition pattern for prescribed fires, in which the fire is ignited in a ring to produce strong central convection, inward spread and high fire intensity; variant of a centerfire.

- **rossby waves**

A series of troughs and ridges on quasi-horizontal surfaces in the major belt of upper tropospheric westerlies. The waves are thousands of kilometers long and have significant latitudinal amplitude.

- **rotor cloud**

A turbulent altocumulus or cumulus cloud formation found in the lee of some mountain barriers when winds cross the barrier at high speed. The air in the cloud rotates around an axis parallel to the range.

S

- **sampling frequency**

The rate at which sensor data are read or sampled.

- **santa ana wind**

(in southern California) a strong, hot, dust-bearing foehn wind that descends to the Pacific Coast around Los Angeles from inland desert regions.

- **sastrugi**

Ridges of snow formed by wind on a snow field.

- **saturation vapor pressure**

The vapor pressure of a system at a given temperature, wherein the vapor of a substance is in equilibrium with a plane surface of that substance's pure liquid or solid phase.

❒ scattering

The process in which a beam of light is diffused or deflected by collisions with particles suspended in the atmosphere.

❒ sea breeze

A thermally driven wind blowing during the day from over a cool ocean surface onto the adjoining warm land, caused by the difference in the rates of heating of the ocean and land surfaces.

❒ sea breeze convergence zone

The zone at the leading edge of a sea breeze where winds converge. The incoming air rises in this zone, often producing convective clouds.

❒ sea breeze front

The leading edge of a sea breeze, whose passage is often accompanied by showers, a wind shift, or a sudden drop in temperature.

❒ secondary ambient air quality standards

Air quality standards designed to protect human welfare, including the effects on vegetation and fauna, visibility and structures.

❒ secondary pollutant

Pollutants generated by chemical reactions occurring within the atmosphere. Compare primary pollutant.

❒ sensible heat flux

The flux of heat from the earth's surface to the atmosphere that is not associated with phase changes of water; a component of the surface energy budget.

❒ separation eddy

An eddy that forms near the ground on the windward or leeward side of a bluff object or steeply rising hillside; streamlines above this eddy go over the object.

❒ severe thunderstorm

A thunderstorm that produces heavy precipitation, frequent lightning, strong, gusty surface winds or hail. A severe thunderstorm can cause flash floods and wind and hail damage and may spawn tornadoes.

❒ shortwave radiation

A term used to describe the radiant energy emitted by the sun in the visible and near-ultraviolet wavelengths (between about 0.1 and 2 micrometers). Compare longwave radiation.

❒ short-wave ridge

A relatively small-scale ridge that is superimposed on and propagates through the longer wavelength Rossby waves.

❑ short-wave trough

A relatively small-scale trough that is superimposed on and propagates through the longer wavelength Rossby waves.

❑ sleet

In the U.S., a term used to describe tiny ice pellets that are formed when rain or partially melted snowflakes refreeze before reaching the ground. Colloquial usage of the term coincides with British usage, which defines sleet as a mixture of rain and snow.

❑ sling psychrometer

An instrument used to measure the water vapor content of the atmosphere, in which wet and dry bulb thermometers are mounted on a frame connected to a handle at one end by means of a bearing or a length of chain. The psychrometer is whirled by hand to provide the necessary ventilation to evaporate water from the wet bulb.

❑ slope wind system

A closed, thermally driven diurnal mountain wind circulation whose lower branch blows up or down the sloping sidewalls of a valley or mountain. The upper branch blows in the opposite direction, thereby closing the circulation.

❑ smoke management

The use of meteorology, fuel moisture, fuel loading, fire suppression and burn techniques to keep smoke impacts from prescribed fires within acceptable limits.

❑ snow

Precipitation in the form of ice crystals, mainly of intricately branched, hexagonal form and often agglomerated into snowflakes, formed directly from the freezing of the water vapor in the air.

❑ snow cornice

A mass of snow or ice projecting over a mountain ridge.

❑ snow grain

Precipitation consisting of white, opaque ice particles usually less than 1 mm in diameter.

❑ snow pellets

Precipitation, usually of brief duration, consisting of crisp, white, opaque ice particles, round or conical in shape and about 2 to 5 mm in diameter.

❑ snow pillow

A windrow of snow deposited in the immediate lee of a snow fence or ridge.

❑ snowflake

An agglomeration of snow crystals falling as a unit.

❐ sodar

Sound Detection and Ranging; an electronic instrument used to detect and determine the range to distant objects (e.g., hydrometeors) or atmospheric discontinuities in temperature or humidity structure that scatter or reflect sound energy.

❐ solstice

Either of the two times per year when the sun is at its greatest angular distance from the celestial equator: about June 21 (Northern Hemisphere summer solstice), when the sun reaches its northernmost point on the celestial sphere, or about December 22 (Northern Hemisphere winter solstice) when it reaches its southernmost point.

❐ sounding

A set of data measuring the vertical structure of one or more atmospheric parameters (e.g., temperature, humidity, pressure, wind) at a given time.

❐ specific gravity

The ratio of the density of any substance to the density of water.

❐ spot fire

A fire ignited outside the perimeter of the main fire by a firebrand.

❐ spotfire

A firing pattern or fire ignition pattern for prescribed fires, in which a fire is ignited by firing equally spaced spots in rapid succession or using simultaneous ignition. The spacing is chosen so that the fires draw together rather than making individual runs.

❐ spotting

Outbreak of secondary fires as firebrands or other burning materials are carried ahead of the main fire line by winds.

❐ spray block

The geographical area to be sprayed.

❐ stability

The degree of resistance of a layer of air to vertical motion.

❐ stable boundary layer

The stably-stratified layer that forms at the surface and grows upward, usually at night or in winter, as heat is extracted from the atmosphere's base in response to longwave radiative heat loss from the ground. Stable boundary layers can also form when warm air is advected over a cold surface or over melting ice.

❐ stable core

Post-sunrise, elevated remnant of the temperature inversion that builds up overnight within a valley.

▢ standard atmosphere

A hypothetical vertical distribution of temperature, pressure and density which, by international consent, is taken to be representative of the atmosphere for purposes of pressure altimeter calibrations, aircraft performance calculations, aircraft and missile design, ballistic tables, etc.

▢ state implementation plan

A formal air quality management plan, produced by an individual state, specifying how state air resources will be managed to achieve federal and state standards.

▢ station model

A specified pattern for plotting, on a weather map, the meteorological symbols that represent the state of the weather at a particular observing station.

▢ stationary front

A front between warm and cold air masses that is moving very slowly or not at all.

▢ stepped leader

A faint, negatively charged channel that emerges from the base of a thunderstorm and propagates toward the ground in a series of steps of about 1 microsecond duration and 50-100 meters in length, initiating a lightning stroke.

▢ stratiform

(Of a cloud) having predominantly horizontal development.

▢ stratocumulus

A cloud of a class characterized by large dark, rounded masses, usually in groups, lines, or waves, the individual elements being larger than those in altocumulus and the whole being at a lower altitude, usually below 8000 ft (2400 m).

▢ stratopause

The boundary between the stratosphere and mesosphere.

▢ stratosphere

The atmospheric shell above the troposphere and below the mesosphere. It extends from the tropopause, usually 10-25 km high, to the height where temperature begins to decrease in the 20 to 25 km region.

▢ stratus

A cloud of a class characterized by a gray, horizontal layer with a uniform base, found at a lower altitude than altostratus, usually below 8000 ft (2400 m).

▢ streamline

The path of an air parcel that flows steadily over or around an obstacle.

▢ subsidence

A descending motion of air in the atmosphere occurring over a rather broad area.

❐ subsidence inversion

A temperature inversion that develops aloft as a result of air gradually sinking over a wide area and being warmed by adiabatic compression, usually associated with subtropical high pressure areas.

❐ subtropical jet

A strong, generally westerly wind concentrated in a relatively narrow and shallow stream in the subtropical upper troposphere; a feature of the general circulation.

❐ supercooled

A liquid cooled below its freezing point without solidification or crystallization.

❐ surface energy budget

The energy or heat budget at the earth's surface, considered in terms of the fluxes through a plane at the earth-atmosphere interface. The energy budget includes radiative, sensible, latent and ground heat fluxes.

❐ surface weather chart

An analyzed synoptic chart of surface weather observations. A surface chart shows the distribution of sea-level pressure (therefore, the position of highs, lows, ridges and troughs) and the location and nature of fronts and air masses. Often added to this are symbols for occurring weather phenomena. Although the pressure is referred to mean sea level, all other elements on this chart are presented as they occur at the surface point of observation.

❐ swath width

The width of the spray swath parallel to the path of the aerial or ground application vehicle over which a desired spray deposition is achieved.

❐ synoptic-scale

Any phenomenon occurring on the spatial scale of the migratory high and low pressure systems of the lower troposphere, with wavelengths of 1000 to 2500 km and on time scales exceeding 12 hours.

T

❐ teleconnection

Linkage between changes in atmospheric circulation occurring in widely separated parts of the globe.

❐ temperature inversion (surface-based or elevated)

A layer of the atmosphere in which air temperature increases with height. When the layer's base is at the surface, the layer is called a surface-based temperature inversion; when the base of the layer is above the surface, the layer is called an elevated temperature inversion.

❐ terpene

Any of a class of monocyclic hydrocarbons of the formula $C_{10}H_{16}$, obtained from plants.

❐ terrain-forced flow

An airflow that is modified or channeled as it passes over or around mountains or through gaps in a mountain barrier.

❐ thermal belt

A zone of high nighttime temperature and relatively low humidity that is often experienced within a narrow altitude range on valley sidewalls, especially evident during clear weather with light winds.

❐ thermally driven circulation

A diurnally reversing closed cellular wind current resulting from horizontal temperature contrasts caused by different rates of heating or cooling over adjacent surfaces; includes slope, cross-valley, along-valley, mountain-plain and sea breeze circulations.

❐ thermistor

A resistor whose resistance changes with temperature. Because of the known dependence of resistance on temperature, the resistor can be used as a temperature sensor.

❐ thermosphere

The atmospheric shell extending from the top of the mesosphere to outer space. It is a region of more or less steadily increasing temperature with height, starting at 70 or 80 km.

❐ thunder

The sound caused by rapidly expanding gases in a lightning discharge.

❐ thunderstorm

A local storm produced by a cumulonimbus cloud and accompanied by lightning and thunder.

❐ time constant

The time required for a measuring instrument to respond to 63.2% of a stepwise change in a measured quantity.

❐ time lag

1. Same as time constant.
2. The time needed under specific atmospheric conditions for a fuel particle to lose about 63 percent of the difference between its initial moisture and its equilibrium moisture contents.

❐ towering cumulus

A tall cumulus cloud, extending through low and middle cloud levels.

❐ trade winds

Any of the nearly constant easterly winds that dominate most of the tropics and subtropics, blowing mainly from the northeast in the Northern Hemisphere and from the southeast in the Southern Hemisphere.

❑ transpiration

The passage of water vapor into the atmosphere through the vascular system of plants.

❑ transport wind

For smoke management computations, the average wind speed and direction through the depth of the mixed layer.

❑ trapper

A valley or basin in which cold air becomes trapped or pooled.

❑ tropopause

The boundary between the troposphere and stratosphere, characterized by an abrupt change in temperature lapse rate (temperatures decrease with height in the troposphere, but increase or remain constant with height in the stratosphere).

❑ troposphere

The portion of the earth's atmosphere from the surface to the tropopause; that is, the lowest 10-20 km of the atmosphere. The troposphere is characterized by decreasing temperature with height, and is the layer of the atmosphere containing most clouds and other common weather phenomena.

❑ trough

On a weather chart, a narrow, elongated area of relatively low pressure.

❑ troughing

The deepening or intensification of a trough or low pressure center.

❑ turbulence

Irregular motion of the atmosphere, as indicated by gusts and lulls in the wind.

❑ twenty-foot wind

Wind measured or estimated over a 2- to 10-minute period at a standard 20-foot level above the vegetative surface or continuous tree canopy and used for input into fire danger models or for fire suppression activities.

U

❑ undersun

An optical effect seen by an observer above a cloud deck when looking toward the sun, as sunlight is reflected upwards off the faces of ice crystals in the cloud deck.

❑ upper-air weather chart

Weather maps that are produced for the portion of the atmosphere above the lower troposphere, generally at and above 850 mb. Isolines on these maps usually represent the heights of a constant pressure surface, such as the 500 mb surface.

❑ upslope fog

A fog that forms when moist, stable air is cooled when it is lifted up a mountain slope.

❑ upslope wind

A diurnal thermally driven flow directed up a mountain slope, and usually occurring during daytime; part of the slope wind system.

❑ up-valley wind

A diurnal thermally driven flow directed up a valley's axis, usually occurring during daytime; part of the along-valley wind system.

❑ upwelling

The process by which warm, less dense surface water is drawn away from a shoreline by offshore currents and replaced by cold, denser water brought up from the subsurface.

V

❑ valley exit jet

A strong elevated down-valley air current issuing from a valley above its intersection with the adjacent plain.

❑ valley volume effect

The reduction in volume of a valley (or basin) as compared to an equal depth volume with a horizontal floor. Because the valley volume is smaller, equivalent heat fluxes cause larger changes in temperature in the valley volume than in the flat-floor volume.

❑ ventilation index

Product of mixing depth and transport wind speed, a measure of the potential of the atmosphere to disperse airborne pollutants from a stationary source.

❑ venturi effect

The speedup of air through a constriction due to the pressure rise on the upwind side of the constriction and the pressure drop on the downwind side as the air diverges to leave the constriction.

❑ virga

Streaks of water droplets or ice particles falling out of a cloud and evaporating before reaching the ground.

❑ visibility

The distance at which a given standard object can be seen and identified with the unaided eye.

❑ visibility protection program

The program specified by the Clean Air Act to achieve a national goal of remedying existing impairments to visibility and preventing future visibility impairment throughout the United States.

❑ visible spectrum

The portion of the electromagnetic spectrum to which the eye is sensitive, i.e., light with wavelengths between 0.4 and 0.7 micrometers.

❑ volatile fraction

Mass fraction of a sprayed liquid that could evaporate in the atmosphere before depositing.

❑ vortex

A whirling mass of air in the form of a column or spiral. It need not be oriented vertically but, for example, could be rotating around a horizontal axis.

W

❑ wake

The region of turbulence immediately behind a solid body caused by the flow of air over or around the body.

❑ warm front

A transition zone between a mass of warm air and the colder air it is replacing.

❑ warm occlusion

A frontal zone formed when a cold front overtakes a warm front and, finding colder air ahead of the warm front, leaves the ground and rises up and over this denser air.

❑ warm sector

The region of warm air within a low pressure center located between an advancing cold front and a retreating warm front.

❑ wasatch wind

A strong easterly wind blowing out of the mouths of the canyons of the Wasatch Mountains onto the plains of Utah.

❑ weather

The state of the atmosphere with respect to wind, temperature, cloudiness, moisture, pressure, etc.

❑ westerlies

The prevailing winds that blow from the west in the mid-latitudes of both the Northern and Southern Hemspipheres.

❑ wildfire

An unwanted fire that requires measures of control.

❑ wind chill equivalent temperature

The apparent temperature felt on the exposed human body owing to the combination of temperature and wind speed.

❑ wind field

The three-dimensional spatial pattern ot winds.

❑ wind rose

A diagram, for a given locality or area, showing the frequency and strength of the wind from various directions.

❑ wind shear

The rate at which wind velocity changes from point to point in a given direction (as, vertically). The shear can be speed shear (where speed changes between the two points, but not direction), direction shear (where direction changes between the two points, but not speed) or a combination of the two.

❑ wind-driven fire

A fire whose behavior is governed primarily by a strong consistent wind, rather than by the convective circulation produced by the fire itself.

❑ winds of most concern to firefighters

Winds that dominate the fire environment, rendering fire suppression activities dangerous or ineffective.

❑ windthrow

A tree or stand of timber that has been blown down by the wind.

❑ windward

The side toward the wind.

Pollution Control

A

❐ abandon

Used in reference to wells, abandonment is a process by which a well is filled and sealed from bottom to top.

❐ absorb

To take up or receive by chemical or molecular action.

❐ acclimation

Response by an animal that enables it to tolerate a change in a single factor (for example, temperature) in its environment.

❐ acenaphthene

A white to light yellowish solid, used to manufacture plastics, and as an intermediate in the manufacture of dyes, pharmaceuticals, insecticide, fungicide, herbicide and plant growth hormones. Chemical formula = $C_{12}H_{10}$. Molecular weight = 154.21 g/mol.

❐ acenaphthylene

A polycyclic aromatic hydrocarbon (PAHs) that is an odiforous chemical component of crude oil and coal tar. Chemical formula = $C_{12}H_8$. Molecular weight = 152.20 g/mol.

❐ acetaldehyde

A colorless, volatile, water-soluble liquid aldehyde used primarily in organic synthesis. Chemical formula = CH_3CHO. Molecular weight = 44.06 g/mol.

❐ acetamide

A white, crystalline amide of acetic acid used primarily as a solvent and also in organic synthesis. Chemical formula = C_2H_5NO. Molecular weight = 59.07 g/mol.

❐ acetone

A volatile, fragrant, flammable liquid ketone used primarily as a solvent and also in organic synthesis. Chemical formula = C_3H_6O. Molecular weight = 58.08 g/mol.

❐ acetonitrile

The colorless liquid nitrile of acetic acid used primarily in organic synthesis and as a solvent. Chemical formula = C_2H_3N. Molecular weight = 41.05 g/mol.

acetophenone

A clear to light yellow liquid used as a fragrance ingredient in soaps, detergents, creams, lotions, and perfumes. It is also used as a flavoring agent in foods and beverages, and as a specialty solvent for plastics and resins. Chemical formula = C_8H_8O. Molecular weight = 120.15 g/mol.

acid

A material with pH of less than 7.0.

acid rain

- The term 'acid rain' was first coined in 1872, after a link was made between sulfur dioxide emissions from the burning of coal in Manchester, England, to the acidification of nearby rainfall.
- The term now tends to be used indiscriminately to describe all forms of pollution associated with the burning of fossil fuels. Not only rain, but also snow, sleet, and mist, collectively known as 'wet deposition'; and gases, dusts, and smog, collectively known as 'dry deposition' are now included in the definition of acid rain.

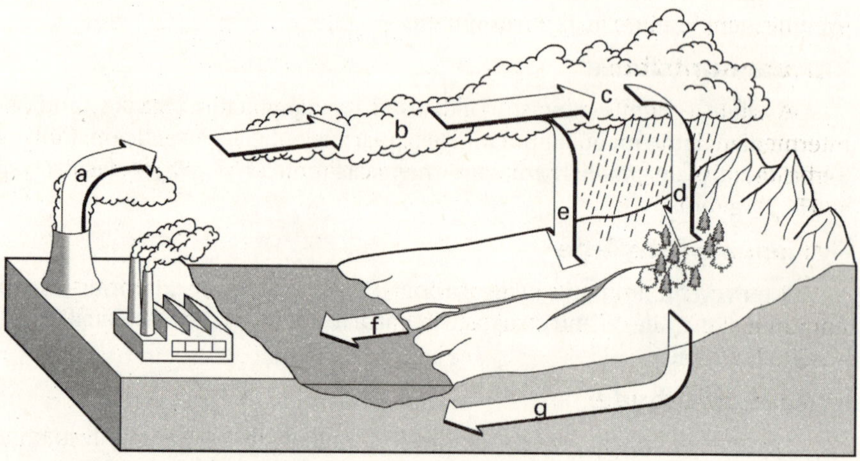

a Acid rain is produced by the release of sulfur dioxide (SO_2) and nitrogen oxides (NO_x) into the atmosphere from factories and power stations
b Gases combine with water to make sulfuric and nitric acids
c Acidic clouds can travel for long distances from the source
d Acid rain and snow fall on trees, killing them
e Acid rain runs into river, killing fish and other organisms
f Acidified water runs into lake, killing fish
g Acidified water leaches nutrients and heavy metals out of soil. These wash into the lake, polluting it

acid rain 2: the global picture

- In the 1980s, 'acid rain' was identified as a major international environmental problem, spilling over from densely populated and heavily industrialized areas of Europe and North America into prime agricultural areas.
- Acid deposition pollution can travel hundreds of miles across state and national boundaries.

- Because their higher rainfall increases the volume of acid deposition, and their often thin soils cannot neutralize the acid, mountain regions suffer most from acid deposition.
- Lakes and streams in parts of Scandinavia and Scotland far from industrial centres have become acidified, with considerable fish stock depletion.
- The `black triangle' bordering Germany, the Czech Republic and Poland has been the scene of the most intense fallout.
- Acid rain has also been the major cause of trees dying (Waldsterben), a particular problem in Central and Eastern Europe.

Global density of industrial facilities at the end of the 20th century

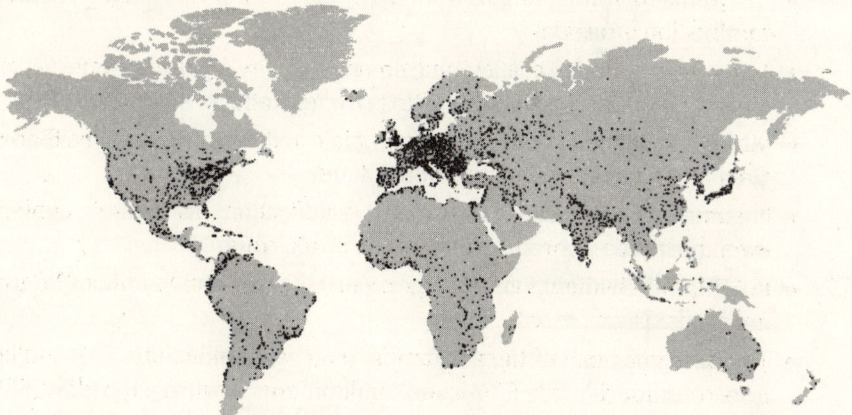

Acid rain levels and extent of forest damage in Europe

percent forest damaged

- high (over 25 percent)
- moderate (15–25 percent)
- light (under 15 percent)

a Great Britain
b Sweden
c Netherlands
d Belgium
e Germany
f Luxembourg
g France
h Switzerland
i Poland
j Czech Rep./Slovakia
k Austria
l Hungary

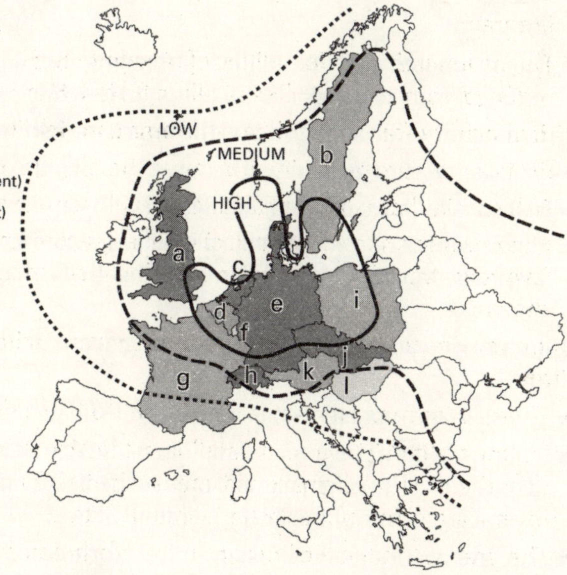

❑ acid rain 3: eastern north america

- Acid rain has widely affected North America for several decades, leading to the acidification of lakes and streams, causing respiratory problems in people, thinning forests, and corroding monuments and buildings.
- Human activity is responsible for over 90 percent of the SO_2 and NO_x present in the atmosphere over eastern North America.
- Coal-, and to a lesser extent, oil-burning electric power plants, and industries such as ore smelting are the major sources of sulfur emissions.
- About 40 percent of NO_x emissions come from the transportation sector, (cars, trucks, planes, etc.).
- The remainder are emitted from fossil fuel-fired power plants and other combustion processes.
- Acid rain can diminish lakes' and streams' ability to sustain some fish and other aquatic species' capability to grow, reproduce, or survive.
- More than 90 percent of streams are acidic in the New Jersey Pine Barrens area, the highest rate in the United States.
- In some sensitive lakes and streams, acidification has severely depleted even the hardiest species, such as the brook trout.
- In 1990, the United States Congress established the Acid Rain Program under the Clean Air Act.
- The principal goal of the program is to achieve reductions of 10 million tons of sulfur dioxide (SO_2) and 2 million tons of nitrogen oxides (NO_x), the primary components of acid rain, by 2010.
- Emissions of SO_2 and NO_x are declining, mainly as a result of the Acid Rain Program.
- The estimated costs to utilities of meeting the SO_2 2010 target has been reduced from an original $4.6 billion to less than $1 billion.

Critical acidity for selected North American fish in lakes and streams

- PH is a measure of acidity. The lower the pH, the more acidic the water.
- Fish species have different abilities to withstand excess acidity.
- Black symbols show favorable pH ranges; white symbols are placed in less favorable ranges. No symbol means that that fish cannot survive in acidity shown.

Reduction in acid rain (SOx) resulting from Acid Rain Program 1995 legislation

- These maps represent snapshots of wet sulfate deposition over time.
- Following the 1995 implementation of the Acid Rain Program, total sulfate deposition fell in a dramatic and unprecedented reduction of up to 25 percent over a large area of the Eastern United States.
- The greatest reductions occur in the Northeastern United States, where many of the most acid sensitive ecosystems are located.

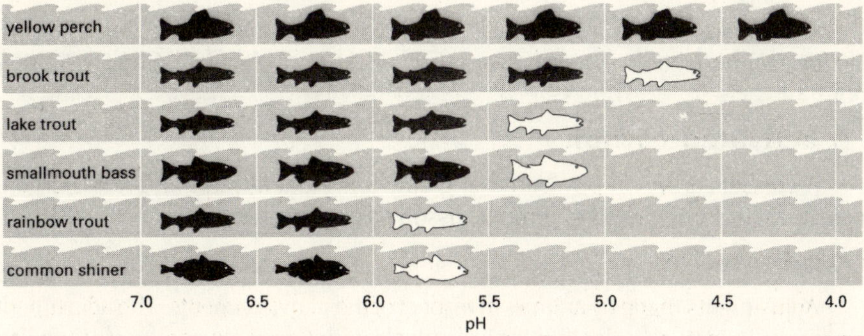

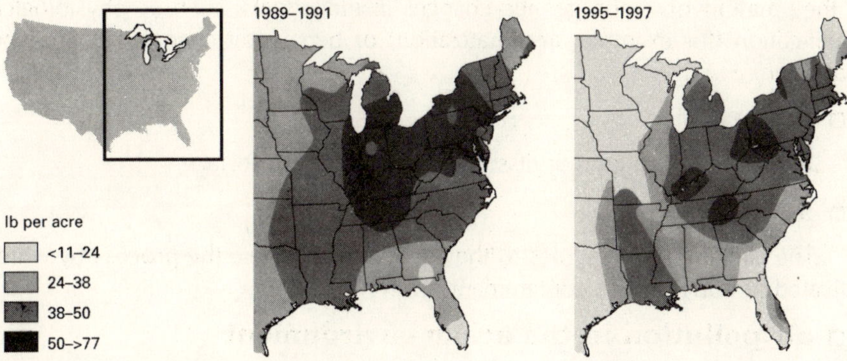

❒ acid rain

Rain with a higher than normal acid range. Caused when polluted air mixes with cloud moisture. The "acid" in acid rain comes from sulfur oxides and nitrogen oxides, products of burning coal and other fuels and from certain industrial processes. Acid rain can make lakes devoid of fish and damage human health and property.

❒ acrolein

A colorless, irritant, pungent liquid aldehyde used primarily in organic synthesis. Chemical formula = C_3H_4O. Molecular weight = 56.06 g/mol.

❒ acrylamide

An amide that is derived from acrylic acid, that polymerizes easily, and is used in the manufacture of synthetic textile fibers. Chemical formula = C_3H_5NO. Molecular weight = 71.08 g/mol.

❒ acrylic acid

CAS Number: 79-10-7. An unsaturated liquid acid that is used as a constituent for varnishes and lacquers. Chemical formula = $C_3H_4O_2$. Molecular weight = 72.06 g/mol.

❑ acrylonitrile

A colorless, volatile, flammable liquid nitrile used primarily in organic synthesis, and for polymerization. Chemical formulat = C_3H_3N. Molecular weight = 53.06 g/mol.

❑ activated carbon

A special carbon in powdered or grainy form that attracts and holds organic molecules. It is often used to remove dissolved contaminants.

❑ adaptation

Adjustments made by animals in respect of their environments. The adjustments may occur by natural selection, as individuals with favorable genetically acquired traits breed more prolifically than those lacking these traits (genotypic adaptation), or they may involve non-genetic changes in individuals, such as physiological modification (for example, acclimatization) or behavioral changes (phenotypic adaptation).

❑ adsorb

To gather a gas, liquid or dissolved substance on a surface.

❑ adsorption

The clinging of molecules to the surface of particles; the process by which activated carbon removes contaminants from water.

❑ air pollution in the urban environment

- Approximately half of the world's population now lives in urban areas, and half of them are exposed to potentially harmful amounts of particulate matter in smogs, sulfur dioxide (SO_2), and ozone.
- Winter smogs largely arise from burning coal to heat buildings during cold weather. When the coal smoke and SO_2 combine with fog in windless weather they create a pollution cap that the sun is not strong enough to clear.
- China's winter smogs cause more than 50,000 premature deaths and 400,000 new cases of chronic bronchitis a year in 11 of its largest cities alone.
- Conditions are worst in thin air at higher altitudes and if the air is trapped inside a valley or under a layer of warm air ('temperature inversion').
- Summer smogs, first reported in Los Angeles, consist of pollutants – mainly from vehicle exhausts – that undergo photochemical changes in bright sunlight, creating substances such as ozone, a gas that can trigger asthma attacks.
- Both situations apply in Mexico City, the world's second largest urban agglomeration, where smog alerts close factories and force cars off city streets several times a year.
- Globally, some 50 percent of cases of chronic respiratory illness are now thought to be associated with air pollution.

- There have also been some improvements in local and regional air quality over the past decade, although significant problems remain.
- In the United States, except for increased NO_x emissions of about 14 percent, emissions of CO_2, volatile organic compounds (VOCs), particulates and SO_2 decreased between 1970 and 1994, because of the 1990 Clean Air Act. Lead emissions had the most spectacular decline (98 percent over the same period), due to the adoption of unleaded fuels.
- In a recent analysis of ozone health impacts in 13 cities, the American Lung Association estimated that high ozone levels were responsible for 10,000 to 15,000 extra hospital admissions, and 30,000 to 50,000 additional emergency visits to hospitals.
- Over the next 10 years, air quality may improve in some cities but is likely to decline further in others, particularly those with growing populations and increased automobile use.

Temperature inversion

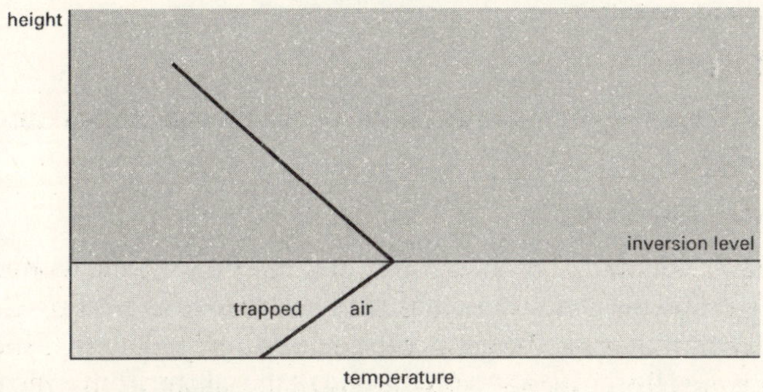

Relative importance of urban pollutants

effects on	health	visual amenity	buildings	flora and fauna	perceived wellbeing
CO	low	low	low	low	low
NO_x	medium	medium	low	low	medium
SO_x	medium	medium	high	low	high
VOCs	high	high	medium	low	medium
particles	high	high	medium	medium	high
lead	high	low	low	low	medium

❑ air quality standards

Federal and state government- prescribed levels of a pollutant in the outside air that cannot be exceeded during a specified period of time in a specified geographical area.

❐ air stripping

Removing contaminants from water by enlarging the water's surface area and forcing air over it to evaporate volatile contaminants.

❐ air toxics

A category of substances in the air that are known or suspected of causing cancer or other health problems in humans, and for which a National Ambient Air Quality standard (NAAQS) does not exist (i.e. excluding ozone, carbon monoxide, PM-10, sulfur dioxide, nitrogen oxide).

❐ aldrin

An exceedingly poisonous cyclodiene insecticide. Chemical formula = $C_{12}H_8Cl_6$. Molecular weight = 344.372 g/mol.

❐ algae

Simple rootless plants that grow in bodies of water in relative proportion to the amount of nutrients available.

❐ algal bloom

An unusual, sudden or excessive abundance of algae. Algal blooms can adversely affect water quality.

❐ aliphatic aldehydes

The aliphatic aldehydes are a homologous series of organic compounds of carbon, hydrogen and oxygen, where the aldehyde functional group, -C=O.H , is contained within the molecule. The general formula is $CnH_2n+2.CHO$. The aliphatic aldehydes are named by replacing "-e" at the end of the alkane from which they are derived with "-al".

❐ alkaline

Capacity of a lake to neutralize acid.

❐ alkalinity

Alkalinity is measured in mg/l as calcium carbonate ($CaCO_3$). It represents a measure of a solution's ability to buffer or neutralize acids. Lakes located in areas of calcareous glacial till (common throughout central and southern Minnesota) will have higher alkalinity than lakes formed on non-calcareous bedrock (common in northeastern Minnesota). Water with alkalinity less than about 75 mg/L could be considered soft, 76-150 moderately hard, 151-300 hard, and greater than 300 very hard. Alkalinity has also been used as a basis for estimating sensitivity to acid precipitation. For this purpose, lakes with alkalinity values less than 5 to 10 mg/L could be considered potentially sensitive to acid precipitation based on current levels of deposition across Minnesota.

❐ allyl chloride

Also known as: 3-Chloropropene. A clear, flammable, toxic liquid that is used to make epichlorohydrin and glycerin, and in production of allyl compounds. Also used in production of varnish, plastics, adhesvies, perfumes, pharmaceuticals and insecticides. Chemical formula = C_3H_5Cl. Molecular weight = 76.53 g/mol.

❑ aluminum

A lightweight, bluish, silver-white, malleable metallic element that has good electrical and thermal conductivity, high reflectivity, and resistance to oxidation and is the most abundant metal in the earth's crust. Chemical symbol = Al. Molecular weight = 26.982 g/mol.

❑ ambient air quality

The state of quality of the air in the surrounding environment.

❑ ambient monitoring

Monitoring within natural systems (lakes, rivers, estuaries, and wetlands) to determine existing conditions.

❑ american rivers at risk

Endangered species

This map shows the 13 rivers classified by the American Rivers organization as endangered in 2001.

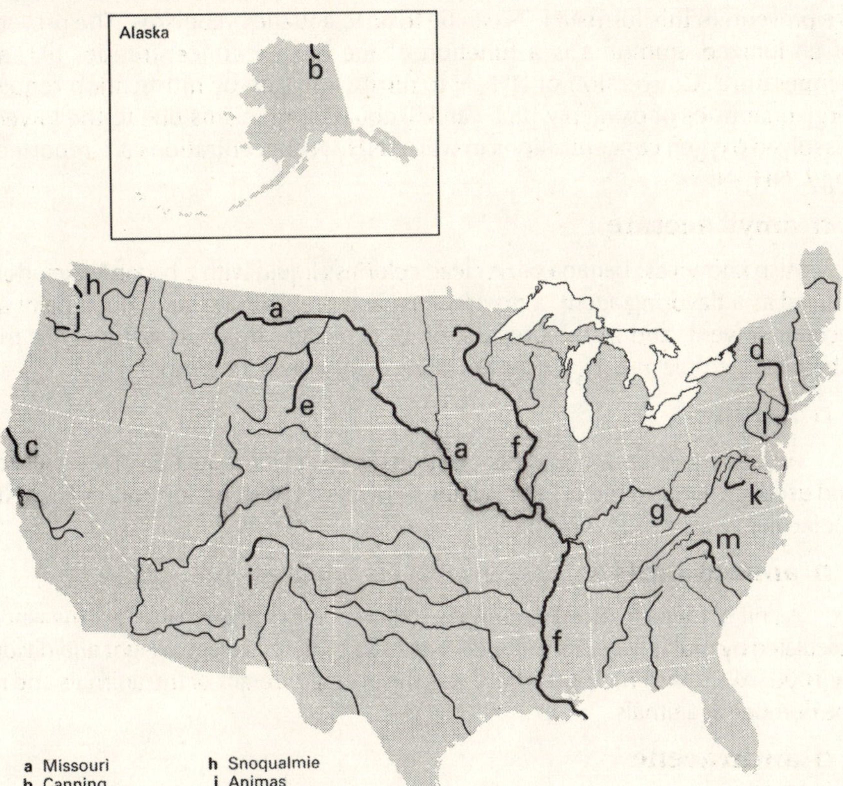

a Missouri
b Canning
c Eel
d Hudson
e Powder
f Mississippi
g Big Sandy
h Snoqualmie
i Animas
j East Fork Lewis
k Paine Run
l Hackensack
m Catawba

- Although pollution levels have fallen in recent years, many American rivers are still at risk.
- The organization American Rivers, named America's longest river, the Missouri, as the most endangered in 2001.
- Of the 13 rivers named as endangered, nearly half are threatened by energy development.
- Dams have reduced the flow of many rivers, threatening wildlife.
- Other rivers are endangered by mining operations.
- The most polluted United States river, by far, is the Mississippi.

❑ ammonia (nh3)

A pungent, colorless, gaseous alkaline compound of nitrogen and hydrogen that is very soluble in water and can easily be condensed to a liquid by cold and pressure. Chemical formula = NH_3. Molecular weight = 17.03 g/mol.

❑ ammonia (nh3-n)

An inorganic form of nitrogen, is contained in fertilizers, septic system effluent, and animal wastes. It is also a product of bacterial decomposition of organic matter. NH_3-N becomes a concern if high levels of the un-ionized form are present. In this form NH_3-N can be toxic to aquatic organisms. The presence of un-ionized ammonia is a function of the NH_3-N concentration, pH, and temperature. Conversion of NH_3-N to nitrite nitrogen by nitrification requires large quantities of oxygen which can kill aquatic organisms due to the lowered dissolved oxygen concentrations in water. NH_3-N concentrations are reported in mg/L NH_3-N.

❑ amyl acetate

Also known as: banana oil. A clear, colorless liquid with a banana-like odor. It is used as a flavoring agent, a dry cleaning and pre-spotting solution, a paint and lacquer solvent, and in the preparation of penicillin. It occurs naturally in fruit. Chemical formula = $C_7H_{14}O_2$. Molecular weight = 130.19 g/mol.

❑ aniline

An oily, liquid, poisonous amine obtained through the reduction of nitrobenzene and used primarily in the organic synthesis of dyes. Chemical formula = $C_6H_5NH_2$. Molecular weight = 93.12 g/mol.

❑ animal units

A unit of measure used to compare different types of animals. This measure is calculated by multiplying the number of animals by their average weight and dividing by 1000: AU = (w * n)/1000 where w is the average weight of the animals and n is the number of animals.

❑ anthracene

A crystalline, aromatic hydrocarbon obtained from coal-tar distillation. Chemical formula = $C_{14}H_{10}$. Molecular weight = 178.23 g/mol.

❐ antimony (sb)

A metalloid element that is commonly metallic silvery white, crystalline, and brittle. It is used especially as a constituent of alloys and semiconductors. Chemical symbol = Sb. Molecular weight: 121.75 g/mol.

❐ aquatic assemblage

An organism group of interacting populations in a given water body, for example, fish assemblage or a benthic macroinvertebrate assemblage.

❐ aquatic biota

Collective term describing the organisms living in or depending on the aquatic environment.

❐ aquatic community

Association of interacting assemblages in a given waterbody, the biotic component of an ecosystem.

❐ aquatic life

Plants or animals living in water.

❐ aquatic life use

A beneficial use designation in which the water body provides suitable habitat for survival and reproduction of desirable fish, shellfish, and other aquatic organisms.

❐ aquatic recreation

Formerly referred to as "swimmable use " support, this use support classification considers not only swimming, but also wading, aesthetics, and other related uses. The various classes of full, partial and non support are defined in the text of this report.

❐ aquifer

An underground layer of sand, gravel or rock that stores or conveys water below the surface of the soil.

❐ aquitard

An underground bed or layer of soil, rock or clay that is too dense to allow easy passage of water.

❐ arar

Applicable, relevant and appropriate requirements -- federal, state and local regulations that apply to cleanup options.

❐ area sources

Area sources are smaller sources of air emissions that are not required to submit criteria pollutant data to the MPCA. Examples of area sources include gas stations, dry cleaners and auto body shops. Individually, these sources do not emit much air pollution, but collectively release large amounts of one or more toxic air pollutants.

❐ arsenic

A solid, poisonous element that is commonly metallic steel-gray, crystalline, and brittle. Chemical symbol = As. Molecular weight = 74.92 g/mol.

❐ arsenic trioxide

An odorless, tasteless, white or transparent solid, produced from by-products of copper smelting. It is then used in the manufacture of other arsenic compounds which are used as wood preservatives, insecticides and herbicides. It is also used in metallurgical process and in the manufacture of glass and ceramics. Chemical formula = As_2O_3. Molecular weight = 197.84 g/mol.

❐ arsine

A colorless, flammable, extremely poisonous gas with an odor like garlic. Chemical formula = AsH_3. Molecular weight = 77.95 g/mol.

❐ artesian

Ground water under sufficient pressure to rise above the top of the aquifer containing it.

❐ asbestos

An incombustible, chemical-resistant, fibrous mineral form of impure magnesium silicate. Prolonged, inhalation of asbestos can lead to asbestosis -- a chronic, progressive lung disease.

❐ ash

In combustion, solid residue that remains after waste has been completely burned.

❐ assemblage

An association of interacting populations of organisms in a given water body. Examples of assemblages used for biological assessments include: algae, amphibians, birds, fish, reptiles and amphibians, macro invertebrates (insects, crayfish, clams, snails, etc.), and vascular plants.

❐ atmosphere

The body of air surrounding the earth.

❐ atrazine

A photosynthesis-inhibiting, persistent herbicide used especially to kill annual weeds and quack grass. Chemical formula = $C_8H_{14}ClN_5$. Molecular weight = 215.72.

❐ attainment area

A geographic area in which levels of a criteria air pollutant meet the health-based primary standard (national ambient air quality standard, or NAAQS) for the pollutant. An area may have an acceptable level for one criteria air pollutant, but may have unacceptable levels for others. Thus, an area could be both attainment and nonattainment at the same time. Attainment areas are defined using federal pollutant limits set by the U.S. Environmental Protection Agency.

▫ attribute

A measurable component of a biological system.

B

▫ bacteria

Bacteria are a group of microscopic, single-celled organisms. They may have spherical, rodlike, or spiral shapes. They inhabit virtually all environments, including soil, water, organic matter, and the bodies of multicellular animals.

▫ bacteria, fecal coliform and fecal streptococcus

Rod-shaped bacteria usually found in the intestinal tracts of animals, including humans. Their presence in water indicates recent contamination by human or animal feces. These bacteria are obviously not harmful to animals or humans, but they do indicate that sewage contamination has occurred and suggests the presence of disease-causing bacteria and viruses such as E. Coli, salmonella, and cryptosporidium. The presence of fecal coliform and fecal streptococcus bacteria in water is affected by several environmental factors, including sunlight, nutrient levels, temperature, amount and character of sediment, predation by other organisms, flow, and amount of runoff.

▫ barium

A silvery-white metal found in nature. Barium enters the environment during the mining, refining, and production of barium compounds, and from the burning of coal and oil. Barium compounds are used by the oil and gas industries to make drilling muds. Barium sulfate is sometimes used by doctors to perform medical tests and to take barium-rays of the stomach. Chemical formula = Ba. Molecular weight = 137.34 g/mol.

▫ barrier wells

Wells designed to interrupt the natural movement of ground water or to contain the spread of contaminants on an aquifer.

▫ baseflow

The stream discharge composed of ground water drainage and delayed surface drainage.

▫ basin

A tract of land in which the ground is broadly tilted toward a common point. Water that falls onto any portion of the basin is carried toward the common point by a single river system.

▫ bedrock

The solid rock found on the surface of the land or just below the soil.

❐ benthos

Animals without backbones, living in or on the sediments, a size large enough to be seen by the unaided eye, and which can be retained by a U.S. Standard No. 30 sieve (28 openings/inch, 0.595-mm openings). Also referred to as benthic macroinvertebrates, infauna, or macrobenthos.

❐ benzaldehyde

A colorless, nontoxic, aromatic liquid found in essential oils (as in peach kernels) and used in flavoring and perfumery, in pharmaceuticals, and in synthesis of dyes. Chemical formula: C_6H_5CHO. Molecular weight: 106.12 g/mol.

❐ benzene

A colorless, volatile, flammable and toxic liquid hydrocarbon used in organic synthesis, as a solvent, and as a motor fuel. Chemical formula = C_6H_6. Molecular weight = 78.11 g/mol.

❐ benzidene

A crystalline base prepared from nitrobenzene and used especially in making dyes. Chemical formula = $C_{12}H_{12}N_2$. Molecular weight = 184.2 g/mol.

❐ benzo(a)anthracene

One of the class of chemicals called Polycyclic Aromatic Hydrocarbons (PAHs). Chemical formula = $C_{18}H_{12}$. Molecular weight = 228.29 g/mol.

❐ benzo(a)pyrene

A polycyclic aromatic hydrocarbon (PAH) that is a by-product of combustion. The combustion sources are numerous, including natural sources such as wildfires, industrial processes, transportation, energy production and use, food preparation, smoking tobacco, and disposal activities such a as open trash burning. Chemical formula = $C_{20}H_{12}$. Molecular weight = 252.32 g/mol.

❐ benzo(b)fluoranthene

A polycyclic aromatic hydrocarbon (PAH) formed during the incomplete burning of fossil fuels, garbage or other organic matter. Chemical formula = $C_{20}H_{12}$. Molecular Weight = 252.32 g/mol.

❐ benzo(ghi)perylene

A polycyclic aromatic compound (PAH) that is a component of crude oil and a product of combustion. Chemical formula = $C_{22}H_{12}$. Molecular weight = 276.34 g/mol.

❐ benzo(k)fluoranthene

A polycyclic aromatic hydrocarbons (PAH) that results from the incomplete burning of orrganic matter, especially fossil fuels and tobacco. Chemical formula = $C_{20}H_{12}$. Molecular weight = 252.32 g/mol.

❐ benzyl chloride

A colorless liquid with a pungent odor. Used as a chemical intermediate in the manufacture of dyes, pharmaceuticals, perfumes and flavor products. Also used as a photographic developer, and as a gumming inhibitor in gasoline. Chemical formula = C_7H_7Cl. Molecular weight = 126.59 g/mol.

❐ beryllium

Beryllium is a hard, grayish metal naturally found in mineral rocks, coal, soil, and volcanic dust. Beryllium compounds are commercially mined, and the beryllium is purified for use in nuclear weapons and reactors, aircraft and space vehicle structures, instruments, x-ray machines, and mirrors. Beryllium ores are used to make specialty ceramics for electrical and high-technology applications. Beryllium alloys are used in automobiles, computers, sports equipment (golf clubs and bicycle frames), and dental bridges. Chemical formula = Be. Molecular weight = 9.01 g/mol.

❐ bioaccumulation 1: clear lake, california

- In the 1940s and 1950s, some 120,000 gallons of the pesticides DDD and methyl parathion were applied to Clear Lake in California in an attempt to eradicate the Clear Lake gnat (Chaoborus astictopus).
- This had devastating effects on Clear Lake's bird populations, especially the western grebe (Aechmophorus occidentalis).
- The grebe was affected by eating lake fish contaminated by the pesticides.
- Five years after the initial DDD applications, 100 grebes were found dead in one season.
- Analysis of their fat tissues revealed extremely high concentrations of DDD (1,600 parts per million).
- Clear Lake fish were also found to have excessively high DDD concentrations (40 to 2,500 parts per million).

The western grebe provided the first identification of the process of bioaccumulation in food webs

The life cycle of the Clear Lake gnat

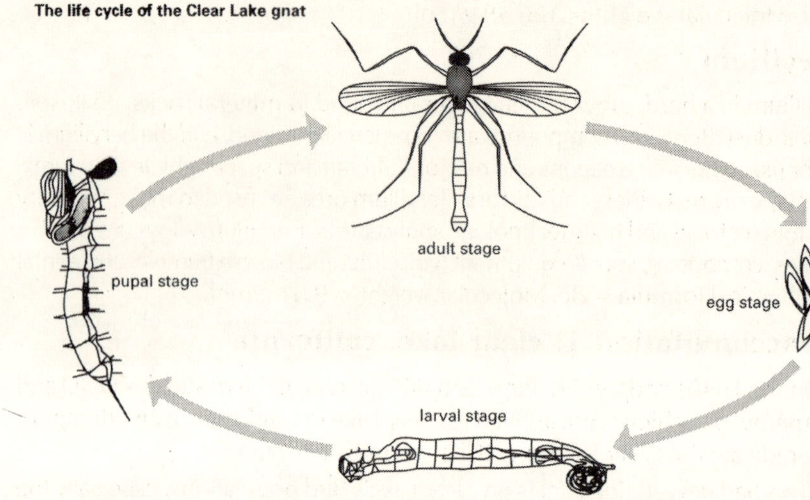

- The western grebe population crashed from over1,000 to 25 breeding pairs in ten years.
- The Clear Lake pollution was described by Rachel Carson in her book, Silent Spring (1962).

❒ bioaccumulation 2: kaneohe bay, hawaii

Algae and coral damage

Kaneohe Bay on the island of Oahu, Hawaii, was once known for its thriving, highly-diverse coral reefs. After 1945, the human population increased tenfold, and sewage discharged into the bay led to a huge increase in the nutrient content of the water. As a result, the green 'bubble algae' (Dictyosphaeria cavernosa) rapidly spread and grew over the coral causing extensive mortality. Filter-feeding organisms multiplied as a result of the increased plankton.

The bubble algae *Dictyosphaeria cavernosa*

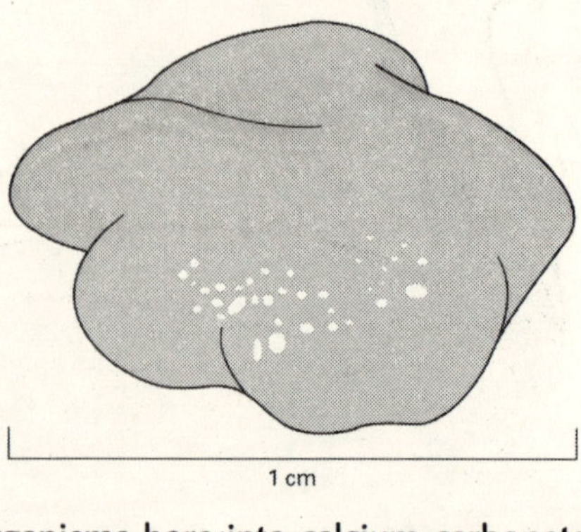

Many of these organisms bore into calcium carbonate, and their increase accelerated the degradation and breakdown of the coral reef structure. The overall result was the massive mortality of coral and associated organisms in the south-east bay. Fortunately, many areas have recovered now that the sewage is being diverted to an offshore discharge site.

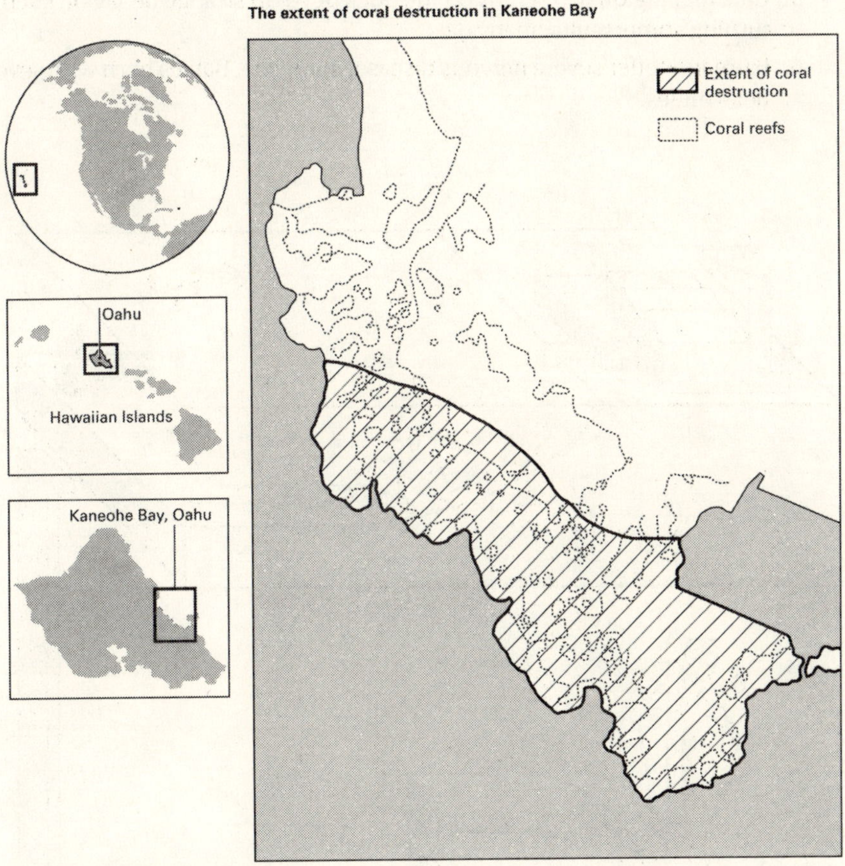

▢ bioaccumulation

Build up of toxic substances in fish flesh. Toxic effects may be passed on to humans eating the fish.

▢ bioaccumulation 3: minamata disease, japan

The inhabitants of the Japanese city of Minamata suffered irreparable damage to their health as a result of industrial mercury pollution. The first cases of Minamata disease – a severe and sometimes fatal affliction of the nervous system – were found in 1953. Over 12,000 people have been officially recognised as affected and were compensated, and over 1,200 of those died. Minamata Bay was only declared safe in 1997.

Stages in the pollution process

1. Chisso Corporation chemical plant discharged methyl mercury waste into Minamata Bay.

2. Methyl mercury taken in by plankton.
3. Fish eat plankton; methyl mercury accumulates in fish tissues.
4. Fishermen catch fish, an important part of the diet of Minamata's citizens.
5. Cats feeding off fish scraps on the dock develop strange behavior, such as turning somersaults; many die.
6. Humans suffer severe nervous disease; some die. Babies born with severe deformities.

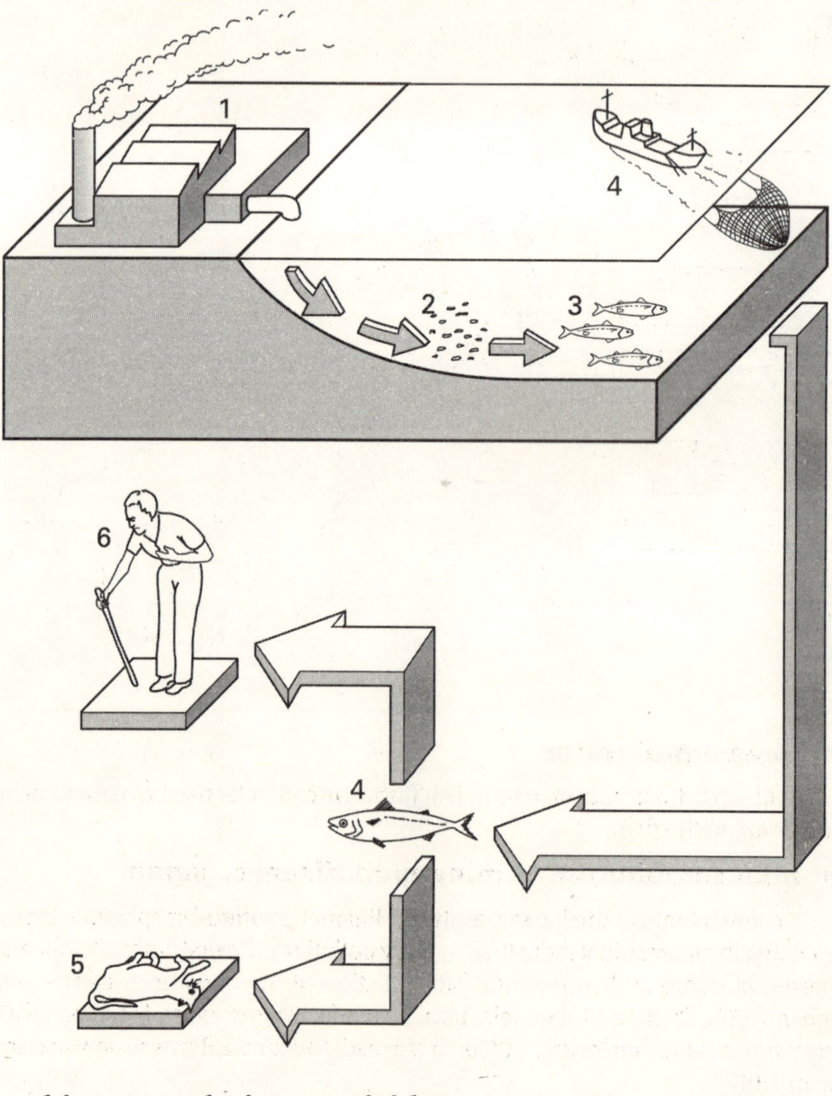

▢ bioaccumulative pesticides

Many organochlorine pesticides phased out since substantial use began in the 1960's have properties that result in their increased accumulation up the food chain (bioaccumulation). Dichlorodiphenyl trichlororethane (DDT), dieldrin, and toxaphene are bioaccumulative pesticides that are no longer in use in the United States, but persist and are still detected in the environment. Exposure to bioaccumulative pesticides can affect nervous, reproductive, and immune system function. In addition, the U. S. Environmental Protection Agency (EPA) classifies most organochlorine pesticides as probable human carcinogens. Minnesota has Water Quality Standards for DDT, dieldrin, and toxaphene, which are measured in water in units ranging from nanograms (one billionth of a gram) per liter (ng/L) to picograms (one trillionth of a gram) per liter (pg/L).

❐ bioavailability

Degree to which chemicals can be taken up by organisms.

❐ biodegradable

Able to be broken down into simpler products by microscopic plants and animals.

❐ biodiversity

Refers to the variety and variability among living organisms and the ecological complexes in which they occur. Diversity can be defined as the number of different items and their relative frequencies. For biological diversity, these items are organized at many levels, ranging from complete ecosystems to the biochemical structures that are the molecular basis of heredity. Thus, the term encompasses different ecosystems, species, and genes.

❐ biological assessment (bioassessment)

Using biomonitoring data biological surveys and other direct measurements of resident biota in surface waters to evaluate the biological condition or health of a place (such as a stream, wetland, or woodlot).

❐ biological criteria or biocriteria

Narrative or numeric expressions that describe the biological condition (structure and function) of aquatic communities inhabiting waters of a designated aquatic life use. Biocriteria are based on the numbers and kinds of organisms present and are regulatory-based biological measurements.

❐ biological integrity

The ability of an aquatic ecosystem to support and maintain a balanced, adaptive community of organisms having a species composition, diversity, and functional organization comparable to that of natural habitats within a region.

❐ biological monitoring or biomonitoring

Sampling the biota of a place (such as a stream, a woodlot, or a wetland); use of a biological entity as a detector and its response as a measure to determine environmental conditions. Toxicity tests and ambient biological surveys are common biological monitoring methods.

❑ biological oxygen demand

The amount of oxygen required by aerobic microorganisms to decompose the organic matter in sample of water. Used as a measure of the degree of water pollution.

❑ biological oxygen demand, five-day (bod5)

The amount of oxygen required by aerobic microorganisms to decompose organic matter in a water sample, based on the maximum rate of O_2 consumption in a water sample over a five-day period in the dark at 20 degrees C. It is used to estimate the total amount of "biodegradable" organic matter in the system and therefore serves as a measure of the degree of water pollution.

❑ biological survey or biosurvey

Collecting, processing, and analyzing a representative portion of the resident aquatic community to determine its structural and/or functional characteristics.

❑ bioregion

Any geographical region characterized by a distinctive flora and fauna.

❑ biota

The plants and animals living in a habitat.

❑ biphenyl

A white, crystalline hydrocarbon used primarily as a heat-transfer medium. Chemical formula = $C_6H_5{\cdot}C_6H_5$. Molecular weight = 154.20 g/mol.

❑ bis(2-ethylhexyl)phthalate (dehp)

Also known as Di(2-ethylhexyl)phthalate and DEHP. A clear, oily liquid that is practically insoluble in water, has a slight odor and is primarily used in the plastics industry as a plasticizer. Chemical formula = $C_{24}H_{38}O_4$. Molecular weight = 390.57 g/mol.

❑ boron

A metalloid element used in metallurgy and in composite structural materials. Chemical symbol = B. Molecular weight = 10.811 g/mol.

❑ bromine

Bromine is the only liquid nonmetallic element. It is a heavy, volatile, mobile, dangerous reddish-brown liquid. The red vapor has a strong unpleasant odor and the vapor irritates the eyes and throat. It causes severe burns upon contact with the skin. In concentrated form, it is a serious health hazard, and maximum safety precautions should be taken when handling it. Bromine is used in agricultural chemicals, dyestuffs, chemical intermediates and flame-retardants. Most is used to prepare 1,2-dibromoethane, which is used as an anti-knock agent in combustion engines. Bromine is obtained by treating brines from wells in Michigan and Arkansas with chlorine, or through extraction from seawater. Chemical formula = Br. Molecular weight = 79.90 g/mol.

❑ bromodichloromethane

Also known as: Chlorodibromomethane and Dibromochloromethane. A colorless to pale yellow liquid, ussed as a fire retardant, solvent, and in laboratories. Chemical formula = $CHBrCl_2$. Molecular weight = 163.83 g/mol.

❑ bromoform

A heavy liquid, with a chloroform-like odor. It was used in the past as a solvent and flame retardants, or to make other chemicals. Now used mainly as a laboratory reagent. Chemical formula = $CHBr_3$. Molecular weight = 252.73 g/mol.

❑ butyl benzyl phthalate

A clear, oily liquid with a slight odor. It is used as a plasticizer and as an intermediate in organis synthesis. The major use is in flooring materials, with minor amounts used in household products. Chemical formula = $C_{19}H_{20}O_4$. Molecular weight = 312.39.

❑ butyraldehyde

Either of two aldehydes, used especially in making synthetic resins. Chemical formula = C_4H_8O. Molecular weight = 72.11 g/mol.

❑ byproduct

A secondary product of a manufacturing process. A waste byproduct is an unwanted byproduct that can either be disposed of or recycled.

C

❑ cadmium

A bluish-white, soft, metallic element used in Nickel-cadmium batteries, steel alloys such as stainless steel, and in protective coatings and plating. Cadmium accumulates in the environment and is thought to be harmful to humans and animal life. Chemical symbol = Cd. Molecular weight = 112.41 g/mol.

❑ calcareous fen

An area of upwelling of calcium-rich ground water that often harbors unique or rare vegetation.

❑ calcium

A silver-white bivalent metallic element of the alkaline-earth group occurring only in combination. Chemical symbol = Ca. Molecular weight = 40.078 g/mol.

❑ cancer inhalation health benchmark (cihb)

A cancer IHB is a concentration in ambient air, at or below which a chemical or defined mixture of chemicals is unlikely to cause an adverse health effect to the general public when exposure occurs daily throughout a person's lifetime. Cancer IHBs are based on unit risk values and the judgment that 1 additional chance in 100,000 of getting cancer in a lifetime is acceptable. For implementation purposes, cancer IHBs are typically compared to an annual average concentration of a chemical or defined mixture of chemicals in air.

❐ carbon dioxide (co_2)

A heavy, colorless, odorless gas, present in the atmosphere or formed by the burning of fuels containing carbon. Carbon dioxide is also exhaled by animals and used by plants in photosynthesis.

❐ carbon disulfide

A colorless, flammable, poisonous liquid used as a solvent for rubber and as an insect fumigant. Chemical formula = CS_2. Molecular weight = 76.14 g/mol.

❐ carbon monoxide (co)

A colorless, toxic gas produced by incomplete burning of carbon-based fuels, including: gasoline, oil and wood. Carbon monoxide is also produced from incomplete combustion of many natural and synthetic products. For instance, cigarette smoke contains carbon monoxide. When carbon monoxide gets into the body, the carbon monoxide combines with chemicals in the blood and prevents the blood from bringing oxygen to cells, tissues and organs. The human body needs oxygen for energy, so high-level exposures to carbon monoxide can cause serious health effects, with death possible from massive exposures. Symptoms of carbon monoxide exposure include vision problems, reduced alertness, and general reduction in mental and physical functions. Carbon monoxide exposures are especially harmful to people with heart, lung and circulatory system diseases.

❐ carbon tetrachloride

A colorless, nonflammable, toxic liquid that has an odor resembling that of chloroform and is used as a solvent and a refrigerant. Chemical formula = CCl_4. Molecular weight = 153.8 g/mol.

❐ carbonyl sulfide

A colorless gas with a typical sulfide odor. It is seed as an intermediate in the production of organic sulfur compounds and alkyl carbonates. Chemical formula = COS. Molecular weight = 60.08 g/mol.

❐ carbonyls

A group of organic chemicals where an oxygen atom is double-bounded to a carbon atom. Examples include: Acetone, Acrolein, Formaldehyde, Methyl Ethyl Ketone.

❐ carcinogen

An cancer causing substance or agent.

❐ catechol

Also known as: Pyrocatechol. Used as a topical antiseptic, antifungal preservative on seed potato pieces, photographic developer, a developer in fur dyes, an intermediate for antioxidants in rubber and lubricating oils, and in pharmaceuticals. Chemical formula = C_6H_6O. Molecular weight = 110.11 g/mol.

▢ causes of ozone depletion

- Ozone (O_3) plays a vital role in the stratosphere, the region of the atmosphere between 6–30 miles above the Earth's surface, by absorbing ultraviolet radiation from the Sun that is harmful to humans.
- Stratospheric ozone is threatened by some of the human-made gases that have been released into the atmosphere, especially those known as chlorofluorocarbons (CFCs).

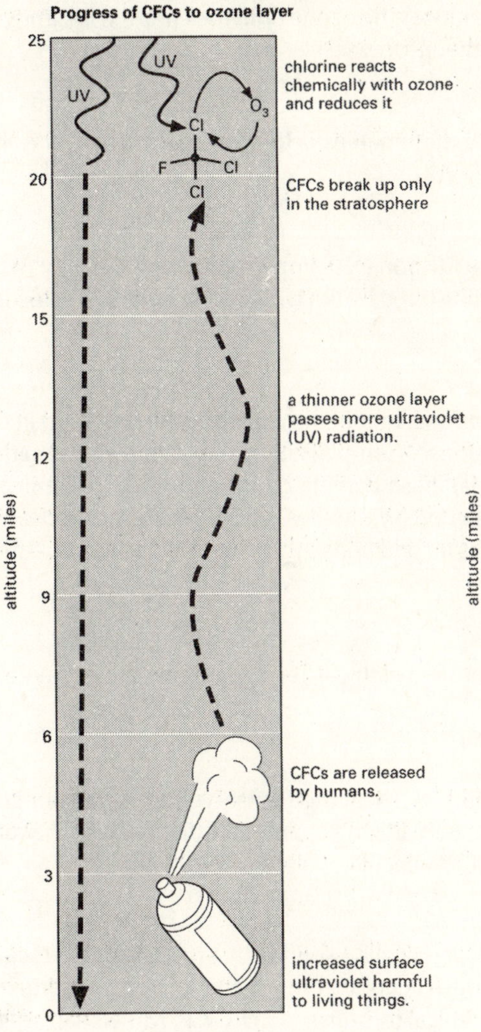

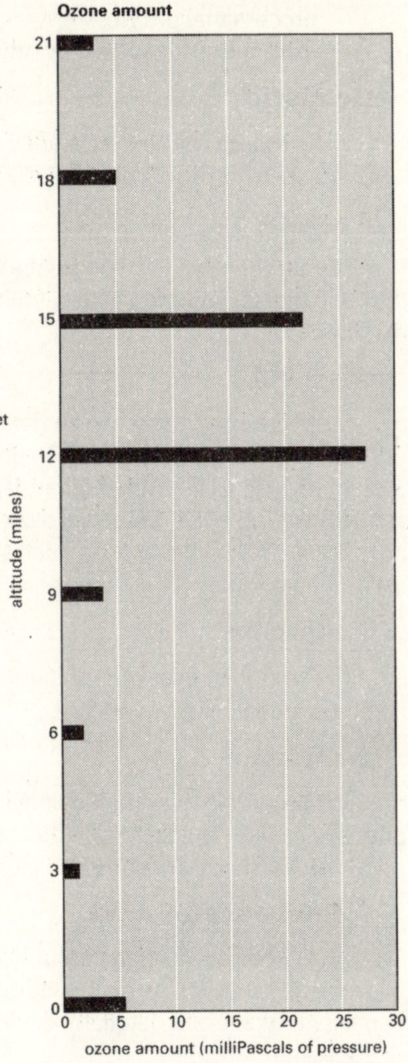

- Widely produced from the 1940s–1970s as propellants in aerosols, refrigerants, electronics cleaners, and constituents of insulating products, CFCs had been hailed as 'wonder chemicals.'

- CFCs mix in the atmosphere all over the world and survive until, after one to two years, they reach the stratosphere and are broken down into their separate elements (chlorine, fluorine and carbon) by ultraviolet radiation.
- The chlorine atoms that are released by ultraviolet react chemically with ozone and break it down.
- In the process of reacting with ozone, more chlorine is produced which then reacts further with the ozone in the stratosphere, in a continuous process that progressively reduces the ozone until the chlorine is removed from the stratosphere by other processes.

❑ caustic

The property of one substance that enables it to burn, corrode or dissolve another substance by chemical reaction.

❑ cercla

Comprehensive Environmental Response, Compensation and Liability Act -- the 1980 federal Superfund law administered by the U.S. Environmental Protection Agency (EPA).

❑ cerclis

Suspected hazardous waste sites throughout the United States are listed in the Comprehensive Environmental Response, Compensation and Liability Information System, or CERCLIS. This federal database contains information on preliminary assessments, potential and actual hazardous waste sites, site inspections, and cleanup activities. CERCLIS sites are candidates for addition to the federal and state Superfund lists.

❑ chloride

Common anionic form of chlorine which carries one net negative charge. A common anion in many waters.

❑ chlorine

A halogen element that is isolated as a heavy, greenish-yellow gas of pungent odor. It is used especially as a bleach, oxidizing agent, and disinfectant in water purification. Chemical formula = Cl_2. Molecular weight = 70.90 g/mol.

❑ chloroacetic acid

A very corrosive, water soluble, crystalline solid with a penetrating, vinegar-like odor. It is used in the manufacture of cellulose ethers, and as a post-emergence contact herbicide and defoliant. Chemical formula = $C_2H_3ClO_2$. Molecular weight = 94.5 g/mol.

❑ chlorobenzene

A colorless, flammable, volatile, toxic liquid used in organic synthesis (as of DDT) and as a solvent. Chemical formula = C_6H_5Cl. Molecular weight = 112.56 g/mol.

❑ chlorodifluoromethane

A colorless gas with a faint, sweetish odor. Used as a propellant, refrigerant and solvent. Chemical formula = $CHClF_2$. Molecular weight = 86.48 g/mol.

❑ chloroethane

Also known as: Ethyl Chloride. A colorless liquid with a pungent odor. It is used as a solvent, refrigerant, and topical anesthetic. Also used in the manufacture of dyes, chemicals, and pharmaceuticals. Was used in the past in the production of tetraethyl lead, an anti-knock additive to leaded gasoline. US government-mandated reduction in the amount of lead additives used in gasoline and a shift to the use of unleaded gasoline has caused a dramatic reduction in the amount of chloroethane required for the production of tetraethyl lead. Chemical formula = C_2H_5Cl. Molecular weight = 64.52 g/mol.

❑ chloroflurocarbons (cfcs)

CFCs are chemicals that have been used in great quantities for refrigeration and air conditioning. When these chemicals are released into the air, they rise into the stratosphere and create chemical reactions that reduce and destroy the Earth's stratospheric ozone layer. The stratospheric ozone layer protects the Earth's surface from the sun's harmful radiation.

❑ chloroform

A colorless, volatile, heavy, toxic liquid with an ether odor used as a solvent or as a veterinary anesthetic. Chemical formula = $CHCl_3$. Molecular weight = 119.38 g/mol.

❑ chlorophyll a (chl-a)

Photosynthetic pigment found in all green plants, and the main pigment in algae. The concentration of chlorophyll a is used to estimate the amount of algae (all phytoplankton in a given area) in surface water. Measured in micrograms per liter (μg/L).While it is normal to find algae in streams and lakes, high levels of algae can lead to low levels of dissolved oxygen in a water body.

❑ chloroprene

A colorless liquid used especially in making neoprene by polymerization. Chemical formula = C_4H_5Cl. Molecular weight = 88.54 g/mol.

❑ chromium

A steel-gray solid with a high melting point. It is used in the production of steel and other metal alloys. It is a naturally occurring element in rocks, soils, animals and plants. Chemical formula = Cr. Molecular weight = 51.996 g/mol.

❑ cis-1,2-dichloroethylene

A colorless, odorless liquid. One of the isomers of 1,2 Dichloroethylene, used as a solvent for waxes and resins and as a refrigerant. Also used in the manufacture of pharmaceuticals and artificial pearls, in the extraction of oils and fats from fish and meat, and the extraction of rubber. Chemical formula = $C_2H_2Cl_2$. Molecular weight = 96.94 g/mol.

❑ cis-1,3-dichloropropene

A colorless liquid with pungent, chloroform-like odor. Used primarily in soil fumigants for nematodes and as a chemical intermediate. Chemical formula = $C_3H_4Cl_2$. Molecular weight = 110.97 g/mol.

❑ clean air act

The original Clean Air Act was passed in 1963, but our national air pollution control program is actually based on the 1970 version of the law. The 1990 Clean Air Act Amendments are the most far- reaching versions of the 1970 law. The 1990 amendments are commonly referred to as the 1990 Clean Air Act.

❑ clean water act (cwa)

An act passed by the U.S. Congress to control water pollution (formerly referred to as the Federal Water Pollution Control Act of 1972).

❑ clean water act section 303(d)

Report to Congress from EPA that identifies those waters for which existing controls are not sufficiently stringent to achieve applicable water quality standards.

❑ clean water act section 305(b)

Biennial reporting requires description of the quality of the Nation's surface waters, evaluation of progress made in maintaining and restoring water quality, and description of the extent of remaining problems by using biological data to make aquatic life use support decisions.

❑ clean water partnership (cwp) program

A program created by the legislature in 1990 to protect and improve ground water and surface water in Minnesota by providing financial and technical assistance to local units of government interested in controlling nonpoint source pollution.

❑ coastal recreation waterway

A designation that includes Great Lakes marine coastal waters and estuaries designated by the Clean Water Act section 303c for use for swimming, bathing, surfing or similar water contact activities. This excludes inland waters, and waters upstream of the mouth of a river or stream that has an impaired natural connection with an open sea.

❑ cobalt

Cobalt is a naturally occurring element found in rocks, soil, water, plants, and animals. There are non radioactive and radioactive forms of cobalt. Non-radioactive cobalt, referred to as stable cobalt, is used to produce metal alloys used in the manufacture of aircraft engines, magnets, grinding and cutting tools, artificial hip and knee joints. Cobalt compounds are also used to color glass, ceramics and paints, and used as a drier for porcelein enamel and paints. Cobalt enters the environment from natural sources and the burning of coal or oil or the production of cobalt alloys. Chemical formula = Co. Molecular weight = 58.93 g/mol.

▫ color (as related to water quality)

Measured against a platinum-cobalt standard (PCU or Pt-Co units), can give an indication of the relative amount of dissolved organic matter in the water. High coloration, or "bog stain" as it is referred to, is usually caused by runoff from wetlands or forested lands. At high values, color may interfere with the expected phosphorus - chlorophyll - Secchi relationship of a lake. In general, values between 0-20 are considered clear, 21-50 are considered moderately colored, 51-100 are considered highly colored, and values greater than 100 can be considered very highly colored. Based on data from the ecoregion reference lakes color begins to influence the phosphorus - chlorophyll - Secchi relationships above about 50 Pt-Co units, and may strongly influence the relationship above 100 Pt-Co units. In highly colored lakes, transparency and chlorophyll values are often lower than expected based on the phosphorus values.

▫ combustor

A facility for controlled burning of municipal solid waste to reduce its volume and weight and, commonly, to produce energy.

▫ commingled

A mixture of any number of recyclable materials, which usually must be separated before they can be recycled.

▫ commissioner's notice

A notice to responsible parties that the MPCA staff intends to recommend that they be issued a Request for Response Action.

▫ community

All the groups of organisms living together in the same area, usually interacting or depending on each other for existence.

▫ composition (structure)

The composition of the taxonomic grouping such as fish, algae, or macroinvertebrates relating primarily to the kinds and number of organisms in the group.

▫ compost

Decomposed organic material resulting from the composting process. Used to enrich or improve the consistency of soil.

▫ condensation

The changing of a gas or vapor to a liquid, as in the formation of water droplets when steam cools.

▫ conductivity

The ability of a substance to conduct or transmit heat, electricity, or sound. Relating to water quality, it is a measure of dissolved minerals and/or ions in the water. Typically measured in micromhos per centimeter (μmhos/cm). If baseline levels of conductivity are established for an area, this measurement can be used to detect possible pollution discharges that have a high ionic strength. This would include manure directly from livestock in streams, land application, or releases from storage basins.

❑ cone of depression

A conical lowering of the water table or pressure surface in an aquifer resulting from water withdrawal, such as pumping.

❑ consent decree

An agreement between MPCA and responsible parties that has been filed with the courts.

❑ consent order

A site cleanup agreement between MPCA and responsible parties.

❑ consequences of ozone depletion

Human health

- Although the ozone layer blocks most of the damaging ultraviolet (UV) radiation received from the Sun, small amounts do reach the Earth's surface. These can cause sunburns and 'suntans.'
- UV radiation is mainly absorbed in the skin and in the outer layers of the eye, and does not penetrate any deeper into the human body.
- The greatest threat to human health from increased UV exposure is an increase in the incidence of skin cancer.
- Each 1 percent loss of total ozone leads to a 3 percent–5 percent increase in skin cancer cases.
- Increased UV exposure almost certainly causes an increase in melanoma.
- Estimating forward to the end of the 21st century, the overall effect of reduction of stratospheric ozone may total 3–5 million additional cases of nonmelanoma skin cancer in the United States.
- In addition, eye disease, including temporary disorders such as 'snow blindness', and more permanent conditions, such as cataracts, could become more common.
- The incidence of cataracts is expected to rise by up to 3 percent.
- UV radiation may also play a role in the suppression of the human immune system.

Marine environment

- There is likely to be a reduction in the numbers of phytoplankton, the photosynthesizing organisms in the sea that form the base of the marine food chain and absorb CO_2 from the atmosphere, so helping to stave off global warming.
- UV also causes direct damage to young fish, shrimp and crab larvae, and other small animals.

- Further up the food chain, this means a reduction in fish stocks, marine animals, and seabirds.

Marine environment **Agriculture and plant life** **Material degradation**

Agriculture and plant life

- Crop yields will suffer, because exposure to high concentrations of UV radiation has been shown to stunt the growth and leaf development of most plant varieties.
- UV-sensitive plants are also more susceptible to pests and disease.

Material degradation

- Many plastics suffer when exposed to UV radiation because photochemical reactions make some plastics brittle, thus shortening their useful lives.

Levels of risk of skin cancer caused by ozone depletion under different restriction estimates

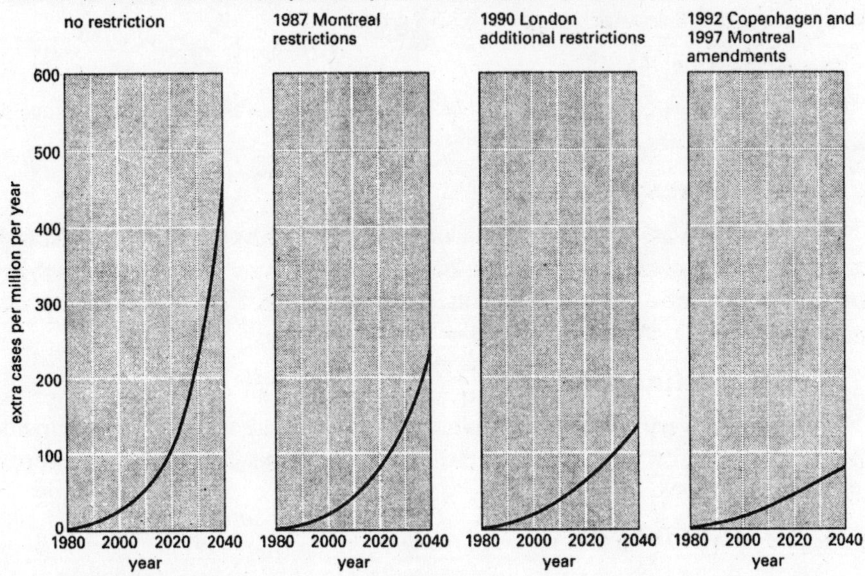

❑ conservation

The protection or wise use of natural resources, such as forests, rivers and fuels, to ensure their continuation.

❑ contaminate

To make impure or unsafe by contact with potentially harmful substances.

❑ continuous emission monotiring systems (cems)

Machines that measure, on a continuous basis, pollutants released by a source. The 1990 Clean Air Act requires continuous emission monitoring systems for certain large sources.

❑ contribution

The relative amount of this pollutant emitted from this facility during this reporting year, expressed as a percentage of the total amount emitted in Minnesota during this reporting year.

❑ copper

Copper is a reddish metal that occurs naturally in rocks, soil, water, and air. It also occurs naturally in plants and animals. Metallic copper is easily molded or shaped, and can be found in the U.S. penny, electrical wiring, and some water pipes. It can also be found in mixtures (called alloys) with other metals such as brass and bronze. Copper is also found as part of other compounds forming salts. Copper salts occur naturally, but are also manufactured. The most common copper salt is copper sulfate. Most copper compounds are blue-green in color. Copper compounds are commonly used in agriculture to treat plant diseases like mildew, for water treatment and, as preservatives for wood, leather, and fabrics. Chemical formula = Cu. Molecular weight = 63.55 g/mol.

❑ corrosive

A substance that is usually highly acidic or alkaline that corrodes materials with which in comes in contact.

❑ cresol/cresylic acid

Also known as: Cresol (mixed isomers). Any of several acids derived from petroleum and coal tar that boil above 204°C and contain varying amounts of xylene and cresol. They are used in disinfectants, solvents, and electrical insulation. Chemical formula = C_7H_8O. Molecular weight = 108.14 g/mol.

❑ criteria air pollutants

A group of very common air pollutants regulated by the U.S. Environmental Protection agency on the basis of criteria (information on health and/or environmental effects of pollution).

❑ crotonaldehyde

A clear to yellow liquid, whose vapor (at high concentrations) can be very irritable to the respiratory system. It is used as an intermediate in the production of other industrial chemicals. It also occurs naturally in foods, and is formed during the combustion of fossil fuels (including engine exhausts), wood and tobacco and in heated cooking oils. Chemical formula = C_4H_6O. Molecular weight = 70.09 g/mol.

❑ cumene

Also known as: Isopropylbenzene. A colorless, flammable liquid with a penetrating, gasoline-like odor. It is used as a thinner for paints, lacquers an enamels, and as a componenet of high octane fuels. It is also used in the production of phenol, acetone, acetophenone, and methylstyrene. Chemical formula = C_9H_{12}. Molecular weight = 120.19 g/mol.

❒ cyanide

A naturally occurring substance found in a number of foods and plants. It found in the air as a result of the combustion of gasoline in automobile engines. It is also emitted during chemical processing and other industries, as well as from incineration of municipal waste. It is also released in cigarette smoke. Chemical formula = CN. Molecular weight = 27.02 g/mol.

❒ cyclohexane

A pungent, saturated, cyclic hydrocarbon found in petroleum or made synthetically. Used as a solvent and in organic synthesis. Chemical formula = C_6H_{12}. Molecular weight = 84.16 g/mol.

❒ cyclohexanone

A watery white to pale yellow liquid with a peppermint or acetone-like odor. Used widely as a solvent and in the preduction of other organic chemicals. Chemical formula = $C_6H_{10}O$. Molecular weight = 98.14 g/mol.

D

❒ data quality

This refers to the i°qualityi± of the data used in the assessment and refers to the number and extent of observations available for the assessment. The terms used were derived from USEPA guidance but actual classifications were developed by MPCA as follows: Poor - < 4 TP observations; Fair ¨C 4 iÜTP< 8, some chlorophyll-a and Secchi; Good ¨C 8<TP< observations. Secchi 12 and chlorophyll-a TP, ¨C Excellent Secchi; some>.

❒ decomposition

The process by which a substance is broekn down into component parts of basic elements. Food and other plant and animal matter decompose under the proper conditions of light, air and moisture.

❒ delisted state superfund sites

Delisted State Superfund Sites, or Delisted PLP (DPLP), sites are those that have been removed from the state Superfund priority list because they have been cleaned up, pose no further risk to public health or the environment, or have been transferred to a more appropriate cleanup program.

❒ density-dependence

Regulation of the size of a population by mechanisms that are themselves controlled by the size of that population (for example, the availability of resources) and whose effectiveness increases as population size increases.

❐ designated use

Classification designated in water quality standards for each water body or segment that defines the optimal purpose for that water body. Examples are drinking water use and aquatic life use.

❐ diatom

Microscopic algae with cell walls made of silicon and have two separating halves.

❐ dibenzofuran

A highly toxic chemical compound that is used in chemical synthesis and as an insecticide. It is a hazardous pollutant in its chlorinated form. Chemical formula = $C_{12}H_8O$. Molecular weight = 168.20 g/mol.

❐ dibromochloromethane

Also known as: Chlorodibromomethane. A colorless to pale yellow liquid used as in the manufacture of fire extinguishing agents, aerosol propellants, refrigerants, and pesticides. Chemical formula = $CHBr_2Cl$. Molecular weight = 208.28 g/mol.

❐ dibutyl phthalate

A colorless, oily ester used chiefly as a solvent, plasticizer, pesticide, and repellent (as for chiggers and mites). Chemical formula = $C_{16}H_{22}O_4$. Molecular weight = 278.35 g/mol.

❐ dichlorobenzenes (mixed isomers)

The group of organic halogen compounds where the two hydrogen atoms in benzene are replaced by chlorine atoms. 1,2-Dichlorobenzene, 1,3-Dichlorobenzene and 1,4-Dichlorobenzene are included in this group. Chemical formula = $C_6H_4Cl_2$. Molecular weight = 147.0 g/mol.

❐ dichlorodifluoromethane (freon-12)

A colorless, odorless gas used in the manufacture of aerosols for cosmetics, pharmaceuticals, insecticides, paints, adhesives, and cleaners. Also used as a refrigerant in home and commercial applications and as a propellant, foaming agent and fumigant. Chemical formula = CCl_2F_2. Molecular weight = 120.91 g/mol.

❐ dichlorotetrafluoroethane (freon 114)

A colorless gas that is used as a solvent, foaming agent, and refrigerant in industrial cooling and air conditioning systems. Chemical formula = $C_2Cl_2F_4$. Molecular weight = 170.92 g/mol.

❐ dicyclopentadiene

A colorless, waxy, flammable solid or liquid used in the production of unsaturated polyester resins, petroleum hydrocarbon resins, pesticides, paints, varnishes, metal derivative and other chemicals. Chemical formula = $C_{10}H_{12}$. Molecular weight = 132.20 g/mol.

❑ **dieldrin**

A white, crystalline, persistent, chlorinated hydrocarbon insecticide that accumulates in and becomes toxic to vertebrates. Chemical formula = $C_{12}H_8Cl_6O$. Molecular weight = 360.371 g/mol.

❑ **diethanolamine**

A colorless powder or liquid with a slight ammonia-like odor. It is used in consumer care products such as soaps, shampoos, pharmaceuticals and cosmetics. Chemical formula = $C_4H_{11}NO_2$. Molecular weight = 105.1 g/mol.

❑ **diethyl sulfate**

A colorless, oily liquid used as an ethylating agent and as a chemical intermediate. Chemical formula = $C_4H_{10}O_4S$. Molecular weight = 154.19 g/mol.

❑ **diethylene glycol**

A clear liquid used in the manufacture of unsaturated polyester resins, polyurethanes and plasticizers. It is also used as a dehydration agent in natural gas processing, as a humectant in the tobacco industry and in the treatment of corks, glue, paper and cellophane. Chemical formula = $O(CH_2CH_2OH)_2$. Molecular weight = 106.12 g/mol.

❑ **diethylene glycol monoethyl ether**

Also known as: 2-(2-Ethoxyethoxy)ethanol. CAS Number: 111-90-0. A colorless liquid with a pleasant, fruity odor. It is used as solvent, and in lacquers, quick-drying varnishes, dyes, and wood stains. Chemical formula = $C_6H_{14}O_3$. Molecular weight = 134.18 g/mol.

❑ **dimethyl phthalate**

A colorless oily liquid with a slightly sweet odor. It is used in solid rocket propellants, lacquers, plastics, safety glasses, rubber coating agents, molding powders, insect repellants, and pesticides. Chemical formula = $C_{10}H_{10}O_4$. Molecular weight = 194.19 g/mol.

❑ **dimethyl sulfate**

A clear, oily liquid with a faint onion odor. Used as a methylating agent in the manufacture of many organic chemicals, dyes and perfumes, for the separation of mineral oils, and for the analysis of auto fluids. Chemical formula = $C_2H_6O_4S$. Molecular weight = 126.13 g/mol.

❑ **di-n-octyl phthalate**

Also known as Dioctyl phthalate and DnOP. A straight chain dioctyl ester of phthalic acid, used primarily as a plasticizer in the manufacture of plastics. Chemical formula = $C_{24}H_{38}O_4$. Molecular weight = 390.56 g/mol.

❑ **dioxin**

Chlorinated dioxins are a class of pollutants created unintentionally and released into the environment during production and combustion of certain organic and chlorinated compounds. Release into air from combustion is the primary route into the environment. In the past, paper mills with bleach-kraft pulping were a significant source of dioxins to some waters in Minnesota. Dioxins are present throughout the environment and increase in concentration up the food chain (bioaccumulation). Tetra-chlorinated, 2, 3, 7, 8-dioxin is the most toxic and is the basis for Minnesota's Dioxin-Water Quality Standards. Dioxins are monitored to help protect the health of humans and wildlife and are measured in femtograms (one quadrillionth of a gram) per liter (fg/L) in water.

❐ disposable

Designed to be thrown away after one use or after a limited period of time.

❐ disposal of hazardous waste

- There are basically four methods of waste disposal: incineration, landfill, dumping at sea, and recycling.

Waste disposal growth 1960–2005

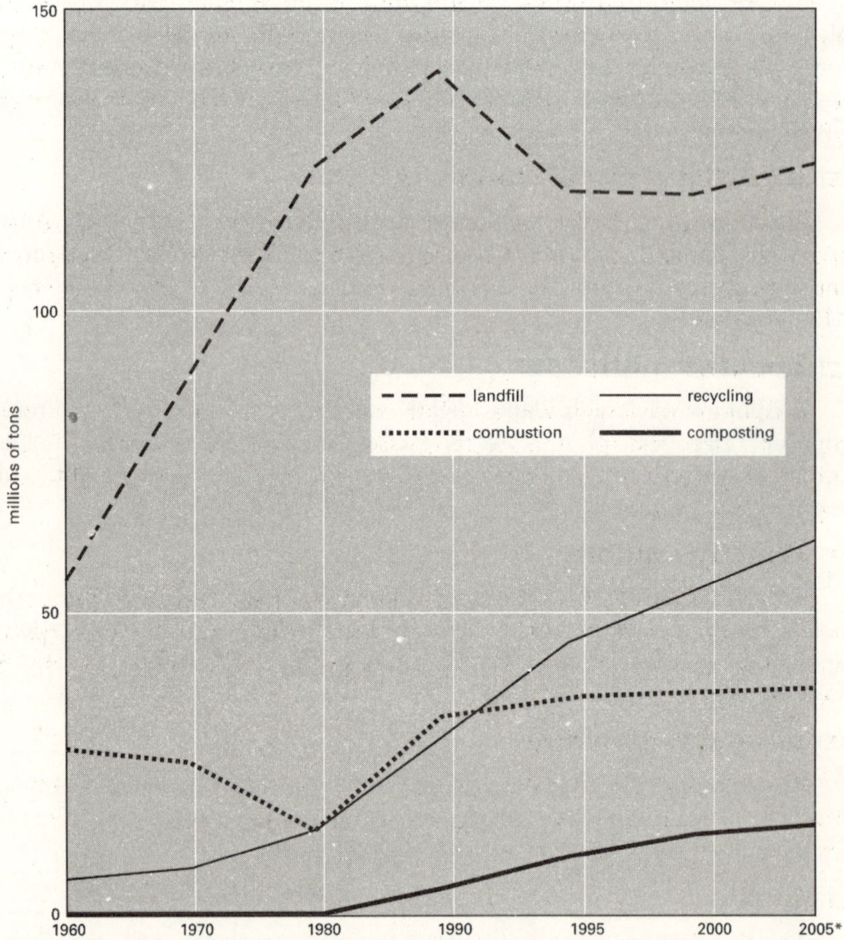

*projected

- Incinerators reduce the volume of solid waste but they do not make the problem of toxic substances in the waste disappear.
- Incinerators emit a wide range of pollutants in their stack gases, ashes, and other residues: for example, they are still the number one source of dioxins worldwide.
- Landfill is not a completely safe disposal strategy for the different types of toxic and radioactive wastes, and those kinds of waste are also capable of producing large quantities of methane, a potent greenhouse gas.
- The environmental disadvantages of burial of nuclear waste include the spread of radioactivity into the surrounding environment. If absorbed into the food chain, it can cause unpredictable genetic damage.
- Obsolete weapons are another source of hazardous waste, particularly nerve gas, chemical weapons stockpiles and nuclear weapons, which were never designed with safe disposal in mind.
- The oceans have long served as a dumping ground for human beings' waste. But direct dumping at sea is one of the fastest ways for toxic compounds to enter the food chain.

❐ dissolved oxygen (do)

The concentration of molecular oxygen (O_2) dissolved in water, usually expressed in milligrams per liter (mg/L), parts per million, or percent of saturation. The DO level represents one of the most important measurements of water quality and is a critical indicator of a water body's ability to support healthy ecosystems. Levels above 5 mg/L are considered optimal, and most fish cannot survive for prolonged periods at levels below 3 mg/L.

❐ dissolved oxygen (do)

The concentration of molecular oxygen (O_2) dissolved in water, usually expressed in milligrams per liter (mg/L), parts per million, or percent of saturation. The DO level represents one of the most important measurements of water quality and is a critical indicator of a water bodies ability to support healthy ecosystems. Levels above 5 milligrams per liter (mg/L) are considered optimal. Most fish cannot survive for prolonged periods at levels below 3 mg/L. Microbial communities in water use oxygen to breakdown organic materials, such as manure, sewage and decomposing algae. Low levels of dissolved oxygen can be a sign that too much organic material is in a water body.

❐ diversity

A combination of the number of taxa and the relative abundance of those taxa. A variety of diversity indexes have been developed to calculate diversity.

❐ dplp sites

DPLP (Delisted Permanent List of Priority) sites are those that have been removed from the state Superfund Program priority list because they have been cleaned up, pose no further risk to public health or the environment, or have been transferred to a more appropriate cleanup program.

❑ drift

Deposits of boulders, gravel, sand, clay or tillmoved and deposted by a glacier or the water from a melting glacier.

E

❑ e. coli (ecoli)

Escherichia coli, a subgroup of fecal coliform bacteria that is present in the intestinal tracts and feces of warm-blooded animals. It is used as an indicator of the potential presence of pathogens. There are many different strains of E. coli that are classified into more than 170 serogroups. Although most strains of E. coli are harmless and live in the intestines of healthy humans and animals, the E. coli O157:H7 strain produces a powerful toxin and can cause severe illness.

❑ ecological integrity

The condition of an unimpaired ecosystem as measured by combined chemical, physical (including physical habitat), and biological attributes.

❑ ecoregion

Areas of relative homogeneity characterized by distinctive regional ecological factors, including land use, soils, topography and potential natural vegetation. There are seven such Ecoregions in the state of Minnesota: NLF = Northern Lakes and Forests CHF = North Central Hardwood Forests NGP = Northern Glaciated Plains WCP= Western Corn Belt Plains RRV = Red River Valley DA = Driftless Area NMW = Northern Minnesota Wetlands

❑ ecosystem

A community of interaction among animals, plants, and microorganisms, and the physical and chemical environment in which they live.

❑ effluent

Liquid flowing out of a system, such as a discharge of liquid waste from a factory or water leaving a sewage treatment plant.

❑ emission

Substances (usually pollutants) discharge into the air.

❑ environmental impact assessment

Economic, social, and environmental change are inherent in the development of nations and societies. Whilst development aims to bring about positive change it can lead to conflicts. In the past, the promotion of economic growth as the motor for increased well-being was the main development thrust with little sensitivity to adverse social or environmental impacts. The need to avoid adverse impacts and to ensure long term benefits led to the concept of sustainability. This has become accepted as an essential feature of development if the aim of increased well-being and greater equity in fulfilling basic needs is to be met for this and future generations.

- To predict environmental impacts of any development activity and to provide an opportunity to mitigate against negative impacts and enhance positive impacts, the environmental impact assessment (EIA) procedure was developed in the 1970s.
- An EIA may be defined as a formal process to predict the environmental consequences of human development activities and to plan appropriate measures to eliminate or reduce adverse effects and to augment positive effects.

Environmental impact
This flowchart for an Environmental Impact Assessment of macroeconomic reforms, devised by the World Wide Fund for Nature (WWF), shows the environmental impact of certain fiscal reforms.

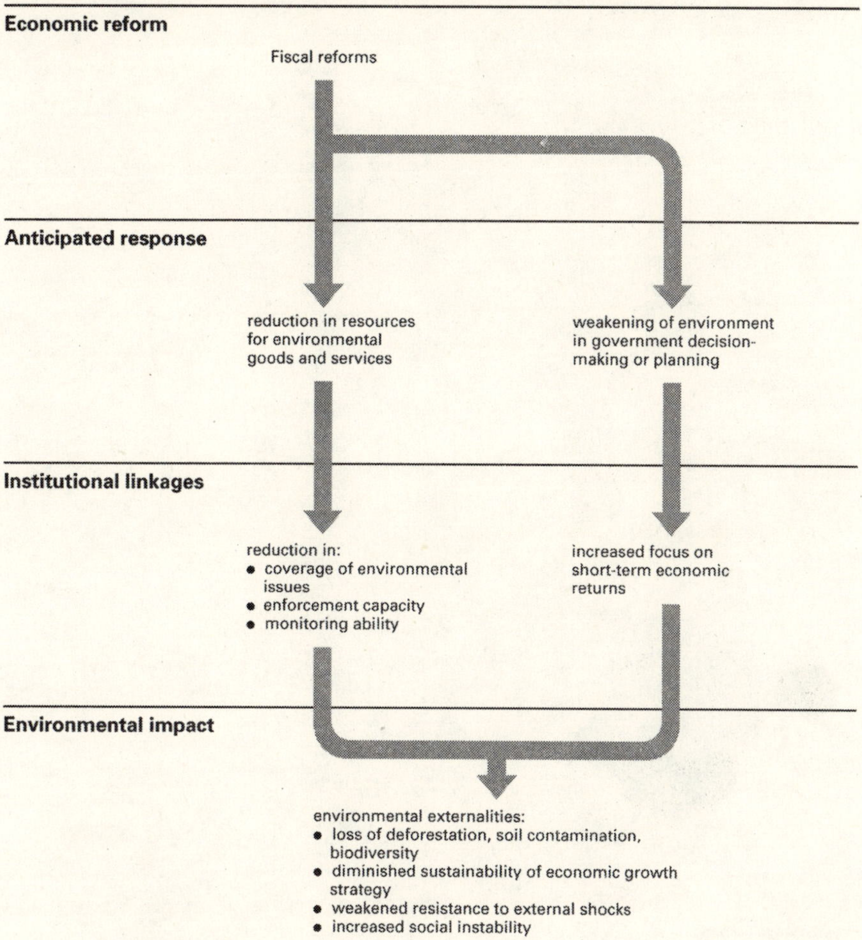

- EIA thus has three main functions:
  1. to predict problems,
  2. to find ways to avoid them, and
  3. to enhance positive effects.

❑ environmental impacts of energy production

- Energy-related emissions account for more than 80 percent of the carbon dioxide released into the atmosphere each year.
- If all economically recoverable conventional fossil fuel reserves were burnt over the next 100 years, this would lead to long-term global warming of well over 3.5°C.
- On a per capita basis, the coal mining industry produces about 22 tons of overburden for every man, woman and child in the United States every year.

Carbon emissions *per capita* 2000

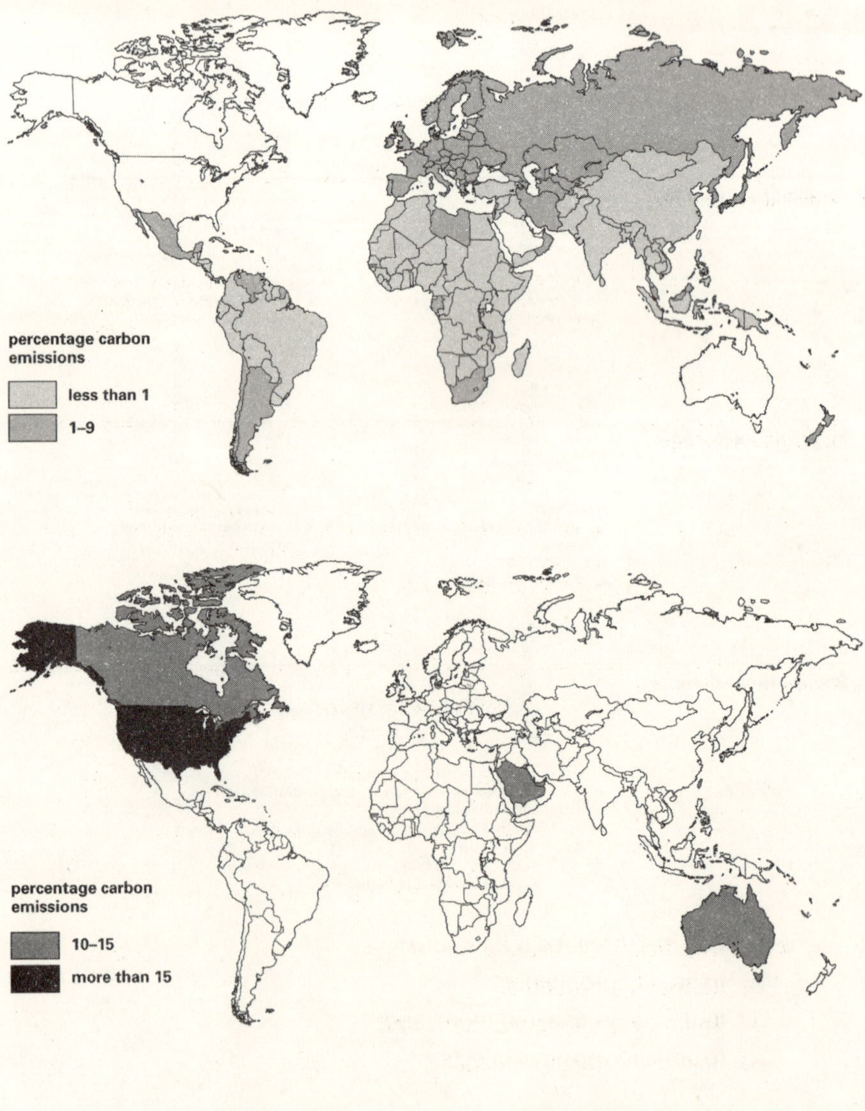

- The per capita share of carbon dioxide emissions in the United States is only slightly lower, at 21 tons per person per year.

☐ environmental impacts of industrial processes

- Since the Industrial Revolution, textiles, steel, glass and soap manufacture are all dependent on the ready availability of basic chemicals like sulfuric acid and the alkali sodium carbonate.
- Our present standard of living would be impossible without power generation and industries such as steel, non-ferrous metals, and chemicals manufacture.
- These processes have also had a profound effect on our environment. Our bodies and our surroundings are contaminated by their wastes.

Heavy metal production ranked in order of increasing toxicity

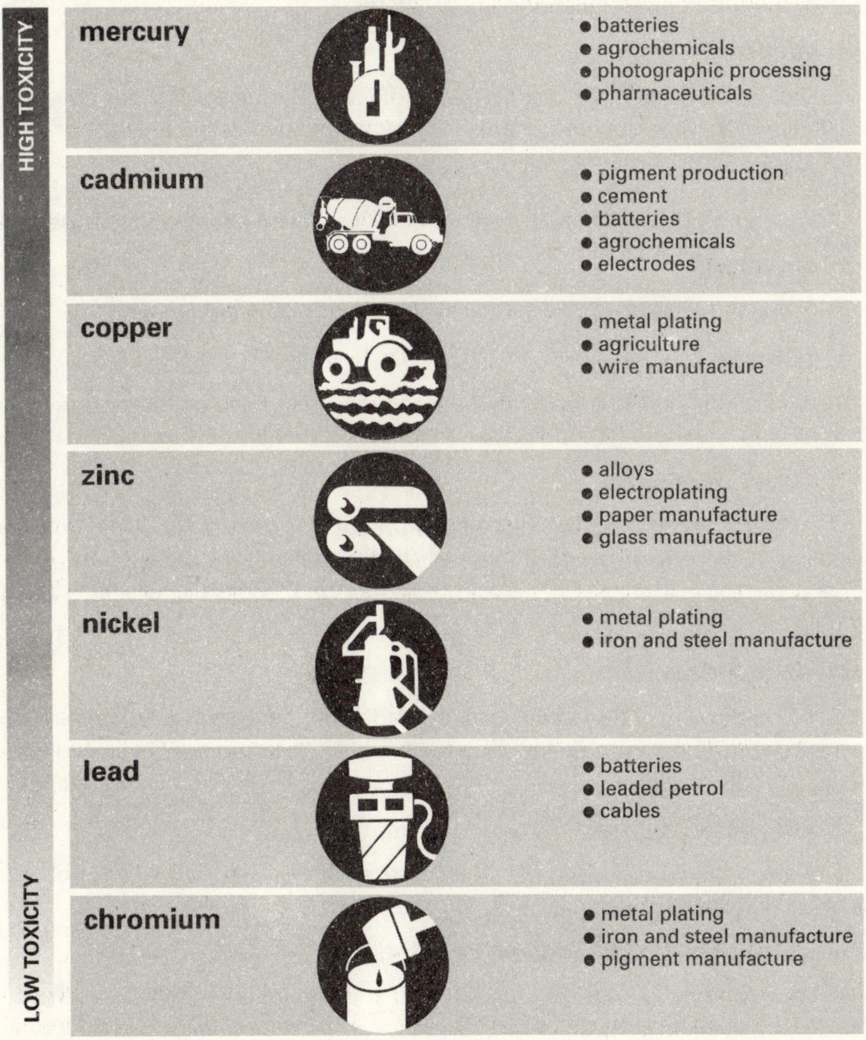

- The Earth's soil, atmosphere and water are permeated with waste metals and organic chemicals which reach us through our food, drinking water and the air we breathe.
- One chemical that has had a particularly strong environmental impact is chlorine.
- The chlorine industry grew rapidly following World War II, producing products such as pesticides, solvents, dry-cleaning fluids and PVC plastic.
- In the 1960s and 1970s it was recognized that many chlorine compounds were toxic and environmentally persistent.

❐ epa

U.S. Environmental Protection Agency -- the federal agency that administers federal environmental regulations and programs.

❐ epilimnion

Most lakes form three distinct layers of water during summertime weather. The epilimnion is the upper layer and is characterized by warmer and lighter water.

❐ erla

Environmental Responsibility and Liability Act. The Minnesota State Superfund.

❐ erosion

The wearing away of land surfaces by the action of wind or water.

❐ ethane

A colorless, odorless, gaseous alkane found in natural gas. It is used as a fuel. Chemical formula = C_2H_6. Molecular weight = 30.069 g/mol.

❐ ethyl acrylate

A clear liquid that is slightly soluble in water. It is used in the manufacture of water-based latex paints and adhesives, textile and paper coatings, and in the production of acrylic fibers. Chemical formula = $C_5H_8O_2$. Molecular weight = 100.11 g/mol.

❐ ethylbenzene

A liquid hydrocarbon that is usually made from benzene and ethylene and is used primarily in the manufacture of styrene. Chemical formula = C_8H_{10}. Molecular weight = 106.16 g/mol.

❐ ethylene glycol

A thick, liquid diol used primarily as an antifreeze and in making polyester fibers. Chemical formula = $C_2H_6O_2$. Molecular weight = 62.07 g/mol.

❐ ethylene glycol monobutyl ether

Also known as: 2-Butoxyethanol. A clear liquid with an ether-like odor. It is used as a solvent in spray lacquers, enamels, varnishes, and latex paints. It is also found in paint thinners, strippers, varnish removers, agricultural chemicals, herbicides, silicone caulks, cutting oils and hydraulic fluids, solvents, dyes and cleaners. Chemical formula = $C_6H_{14}O_2$. Molecular weight = 118.2 g/mol.

❑ ethylene glycol monopropyl ether

Also known as: 2-Propoxyethanol. An oily, colorless, odorless liquid. It is used as a solvent for inks and varnishes, in hydraulic fluids, and as an antiicing agent. Chemical formula = $C_5H_{12}O_2$. Molecular weight = 104.17 g/mol.

❑ ethylene oxide

A colorless, flammable, toxic compound used especially in synthesis of other chemicals, and in sterilization and fumigation. Chemical formula = C_2H_4O. Molecular weight = 44.1 g/mol.

❑ eutrophic

A lake that has a high level of plant nutrients and biological productivity and a low oxygen content.

❑ eutrophic lake

A nutrient-rich lake -- usually shallow, "green" and with limited oxygen in the bottom layer of water.

❑ eutrophication

The aging process by which lakes are fertilized with nutrients. Natural eutrophication will very gradually change the character of a lake. Cultural eutrophication is the accelerated aging of a lake as a result of human activities.

- Eutrophication describes the state of a water body which becomes so rich in nutrients that the natural wildlife is unable to survive.
- The area becomes swamped with algae and foreign species starving the water body of oxygen.
- Common causes are intensive farming and sewage disposal.
- Inorganic fertilizers, used widely in intensive farming, are rich in very soluble nitrates, as well as phosphates and potassium.
- When it rains these chemicals are washed over and through the soil into nearby waterways.
- The water body is unable to cope with this excess and the wildlife supported by it begins to die out.
- New less-sophisticated life, such as algae, takes over.
- It can be avoided by the use of more organic fertilizers (manure) which are high in nutrients but are natural to the environment.
- Untreated human sewage is a massive form of food supply for decomposing organisms.
- When sewage is directly pumped into a river or lake it becomes an ideal haven for them.

- These decomposers use up much of the available oxygen to breakdown the sewage which results in eutrophication taking place.

Water rich in oxygen
Water with few nutrients is rich in oxygen and supports a variety of animal and plant life.

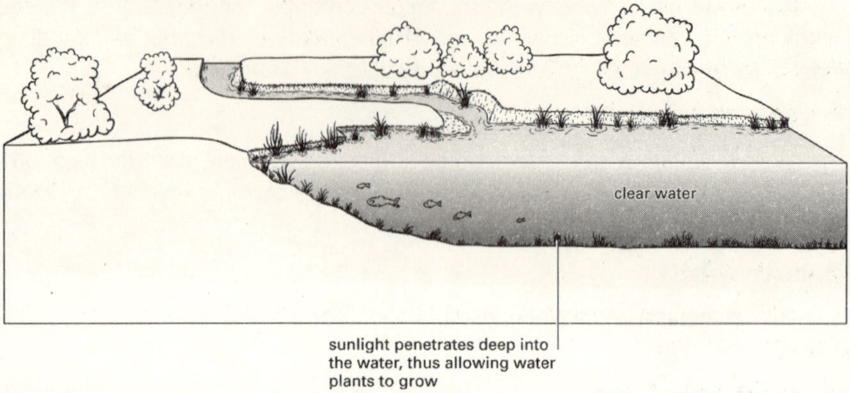

Water low in oxygen
Water with high concentrations of nutrients is low in oxygen, and few animals or fish can live in it.

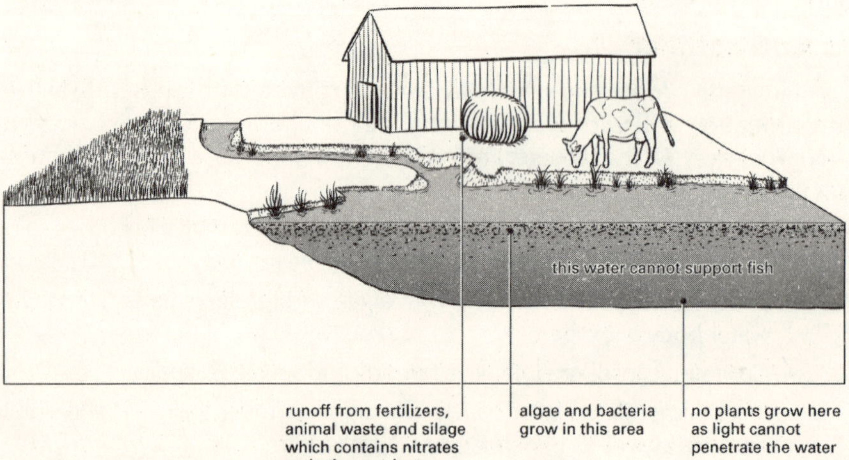

- Proper treatment and breakdown of sewage before it is discharged into a river could prevent eutrophication taking place.

F

❑ fall turnover

Cooling surface waters, activated by wind action, sink to mix with lower levels of water. As in spring turnover, all water is now at the same temperature.

❑ feasibility study

An investigation to determine which course of action should be taken, based on which is the most cost effective. Often preceded by a remedial investigation.

▫ fecal coliform bacteria (fc)

A group of bacteria found in the intestinal tract of humans and animals, and also found in soil. While harmless in themselves, coliform bacteria are commonly used as indicators of the presence of pathogenic organisms and other disease-causing bacteria, such as those that cause typhoid, dysentery, hepatitis A and cholera. Measured in number of bacteria per 100 milliliters of water. Failing septic systems and runoff from feedlots are common sources of fecal coliform in water samples.

▫ federal superfund sites

Federal Superfund sites, also referred to as National Priority List (NPL) sites, are those considered high priority for cleanup under the federal Superfund Program.

▫ feedlot

A lot or building or a group of lots or buildings used for the confined feeding, breeding or holding of animals. This definition includes areas specifically designed for confinement in which manure may accumulate or any area where the concentration of animals is such that a vegetative cover cannot be maintained. Lots used to feed and raise poultry are considered to be feedlots. Pastures are not animal feedlots.

▫ field ph (ph)

A measure of the relative acidity or alkalinity of water. The pH scale ranges from 1 to 14, with 1 being the most acidic and 14 the most alkaline. Pure water is neutral with a pH of 7.

▫ field turbidity (turb)

Indicates the degree to which light is scattered in water by suspended particulate material and soluble colored compounds. It provides an estimate of the muddiness or cloudiness of the water due to clay, silt, finely divided organic and inorganic matter, soluble colored organic compounds, plankton, and microscopic organisms.

▫ field turbidity (turb)

A measurement that provides an estimate of the muddiness or cloudiness of the water due to clay, silt, fine organic and inorganic matter, soluble colored organic compounds, plankton, and microscopic organisms. It is an indication of the degree to which light is scattered in water by suspended particles and soluble colored compounds. Measured in Nephelometric Turbidity Units (NTU) or Formazin Turbidity Units (FTU).

▫ fine particulate matter (pm2.5)

Fine particulate matter is a complex mixture of very small liquid droplets or solid particles in the air. Major sources are cars, trucks, construction equipment, coal-fired power plants, wood burning, vegetation and livestock. These particles can be directly released when coal, gasoline, diesel fuels and wood are burned. Many fine particles are also formed in the atmosphere from chemical reactions of nitrogen oxides, sulfur oxides, organic compounds and ammonia. Fine particulates are associated with increased hospitalizations and deaths due to respiratory and heart disease and can worsen the symptoms of asthma. People with respiratory or heart disease, the elderly and children are the groups most at risk. Fine particles are also major contributors to reduced visibility (haze).

❐ fluoranthene

One of the class of chemicals called Polycyclic Aromatic Hydrocarbons (PAHs). A yellow, crystalline hydrocarbon found in coal tar. It is used as an intermediate for dyes, pharmaceuticals and agrochemicals. Chemical formula = $C_{16}H_{10}$. Molecular weight = 202.26 g/mol.

❐ formaldehyde

CAS Number: 50-00-0. A colorless, pungent, irritating gas used primarily in aqueous solution as a disinfectant and preservative, and in chemical synthesis. Chemical formula = CH_2O. Molecular weight = 30.03 g/mol.

❐ fossil fuels

Fuels derived from organic substances such as coal, petroleum and natural gas.

G

❐ gallium

Gallium is the only metal, except for mercury, caesium, and rubidium, which can be liquid near room temperatures, which makes possible its use in high-temperature thermometers. Ultra-pure gallium has a beautiful, silvery appearance. Gallium compounds are also used to create semiconductors and microwave equipment. Chemical formula = Ga. Molecular weight = 69.72 g/mol.

❐ gas station

Any facility that dispenses gasoline to a vehicle for consumption in that vehicle. Bulk plants, petroleum distribution terminals, and refineries are not gas stations.

❐ generation of hazardous waste

- Precise definitions of hazardous waste vary.
- But there is general agreement that hazardouswaste includes wastes that are: toxic, poisonous, explosive, corrosive, flammable, infectious, or toxic to the environment.
- Worldwide, some 440 million tons of hazardous waste are generated each year, three-quarters by OECD countries.
- Hazardous wastes mainly arise from chemical production, energy production, pulp and paper factories, mining industries, and leather and tanning processes.
- About 75 percent of pesticide use and hazardous waste generation occurs in developed countries.
- The top ten hazardous substances are: arsenic, lead, mercury, vinyl chloride, benzene, polychlorinated biphenyls (PCBs), cadmium, benzo(a)pyrene, polycyclic aromatic hydrocarbons, benzo(b)fluoranthene.

- Since the 1960s, more than 200,000 tons of spent nuclear fuel have been produced by 400 reactors in 30 countries, and every year 10,000 tons are added.

Increase in dangerous chemicals at end of 20th century
The production and imports of dangerous chemicals relative to GDP are on the increase in the European Union.

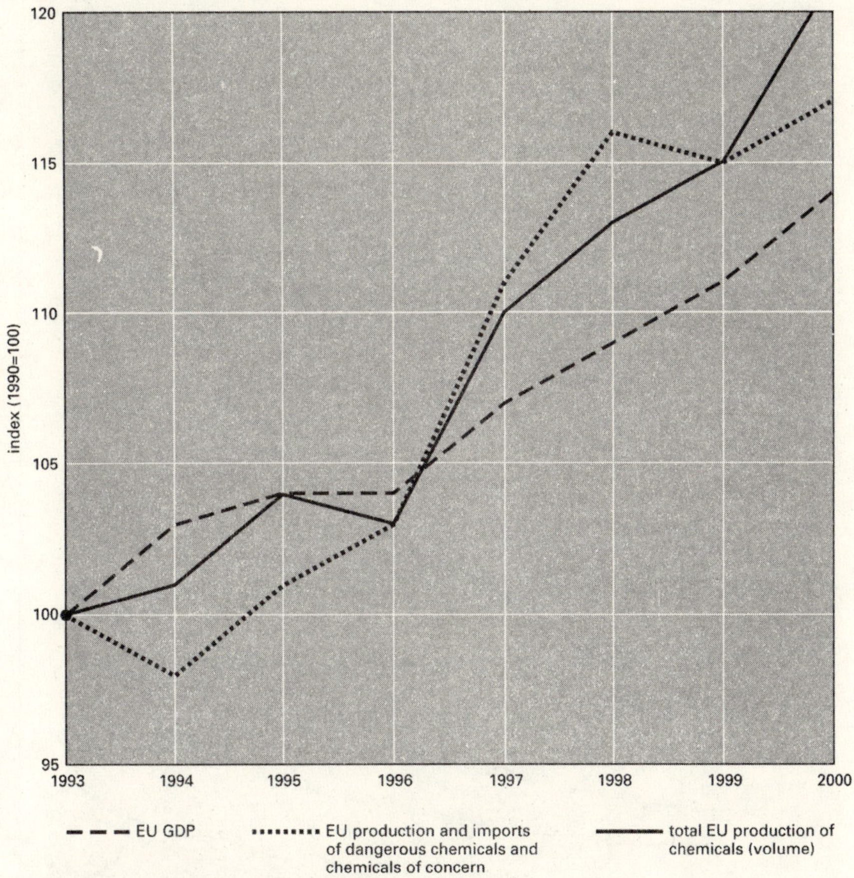

❐ global methane emissions

- Since 1700, the atmospheric concentration of methane has increased by about 145 percent.
- Methane accounts for some 20 percent of long-lived greenhouse gases.
- Atmospheric methane is produced both from natural sources (e.g., wetlands) and from human activities such as: flaring and venting of natural gas, coal mining, biomass burning, livestock farming, rice farming and landfills (garbage burial).

150-year record of global methane emissions from human activity

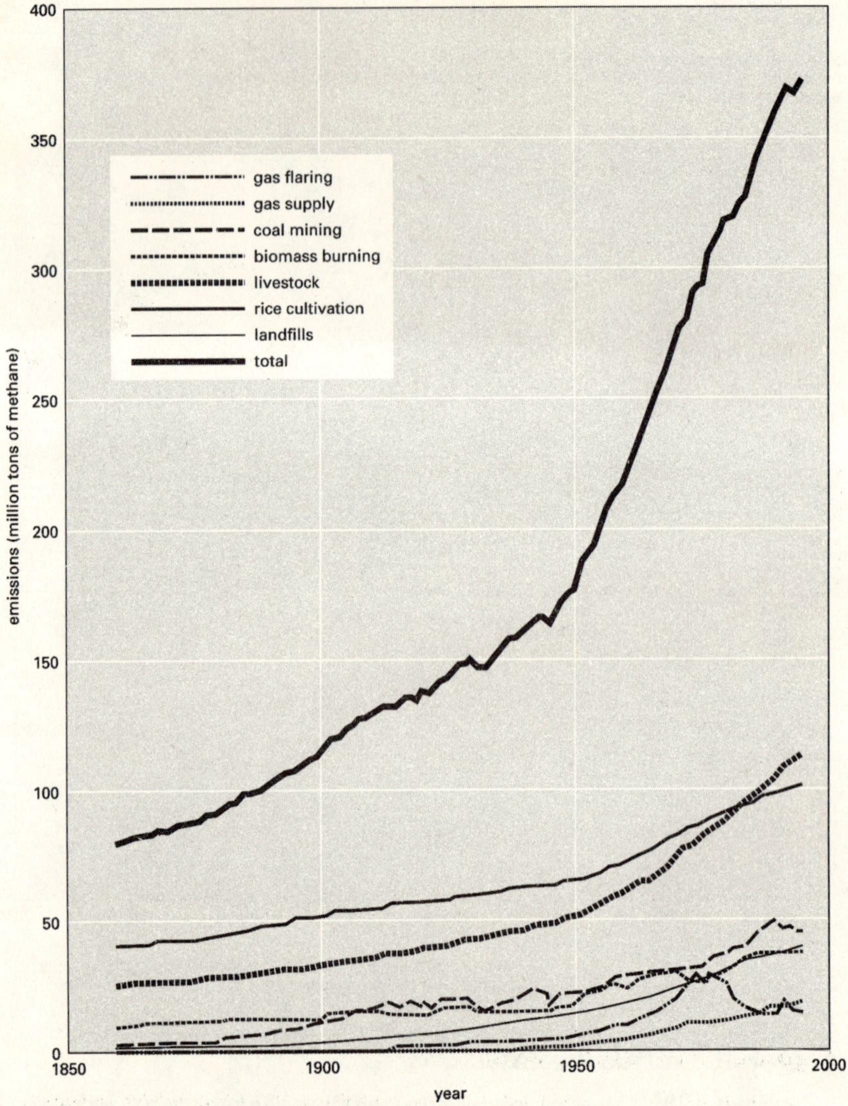

❏ global warming 1: sources

- Global warming refers to an expected rise in global average temperature due to the continued emission of greenhouse gases produced by industry and agriculture which trap heat in the atmosphere.
- Changing patterns of precipitation frequency and intensity, changes in soil moisture, and a rise of the global sea level are expected to be accompanied by higher temperatures.

- Carbon dioxide, the most important human-made greenhouse gas, is released primarily by the burning of fossil fuels like coal, oil, and natural gas.
- Its concentration in the atmosphere has risen by 30 percent over its value in pre-industrial times.

Comparison of historical carbon dioxide emissions from fossil fuel consumption in 10 world regions over 100-year period 1900–2000

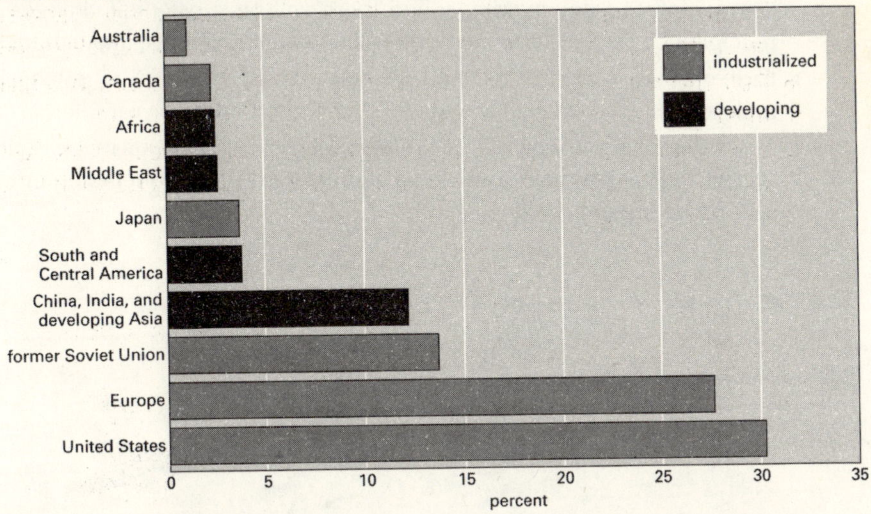

Contribution of different gases to global warming

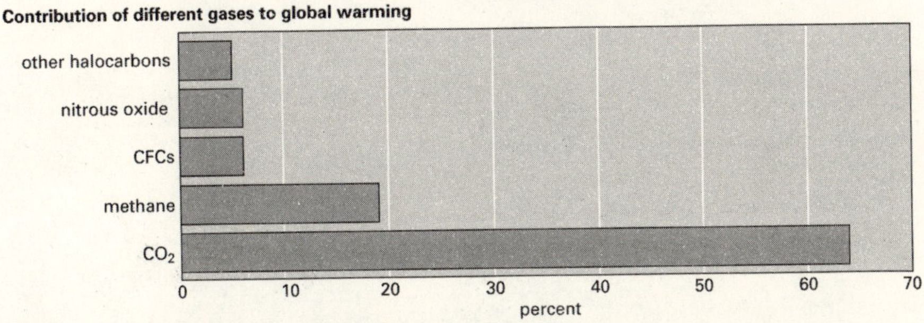

Contribution of human activities to CO_2 emissions

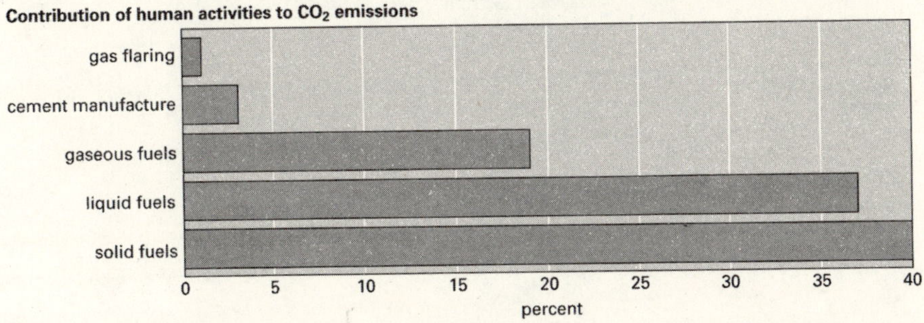

- Concentrations of other greenhouse gases have also risen; methane levels have more than doubled and nitrous oxide levels are increasing, too.

❐ global warming 2: effects

- Climate scientists all over the world now generally agree that global average temperature has risen by about 1°F in the past 140 years, and that Earth's temperature will rise further over the next century if the release of greenhouse gases from human activity continues to grow.
- Land areas will warm twice as much as the oceans; high latitudes will warm more quickly in winter; and there will be substantial changes in precipitation, especially in the tropics, according to current climate models.
- Earth may warm by 2.5°F to 10.4°F by the end of the 21st century, potentially making it warmer than at any time in the last 140 million years.
- The risk of extreme weather, including intense El Niño events in the Pacific Ocean, hurricanes in coastal areas and droughts in continental interiors will increase significantly.

Precipitation anomalies

unusually wet

unusually dry

Outbreaks of climate-related diseases

Rift Valley fever
malaria
hantavirus
encephalitis
dengue
cholera

Heatwaves, fire and smoke

- Sea levels will rise as glaciers and ice sheets melt, and the oceans expand.
- Possible ecological effects include the destruction of most of the Amazon rainforest (from warming and drying) by the end of the 21st century.
- The Asian summer monsoon will become more variable.
- Warming is also likely to spread pests and diseases to new regions.
- The loss of forests globally will further accelerate the emissions of CO_2 into the air, worsening climate change.
- Nearly 70 percent of the world's population could be living in malaria transmission zones within a century.
- Warming might reduce the strength of the North Atlantic ocean circulation through the 21st century, with possible collapse of the Gulf Stream and cooling of Western Europe in the 22nd century.
- Declining rainfall and a low technical capacity to adapt are likely to cause falling crop yields in much of Africa and India.

❑ **global warming**

An increase in the Earth's temperature caused by human activities, such as burning coal, oil and natural gas. This releases carbon dioxide, methane, and other greenhouse gases into the atmosphere. Greenhouse gases form a blanket around the Earth, trapping heat and raising temperatures on the ground. This is steadily changing our climate.

❑ **glycol monobutyl ether acetate**

Also known as: 2-butoxyethyl acetate. A colorless liquid with a fruity odor. It is used as a slow-evaporating solvent for lacquers, varnishes, enamels and epoxy resins. Chemical formula = $C_8H_{16}O_3$. Molecular weight = 160.2 g/mol.

❑ **gpm**

Gallons per minute.

❑ **grazing capacity**

The maximum animal stocking rate possible without inducing damage to vegetation, water or related resources.

❑ **greenhouse effect**

An increase in the earth's temperature resulting from the accumulation of carbon dioxide and water vapor in the atmosphere. Excess amounts of these gases trap heat in the upper atmosphere and reflect it back to earth.

❑ **greenhouse gas**

A greenhouse gas is a gas, either naturally occurring or man-made, that absorbs heat in the atmosphere. Some greenhouse gases occur naturally in the atmosphere, while others result from human activities. Naturally occurring greenhouse gases include water vapor, carbon dioxide, methane, nitrous oxide, and ozone. Certain human activities, however, add to the levels of most of these naturally occurring gases: - Carbon dioxide is released to the atmosphere when solid waste, fossil fuels (oil, natural gas, and coal), and wood and wood products are burned. - Methane is emitted during the production and transport of coal, natural gas, and oil. Methane emissions also result from the decomposition of organic wastes in municipal solid waste landfills, and the raising of livestock. - Nitrous oxide is emitted during agricultural and industrial activities, as well as during combustion of solid waste and fossil fuels. Very powerful greenhouse gases that are not naturally occurring include hydrofluorocarbons (HFCs), perfluorocarbons (PFCs), and sulfur hexafluoride (SF6), which are generated in a variety of industrial processes. Each greenhouse gas differs in its ability to absorb heat in the atmosphere. HFCs and PFCs are the most heat-absorbent. Methane traps over 21 times more heat per molecule than carbon dioxide, and nitrous oxide absorbs 270 times more heat per molecule than carbon dioxide. Often, estimates of greenhouse gas emissions are presented in units of millions of metric tons of carbon equivalents (MMTCE), which weights each gas by its GWP value, or Global Warming Potential.

❑ greenhouse gas estimates

- CO_2 emissions have quadrupled between 1950 and the present day.
- Preindustrial atmospheric CO_2 concentrations were approximately 275 parts per million volume (ppmv).
- Today concentrations are approximately 367 ppmv, an increase of over 30 percent.
- According to global carbon emission estimates, even with the implementation of the Kyoto Protocol, emission levels in 2010 are still expected to be more than 30 percent higher than 1990 levels, largely due to increases from rapidly growing developing countries.

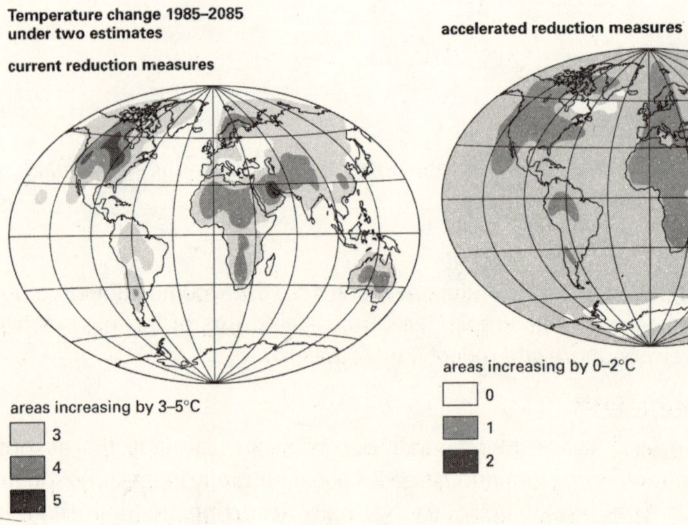

- It will be virtually impossible to restore the atmosphere to its pre-industrial level of CO_2.

- Each stabilization scenario so far evolved represents a huge decrease in annual emissions of CO_2.

Four greenhouse gas emission stabilization estimates

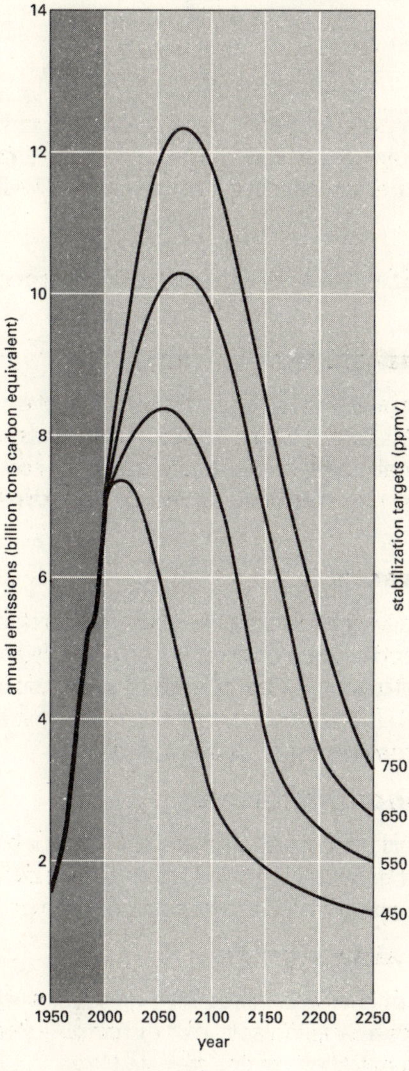

❐ ground water

Underground water in an aquifer, used for drinking water in 75 percent of Minnesota households. The subsurface water supply in the saturated zone below the level of the water table.

❐ ground water pumpout

A cleanup action for contaminated ground water consisting of a well or series of wells pumping continuously to remove ground water from the aquifer.

❐ grout curtain

A wall of bentonite or similar material injected underground to retard horizontal flow of ground water in order to isolate or control contamination; a slurry wall.

H

❐ habitat

The sum of the physical, chemical, and biological environment occupied by individuals of a particular species, population, or community, including the food, cover, and space resources needed for plant and animal livelihood.

❐ half-life

The time required for one-half of a specified substance to degrade or become inert.

❐ hazardous air pollutant (hap)

A hazardous air pollutant is a substance that is defined as hazardous by the 1990 amendments of the Clean Air Act. These substances include certain volatile organic chemicals, pesticides, herbicides, and radionuclides that present tangible hazard, based on scientific studies of exposure to humans and other mammals.

❐ hazardous waste

A waste that: is easily ignitable under ordinary temperature and pressure; readily supplies oxygen or reactive gas to a fire; is corrosive (highly acidic or caustic); is explosive or generates toxic gas; is acutely toxic to animals if it comes into contact with skin or is inhaled, eaten or drunk; or contains toxic chemicals that can be dissolved in an acidic environment, such as a landfill.

❐ health benchmarks for air

A level of chemical concentration in ambient air, at or below which a chemical or defined mixture of chemicals is not likely to cause an adverse health effect to the general public when exposure occurs over a prescribed period of time.

❐ health risk values (hrv)

Concentrations of chemicals or defined mixtures of chemicals emitted to air that are unlikely to pose a significant risk of harmful effects when humans are exposed to those concentrations over a specified time.

❐ heavy metals

Metals with high molecular weights that are of concern because they are generally toxic to animal life and human health if naturally occurring concentrations are exceeded. Examples include, arsenic, chromium, lead and mercury.

❐ herbicides

Chemicals used to kill undesirable vegetation.

❒ herptiles

Reptiles and amphibians.

❒ hexachlorobenzene

A white, crystalline solid which is not very soluble in water. It is formed as a by-product in the manufacture of other chemicals. Hexachlorobenzene was also used as a pesticide until 1965. It was also used in fireworks, ammunitions and in the production of rubber. There are no current commercial uses of the substance. Chemical formula = C_6Cl_6. Molecular weight = 284.79 g/mol.

❒ hexachlorobutadiene

A colorless liquid with a turpentine-like odor. It is used mainly as an intermediate in the manufacture of rubber compounds. Chemical formula = C_4Cl_6. Molecular weight = 260.76 g/mol.

❒ hexamethylene diisocyanate

A colorless liquid with an irritating odor. It is used as a chemical intermediate in the production of light-stable, weather-resistant urethane coatings and elastomers. Chemical Formula - $C_8H_{12}N_2O_2$. Molecular weight = 168.22 g/mol.

❒ hexanaldehyde

Also known as: Hexanal. A colorless liquid with a soft, fruity odor when diluted; a strong, green grass odor when not diluted. It is used as a flavoring additive in food, and in the creation of plasticizers, rubber, dyes, plastic resins and insecticides. Chemical formula = $C_6H_{12}O$. Molecular weight = 100.18 g/mol.

❒ hexylene glycol

A clear liquid with a mild, sweetish odor. It is a component of industrial coatings and used in the production of other industrial and consumer products. Chemical formula = $C_6H_{14}O_2$. Molecular weight = 118.2 g/mol.

❒ historical data

Data sets from previous studies, which can range from handwritten field notes to published journal articles.

❒ hrs

Hazard Ranking System Score. A mathematical score given to each Superfund site, used to set priorities for cleanup of sites. Also called an HRS score.

❒ hydrazine

CAS Number: 302-01-2. A colorless, fuming, corrosive, strongly reducing liquid base, used primarily in fuels for rocket and jet engines. Chemical formula = N_2H_4. Molecular weight = 32.05 g/mol.

▫ hydrochloric acid (hcl)

An aqueous solution of hydrogen chloride that is a strong corrosive irritating acid. It is normally present in dilute form in gastric juice, and is widely used in industry and in the laboratory. Chemical formula = HCl. Molecular weight = 36.47 g/mol.

▫ hydrogen cyanide

A colorless gas with a faint, bitter, almond-like odor. It is used in electroplating and metallury, in the production of other chemicals, and in some cases, fumigation. Chemical formula = HCN. Molecular weight = 27.03 g/mol.

▫ hydrogen fluoride (hf)

Also known as: Hydrofluoric acid. An aqueous solution that is a weak poisonous acid. It resembles hydrochloric acid chemically but attacks silica and silicates, and therefore is used often in finishing and etching glass. Chemical formula = HF. Molecular weight = 20.01 g/mol.

▫ hydrogen sulfide (h_2s)

A flammable, poisonous gas that has an odor suggestive of rotten eggs. It is found in many mineral waters and in decaying organic matter. Chemical formula = H_2S. Molecular weight = 34.08 g/mol.

▫ hydrogeologic study

A study of an area's geology, ground water and ground water movement.

▫ hydrogeomorphic (hgm) approach

A functional assessment method that compares a wetland's condition to similar wetland types (as defined by HGM classification) that are relatively unaltered. HGM functions normally fall into one of three major categories: (1) hydrologic (storage of surface water), (2) biogeochemical (removal of elements and compounds), and (3) habitat (maintenance of plant and animal communities). Hydrogeomorphic classes include riverine, depressional, slope, mineral soil flats, organic soil flats, estuarine fringe, and lacustrine fringe.

▫ hydrologic unit code (huc)

The 8-character federal code identifying the Cataloguing Unit, the smallest of the different hydrologic units. The coding is hierarchical with a 2-character region, a 4-character subregion, a 6-character accounting unit and an 8-character cataloguing unit.

▫ hydrology

The science of dealing with the properties, distribution, and circulation of water both on the surface and under the earth.

▫ hydroquinone

A white, crystalline, strongly reducing phenol used as a photographic developer and as an antioxidant and stabilizer. Chemical formula = $C_6H_6O_2$. Molecular weight = 110.06 g/mol.

❐ hypereutrophic

A very nutrient-rich lake characterized by frequent and severe nuisance algal blooms and low transparency.

❐ hypolimnion

The bottom layer of lake water during the summer months. The water in the hypolimnion is denser and much colder than the water in the upper two layers.

I

❐ impaired water

The Clean Water Act requires states to publish, every two years, an updated list of streams and lakes that are not meeting their designated uses because of excess pollutants.

❐ incinerator

A furnace that burns under controlled conditions.

❐ indeno(1,2,3-cd)pyrene (ip)

A polycyclic aromatic hydrocarbon (PAH) that is a by-product of the combustion of coal, wood, gasoline, municipal waste, and cigarette smoke. IP has also been found in gasoline, fresh and used motor oil, and road runoff. Chemical formula = $C_{22}H_{12}$. Molecular weight = 276.34 g/mol.

❐ index of biological integrity (ibi)

A method for describing water quality using characteristics of aquatic communities, such as the types of fish and invertebrates found in the water body. It is expressed as a numerical value between 0 (lowest quality) to 100 (highest quality).

❐ indium

Indium is a very soft, silvery-white metal with a brilliant lustre. The pure metal gives a high-pitched "scream" when bent. It is useful for making low-melting alloys. When plated or evaporated onto glass, it forms a mirror as good as that formed by silver, but with superior resistance to atmospheric corrosion. Indium is a byproduct of the formation of lead and zinc. Indium metal is isolated by the electrolysis of indium salts in water. Chemical formula = In. Molecular weight = 114.82 g/mol.

❐ indoor air pollutants

- In industrialized countries, there are many possible sources of air pollutants in the home.
- DIY work may lead to a temporary increase in indoor pollutants such as volatile organic compounds (VOCs), during painting or stripping in enclosed spaces, or laying loft insulation.
- In some parts of the world, the radioactive gas radon can seep into the house from the underlying geology, and accumulate indoors if ventilation is poor.

- If a gas or kerosene appliance is faulty, incomplete combustion may result in the release of carbon monoxide.
- This gas also builds up when people smoke cigarettes indoors.
- Over half the world's population still cooks with wood, coal, dung, crop residues and charcoal.
- Indoor air pollution from these fuels causes some 3 million deaths every year, according to the World Health Organization (WHO).
- In Latin American countries, where one quarter of households use solid fuels, an estimated 30,000 people die each year from acute respiratory infections attributable to indoor air quality.

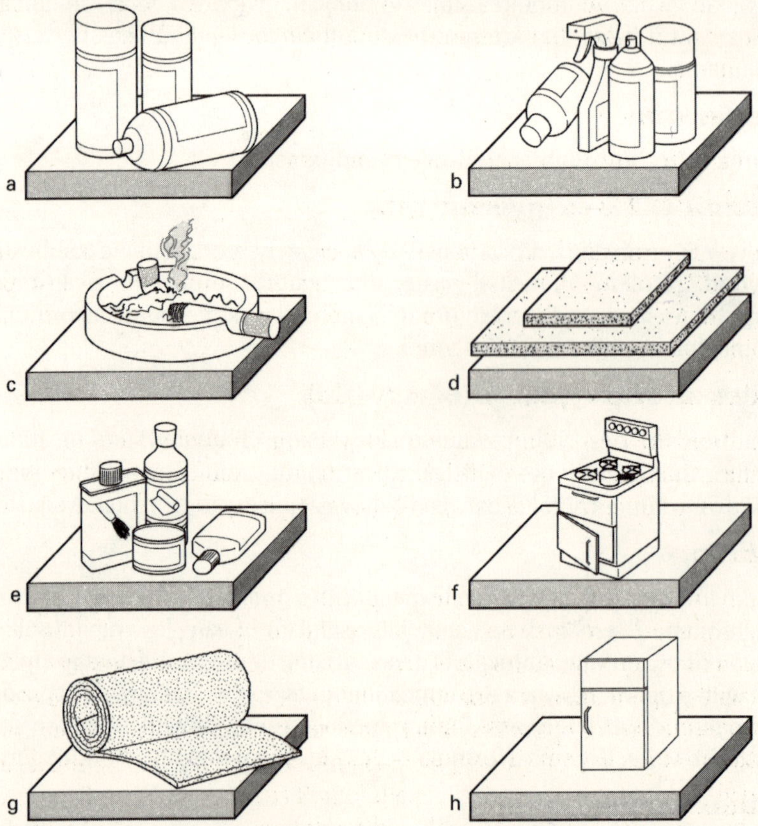

a Air fresheners, moth repellents, and dry cleaning fluids may contain para-dichlorobenzene or tetrachloroethylene, both potential carcinogens.
b Insecticides contain toxic chemicals which can accumulate inside the body.
c Cigarettes produce toxic gases such as benzopyrene and carcinogens.
d Wood board and other insulating materials emit formaldehyde which causes eye, skin, throat, and lung irritations.
e Paint thinners and strippers contain poisonous solvents which persist in the air.
f Badly ventilated gas and kerosene cookers release carbon monoxide and nitrogen oxides.
g Some carpets and plastics can emit styrene which can cause kidney damage.
h Old refrigerators must be disposed of carefully because of the chlorofluorocarbon (CFC) contained in their coolant systems.

- In India, where 80 percent of households use solid fuel, an estimated half million children die annually from indoor air pollution. The figure for sub-Saharan Africa is roughly the same.

❒ industrial and hazardous waste production

- Five tons of solid waste is produced every year for every European, far more for each citizen of the United States of America.
- It is increasing, rising by 10 percent in Europe over the first five years of the 1990s.
- No one knows how much of this waste is hazardous; records are incomplete and national definitions of it vary.

Industrial waste in OECD countries 2000

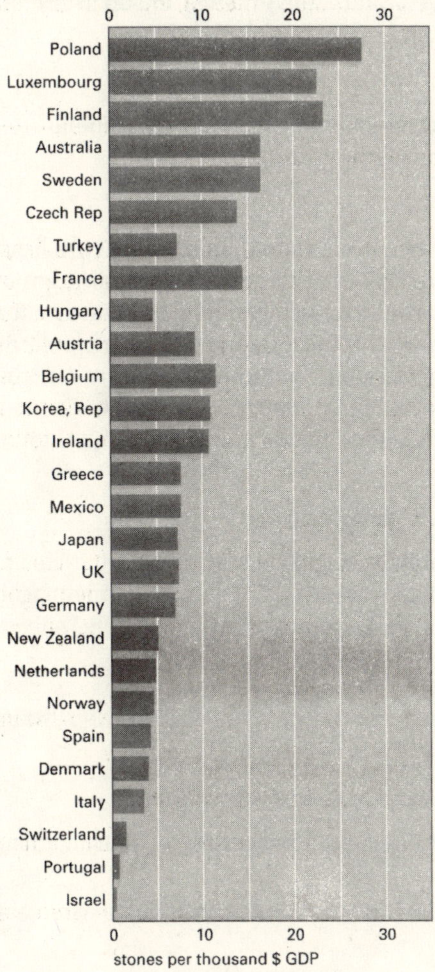

Nuclear waste from spent fuel in selected OECD countries 2000

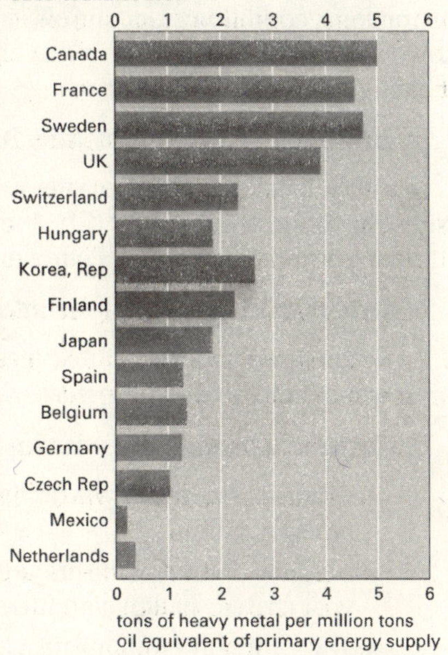

- The European Environment Agency reports that, in the early 1990s, Western Europe produced 46 million tons of hazardous waste, Russia some 22 million tons, and Eastern Europe some 11 million tons.
- The two most common methods of disposal, landfill and incineration, both pose serious risks if not carried out to the highest standards.
- Some wastes – such as polychlorinated biphenyls (PCBs) and those from the nuclear industry – take many years to become harmful.
- In many countries of the developing world there are dumps of dangerous and obsolete pesticides, which frequently expose local communities to oisons in their air, food and water.
- To date just 3,500 tons have been removed from Africa and the Near East at a cost of $24 million. But estimates suggest that a further 20,000 tons remain in Africa, 80,000 tons in Asia and Latin America, and up to 200,000 tons in Central and Eastern Europe.

❑ inorganics

Matter which does not contain carbon and hydrogen atoms. Minerals (e.g. Silica), and metals (e.g. Lead) are included in this group.

❑ institutional controls (ic)

Institutional controls are defined by Minnesota Statute, subdivision 9a, as legally enforceable restrictions, conditions, or controls on the use of real property, ground water, or surface water located at or adjacent to a facility where response actions are taken that are reasonably required to assure that the response actions are protective of public health or welfare or the environment. Institutional controls include restrictions, conditions, or controls enforceable by contract, easement, restrictive covenant, statute, ordinance, or rule, including official controls such as zoning, building codes, and official maps.

❑ interest type (ust/ast/lust programs)

This indicates the type of site. It could be a tank site (TS), leak site (LS), spill site (SS), composting facility (CF), thermal processor (TP), land treatment facility (LTF) or a non-tank Voluntary Petroleum Investigation and Cleanup (VPIC) site.

❑ interim remedial measure

Action taken to protect public health or remove or continue an obvious source of contamination before a remedial investigation is complete.

❑ international movement of hazardous waste

- Trade in hazardous wastes has become a global problem, and it demands global solutions.
- About a tenth of all the hazardous wastes generated around the world each year crosses national frontiers.
- This probably amounts to at least 50 million tons.
- A large proportion of this has gone, and is still going, from industrialized countries to developing ones, where disposal costs are lower.

- Officially, fewer than 1,000 tons a year are traded to developing countries.
- But illegal traffic in hazardous waste poses a serious threat to the environment and human health.
- There are cases involving large-scale illegal traffic of hazardous wastes, usually involving moneylaundering activities, and sometimes linked to illegal trade in arms.
- There have also been cases of ships containing waste cargo wandering the world's oceans, seeking ports to discharge their waste cargos.
- Such 'toxic trade' led to the creation of the UN Convention for the Control of Transboundary Movements of Hazardous Waste and Their Disposal, known as the Basel Convention, which came into force in 1992.
- Two groups of hazardous chemicals – heavy metals and persistent organic pollutants (POPs) – are currently receiving particular attention.
- Although the emissions of some of these substances are falling, concentrations in the environment are of concern.

❏ ion

An electrically charged atom or group of atoms.

❏ iron

A heavy, malleable, magnetic, silver-white metallic element that readily rusts in moist air, occurs native in meteorites and combined in most igneous rocks. It is the most used of metals, and is vital to biological processes. Chemical symbol = Fe. Molecular weight = 55.845 g/mol.

❏ isophorone

A watery white liquid with a peppermint-like odor. It is used as a solvent, in some herbicide and pesticide formulations and in adhesives for plastics, polyvinylchloride, and polystyrene materials. Chemical Formula = $C_9H_{14}O$. Molecular weight = 138.21 g/mol.

J

❏ junkyard

A lot or property where worn-out or discarded items, metal or other scrap material is stored for possible resale.

K

❏ karst

Topography characterized by depressions without external drainage, such as sinkholes, underground caverns and solution channels.

L

❑ lactol spirits

Also known as: Light aliphatic solvent naptha. A light-colored liquid with a mild, petroleum-like odor. It is a mixture of liquid hydrocarbons that is used as a solvent, and in a wide array of industrial and commercial products, including stains, varnishes, pesticides, and cleaners.

❑ lake id number

Minnesota Department of Natural Resources lake identification number. Lake ID numbers are taken from the Minnesota Department of Natural Resources, 1968 publication "Bulletin 25: An Inventory of Minnesota Lakes." It is a six-digit code with the first two digits indicating the county and the next four the lake number. These numbers are used for storing/retrieving data in STORET. Lakes with a 01, 02, etc., suffix indicates that the lake has been divided into "bays" for data storage and assessment purposes. The subsequent area is an estimate for that bay. (Example: Lake ID = 01-0123-01 indicates - Anoka County, Cedar Lake, main basin.)

❑ lake management

A process that involves study, assessment of problems, and decisions on how to maintain a lake as a thriving ecosystem.

❑ lake restoration

Actions directed toward improving the quality of a lake.

❑ lake stewardship

An attitude that recognizes the vulnerability of lakes and the need for citizens, both individually and collectively, to assumes responsibility for their care.

❑ landfill (sanitary landfill)

A specially engineered site for disposing of solid waste on land, constructed so that it will reduce hazard to public health and safety. Some qualities include: an impermeable lower layer to block the movement of leachate into ground water; a leachate collection system; gravel layers permitting the control of methane; and daily covering of garbage with soil.

❑ lanthanum

A soft, silvery-white metal that rapidly tarnishes in air and burns easily. It reacts with water to give off hydrogen gas. Used in optical glass and for flints. Chemical formula = La. Molecular weight = 138.9055 g/mol.

❑ large quantity generator (lqg)

A hazardous waste generator that generates over 1,000 kg (2,200 lbs.) per month of hazardous waste. LQGs must comply with general federal and state requirements for managing hazardous waste. LQGs need an EPA identification number that is used to monitor and track hazardous waste activities.

❒ leachate

Liquids that have percolated through a soil that carry substances in solution or suspension. Liquids can be rain water or snow melt that enter soil and carry contaminates from buried wastes. Leachate can seep into ground or surface water, or pool on or around a landfill.

❒ lead

Lead is a metal found naturally in the environment as well as in manufactured products. The major sources of lead emissions have historically been motor vehicles (such as cars and trucks) and industrial sources. Due to the phase out of leaded gasoline, metals processing is the major source of lead emissions to the air today. The highest levels of lead in air are generally found near lead smelters. Other stationary sources are waste incinerators, utilities, and lead-acid battery manufacturers. In the ambient air, lead exists primarily as inhalable-size particulate matter. Lead occurs naturally in small quantities in soil, water and air.

❒ lead

A bluish-white, soft, heavy, metallic element found mostly in combination with other metals, and used especially in pipes, cable sheaths, batteries, solder, and shields against radioactivity. Chemical symbol = Pb. Molecular weight = 207.2 g/mol.

❒ lead chromate

A yellow solid that is used to produce dyes, inks, paints and plastics. Chemical formula = $PbCrO_4$. Molecular weight = 323.2 g/mol.

❒ lead emissions

- On a global basis, lead in gasoline has been estimated to contribute 95 percent of the lead air pollution found in the world's major cities.
- Alkyl-lead is used primarily as a fuel additive to reduce `knock´ in combustion engines.
- Unlike metallic forms of lead, alkyl-lead is easily absorbed through the skin.
- Lead interferes with normal cell function and with a number of physiologic processes once it enters the body.
- Lead primarily affects the peripheral and central nervous systems, the blood cells, and metabolism of vitamin D and calcium, and also causes reproductive toxicity.
- Soon after their anti-knock properties were first discovered in 1921, alkylated lead compounds were included in gasoline.
- In 1990, the Clean Air Act Amendments (CAAA) completely prohibited the use of leaded gasoline as fuel for on-road automotive use in the United States.
- As a result, the production of leaded gasoline and its use in on-road vehicles has declined dramatically, as have estimates of lead emissions resulting from onroad vehicles.

Gradual phaseout of lead in gasoline correlated with children's blood-lead concentration

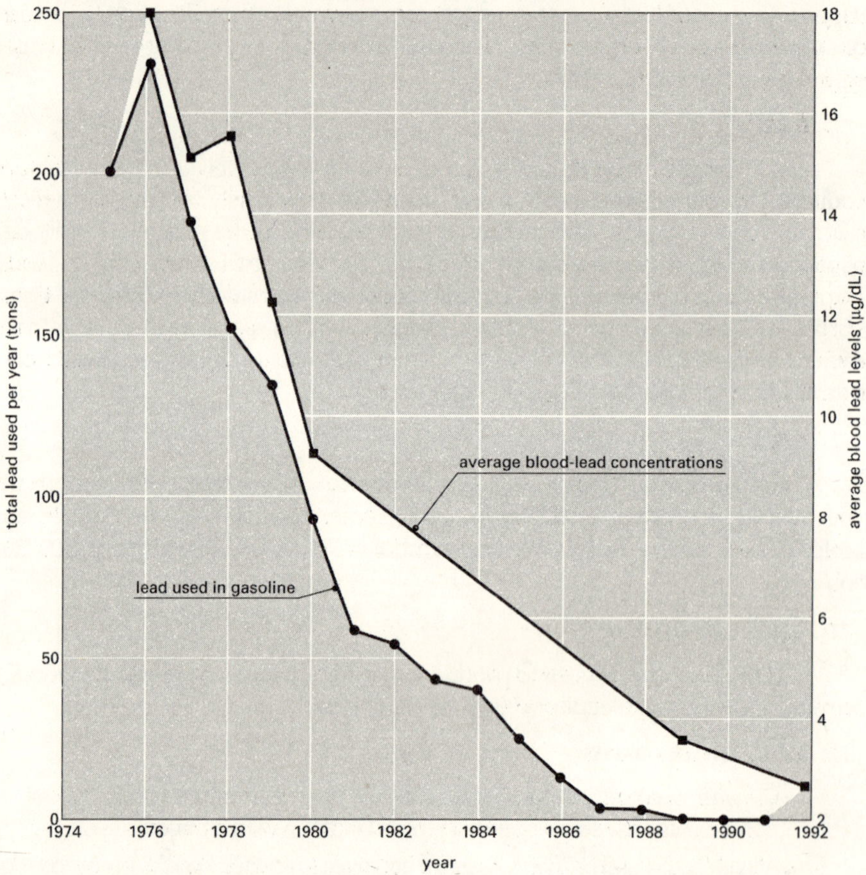

- The production of leaded gasoline decreased from 77.5 billion gallons in 1967 to 3.1 billion gallons in 1991.

❐ liability limit

Set by the Minnesota Superfund law, it is a limit of what political subdivisions -- such as counties, cities, townships or joint powers boards -- can be required to pay for cleanup as a responsible party.

❐ limnetic community

The area of open water in a lake providing the habitat for phytoplankton, zooplankton and fish.

❐ littoral community

The shallow areas around a lake's shoreline, dominated by aquatic plants. The plants produce oxygen and provide food and shelter for animal life.

M

❒ macroinvertebrates

Animals without backbones that can be seen with the naked eye. Includes insects, crayfish, snails, mussels, clams, fairy shrimp.

❒ magnesium

A lightweight, silver-white, malleable metallic element that occurs abundantly in nature and is used in metallurgical and chemical processes, in photography, signaling, and pyrotechnics because of the intense white light it produces on burning, and in construction especially in the form of light alloys. Chemical symbol = Mg. Molecular weight = 24.305 g/mol.

❒ major air pollutants: 1

Carbon Dioxide (CO_2)

- A colorless, odorless gas produced by burning large quantities of carbon rich, fossil fuels, and by the burning of timber and scrub.
- Increased quantities in the atmosphere contribute to the 'greenhouse effect' and global warming.
- High levels can also contribute to acid rain by producing carbonic acid (H_2CO_3) when it dissolves in atmospheric water droplets.

Carbon Monoxide (CO)

- A colorless, odorless, poisonous gas produced by the incomplete combustion of fossil fuels and organic compounds.
- Levels of just 0.1 percent in the atmosphere can begin poisoning vertebrate animals.
- Sources include fossil-fuelled power stations, petrol engines, forest fires, and tobacco smoking.

Chlorofluorocarbons (CFCs)

- Gases manufactured by replacing the carbon atoms in hydrocarbon compounds with those of elements from the halogen group: usually fluorine and chlorine; sometimes bromine.
- CFCs decompose slowly in the stratosphere under ultraviolet light, producing chlorine atoms that oxidize with ozone molecules, so depleting the protective ozonosphere.

Dioxins

- Highly toxic, complex compounds produced as byproducts in some chemical processes.
- Herbicide production, paper bleaching, and other chemical reactions produce dioxins.
- They can enter the atmosphere carried upward by exhaust gas thermals from incinerators, from where they descend to pollute the environment.

Methane (CH_4)

- A naturally-occurring substance produced by the decomposition of organic materials.
- It is the chief constituent of natural gas and firedamp.
- It is also produced by the digestive tracts of animals – particularly herbivores – and by the breakdown of organic waste.

Nitric Oxide (NO)

- A product of industry formed as a byproduct when nitric acid is reduced by chemical reaction.
- As an air pollutant nitric oxide produces nitrogen dioxide (see below).
- It can also react with ozone so acting as a depleter of the ozone layer or ozonosphere and contributing to the greenhouse effect and global warming.

Nitrogen Dioxide (NO_2)

- A reddish-brown, poisonous gas produced when various metals are dissolved in nitric acid (HNO_3) during industrial processes.
- Another important source is from the exhausts of petrol engines.
- It is one of the chemicals responsible for acid rain.

Sulfur Dioxide (SO_2)

- A poisonous gas with a pungent odor.
- It is manufactured industrially by burning sulfur in air, so that it oxidizes, and is used in various chemical processes. It is one of the chemicals responsible for acid rain.

Ozone (O_3)

- An unstable, poisonous, bluish gas with a pungent smell, often associated with electrical motors.
- In the atmosphere ozone readily decomposes into molecular oxygen, but in the stratosphere it forms the protective ozone layer or ozonosphere.

Ozone Depleters

- A group of gaseous chemicals that cause ozone depletion in the stratosphere by reacting with ozone to form oxides.
- Ozone depleters include: chlorofluorocarbons (CFCs), methyl chloroform, carbon tetrachloride, methyl bromine and nitric oxide.

Vapors

- Evaporated liquids, including gasoline, acetone, methanol, alcohol, turpentine and water, that enter the atmosphere as air pollutants.

Particulates

- Tiny particles from burning or internal combustion engines are carried by convection and dispersed into the environment.
- Particulates affect human health more than any other group of pollution.
- Smog is a mixture of smoke and fog.

❒ major air pollutants: 2

Sources of main air pollutant gases

CFCs, which were in widespread use in refrigeration and aerosols until about 1990, have now been largely replaced by less damaging chemicals following the signing of the Montreal Protocol in 1987.

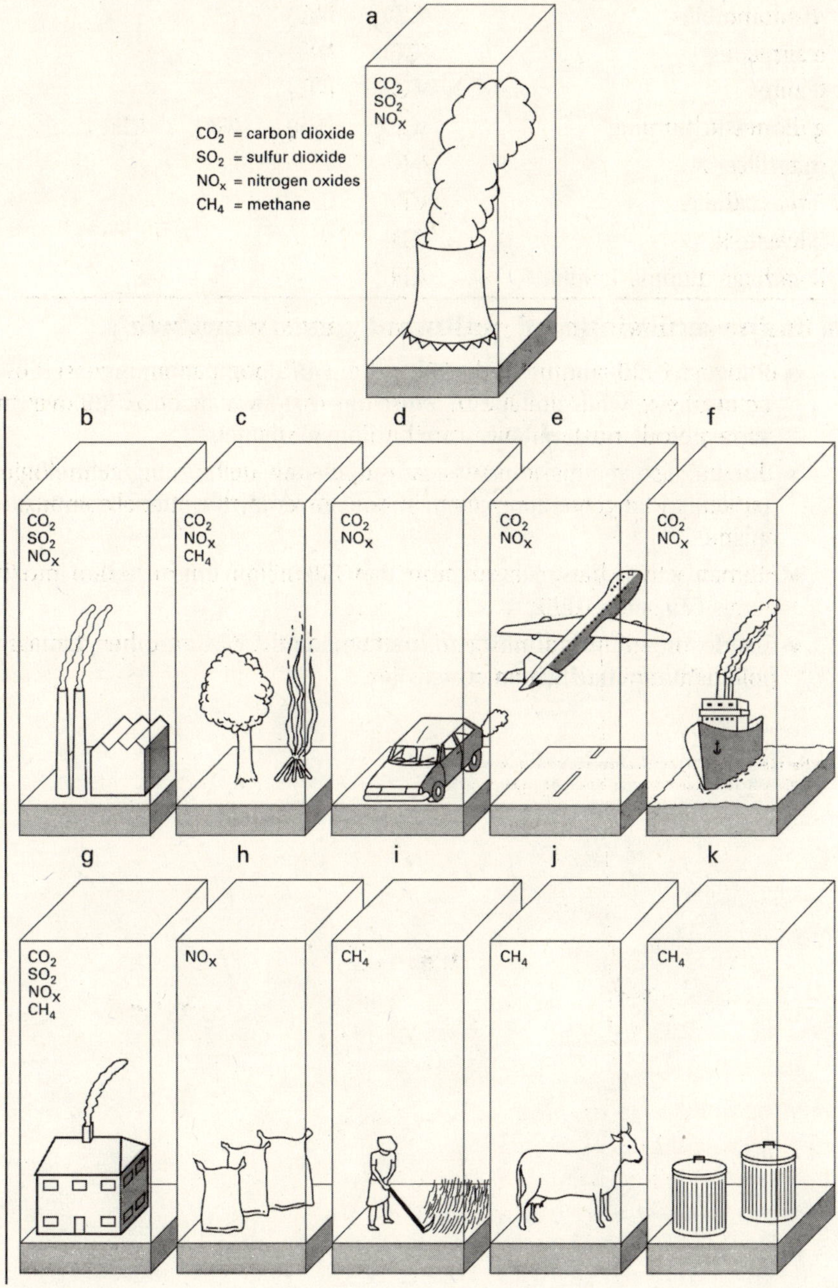

sources	pollutant gases			
a power stations	CO_2	SO_2	NO_x	
b factories	CO_2	O_2	NO_x	
c burning tropical forests	CO_2	NO_x	CH_4	
d automobiles	CO_2	NO_x		
e airplanes	CO_2	NO_x		
f ships	CO_2	NO_x		
g domestic burning	CO_2	NO_x	CO_2	CH_4
h fertilizers	NO_x			
i rice paddies	CH_4			
j livestock	CH_4			
k garbage dumps, landfill	CH_4			

❒ major emissions of pollutant gases worldwide

- Between 1800 and the mid-1990s, the world population increased by a factor of six, while global CO_2 emissions rose by a factor of 800 over the same period, most notably from burning fossil fuels.
- The increase in emissions was caused by new fuelburning technologies, particularly for generating electricity and powering the internal combustion engine.
- Human activity has released more than 350 billion tons of carbon into the air as CO_2 since 1850.
- Oxides of sulfur and nitrogen, methane and CFCs are other damaging pollutants emitted by human activities.

Highest levels of CO_2 emissions around the world 1995
2,500–35,000 lb of carbon equivalent per square mile

Twenty-year survey of change in *per capita* emissions 1975–1995

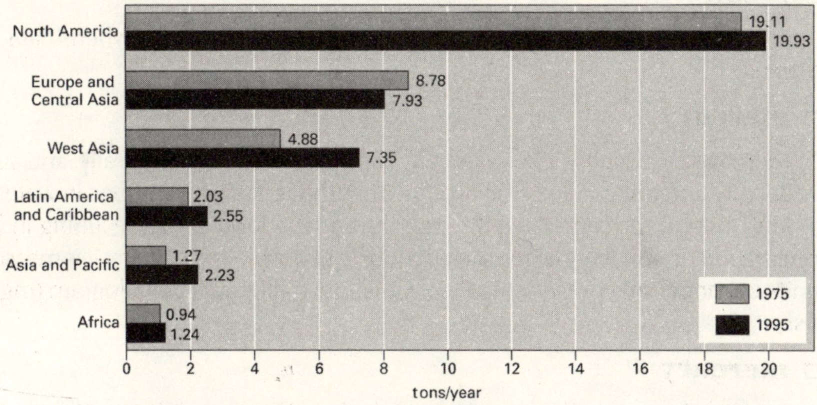

- Despite up to 50 percent of emissions being currently absorbed by oceans or terrestrial ecosystems, CO_2 concentrations in the air have been raised by 30 percent since 1850.
- Only in Europe have per capita emissions of carbon dioxide declined over the past 20 years. Emissions are much higher in North America than in other regions.
- Because of increasing industrialization in the developing world, emissions on a global scale are still on the rise.

❐ maleic anhydride

A caustic, crystalline, cyclic anhydride used especially in making resins. Chemical formula = $C_4H_2O_3$. Molecular weight = 98.06 g/mol.

❐ manganese

A gray-white metal, resembling iron, but is harder and very brittle. It is reactive chemically, and decomposes in cold water slowly. Manganese is used to form many important alloys. In steel, manganese improves rolling and forging qualities, strength, toughness, stiffness, wear resistance, and hardness. Manganese colors glass an amethyst color, and is responsible for the color of true amethyst. Chemical formula = Mn. Molecular weight = 54.938 g/mol. Manganese dioxide is used in the preparation of oxygen and chlorine, and in drying black paints. Manganese permanganate is a powerful oxidizing agent and is used in quantitative analysis and in medicine. Manganese is also widely distributed throughout the animal kingdom. It is an important trace element and may be essential for utilization of vitamin B1.

❐ materials recovery facility

A system that separates collected mixed residential recyclables by type so that they can be recycled into new products available for market.

❑ maximum

Highest observation.

❑ mean

TP, chlorophyll-a, and Secchi was calculated for each lake where data was available.

❑ mercury

Mercury is a highly toxic element that is found both naturally and as an introduced contaminant in the environment. Although concentrations in water are very low, mercury accumulates through the aquatic food chain, resulting in high concentrations in fish that can threaten the health of people and wildlife. It is measured in units of nanograms per liter (ng/L) in water and milligrams per kilogram (mg/Kg) in fish.

❑ mercury

Mercury is a naturally occurring metal which has several forms. The metallic mercury is a shiny, silver-white, odorless liquid. If heated, it is a colorless, odorless gas. Mercury is known to be harmful to the central nervous system, and is readily absorbed through the respiratory tract, the gastrointestinal tract, and through unbroken skin. Metallic mercury is used to produce chlorine gas and caustic soda, and is used in thermometers, dental fillings, mercury vapor lamps, electrical switches, and batteries. Salts of mercury are sometimes used in skin lightening creams and as antiseptic creams and ointments. Chemical formula = Hg. Molecular weight = 200.59 g/mol.

❑ mesotrophic

Describes a lake of moderate photosynthetic productivity.

❑ mesotrophic lake

Midway in nutrient levels between the eutrophic and oligotrophic lakes.

❑ metal concentrations: the river rhine

- At 825 miles (1,320 kilometers) the Rhine is the longest river in Europe.
- It rises in the Swiss Alps, flows generally north, passing through or bordering on Switzerland, Liechtenstein, Austria, Germany, France, and the Netherlands before emptying into the North Sea at Rotterdam.
- Its important tributaries are the Aare, Neckar, Main, Moselle, and Ruhr rivers. The population of the Rhine River Basin is 50 million.
- The River Rhine and its tributaries are polluted by five heavy metals: cadmium, copper, mercury, lead and zinc.
- Rhine pollution comes from towns, cities and industries on its banks in Switzerland, Germany, France, and the Netherlands.
- Between 1983–1987 and 1993–1997, total heavy metal loads decreased, mainly because of reduced industrial emissions in Germany.

- This follows the introduction of the Rhine Action Plan Against Chemical Pollution adopted by the European Community in 1987.
- Apart from metal pollution, the Rhine is also polluted by fertilizers, which have been responsible for high levels of nitrogen and phosphorous in the water.

Sources of metal pollution in the Rhine at the end of the 20th century **Tons per annum of metals**

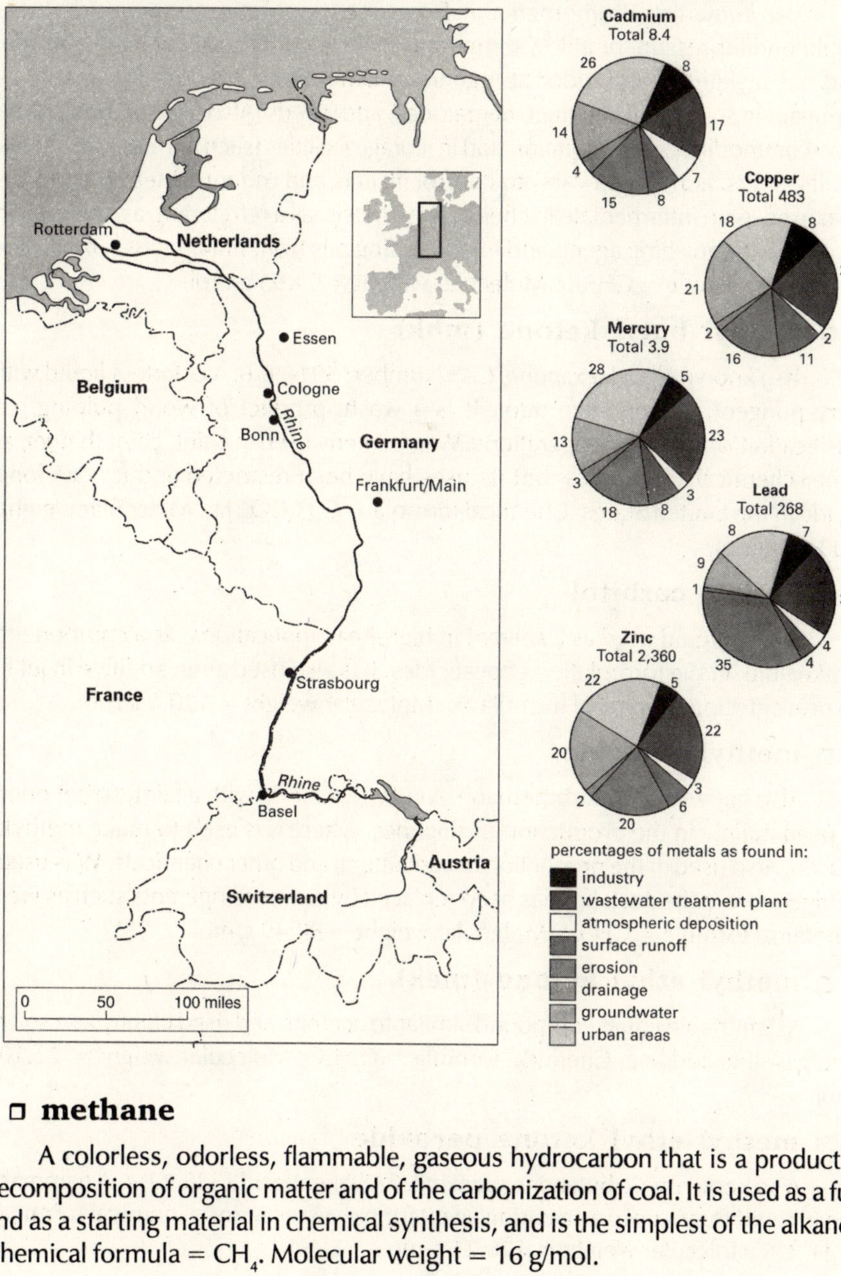

▢ methane

A colorless, odorless, flammable, gaseous hydrocarbon that is a product of decomposition of organic matter and of the carbonization of coal. It is used as a fuel and as a starting material in chemical synthesis, and is the simplest of the alkanes. Chemical formula = CH_4. Molecular weight = 16 g/mol.

❒ methanol

A lightweight, volatile, flammable, poisonous, liquid alcohol used especially as a solvent, antifreeze, or denaturant to create ethyl alcohol. Also used in the synthesis of other chemicals. Chemical formula = CH_3OH. Molecular weight = 32.04 g/mol.

❒ methyl bromide

Also known as: Bromomethane. Exists as a gas at room temperature and as a liquid under pressure or at low temperatures. It is odorless at low concentrations and has a slightly fruity odor at higher concentrations. It is used primarily as a fumigant in soil to control fungi, nematodes, and weeds. Also used in fumigation of food commodities, such as grains, and in storage facilities (such as mills, warehouses, vaults, ships, and freight cars) to control insects and rodents. Methyl bromide is also used as an intermediate in chemical synthesis, as a refrigerant, as an herbicide, as a fire extinguishing agent, and for extracting oils from nuts, seeds, and flowers. Chemical formula = CH3Br. Molecular weight = 94.95 g/mol.

❒ methyl butyl ketone (mbk)

Also known as: 2-Hexanone. CAS Number: 591-78-6. A colorless liquid with a very pungent, acetone-like odor. It is a waste product of wood pulping, coal gasification, and oil shale operations. Was formerly used in paint, paint thinner, and other chemical substances, but its uses have been restricted and it is no longer made in the United States. Chemical formula = $C_4H_9COCH_3$. Molecular weight = 100.16 g/mol.

❒ methyl carbitol

A clear liquid used as a solvent in high-heat applications, as a component in brake fluid, in the formulation of pesticides. It is also used as an additive in jet fuel to prevent icing. Chemical formula = . Molecular weight = 120.1 g/mol.

❒ methyl chloride

Also known as: Chloromethane. A colorless liquid with a faint, sweet odor. It is used mainly in the production of silicones, where it is used to make methylate silicon. Also used in the production of agricultural and other chemicals. Was used in refrigerators in the past, but has been replaced by newer refrigerants such as Freon. Chemical formula = CH_3Cl. Molecular weight = 50.49 g/mol.

❒ methyl ethyl ketone (mek)

A flammable liquid compound similar to acetone and used chiefly as a solvent and gasoline additive. Chemical formula = C_4H_8O. Molecular weight = 72.10 g/mol.

❒ methyl ethyl ketone peroxide

Also known as: 2-butanone peroxide. CAS Number: 12338-12-4. A flammable, clear liquid that is used as a curing agent for polyester resins. Chemical formula = $C_4H_{16}O_4$. Molecular weight = 176.21 g/mol.

❒ methyl iodide

A colorless, nonflammable liquid that has has a pungent odor. It is used as an intermediate in the manufacture of pharmaceuticals and pesticides. Chemical formula = CH3I. Molecular weight = 141.95 g/mol.

❒ methyl isobutyl ketone

Also known as: Hexone. A colorless, flammable liquid with a faint camphor-like odor. It is used as a solvent for gums, resins, paints, varnishes, and lacquers, as an alcohol denaturant, in the extraction of rare metals, and in the creation of synthetic flavorings. Chemical formula = $C_6H_{12}O$. Molecular weight = 100.16.

❒ methyl methacrylate

A volatile, flammable liquid that polymerizes readily and is used as a monomer for resins. Chemical formula = $C_5H_8O_2$. Molecular weight = 100.1 g/mol.

❒ methyl tert-butyl ether (mtbe)

CAS Number: 1634-04-4. A colorless liquid that is used as an octane booster in gasoline. Chemical formula = $C_5H_{12}O$. Molecular weight = 88.15 g/mol.

❒ methylene chloride

Also known as: Dichloromethane. CAS Number: 75-09-2. A nonflammable liquid used especially as a solvent, paint remover, and aerosol propellant. Chemical formula = CH_2Cl_2. Molecular weight = 84.93 g/mol.

❒ methylhydrazine

A clear, corrosive liquid with an ammonia-like odor, Used as a high-energy fuel, rocket propellant and fuel for thrusters, and as fuel for small electrical power generating units. Chemical formula = CH_3NHNH_2. Molecular weight = 46.1 g/mol.

❒ metric

A calculated term or enumeration representing some aspect of biological assemblage, function, or other measurable aspect and is a characteristic of the biota that changes in some predictable way with increased human influence. A multimetric approach involves combinations of metrics to provide an integrative assessment of the status of aquatic resources.

❒ microinvertebrates

Animals without backbones that are not large enough to be seen by the unaided eye; they will not be retained by a U.S. Standard No. 30 sieve (28 meshes per inch, 0.595 mm openings).

❒ minimally impaired

Sites or conditions with slight anthropogenic perturbation relative to the overall region of the study.

❐ minimum

Lowest observation.

❐ mobile source

A moving source of pollution, such as a car or truck.

❐ molybdenum

A soft, silvery metal used in steel and other metal alloys, electrodes, and catalysts. Chemical formula = Mo. Molecular weight = 95.94 g/mol.

❐ monitored or evaluated

Any lake with total phosphorus (TP), chlorophyll-a, or Secchi transparency data collected between 1970-1998, and stored under MPCA's agency code 21 MINNL in STORET is included in this assessment. Lakes were classified into one of two categories for this data assessment. "M" implies monitored (recent) and indicates that summer (June through September) data collected between 1989 and 1998 was available for that lake. "E" implies evaluated (old). This indicates that data for this lake was collected between 1970-1988. Non-summer total phosphorus measurements may also be included in this category. Overall, approximately 2,245 lakes are included in this assessment - 1,539 with "monitored" data and 706 with "evaluated" data. Of the "monitored" lakes 896 (58%) had TP data while for the "evaluated" lakes 623 (88 %) had TP data.

❐ monitoring well

A well installed at a site to determine the direction of ground water flow, levels of contamination and other ground water characteristics.

❐ multi-aquifer well

A well that is open to more than one aquifer. Such wells can be pathways for contaminants to reach deep ground water.

❐ multimetric

Analysis techniques using several measurable characteristics of a biological assemblage.

❐ multivariate community analysis

Statistical methods (such as ordination or discriminant analysis) for analyzing physical and biological community data using multiple variables.

❐ municipal solid waste stream

Waste generated in households, commercial establishments, institutions and light industries. Industrial process, agricultural and mining wastes and sewage sludge are not municipal solid waste.

❐ mutagen

A substance that causes genetic changes in humans or animals.

□ **m-xylene**

A clear liquid with a sweet odor. One of the three isomers of xylene. Used as a solvent, as an intermediate for dyes & organic synthesis, in insecticides and aviation fuel. Chemical formula = C_8H_{10}. Molecular weight = 106.17 g/mol.

N

□ **n,n-dimethylformamide**

A clear liquid that is used as an industrial solvent. Chemical formula = C_3H_7NO. Molecular weight = 73.09 g/mol.

□ **napthalene**

A crystalline, aromatic hydrocarbon usually obtained by distillation of coal tar and used primarily in organic synthesis. Chemical formula = $C_{10}H_8$. Molecular weight = 128.6 g/mol.

□ **narrative biological criteria**

General statements of attainable or attained conditions of biological integrity and water quality for a given designated aquatic life use.

□ **national priorities list (npl)**

The federal list of Superfund sites nationwide.

□ **natural cycle**

A series of events in nature that repeat over and over, usually involving a complete process of growth or action. Examples include: the oxygen, nutrient, hydrologic and mineral cycles.

□ **n-butane**

A colorless gas with a gasoline or natural gas-like odor. It is used in the manufacture of synthetic rubber, various organic compounds. It is also used as a food additive, as a solvent, a fuel and a propellant. Chemical formula = C_4H_{10}. Molecular weight = 58.1 g/mol.

□ **n-butanol**

Either of two flammable four-carbon alcohols derived from normal butane. Chemical formula = C_4H_9OH. Molecular weight = 74.122 g/mol.

□ **n-butyl acetate**

A clear, colorless liquid with a mild, fruity odor. It is used as an industrial solvent, in the manufacture of lacquer, and in artificial perfumes and flavoring agents. Chemical formula = $C_6H_{12}O_2$. Molecular weight = 116.16 g/mol.

□ **nephelometric turbidity units (ntu)**

The unit used to describe turbidity. Nephelometric refers to the way the instrument, a nephelometer, measures how much light is scattered by suspended particles in the water. The greater the scattering, the higher the turbidity. Therefore, low NTU values indicate high water clarity, while high NTU values indicate low water clarity.

❐ n-heptane

One of the isomers of heptane, used especially as a solvent in glues, varnishes, cements and inks. Also used as a standard in gasoline octane ratings. Chemical formula = C_7H_{16}. Molecular weight = 100.2 g/mol.

❐ n-hexane

Any of the several isomeric volatile liquid alkanes found in petroleum. Chemical formula = C_6H_{14}. Molecular weight = 86.17 g/mol.

❐ nickel

A silver-white, hard, yet malleable metallic element capable of a high polish and resistant to corrosion that is used chiefly in alloys and as a catalyst. Chemical symbol = Ni. Molecular weight = 58.71 g/mol.

❐ nitric oxide (no)

A colorless, poisonous gas formed by oxidation of nitrogen or ammonia. Chemical formula = NO. Molecular weight = 30.00614 g/mol.

❐ nitrobenzene

A poisonous, yellow, insoluble oil with an almond odor that is used as a solvent, as a mild oxidizing agent, and in making aniline. Chemical formula = $C_6H_5NO_2$. Molecular weight = 123.11 g/mol.

❐ nitrogen dioxide (no_2)

A toxic, reddish-brown gas that is a strong oxidizing agent. It is produced by combustion (as of fossil fuels), and is an atmospheric pollutant. Chemical formula = NO_2. Molecular weight = 46.0055 g/mol.

❐ nitrogen oxides (nox) emissions in selected countries-308 (at)

❐ nitrogen oxides (nox)

Nitrogen oxides, or NOx, is the generic term for a group of highly reactive gases, all of which contain nitrogen and oxygen in varying amounts. Many of the nitrogen oxides are colorless and odorless. However, one common pollutant, nitrogen dioxide (NO2) along with particles in the air can often be seen as a reddish-brown layer over many urban areas. Nitrogen oxides form when fuel is burned at high temperatures, as in a combustion process. The primary sources of NOx are motor vehicles, electric utilities, and other industrial, commercial, and residential sources that burn fuels.

❐ nitrous oxide

Also known as: Laughing gas and Dinitrogen monoxide. A colorless, nonflammable gas with a slightly sweet odor. Used mainly as an anesthetic in minor and major surgeries. Chemical formula = N_2O. Molecular weight = 44.02 g/mol.

❒ noncancer chronic inhalation health benchmark (ihb)

A chronic noncancer IHB is a concentration in ambient air, at or below which a chemical or defined mixture of chemicals is unlikely to cause an adverse health effect to the general public when exposure occurs daily throughout a person's lifetime. Noncancer IHBs are used for effects other than cancer. For implementation purposes, chronic IHBs are typically compared to an annual average concentration of a chemical or defined mixture of chemicals in air.

❒ nonpoint source

Polluted runoff -- nutrients and pollution sources not discharged form a single point: e.g., runoff from agricultural fields or feedlots.

❒ nonrenewable resources

Natural materials that are considered finite, or exhaustible because of their scarcity, the great length of time for their formation or their rapid depletion.

❒ nontoxic

Not containing ingredients that are harmful, poisonous or destructive.

❒ npdes

National Pollutant Discharge Elimination System.

❒ n-pentane

A clear, colorless solution with a mild, gasoline-like odor. It is used in natural gas, lighter fluid, blowtorch fuels, and aerosol propellants. Chemical formula = C_5H_{12}. Molecular weight = 72.15 g/mol.

❒ n-propyl acetate

Also known as: Acetic acid. A clear, colorless liquid with a pungent odor. It is used as a solvent, and in the production of pharmaceuticals, dyes, inks and insecticides. Chemical formula = $C_5H_{10}O_2$. Molecular weight = 102.15 g/mol.

❒ nuclear waste disposal at sea

- During the 48-year history of the sea disposal of nuclear waste, 14 countries used more than 80 sites to dispose of some 85 PBq of radioactive waste.
- Five countries used the sea disposal option regularly for large quantities of waste: Belgium, the former Soviet Union, Switzerland, United Kingdom, USA.
- Some 53 percent of the radioactivity of the dumped waste is lowlevel packaged waste, most of it dumped in the Northeast Atlantic by the UK.
- Around 43 percent is associated with the dumping of reactors with spent nuclear fuel by the former Soviet Union in the Kara Sea.
- The disposal of nuclear waste material at sea was finally banned by a resolution under the London Convention on Dumping in 1993.
- No further waste was dumped after the London Convention took place.

Radioactive waste dumped at sea 1946–1993
(PBq=Petabecquerel [Peta=10 to the power of 15])

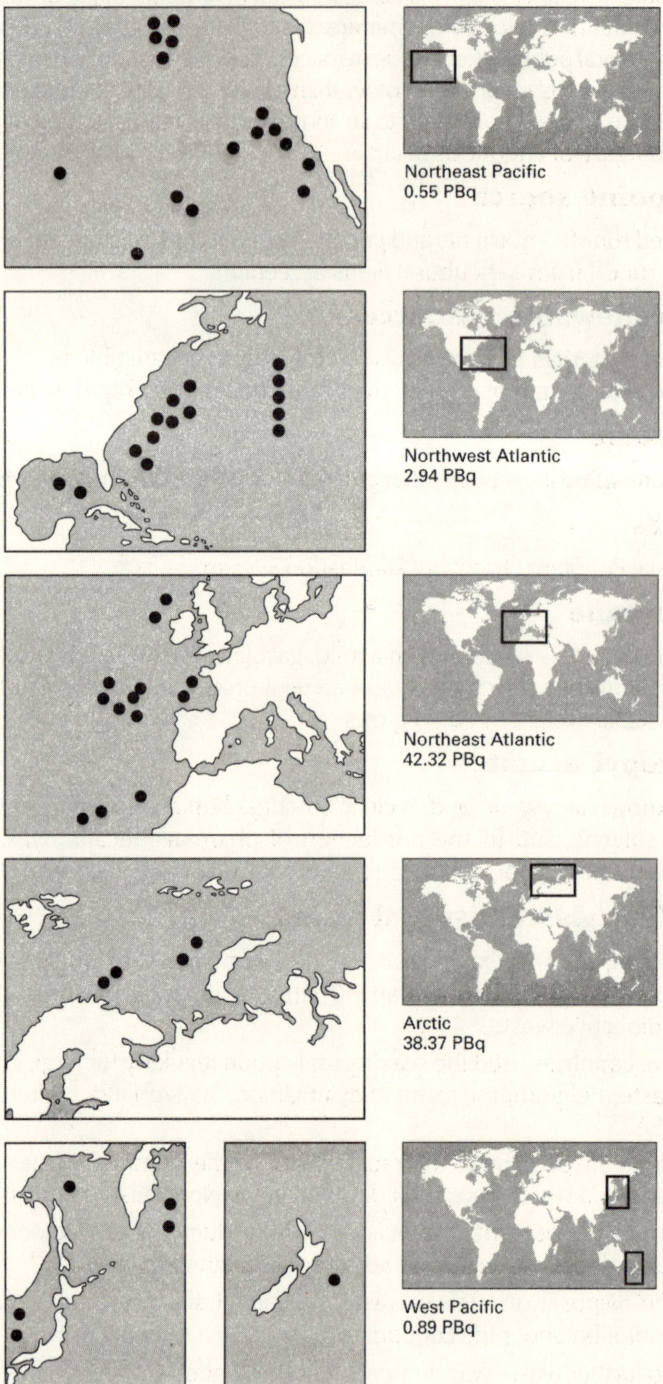

❑ nuclear waste disposal on land

- Since the 1960s, more than 200,000 tons of spent fuel have been produced by 400 reactors in 30 countries, and every year 10,000 tons are added.
- By 2050, production is predicted to rise to 450,000 tons.
- One much-debated method of nuclear waste disposal is burial.
- The most widely accepted plans are to bury vitrified high-level wastes sealed into stainless steel canisters, or to encapsulate spent fuel rods in corrosion-resistant metals such as copper or stainless steel, and for these to be buried in stable rock structures deep underground.
- The environmental disadvantages of burial of nuclear wastes include the possible spread of radioactivity into the surrounding environment.
- If absorbed into the food chain, it can cause unpredictable genetic damage.
- Furthermore, some elements have a half-life of up to 100,000 years, and so the effects of nuclear contamination would be permanent and almost irreversible.

Route nuclear fuel follows when used in the nuclear industry

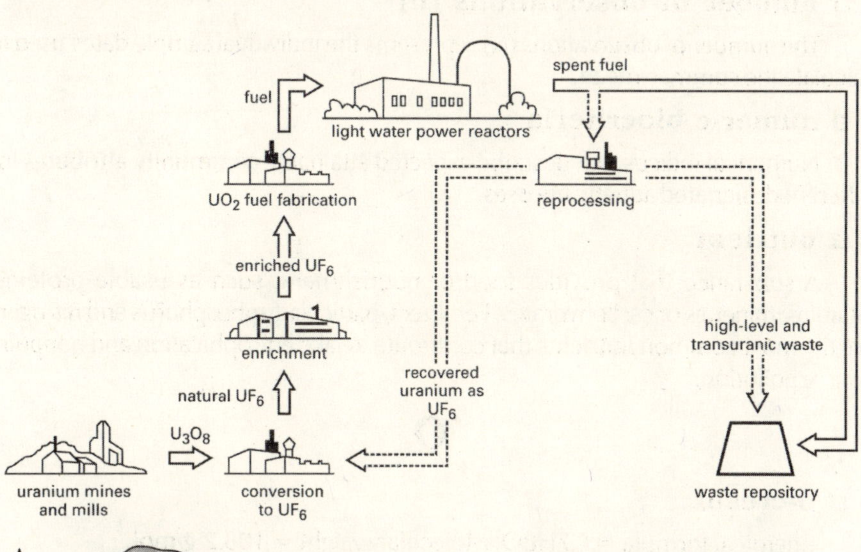

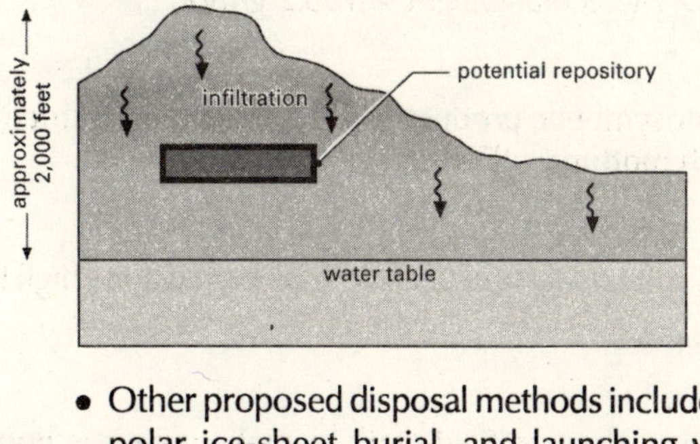

Proposed Yucca Mountain Repository

- The water table under Yucca Mountain is extremely deep.
- This makes it possible to put a repository about 1,000 feet underground and still have it about 800 feet above the water table.
- This significantly limits the risk of water reaching the corroding waste containers, and carrying radioactive material away from a repository.

- Other proposed disposal methods include hot rock disposal, seabed burial, polar ice-sheet burial, and launching waste into space. They all have disadvantages.

- In 'hot rock' disposal, the material is sent into deep boreholes, melting the rock and mixing to produce a self-sealing rock mass.
- Several countries have tried to find safe burial sites, but none has so far come into use.
- Cases in point include Yucca Mountain in Nevada in the United States, Gorleben in Germany, and Cumbria in Britain.
- In all these cases, the countries were faced with more geological complexities and political opposition than expected.
- Yucca Mountain is the US Government's potential geologic repository designed to store and dispose of spent nuclear fuel and high-level radioactive waste.
- The site is located in Nye County, Nevada, about 100 miles northwest of Las Vegas, near the Nevada Test Site.
- The repository would be approximately 1,000 feet below the top of the mountain, and 1,000 feet above the ground water.
- It would be able to store some 70,000 tons of nuclear waste.

❑ number of observations (n)

The number of observations (N) represents the individual sample dates used to calculate the summer-mean.

❑ numeric biocriteria

Numerical indices that describe expected attainable community attributes for different designated aquatic life uses.

❑ nutrient

A substance that provides food or nourishment, such as usable proteins, vitamins, minerals or carbohydrates. Fertilizers, particularly phosphorus and nitrogen, are the most common nutrients that contribute to lake eutrophication and nonpoint source pollution.

O

❑ o-cresol

Chemical formula = C_7H_8O. Molecular weight = 108.2 g/mol.

❑ oligotrophic

Describes a lake of low photosynthetic productivity. As a relatively nutrient-poor lake, it is clear and deep with bottom waters high in dissolved oxygen.

❑ oligotrophic lake

A relatively nutrient-poor lake, it is clear and deep with bottom waters high in dissolved oxygen.

❑ on-site inspector

The MPCA staff member who oversees field work to make sure it is done correctly.

❐ open dump

An uncovered space once used for depositing municipal refuse, which sometimes causes health and sanitation problems. These have been replaced with modern sanitary landfills with designs that contain waste and prevent contamination of surrounding areas.

❐ organism

A living body made up of cells, tissues and organs.

❐ other organics

Organic chemicals are those that contain carbon and hydrogen atoms. This group of organics does not include Carbonyls or Polynuclear Aromatic Hydrocarbons (PAHs). Examples of other organics are: Benzene, Vinyl Chloride, and 1,3-Butadiene

❐ overfill protection (none indicated)

This means that the tank has no overfill protection.

❐ overfill protection alarm

This is a device that is connected to a probe inside an underground storage tank. A float on the probe detects product level inside the tank and triggers a bell or whistle that alerts a delivery person that the tank is 90 percent full.

❐ overfill protection automatic shutoff device

This is a device that is installed in the fill pipe drop tube of an underground storage tank. It is usually set at 95 percent of tank capacity. When product reaches that level, a float arm shuts off the fill pipe to additional product. It is also called a "flapper valve".

❐ overfill protection ball float

This device is installed at the connection of the vent pipe to the tank. It extends into the tank and should be set at the 90 percent full level. Ball float valves work via a ball inside a wire cage that floats up with the rising product level which blocks off the vent line. When this occurs the delivery person is alerted to shut down the delivery.

❐ overfill protection type unknown

This signifies that the 30-day notification form submitted by the tank owner indicated that a tank has a form of overfill protection, but the specific type is unknown. On older MPCA notification forms, the type of overfill protection did not have to be indicated.

❐ o-xylene

A clear liquid with a sweet odor. One of the three isomers of xylene. Used as a raw material in the production of plasticizers, polyesters, dyes, insecticides and gasoline and other motor fuels. Chemical fromula = C_8H_{10}. Molecular weight = 106.16 g/mol.

❐ ozone

Ground-level ozone is formed in the atmosphere when nitrogen oxides and volatile organic compounds react in the presence of heat and sunlight. Cars, trucks, power plants, and solvents contribute to the formation of ozone, which is a major component of smog. Ozone can be transported into an area from sources hundreds of miles upwind. It is irritating to the eyes, nose, throat and lungs, and it can worsen the symptoms of asthma. The elderly, children, and people with respiratory illnesses are most at risk. Ozone can also damage plants, including crops and trees.

❐ ozone

A triatomic, very reactive form of oxygen that is a bluish, irritating gas of pungent odor. It is formed naturally in the atmosphere by a photochemical reaction. It is a major air pollutant in the lower atmosphere but a beneficial component of the upper atmosphere. Ozone is used for oxidizing, bleaching, disinfecting, and deodorizing. Chemical formula = O_3. Molecular weight = 47.998 g/mol.

P

❐ palladium

A moldable, silver-white metal used extensively in industry as an alloy with gold, silver and copper. Chemical formula = Pd. Molecular weight = 106.40 g/mol.

❐ particulate matter

Particulate matter, or PM, is the term for particles found in the air, including dust, dirt, soot, smoke, and liquid droplets. Particles can be suspended in the air for long periods of time. Some particles are large or dark enough to be seen as soot or smoke. Others are so small that individually they can only be detected with an electron microscope. Some particles are directly emitted into the air. They come from a variety of sources such as cars, trucks, buses, factories, construction sites, tilled fields, unpaved roads, stone crushing, and burning of wood.

❐ particulate matter, coarse (pm10)

In July 1987, EPA began using a new indicator, PM-10, which includes only those particles with aerodynamic diameter smaller than 10 um. Ten microns is approximately one seventh the diameter of a human hair. This fraction of TSP is responsible for most of the adverse human health effects of particulate matter because of the particles' ability to reach the lower regions of the respiratory tract.

❐ particulate matter, fine (pm2.5)

Fine particulate matter is a complex mixture of very small liquid droplets or solid particles in the air. Major sources are cars, trucks, construction equipment, coal-fired power plants, wood burning, vegetation and livestock. These particles can be directly released when coal, gasoline, diesel fuels and wood are burned. Many fine particles are also formed in the atmosphere from chemical reactions of nitrogen oxides, sulfur oxides, organic compounds and ammonia. Fine particulates are associated with increased hospitalizations and deaths due to respiratory and heart disease and can worsen the symptoms of asthma. People with respiratory or heart disease, the elderly and children are the groups most at risk.

❐ pathogens

Disease-causing organisms.

❐ pcbs

Polychlorinated biphenyls (PCBs) are a class of industrial chemicals manufactured from 1930 to 1977 for use in electrical and hydraulic products. PCBs are still present in the environment because of their persistence and accumulation at higher concentrations up the food chain (bioaccumulation), but have been showing a trend of decreasing concentrations. PCBs are a concern to human and ecological health, because they can affect neurodevelopment and reproduction and are classified by the U. S. Environmental Protection Agency (EPA) as probable human carcinogens. In Minnesota, PCBs are assessed in fish and in some surface waters. Minnesota's Water Quality Standards for PCBs are based on the total concentration of PCBs and are measured in nanograms (one billionth of a gram) per liter (ng/L) to picograms (one trillionth of a gram) per liter (pg/L) in water. In fish, PCBs are measured in units of nanogram (one billionth of a gram) per gram (ng/g).

❐ p-cresol

Chemical formula = $CH_3C_6H_4OH$. Molecular weight = 108.14 g/mol.

❐ pentachlorophenol

A crystalline compound used as a wood preservative, fungicide and disinfectant. Chemical formula = C_6HCl_5O. Molecular weight = 266.35 g/mol.

❐ perchlorate

Waste perchlorates are regulated as hazardous waste because they exhibit the characteristic of an oxidizer. (MN Rule. 7045.0131, sub.3). Ammonium perchlorate if in solution and stored in an aboveground storage tank would be regulated and if stored in an underground storage tank would be regulated as a hazardous substance. However, currently there are no risk-based soil, ground water, or surface water criteria developed at least at the state level. Some states, however (i.e. California has nonpromulgated Advisory Level [AL] of 4 ug/L for perchlorate), and USEPA has received peer review comments on their draft toxicological review and assessment document.

❐ perchloroethylene

Also known as: Tetrachloroethylene. A colorless, nonflammable liquid used often as a solvent in dry cleaning and for removal of grease from metals. Chemical formula = C_2Cl_4. Molecular weight = 165.83 g/mol.

❐ percolation

Movement of water through soil layers of materials.

❐ permanent list of priorities (plp)

Permanent List of Priorities (PLP) sites are those the state intends to investigate and cleanup under the Minneso t a Superfund Program. Known or suspected environmental contamination threatens public health, welfare or the environment at PLP sites.

▫ permitted solid waste sites

Permitted Solid Waste (SW) sites include facilities that manage solid waste (household or business garbage). Among the types of facilities included are landfills, transfer stations, demolition landfills, composting facilities and solid waste incinerators.

▫ persistence time

The time required for a pesticide to become inert. Arbitrarily assumed to equal four half-lives when measured persistence time is not available.

▫ personal exposure

The sum of exposure from each microenvironment where a person spends time. Examples of microenvironments include home, car, office, etc. Personal exposure can be measured using a sampling device worn by a study participant throughout a normal day.

▫ pesticide

A chemical substance used to kill or control pests, such as weeds, insects, fungus, mites, algae, rodents and other undesirable agents.

▫ ph scale

A measure of acidity, with 7 being neutral. Numbers under 7 are acidic and numbers over 7 are alkaline.

▫ phenanthrene

A crystalline, aromatic hydrocarbon of coal tar. Chemical formula = $C_{14}H_{10}$. Molecular weight = 178.22 g/mol.

▫ phenol

A corrosive, poisonous, crystalline, acidic compound present in coal tar and wood tar. In dilute solution, it is used as a disinfectant. Chemical formula = C_6H_5OH. Molecular weight = 94.1 g/mol.

▫ pheophytin-a (pheo)

A degradation product of chlorophyll-a that interferes with the measurement of chlorophyll-a, resulting in its over- or under-estimation. Pheophytin-a measurements help to determine a more accurate measure of chlorophyll-a.

▫ phosphate

An essential nutrient containing phosphorus and oxygen. Phosphate is often a critical nutrient in lake eutrophication management.

▫ phosphine

A colorless, poisonous, flammable gas that is a weaker base than ammonia. It is used to fumigate stored grain. Chemical formula = PH_3. Molecular weight = 34.0 g/mol.

▢ phosphorus

A nonmetallic element of the nitrogen family that occurs widely especially as phosphates. Chemical symbol = P. Molecular weight = 30.973761 g/mol.

▢ photosynthesis

The process by which green plants produce oxygen from sunlight, water and carbon dioxide.

▢ phthalic anhydride

A crystalline, cyclic, acid anhydride used especially in making alkyd resins. Chemical formula = $C_8H_4O_3$. Molecular weight = 148.12 g/mol.

▢ phytoplankton

Algae -- the base of the lake's food chain, it also produces oxygen.

▢ piping release detection annual tightness test

This is a precision or integrity test used to determine if a pipe is leaking at a volume of at least 0.1 gallons per hour.

▢ piping release detection automatic leak detec

This equipment is installed on pressurized piping lines. Line leak detectors (LLDs) must be able to detect large leaks of three (3) gallons per hour (gph). Mechanical line leak detectors are installed at the submersible pump head and usually work by restricting the flow of product. Electronic line leak detectors can be installed either at the pump head or at the dispenser and work by detecting pressure increases or decreases that may indicate leaks. Some electronic units are capable of performing 0.1 gph line tightness tests as well. Annual line tightness tests are required if the line leak detectors do not have this capability. All line leak detectors must be checked for proper functioning at least annually. Records of these function checks must be kept for at least ten (10) years.

▢ piping release detection european suction

This is a leak detection system that uses a pump at the dispenser to draw product from the tank to the dispenser. Also, this system is similar to a safe suction system with enough slope so that the product in the piping can drain back into the tank if suction is released.

▢ piping release detection european suction

This is a leak detection system that uses a pump at the dispenser to draw product from the tank to the dispenser. Also, this system is similar to a safe suction system with enough slope so that the product in the piping can drain back into the tank if suction is released.

▢ piping release detection ground water monitoring

This is a method of external leak detection that involves the use of one or more monitoring wells placed close to the piping that can be monitored visually or by electronic sensors. The wells are checked monthly for the presence of product that has leaked from the piping and is floating on to the groundwater. The wells are checked monthly for the presence of product that has leaked from the piping and is floating on to the groundwater. Also, there are a number of criteria that must be met to use this as a leak detection option. . If this method is used, it is normally used for both the tank and the piping.

❑ piping release detection interstitial monitoring

A leak detection method used to check the area between the pipe and the secondary containment for leaks and alert the operator if a leak is suspected.

❑ piping release detection other

This is any other form of leak detection used for piping that is not specifically mentioned in the UST or AST rules.

❑ piping release detection three-year tightness test

This is a periodic precision or integrity test for some suction piping systems used to determine if piping is leaking.

❑ piping release detection vapor monitoring

This method involves sampling monthly for vapors that have been released from product contained in the piping. The detection and measurement of fumes using this method indicate a possible leak. Also, there are a number of criteria that must be met to use this as a leak detection option. If this method is used, it is normally utilized for both the tank and the piping.

❑ plume

The path and form taken by contaminated ground water as it moves from the source.

❑ point source pollution

Pollution arising from a well- defined origin, such as a discharge from an industrial plant.

❑ poisoned land/sea food chains

- Species at the top of food chains are particularly vulnerable to pesticides and other poisons, as a result of a process called 'biomagnification'.
- Spraying a marsh to control mosquitoes will cause trace amounts of pesticide to accumulate in the cells of microscopic aquatic organisms, the plankton, in the marsh.
- In feeding on the plankton, filter-feeders, like clams and some fish, harvest DDT as well as food. (Concentrations of DDT ten times greater than those in the plankton have been measured in clams.)
- The process of concentration goes right on up the food chain from one trophic level to the next.
- Gulls, which feed on clams, may accumulate DDT to 40 or more times the concentration in their prey. This represents a 400-fold increase in concentration along the length of this short food chain.

- There is abundant evidence that some carnivores at the ends of longer food chains (e.g., ospreys, pelicans, falcons, and eagles) suffered serious declines in fecundity and hence in population size because of this phenomenon in the years before use of DDT was banned (1972) in the United States.
- Humans can be affected too, particularly people like the Inuit who depend on sea mammals for their food.
- The seals, whales, and walruses that are the staples of the Inuit diet have become deposits for the world's 12 most toxic chemicals, persistent organic pollutants (POPs) that collect in the animals' fat and are passed on to the Inuit as they eat, or through breast milk.

DDT in the food chain

The diagram shows how DDT becomes concentrated in the tissues of organisms representing four successive trophic levels in a food chain. The concentration effect occurs because DDT is metabolized and excreted much more slowly than the nutrients that are passed from one trophic level to the next. So DDT accumulates in the bodies (especially in fat).

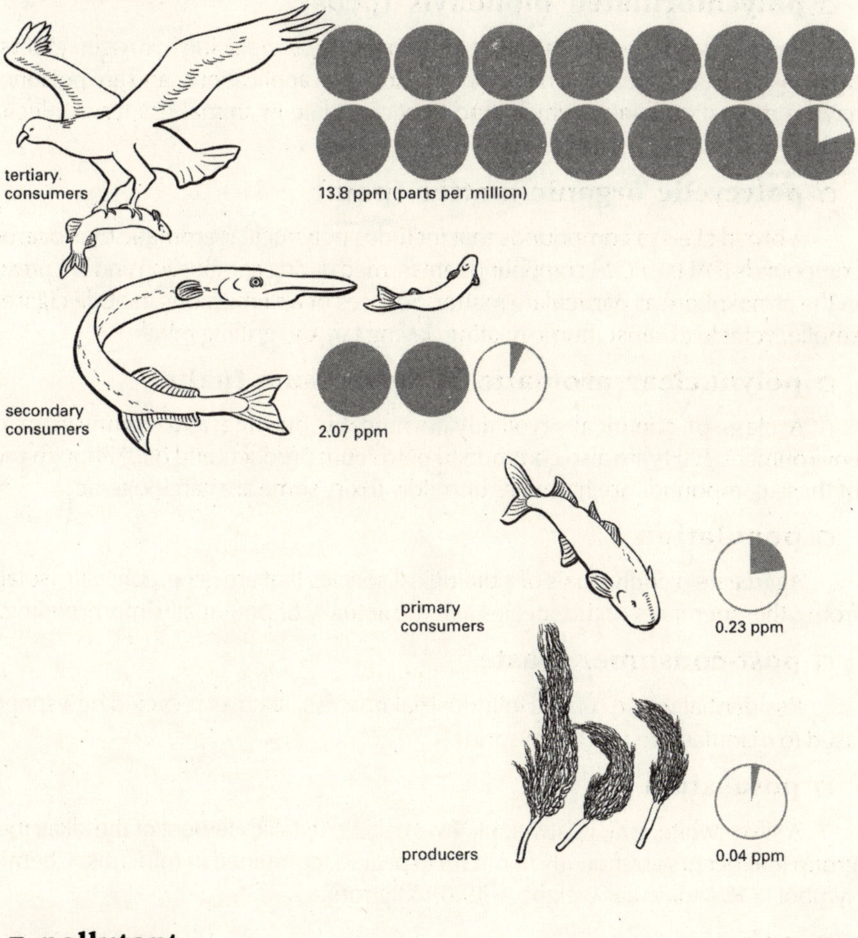

▫ pollutant

A substance that pollutes air, water or land.

❐ pollution

The contamination of soil, water or air by the discharge of potentially harmful substances.

❐ pollution control technologies

Equipment designed to reduce pollution and the resulting adverse environmental and health effects from waste combustion or disposal, or private or industrial emissions. Filters are often used to control emissions of particles, and scrubbers can control emissions of acidic gases, such as sulfur dioxide and hydrogen chloride.

❐ pollution prevention

Eliminating or reducing at the source the use, generation or release of toxic pollutants, hazardous substances and hazardous wastes.

❐ polychlorinated biphenyls (pcbs)

Any of several compounds that are produced by replacing hydrogen atoms in biphenyl with chlorine. They have various industrial applications, and are poisonous environmental pollutants which tend to accumulate in animal tissues. Molecular weight = 291.98 - 360.86 g/mol.

❐ polycyclic organic matter (pom)

A broad class of compounds that includes polynuclear aromatic hydrocarbon compounds (PAHs). POM compounds are formed during combustion and are present in the atmosphere as particulate matter. Sources of air emissions include cigarette smoke, vehicle exhaust, home heating, laying tar, and grilling meat.

❐ polynuclear aromatic hydrocarbons (pahs)

A class of chemicals typically formed by burning and common in the environment. PAHs are also common to petroleum products and oil. Although most of these compounds are harmless or mildly toxic, some are carcinogenic.

❐ population

Aggregate of individuals of a biological species that are geographically isolated from other members of the species and are actually or potentially interbreeding.

❐ post-consumer waste

Residential waste reused in industrial process, such as recycled newspapers used to manufacture more newsprint.

❐ potassium

A silver-white, soft, lightweight, low-melting metallic element of the alkali metal group that occurs abundantly in nature especially combined in minerals. Chemical symbol = K. Molecular weight = 39.0983 g/mol.

❐ ppb

Parts per billion. Ten ppb means ten parts of a contaminant to one billion parts of water.

❑ p-phenylenediamine

An off-white crystal or powder. Most commonly used in hair dyes, and as a dye intermediate. It is also used as a photographic developing agent and a chemical intermediate and as a vulcanization accelerator and as an antioxidant in rubber compounds. Prlonged skin contact is not recommended because the compound is known to cause skin irriations in humans. Chemical Formula = C_6H_8N. Molecular weight = 108.4 g/mol.

❑ primary standard

"Primary ambient air quality standards" or "primary standards" mean levels established to protect the public health from adverse effects. The adverse effects that the standards should protect against include acute or chronic subjective symptoms and physiological changes that are likely to interfere with normal activity in healthy or sensitive individuals or to interfere unreasonably with the enjoyment of life or property.

❑ profundal community

The area below the limnetic zone where light does not penetrate. This area roughly corresponds to the hypolimnion layer of water and is home to organisms that break down or consume organic matter.

❑ propane

A colorless, odorless, flammable gas used as a propellant in aerosol products, as a fuel for cooking (as in gas grills), and in cigarette lighters. A foul-smelling odorant is often added when propane is used for fuel purposes. Chemical formula = C3H8. Molecular weight = 44.1 g/mol.

❑ propene

Also known as: Propylene. A flammable, gaseous hydrocarbon obtained by cracking petroleum hydrocarbons. It is used chiefly in organic synthesis. Chemical formula = C_3H_6. Molecular weight = 42.08 g/mol.

❑ propionaldehyde

A clear liquid with an intense, suffocating, slighlty fruity odor. It is used as an intermediate in the manufacture of polyvinyl & other plastics, in the synthesis of rubber chemicals, as a disinfectant and as as preservative. Chemical formula = C_3H_6O. Molecular weight = 58.08 g/mol.

❑ propylene glycol methyl ether acetate (pgmea)

A colorless liquid with a mild, fruity odor. Used as a solvent, flow improver, and coalescent in coating and printing ink industries. Chemical Formula = $C_6H_{12}O_3$. Molecular weight = 132.159 g/mol.

❑ propylene oxide

A colorless, highly flammable liquid that is soluble in water and has a sweet, alcoholic or ether-like odor. It is used primarily in the production of polyethers and propylene glycol. Also used in the fumigation of foodstuffs and plastic medical instruments, and in the preparation of ubricants, surfactants, and oil demulsifiers. Chemical formula = C_3H_6O. Molecular weight = 58.08 g/mol.

❐ p-xylene

A clear liquid with a sweet odor. Used in pharmaceutical synthesis, insecticides, and frequently used for paints and in the printing trade. Chemical formula = C_8H_{10}. Molecular weight = 106.17 g/mol..

Q

❐ quinoline

A pungent, oily, nitrogenous base obtained usually by distillation of coal tar or by synthesis from aniline. It is the parent compound of many alkaloids, drugs, and dyes. Chemical formula = C_9H_7N. Molecular weight = 129.15 g/mol.

❐ quinone

Also known as: p-Benzoquinone. CAS Number: 106-51-4. Exists as yellow prisms, with a strong odor resembling chlorine. Quinone is used as a chemical intermediate, a polymerization inhibitor, an oxidizing agent, a photographic chemical, a tanning agent, and a chemical reagent. Chemical formula = $C_6H_4O_2$. Molecular weight = 108.1 g/mol.

R

❐ rank

The ranking of this facility's emissions compared to other point sources that emitted this pollutant during this reporting year.

❐ raw materials

Unprocess natural substances, such as wood and metals, used in the manufacture of products.

❐ rcra investigation/cleanup sites

Sites where RCRA (Resource Conservation and Recovery Act) hazardous waste generators had an actual or potential release requiring investigation and/or cleanup. These generators fall into the very small, small and large quantity generator classes.

❐ rcra tsd facilities

RCRA TSD (Resource Conservation and Recovery Act Treatment, Storage and Dísposal) facilities are those permitted to treat, store and dispose of hazardous wastes.

❐ recyclable

Material that still has useful physical or chemical properties after serving its original purpose and can, therefore, be reused or remanufactured into additional products. Plastic, paper, glass, used oil, tin and aluminum cans are examples of recyclable materials.

❒ recycle

The process of collecting materials from the waste stream and separating them by type, remaking them into new products, and marketing and reusing the materials as new products.

❒ regionalization or ecoregionalization

Procedure for subdividing a geographic area into regions of relative homogeneity in ecological systems or in relationship between organisms and their environment.

❒ regions at risk from sea-level rise

- Ice is essential to global climatic and ecological stability. Several large-scale effects have been predicted if large areas of the world's ice melt.

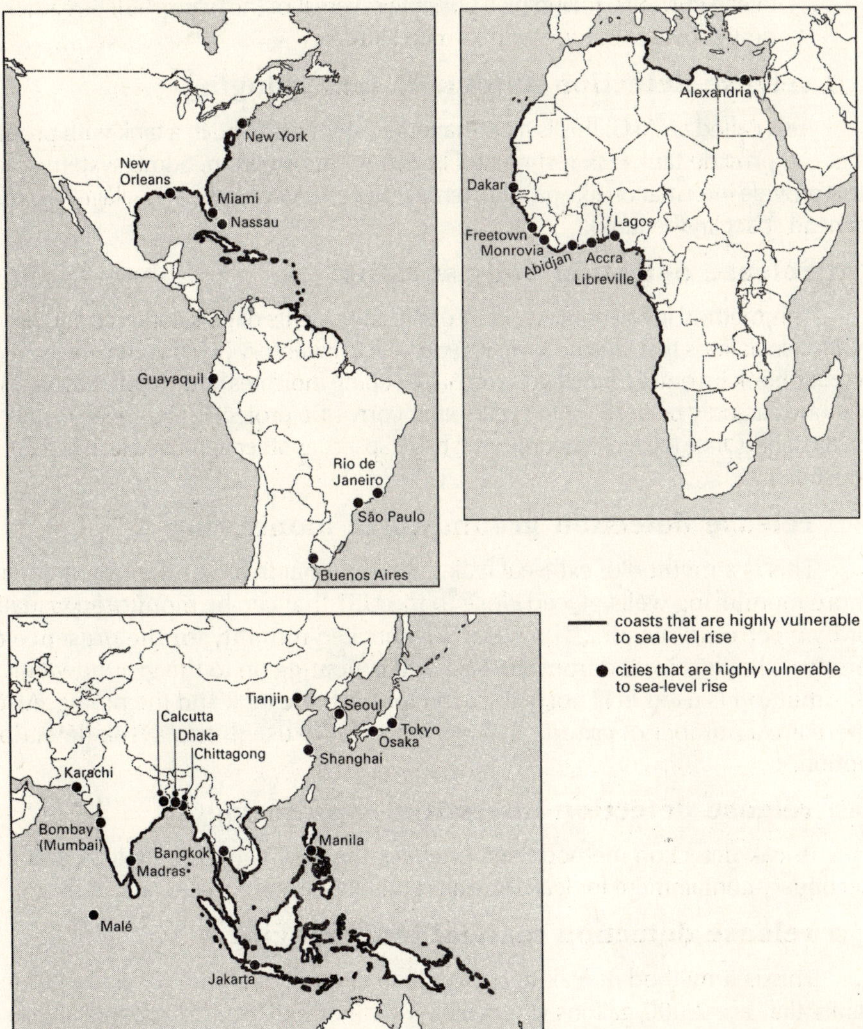

- First, as land-based ice on Greenland and Antarctic melts and drains into the oceans, sea-levels will rise – although the larger cause of sea-level rise is the expansion of water as it warms.
- Higher water combined with storm surges could inundate many coastal cities and ecosystems.
- Second, as glaciers and snow in the mountains melt and increase runoff, human settlements in major river valleys downstream from the mountains (such as the Ganges and Jiang [Pearl] river basins below the Himalayas) will become increasingly vulnerable to floods.
- Hundreds of millions of people could be displaced by floods or altered coastlines within decades.
- Climate models also suggest that melting could cause major alterations in ocean currents, including a possible reversal of the warm Gulf Stream that now provides Europe with its mild climate.

❑ release detection automatic tank gauging

Also called an ATG, this is a permanent system installed in a tank with probes that can provide tank leak testing and inventory information. Some systems may also provide information like product temperature, water levels, and ullage (amount of head space in the tank).

❑ release detection daily sticking

This method involves taking daily of measurements of tank contents and using math calculations that let you compare your "stick" inventory (what you've measured) to your book inventory (what your record-keeping indicates you should have). This method can only be used for ten years after corrosion protection has been installed on the tank. Once the ten-year time limit has elapsed, another approved leak detection must be used.

❑ release detection groundwater monitoring

This is a method of external leak detection that involves the use of one or more monitoring wells placed close to the UST that can be monitored visually or by electronic sensors. The wells are checked monthly for the presence of product that has leaked from the UST and is floating on to the groundwater. If this method is used, it is normally used for both the tank and the piping. Also, there are a number of criteria that must be met to use this as a leak detection option.

❑ release detection interstitial monitoring

A leak detection method used to check the area between the tank and the secondary containment for leaks and alert the operator if a leak is suspected.

❑ release detection manual tank gauging

This is a method of leak detection than can only be used for underground tanks that are 2,000 gallons or smaller. This method requires monitoring on a weekly basis.

❐ release detection other

This indicates an alternative form of leak detection that is not specifically defined in the UST or AST rules.

❐ release detection other

This indicates an alternative form of leak detection that is not specifically defined in the in the UST or AST rules.

❐ release detection soil vapor monitoring

This method involves sampling monthly for vapors that have been released from product contained in the tank. If this method is used, it is normally utilized for both the tank and piping. The detection and measurement of fumes using this method indicate a possible leak. Also, there are a number of criteria that must be met to use this as a leak detection option.

❐ release detection statistical inventory reconciliation (sir)

This is the date that the annual leak SIR report is due to MPCA. The date is February 15th.

❐ release detection tightness test

This is a precision or integrity test used to determine if a tank is leaking.

❐ remedial action

Cleanup actions taken, such as at Superfund sites. Also referred to as RA.

❐ remedial design

The design of a cleanup system. It may include developing engineering plans and specifications, and bidding documents for construction of the system.

❐ remedial investigation

A soil, ground water, surface water and/or air investigation to determine the extent of contamination present at a site.

❐ renewable resource

A naturally occurring raw material or from of energy derived from an endless or cyclical source, such as the sun, wind, falling water, biofuels and trees. With proper management and wise use, the consumption of these resources can be approximately equal to replacement by natural or human- assisted systems.

❐ request for response action (rfra)

A formal MPCA Citizens Board request to responsible parties at Superfund sites asking them to take specific actions at the site leading to cleanup.

❐ residue

Something that remains, or is left over. Residues of toxic materials left in empty containers can pose a threat to the environment.

❒ resource conservation and recovery act (rcra)

RCRA was enacted by Congress in 1976, as an amendment to the 1965 Solid Waste Disposal Act. The goals of RCRA are to: Protect human health and the environment from the hazards posed by waste disposal Conserve energy and natural resources through waste recycling and recovery Reduce or eliminate, as expeditiously as possible, the amount of waste generated, including hazardous waste Ensure that wastes are managed in a manner that is protective of human health and the environment.

❒ respiration

Oxygen consumption.

❒ responsible parties (rp)

Businesses, local governments and individuals responsible, as defined by Superfund law, for contamination at a site because of past or current activities. Also call PRPs for potentially responsible parties.

❒ reuse

The use of a product more than once in its same form for the same or different purpose.

❒ rise in carbon emissions

- Atmospheric CO_2 concentrations have increased by 30 percent since 1750, and are set to continue rising steeply.
- CO_2 emissions from fossil fuel combustion worldwide increased from 3,811 million tons of carbon equivalent in 1970 to 5,821 million in 1990, an average annual growth rate of 2.1 percent.
- Because of economic recession and political and economic upheaval in Eastern Europe between 1990 and 1999, the growth in carbon dioxide emissions slowed to an average annual rate of 0.5 percent.
- World carbon dioxide emissions are projected to reach 9,762 million tons carbon equivalent in 2020, an increase of 3,671 million tons over 1999 emissions.
- Approximately 67 percent of the growth in emissions between 1999 and 2020 is projected to come from developing countries.
- Population growth, rising personal incomes, rising standards of living, and further industrialization in the developing countries are expected to have a much greater influence on levels of energy consumption than in industrialized countries.
- The industrialized nations are expected to account for 25 percent of the total increase in emissions, while Eastern Europe and the former Soviet Union will account for 8 percent.

Atmospheric CO_2 concentrations 1750–2000

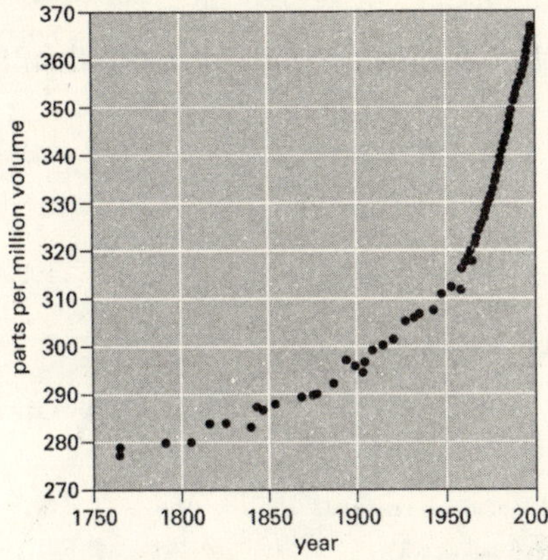

World CO_2 emissions by region 1970–2020

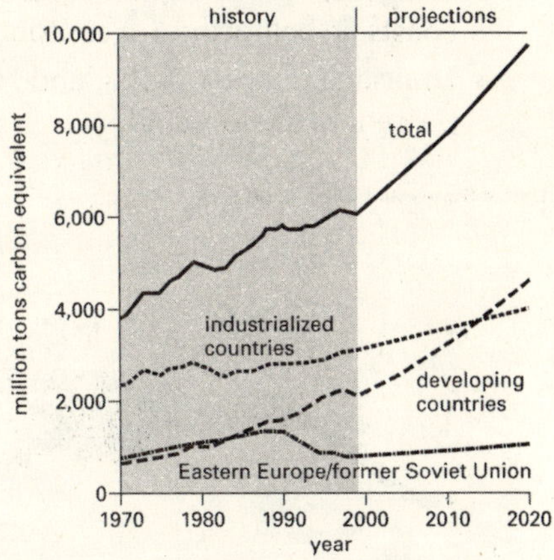

Fifty-year period cumulative carbon emissions country-by-country as a percentage of global output

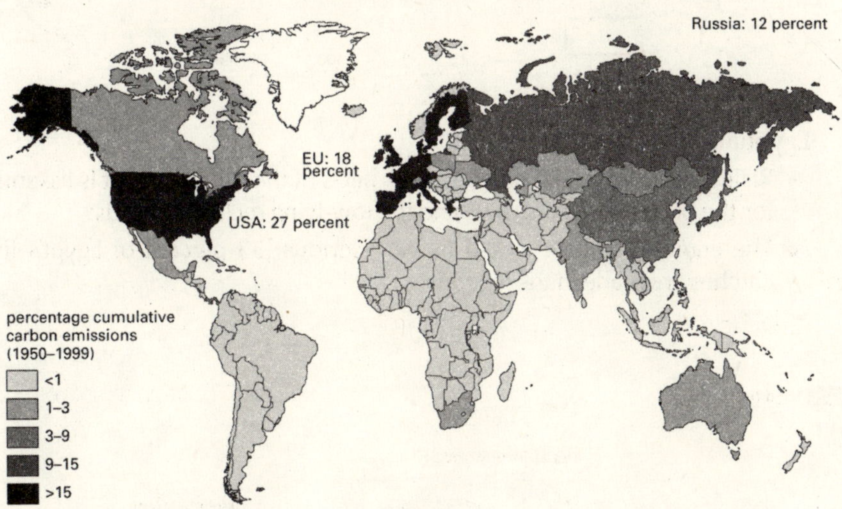

▫ risk in close up 1: united states gulf coast and egyptian nile delta

- Over the next century, sea level is most likely to rise nearly 2 ft along most of the US Atlantic and Gulf Coasts.
- The 11 ft-6 in contour roughly illustrates an area that might be flooded over a period of several centuries.
- The 5-ft contour roughly represents the area that would be inundated during spring high water with a 27 in rise in sea level.

- Such a rise appears to be most likely to occur in the next 120 years, but has a 1 percent chance of occurring in the next 60 years.
- Altogether, some 22,400 square miles) of land along the Atlantic and Gulf coasts lie below the 5-ft contour.
- Louisiana, Florida, Texas, and North Carolina account for more than 80 percent of the low land.

United States Gulf Coast flood area

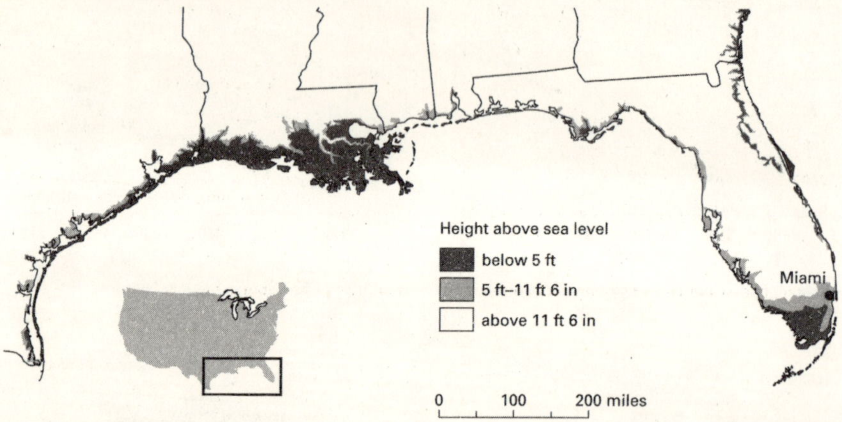

Egyptian Nile Delta flooding

- Rising sea level would destroy weak parts of the sand belt that is essential for the protection of lagoons and the low-lying reclaimed lands.
- The economic effect would be very serious: 33 percent of Egypt's fish catches are made in the lagoons.

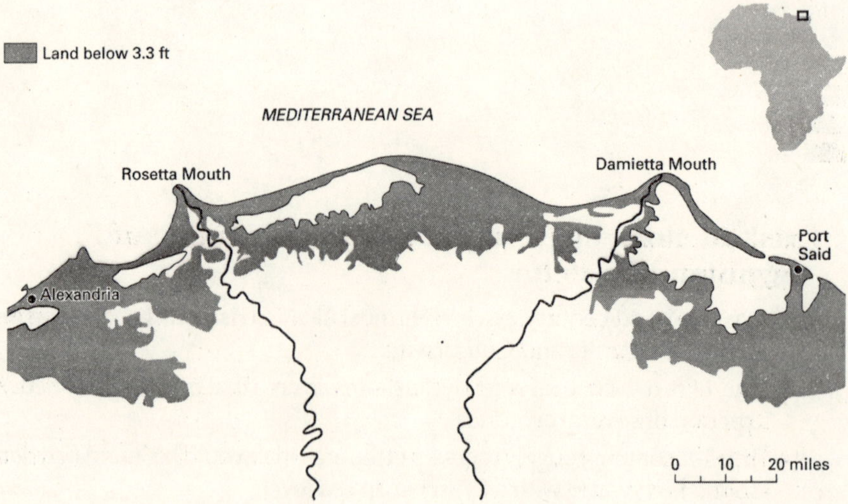

- Sea level rise would change the water quality and affect most freshwater fish. Valuable agricultural land would be inundated.
- Vital, low-lying installations in Alexandria and Port Said would be threatened.
- A 3.3 ft rise would affect a population of 6.1 millioN and 1,740 sq miles of cropland.

❐ risk in close up 2: bangladesh and the netherlands

- Bangladesh, one of the world's poorest nations, is also the country most vulnerable to sealevel rise.
- The population is already severely affected by storm surges: catastrophic events in the past have caused damage up to 62 miles inland.
- A 5-ft rise in sea level would inundate 17 percent of Bangladesh (8,500 square miles), displacing 34 million people (15 percent of the projected population in 2030).
- A 10-ft rise would inundate the capital, Dhaka.

Land below sea level in Bangladesh

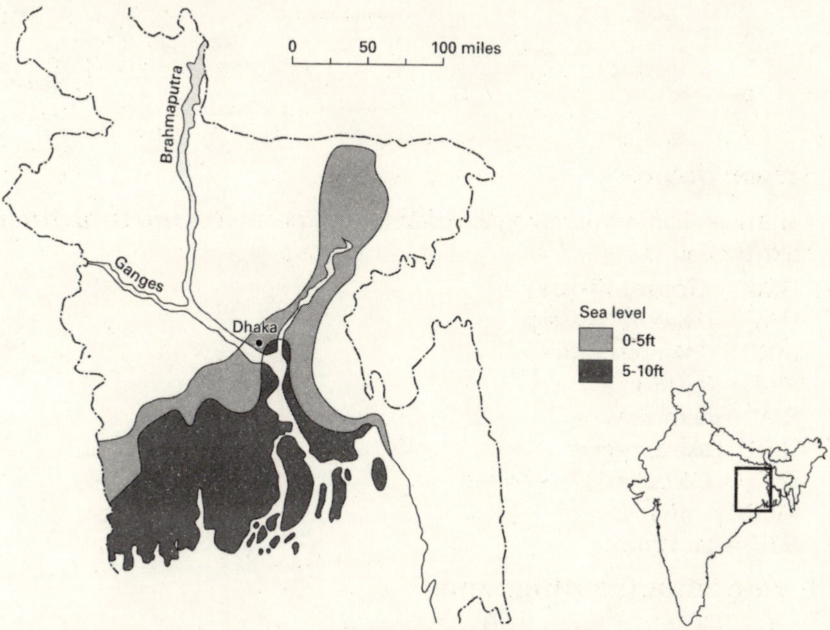

- A huge area of the Netherlands (one quarter of the land area), including the city of Amsterdam, is below sea level, protected from the sea by dykes.
- The Netherlands is evolving plans to solve the problems arising from rising sea levels caused by global warming and greater river discharges.
- Anticipating these developments now will make it possible to limit flooding in the future.
- The government is allocating EUR 1.27 billion (over $1 billion) up to 2020 for the implementation of these plans.

❒ river basin

The river than surrounding land drains to. Minnesota is characterized by nine major River Basins.

UMB = Upper Mississippi
LMB = Lower Mississippi
MNB = Minnesota River
RRB = Red River
RAB = Rainy River
LSB = Lake Superior
CDB = Cedar and Des Moines
MOB = Missouri
SCB = St. Croix

❒ root zone (rooting zone)

The depth of soil penetrated by plant roots.

❒ rubidium

A soft, silvery-white, highly reactive metal. It is used as a catalyst; and in photocells and vacuum- and cathode-ray tubes. Chemical formula = Rb. Molecular weight = 85.4678 g/mol.

❒ runoff

That portion of precipitation or irrigation water that flows off a field or paved area and enters surface water.

S

❐ sample depth (depth)

Sample collection depth measured from the surface of a lake.

❐ scrap

Discarded or rejected industrial waste material often suitable for recycling.

❐ scrubber

Common antipollution device that uses a liquid or slurry spray to remove acidic gases and particles from municipal waste combustion facility flue gas.

❐ sea-level rise estimates

- The flood risk to coastal populations will significantly increase if sea levels rise.
- The southern Mediterranean, Africa, southern Asia and Southeast Asia are most vulnerable to the impacts of flooding.
- Various studies have been prepared to show the possible impact of reducing emission levels.
- If emissions do not drop below their current levels, by the 2080s sea level will be about 16 inches higher than today. It is estimated this will increase annual flood victims from 13 million to 94 million.
- 60 percent of this increase will occur in southern Asia (along coasts from Pakistan, through India, Sri Lanka and Bangladesh to Burma), and 20 percent will occur in Southeast Asia (from Thailand to Vietnam including Indonesia and the Philippines).
- In the emissions estimates shown below the flood impacts of sea-level rise are reduced as levels of CO_2 stabilize.
- By the 2080s, the annual number of people flooded could be 34 million under the 750 parts per million (ppm) CO_2 emissions estimate and 19 million under the 550 ppm estimate. Again, most of those flooded will be in southern and Southeast Asia.

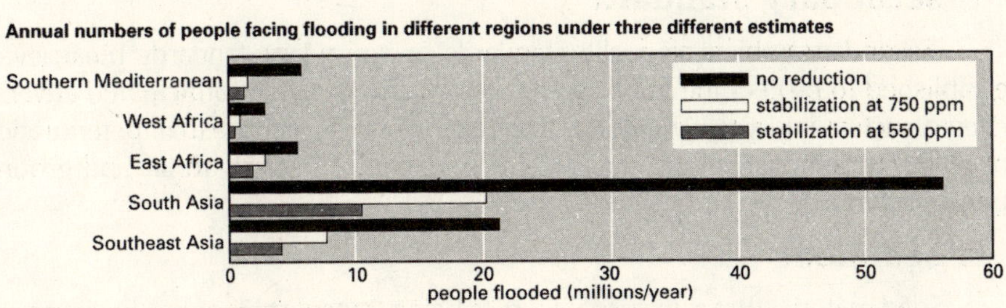

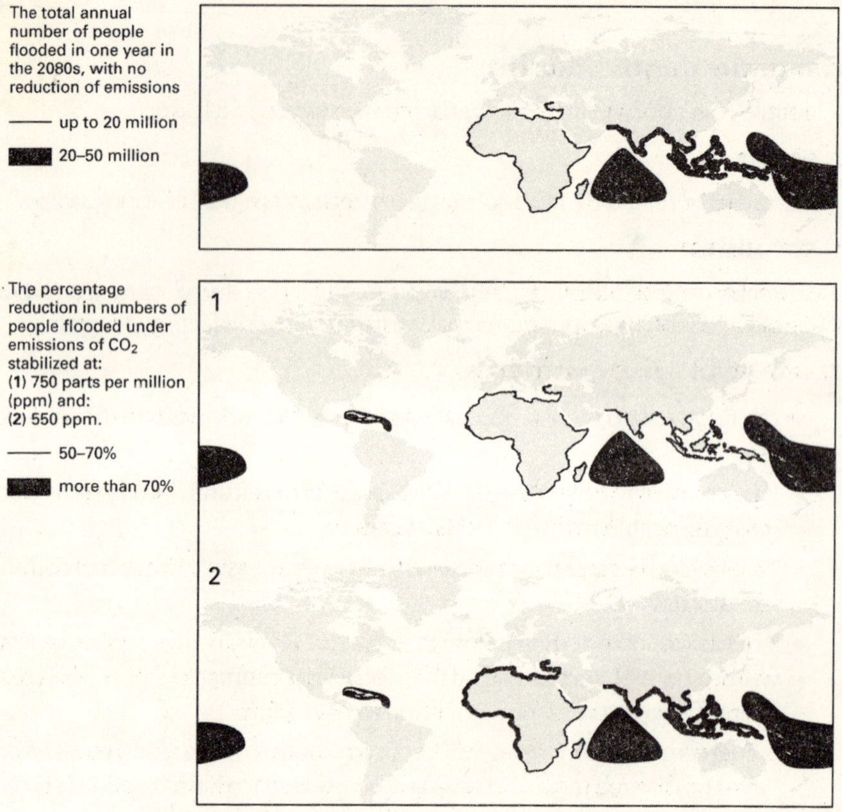

❐ secchi depth (secchi)

The depth in a lake, measured in meters, to which a Secchi disk can be observed, as a measure of light penetration in water. The Secchi disk is lowered into a section of shaded water until it can no longer be seen and then lifted back up until it can be seen once again. Averaging the two depths gives the clarity of the water.

❐ secchi disk

A device measuring the depth of light penetration in water.

❐ secondary standard

"Secondary ambient air quality standards" or "secondary standards" mean levels established to protect the public welfare from any known or anticipated adverse effects, such as injury to agricultural crops and livestock, damage to or deterioration of property, annoyance and nuisance of persons, or hazards to air and ground transportation.

❐ sediment

Solid material that is in suspension, is being transported, or has been moved from its original location by air, water, gravity or ice.

❒ sediment yield

The quantity of sediment arriving at a specific location.

❒ sedimentation

The addition of soils to lakes, a part of the natural aging process, making lakes shallower. The process can be greatly accelerated by human activities.

❒ seepage

Percolation of water through the soil from unlined canals, ditches, laterals, watercourses, or water storage facilities.

❒ seepage and leakage

Occurrence of seepage and leakage
A cross-section of a clay-lined landfill site on a slope leading down to a river shows the way in which both seepage and leakage can occur,

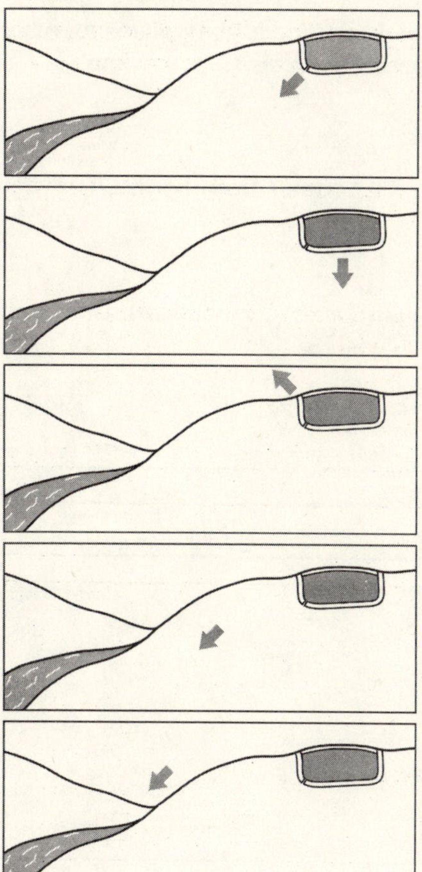

Leakage of pollutants can occur when the container has holes or cracks in it so that the pollutant is able to escape into the environment. This can also happen when the container simply overflows.

Seepage takes place when pollutants escape because the lining material of a container is porous or permeable, allowing the passage of pollutants even though there are no visible holes.

Gaseous as well as liquid pollutants can seep or leak into the environment. They behave in a similar way to liquids, though they may be heavier or lighter than air.

Beneath-ground pollutants can continue to permeate bedrock. They will seep or leak depending on the nature and structure of the rock. Additionally they may dissolve into the groundwater and find their way into the land's natural drainage system.

Above-ground liquid pollutants may contaminate surface runoff water and drain into the river system. Gaseous pollutants may mix with the air or rise through it to join the layer of greenhouse gases in the upper atmosphere, depending on their weight.

Seepage and leakage can occur from both urban and rural sites

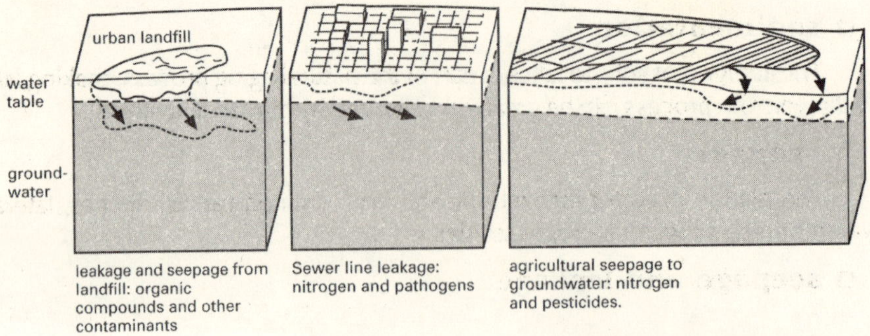

❐ selenium

A metallic gray solid used heavily in the production of glass, pigments, rubber, metal alloys, petrochemicals and photographic chemicals. Chemical formula = Se. Molecular weight = 78.996 g/mol +/- 3 g.

❐ sewage

The organic waste and wastewater that comes from homes, farms and businesses.

❐ sewage pollution

- Sewage is defined as waste matter from domestic or industrial establishments that is carried away in sewers or drains.

Main constituents of sewage

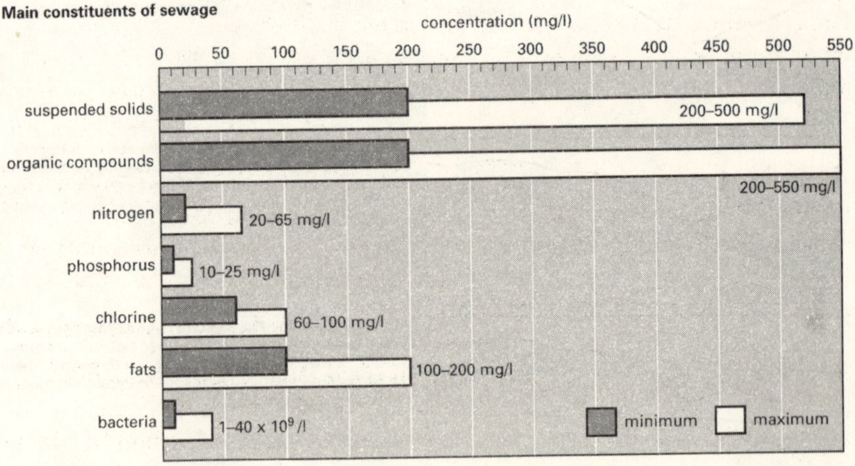

- Sewage is still discharged untreated into natural waters from industrial and domestic outfalls, in addition to the wastes produced by animals which contaminate water courses and eventually the marine environment.

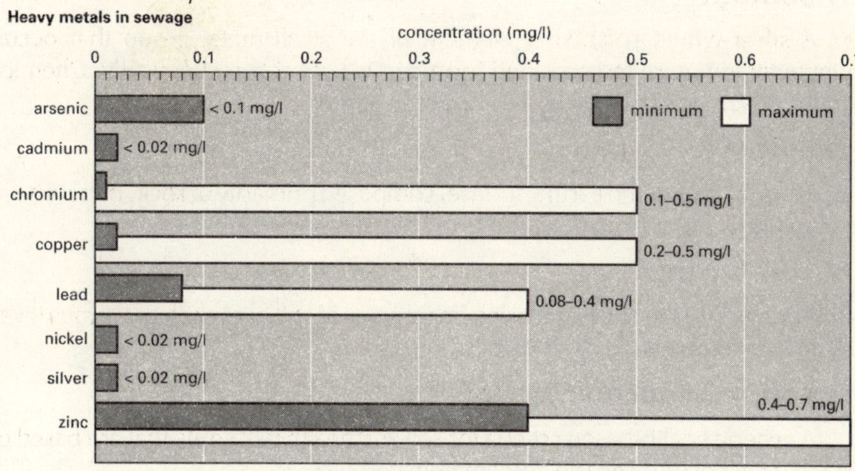

- Low levels of fecal material contain organic and nutrient components with which the natural system has inbuilt mechanisms to cope.
- But high concentrations of industrial and domestic waste entering the marine environment can disrupt marine ecosystems, and toxins and pathogens are introduced which may also affect people.
- Sewage is 99 percent water with 1 percent impurities amounting to about about 1,000 mg/l.
- The 60 million population of the UK alone produces 6 billion gallons of sewage everyday in the UK, of which 300,000 gallons is discharged untreated into the sea.

❑ sheet erosion

The removal of a fairly uniform layer of soil material from the land surface by the action of rainfall and surface runoff.

❑ silver

A white metallic element that is very malleable, capable of a high degree of polish, and has the highest thermal and electric conductivity of any substance. Chemical symbol = Ag. Molecular weight = 107.8682 g/mol.

❑ sinkhole

A depression in the landscape where limestone has been dissolved.

❑ small quantity generator (sqg)

A hazardous waste generator that generates between 220 and 2,200 lbs (100 and 1,000 kg) per month of hazardous waste. SQGs must comply with general federal and state requirements for managing hazardous waste. SQGs need an EPA identification number that is used to monitor and track hazardous waste activities.

❐ sodium

A silver-white, soft, waxy element of the alkali metal group that occurs abundantly in nature in combined form and is very active chemically. Chemical symbol = Na. Molecular weight = 22.98977 g/mol.

❐ soil

The top layer of the earth's surface, composed primarily of rock, minerals and decomposed matter from dead plants or animals.

❐ soil borings

Borings into the earth that bring up samples of soil or rock from various levels to study for composition or contamination.

❐ soil reference value (srv)

A generic health-based criteria for soil and health risk limits that are based on a standard exposure scenario for contaminated sites.

❐ solid waste

Trash and garbage without enough liquid to flow freely.

❐ source reduction

The design, manufacture and use of products which reduces their amount and toxicity. Source reduction measures include: reusing items, minimizing the amount of products used, extending the useful life of a product, and reducing the amount of excessive product packaging.

❐ specific conductance (spec cond)

A temperature-adjusted measurement of conductivity. Typically measured in micromhos per centimeter (μmhos/cm). Conductivity varies depending on water temperature. Specific conductance takes into account the water sample temperature to yield a conductivity value as if the sample were at 25 degrees Celsius.

❐ spending on pollution control in usa

Capital expenditure of United States refining industries on pollution-abatement measures have accounted for a varying share of total capital investment which was highest after the Clean Air Act amendments of 1990.

In three years –1992, 1993, and 1994 – they accounted for over half of all capital expenditure and reached nearly $3,000 million.

Acidification:	13.74%
Areas under protection:	3,481
(per capita):	11.7707 per 1 million people
Biosphere reserves area:	31,570 thousand hectares
(per capita):	106.751 thousand hectares per 1
Breeding birds threatened:	8.31%
Carbon efficiency:	1.77 CO2 emissions/$ GDP

Latest available figures for United States refining industry capital expenditure and pollution abatement spending

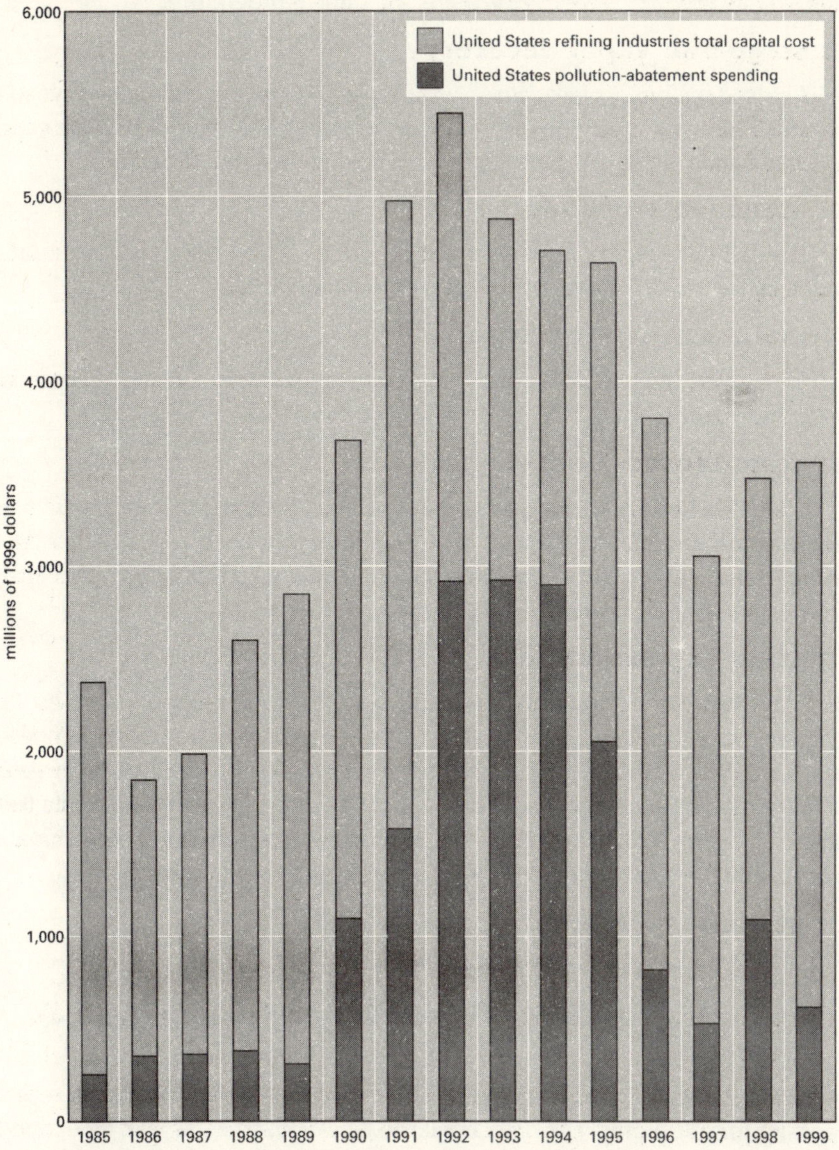

❐ spill containment

This indicates that a tank has a catchment basin or spill bucket that is sealed around the fill pipe to prevent spills that may occur when the delivery hose is disconnected from the fill pipe. This basin or bucket must be able to contain the contents of the delivery hose.

❐ spring turnover

After ice melts in spring, warming surface water sinks to mix with deeper water. At this time of the year, all water is the same temperature.

❐ stage-one vapor recovery

A closed system of pipes, hoses, and fittings, connecting a gasoline-unloading tank and a gasoline-receiving tank at a gasoline service station so that the vapors displaced from the receiving tank are transferred to the unloading tank.

❐ standard error (se)

The standard error of the mean (SE) is calculated as the standard deviation divided by the square root of the number of observations.

❐ state assessment sites

Sites where the MPCA has spent federal or state funds to investigate potentially contaminated properties.

❐ state closed landfill sites

State Closed Landfill Program sites are closed municipal waste landfills that were formerly permitted by the state. After a binding agreement is negotiated between the MPCA and the site owner/operator, the state takes responsibility for monitoring and cleanup systems needed for long-term care.

❐ state superfund sites

State Superfund sites, also referred to as or Permanent List of Priorities (PLP) sites, are those the state intends to investigate and cleanup under the Minnesota Superfund Program. Known or suspected environmental contamination threatens public health, welfare or the environment at PLP sites. These sites include those addressed by the MPCA, as well as sites with agricultural chemical contamination which are addressed by the Minnesota Department of Agriculture.

❐ stationary source

A nonmoving source of pollution.

❐ statistical inventory reconciliation (sir) piping leak detection

This is indicated if the tank is using SIR for leak detection. There is not separate reporting for SIR piping leak detection. The tank SIR serves as leak detection for both the tank and piping.

❐ statistical inventory reconciliation (sir) tank leak detection

This method involves taking daily measurements of tank contents, deliveries, and sales. These figures are sent to a SIR vendor (a third party) for statistical analysis that checks if the tank system is leaking.

❏ stormwater pollution prevention plan (swppp)

A plan for stormwater discharge that includes erosion prevention measures and sediment controls that, when implemented, will decrease soil erosion on a parcel of land decrease off-site nonpoint pollution.

❏ stream transparency (stream tra)

The clarity of stream water (in units of centimeters) as measured in a Citizen Stream Monitoring Program transparency tube. A greater transparency reading in centimeters reflects higher water clarity. Stream water clarity is an indicator of how much sediment, algae, and other materials are suspended in the water. The transparency tube is a clear, 60 cm-long tube with a colored disk on the bottom for measuring the depth at which the disk is visible. To measure water clarity, the tube is filled with water collected from a stream or river. Looking down into the tube, water is released through a valve until the colored disk is visible. The depth of the water when the disk becomes visible is recorded in centimeters, which are marked on the side of the tube. If the disk is visible when the tube is full, the transparency reading is ">60 centimeters."

❏ strontium

A silvery-white metal that rapidly turns yellow when it oxidizes. It is used in fireworks, signal flares and in tracer bullets. Also added to alloys of tin and lead to add hardness and durability. Strontium usually occurs as a mixture of four stable isotopes. 26 other less common, yet unstable isotopes exist, the most harmful of which is 90Sr. 90Sr is the product of nuclear fallout and can cause health problems. Chemical formula = Sr. Molecular weight = 87.62 g/mol.

❏ strontium chromate

A yellow powder that is used in metal coatings, anti-corrosive paints, pigments, ceramics, and glass. Chemical formula = $SrCrO_4$. Molecular weight = 203.62 g/mol.

❏ styrene

Also known as: Ethylbenzene. A fragrant, liquid, unsaturated hydrocarbon used chiefly in making synthetic rubber, resins, and plastics and in improving drying oils. Chemical formula = C_8H_8. Molecular weight = 104.16 g/mol.

❏ sulfate (so_4)

One of several minerals containing positive sulfur ions bonded to negative oxygen ions. Chemical formula = SO_4. Molecular weight = 96.058 g/mol.

❏ sulfur

Tends to appear as a yellowish solid. Pure sulfur has no odor, but traces of impurity may create a rotten egg odor. Sulfur is used widely in the creation of other chemicals and in industrial and manufacturing applications, and is therefore rarely found in pure form. Most sulfur is used to create sulfuric acid. Chemical formula = S. Molecular weight = 32.065 g/mol.

❑ sulfur dioxide (so_2)

Sulfur dioxide is a heavy, pungent, colorless gas formed primarily by the combustion of coal, oil, and diesel fuels. Elevated levels can impair breathing, lead to other respiratory symptoms, and at very high levels aggravate heart disease. People with asthma are most at risk. Sulfur dioxide also contributes to acid rain, which can damage plants, lakes and buildings.

❑ sulfur dioxide (so_2) emissions: changing patterns

- As more countries industrialize, acidification of the environment is becoming a global problem.
- Asian emissions of sulfur dioxide (SO_2) now exceed those of Europe and North America combined.

SO_2 emissions from fossil fuel burning

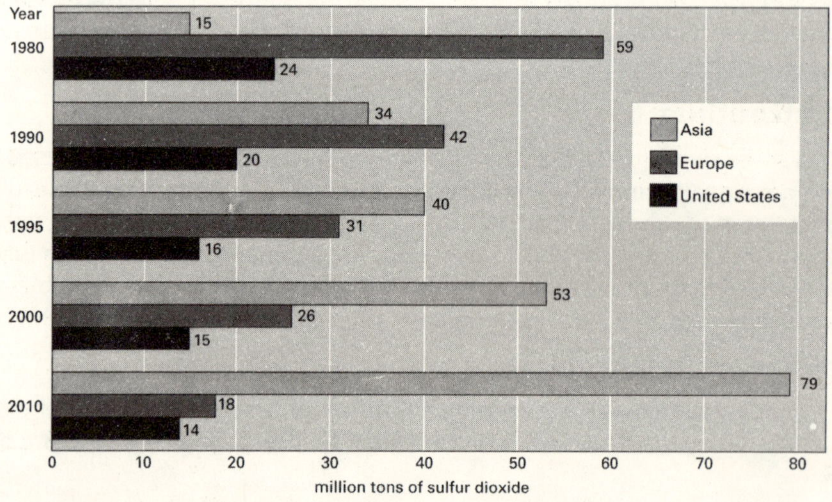

Latest available figures for SO_2 concentrations in eight Chinese cities

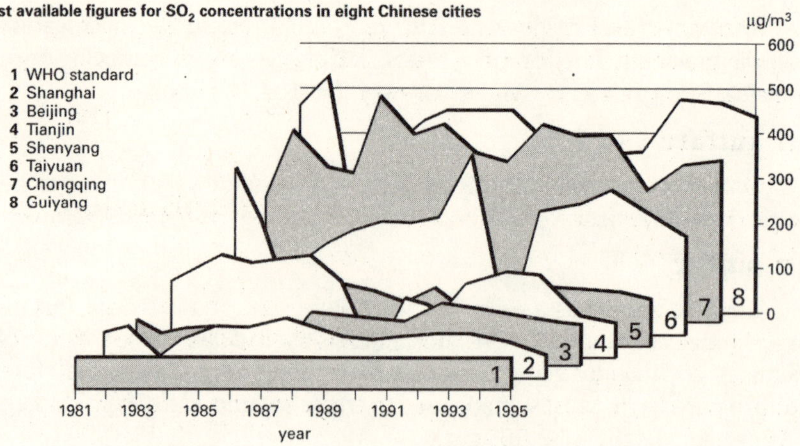

- The largest source is China, which emits 18 million tons of SO_2 a year.
- China's losses to crops and forests from acid deposition stand at US$5 billion a year.
- After investing heavily to clean up its own emissions, Japan is now suffering cross-border pollution from its neighbors.
- Acid fallout over large areas of China will, without a clean-up, exceed by 2020 the levels reached in the 1970s in Central Europe.
- China has adopted a number of measures to reduce acid rain.
- Clean coal, desulfurization technologies, improved efficiency of energy use, and the imposition of a levy on sulfur dioxide emissions are among strategies currently employed by China.
- Future levels of atmospheric pollution will be governed largely by the use of fossil fuel energy.
- In Europe and the United States, SO_2 emissions have declined since 1980 despite increasing energy consumption during 1980–90.

❑ sulfur dioxide (so_2)

A heavy, pungent, toxic gas that is easily condensed to a colorless liquid. It is used in making sulfuric acid, in bleaching, as a preservative, as a refrigerant. It is a major air pollutant especially in industrial areas. Chemical formula = SO_2. Molecular weight = 64.1 g/mol.

❑ sulfur emissions

- Most sulfur emissions are generated by the burning of hard coal, brown coal and oil.
- Sulfur emissions are an important constituent of acid rain and a contributor to global climate change.
- In 1900, the largest coal producing country was Great Britain, followed by the US and Germany.
- By 1960, the three major coal-producing countries were China, the US, and the USSR.
- By 1990, China predominated in coal production, with an output of 1.1 Gigatons per year.
- In 1990 the United States, the USSR, and China were the main sulfur emitters (i.e. approximately 50 percent of the total) in the world.
- Although the USSR and the US appear to have stabilized their sulfur emissions over the past 20 years, recent increases in global emissions have been linked to the rapid industrialization in China.
- Switching from high- to low-sulfur coals has reduced sulfur emissions in some cases.
- In a few countries, such as Germany, flue gas desulfurization (FGD) has made important contributions to emission reductions.

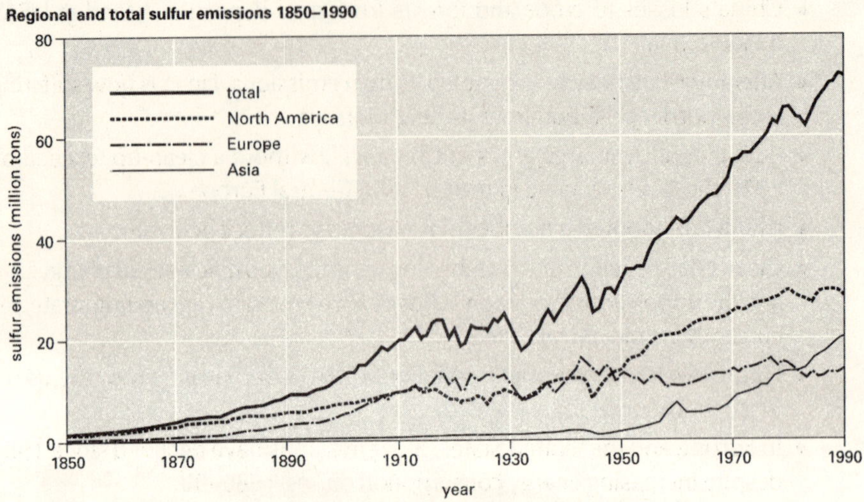

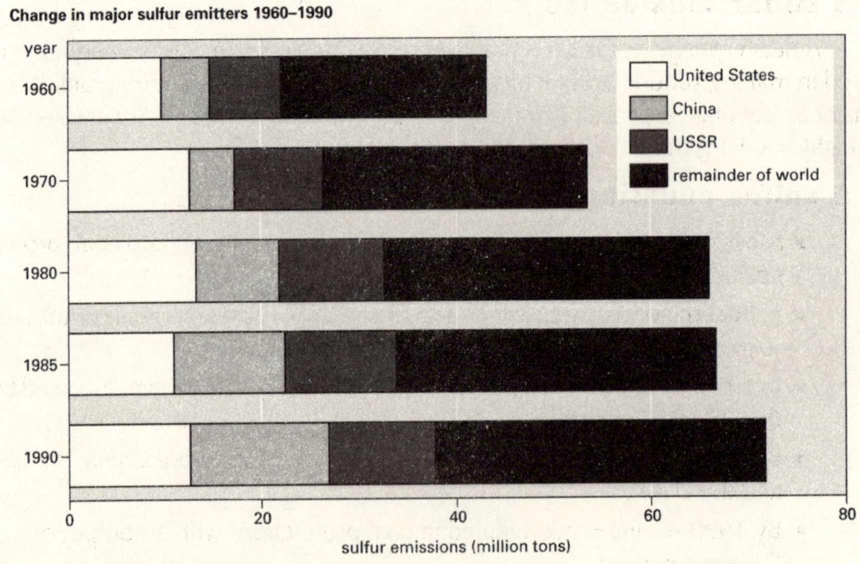

❑ superfund

Federal and state laws designed to clean up sites where past, improper disposal of hazardous substances caused soil and ground water contamination.

❑ superfund amendments and reauthorization act (sara

The federal law passed in 1986 which amended CERCLA and provided additional money and further direction for the federal Superfund program.

❐ surface water

Aboveground water, such as streams, rivers and lakes.

❐ sustainable energy

- Energy-related emissions account for more than 80 percent of the CO_2 released into the atmosphere each year.
- Burning all economically recoverable fossil fuel reserves over the next century would lead to longterm global warming of well over 3.5°C.
- Renewable energy sources supply 14 percent of the total world energy demand, but this is often with high levels of household and local air pollution.
- 'New' renewable energy sources, such as geothermal, wind, solar, marine and small-scale hydroelectric, currently contribute some 2 percent of world energy consumption.
- Although renewable energy sources are the fastest growing energy source, they will probably still only provide 3 percent of all energy by 2020.

Geothermal

- Geothermal energy is contained in the rock and the fluid in the Earth's crust.
- Such resources occur in five forms: hydrothermal fluids, hot dry rock, geopressured brines, magma, and ambient ground heat.
- Of these, only hydrothermal fluids have been developed commercially for power generation.
- Geothermal energy is the largest source of nonwater- based renewable energy in the United States.
- In a typical system, boreholes are drilled to a depth of around 2.4 miles, where water temperatures are 165°C.
- Water is injected into one borehole and energy is extracted from the other.

Wind

- The price for wind-generated electricity has fallen by an average of 3 percent per annum since the 1980s, mainly because of increasing turbine capacities.
- Wind energy now contributes 13 percent of national energy consumption in Denmark, the highest proportion of any country in the world.
- New wind farms can generate electricity at $0.04/kWh, a price competitive with many conventional technologies, and electricity from one new US wind farm will cost only $0.025/kWh.

Solar (Photovoltaic)

- Semiconductors which convert solar energy directly into electricity are called photovoltaic (PV) devices or solar cells.
- There are about 400,000 photovoltaic systems worldwide, generating a cumulative capacity of over 800 MW.
- Solar energy is also used for direct heating of domestic water and thermal generation of electricity, generating some 400 MW worldwide.
- PV-installed capacity is growing at the rate of 30 percent a year.

Marine

- A study of potential sites around Europe estimated that marine current turbines could generate around 12,500 MW of electricity.
- Marine energy can be harnessed by tidal mills or barrages, or by using a marine current turbine to exploit the energy of flowing water.

Biomass

- Energy crops, such as sugar cane or corn, can be converted into ethanol for fuel.
- Domestic, agricultural and industrial wastes and crop residues can be burned or gasified to produce energy.
- About 7,000 MW of biomass-fired power generation is currently fed into the US national electricity grid.
- The largest and most successful use of energy crops are the US program to produce ethanol from corn (1.05 billion gallons in 1999) and the Brazilian industry to produce ethanol from sugar cane (3.7 billion gallons in 1999).

Hydrogen

- Hydrogen is one of the most promising energy carriers for the future.
- Hydrogen can also be combined with gasoline, ethanol (alcohol), methanol, and natural gas to increase performance and reduce pollution.
- It is a high-efficiency, low polluting fuel that can be used for transportation, heating and power generation in places where it is difficult to use electricity.
- Hydrogen fuel cells combine hydrogen with oxygen and use the release of chemical energy to create an electric voltage.
- Iceland plans to replace fossil fuels with hydrogen over the next 30 years and create the first hydrogen economy.

T

❐ taxa

A grouping of organisms given a formal taxonomic name such as species, genus, family.

❐ taxa richness

The number of distinct species or taxa that are found in an assemblage, community, or sample.

❐ temperature

A specific degree of hotness or coldness as indicated on or referred to a standard scale.

❐ teratogen

A substance that causes defective development in fetuses (birth defects).

❐ tetrahydrofuran

A flammable liquid ether that is derived from furan and used as a solvent and as an intermediate in organic synthesis. Chemical formula = C_4H_8O. Molecular weight = 72.1 g/mol.

❐ the carbon budget

- Human burning of fossil fuels and changes in landuse continue to disturb the carbon cycle, increasing the atmospheric concentration of carbon dioxide.
- The terrestrial biosphere has historically been a source of carbon to the atmosphere; it is currently a net sink. The current terrestrial carbon sink is caused by land management practices, higher carbon dioxide, nitrogen deposition and possibly recent changes in climate.
- The oceans, vegetation, and soils actively exchange CO_2 with the atmosphere and are significant reservoirs of carbon.
- Oceans contain about 50 times as much carbon as the atmosphere, predominantly in the form of dissolved inorganic carbon.

The global carbon cycle

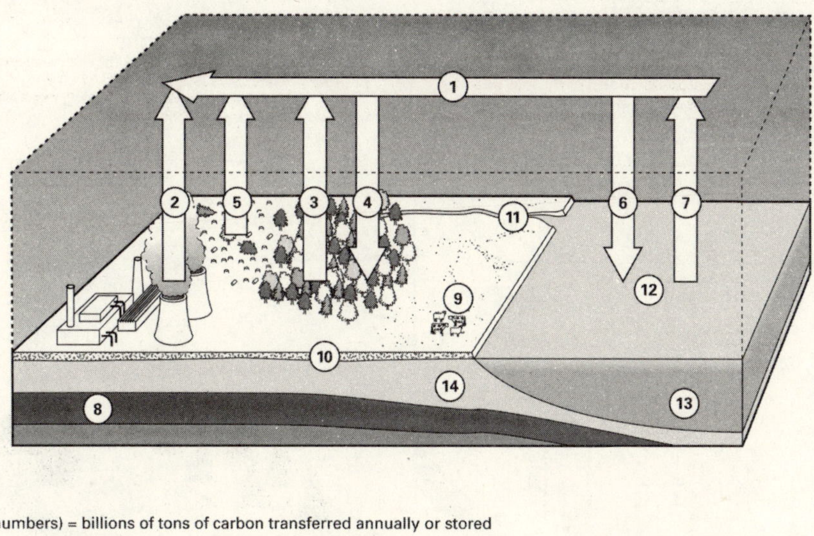

(numbers) = billions of tons of carbon transferred annually or stored

1 atmosphere (750)
2 fossil fuel burning (6)
3 respiration (60)
4 photosynthesis (62)
5 deforestation (1)
6 rainfall (92)
7 evaporation (90)
8 fossil fuels (5,000)
9 plants and animals (610)
10 soil and plant debris (1,600)
11 rivers (1)
12 surface ocean (1,000)
13 intermediate/deep ocean (38,000)
14 rocks (75,000,000)

- Terrestrial vegetation and soils contain about three and a half times as much carbon as the atmosphere; the exchange is controlled by photosynthesis and respiration.
- Ocean uptake of carbon is limited by the solubility of CO_2 in seawater and the slow rate of mixing between surface and deep-ocean waters.

❐ the greenhouse effect

Causes: An accumulation of greenhouse gases in the upper atmosphere is the cause of the greenhouse effect. These gases include carbon dioxide, methane, CFCs and water vapor, which form a layer that acts in a similar way to the glass of a real greenhouse.

Sources of the greenhouse effect

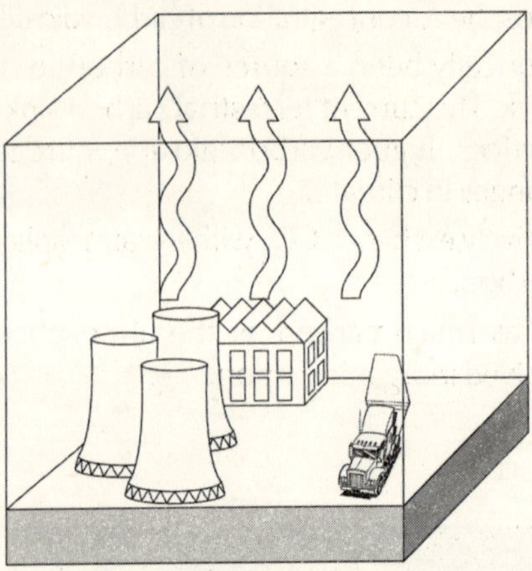

a The burning of organic and fossil fuels produces large amounts of carbon dioxide. Wood, coal, oil and petroleum based fuels all produce CO_2 when they combust.

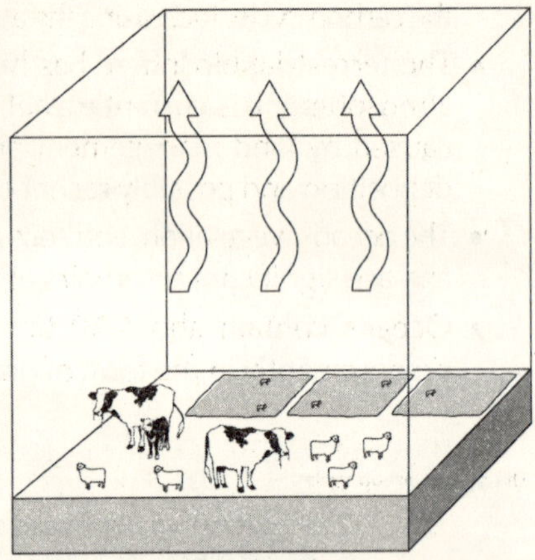

b Methane gas (CH_4) is a product of the decomposition of organic material. It is produced within the guts of herbivorous livestock and by rotting vegetation in marshes and rice paddies.

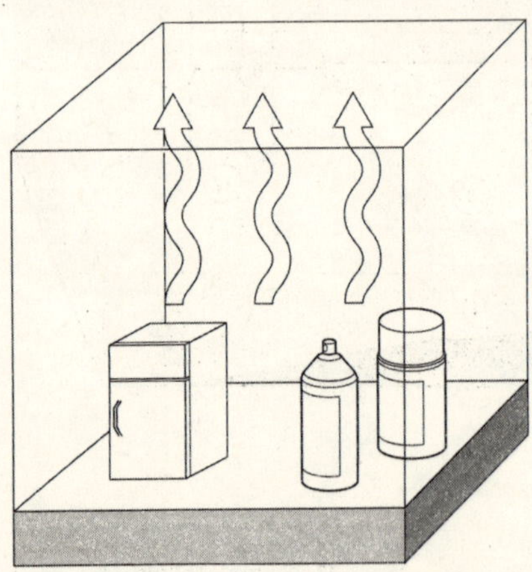

c The gases known as CFCs (chlorofluorocarbons) were used until recently as coolants in refrigerators and propellants in aerosols. They readily escape into the atmosphere.

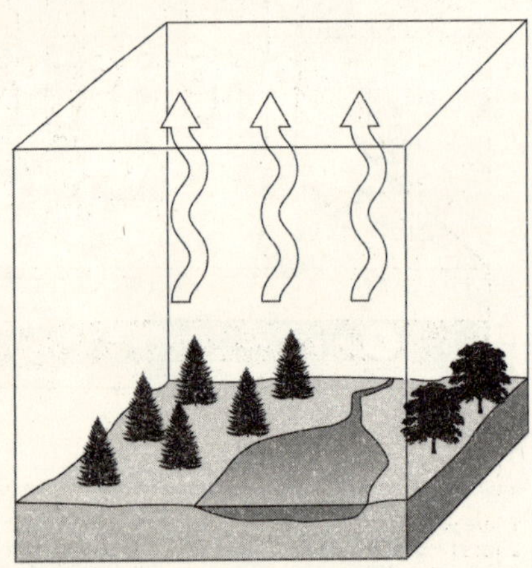

d Water vapor (H_2O) is produced by water evaporating into the atmosphere. Rivers, lakes, oceans, and plants all generate vast amounts of water vapor.

Effect: Normally, infrared radiation in the Sun's rays bounces off the Earth back into space. The greenhouse gases act as a barrier to the infrared radiation and reflect it back down onto the surface of the Earth.

Consequences: The infrared radiation becomes trapped between the layer of greenhouse gases and the Earth's surface, bouncing back and forth. The consequence is that the energy in the radiation has to be absorbed by the planet and its atmosphere, the phenomenon known as global warming.

❑ the us toxics release inventory (tri)

Total toxic releases by industry in 2000

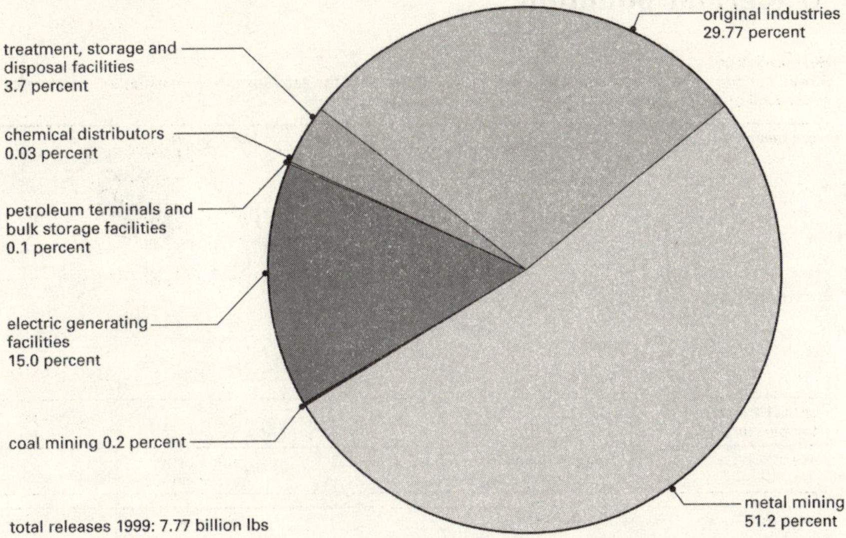

Distribution of Toxic Release Inventory on- and off-site releases 1988–2000

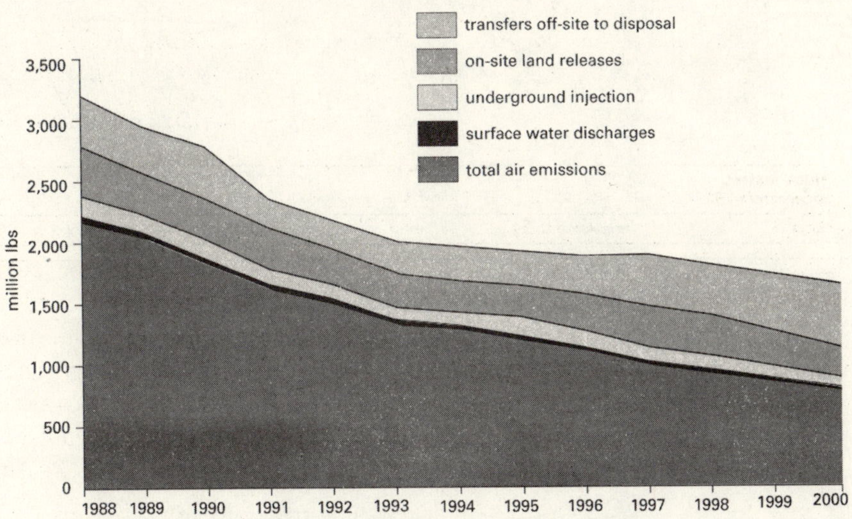

- The Toxics Release Inventory (TRI) is a publiclyavailable database containing information on toxicchemical releases and other waste-management activities that are reported annually by manufacturing facilities.

- There are now nearly 650 toxic chemicals and chemical compounds on the list of chemicals that must be reported to the EPA.
- Facilities must report the quantity of each chemical that they release into the air, the water, and onto the land each year.
- Since the inventory began in 1988, most releases have decreased.

❐ thermal pollution

Heat discharge

Power stations are cooled by water that may be discharged into the sea. Sharp rises in water temperature can directly kill plants and animals or damage fragile ecosystems.

organism	algae	mangroves
	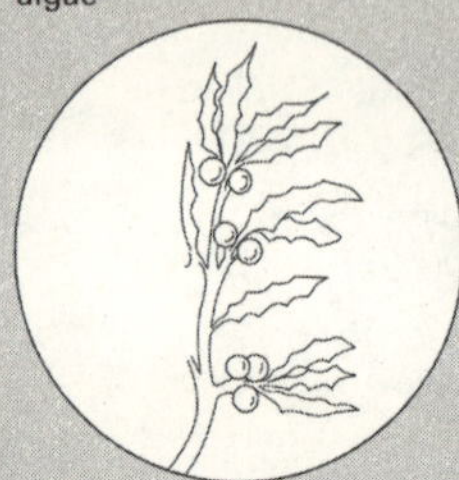	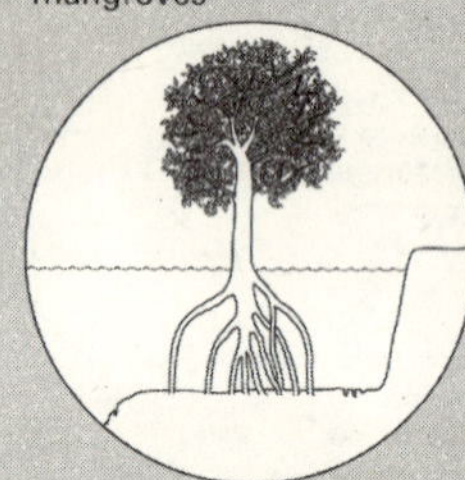
critical water temperature (°C)	34	38
effect	major change in community structure	reduction in photosynthesis

organism	copepods	corals
	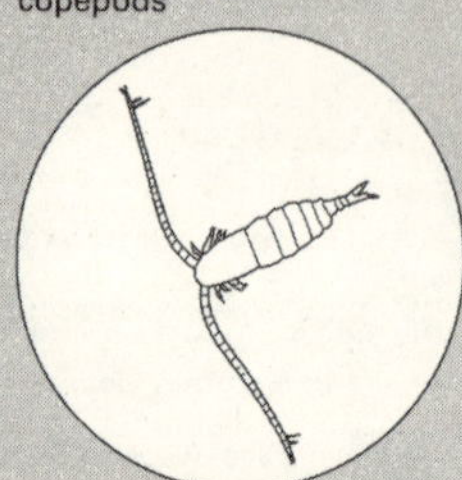	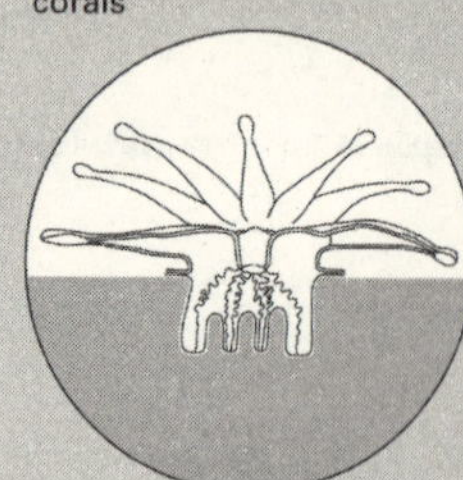
critical water temperature (°C)	35–40	35–40
effect	mass mortality	high mortality

organism	fish
	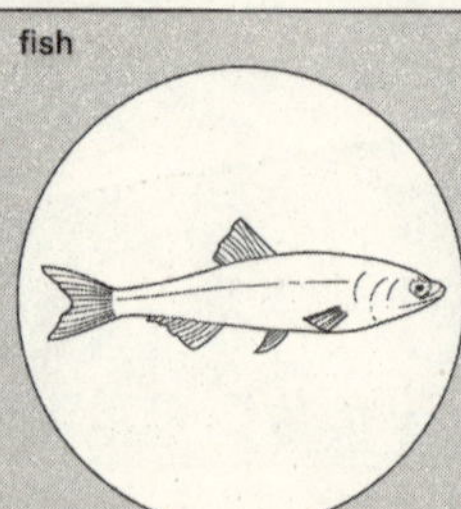
critical water temperature (°C)	34–38
effect	death

- The main cause of thermal pollution is the use of water as a cooling agent in nuclear power plants and other industrial facilities.
- The deforestation of shorelines and soil erosion also contribute to the problem.
- Soil erosion makes the water muddy, which increases the light absorbed.
- Deforestation increases soil erosion and increases the amount of sunlight striking the water.
- As seawater gets warmer, the amount of oxygen dissolved in the water is decreased, and rates of photosynthesis and plant growth are increased.
- Higher water temperatures can destroy coral reefs through a phenomenon known as bleaching.

❐ thermocline

During summertime, the middle layer of lake water. Lying below the epilimnion, this water rapidly loses warmth.

❐ till

Unstratified sediment deposited by a glacier; generally does not yield a good supply of water to a well.

❐ tin

A soft, faintly bluish-white, lustrous, low- melting, crystalline metallic element that is malleable and ductile at ordinary temperatures and that is used as a protective coating, in tinfoil, and in soft solders and alloys. Chemical symbol = Sn. Molecular weight = 118.71 g/mol.

❐ titanium

A shiny, dark gray metal that is used extensively in industrial applications and in steel and aluminum alloys. Titanium is probably best known for its use in surgical implants. It is also used for treatment of some skin disorders. Chemical formula = Ti. Molecular weight = 47.90 g/mol.

❐ toluene

Also known as: Methylbenzene. A liquid aromatic hydrocarbon that resembles benzene but is less volatile, flammable, and toxic. It is used as a solvent, in organic synthesis, and as an antiknock agent for gasoline. Chemical formula = $C_6H_5CH_3$. Molecular weight = 92.15 g/mol.

❐ total kjeldahl nitrogen (tkn)

The sum of organic nitrogen and ammonia in a water body. Measured in milligrams per liter (mg/L). High measurements of TKN typically results from sewage and manure discharges to water bodies.

❐ total maximum daily loads (tmdls)

Calculation of the maximum amount of a pollutant that a water body can receive and still meet water quality standards and an allocation of that amount to the pollutants source.

❐ total nitrate and nitrite nitrogen (tnnn)

Nitrate (NO_3) plus nitrite (NO_2) as nitrogen. In lakes, most nitrate/nitrogen is in NO_3 form. It is measured in milligrams per liter (mg/L). Elevated levels of nitrates/nitrogen are often caused by over application of fertilizers that leach into water bodies.

❐ total nitrogen

Total Nitrogen is the sum of nitrate (NO_3), nitrite (NO_2), organic nitrogen and ammonia (all expressed as N). Note that for laboratory analysis purposes, Total Kjeldahl Nitrogen (TKN) is a test performed that is made up of both organic nitrogen and ammonia.

❐ total phosphorus (tp)

A nutrient essential to the growth of organisms, and is commonly the limiting factor in the primary productivity of surface water bodies. Total phosphorus includes the amount of phosphorus in solution (reactive) and in particle form. Agricultural drainage, wastewater, and certain industrial discharges are typical sources of phosphorus, and can contribute to the eutrophication of surface water bodies.

❐ total suspended particulates (tsp)

The total number of particles of solid or liquid matter - such as soot, dust, aerosols, fumes and mist - found in a sample of ambient air.

❐ total suspended solids

Very small particles remaining dispersed in a liquid due to turbulent mixing that can create turbid or cloudy conditions. Measured in milligrams per liter (mg/L). 1. A measure of the material suspended in wastewater. Total suspended solids (TSS) cause: a) interference with light penetration, b) buildup of sediment and c) potential reduction in aquatic habitat. Solids also carry nutrients that cause algal blooms and other toxic pollutants that are harmful to fish. 2.Very small particles remaining dispersed in a liquid due to turbulent mixing exceeding gravitational sinking that can create turbid or cloudy conditions.

❐ toxic

Producing or containing a poisonous substance that may be harmful or deadly.

❐ toxic pollutants in the great lakes

Pollution factors

- The Great Lakes – Superior, Michigan, Huron, Erie and Ontario – and their connecting channels form the largest fresh surface water system on Earth.
- Covering more than 94,000 square miles and draining more than twice as much land, they hold an estimated 6 quadrillion gallons of water, about one-fifth of the world's fresh surface water supply.
- Ways that humans have affected the quality of the Great Lakes' water over the centuries include sewage disposal, toxic contamination through heavy

metals and pesticides, overdevelopment of the water's edge, runoff from agriculture and urbanization, and air pollution.

- Toxic pollutants include human-made organic chemicals and heavy metals that can be acutely toxic in relatively small amounts, and injurious through long-term exposure in minute concentrations.
- Pollutants tend to accumulate as the Lakes are a closed system.
- Creatures higher up the food chain have a much higher concentration of pollutants as a result of bioaccumulation.
- A person who eats one meal of lake trout from Lake Michigan will be exposed to more PCBs in one meal than in a lifetime of drinking water from the lake.
- However, pollution levels generally have fallen in recent years.

Phosphorous concentrations
This map shows phosphorus concentrations arising from pollution by detergents, sewage treatment, and fertilizer runoff.

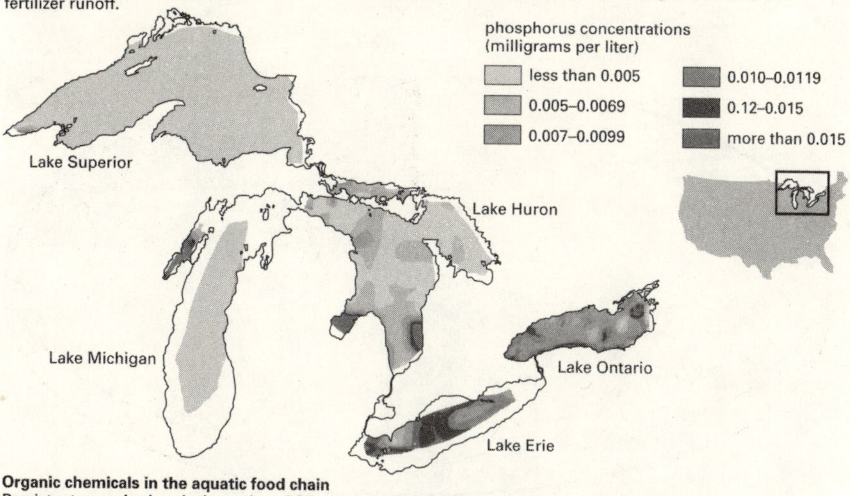

Organic chemicals in the aquatic food chain
Persistent organic chemicals, such as PCBs, bioaccumulate. This diagram shows the degree of concentration in each level of the Great Lakes aquatic food chain for PCBs (in parts per million, or ppm). The highest levels are reached in the eggs of fish-eating birds such as herring gulls.

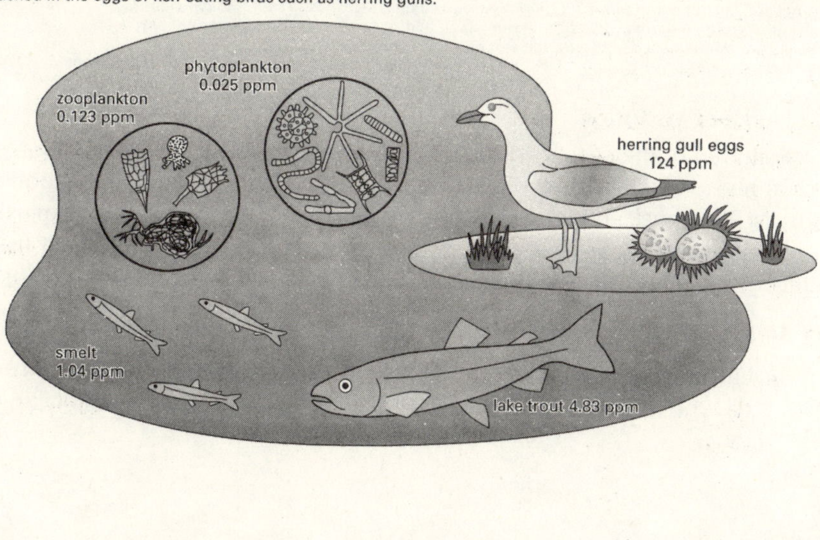

❑ toxic waste dumping at sea

- Large quantities of oil, nuclear waste, plastics and other debris, and a vast variety of chemical contaminants, are dumped or discharged at sea.
- Many directly poison marine life or cause chronic disease, reproductive failure, or deformities.
- The sources of marine pollution include commercial, military and recreational shipping and boating; run-off from urban streets and agricultural fields; oil drilling installations; and industries and sewage treatment plants.
- The discharge of nutrients and other substances – for example, in human wastes in sewage; fertilizers and animal waste from farm runoff, and air emissions from coal- and oil-burning industries – is polluting coastal waters with excess nutrients.
- These substances cause oxygen depletion of nearshore waters, harmful algal blooms, and dramatic reductions in the richness of sea-life communities.

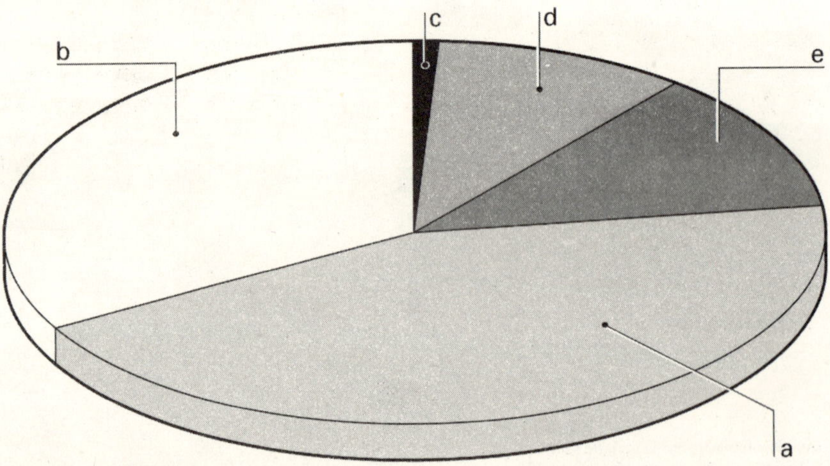

Sources of marine pollution

a nutrient pollution and land-derived pollutants, including industrial effluent and sewage – 42 percent
b air-based pollution, including sulfur and nitrogen oxides returned in rain – 35 percent
c pollution from offshore production, including oil and gas extraction and mining – 1 percent
d deliberate dumping at sea of garbage and industrial waste – 10 percent
e transportation-derived garbage, including deliberate and accidental discharge of oil – 12 percent

Nutrient pollution

Pollution of the seas by nutrients is a serious problem. Dissolved inorganic nitrates and phosphates from municipal sewage plants, agricultural runoff and industrial effluent all enter coastal waters, where they can cause sudden explosions in populations of algae. The algae in turn can produce toxins that affect fish or shellfish, and reduce the oxygen content of water, stressing marine organisms.

❑ toxic waste dumping on land

- Europe produces more than 2.5 billion tons of solid waste a year, and every day the inhabitants of New York City throw away approximately 26,000 tons.

- Land pollution occurs because of:
  Deposition of solid waste
  Accumulation of non-biodegradable materials
  Toxification of chemicals into poisons
  Alteration of soil chemical composition
- In most countries, most garbage is still buried in landfill sites.
- Landfill sites are a possible source of toxic chemicals and produce large quantities of methane gas.
- They must be managed so that pollutants do not seep into groundwater and should therefore be kept dry, but this slows down the rate of decomposition.
- Tyres constitute another problem, as they are virtually non-degradable and spread noxious fumes when burnt.
- Western Europe, the United States and Japancombined produce around 580 million tyres a year.

Cross-section of typical landfill site
- The compacted garbage (between soil cover) will biodegrade naturally if it has been properly graded.
- The leachate pump contains a mixture of garbage liquid and rainwater which is pumped out and sent to a sewage treatment plant.

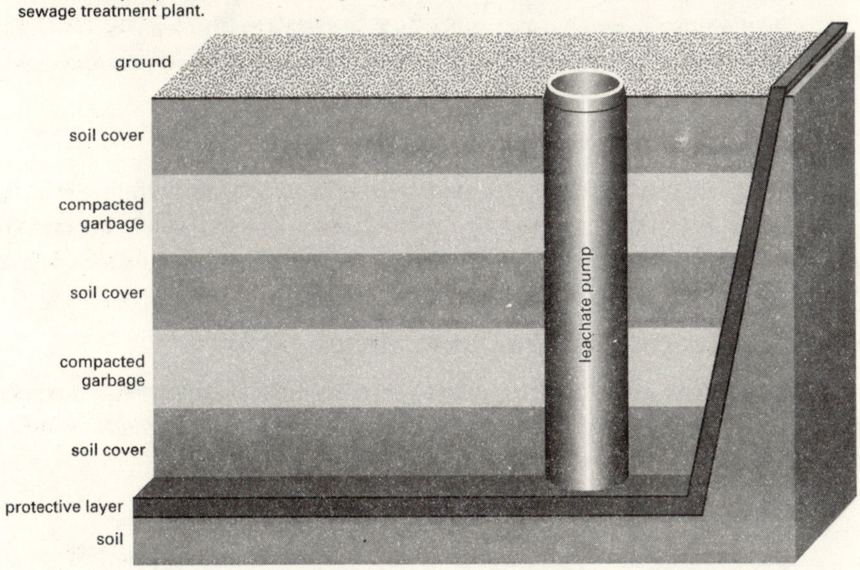

☐ toxicity

The extent to which a substance is poisonous to plants or animals.

☐ trans-1,2-dichloroethylene

A colorless, odorless liquid. One of the isomers of 1,2 Dichloroethylene, used as a solvent for waxes and resins and as a refrigerant. Also used in the manufacture of pharmaceuticals and artificial pearls, in the extraction of oils and fats from fish and meat, and the extraction of rubber.

Chemical formula = $C_2H_2Cl_2$. Molecular weight = 96.94 g/mol.

❒ trans-1,3-dichloropropene

A colorless liquid with pungent, chloroform-like odor. Used primarily in soil fumigants for nematodes and as a chemical intermediate. Chemical formula = $C_3H_4Cl_2$. Molecular weight = 110.97 g/mol.

❒ transparency (trans)

The transparency of stream water (in units of centimeters) as measured in a Citizen Stream Monitoring Program transparency tube. The transparency tube is a clear, 60 cm-long tube with colored disk on the bottom for measuring depth at which disk is visible. The clarity of stream water is an indicator of how much sediment, algae, and other materials are suspended in the water.

❒ trash

Material considered worthless, unnecessary or offensive that is usually thrown away. Garbage.

❒ trichloroethylene

A nonflammable liquid used especially as a solvent in dry cleaning, and for the removal of grease from metal. Chemical formula = C_2HCl_3. Molecular weight = 131.40 g/mol.

❒ trichlorofluoromethane (freon 11)

A colorless liquid that has little odor at less than 20% concentration in air, but has a slightly sweet odor at higher concentrations. It is used widely in consumer products as a propellant. Also used as a refrigerant, and a bronchial dilator. Chemical formula = CCl_3F. Molecular weight = 137.38 g/mol.

❒ trichlorotrifluoromethane (freon 113)

A colorless, volatile liquid. It is used as a solvent in degreasing metal equipment, a dry cleaner for textiles of many types, and as a refrigerant. Chemical formula = $C_2Cl_3F_3$. Molecular weight = 187.38 g/mol.

❒ triethylamine (tea)

A colorless, flammable liquid that is slightly soluble in water and has a strong fishy ammonia-like odor. It is used in curing and hardening of polymers, as a corrosion inhibitor, and in propellants, pesticides, pharmaceuticals, and paints/coatings. Chemical formula = $C_6H_{15}N$. Molecular weight = 101.19 g/mol.

❒ trifluralin

A yellow-orange crystalline solid, used as an herbicide to control annual grasses and some broadleaf annual weeds on a variety of crops, shrubs, and flowers. It is used mostly on cotton, as well as on soybeans and some fruits and vegetables. Chemical formula = $C_{13}H_{16}F_3N_3O_4$. Molecular weight = 335.3 g/mol.

❐ trimethylbenzene (mixed isomers)

Any of three isomers of trimethylbenzene, including: 1,2,3-Trimethylbenzene, 1,2,4-Trimethylbenzene, and 1,3,5-Trimethylbenzene. Chemical formula = C_9H_{12}. Molecular weight = 120.2 g/mol.

❐ trophic status

The level of growth or productivity of a lake as measured by phosphorus content, algae abundance, and depth of light penetration.

❐ tsi rankings

All trophic status rankings are conducted by ecoregion and expressed as a percentile. In general, as trophic state increases, moving from oligotrophy to eutrophy (e.g., phosphorus and chlorophyll increase and Secchi transparency decreases), percentile ranking decreases.

❐ turbidity

"Turbidity in water is caused by suspended matter, such as clay, silt, finely divided organic and inorganic matter, soluble colored organic compounds, and plankton and other microscopic organisms... Turbidity is an expression of the optical property that causes light to be scattered and absorbed rather than transmitted in straight lines through the sample." In streams, a major cause of elevated turbidity are disturbed and eroding soils carried by storm run-off to streams. Once in the stream system, elevated turbidity reduces the depth of photosynthesis and the feeding ability of aquatic organisms. When soils settle out in downstream reaches with slower flow, bed substrate becomes embedded, removing essential habitat for aquatic insects and other organisms.

U

❐ u.s. environmental protection agency (epa)

The federal agency that enforces federal regulations and administers federal programs, such as Superfund.

❐ unit risk

The upper-bound additional lifetime cancer risk estimated to result from continuous exposure to an agent at a concentration of 1 ìg/m3 in air.

❐ unpermitted dump sites

The list of Unpermitted Dump Sites was created in the 1980's and was originally called the Open Dump Inventory (ODI). Sites maintained in this list were later referred to as Outstate Dump Inventory (ODI) and Metro Dump Inventory (MDI) sites. Because these sites are no longer open the name has been changed to Unpermitted Dump Sites and contains dump sites for the entire state. Included in the list are abandoned dumps, demolition sites, tree disposal sites, industrial dumps and other dumps. Most of these sites existed prior to the creation of the MPCA in 1967, and detailed information about them is not generally available.

❐ use attainability analysis (uaa)

Analysis that describes factors limiting designated use of water bodies.

❐ usgs topographic 7.5-minute quadrangle map

A U.S. Geological Survey (USGS) topographic map use contours to show shape and elevation of terrain. These maps divide the United States into quadrangles bounded by two lines of latitude and two lines of longitude. Individual topographic maps are commonly referred to as quadrangles or quads. The smallest topographic quadrangle commonly published by the USGS are 7.5 minute quadrangles, which measure 7.5 minutes of latitude and longitude.

V

❐ vanadium

A grayish, malleable metallic element found combined in minerals and used especially to form alloys (as vanadium steel). Chemical symbol = V. Molecular weight = 50.9415 g/mol.

❐ vinyl acetate

A clear, flammable liquid that easily dissolves in water. It is used as an intermediate in the production of polyvinyl acetate, polyvinyl alcohol, and other industrial chemicals. Chemical formula = $C_4H_6O_2$. Molecular weight = 86.09 g/mol.

❐ vinyl bromide

A colorless gas that is used primarily in the manufacture of flame retardant synthetic fibers. Chemical formula = C_2H_3Br. Molecular weight = 106.9 g/mol.

❐ vinyl chloride

A flammable, gaseous, carcinogenic compound that is used to make vinyl resins. Chemical formula = $C2H_3Cl$. Molecular weight = 62.5 g/mol.

❐ vinylidene chloride

Also known as: 1,1-Dichloroethylene. A colorless liquid with a mild, sweet, chloroform-like odor. Used as an intermediate in chemical synthesis, and to produce polyvinylidene chloride copolymers. Chemical formula = $C_2H_2Cl_2$. Molecular weight = 96.95 g/mol.

❐ volatile organic compounds (voc)

Characteristics: Volatile organic compounds (VOC) are a principal component in atmospheric reactions that form ozone and other photochemical oxidants. VOCs are emitted from diverse sources, including automobiles, chemical manufacturing facilities, drycleaners, paint shops and other commercial and residential sources that use solvent and paint. The term, volatile organic compound is defined in federal rules as a chemical that participates in forming ozone. Methane, a nonreactive compound is not a VOC, nor are other oganic chemicals with negligible photochemical reactivity.

❒ voluntary investigation and cleanup (vic) program

The Voluntary Investigation and Cleanup (VIC) Program allows buyers, sellers, developers or local governments to voluntarily investigate and, if necessary, clean up contaminated land to facilitate its sale, financing or redevelopment. Voluntary parties that complete investigation and/or cleanup activities under MPCA oversight can receive liability assurances that protect them from future Superfund liability.

W

❒ waste

Material that has been discarded because it has worn out, is used up, or is no longer needed.

❒ water body type (wbt)

Refers to water body type as indicated in MDNR's PWI. PWI serves to document all public waters (lakes and wetlands) and assign individual ID numbers. (consistent with Bulletin 25). However some lakes in Bulletin 25 are really wetlands and are coded as "LW" whereas lakes have the codes "LP" or "L" in the PWI. Lakes that are "wetlands" based on the PWI were not assessed for the 303(d) list, however this differentiation was not made for the 305(b) assessments noted herein.

❒ water density

Water is most dense at 39 degrees Fahrenheit and expands (becomes less dense) at both higher and lower temperatures.

❒ water quality standard

A legally established state regulation consisting of three parts: (1) designated uses, (2) criteria, and (3) antidegradation policy.

❒ water table

The water surface in an unconfined aquifer at which the pore water pressure is at equilibrium with atmospheric pressure.

❒ waters of the state

All streams, lakes, ponds, watercourses, waterways, wells, springs, reservoirs, aquifers, irrigation systems, and all other bodies or accumulations of water, surface or underground, natural or artificial, public or private.

❒ watershed

The surrounding land area that drains into a lake, river or river system.

❒ wetland

Those areas that are inundated or saturated by surface or groundwater at a frequency and duration sufficient to support, and that under normal circumstances do support, a prevalence of vegetation typically adapted for life in saturated soil conditions. Wetlands generally include swamps, marshes, bogs, and similar areas.

❐ wetlands

A lowland area, such as a marsh or swamp, saturated with water. Wetlands are crucial wildlife habitat, and important for flood control and maintaining the health of surrounding ecosystems.

❐ world energy production

- Global energy use has increased nearly 70 percent since 1971, and is projected to increase at more than two percent annually for the next 15 years.
- This increase will raise greenhouse gas emissions about 50 percent higher than current levels, unless a concerted effort takes place to increase energy efficiency and move away from today's heavy reliance on fossil fuels.

World primary energy production by source in 2000

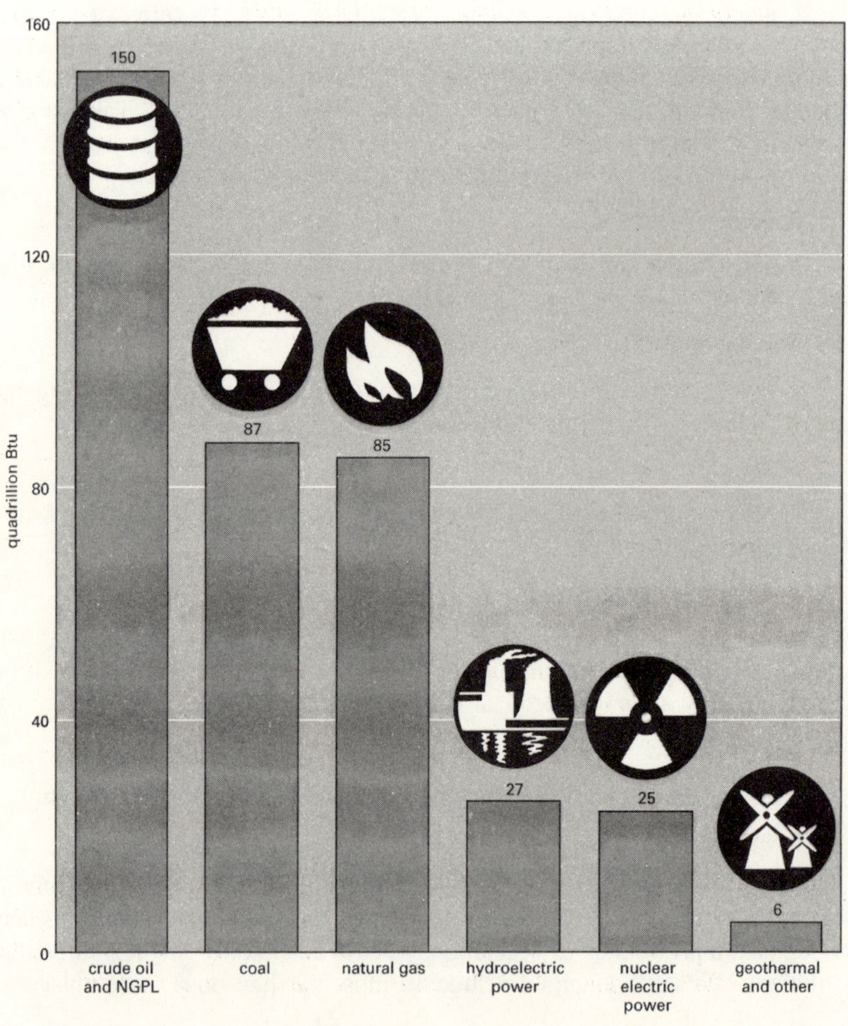

- Fossil fuels account for some 80 percent of the world's primary energy consumption, and supply roughly 90 percent of the world's commercial energy.

X

❐ xylene(s)

Any of three toxic, flammable, oily aromatic hydrocarbons (o-xylene, m-xylene, p-xylene) that are di-methyl homologues of benzene. They are usually obtained from petroleum or natural gas distillates. Chemical formula = C_8H_{10}. Molecular weight = 106.16 g/mol for mixed xylene.

Y

❐ yttrium

A highly crystalline, dark gray metal. Yttrium is largely used in its oxide form, yttria (YxO_3), which is used to make red phosphors for color television picture tubes. It is also used in small amounts as a metal alloy. Chemical formula = Y. Molecular weight = 88.906 g/mol.

Z

❐ zinc

A bluish, pale gray metal that is an essential nutrient in mammals. Thin layers of zinc are used to galvanize iron or steel, which protect the underlying metal from rusting. Zinc is also used in brass, bronze and nickel alloys, and can be found in solder, cosmetics and pigments. Chemical formula = Zn. Molecular weight = 65.409 g/mol.

❐ zirconium

A grayish-white, shiny metal that is used in the manufacture of cast iron and steel, as well as the production or surgical appliances and superconductive magnets. The oxide form of Zirconium (zircon) has a high index of refraction and is therefore used as a gem. Chemical formula = Zr. Molecular weight = 91.22 g/mol.

❐ zooplankton

Microscopic animals.

Recycling Terminology

A

❒ **abatement**

The reduction in landfill pollution by source reduction and waste recycling.

❒ **abc**

Aluminum Beverage cans.

❒ **abs**

Acrylonitrile butiadene styrene.

❒ **ac**

Aerosol cans.

❒ **aeration**

The process of exposing compost material to air.

❒ **al**

Aluminum.

❒ **ap**

Auto parts.

❒ **asb**

Asbestos.

❒ **asph**

Asphalt.

B

❒ **ba**

Brass.

❒ **back-end system**

Any of several processes for recovering resources from the organic portion of the waste stream (Ex.: Fluid bed incineration; Pyrolysis; Composting; Combustion; Fiber reclamation).

❒ **bale**

A compacted and bound cube of recycled material.

❐ **baler**

Equipment that compacts and binds recyclable materials to reduce volume and transportation costs.

❐ **benefication**

The mechanical processing of waste glass to decontaminate it and crush it to a more uniform size.

❐ **bias ply construction**

A tire building method consisting of a casing built using multiple layers of textile laid diagonally across the run of the tire.

❐ **bimetal can (bi-metal)**

A food or beverage can with a steel body and an aluminum lid: it is 100% recyclable by the steel industry.

❐ **biodegradable**

Able to break down or decompose rapidly under natural conditions and processes.

❐ **biological oxygen demand (bod)**

The amount of oxygen needed for aerobic microorganisms to function in organic-rich water such as sewage.

❐ **biological treatment**

Technology that uses bacteria to consume waste/ organic materials.

❐ **bleaching**

Method used in the paper-making process to produce a bright white paper sheet. It produces dioxins.

❐ **bo**

Bronze.

❐ **bottle bank**

Containers meant to receive beverage bottles which people drop off to be recycled.

❐ **bottle bill**

A law requiring deposits on beverage containers.

❐ **boxboard**

Paperboard used for fabricating boxes. Different boxboard grades are classified as to the composition of the top liner, filler (middle layer), and back liner. It includes folding b-- (cereal boxes), setup b-- (shoe boxes), and foodboard (milk cartons).

❐ **brgl**

Brown glass.

❐ **brk**

Bricks.

❐ **broke**

Paper that has been discarded anywhere in the process of manufacture in the paper mill.

❐ **brown goods**

Obsolete electronic products, such as radios and televisions.

❐ **bt**

Automotive batteries.

❐ **buy-back center**

A recycling facility that purchases small amounts of secondary materials from the public.

C

❐ **capture rate**

The percentage of generated secondary materials actually recovered from a household or business.

❐ **carcass**

The foundation structure of a tire, including sidewalls, bead and cord.

❐ **cc**

Catalytic converters.

❐ **cellulose insulation**

Insulation commonly used in construction and manufactured from waste paper (primarily newsprint) processing, with the addition of chemicals acting as retardants to fire or fibre breakdown.

❐ **chbd**

Chipboard.

❐ **chlorofluorocarbons (cfcs)**

Family of inert, nontoxic and easily liquified chemicals manufactured for use as coolants, cleaning solvents, plastic, aerosol propellants and foam insulation.

❐ **clgl**

Clear glass.

❐ **co**

Copper.

❒ **co-collection**

The collection of bagged recyclables together with other municipal garbage, separated later for recycling or disposal.

❒ **commingled**

Mixed recyclables that are collected or processed together.

❒ **comminution**

Mechanical shredding or pulverizing of waste; used in solid and water waste treatment.

❒ **compactor**

Equipment that densifies recyclable material and contains it under pressure, not allowing it to expand until it is unloaded.

❒ **compost**

A mixture of garbage, degradable trash and soil in which bacteria in the soil break down the mixture into a soil conditioner (not a fertilizer). It has high organic content but low nitrogen.

❒ **conc**

Concrete.

❒ **containerboard**

The component materials used in the fabrication of corrugated cardboard.

❒ **corrugated cardboard, old (occ)**

Unbleached, unwaxed kraft paper with ruffled inner liner. A recyclable material used to manufacture cardboard boxes.

❒ **cos**

Computer scrap.

❒ **cpo**

Computer print out paper.

v **cpt**

Carpet.

❒ **cptf**

Carpet, foam padding.

❒ **cradle-to-grave**

A system that handles solid waste from creation through disposal.

❒ **crumb rubber**

Ground or shredded rubber.

❐ crusher

A mechanical device used to break secondary materials into smaller pieces.

❐ cryogenic size reduction

Process in which flexible substances are made brittle by cooling to extremely low temperatures, using liquid nitrogen and ground rubber.

❐ cspe-r

Chlorosulforated polyethylene.

❐ cullet

Crushed glass which can be added to a batch of new materials in the manufacturing of new glass products. It increases the rate of heat gain by batch and reduces fuel costs. Domestic c-- if produced in house during the manufacturing process. Foreign c-- if it comes from an external source.

D

❐ degradability

Ability of materials to break down, by bacterial (biodegradable) or ultraviolet (photodegradable) action.

❐ deinking

A process that removes inks, dyes or other contaminants from collected wastepaper.

❐ detinner

A company that buys tin mill products, removes the tin through appropriate processes and sells the detinned steel, to steel mills and foundries and the recovered tin on the tin market.

❐ dioxin

Chlorinated organic compound: a by-product of the paper-making process that uses chlorine as a bleaching agent. Dioxins can be released into the atmosphere through the incineration of chlorinated paper. They are believed to be highly toxic to humans,

❐ diversion credits

A financial incentive provided to municipalities or private recycling operations based on the tonnage diverted from the waste stream.

E

❐ elastomer

A material which can be stretched at room temperature to at least twice its size and upon release of the stress, will return to approximately its original lenght.

❐ energy from waste

A recognized alternative process to reduction or recovery of recyclable materials which are not currently economical.

❐ exothermics

Materials used to generate heat in chemical or other processes.

F

❐ fd

Fiber drums.

❐ ferrous metals (fe)

magnetic metals which are predominantly composed of iron.

❐ flb

Fluorescent light bulbs.

❐ flow control law

Local ordinance controlling, or giving a municipal official authority to control, the collection and/or disposal of municipal solid waste produced in a specific geographical area.

❐ fly ash

A fine residue, left after trash is burned in an incinerator, which can be carried in the air. It can contain harmful or toxic substances such as dioxins, lead and mercury.

❐ forced deposit

Requires a monetary deposit on beverage containers which is returned to the consumer when the containers are returned to the retailer.

❐ front end separation

A system in which certain materials removed from the waste stream are directed toward a specific recovery system such as recycling or waste to energy incineration.

❐ fsb

Ferrous steel beams only.

G

❐ gaylord container

The trade name for a large, reusable corrugated container used for shipping materials.

❐ gl

Glass, all grades.

❐ **glasphalt**

A trade name for highway paving material in which recovered ground glass replaces some of the gravel normally used in asphalt.

❐ **grading**

Rating of waste or recyclable materials into homogenous categories by type and quality.

❐ **grgl**

Green glass.

❐ **grinding**

A size reduction technique that usually utilizes a series of rollers.

❐ **gyp**

Gypsum, wallboard, drywall.

H

❐ **hbt**

Household batteries.

❐ **heavy metals**

Elements, including cadmium, mercury, lead and arsenic which may be found in the waste stream as part of discarded items (batteries, lighting fixtures, colorants, ink).

❐ **high density polyethylene (hdpe)**

Used to make plastic bottles, milk cartons and other products. It produces toxic fumes when burned. Often referred to as No.2 Plastic.

❐ **high grade paper (hg)**

Relatively valuable types of paper such as computer printout, white ledger and tab cards.

❐ **humus**

The substance which results from decay of plant or animal matter. Biodegradable matters form humus as they decompose.

❐ **hzw**

Hazardous waste.

I

❐ **integrated solid waste management**

U.S. Environmental Protection Agency recommendation that promotes solid waste management through an integrated system that uses resource reduction, recycling, waste to energy incineration and land filling to manage the reclamation, reuse or disposal of plastics in the waste stream.

- **intensive recycling**

A concept promoted by opponents of waste to energy systems, whereby municipal recycling efforts target all recyclables in the waste stream.

- **intermediate processing center (ipc)**

Facility that separates, cleans and bails or packages materials for sale to manufacturers or brokers.

- **kft**

Kraft paper.

L

- **landfill**

A private or municipal site where non-hazardous solid or municipal waste is buried.

- **lb**

Light bulbs.

- **ld**

Lead.

- **leaching**

Process by which soluble materials are dissolved and carried through the soil by a percolating liquid.

- **liner**

Barrier designed to prevent the leaching of contents from a landfill. Commonly comprised of plastic or dense clay.

- **linerboard**

A paperboard used as the facing material in the production of corrugated shipping containers.

- **lldpe**

Linear low density polyethylene.

- **low density polyethylene (ldpe)**

Often referred as No.4 Plastic.

- **low grade paper (lg)**

Less valuable types of paper such as mixed office paper, corrugated cardboard and newspaper.

M

❒ materials recovery facility (mrf)

A recycling facility that sorts and processes collected mixed recyclables into individual streams for market. Also known as an intermediate processing center (IPC).

❒ mdpe

Medium density polyethylene.

❒ midden

A pit in which trash or garbage is buried. Usually only organic garbage.

❒ mill broke

Waste paper produced by mills during the paper making process that has traditionally been reused in manufacturing paper. It cannot be considered in the recycled content according to U.S. EPA guidelines.

❒ minimum recycled content laws

Laws requiring a product or type of packaging to contain a certain percentage of recycled material.

❒ mjc

Milk, juice cartons.

❒ mlr

Multi layered resins.

❒ mulching

The natural and gradual decomposition of dead organic matter that has been evenly distributed in a thin layer on the ground.

❒ municipal solid waste (msw)

Residential and commercial trash and/or garbage generated by a particular municipal area.

❒ mw

Medical waste.

❒ mwp

Mixed waste paper.

N

❒ new scrap

Material which is discarded during a manufacturing or processing operation and which cannot be directly fed back to that operation.

- **newsprint**

Alternate term for the low grade paper used to make newspaper.

- **ni**

Nickel.

- **non ferrous scrap metals (nf)**

Metals which contain no iron, such as aluminum, copper, brass and bronze.

- **noryl**

Plastics noryl.

- **npc**

Nylon polycarbonate.

O

- **of**

Oil filters.

- **office paper**

Waste paper generated by offices, including stationery, photocopy and computer paper.

- **omg**

Old magazines.

- **onp**

Old newspapers.

- **owm**

Oil water mixture.

- **ozone depletion**

Destruction of the stratospheric ozone layer of the earth's atmosphere due to the release of chlorofluorocarbons, or CFCs, into the environment.

P

- **pa**

Paper, all grades.

- **paper stock**

Scrap or waste papers that have been sorted and baled into specific grades. It is commonly used interchangeably with the term waste paper.

❐ paperboard

General term for heavyweight grades of paper that are used for containers, boxes, cartons and packaging materials. It is divided into: Containerboard, Boxboard and Other Paperboard.

❐ pc

Polycarbonate.

❐ pd

Plastic drums.

❐ pe

Polyethylene.

❐ perforator/flattener

Equipment that perforates and flattens material, then ejects it into a receptacle or processor. Used to prevent plastic bottles from expanding after flattening.

❐ pg

Plate glass.

❐ photodegradable

A process where ultraviolet radiation degrades the chemical bond or link in the polymer or chemical structure of a plastic.

❐ pl

Plastic, all grades.

❐ pm

Precious metals.

❐ polyethylene teraphthalate (pet)

A type of plastic that is clear or colored transparent with high gloss. It is used for carbonated beverage bottles and some household cleanser containers. Often referred to as No. 1 Plastic.

❐ polypropilene (pp)

Plastic with a smooth surface that cracks easily when bent and is difficult to scratch. Typical uses are: battery cases, dairy tubs, jar lids, straws and syrup bottles. It is hard to collect in marketable quantities for recycling and has limited uses in its recycled form. Often referred to as No. 5 Plastic.

❐ polystyrene (ps)

Plastic with a smooth surface that cracks easily when bent. Used for fast food packaging, styrofoam cups and packing peanuts, it takes up a large part of landfill space because of its bulk. Often referred to as No. 6 Plastic.

❑ **polyvinyl chloride (pvc)**

Environmentally indestructible plastic that releases toxic hydrocloric acid when burned. It is used for food wraps and containers for personal care products. Often referred to as V-3 or No. 3 Plastic.

❑ **post consumer material**

Any household or commercial product which has served its original, intended use.

❑ **post industrial material**

Industrial manufacturing scrap.

❑ **pre-consumer waste**

Waste produced by converting or printing operations and includes any scraps, trimmings, overruns, etc.

❑ **precycle**

Source reduction option whereby evaluation and selection of items for purchase is dependent upon method of manufacture, product content and recyclability of product after consumer use.

❑ **pw**

Paper mill waste.

❑ **pyrolysis**

A process of producing fuels from waste by heating it in an oxygen-deficient atmosphere.

R

❑ **rb**

Rubber, all grades.

❑ **rbc**

Rubber crumbs.

❑ **rbf**

Rubber buffing.

❑ **rbit**

Rubber inner tubes.

❑ **rbs**

Rubber, shredded tires.

❑ **rbt**

Rubber, used tires.

❐ **rd**

Radiators.

❐ **reclaimer**

A company that processes post-consumer materials into new products.

❐ **recycling**

Process by which materials that would otherwise become solid waste are collected, separated or processed and returned to the economic mainstream to be reused in the form of raw materials or finished goods.

❐ **redemption**

To turn in a secondary material to the original supplier.

❐ **refuse derived fuel (rdf)**

Fuel produced by separating, shredding and processing mixed wastes.

❐ **refuse reclamation**

Transformation of solid waste into useful products, such as soil conditioners or recycled materials.

❐ **regrind**

Ground up recyclable plastics.

❐ **remanufacturing**

The process of disassembly and reassembly of products during which time parts are cleaned, repaired or replaced.

❐ **reprocessing**

Operation of reforming reclaimed materials into new products.

❐ **resource recovery**

Any process of obtaining matter or energy from materials formerly discarded.

❐ **retreading**

The process of removing a worn tread and placing a new tread on an undamaged tire casing.

❐ **reuse**

To use a product repeteadly in the same form (e.g.: glass bottles, cloth diapers).

❐ **rs**

Rail scrap.

❐ **rubber asphalt**

A product that combines ground-up scrap tires and asphalt. It is primarily used in highway, airport runway and street projects as a stress-absorbing membrane interlayer.

❐ rubbish

Solid waste that does not contain food waste.

S

❐ salvage

The act of obtaining a secondary material through collection, sorting, etc.

❐ sc

Steel cans.

❐ scrap

That portion of solid waste which can be economically recycled.

❐ sd

Steel drums.

❐ sg

Specialty grades, paper.

❐ shredding

Size reduction by shearing action.

❐ sht

Stainless, high temperature steel.

❐ sm

Special metals.

❐ solid waste

Nonsoluble, discarded solid materials, including sewage sludge, municipal garbage, industrial wastes, agricultural refuse, demolition wastes and mining residues.

❐ solid waste management

The handling of activities which provide for the collection, separation, storage, transport, transfer, processing, recycling, incineration, treatment and disposal of solid waste.

❐ source reduction

Reducing the quantity of waste which in turn lessens the amount of material that enters the waste stream.

❐ source separation

The sorting of specific waste materials prior to their collection or deposition into a collection container.

❐ special wastes

Any waste requiring special handling such as scrap tires, used motor oil, hospital wastes or household hazardous wastes.

❑ **steel can**

A rigid container made exclusively or primarily of steel. It is 100% recyclable.

❑ **sweating furnaces**

Furnace in which metals are separated by melting at different temperatures.

T

❑ **tb**

Telephone books.

❑ **test linerboard**

Linerboard that is made exclusively out of recycled materials such as double-lined kraft cuttings and old corrugated containers.

❑ **tf**

Tin foil.

❑ **thermoplastics (tpo)**

Plastics which can be reformed repeatedly by application of heat and pressure.

❑ **thermosets**

Plastics which once molded cannot be reformed using heat or pressure.

❑ **ti**

Titanium.

❑ **tin can**

A term sometimes used to describe a steel food can. Technological developments have allowed for the tin coating on a can to become progressively thinner, to the point that tin now represents less than one percent of the weight of a steel can.

❑ **tipping fee**

Charge for the unloading or dumping of waste at a recycling facility, composting facility, landfill, transfer station or waste to energy facility.

❑ **tp**

Truck parts.

❑ **ts**

Tin (metal).

❑ **tx**

Textiles, all grades.

❑ **txc**

Textiles, cotton.

❑ **txj**

Textiles, jute.

❑ **txn**

Textiles, nonwoven.

❑ **txr**

Textiles, wiping rags.

❑ **txs**

Textiles, synthetic.

❑ **txu**

Textiles, used clothing.

❑ **txw**

Textiles, wool.

U

❑ **ubc**

Used beverage containers.

❑ **usbc**

Used steel beverage containers.

❑ **ust**

Underground storage tanks.

V

❑ **volume reduction**

Processing waste materials to decrease the amount of space the materials occupy. It is accomplished by mechanical, thermal or biological means.

❑ **vulcanization of rubber**

A chemical reaction of sulphur with rubber to cause cross-linking of the polymer chain which increases strength and resiliency of rubber.

W

❑ **waste**

Unwanted materials remaining from manufacturing processes, or refuse from humans and animals.

❑ **waste exchange**

The use by one company of an industrial waste generated by another firm.

❐ **waste recycling**

A method of recovering waste as resources. It includes the reuse of wastes or the collection and treatment of a waste product for use as a replacement of all or part of the raw material in the manufacturing process.

❐ **waste reduction**

The prevention of waste at its source.

❐ **waste stream**

The flow of waste material from generation to disposal.

❐ **waste to energy incineration**

An alternative process to reduction or recovery of recyclable materials which are not currently economical.

❐ **white goods (wga)**

A generic term for obsolete major appliances such as refrigerators, water heaters and stoves.

❐ **windrowing**

The placement and management of compostable material in piled rows, where micro-organisms break down organic material into a finished compost product.

❐ **wo**

Waste oil.

❐ **wp**

Waste wood products.

❐ **ww**

Waste water.

Y

❐ **yard waste (yw)**

Leaves, grass clippings and other organic wastes produced as part of yard and garden development and maintenance.

Z

❐ **zi**

Zinc.

Snow Terminology

A

❒ **ablation**

The process of being removed. Snow ablation usually refers to removal by melting.

❒ **accretion**

Growth of precipitation particles by collision of ice crystals with super cooled liquid droplets which freeze on impact.

B

❒ **blizzard**

Winds of at least 35 miles per hour along with considerable falling and/or blowing snow reducing visibility to less than one-quarter mile for a period of at least three hours.

C

❒ **climatology**

The science and study of climate; the aggregate of all weather conditions at a point over a long period of time.

❒ **condensation**

The process in which water vapor becomes liquid.

D

❒ **dendrite**

Hexagonal ice crystals with complex and often fernlike branches.

❒ **depth hoar**

Large (one to several millimeters in diameter), cohesionless, coarse, faceted snow crystals which result from the presence of strong temperature gradients within the snow pack.

E

❒ **equivalent**

The water content obtained from melting.

evaporation (water)

The physical process in which liquid water changes into a gas.

F

firn

Rounded, well-bonded snow that is older than one year.

G

graupel

Snowflakes that become rounded pellets due to riming. Typical sizes are two to five millimeters in diameter (0.1 to 0.2 inches). Graupel is sometimes mistaken for hail.

M

metamorphism

Changes in the structure and texture of snow grains which results from variations in temperature, migration of liquid water and water vapor, and pressure within the snow cover.

P

polycrystal

A snowflake composed of many individual ice crystals.

precipitation

The accumulated depth of rain or drizzle and also the melted water content of snow or other forms of frozen precipitation, including hail.

R

rime

A deposit of ice formed when supercooled water droplets freeze on contact with an object.

S

saturation vapor pressure (water)

The maximum amount of water vapor necessary to keep moist air in equilibrium with a surface of pure water. This is the maximum water vapor the air can hold for any given combination of temperature and pressure.

❐ snow core

A sample of snow, either just the freshly fallen snow or the combined old and new snow on the ground, obtained by pushing a cylinder down through the snow layer and extracting it.

❐ snow density

The mass of snow per unit volume which is equal to the water content of snow divided by its depth.

❐ snow depth

The combined total depth of both old and new snow on the ground.

❐ snow flurries

Snow that falls for short durations and which often changes in intensity. Flurries usually produce little accumulation.

❐ snow load

The downward force on an object or structure caused by the weight of accumulated snow.

❐ snow squall

A brief, but intense fall of snow that greatly reduces visibility and which is often accompanied by strong winds.

❐ snowboard

A solid, flat, white material, such as painted plywood, approximately two feet on each side, that is laid on the ground or on the surface of the snow by weather observers to obtain more accurate measurements of snowfall and water content.

❐ snowbursts

Very intense showers of snow, often of short duration, that greatly restrict visibility and produce periods of rapid snow accumulation.

❐ snowfall

The depth of new snow that has accumulated since the previous day or since the previous observation.

❐ snowflake

A cluster of ice crystals that falls from a cloud.

❐ snowpack

The total snow and ice on the ground, including both new snow and the previous snow and ice which has not melted.

❐ sublimation

The process in which ice changes directly to water vapor without melting, but also in meteorology the opposite process in which water vapor is transformed to ice (also called deposition).

❒ supercooled

The condition when a liquid remains in the liquid state even through its temperature is below its freezing point.

❒ supersaturation

The condition which occurs in the atmosphere when the relative humidity is greater than 100 percent.

❒ surface hoar

The deposition (sublimation) of ice crystals on a surface which occurs when the temperature of the surface is colder than the air above and colder than the frost point of that air.

V

❒ vapor pressure

The pressure exerted by water vapor molecules in a given volume of air.

Toxic Substances

A

❐ absorption

The process of taking in. For a person or an animal, absorption is the process of a substance getting into the body through the eyes, skin, stomach, intestines, or lungs.

❐ acute

Occurring over a short time.

❐ acute exposure

Contact with a substance that occurs once or for only a short time (up to 14 days).

❐ additive effect

A biologic response to exposure to multiple substances that equals the sum of responses of all the individual substances added together.

❐ adverse health effect

A change in body function or cell structure that might lead to disease or health problems

❐ aerobic

Requiring oxygen.

❐ ambient

Surrounding (for example, ambient air).

❐ anaerobic

Requiring the absence of oxygen.

❐ analyte

A substance measured in the laboratory. A chemical for which a sample (such as water, air, or blood) is tested in a laboratory. For example, if the analyte is mercury, the laboratory test will determine the amount of mercury in the sample.

❐ analytic epidemiologic study

A study that evaluates the association between exposure to hazardous substances and disease by testing scientific hypotheses.

❐ antagonistic effect

A biologic response to exposure to multiple substances that is less than would be expected if the known effects of the individual substances were added together.

B

❐ background level

An average or expected amount of a substance or radioactive material in a specific environment, or typical amounts of substances that occur naturally in an environment.

❐ biodegradation

Decomposition or breakdown of a substance through the action of microorganisms (such as bacteria or fungi) or other natural physical processes (such as sunlight).

❐ biologic indicators of exposure study

A study that uses (a) biomedical testing or (b) the measurement of a substance [an analyte], its metabolite, or another marker of exposure in human body fluids or tissues to confirm human exposure to a hazardous substance.

❐ biologic monitoring

Measuring hazardous substances in biologic materials (such as blood, hair, urine, or breath) to determine whether exposure has occurred. A blood test for lead is an example of biologic monitoring.

❐ biologic uptake

The transfer of substances from the environment to plants, animals, and humans.

❐ biomedical testing

Testing of persons to find out whether a change in a body function might have occurred because of exposure to a hazardous substance.

❐ biota

Plants and animals in an environment. Some of these plants and animals might be sources of food, clothing, or medicines for people.

❐ body burden

The total amount of a substance in the body. Some substances build up in the body because they are stored in fat or bone or because they leave the body very slowly.

C

❐ cancer

Any one of a group of diseases that occur when cells in the body become abnormal and grow or multiply out of control.

❐ cancer risk

A theoretical risk for getting cancer if exposed to a substance every day for 70 years (a lifetime exposure). The true risk might be lower.

❐ carcinogen

A substance that causes cancer.

❐ case study

A medical or epidemiologic evaluation of one person or a small group of people to gather information about specific health conditions and past exposures.

❐ case-control study

A study that compares exposures of people who have a disease or condition (cases) with people who do not have the disease or condition (controls). Exposures that are more common among the cases may be considered as possible risk factors for the disease.

❐ central nervous system

The part of the nervous system that consists of the brain and the spinal cord.

❐ chronic

Occurring over a long time.

❐ chronic exposure

Contact with a substance that occurs over a long time (more than 1 year).

❐ cluster investigation

A review of an unusual number, real or perceived, of health events (for example, reports of cancer) grouped together in time and location. Cluster investigations are designed to confirm case reports; determine whether they represent an unusual disease occurrence; and, if possible, explore possible causes and contributing environmental factors.

❐ community assistance panel (cap)

A group of people from a community and from health and environmental agencies who work with ATSDR to resolve issues and problems related to hazardous substances in the community. CAP members work with ATSDR to gather and review community health concerns, provide information on how people might have been or might now be exposed to hazardous substances, and inform ATSDR on ways to involve the community in its activities.

❐ comparison value (cv)

Calculated concentration of a substance in air, water, food, or soil that is unlikely to cause harmful (adverse) health effects in exposed people. The CV is used as a screening level during the public health assessment process. Substances found in amounts greater than their CVs might be selected for further evaluation in the public health assessment process.

❒ comprehensive environmental response, compensation, and liability act of 1980 (cercla)

CERCLA, also known as Superfund, is the federal law that concerns the removal or cleanup of hazardous substances in the environment and at hazardous waste sites. ATSDR, which was created by CERCLA, is responsible for assessing health issues and supporting public health activities related to hazardous waste sites or other environmental releases of hazardous substances.

❒ concentration

The amount of a substance present in a certain amount of soil, water, air, food, blood, hair, urine, breath, or any other media.

❒ contaminant

A substance that is either present in an environment where it does not belong or is present at levels that might cause harmful (adverse) health effects.

D

❒ delayed health effect

A disease or an injury that happens as a result of exposures that might have occurred in the past.

❒ dermal

Referring to the skin. For example, dermal absorption means passing through the skin.

❒ dermal contact

Contact with (touching) the skin.

❒ descriptive epidemiology

The study of the amount and distribution of a disease in a specified population by person, place, and time.

❒ detection limit

The lowest concentration of a chemical that can reliably be distinguished from a zero concentration.

❒ disease prevention

Measures used to prevent a disease or reduce its severity.

❒ disease registry

A system of ongoing registration of all cases of a particular disease or health condition in a defined population.

❒ dod

United States Department of Defense.

❐ doe

United States Department of Energy.

❐ dose (for chemicals that are not radioactive)

The amount of a substance to which a person is exposed over some time period. Dose is a measurement of exposure. Dose is often expressed as milligram (amount) per kilogram (a measure of body weight) per day (a measure of time) when people eat or drink contaminated water, food, or soil. In general, the greater the dose, the greater the likelihood of an effect. An "exposure dose" is how much of a substance is encountered in the environment. An "absorbed dose" is the amount of a substance that actually got into the body through the eyes, skin, stomach, intestines, or lungs.

❐ dose (for radioactive chemicals)

The radiation dose is the amount of energy from radiation that is actually absorbed by the body. This is not the same as measurements of the amount of radiation in the environment.

❐ dose-response relationship

The relationship between the amount of exposure to a substance and the resulting changes in body function or health (response).

E

❐ environmental media

Soil, water, air, biota (plants and animals), or any other parts of the environment that can contain contaminants.

❐ environmental media and transport mechanism

Environmental media include water, air, soil, and biota (plants and animals). Transport mechanisms move contaminants from the source to points where human exposure can occur. The environmental media and transport mechanism is the second part of an exposure pathway.

❐ epidemiology

The study of the distribution and determinants of disease or health status in a population; the study of the occurrence and causes of health effects in humans.

❐ exposure

Contact with a substance by swallowing, breathing, or touching the skin or eyes. Exposure may be short-term [acute exposure], of intermediate duration, or long-term.

❐ exposure assessment

The process of finding out how people come into contact with a hazardous substance, how often and for how long they are in contact with the substance, and how much of the substance they are in contact with.

❑ exposure investigation

The collection and analysis of site-specific information and biologic tests (when appropriate) to determine whether people have been exposed to hazardous substances.

❑ exposure pathway

The route a substance takes from its source (where it began) to its end point (where it ends), and how people can come into contact with (or get exposed to) it. An exposure pathway has five parts: a source of contamination (such as an abandoned business); an environmental media and transport mechanism (such as movement through groundwater); a point of exposure (such as a private well); a route of exposure (eating, drinking, breathing, or touching), and a receptor population (people potentially or actually exposed). When all five parts are present, the exposure pathway is termed a completed exposure pathway.

❑ exposure registry

A system of ongoing follow up of people who have had documented environmental exposures.

❑ exposure-dose reconstruction

A method of estimating the amount of people's past exposure to hazardous substances. Computer and approximation methods are used when past information is limited, not available, or missing.

F

❑ feasibility study

A study by EPA to determine the best way to clean up environmental contamination. A number of factors are considered, including health risk, costs, and what methods will work well.

G

❑ geographic information system (gis)

A mapping system that uses computers to collect, store, manipulate, analyze, and display data. For example, GIS can show the concentration of a contaminant within a community in relation to points of reference such as streets and homes.

❑ grand rounds

Training sessions for physicians and other health care providers about health topics.

❑ groundwater

Water beneath the earth's surface in the spaces between soil particles and between rock surfaces.

H

❐ half-life (t½)

The time it takes for half the original amount of a substance to disappear. In the environment, the half-life is the time it takes for half the original amount of a substance to disappear when it is changed to another chemical by bacteria, fungi, sunlight, or other chemical processes. In the human body, the half-life is the time it takes for half the original amount of the substance to disappear, either by being changed to another substance or by leaving the body. In the case of radioactive material, the half life is the amount of time necessary for one half the initial number of radioactive atoms to change or transform into another atom (that is normally not radioactive). After two half lives, 25% of the original number of radioactive atoms remain.

❐ hazard

A source of potential harm from past, current, or future exposures.

❐ hazardous substance release and health effects database (hazdat)

The scientific and administrative database system developed by ATSDR to manage data collection, retrieval, and analysis of site-specific information on hazardous substances, community health concerns, and public health activities.

❐ hazardous waste

Potentially harmful substances that have been released or discarded into the environment.

❐ health consultation

A review of available information or collection of new data to respond to a specific health question or request for information about a potential environmental hazard. Health consultations are focused on a specific exposure issue. Health consultations are therefore more limited than a public health assessment, which reviews the exposure potential of each pathway and chemical.

❐ health education

Programs designed with a community to help it know about health risks and how to reduce these risks.

❐ health investigation

The collection and evaluation of information about the health of community residents. This information is used to describe or count the occurrence of a disease, symptom, or clinical measure and to evaluate the possible association between the occurrence and exposure to hazardous substances.

❐ health promotion

The process of enabling people to increase control over, and to improve, their health.

❒ health statistics review

The analysis of existing health information (i.e., from death certificates, birth defects registries, and cancer registries) to determine if there is excess disease in a specific population, geographic area, and time period. A health statistics review is a descriptive epidemiologic study.

I

❒ in vitro

In an artificial environment outside a living organism or body. For example, some toxicity testing is done on cell cultures or slices of tissue grown in the laboratory, rather than on a living animal.

❒ in vivo

Within a living organism or body. For example, some toxicity testing is done on whole animals, such as rats or mice.

❒ incidence

The number of new cases of disease in a defined population over a specific time period.

❒ indeterminate public health hazard

The category used in ATSDR 's public health assessment documents when a professional judgment about the level of health hazard cannot be made because information critical to such a decision is lacking.

❒ ingestion

The act of swallowing something through eating, drinking, or mouthing objects. A hazardous substance can enter the body this way.

❒ inhalation

The act of breathing. A hazardous substance can enter the body this way.

❒ intermediate duration exposure

Contact with a substance that occurs for more than 14 days and less than a year.

L

❒ lowest-observed-adverse-effect level (loael)

The lowest tested dose of a substance that has been reported to cause harmful (adverse) health effects in people or animals.

M

❒ medical monitoring

A set of medical tests and physical exams specifically designed to evaluate whether an individual's exposure could negatively affect that person's health.

❐ metabolism

The conversion or breakdown of a substance from one form to another by a living organism.

❐ mg/cm2

Milligram per square centimeter (of a surface).

❐ mg/kg

Milligram per kilogram.

❐ mg/m3

Milligram per cubic meter; a measure of the concentration of a chemical in a known volume (a cubic meter) of air, soil, or water.

❐ migration

Moving from one location to another.

❐ minimal risk level (mrl)

An ATSDR estimate of daily human exposure to a hazardous substance at or below which that substance is unlikely to pose a measurable risk of harmful (adverse), noncancerous effects. MRLs are calculated for a route of exposure (inhalation or oral) over a specified time period (acute, intermediate, or chronic). MRLs should not be used as predictors of harmful (adverse) health effects.

❐ morbidity

State of being ill or diseased. Morbidity is the occurrence of a disease or condition that alters health and quality of life.

❐ mortality

Death. Usually the cause (a specific disease, a condition, or an injury) is stated.

❐ mutagen

A substance that causes mutations (genetic damage).

❐ mutation

A change (damage) to the DNA, genes, or chromosomes of living organisms.

N

❐ national priorities list for uncontrolled hazardous waste sites (national priorities list or npl)

EPA's list of the most serious uncontrolled or abandoned hazardous waste sites in the United States. The NPL is updated on a regular basis.

❐ national toxicology program (ntp)

Part of the Department of Health and Human Services. NTP develops and carries out tests to predict whether a chemical will cause harm to humans.

❐ no apparent public health hazard

A category used in ATSDR 's public health assessments for sites where human exposure to contaminated media might be occurring, might have occurred in the past, or might occur in the future, but where the exposure is not expected to cause any harmful health effects.

❐ no public health hazard

A category used in ATSDR 's public health assessment documents for sites where people have never and will never come into contact with harmful amounts of site-related substances.

❐ no-observed-adverse-effect level (noael)

The highest tested dose of a substance that has been reported to have no harmful (adverse) health effects on people or animals.

P

❐ physiologically based pharmacokinetic model (pbpk model)

A computer model that describes what happens to a chemical in the body. This model describes how the chemical gets into the body, where it goes in the body, how it is changed by the body, and how it leaves the body.

❐ pica

A craving to eat nonfood items, such as dirt, paint chips, and clay. Some children exhibit pica-related behavior.

❐ plume

A volume of a substance that moves from its source to places farther away from the source. Plumes can be described by the volume of air or water they occupy and the direction they move. For example, a plume can be a column of smoke from a chimney or a substance moving with groundwater.

❐ point of exposure

The place where someone can come into contact with a substance present in the environment.

❐ population

A group or number of people living within a specified area or sharing similar characteristics (such as occupation or age).

❐ potentially responsible party (prp)

A company, government, or person legally responsible for cleaning up the pollution at a hazardous waste site under Superfund. There may be more than one PRP for a particular site.

❐ ppb

Parts per billion.

❐ ppm

Parts per million.

❐ prevalence

The number of existing disease cases in a defined population during a specific time period.

❐ prevalence survey

The measure of the current level of disease(s) or symptoms and exposures through a questionnaire that collects self-reported information from a defined population.

❐ prevention

Actions that reduce exposure or other risks, keep people from getting sick, or keep disease from getting worse.

❐ public availability session

An informal, drop-by meeting at which community members can meet one-on-one with ATSDR staff members to discuss health and site-related concerns.

❐ public comment period

An opportunity for the public to comment on agency findings or proposed activities contained in draft reports or documents. The public comment period is a limited time period during which comments will be accepted.

❐ public health action

A list of steps to protect public health.

❐ public health advisory

A statement made by ATSDR to EPA or a state regulatory agency that a release of hazardous substances poses an immediate threat to human health. The advisory includes recommended measures to reduce exposure and reduce the threat to human health.

❐ public health assessment (pha)

An ATSDR document that examines hazardous substances, health outcomes, and community concerns at a hazardous waste site to determine whether people could be harmed from coming into contact with those substances. The PHA also lists actions that need to be taken to protect public health.

❐ public health hazard

A category used in ATSDR's public health assessments for sites that pose a public health hazard because of long-term exposures (greater than 1 year) to sufficiently high levels of hazardous substances or radionuclides that could result in harmful health effects.

❒ public health hazard categories

Public health hazard categories are statements about whether people could be harmed by conditions present at the site in the past, present, or future. One or more hazard categories might be appropriate for each site. The five public health hazard categories are no public health hazard, no apparent public health hazard, indeterminate public health hazard, public health hazard, and urgent public health hazard.

❒ public health statement

The first chapter of an ATSDR toxicological profile. The public health statement is a summary written in words that are easy to understand. The public health statement explains how people might be exposed to a specific substance and describes the known health effects of that substance.

❒ public health surveillance

The ongoing, systematic collection, analysis, and interpretation of health data. This activity also involves timely dissemination of the data and use for public health programs.

❒ public meeting

A public forum with community members for communication about a site.

R

❒ radioisotope

An unstable or radioactive isotope (form) of an element that can change into another element by giving off radiation.

❒ radionuclide

Any radioactive isotope (form) of any element.

❒ receptor population

People who could come into contact with hazardous substances.

❒ reference dose (rfd)

An EPA estimate, with uncertainty or safety factors built in, of the daily lifetime dose of a substance that is unlikely to cause harm in humans.

❒ registry

A systematic collection of information on persons exposed to a specific substance or having specific diseases.

❒ remedial investigation

The CERCLA process of determining the type and extent of hazardous material contamination at a site.

□ resource conservation and recovery act (1976, 1984) (rcra)

This Act regulates management and disposal of hazardous wastes currently generated, treated, stored, disposed of, or distributed.

□ rfa

RCRA Facility Assessment. An assessment required by RCRA to identify potential and actual releases of hazardous chemicals.

□ risk

The probability that something will cause injury or harm.

□ risk communication

The exchange of information to increase understanding of health risks.

□ risk reduction

Actions that can decrease the likelihood that individuals, groups, or communities will experience disease or other health conditions.

□ route of exposure

The way people come into contact with a hazardous substance. Three routes of exposure are breathing [inhalation], eating or drinking [ingestion], or contact with the skin [dermal contact].

S

□ sample

A portion or piece of a whole. A selected subset of a population or subset of whatever is being studied. For example, in a study of people the sample is a number of people chosen from a larger population. An environmental sample (for example, a small amount of soil or water) might be collected to measure contamination in the environment at a specific location.

□ sample size

The number of units chosen from a population or an environment.

□ solvent

A liquid capable of dissolving or dispersing another substance (for example, acetone or mineral spirits).

□ source of contamination

The place where a hazardous substance comes from, such as a landfill, waste pond, incinerator, storage tank, or drum. A source of contamination is the first part of an exposure pathway.

□ special populations

People who might be more sensitive or susceptible to exposure to hazardous substances because of factors such as age, occupation, sex, or behaviors (for example, cigarette smoking). Children, pregnant women, and older people are often considered special populations.

❑ stakeholder

A person, group, or community who has an interest in activities at a hazardous waste site.

❑ statistics

A branch of mathematics that deals with collecting, reviewing, summarizing, and interpreting data or information. Statistics are used to determine whether differences between study groups are meaningful.

❑ substance

A chemical.

❑ substance-specific applied research

A program of research designed to fill important data needs for specific hazardous substances identified in ATSDR 's toxicological profiles. Filling these data needs would allow more accurate assessment of human risks from specific substances contaminating the environment. This research might include human studies or laboratory experiments to determine health effects resulting from exposure to a given hazardous substance.

❑ superfund amendments and reauthorization act (sara)

In 1986, SARA amended the Comprehensive Environmental Response, Compensation, and Liability Act of 1980 (CERCLA) and expanded the health-related responsibilities of ATSDR. CERCLA and SARA direct ATSDR to look into the health effects from substance exposures at hazardous waste sites and to perform activities including health education, health studies, surveillance, health consultations, and toxicological profiles.

❑ surface water

Water on the surface of the earth, such as in lakes, rivers, streams, ponds, and springs.

❑ survey

A systematic collection of information or data. A survey can be conducted to collect information from a group of people or from the environment. Surveys of a group of people can be conducted by telephone, by mail, or in person. Some surveys are done by interviewing a group of people.

❑ synergistic effect

A biologic response to multiple substances where one substance worsens the effect of another substance. The combined effect of the substances acting together is greater than the sum of the effects of the substances acting by themselves.

T

❐ teratogen

A substance that causes defects in development between conception and birth. A teratogen is a substance that causes a structural or functional birth defect.

❐ toxic agent

Chemical or physical (for example, radiation, heat, cold, microwaves) agents that, under certain circumstances of exposure, can cause harmful effects to living organisms.

❐ toxicological profile

An ATSDR document that examines, summarizes, and interprets information about a hazardous substance to determine harmful levels of exposure and associated health effects. A toxicological profile also identifies significant gaps in knowledge on the substance and describes areas where further research is needed.

❐ toxicology

The study of the harmful effects of substances on humans or animals.

❐ tumor

An abnormal mass of tissue that results from excessive cell division that is uncontrolled and progressive. Tumors perform no useful body function. Tumors can be either benign (not cancer) or malignant (cancer).

❐ uncertainty factor

Mathematical adjustments for reasons of safety when knowledge is incomplete. For example, factors used in the calculation of doses that are not harmful (adverse) to people. These factors are applied to the lowest-observed-adverse-effect-level (LOAEL) or the no-observed-adverse-effect-level (NOAEL) to derive a minimal risk level (MRL). Uncertainty factors are used to account for variations in people's sensitivity, for differences between animals and humans, and for differences between a LOAEL and a NOAEL. Scientists use uncertainty factors when they have some, but not all, the information from animal or human studies to decide whether an exposure will cause harm to people [also sometimes called a safety factor].

❐ urgent public health hazard

A category used in ATSDR's public health assessments for sites where short-term exposures (less than 1 year) to hazardous substances or conditions could result in harmful health effects that require rapid intervention.

V

❐ volatile organic compounds (vocs)

Organic compounds that evaporate readily into the air. VOCs include substances such as benzene, toluene, methylene chloride, and methyl chloroform.

Water Quality and Waste Management

A

- **activated carbon filter**

  Water treatment process to remove taste, odor, some organic compounds, and radon.

- **adsorption**

  Retention of a substance by soil particles.

- **aerobic**

  In the presence of or requiring oxygen.

- **agri-chemical**

  Chemicals (fertilizers and pesticides) used in agricultural production.

- **algal bloom**

  Large, visible masses of algae found in bodies of water during warm water.

- **alkalinity**

  Capacity of water to neutralize acids by its content of bicarbonates, carbonates, or hydroxides.

- **ambient monitoring**

  Performed to determine existing environmental conditions or contaminant levels in the environment, against which future conditions can be compared.

- **anaerobic (anoxic)**

  In the absence of oxygen.

- **aquifer**

  Water-bearing formation of rock or soil that will yield useable supplies of water. May be classified as confined or unconfined.

- **artesian (flowing) aquifer**

  Aquifer in which water is held under pressure by confining layers, forcing water to rise in wells above the top of the aquifer.

- **assimilative capacity**

  Natural ability of soil and water to use and decompose potential pollutants without harmful effects to the environment.

❑ **available nitrogen**

Amount of nitrogen present as either nitrate or ammonium, forms which can be readily taken up by plants.

❑ **available water**

The portion of water in soil that can be readily absorbed by plant roots.

B

❑ **background level**

Amount of a substance which occurs naturally in the environment.

❑ **bacteria**

Microscopic one-celled organisms which live everywhere and perform a variety of functions. While decomposing organic matter in water, bacteria can greatly reduce the amount of oxygen in the water.

❑ **baler**

Machine used to compress and bind recyclables, such as aluminum, paper, corrugated cardboard and plastics.

❑ **bentonite**

Highly plastic clay consisting of the minerals montmorillonite and beidellite that swells when wet and is often used as a lining material to seal landfills and lagoons.

❑ **best management practice (bmp)**

Structural or managerial technique recognized as the most effective and practical means of controlling pollution for an agricultural, urban, forested, or mining area.

❑ **biochemical oxygen demand (bod)**

Laboratory measurement of the amount of oxygen consumed by microorganisms while decomposing organic matter in a product. Bod levels are indicative of the effect of the waste on fish or other aquatic life which require oxygen to live, and though not a specific compound, it is defined as a conventional pollutant under the federal clean water act.

❑ **biodegradeable**

Capable of being broken down (decomposed) by microorganisms.

❑ **black water**

Liquid and solid human body waste and the carriage water generated by toilet use.

❑ **bottle bill**

Law requiring deposits on beverage containers (see container deposit legislation).

❒ btx

A test for benzene, toluene, and xylene, three organic compounds characteristically present in gasoline.

❒ buffer zone

Neutral area which acts as a protective barrier separating two conflicting forces. An area which acts to minimize the impact of pollutants on the environment or public welfare. For example, a buffer zone is established between a composting facility and neighboring residents to minimize odor problems.

❒ bulky waste

Large items of refuse including, but not limited to, appliances, furniture, large auto parts, non-hazardous construction and demolition material, trees, branches and stumps which cannot be handled by normal solid waste processing, collection and disposal methods.

❒ buy-back center

Facility where individuals bring recyclables in exchange for payment.

C

❒ carcinogen

Substance which causes cancer.

❒ chemical oxygen demand (cod)

Laboratory measurement of the amount of oxygen used in chemical reactions that occur in water as a result of the addition of wastes. A major objective of conventional wastewater treatment is to reduce the chemical and biochemical oxygen demand.

❒ chlorination

Addition of chlorine as a means of disinfecting drinking water or wastewater.

❒ co-compositing

Simultaneous composting of two or more diverse waste streams.

❒ coliform bacteria

Microorganisms which typically inhabit the intestines of warm-blooded animals. They are commonly measured in drinking water analyses to indicate pollution by human or animal waste.

❒ commercial waste

Materials originating in wholesale, retail, institutional or service establishments such as offices, stores, markets, theaters, hotels and warehouses.

❒ commingled recyclables

Mixture of several recyclable materials in one container.

- **compactor**

Power-driven device used to compress materials to smaller volume.

- **composting**

Controlled microbial degradation of organic waste yielding an environmentally sound product with value as a soil amendment.

- **confined aquifer**

Water-bearing formation whose upper boundary is a layer which does not transmit water readily.

- **contaminant**

Any physical, chemical, biological, or radiological substance causing an impurity in the environment.

- **corrosive**

Capable of eating away materials and destroying living tissue on contact.

- **corrugated**

Structural paper or cardboard shaped in parallel furrows and ridges for rigidity.

- **curbside collection**

Program where recyclable materials are collected at the curb, often from special containers, to be taken to various processing facilities.

D

- **deactivation**

Process in which a pesticide adheres to a soil particle or some organic material so tightly that it is no longer biologically available.

- **decomposition**

Breaking down into component parts or basic elements.

- **decomposition**

Breaking down into component parts or basic elements.

- **decomposition gases**

Produced in the breakdown of garbage or other material. Some, such as methane, are flammable.

- **degradable**

Capable of being chemically reduced or broken down.

- **denitrification**

Biochemical conversion of nitrate (no3) to nitrite (no2), ammonia (nh3), and free nitrogen (n), as in soil by microorganisms.

❑ **dioxins**

Heterocyclic hydrocarbons that occur as toxic impurities, especially in pesticides.

❑ **discharge**

Flow of surface water in a stream or the flow of ground water from a spring, ditch, or flowing artesion well.

❑ **disposable**

Manufactured to be used for a short time and then thrown away; not durable or repairable.

❑ **dissolved oxygen (do)**

Oxygen dissolved in water and readily available to fish and other aquatic organisms.

❑ **diversion rate**

Measure of the amount of waste material being diverted for recycling compared with the total amount previously thrown away.

❑ **drawdown**

Vertical drop of the water level in a well during pumping.

❑ **drop-off center**

Method of collecting recyclable or compatible materials in which materials are taken by individuals to collection sites and deposited into designated containers.

E

❑ **ecosystem**

Community of animals and plants and the physical environment in which they live.

❑ **effluent**

Discharge or emission of a liquid or gas.

❑ **energy recovery**

Conversion of waste energy, generally through the combustion of processed or raw refuse (incineration), to produce steam.

❑ **erosion**

Natural breakdown and movement of soil and rock by water, wind, or ice. The process may be accelerated by human activities.

❑ **escherichia coli (e. coli)**

Species of coliform bacteria that inhabit intestines of people and animals.

❐ **eutrophication**

Degradation of water quality due enrichment by nutrients, primarily nitrogen (n) and phosphorus (p), which results in excessive plant (principally algae) growth and decay. Low dissolved oxygen (do) in the water is a common consequence.

❐ **evapotranspiration (et)**

Loss of water to the atmosphere from the earth's surface by evaporation and by transpiration through plants.

❐ **explosive / reactive**

Capable of causing an explosion or releasing poisonous fumes when exposed to air, water, or other chemicals.

F

❐ **formulation**

The combination of active and inactive (inert) ingredients which make up a pesticide.

❐ **fumigant**

Gaseous material used to destroy insects, pathogens, or other pests in soil or grain bins.

G

❐ **garbage**

Waste food that is thrown away, generally defined as wet food waste. The term is also used to describe all products discarded, regardless of their reusability or recyclability.

❐ **geographic information system**

Computerized database system containing natural resources and land use data that can be used to analyze and display information in spatial, or map, format.

❐ **giardiasis**

Presence of the giardia lamblia protozoan in the human small intestine which can cause diarrhea.

❐ **grey water**

Wastewater other than sewage, such as sink or washing machine drainage.

H

❐ **half-life**

Time required for one-half of a specified substance to decompose.

▫ hammermill

Type of crusher or shredder used to break up waste materials into smaller pieces.

▫ hardness

Characteristic of water which describes the presence of dissolved minerals. Carbonate hardness is caused by calcium and magnesium bicarbonate; noncarbonate hardness is caused by calcium sulfate, calcium chloride, magnesium sulfate, and magnesium chloride.

▫ hazardous waste

Solid, liquid, or gaseous substance which, because of its source or measurable characteristics, is classified under state or federal law as potentially dangerous and is subject to special handling, shipping, and disposal requirements.

▫ head

The height of a column of water above a standard datum such as mean sea level.

▫ health advisory (ha)

Non-regulatory, health-based reference level of drinking water contaminants at which adverse health effects are believed to be minimal. Ha levels are established for 1-day, 10-day, longer-term, and lifetime exposure periods, and they include large safety margins.

▫ heavy metals

Those metals (elements with high density, malleability, and electrical and thermal conductivity) that have high specific gravity and high atomic mass, such as lead, cadmium, zinc, copper, silver, and mercury. These may be found in the waste stream as part of discarded items such as batteries, lighting fixtures, colorants and inks.

▫ herbicide

Chemical used to destroy or inhibit undesirable plant growth.

▫ high grade paper

Relatively valuable paper such as computer printout, white ledger, and tab cards. Also used to refer to industrial trimmings at paper mills that are recycled.

▫ household hazardous waste

Discarded or used portions of home cleaning products, workshop and outdoor chemicals, automotive fluids, and personal care products that contain toxic chemicals. Products labeled warning, caution, poisonous, toxic, flammable, reactive, or explosive are considered hazardous.

▫ humus

Organic materials resulting from decay or plant or animal matter. Also referred to as compost.

❐ hydrologic cycle

The movement of water in and on the earth and atmosphere through processes such as precipitation, evaporation, runoff, and infiltration.

❐ hydrology

Science dealing with the properties, distribution, and flow of water on or in the earth.

❐ hydrolysis

Reaction of a water molecule with another larger molecule, resulting in the splitting of the larger molecule.

I

❐ ignitable

Capable of burning or causing a fire.

❐ incinerator

Facility in which the combustion of solid waste takes place.

❐ industrial waste

Materials discarded from industrial operations or derived from manufacturing processes.

❐ infiltration

Entry of water from precipitation, irrigation, or runoff into the soil profile.

❐ inorganic chemicals

Natural or synthetic chemicals that contain no carbon.

❐ insecticide

Substance that kills insects.

❐ institutional waste

Material originating in schools, hospitals, prisons, research institutions, and other public buildings.

❐ integrated solid waste management

Practice of using several alternative waste management techniques to manage and dispose of specific components of the municipal waste stream. Waste management alternatives include source reduction, recycling, composting, energy recovery, and land filling.

❐ intermediate processing center (ipc)

Type of materials recovery facility (mrf) that processes residentially collected mixed recyclables into new products available for market; often used interchangeably with mrf.

□ in-vessel composting

Method in which the compost material is continuously and mechanically mixed and aerated in a large, contained area.

L

□ leaching

Movement through soil of dissolved or suspended substances in water.

□ lethal dose (ld)

Amount of a substance required to cause death in an organism.

□ loading

Amount of a substance entering the environment (soil, water, or air).

M

□ manual separation

Sorting of recyclables or compositable materials from waste by hand sorting.

□ mass burn

Municipal waste combustion technology in which solid waste is burned in a controlled system without prior sorting or processing.

□ materials market

Combination of manufacturing interests which buy recyclable materials and process them for reuse. The demand for goods made of recycled materials determines the economic feasibility of recycling and resource recovery.

□ materials recovery facility (mrf)

Facility that separates and processes recyclable materials for sale to an end user.

□ maximum contaminant level (mcl)

Enforceable epa standard for the maximum permissible concentration of a contaminant in public water supplies. An mcl is set after considering health effects as well as the feasibility and cost of analysis and treatment of the regulated contaminant.

□ maximum contaminant level goal (mclg)

Preliminary standard based entirely on health effects which is used by epa to establish the mcl for a contaminant. For a chemical believed to cause cancer, the mclg is zero.

□ mechanical separation

Sorting of waste into various components using mechanical means, such as cyclones, trommels, and screens.

❐ metabolites

Breakdown chemical products resulting when a pesticide passes through a biological system.

❐ methane

Odorless, colorless, flammable and explosive gas produced by municipal solid waste undergoing anaerobic decomposition. Methane is emitted from municipal solid waste landfills.

❐ mineral water

Contains large amounts of dissolved minerals such as calcium, sodium, magnesium, and iron. Some tap waters contain as many or more minerals than some commercial mineral waters. There is no scientific evidence that either high or low mineral content water is beneficial to humans.

❐ mineralization

Microbial conversion of an element from an organic (containing carbon) to an inorganic (not containing carbon) state.

❐ modular incinerator

Small-scale waste combustion units prefabricated at a manufacturing facility and transported to the mwc facility site.

❐ most probable number (mpn)

Statistical expression for estimating the number of microorganisms in a culture or a volume of water.

❐ mulch

Natural or artificial layer of plant residue or proper material covering the land surface which conserves soil moisture, holds soil in place, aids in establishing plant cover, and minimizes temperature fluctuations.

❐ municipal solid waste (msw)

Non-hazardous discarded material generated in residential, commercial, institutional, and light industrial settings. It is defined by local governments, and in general does not include automobile oil, tires, lead-acid batteries, hazardous or infectious wastes, demolition debris, etc.

N

❐ nimby

Acronym for "not in my back yard" which is an expression of resident opposition to the sitting of a solid waste facility based on the particular location proposed.

❐ nitrification

Biochemical oxidation of ammonia (nh_3), ammonium (nh_4), or atmospheric nitrogen (n) to nitrate (no_3) or nitrite (no_2).

❒ no observeable adverse effect level (noael)

Chemical exposure dose or level producing no observeable adverse effect in long-term toxicity studies. This level is used to establish a tolerance for human consumption.

❒ nondischarge systems

Wastewater disposal systems that do not discharge to surface waters, such as spray irrigation, land application, or conventional septic systems.

❒ nonpoint source (nps) contamination

Water contamination derived from diffuse sources such as construction sites, agricultural fields, and urban runoff.

❒ nuisance contaminant

Constituents in water which are not normally harmful to health but may cause offensive taste, odor, color, corrosion, foaming, or staining.

❒ nutrient

Element essential for plant or animal growth. Major nutrients include nitrogen, phosphorus, carbon, oxygen, sulfur, and potassium.

O

❒ organic compound

Any carbon-based substance, including some petroleum products, solvents, pesticides, and halomethanes. Volatile organic compounds (vocs) are those which are readily vaporized; a number of these are known or probable carcinogens.

❒ oxygen demand

Materials such as food waste and dead plant or animal tissue that use up dissolved oxygen in the water when they are degraded through chemical or biological processes. Chemical and biochemical oxygen demand (cod and bod) are measures of the amount of oxygen consumed when a substance degrades.

P

❒ package treatment plant

Prefabricated, small-scale wastewater treatment system used in subdivisions or trailer parks.

❒ pathogen

Disease-causing biological agent such as a bacterium, virus, or fungus.

❒ percolation

Movement of water through soil or rock.

❒ permeability

Capacity of soil, sediment, or porous rock to transmit water.

❒ persistence

Resistance to degradation as measured by the period of time required for complete breakdown of a material. Depends on temperature, ph, soil type, light intensity, etc.

❒ pesticide

Substance used for controlling, destroying, or repelling a specific pest. Includes fungicides, herbicides, insecticides, nematicides, rodenticides, defoliants, and plant growth regulators.

❒ ph

Numerical measure of acidity, with a scale of 0 to 14. Neutral is ph 7, values below 7 are acidic, and values above 7 are alkaline.

❒ photodegradeable

Capable of being broken down (decomposed) by a chemical reaction initiated by direct exposure to the sun's ultraviolet radiation.

❒ point source contamination

Water contamination from specific sources such as leaking underground storage tanks, landfills, industrial waste discharge points, or chemical mixing sites.

❒ point-of-entry(poe)

Water treatment system located at the entry point to the home which treats all water used in the home.

❒ point-of-use

Water treatment system located at the tap which treats only water used from the tap.

❒ pollution

Presence of a contaminant to such a degree that the environment (land, water, or air) is not suitable for a particular use.

❒ polychlorinated byphenyl (pcb)

Hazardous compound (suspected carcinogen) used for electrical insulation and heating/cooling equipment which has been found in air, soil, water, and fish across the country.

❒ polyethylene terephthalate (pet)

Recyclable plastic used to make bottles such as soda bottles. Recycled pet is used in car bumpers, furniture, skis, surfboards, carpet yarn, polyester fiber, films and sheets, and molded parts.

❑ **polyvinyl chloride (pvc)**

Common plastic material which releases hydrochloric acid when burned.

❑ **post-consumer recycling**

Reuse of materials generated from residential and commercial waste; excludes recycling of material from industrial processes that has not reached the consumer, such as glass broken in the manufacturing process.

❑ **post-consumer waste**

Material discarded by a business or residence that has fulfilled its useful life.

❑ **primary drinking water standards**

Enforceable epa standards which establish mcls for drinking water contaminants after considering health effects and the feasibility and cost of analysis and treatment of regulated contaminants.

❑ **publicly owned treatment works (potw)**

Wastewater treatment facility supported by public funding.

❑ **pyrolysis**

Chemical decomposition of a material by heat in the absence of oxygen.

R

❑ **radon**

Colorless, odorless, tasteless, radioactive gas.

❑ **receiving waters**

Bodies of water that receive runoff or wastewater discharges, such as rivers, streams, lakes, estuaries, and ground water.

❑ **recharge**

Downward movement of water through soil to ground water.

❑ **recharge area**

Land area over which precipitation infiltrates into soil and percolates downward to replenish an aquifer.

❑ **recyclables**

Materials that still have useful physical or chemical properties after serving their original purpose and that can be reused or remanufactured into additional products, thereby serving as substitutes for raw materials.

❑ **recycling**

Process by which materials otherwise destined for disposal are collected, reprocessed or remanufactured, and reused. Mandatory recycling programs require by law that consumers separate trash so that some or all recyclable materials are not burned or dumped in landfills.

❐ refractory

Material able to withstand dramatic heat variations which may be used to construct conventional combustion chambers in incinerators.

❐ refuse derived fuel (rdf)

Product of a mixed waste processing system in which certain recyclable and non-combustible materials are removed, and the remaining combustible material is converted for use as a fuel to create energy. Densified refuse derived fuel (d-rdf) results when the fuel is processed to form briquettes, pellets, or cubes.

❐ residential waste

Materials generated in single and multiple-family homes.

❐ residue

Materials remaining after processing, incineration, composting, or recycling have been completed; normally disposed of in landfills.

❐ resource conservation and recovery act (rcra)

Federal legislation related to hazardous waste (subtitle c), solid, non-hazardous waste (subtitle d), and the recovery and use of recycled materials and energy (subtitle f).

❐ resource recovery

Extraction and utilization of materials and energy from the waste stream.

❐ reuse

Use of a product, such as a soft drink bottle, in its original form more than once for the same purpose.

❐ reverse osmosis (ro)

Water treatment process in which contaminants are removed by forcing water through a membrane having microscopic holes that allow water molecules, but not larger compounds, to pass through. Ro units do not remove all chemicals, and they generally discharge more than half of the total water as waste.

❐ roll-off container

Large waste container that fits onto a tractor trailer and can be loaded and unloaded hydraulically.

❐ runoff

The portion of precipitation, snow melt, or irrigation which flows over and through soil, eventually reaching surface water (streams, rivers, lakes).

S

❐ safe drinking water act(sdwa)

Passed by congress in 1974, and amended in 1986, to insure safe drinking water. It directs the epa to establish and enforce water quality standards to protect public health.

❐ salinity

Quality of water based on its salt content; seawater contains approximately 18,000 parts per million of salt.

❐ sanitary landfill

Solid waste disposal site where waste is spread in layers, compacted, and covered with soil or other cover materials each day to minimize pest, aesthetic, disease, air pollution, and water pollution problems. Modern sanitary landfills are equipped with leachate collection and monitoring systems and methane gas controls and are operated in accordance with environmental protection standards.

❐ saturated zone

Portion of the soil or rock profile in which all pores are filled with water.

❐ scavenger

One who illegally removes materials at any point in the solid waste management system.

❐ scrap

Discarded or rejected industrial waste material often suitable for recycling.

❐ scrubber

Anti-pollution device that uses a liquid or slurry spray to remove acid gases and particulates from municipal waste combustion facility flue gases.

❐ secondary drinking water standards

Epa guidelines for establishing secondary maximum contaminant levels, non-enforceable standards for nuisance contaminants that cause offensive taste, odor, color, corrosion, foaming, and staining.

❐ secondary material

Used in place of a primary or raw material in manufacturing a product.

❐ sediment

Eroded soil and rock material, and plant debris, transported and deposited by water.

❐ septic tank

Sewage disposal tank in which a continuous flow of waste material is decomposed by anaerobic (in the absence of oxygen) bacteria.

❐ signal word

Warning required by the federal hazardous substances act of 1960 to be used on the label of a hazardous substance. Examples include danger, warning, caution, and poison.

❐ sludge

Heavy, slimy residue remaining from the treatment of municipal and industrial water and wastewater. Digested sewage sludge remains after decomposition under controlled temperature, ph, and mixing in a digester tank.

❐ softening

Process of removing hardness caused by calcium and magnesium minerals from water.

❐ soil liner

Landfill liner composed of compacted soil or synthetic material designed to assist in containment of leachate.

❐ solid waste management

Related to storage, collection, transportation, treatment, utilization, processing, and final disposal of solid waste or resource recovery, and facilities necessary for such activities.

❐ solubility

Amount of a substance that will dissolve in a given amount of another substance, typically water.

❐ soluble

Capable of being dissolved easily.

❐ solvent

Liquid capable of dissolving another substance.

❐ source reduction

Design, manufacture, acquisition and reuse of materials so as to minimize the quantity and/or toxicity of waste produced. Waste is eliminated by redesigning products or by otherwise changing societal patterns of consumption, use, and waste generation.

❐ source separation

Segregation of specific recyclable materials at the point of generation for separate collection; often part of a curbside recycling program.

❐ special waste

Items that require special or separate handling, such as household hazardous waste, bulky waste, tires, and used oil.

❒ **static water level**

Water level in a well before pumping.

❒ **superfund**

Common name for the comprehensive environmental response, compensation and liability act (cercla) designed to clean up abandoned or inactive hazardous waste dump sites.

T

❒ **tipping fee**

Charge, usually in dollars per ton, for the unloading or dumping of waste at a landfill, transfer station, recycling center, or waste-to-energy facility, also called a disposal or service fee.

❒ **total dissolved solids (tds)**

Concentration of all substances dissolved in water (solids remaining after evaporation of a water sample).

❒ **toxic substance (toxicant)**

Harmful to plant or animal life, either immediately (acute toxicity) or over a long time period (chronic toxicity).

❒ **transfer station**

Site where waste materials are taken and temporarily stored after collection, pending shipment to a disposal site or resource recovery facility. Recycling and some processing may also take place at transfer stations.

❒ **transmissivity**

Rate at which water passes through a unit width of an aquifer.

❒ **trash**

Material considered worthless, unnecessary or offensive that is usually thrown away. In common usage, it is a synonym for garbage, rubbish or refuge.

❒ **tub grinder**

Machine to grind or chip wood wastes for mulching, composting or size reduction.

❒ **turbidity**

Measure of water cloudiness due to suspended solids.

U

❒ **unconfined (water table) aquifer**

Water-bearing formation whose upper boundary is the water table (as opposed to a confining layer).

❐ unsaturated zone

Portion of the soil profile which contains both air and water. Water in this zone cannot enter a well.

V

❐ vector

A carrier, typically an insect or rodent, capable of transmitting a disease.

❐ virgin material

Raw materials which have never been processed in a manufacturing system, usually requiring more energy to produce than when substituted for by recyclable materials.

❐ volatilization

Conversion of substance to gaseous form.

❐ volume reduction

The processing of waste materials so as to decrease the amount of space the materials occupy, usually by compacting or shredding (mechanical), incineration (thermal), or composting (biological).

W

❐ waste exchange

A computer and catalog network that redirects waste materials back into the manufacturing or reuse process by matching companies generating specific waste with companies that use those wastes as manufacturing inputs.

❐ waste stream

The total waste generated by all contributors (households, industry, government) in a particular area (city, county, state).

❐ wastewater treatment plant (wwtp)

Facility that treats wastewater (and sometimes runoff) from domestic and/or industrial sources by a combination of physical, chemical, and biological processes.

❐ water table

Top of an unconfined aquifer, below which the pore spaces are saturated with water.

❐ watershed (drainage basin)

All land and water that drains runoff to a stream or other surface water body.

❐ waterwall incinerator

Waste combustion facility using lined steel tubes filled with circulating water for cooling. Heat from the combustion gases is transferred to the water, and the resultant steam is sold or used to generate electricity.

❐ wetlands

Areas that are regularly wet or flooded and have a water table that stands at or above the land surface for at least part of the year. Coastal wetlands extend back from estuaries and include salt marshes, tidal basins, marshes, and mangrove swamps. Inland freshwater wetlands consist of swamps, marshes, and bogs.

❐ white goods

Large household appliances such as refrigerators, stoves, air conditioners and washing machines.

❐ windrow

A large, elongated pile of composting material.

Y

❐ yard waste

Leaves, grass clippings, prunings, and other natural organic matter discarded from yards and gardens.

Z

❐ zoning

Designation by ordinances of areas of land reserved and regulated for different land uses; a type of regulatory ordinance based on a land use plan.

Water Quality Terminology

A

❐ **absorbed dose**

The amount of a chemical that enters the body of an exposed organism.

❐ **absorption**

The uptake of water or dissolved chemicals by a cell or an organism (as tree roots absorb dissolved nutrients in the soil).

❐ **absorption factor**

The fraction of a chemical making contact with an organism that is absorbed by the organism.

❐ **acceptable daily intake (adi)**

Estimate of the largest amount of chemical to which a person can be exposed on a daily basis that is not anticipated to result in adverse effects (usually expressed in mg/kg/day).

❐ **accuracy**

How closely an instrument measures the true or actual value of the process variable being measured or sensed.

❐ **acid mine drainage**

Drainage of water from areas that have been mined for coal of other mineral ores; the water has low pH, sometimes less than 2.0 (is acid), because of its contact with sulfur-bearing material; acid drainage is harmful because it often kills aquatic organisms.

❐ **acid rain**

Precipitation which has been rendered (made) acidic by airborne pollutants.

❐ **acidic (uh-sid-ick)**

The condition of water or soil which contains a sufficient amount of acid substances to lower the pH below 7.0.

❐ **acidified (uh-sid-uh-fie-d)**

The addition of an acid (usually nitric or sulfuric) to a sample to lower the pH below 2.0. The purpose of acidification is to "fix" a sample so it won't change until it is analyzed.

❐ acre-foot

A volume of water that covers one acre to a depth of one foot, or 43,560 cubic feet (1233.5 cubic meters).

❐ action level

The concentration of lead or copper in water specified at Code of Federal Regulations 141.80(c) which determines, in some cases, the treatment requirements contained in subpart I of this part that a water system is required to complete.

❐ activated carbon

Adsorptive particles or granules of carbon usually obtained by heating carbon (such as wood). These particles or granules have a high capacity to selectively remove certain trace and soluble materials from water.

❐ active transport

An energy-expending mechanism by which a cell moves a chemical across the cell membrane from a point of lower concentration to a point of higher concentration, against the diffusion gradient.

❐ acute

Occurring over a short period of time; used to describe brief exposures and effects which appear promptly after exposure.

❐ acute exposure

A single exposure to a toxic substance which results in severe biological harm or death. Acute exposures are usually characterized as lasting no longer than a day.

❐ acute toxicity

The ability of a substance to cause poisonous effects resulting in severe biological harm or death soon after a single exposure or dose. Also, any severe poisonous effect resulting from a single short-term exposure to a toxic substance.

❐ additive effect

Combined effect of two or more chemicals equal to the sum of their individual effects.

❐ adsorbate (add-sore-bait)

The material being removed by the adsorption process.

❐ adsorbent (add-sore-bent)

The material (activated carbon) that is responsible for removing the undesirable substance in the adsorption process.

❐ adsorption

The process by which chemicals are held on the surface of a mineral or soil particle (compare with Absorption).

▫ aeration (air-a-shun)

The process of adding air to water. Air can be added to water by either passing air through water or passing water through air.

▫ aerobic (air-0-bick)

A condition in which free" (atmospheric) or dissolved oxygen is present in the water.

▫ age tank

A tank used to store a chemical solution of known concentration for feed to a chemical feeder. Also called a day tank.

▫ aggregate

A mass or cluster of soil particles, often having a characteristic shape.

▫ agrochemical

Synthetic chemicals (pesticide and fertilizers) used in agricultural production.

▫ air binding

A situation where air enters the filter media. Air is harmful to both the filtration and backwash processes. Air can prevent the passage of water during the filtration process and can cause the loss of filter media during the backwash process.

▫ air gap

An open vertical drop, or vertical empty space, that separates a drinking (potable) water supply to be protected from another water system in a water treatment plant or other location. This open gap prevents the contamination of drinking water by backsiphonage or backflow because there is no way raw water or any other water can reach the drinking water.

▫ air padding

Pumping dry air into a container to assist with the withdrawal of a liquid or to force a liquefied gas such as chlorine out of a container.

▫ air stripping

A treatment process used to remove dissolved gases and volatile substances from water. Large volumes of air are bubbled through the water being treated to remove (strip out) the dissolved gases and volatile substances.

▫ alarm contact

A switch that operates when some pre-set low, high or abnormal condition exists.

▫ algae

Microscopic plants which contain chlorophyll and live floating or suspended in water. They also may be attached to structures, rocks or other submerged surfaces. They are food for fish and small aquatic animals. Excess algal growths can impart tastes and odors to potable water. Algae produce oxygen during sunlight hours and use oxygen during the night hours. Their biological activities appreciably affect the pH and dissolved oxygen of the water.

❐ **algal bloom (al-gull)**

Sudden, massive growths of microscopic and macroscopic plant life, such as green or blue green algae, which develop in lakes and reservoirs.

❐ **algicide (al-gi-side)**

Any substance or chemical specifically formulated to kill or control algae.

❐ **aliphatic hydroxy acids (al-uh-fat-ick)**

Organic acids with carbon atoms arranged in branched or unbranched open chains rather than in rings.

❐ **aliquot (al-li-kwot)**

Portion of a sample.

❐ **alkali (al-ka-lie)**

Various soluble salts, principally of sodium, potassium, magnesium, and calcium, that have the property of combining with acids to form neutral salts and may be used in chemical water treatment processes.

❐ **alkaline (al-ka-line)**

The condition of water or soil which contains a sufficient amount of alkali substances to raise the pH above 7.0.

❐ **alkalinity (al-ka-lin-it-tee)**

The capacity of water to neutralize acids. This capacity is caused by the water's content of carbonate, bicarbonate, hydroxide and occasionally borate, silicate, and phosphate. Alkalinity is expressed in milligrams per liter of equivalent calcium carbonate. Alkalinity is not the same as pH because water does not have to be strongly basic (high pH) to have a high alkalinity. Alkalinity is a measure of how much acid can be added to a liquid without causing a great change in pH.

❐ **alluvial (uh-lou-vee-ul)**

Relating to mud and/or sand deposited by flowing water. Alluvial deposits may occur after a heavy rain storm.

❐ **alternating current (ac)**

An electric current that reverses its direction (positive/negative values) at regular intervals.

❐ **ambient**

Environmental or surrounding conditions.

❐ **ambient temperature (am-bee-ent)**

Temperature of the surrounding air (or other medium). For example, temperature of the room where a gas chlorinator is installed.

❒ **ammonium**

One form of nitrogen that is usable by plants.

❒ **amperage (am-purr-age)**

The strength of an electric current measured in amperes. The amount of electric current flow, similar to the flow of water in gallons per minute.

❒ **ampere (am-peer)**

The unit used to measure current strength. The current produced by electromotive force of one volt acting through a resistance of one ohm.

❒ **amperometric (am-purr-o-met-rick)**

Based on the electric current that flows between two electrodes in a solution.

❒ **amperometric titration**

A means of measuring concentrations of certain substances in water (such as strong oxidizers) based on the electric current that flows during a chemical reaction.

❒ **anaerobic (an-air-o-bick)**

A condition in which "free" (atmospheric) or dissolved oxygen is NOT present in water.

❒ **analog**

The readout of an instrument by a pointer (or other indicating means) against a dial or scale.

❒ **analyzer**

A device which conducts periodic or continuous measurement of some factor such as chlorine, fluoride or turgidity. Analyzers operate by any of several methods including photocells, conductivity or complex instrumentation.

❒ **animal studies**

Investigations using animals as surrogates for humans, on the expectation that results in animals are pertinent to humans.

❒ **anion (an-eye-en)**

A negatively charged ion in an electrolyte solution, attracted to the anode under the influence of a difference in electrical potential. Chloride (Cl-) is an anion.

❒ **anionic polymer (an-eye-on-ick)**

A polymer having negatively charged groups of ions; often used as a filter aid and for dewatering sludges.

❒ **annular space (an-you-ler)**

A ring-shaped space located between two circular objects, such as two pipes.

❐ anode (an-0-d)

The positive pole or electrode of an electrolytic system, such as a battery. The anode attracts negatively charged particles or ions (anions).

❐ antagonism

Interference or inhibition of the effect of one chemical by the action of another chemical.

❐ appropriative

Water rights to or ownership of a water supply which is acquired for the beneficial use of water by following a specific legal procedure.

❐ appurtenance (uh-purr-ten-nans)

Machinery, appliances, structures and other parts of the main structure necessary to allow it to operate as intended, but not considered part of the main structure.

❐ aquatic

Plants of animal life living in, growing in, or adapted to water.

❐ aqueous (a-kwee-us)

Something made up of, similar to, or containing water; watery.

❐ aquifer (ack-wi-fer)

A natural underground layer of porous, water-bearing materials (sand, gravel) usually capable of yielding a large amount or supply of water.

❐ artesian (are-tee-zhun - aquifer or well)

Water held under pressure in porous rock or soil confined by impermeable geologic formations. An artesian well is free flowing.

❐ aseptic (a-sep-tick)

Free from the living germs of disease, fermentation or putrefaction. Sterile.

❐ assay

A test for a particular chemical or effect.

❐ association of boards of certification

An international organization representing over 150 boards which certify the operators of waterworks and waste water facilities.

❐ asymmetric (a-see-met-rick)

Not similar in size, shape, form or arrangement of parts on opposite sides of a line, point or plane.

❐ atom

The smallest unit of a chemical element; composed of protons, neutrons and electrons.

❑ available chlorine

A measure of the amount of chlorine available in chlorinated lime, hypochlorite compounds, and other materials that are used as a source of chlorine when compared with that of elemental (liquid or gaseous) chlorine.

❑ available expansion

The vertical distance from the sand surface to the underside of a trough in a sand filter. This distance is also called FREEBOARD.

❑ axial to impeller

The direction in which material being pumped flows around the impeller or flow parallel to the impeller shaft.

❑ axis of impeller

An imaginary line running along the center of a shaft (such as an impeller shaft).

B

❑ back pressure

A pressure that can cause water to backflow into the water supply when a user's water system is at a higher pressure than the public water system.

❑ backflow

A reverse flow condition, created by a difference in water pressures, which causes water to flow back into the distribution pipes of a potable water supply from any source or sources other than an intended source.

❑ background level

In toxic substances monitoring, the average presence of a substance in the environment, originally referring to naturally occurring phenomena.

❑ backsiphonage

A form of backflow caused by a negative or below atmospheric pressure within a water system.

❑ backwashing

The process of reversing the flow of water back through the filter media to remove the entrapped solids.

❑ bacteria (back-teer-e-uh)

Singular: bacterium. Microscopic living organisms usually consisting of a single cell. Bacteria can aid in pollution control by consuming or breaking down organic matter in sewage, or by similarly acting on oil spills or other water pollutants. Some bacteria in soil, water or air may also cause human, animal and plant health problems.

❒ baffle

A flat board or plate, deflector, guide or similar device constructed or placed in flowing water or slurry systems to cause more uniform flow velocities, to absorb energy, and to divert, guide, or agitate liquids (water, chemical solutions, slurry).

❒ bailer (bay-ler)

A 10- to 20-foot-long pipe equipped with a valve at the lower end. A bailer is used to remove slurry from the bottom or the side of a well as it is being drilled.

❒ base metal

A metal (such as iron) which reacts with dilute hydrochloric acid to form hydrogen.

❒ batch process

A treatment process in which a tank or reactor is filled, the water is treated or a chemical solution is prepared, and the tank is emptied. The tank may then be filled and the process repeated.

❒ best available technology (bat)

The best technology treatment techniques, or other means which the Administrator finds, after examination for efficacy under field conditions and not solely under laboratory conditions, are available (taking cost into consideration). For the purposes of setting MCLs for synthetic organic chemicals, any BAT must be at least as effective as granular activated carbon.

❒ best management practices (bmps)

Structural, nonstructural and managerial techniques that are recognized to be the most effective and practical means to control nonpoint source pollutants yet are compatible with the productive use of the resource to which they are applied. BMPs are used in both urban and agricultural areas.

❒ bias

An inadequacy in experimental design that leads to results or conclusions not representative of the population under study.

❒ bioaccumulation

The retention and concentration of a substance by an organism.

❒ bioassay

Test which determines the effect of a chemical on a living organism.

❒ biochemical oxygen demand (bod)

The amount of oxygen consumed by microorganisms (mainly bacteria) and by chemical reactions in the biodegradation of organic matter.

❐ **bioconcentration**

The accumulation of a chemical in tissues of an organism (such as fish) to levels that are greater than the level in the medium (such as water) in which the organism resides (see bioaccumulation).

❐ **biodegradation**

Decomposition of a substance into more elementary compounds by the action of microorganisms such as bacteria.

❐ **biological growth**

The activity and growth of any and all living organisms.

❐ **bioremediation**

A process of adding nutrient to ground water to speed up the natural process in which bacteria break down gasoline into harmless compounds.

❐ **biotransformation**

Conversion of a substance into other compounds by organisms; includes biodegradation.

❐ **black water**

Liquid and solid human body waste and the carriage water generated through toilet usage.

❐ **blank**

A bottle containing only dilution water or distilled water; the sample being tested is not added. Tests are frequently run on a SAMPLE and a BLANK and the differences are compared.

❐ **bonnet (bon-it)**

The cover on a gate valve.

❐ **brackish**

Mixed fresh and salt waters.

❐ **brake horsepower**

1) The horsepower required at the top or end of a pump shaft (input to a pump).
2) The energy provided by a motor or other power source.

❐ **breakpoint chlorination**

Addition of chlorine to water until the chlorine demand has been satisfied. At this point, further additions of chlorine will result in a free residual chlorine that is directly proportional to the amount of chlorine added beyond the breakpoint.

❐ **breakthrough**

A crack or break in a filter bed allowing the passage of floc or particulate matter through a filter. This will cause an increase in filter effluent turbidity. A breakthrough can occur: 1) when a filter is first placed in service, 2) when the effluent valve suddenly opens or closes, and 3) during periods of excessive head loss through the filter (including when the filter is exposed to negative heads).

❑ brinelling (bruh-nel-ing)

Tiny indentations (dents) high on the shoulder of the bearing race or bearing. A type of bearing failure.

❑ buffer

A solution or liquid whose chemical makeup neutralizes acids or bases without a great change in pH.

❑ buffer capacity

A measure of the capacity of a solution or liquid to neutralize acids or bases. This is a measure of the capacity of water for offering a resistance to changes in pH.

❑ buffer strips

Strips of grass or other close-growing vegetation that separate a waterway (ditch, stream, creek) from an intensive land use area (subdivision, farm); also referred to as filter strips, vegetated filter strips, and grassed buffers.

C

❑ c factor

A factor of value used to indicate the smoothness of the interior of a pipe. The higher the C Factor, the smoother the pipe, the greater the carrying capacity, and the smaller the friction or energy losses from water flowing in the pipe. To calculate the C Factor, measure the flow, pipe diameter, distance between two pressure gages, and the friction or energy loss of the water between the gages. C Factor = Flow (GPM)/193.75 (Diameter, ft)2.63 (Slope)0.54

❑ caisson (kay-sawn)

A structure or chamber which is usually sunk or lowered by digging from the inside. Used to gain access to the bottom of a stream or other body of water.

❑ calcium carbonate (caco3) equivalent

An expression of the concentration of specified constituents in water in terms of their equivalent value to calcium carbonate. For example, the hardness in water which is caused by calcium, magnesium and other ions is usually described as calcium carbonate equivalent.

❑ calibration

A procedure which checks or adjusts an instrument's accuracy by comparison with a standard or reference.

❑ cancer

A disease characterized by the rapid and uncontrolled growth of aberrant cells into malignant tumors.

❐ capillary action

The movement of water through very small spaces due to molecular forces.

❐ capillary forces

The molecular forces which cause the movement of water through very small spaces.

❐ capillary fringe

The porous material just above the water table which may hold water by capillarity (a property of surface tension that draws water upwards) in the smaller void spaces.

❐ capital costs

Costs (usually long-term debt) of financing construction and equipment. Capital costs are usually fixed, one-time expenses which are independent of the amount of water produced.

❐ carcinogen (car-sin-o-jen)

Any substance which tends to produce cancer in an organism.

❐ carcinogenic

Cancer-producing.

❐ cas registration number

A number assigned by the Chemical Abstracts Service to identify a chemical.

❐ catalyst (cat-uh-list)

A substance that changes the speed or yield of a chemical reaction without being consumed or chemically changed by the chemical reaction.

❐ catalyze (cat-uh-lize)

To act as a catalyst. Or, to speed up a chemical reaction.

❐ catalyzed (cat-uh-lized)

To be acted upon by a catalyst.

❐ cathode (ka-thow-d)

The negative pole or electrode of an electrolytic cell or system. The cathode attracts positively charged particles or ions (cations).

❐ cathodic protection (ca-thod-ick)

An electrical system for prevention of rust, corrosion, and pitting of metal surfaces which are in contact with water or soil. A low- voltage current is made to flow through a liquid (water) or a soil in contact with the metal in such a manner that the external electromotive force renders the metal structure cathodic. This concentrates corrosion on auxiliary anodic parts which are deliberately allowed to corrode instead of letting the structure corrode.

❒ cation (cat-eye-en)

A positively charged ion in an electrolyte solution, attracted to the cathode under the influence of a difference in electrical potential. Sodium ion ($Na+$) is a cation.

❒ cationic polymer

A polymer having positively charged groups of ions; often used as a coagulant aid.

❒ cavitation (cav-uh-tay-shun)

The formation and collapse of a gas pocket or bubble on the blade of an impeller or the gate of a valve. The collapse of this gas pocket or bubble drives water into the impeller or gate with a terrific force that can cause pitting on the impeller or gate surface. Cavitation is accompanied by loud noises that sound like someone is pounding on the impeller or gate with a hammer.

❒ central nervous system

Portion of the nervous system which consists of the brain and spinal cord; CNS.

❒ centrate

The water leaving a centrifugal after most of the solids have been removed.

❒ centrifugal pump (sen-trif-h-gull)

A pump consisting of an impeller fixed on a rotating shaft that is enclosed in a casing, and having an inlet and discharge connection. As the rotating impeller whirls the water around, centrifugal force builds up enough pressure to force the water through the discharge outlet.

❒ centrifuge

A mechanical device that uses centrifugal or rotational forces to separate solids from liquids.

❒ check valve

A special valve with a hinged disc or flap that opens in the direction of normal flow and is forced shut when flows attempt to go in the reverse or opposite direction of normal flow.

❒ chelation (key-lay-shun)

A chemical complexing (forming or joining together) of metallic cations (such as copper) with certain organic compounds, such as EDTA (ethylene diamine tetracetic acid). Chelation is used to prevent the precipitation of metals (copper).

❒ chemical oxygen demand (cod)

An indirect measure of the amount of oxygen used by inorganic and organic matter in water. The measure is a laboratory test based on a chemical oxidant and therefore does not necessarily correlate with biochemical oxygen demand.

❑ chisel plowing

Cropland preparation by a special implement (chisel) that avoids complete inversion of the soil (as occurs with conventional moldboard plowing). Chisel plowing can leave a protective cover of crop residues on the soil surface that helps prevent erosion and improve infiltration.

❑ chloramines (klor-uh-means)

Compounds formed by the reaction of hypochlorous acid (or aqueous chlorine) with ammonia.

❑ chlorination (klor-uh-nay-shun)

The application of chlorine to water, generally for the purpose of disinfection, but frequently for accomplishing other biological or chemical results (aiding coagulation and controlling tastes and odors).

❑ chlorinator (klor-uh-nay-ter)

A metering device which is used to add chlorine to water.

❑ chlorine demand

Chlorine demand is the difference between the amount of chlorine added to water and the amount of residual chlorine remaining after a given contact time. Chlorine demand may change with dosage, time, temperature, pH, and nature and amount of the impurities in the water. Chlorine Demand, mg/L = Chlorine Applied, mg/L - Residual, mg/L

❑ chlorine requirement

The amount of chlorine which is needed for a particular purpose. Some reasons for adding chlorine are reducing the number of coliform bacteria (Most Probable Number), obtaining a particular chlorine residual, or oxidizing some substance in the water. In each case a definite dosage of chlorine will be necessary. This dosage is the chlorine requirement.

❑ chlorine-contact chamber

That part of a water treatment plant where effluent is disinfected by chlorine.

❑ chlorophenolic (klor-o-fee-no-lick)

Chlorophenolic compounds are phenolic compounds (carbolic acid) combined with chlorine.

❑ chlorophenoxy (klor-o-fuh-knox-ee)

A class of herbicides that may be found in domestic water supplies and cause adverse health effects. Two widely used chlorophenoxy herbicides are 2,4-D (2,4-Dichlorophenoxy acetic acid) and 2,4,5-TP (2,4,5-Trichlorophenoxy propionic acid (silvex)).

❑ chlororganic (klor-or-gan-nick)

Organic compounds combined with chlorine. These compounds generally originate from, or are associated with, life processes such as those of algae in water.

❑ chronic

Occurring over a long period of time, either continuously or intermittently; used to describe ongoing exposures and effects that develop only after a long exposure.

❑ chronic exposure

Long-term, low-level exposure to a toxic chemical.

❑ circle of influence

The circular outer edge of a depression produced in the water table by the pumping of water from a well. Also see cone of influence and cone of depression.

❑ circuit

The complete path of an electric current, including the generating apparatus or other source; or, a specific segment or section of the complete path.

❑ circuit breaker

A safety device in an electrical circuit that automatically shuts off the circuit when it becomes overloaded. The device can be manually reset.

❑ cistern (sis-turn)

A small tank (usually covered) or a storage facility used to store water for a home or farm.

❑ clarifer (klair-uh-fire)

A large circular or rectangular tank or basin in which water is held for a period of time, during which the heavier suspended solids settle to the bottom. Clarifiers are also called SETTLING BASINS and SEDIMENTATION BASINS.

❑ class (pipe and fittings)

The working pressure rating of a specific pipe for use in water distribution systems which includes allowances for surges. This term is used for cast iron, ductile iron, asbestos cement and some plastic pipe.

❑ clay

One type of soil particle with a diameter of approximately one ten-thousandth of an inch.

❑ clay soil

A soil containing more than 40 percent clay, but less than 45 percent sand, and less than 40 percent silt.

❑ clear well

A reservoir for the storage of filtered water of sufficient capacity to prevent the need to vary the filtration rate with variations in demand. Also used to provide chlorine contact time for disinfection.

▫ clinical studies

Studies of humans suffering from symptoms induced by chemical exposure.

▫ coagulant aid

Any chemical or substance used to assist or modify coagulation.

▫ coagulants (co-agg-you-lents)

Chemicals that cause very fine particles to clump together into larger particles. This makes it easier to separate the solids from the water by settling, skimming, draining or filtering.

▫ coagulation (co-agg-yoo-lay-shun)

The clumping together of very fine particles into larger particles caused by the use of chemicals (coagulants). The chemicals neutralize the electrical charges of the fine particles and cause destabilization of the particles. This clumping together makes it easier to separate the solids from the water by settling, skimming, draining, or filtering.

▫ cohesion

Molecular attraction which holds two particles together.

▫ coliform (coal-i-form)

A group of bacteria found in the intestines of warm-blooded animals (including humans) also in plants, soil, air and water. Fecal coliforms are a specific class of bacteria which only inhibit the intestines of warm-blooded animals. The presence of coliform a is an indication that the water is polluted and may contain pathogenic organisms.

▫ coliform organism

Microorganisms found in the intestinal tract of humans and animals. Their presence in water indicates fecal pollution and potentially dangerous bacterial contamination by disease-causing microorganisms.

▫ colloids (call-loids)

Very small, finely divided solids (particles that do not dissolve) that remain dispersed in a liquid for a long time due to their small size and electrical charge. When most of the particles in water have a negative electrical charge, they tend to repel each other. This repulsion prevents the particles from clumping together, becoming heavier, and settling out.

▫ colorimetric measurement

A means of measuring unknown chemical concentrations in water by measuring a sample's color intensity. The specific color of the sample developed by addition of chemical reagents, is measured with a photoelectric colorimeter or is compared with "color standards" using, or corresponding with, known concentrations of the chemical.

❒ combined available residual chlorine

The concentration of residual chlorine which is combined with ammonia (NH_3) and/or organic nitrogen in water as a chloramine (or other chloro derivative) yet is still available to oxidize organic matter and utilize its bactericidal properties.

❒ combined residual chlorination

The application of chlorine to water to produce combined available residual chlorine. This residual can be made up of monochloramines, dichloramines, and nitrogen trichloride.

❒ combined sewer

A sewer that transports surface runoff and human domestic wastes (sewage), and sometimes industrial wastes. Wastewater and runoff in a combined sewer may occur in excess of the sewer capacity and cannot be treated immediately. The excess is frequently discharged directly to a receiving stream without treatment, or to a holding basin for subsequent treatment and disposal.

❒ community water system

A public water system which serves at least 15 service connections used by year- round residents or regularly serves at least 25 year-round residents. Also see non-community water system, transient water system and non-transient non-community water system.

❒ complete treatment

A method of treating water which consists of the addition of coagulant chemicals, flash mixing, coagulation - flocculation, sedimentation and filtration.

❒ compliance cycle

Thee nine-year calendar year cycle during which public water systems must monitor. Each compliance cycle consists of three three-year compliance periods. The first calendar year cycle begins January 1, 1993 and ends December 31, 2001; the second begins January 1, 2002 and ends December 31, 2010; the third from January 1, 2011 to December 31, 2019, etc.

❒ compliance period

A three year calendar period within a compliance cycle. Each compliance cycle has three three year compliance periods. Within the first compliance cycle, the first compliance period runs from January 1, 1993 to December 31, 1995; the second from January 1, 1996 to December 31, 1998; the third from January 1, 1999 to December 31, 2001.

❒ composite (proportional) samples (come-pah-zit) the sampling period

A composite sample is a collection of individual samples obtained at regular intervals, usually every one or two hours during a 24-hour time span. Each individual sample is combined with the others in proportion to the rate of flow when the sample was collected The resulting mixture (composite sample) forms a representative sample.

❐ composting

A controlled microbial degradation of organic waste yield an environmentally sound, nuisance-free product of potential value as a soil conditioner.

❐ compound

A substance composed of two or more elements whose composition is constant. For example, table salt (sodium chloride - NACl) is a compound.

❐ concentration polarization

1) The ratio of the salt concentration in the membrane boundary layer to the salt concentration in the bulk stream. The most common and serious problem resulting from concentration polarization is the increasing tendency for precipitation of sparingly soluble salts and the deposition of particulate matter on the membrane surface. 2) Used in corrosion studies to indicate a depletion of ions near an electrode. 3) The basis for chemical analysis by a polarograph.

❐ conductance

A rapid method of estimating the dissolved solids content of a water supply. The measurement indicates the capacity of a sample of water to carry an electrical current, which is related to the concentration of ionized substances in the water. Also called SPECIFIC CONDUCTANCE.

❐ conductivity

A measure of the ability of a solution (water) to carry an electric current.

❐ conductor

A substance, body, device or wire that readily conducts or carries electrical current.

❐ cone of depression

The depression, roughly conical in shape, produced in the water table by the pumping of water from a well. Also see circle of influence and cone of influence.

❐ cone of influence

The depression, roughly conical in shape, produced in the water table by the pumping of water from a well. Also see circle of influence and cone of depression

❐ confined aquifer

An aquifer in which ground water is confined under pressure which is significantly greater than atmospheric pressure.

❐ confluent growth

A continuous bacterial growth covering the entire filtration area of a membrane filter, or a portion thereof, in which bacterial colonies are not discrete.

❐ confounding factors

Variables other than chemical exposure level which can affect the incidence or degree of a parameter being measured.

❑ consumptive use

Water removed from available supplies without direct return to a water resource system for uses such as manufacturing, agriculture, and food preparation.

❑ contactor

An electrical switch, usually magnetically operated.

❑ contaminant

Any physical, chemical, biological, or radio- logical substance or matter that has an adverse effect on air, water, or soil.

❑ contamination

The introduction into water of microorganisms, chemicals, toxic substances, wastes, or wastewater in a concentration that makes the water unfit for its next intended use.

❑ continuous sample

A flow of water from a particular place in a plant to the location where samples are collected for testing. This continuous stream may be used to obtain grab or composite samples. Frequently, several taps (faucets) will flow continuously in the laboratory to provide test samples from various places in a water treatment plant.

❑ contour farming

A conservation-based method of farming in which all farming operations (for example, tillage and planting) are performed across (rather than up and down) the slope. Ideally, each crop row is planted at right angles to the ground slope.

❑ contour strip farming

A kind of contour farming in which row crops are planted in strips, between alternating strips of close-growing, erosion resistant forage (grass, grain, hay) crops.

❑ control loop

The path through the control system between the sensor, which measures a process variable, and the controller, which controls or adjusts the process variable.

❑ control system

A system which senses and controls its own operation on a close, continuous basis in what is called proportional (or modulating) control.

❑ controller

A device which controls the starting, stopping, or operation of a device or piece of equipment.

❑ conventional filtration

A method of treating water to remove particulates. The method consists of the addition of coagulant chemicals, flash mixing, coagulation - flocculation, sedimentation and filtration.

❑ conventional filtration treatment

A series of processes including coagulation, flocculation, sedimentation, and filtration resulting in substantial particulate removal.

❑ conventional tillage

The traditional method of farming in which soil is prepared for planting by completely inverting it with a moldboard plow. Subsequent working of the soil with other implements is usually performed to smooth the soil surface. Bare soil is exposed to the weather for some varying length of time depending on soil and climatic conditions.

❑ conveyance loss

Water lost in conveyance (pipe, channel, conduit, ditch) by leakage or evaporation.

❑ corporation stop

A water service shutoff valve located at a street water main. This valve cannot be operated from the ground surface because it is buried and there is no valve box.

❑ corrosion

The gradual decomposition or destruction of a material by chemical action, often due to an electrochemical reaction. Corrosion may be caused by: 1) stray current electrolysis, 2) galvanic corrosion caused by dissimilar metals, or 3) differential concentration cells. Corrosion starts at the surface of a material and moves inward.

❑ corrosion inhibitor

A substances that slows the rate of corrosion of metal plumbing materials by water, especially lead and copper materials, by forming a protective film on the interior surface of those materials.

❑ corrosivity

An indication of the corrosiveness of a water. The corrosiveness of a water is described by the water's pH, alkalinity, hardness, temperature, total dissolved solids, dissolved oxygen concentration, and the Langelier Index.

❑ cost sharing

A publicly financed program through which society, as the beneficiary of environment protection, shares part of the cost of pollution control with those who must actually install the controls.

❑ cost/beneflt analysis

A quantitative evaluation of the costs which would be incurred versus the overall benefits to society of a proposed action such as the establishment of an acceptable dose of a toxic chemical.

❐ coulomb (coo-lahm)

A measurement of the amount of electrical charge conveyed in one second by an electric current of one ampere. One coulomb equals about 6.25 x 1018electrons (6,250,000,000,000,000,000 electrons).

❐ coupon

A steel specimen inserted into water to measure the corrosiveness of water. The rate of corrosion is measured as the loss of weight of the coupon (in milligrams) per surface area (in square decimeters) exposed to the water per day. 10 decimeters = 1 meter = 100 centimeters

❐ cover crop

A crop that provides temporary protection for delicate seedlings and/or provides a canopy for seasonal soil protection and improvement between normal crop production periods. Except in orchards where permanent vegetative cover is maintained, cover crops usually are grown for one year of less. When plowed under and incorporated into the soil, cover crops are also referred to as gren manure crops.

❐ crop rotation

A system of farming in which a regular succession of different crops are planted on the same land area, as opposed to growing the same crop time after time (monoculture).

❐ cross connection

Any actual or potential connection between a drinking (potable) water system and an unapproved water supply or other source of contamination. For example, if you have a pump moving nonpotable water and hook into the g water system to supply water for the pump seal, a cross-connection or mixing between the two water systems can occur. This mixing may lead to contamination of the drinking water.

❐ ct or ctcalc

The product of "residual disinfectant concentration" (C) in mg/l determined before or at the first customer, and the corresponding "disinfectant contact time" (T) in minutes, i.e., "C" x "T". If a public water system applies disinfectants at more than one point prior to the first customer, it must determine the CT of each disinfectant sequence before or at the first customer to determine the total percent inactivation or "total inactivation ratio". In determining the total inactivation ratio, the public water system must determine the residual disinfectant concentration of each disinfection sequence and corresponding contact time before any subsequent disinfection application point(s). "CT99.9" is the CT value required for 99.9 Percent (3-log) inactivation of Giardia lamblia cysts. CT99.9 a variety of disinfectants and conditions appear in Tables 1. l- 1.6, 2.1, and 3.1 of section 141.74(b)(3) in the code of Federal Regulations. CT99.9 is the inactivation ratio. The sum of the inactivation ratios, or total inactivation ratio shown as

E = (CT calc) / (CT99.9)

is calculated by adding together the inactivation ratio for each disinfection sequence. A total inactivation ratio equal to or greater than 1.0 is assumed to provide a 3-log inactivation of Giardia lamblia cysts.

❒ cumulative exposure

The summation of exposures of an organism to a chemical over a period of time.

❒ curb stop

A water service shutoff valve located in a water service pipe near the curb and between the water main and the building. This valve is usually operated by a wrench or valve key and is used to start or stop flows in the water service line to a building. Also called a "curb cock."

❒ curie

A measure of radioactivity. One Curie of radioactivity is equivalent to 3.7 x 1010 or 37,000,000,000 nuclear disintegrations per second.

❒ current

A movement or flow of electricity. Water flowing in a pipe is measured in gallons per second past a certain point, not by the number of water molecules going past a point. Electric current is measured by the number of coulombs per second flowing past a certain point in a conductor. A coulomb is equal to about 6.25 x 1018 electrons (6,250,000,000,000,000,000 electrons). A flow of one coulomb per second is called one ampere, the unit of the rate of flow of current.

D

❒ dateometer (day-tom-uh-ter)

A small calendar disc attached to motors and equipment to indicate the year in which the last maintenance service was performed.

❒ day tank

A tank used to store a chemical solution of known concentration for feed to a chemical feeder. A day tank usually stores sufficient chemical solution to properly treat the water being treated for at least one day. Also called an AGE TANK.

❒ dead end

The end of a water main which is not connected to other parts of the distribution system by means of a connecting loop of pipe

❒ decant

To draw off the upper layer of liquid (water) after the heavier material (a solid or another liquid) has settled.

❒ dechlorination (dee-klor-uh-nay-shun) physical process

The deliberate removal of chlorine from water. The partial or complete reduction of residual chlorine by any chemical.

❒ decibel (des-uh-bull)

A unit for expressing the relative intensity of sounds on a scale from zero for the average least perceptible sound to about 130 for the average level at which sound causes pain to humans.

❒ decomposition

The conversion of chemically unstable materials to more stable forms by chemical or biological action. If organic matter decays when there is no oxygen present (anaerobic conditions or putrefaction), undesirable tastes and odors are produced. Decay of organic matter when oxygen is present (aerobic conditions) tends to produce much less objectionable tastes and odors.

❒ defluoridation (de-floor-uh-day-shun)

The removal of excess fluoride in drinking water to prevent the mottling (brown stains) of teeth.

❒ degasification (dee-gas-if-uh-kay-shun)

A water treatment process which removes dissolved gases from the water. The gases may be removed by either mechanical or chemical treatment methods or a combination of both.

❒ degradation

Chemical or biological breakdown of a complex compound into simpler compounds.

❒ demineralization (dee-min-er-al-uh-zay-shun)

A treatment process which removes dissolved minerals (salts) from water.

❒ denitrification

The biochemical conversion of nitrate; and nitrite nitrogen in the soil dissolved in water to gaseous nitrogen.

❒ density (den-sit-tee)

A measure of how heavy a substance (solid, liquid or gas) is for its size. Density is expressed in terms of weight per unit volume, that is, grams per cubic centimeter or pounds per cubic foot. The density of water is 1.0 gram per cubic centimeter or about 62.4 pounds per cubic foot.

❒ dermal exposure

Contact between a chemical and the skin.

❒ desalinization (dee-say-leen-uh-zay-shun)

The removal of dissolved salts (such as sodium chloride, NACI) from water by natural means (leaching) or by specific water treatment processes.

❒ desiccant (dess-uh-kant)

A drying agent which is capable of removing or absorbing moisture from the atmosphere in a small enclosure.

❑ **desiccation (dess-uh-kay-shun)**

A process used to thoroughly dry air; to remove virtually all moisture from air.

❑ **desiccator (dess-uh-kay-tor)**

A closed container into which heated weighing or drying dishes are placed to coot in a dry environment. The dishes may be empty or they may contain a sample. Desiccators contain a substance, such as anhydrous calcium chloride, which absorbs moisture and keeps the relative humidity near zero so that the dish or sample will not gain weight from absorbed moisture.

❑ **destratification (de-strat-uh-fuh-kay-shun)**

The development of vertical mixing within a lake or reservoir to eliminate (either totally or partially) separate layers of temperature, plant, or animal life. This vertical mixing can be caused by mechanical means (pumps) or through the use of forced air diffusers which release air into the lower layers of the reservoir.

❑ **detention lag**

The time period between the moment a change is made and the moment when such a change is finally sensed by the associated measuring instrument.

❑ **detention time**

1) The theoretical (calculated) time required for a small amount of water to pass through a tank at a given rate of flow.
2) The actual time in hours, minutes or seconds that a small amount of water is in a settling basin, flocculating basin or rapid-mix chamber. In storage reservoirs, detention time is the length of time entering water will be held before being drafted for use (several weeks to years, several months being typical).

Detention Time (hr) = Basin Volume (gal.)(24 hr/day)t Flow (gal/day)

❑ **dew point**

The temperature to which air with a given quantity of water vapor must be cooled to cause condensation of the vapor in the air.

❑ **dewater**

1) To remove or separate a portion of the water present in a sludge or slurry. To dry sludge so it can be handled and disposed.
2) To remove or drain the water from a tank or a trench.

❑ **diatomaceous earth**

A fine, siliceous (made of silica) "earth" composed mainly of the skeletal remains of diatoms, a type of free-floating, microscopic plant found in the ocean.

❑ **diatomaceous earth filtration (de filtration)**

A filtration method resulting in substantial particulate removal, that uses a process in which: 1) a "precoat" cake of diatomaceous earth filter media is deposited on a support membrane (septum), and 2) while the water is filtered by passing through the cake on the septum, additional filter media, known as "body feed," is continuously added to the feed water to maintain the permeability of the filter cake.

▫ diffusion

The movement of suspended or dissolved particles from a more concentrated to a less concentrated region as a result of the random movement of individual particles; the process tends to distribute them uniformly throughout the available volume.

▫ digital readout

Use of numbers to indicate the value or measurement of a variable. The readout of an instrument by a direct, numerical reading of the measured value.

▫ dilute solution

A solution that has been made weaker usually by the addition of water.

▫ dimictic (die-mick-tick)

Lakes and reservoirs which freeze over and normally go through two stratification and two mixing cycles within a year.

▫ direct current (dc)

Electrical current flowing in one direction only and essentially free from pulsation.

▫ direct filtration

A filtration method of treating water which consists of the addition of coagulant chemicals, flash mixing, coagulation, minimal flocculation, and filtration. The flocculation facilities may be omitted, but the physical chemical reactions will occur to some extent. The sedimentation process is omitted. Also see conventional filtration and in-line filtration.

▫ direct runoff

Water that flows over the ground surface or through the ground directly into streams, rivers, or lakes.

▫ discharge head

The pressure (in pounds per square inch or psi) measured at the centerline of a pump discharge and very close to the discharge flange, converted into feet.

▫ disinfectant

Any oxidant, including but not limited to chlorine, chlorine dioxide, chloramines, and ozone, that is added to water in any part of the treatment or distribution process and is intended to kill or inactivate pathogenic microorganisms.

▫ disinfectant contact time ("t" in ct calculations)

The time in minutes that it takes for water to move from the point of disinfectant application or the previous point of disinfectant residual measurement to a point before or at the point where residual disinfectant concentration (C) Is measured.

Where only one C is measured. T is the time in minutes that it takes for water to move from the point of disinfectant application to a point before or at where residual disinfectant concentration (C) is measured. Where more than one C is measured, T is (a) for the first measurement of C, the time in minutes that it takes for water to move from the first or only point of disinfectant application to a point before or at the point where the first C+ is measured and (b) for subsequent easurements of C, the time in minutes that it takes for water to move from the previous C measurement point to the C measurement point for which the particular T is being calculated- Disinfectant contact time in pipelines must be calculated based on plug flow by dividing the internal volume of the pipe by the maximum hourly flow rate through that pipe. Disinfectant contact time within mixing basins and storage reservoirs must be determined by tracer studies or an equivalent demonstration.

❑ **disinfection**

The process designed to kill most microorganisms in water, including essentially all pathogenic (disease- causing) bacteria. There are several ways to disinfect, with chlorine being most frequently used in water treatment.

❑ **disinfection by-product**

A compound formed by the reaction of a disinfectant such as chlorine with organic material in the water supply.

❑ **dissolved oxygen (do)**

Measure of water quality indicating free oxygen dissolved in water.

❑ **distillate (dis-tuh-late)**

In the distillation of a sample, a portion is evaporated; the part that is condensed afterwards is the distillate.

❑ **divalent (die-vay-lent)**

Having a valence of two, such as the ferrous ion, Fe_2+.

❑ **diversion**

1) Use of part of a stream flow as a water supply.
2) A structural conveyance (or ditch) constructed across a slope to intercept runoff flowing down a hillside, and divert it to some convenient discharge point.

❑ **domestic or other non-distribution system plumbing problem**

A coliform contamination problem in a public water system with more than one service connection that is limited to the specific service connection from which the coliform positive sample was taken.

❑ **dosage**

The quantity of a chemical administered to an organism.

❑ dose

The actual quantity of a chemical to which an organism is exposed. See absorbed dose.

❑ dose equivalent

The product of the absorbed dose from ionizing radiation and such factors as account for differences In biological effectiveness due to the " of radiation and is distribution in the body as specified by the International Commission on Radiological Units and Measurements (ICRU).

❑ dose-response

A quantitative relationship between the dose of a chemical and an effect caused by the chemical.

❑ dose-response curve

A graphical presentation of the relationship between degree of exposure to a chemical (dose) and observed biological effect or response.

❑ dose-response evaluation

A component of risk assessment that describes the quantitative relationship between the amount of exposure to a substance and the extent of toxic injury or disease.

❑ dose-response relationship

The quantitative relationship between the amount of exposure to a substance and the extent of toxic injury produced.

❑ downgradients

The direction that ground water flows; similar in concept to: downstream for surface water, such as a river.

❑ dpd (pronounce as separate letters)

A method of measuring the chlorine residual in water. The residual may be determined by either titrating or comparing a developed color with color standards. DPD stands for N,N-diethyl-p- phenylene-diamine.

❑ draft

1) The act of drawing or removing water from a tank or reservoir. 2) The water which is drawn or removed from a tank or reservoir.

❑ drainage

A technique to improve the productivity of some agricultural land by removing excess water from the soil; surface drainage is accomplished with open ditches; subsurface drainage uses porous conduits (drain tile) buried beneath the soil surface.

❑ drainage basin

The area of land that drains water, sediment, and dissolved materials to a common outlet at some point along a stream channel.

❐ drawdown

1) The drop in the water table or level of water in the ground when water is being pumped from a well.
2) The amount of water used from a tank or reservoir. 3) The drop in the water level of a tank or reservoir.

❐ dwel (drinking water equivalent level)

Estimated exposure (in mg/L) which is interpreted to be protective for non carcinogenic endpoints of toxicity over a lifetime of exposure. DWEL was developed for chemicals that have a significant carcinogenic potential (Group B). Provides risk manager with evaluation on non-cancer endpoints, but infers that carcinogenicity should be considered the toxic effect of greatest concern.

❐ dynamic pressure

When a pump is operating, the vertical distance (in feet) from a reference point (such as a pump centerline) to the hydraulic grade line is the dynamic head.

E

❐ eductor (e-duck-ter)

A hydraulic device used to create a negative pressure (suction) by forcing a liquid through a restriction, such as a Venturi. An eductor or aspirator (the hydraulic device) may be used in the laboratory in place of a vacuum pump. As an injector, it is used to produce vacuum for chlorinators.

❐ effective corrosion inhibitor residual

A concentration of corrosion inhibitor sufficient to form a protective coating on the interior walls of a pipe, reducing its corrosion.

❐ effective range

That portion of the design range (usually upper 90 percent) in which an instrument has acceptable accuracy.

❐ effective size (es)

The diameter of the particles in a granular sample (filter media) for which 10 percent of the total grains are smaller and 90 percent larger on a weight basis. Effective size is obtained by passing granular material through sieves with varying dimensions of mesh and weighing the material retained by each sieve. The effective size is also approximately the average size of the grains.

❐ effluent (ef-loo-ent)

Water or some other liquid-raw, partially or completely treated-flowing from a reservoir, basin, treatment process or treatment plant.

❐ ejector

A device used to disperse a chemical solution into water being treated.

❑ electrochemical reaction

Chemical changes produced by electricity (electrolysis) or the production of electricity by chemical changes (galvanic action). In corrosion, a chemical reaction is accompanied by the flow of electrons through a metallic path. The electron flow may come from. An external force and cause the reaction, such as electrolysis caused by a D.C. (direct current) electric railway or the electron flow may be caused by a chemical reaction as in the galvanic action of a flashlight dry cell.

❑ electrochemical sries

A list of metals with the standard electrode potentials given in volts. The size and sip of the electrode potential indicates how easily these elements will take on or give up electrons, or corrode. Hydrogen is conventionally assigned a value of zero.

❑ electrolysis (ee-leck-trawl-us-sis)

The decomposition of material by an outside electrical current.

❑ electrolyte (ee-leck-tro-light)

A substance which dissociates (separates) into two or more ions when it is dissolved in water.

❑ electrolytic cell (ee-leck-tro-lit-ick)

A device in which the chemical decomposition of material causes an electric current to flow. Also, a device in which a chemical reaction occurs as a result of the flow of electric current. Chlorine and caustic (NaOH) are made from salt (NACl in electrolytic cells.

❑ electromotive force (emf)

The electrical pressure available to cause a flow of current (amperage) when an electrical circuit is closed.

❑ electromotive series

A list of metals and alloys presented in the order of their tendency to con-ode (or go into solution). Also called the Galvanic Series. This is a practical application of the theoretical ELECTROCHEMICAL SERIES.

❑ electron

An extremely small, negatively charged particle; the part of an atom that determines its chemical properties.

❑ element

A substance which cannot be separated into its constituent parts and still retain its chemical identity. For example, sodium (Na) is an element.

❑ end bells

Devices used to hold the rotor and stator of a motor in position.

❐ end point

Samples are titrated to the end point. This means that a chemical is added, drop by drop, to a sample until a certain color change (blue to clear, for example) occurs. This is called the END POINT of the titration. In addition to a color change, an end point may be reached by the formation of a precipitate or the reaching of a specified pH. An end point may be detected by the use of an electronic device such as a pH meter.

❐ endangerment assessment

A site-specific risk assessment of the actual or potential danger to human health or welfare and the environment from the release of hazardous substances or waste. The endangerment assessment document is prepared in support of enforcement actions under CERCLA or RCRA.

❐ endemic (en-dem-ick)

Something peculiar to a particular people or locality, such as a disease which is always present in the population.

❐ endrin (en-drin)

A pesticide toxic to freshwater and marine aquatic life that produces adverse health effects in domestic water supplies.

❐ energy grade line (egl)

A line that represents the elevation of energy head of water flowing in a pipe, conduit or channel. The line is drawn above the hydraulic grade line (gradient) a distance equal to the velocity head (V2/2g) of the water flowing at each section or point along the pipe or channel. Also see hydraulic gradeline.

❐ enteric

Of intestinal origin, especially applied to wastes or bacteria.

❐ entrain

To trap bubbles in water either mechanically through turbulence or chemically through a reaction.

❐ enzymes (en-zimes)

Organic substances (produced by living organisms) which cause or speed up chemical reactions. Organic catalysts and/or biochemical catalysts.

❐ epidemic

Widespread outbreak of a disease, or a large number of cases of a disease in a single community or relatively small area. Disease may spread from person to person, and/or by the exposure of many persons to a single source, such as a water supply.

❐ epidemiologic study

Study of human populations to identify causes of disease. Such studies often compare the health status of a group of persons who have been exposed to a suspect agent with that of a comparable non-exposed group.

❒ epidemiology (ep-uh-de-me-all-o-gee)

A branch of medicine which studies epidemics (diseases which affect significant numbers of people during the same time period in the same locality). The objective of epidemiology is to determine the factors that cause epidemic diseases and how to prevent them.

❒ epilimnion (ep-ub-lim-knee-on)

The upper layer of water in a thermally stratified lake or reservoir. This layer consists of the warmest water and has a fairly uniform (constant) temperature. The layer is readily mixed by wind action.

❒ erosion

Wearing away of soil by timing water, wind, or ice; erosion is the process by which the earth's surface is shaped and occurs even in remote, uninhabited areas at a slow rate (geologic erosion); of more concern is accelerated erosion caused by people's activities.

❒ ester

A compound formed by the reaction between an acid and an alcohol with the elimination of a molecule of water.

❒ eutrophic (you-tro-fick)

Reservoirs and lakes which are rich in nutrients and very productive in terms of aquatic animal and plant life.

❒ eutrophication (you-tro-fi-kay-shun)

The increase in the nutrient levels of a lake or other body of water; this usually causes an increase in the growth of aquatic animal and plant life.

❒ evaporation

The process by which water or other liquid becomes a gas (water vapor or ammonia vapor). Water from land areas, bodies of water, and all other moist surfaces is absorbed into the atmosphere as a vapor.

❒ evapotranspiration (ee-vap-o-trans-purr-a-shun)

The combined processes of evaporation and transpiration. It can be defined as the sum of water used by vegetation and water lost by evaporation.

❒ exemption

A State with primacy may relieve a public water system from a requirement respecting an MCL, treatment technique or both, by granting an exemption if certain conditions exist. These are: 1) the system cannot comply with a MCL or treatment technique due to compelling factors which may include economic factors; 2) the system was in operation on the effective date of the MCL or treatment technique requirement; and 3) the exemption will not result in an unreasonable public health risk.

❑ exposure

Contact with a chemical or physical agent.

❑ exposure assessment

The determination or estimation (qualitative or quantitative) of the magnitude, frequency, duration, route, and extent (number of people) of exposure to a chemical.

❑ exposure coefficient

Term which combines information on the frequency, mode, and magnitude of contact with contaminated medium to yield a quantitative value of the amount of contaminated medium contacted per day.

❑ exposure level (chemical)

The amount (concentration) of a chemical at the absorptive surfaces of an organism.

❑ exposure scenario

A set of conditions or assumptions about sources, exposure pathways, concentrations of toxic chemicals and populations (numbers, characteristics and habits) which aid the investigator in evaluating and quantifying exposure in a given situation.

❑ extrapolation

Estimation of unknown values by extending or projecting from known values.

F

❑ facultative (fack-ul-tay-tive)

Facultative bacteria can use either molecular (dissolved) oxygen or oxygen obtained from food material such as sulfate or nitrate ions. In other words, facultative bacteria can live under aerobic or anaerobic conditions.

❑ fecal coliform bacteria

Bacteria found in the intestinal tracts of animals. Their presence in water or sludge is an indicator of pollution and possible contamination by pathogens.

❑ feedback

The circulating action between a sensor measuring a process variable and the controller which controls or adjusts the process variable .

❑ filtration

A process for removing particulate matter from water by passage through porous media.

❑ finished water

Water that has passed through a water treatment plant; all the treatment processes are completed or "finished". This water is ready to be delivered to consumers.

❐ first draw sample

A one-liter sample of tap water, collected in accordance with CFR Section 141.86(b)(2), that has been standing in plumbing pipes at least 6 hours and is collected without flushing the tap.

❐ first draw

The water that immediately comes out when a tap is first opened. This water is likely to have the highest level of lead contamination from plumbing materials.

❐ fix, sample

A sample is fixed in the field by adding chemicals that prevent the water quality indicators of interest in the sample from changing before final measurements are performed later in the lab.

❐ flagellates (flaj-el-lates)

Microorganisms that move by the action of tail-like projections.

❐ flame polished

Melted by a flame to smooth out irregularities. Sharp or broken edges of glass (such as the end of a glass tube) are rotated in aflame until the edge melts slightly and becomes smooth.

❐ floc

Clumps of bacteria and particulate impurities that have come together and formed a cluster. Found in flocculation tanks and settling or sedimentation basins.

❐ flocculation

The gathering together of fine particles in water by gentle mixing after the addition of coagulant chemicals to form larger particles.

❐ floodplain

The flat or nearly flat land on the floor of a steam valley or tidal area that is covered by water during floods.

❐ fluidized (flew-id-i-zd)

A mass of solid particles that is made to flow like a liquid by injection of water or gas is said to have been fluidized. In water treatment, a bed of filter media is fluidized by backwashing water through the filter.

❐ fluoridation (floor-uh-day-shun)

The addition of a chemical to increase the concentration of fluoride ions in drinking water to a predetermined optimum limit to reduce the incidence (number) of dental caries (tooth decay) in children. Defluoridation is the removal of excess fluoride in drinking water to prevent the mottling (brown stains) of teeth.

❐ fluorosis

An abnormal condition caused by excessive intake of fluorine, characterized chiefly by mottling of the teeth.

❐ flush

1) To open a cold-water tap to clear out all the water which may have been sitting for a long time in the pipes. m new homes, to flush a system means to send large volumes of water gushing through the unused pipes to remove loose particles of solder and flux. 2) To force large amounts of water through liquid to clean out piping or tubing. storage or process tanks.

❐ flushing

A method used to clean water distribution lines. Hydrants are opened and water with a high velocity flows through the pipes, removes deposits from the pipes, and flows out the hydrants.

❐ foot valve

A special type of check valve located at the bottom end of the suction pipe on a pump. This valve opens when the pump operates to allow water to enter the suction pipe but closes when the pump shuts off to prevent water from flowing out of the suction pipe.

❐ formation

A group of similar consolidation (that is, relatively solid) rocks of unconsolidated (that is, relatively loose) minerals.

❐ free available residual chlorine

That portion of the total available residual chlorine composed of dissolved chlorine gas cl_2), hypochlorous acid (HOCl), and/or hypochlorite ion (OCl-) remaining in water after chlorination. This does not include chlorine that has combined with ammonia, nitrogen, or other compounds.

❐ free residual chlorination

The application of chlorine to water to produce a free available chlorine residual equal to at least 80 percent of the total residual chlorine (sum of free and combined available chlorine residual).

❐ freeboard

1) The vertical distance from the normal water surface to the top of the confining wall.
2) The vertical distance from the sand surface to the underside of a trough in a sand filter. This distance is also called AVAILABLE EXPANSION

❐ fresh water

Water that generally contains less than 1,000 milligrams-per-liter of dissolved solids.

❐ friction losses

The head, pressure or energy (they are the same) lost by water flowing in a pipe or channel as a result of turbulence caused by the velocity of the flowing water and the roughness of the pipe. channel walls, and restrictions caused by fittings. Water flowing in a pipe loses pressure or energy as a result of friction losses.

❏ fungi (fun-ji)

Mushrooms, molds, mildews, rusts, and smuts that are small non-chlorophyll-bearing plants lacking roots, stems and leaves. They occur in natural waters and grow best in the absence of light. Their decomposition may cause objectionable tastes and odors in water.

G

❏ gage pressure

The pressure within a closed container or pipe as measured with a gage. In contrast, absolute pressure is the sum of atmospheric pressure (14.7 lbs/sq in) PLUS pressure within a vessel (as measured by a gage). Most pressure gages read in gage pressure or psig (pounds per square inch gage pressure).

❏ galvanic series

A list of metals and alloys presented in the order of their tendency to corrode (or go into solution). Also called the ELECTROMOTIVE SERIES. This is a practical application of the theoretical ELECTROCHEMICAL SERIES.

❏ galvanize

To coat a metal (especially iron or steel) with zinc. Galvanization is the process of coating a metal with zinc.

❏ galvnic call

An electrolytic cell capable of producing electrical energy by electrochemical action. The decomposition of materials in the cell causes an electric (electron) current to flow from cathode to anode.

❏ garnet (gar-nit)

A group of hard, reddish, glassy, mineral sands made up of silicates of base metals (calcium, magnesium, iron and manganese). Garnet has a higher density than sand.

❏ gastroenteritis

An inflammation of the stomach and intestine resulting in diarrhea, with vomiting and cramps when irritation is excessive. When caused by an infectious agent, it is often associated with fever.

❏ gauge, pipe

A number that defines the thickness of the sheet used to make steel pipe. The larger the number, the thinner the pipe wall.

❏ gavage

Type of exposure in which a substance is administered to an animal through a stomach tube.

❐ geological log

A detailed description of all underground features discovered during the drilling of a well (depth, thickness and type of formations).

❐ geophysical log

A record of the structure and composition of the earth encountered when drilling a well or similar type of test hole or boring.

❐ germicide (germ-uh-side)

A substance formulated to kill germs or microorganisms. The germicidal properties of chlorine make it an effective disinfectant.

❐ giardia lamblia

Flagellate protozoan which is shed during its cyst stage into the feces of man and animals. When water containing these cysts is ingested, the protozoan causes a severe gastrointestinal disease called giardiasis.

❐ giardiasis (gee-are-dye-us-sis)

Intestinal disease caused by an infestation of Giardia flagellates.

❐ glass, pipe and fittings

The working pressure rating of a specific pipe for use in water distribution systems which includes allowances for surges. This term is used for cast iron, ductile iron, asbestos cement and some plastic pipe.

❐ gooseneck

A portion of a service connection between the distribution system water main and a meter. Sometimes called a pigtail.

❐ grab sample

A single sample collected at a particular time and place which represents the composition of the water only at that time and place.

❐ grade

1) The elevation of the invert of the bottom of a pipeline. canal, culvert or similar conduit.
2) The inclination or slope of a pipeline, conduit, stream channel, or natural ground surface; usually expressed in terms of the ratio or percentage of number of units of vertical rise or fall per unit of horizontal distance. A 0.5 percent grade would be a drop of one-half foot per hundred feet of pipe.

❐ gram

A unit of mass equivalent to one milliliter of water at 4 degrees Celsius. 1/454 of a pound.

❐ gravimetric

A means of measuring unknown concentrations of water quality indicators in a sample by WEIGHING a precipitate or residue of the sample.

❐ grey water

Wastewater other than sewage, such as sink drainage or washing machine discharge.

❐ gross alpha particle activity

The total radioactivity due to alpha particle emission as inferred from measurements on a dry sample.

❐ gross beta particle activity

The total radioactivity due to beta particle emission as inferred from measurements on a dry sample.

❐ ground water

The supply of fresh water found beneath the Earth's surface. usually in aquifers. which is often used for supplying wells and springs. Because ground water is a major source of drinking water there is growing concern over areas where leaching agricultural or industrial pollutants or substances from leaking underground storage tanks are contaminating ground water.

❐ ground water under the direct influence (udi) of surface water

Any water beneath the surface of the ground with: 1) significant occurrence of Insects or other macroorganisms algae. or large-diameter pathogens such as Giardia lamblia or, 2) significant and relatively rapid shifts in water characteristics such as turbidity, temperature, conductivity, or pH which closely correlate to climatological or surface water conditions. Direct influence must be determined for individual sources in accordance with criteria established by the State. The State determination of direct influence may be based on site-specific measurements of water quality and/ or documentation of well construction characteristics and geology with field evaluation.

H

❐ half-life

The length of time required for the mass, concentration, or activity of a chemical or physical agent to be reduced by one-half.

❐ halogen

One of the chemical elements chlorine, bromine, or iodine.

❐ hard water

Alkaline water containing dissolved salts that interfere with some industrial processes and prevent soap from lathering. Water may be considered hard if it has a hardness greater than the typical hardness of water from the region. Some textbooks define hard water as water with a hardness of more than 100 mgAL as calcium carbonate.

❐ hardness, water

A characteristic of water caused mainly by the salts of calcium and magnesium, such as bicarbonate, carbonate, sulfate, chloride and nitrate. Excessive hardness in water is undesirable because it causes the formation of soap curds, increased use of soap, deposition of scale in boilers, damage in some industrial processes, and sometimes causes objectionable tastes in drinking water.

❐ hazard evaluation

A component of risk assessment that involves gathering and evaluating data on the types of health injury or disease (e.g., cancer) that may be produced by a chemical and on the conditions of exposure under which injury or disease is produced.

❐ head

The vertical distance (in feet) equal to the pressure (in psi) at a specific point. The pressure head is equal to the pressure in psi times 2.31 ft/psi.

❐ head loss

The head, pressure or energy (they are the same) lost by water flowing in a pipe or channel as a result of turbulence caused by the velocity of the flowing water and the roughness of the pipe, channel walls or restrictions caused by fittings. Water flowing in a pipe loses head, pressure or energy as a result of friction losses.

❐ header

A large pipe to which a series of smaller pipes are connected. Also called a MANIFOLD.

❐ heat sensor

A device that opens and closes a switch in response to changes in the temperature. This device might be a metal contact, or a thermocouple which generates a minute electrical current proportional to the difference in heat, or a variable resistor whose value changes in response to changes in temperature.

❐ heavy metals

Metallic elements with high atomic weights, e.g., mercury, chromium, cadmium, arsenic, and lead. They can damage living things at low concentrations and tend to accumulate in the food chain.

❐ hectare (heck-tar)

A measure of area in the metric system similar to an acre. One hectare is equal to 10,000 square meters and 2.4711 acres.

❐ hematopoiesis

The production of blood and blood cells; hemopoiesis.

❐ hepatitis (hep-uh-tie-this)

Hepatitis is an inflammation of the liver usually caused by an acute viral infection. Yellow jaundice is one symptom of hepatitis.

□ **hepatoma**

A malignant tumor occurring in the liver.

□ **herbicide (herb-uh-side)**

A compound, usually a man- made organic chemical, used to kill or control plant growth.

□ **hertz**

The number of complete electromagnetic cycles or waves in one second of an electrical or electronic circuit. Also called the frequency of the current. Abbreviated Hz.

□ **heterotrophic microorganisms**

Bacteria and other microorganisms that use organic matter synthesized by other organisms for energy and growth.

□ **heterotrophic plate count (hpc)**

The number of colonies of heterotrophic bacteria grown on selected solid media at a given temperature and incubation period, usually expressed in number of bacteria per milliliter of sample.

□ **high-line jumpers**

Pipes or hoses connected to fire hydrants and laid on top of the ground to provide emergency water service for an isolated portion of a distribution system.

□ **high-to-low-dose extrapolation**

The process of prediction of low exposure risks to rodents from the measured high exposure-high risk data.

□ **histology**

The study of the structure of cells and tissues; usually involves microscopic examination of tissue slices.

□ **hose bib**

Faucet. A location in a water line where a hose is connected.

□ **human equivalent dose**

A dose which, when administered to humans, produces an effect equal to that produced by a dose in animals.

□ **human exposure evaluation**

A component of risk assessment that involves describing the nature and size of the population exposed to a substance and the magnitude and duration of their exposure. The evaluation could concern past exposures, current exposures, or anticipated exposures.

□ **human health risk**

The likelihood (or probability) that a given exposure or series of exposures may have or will damage the health of individuals experiencing the exposures.

▫ humus

Organic portion of the soil remaining after prolonged microbial decomposition.

▫ hydrated lime

Limestone that has been burned and treated with water under controlled conditions until the calcium oxide portion has been converted to calcium hydroxide ($Ca(OH)_2$). Hydrated lime is quicklime combined with water. $CaO + H_2O \rightarrow Ca(OH)_2$. Also see quicklime.

▫ hydraulic grade line

The surface or profile of water flowing of hydraulic gradient The slope of the hydraulic grade line.

▫ hydraulic gradient

The slope of the hydraulic grade line This is the slope of the water surface in an open channel, the slope of the water surface of the groundwater table, or the slope of the water pressure for pipes under pressure.

▫ hydrochlorination (hi-poe-klor-uh-nay-shun)

The application of hypochlorite compounds to water for the purpose of disinfection.

▫ hydrochlorinators (hi-poe-klor-uh-nay-tors)

Chlorine pumps, chemical feed pumps or devices used to dispense chlorine solutions made from hypochlorites such as bleach (sodium hypochlorite) or calcium hypochlorite into the water being treated.

▫ hydrogeologic conditions

Conditions stemming from the interaction of ground water and the surrounding soil and rock.

▫ hydrogeologic cycle

The natural process recycling water from the atmosphere down to (and through) the earth and back to the atmosphere again.

▫ hydrogeologist (hi-dro-gee-all-uh-gist)

A person who studies and works with groundwater.

▫ hydrogeology

The geology of ground water, with particular emphasis on the chemistry and movement of water.

▫ hydrograph

A graph of the rate of runoff plotted against time for a point on a channel.

▫ hydrologic cycle (hi-dro-loj-ick)

Movement or exchange of water between the atmosphere and the earth.

□ **hydrology**

The study of the occurrence, distribution and circulation of the natural waters of the earth.

□ **hydrolysis (hi-droll-uh-sis)**

A chemical reaction in which a compound is converted into another compound by taking up water.

□ **hydrophilic (hi-dro-fill-ick)**

Having a strong affinity (liking) for water. The opposite of hydrophobic.

□ **hydrophobic (hi-dro-foe-bick)**

Having a strong aversion (dislike) for water. The opposite of hydrophilic.

□ **hydropneumatic (hi-dro-new-mat-ick)**

A water system, usually small, in which a water pump is automatically controlled (started and stopped) by the air pressure in a compressed-air tank.

□ **hydrostatic pressure (hi-dro-stat-ick)**

1) The pressure at a specific elevation exerted by a body of water at rest or,
2) In the case of groundwater, the pressure at a specific elevation due to the weight of water at higher levels in the same zone of saturation.

□ **hypochlorite (hi-poe-klor-ite)**

Chemical compounds containing available chlorine; used for disinfection. They are available as liquids (bleach) or solids (powder, granules and pellets).

□ **hypolimnion (hi-poe-lim-knee-on)**

The lowest layer in a thermally stratified lake or reservoir. This layer consists of colder, more dense water, has a constant temperature and no mixing occurs.

I

□ **imhoff cone**

A clear. cone-shaped container marked with graduations. The cone is used to measure the volume of settleable solids in a specific volume (usually one liter) of water.

□ **impeller**

A rotating set of vanes in a pump designed to pump or lift water.

□ **impermeable (im-purr-me-uh-bull)**

Not easily penetrated. The property of a material or soil that des not allow, or allows only with great difficulty, the movement or passage of water.

□ **in situ**

In place, the original location, in the natural environment.

▫ in vitro

In glass; a laboratory experiment performed in a test tube or other vessel.

▫ in vitro studies

Studies of-chemical effects conducted in tissues, cells or subcellular extracts from an organism (i.e., not in the living organism).

▫ in vivo

With in a living organism; a laboratory experiment performed in which the substance under study is inserted into a living organism.

▫ in vivo studies

Studies of chemical effects conducted in intact living organisms.

▫ incidence of tumors

Percentage of animals with tumors.

▫ indicator (chemical)

A substance that gives a visible change, usually of color, at a desired point in a chemical reaction, generally at a specified end point.

▫ indicator (instrument)

A device which indicates the result of a measurement. Most indicators in the water utility field use either a fixed scale and movable indicator (pointer) such as a pressure gage or a movable scale and movable indicator like those used on a circular-flow recording chart.

▫ infiltration

1) The gradual flow or movement of water into and through (to percolate or pass through) the pores of the soil. Also see percolation.
2) The penetration of water from the soil into sewer or other pipes through defective joints, connections or manhole walls.

▫ infiltration gallery

A subsurface groundwater collection system, typically shallow in depth, constructed with open jointed or perforated pipes that discharge collected water into a water-tight chamber. From this chamber the water is pumped to treatment facilities and into the distribution system. Infiltration galleries are usually located close to streams or ponds and may be under the direct influence of surface water.

▫ infiltration rate

Quantity of water (usually measured in inches) that will enter a particular type of soil per unit time (usually one hour).

▫ influent (in-flu-ent)

Water or other liquid-raw or partially flowing INTO a reservoir, basin, treatment process or treatment plant.

- **ingestion**

Type of exposure through the mouth.

- **inhalation**

Type of exposure through the lungs.

- **initial compliance period**

The first full three-year compliance period which begins at least 18 months after promulgation.

- **in-line filtration**

The addition of chemical coagulants directly to the filter inlet pipe. The chemicals are mixed by the flowing water. Flocculation and sedimentation facilities are eliminated. This pretreatment method is commonly used in pressure filter installations.

- **inorganic**

Material such as sand, salt, iron, calcium salts and other mineral materials. Inorganic substances are of mineral origin, whereas organic substances are usually of animal or plant origin.

- **input horsepower**

The total power used in operating a pump and motor.

Input HP = (Brake HP)(100%)Motor Efficiency, %)

- **insecticide**

Any substance or chemical formulated to kill or control insects.

- **instream uses**

Water uses that can be carried out without removing the water from its source, as in navigation and recreation.

- **integrated exposure assessment**

A summation over time, in all media, of the magnitude of exposure to a toxic chemical.

- **integrator**

A device or meter that continuously measures and calculates (adds) total flows in gallons, or million cubic feet. or some other unit of volume measurement.

- **interface**

The common boundary layer between two substances such as water and a solid (metal); or between two fluids such as water and a gas (air); or between a liquid (water) and another liquid (oil).

- **interflow**

Lateral movement of water in the upper layer of soil.

❐ interlock circuit exposed to service

An electrical switch, usually magnetically operated. Used to interrupt all (local) power to a panel or device when the door is opened or the

❐ internal friction

Friction within a fluid (water) due to cohesive forces.

❐ interspecies extrapolation model

Model used to extrapolate from results observed in laboratory animals to humans.

❐ interstate carrier

Any vehicle or transport which conveys passengers in interstate commerce.

❐ interstice (in-tur-stuhz)

A very small open space in a rock or granular material. Also called a void or void space.

❐ invert

The lowest point of the channel inside a pipe, conduit, or canal.

❐ ion

An electrically charged atom, radical (such as SO_{42}-), or molecule formed by the loss or gain of one or more electrons.

❐ ionic concentration

The concentration of any ion in solution, usually expressed in moles per liter.

❐ ionization (eye-on-uh-zay-shun)

The splitting or dissociation (separation) of molecules into negatively and positively charged ions.

❐ irreversible effect

Effect characterized by the inability of the body to partially or fully repair injury caused by a toxic agent.

J

❐ jar test

A laboratory procedure that simulates a water treatment plant's coagulation/ flocculation units with differing chemical doses and also energy of rapid mix, energy of slow mix, and settling time. The purpose of this procedure is to ESTIMATE the minimum or ideal coagulant dose required to achieve certain water quality goals. Samples of water to be treated are commonly placed in six jars. Various amounts of chemicals are added to each jar, and the settling of solids is observed. The dose of chemicals that provides satisfactory settling removal of turbidity and/or color is the dose used to treat the water being taken into the plant at that time. When evaluating the results of a jar test, the operator should also consider the floc quality in the flocculation area and the floc loading on the filter.

❐ jogging

The frequent starting and stopping of an electric motor.

❐ joule (jewel)

A measure of energy, work or quantity of heat. One joule is the work done when a force of one newton is displaced a distance of one meter in the direction of force.

K

❐ kilo

1) Kilogram.
2) Kilometer.
3) A prefix meaning "thousand" used in the metric system and other scientific systems of measurement.

❐ kinetic energy

Energy possessed by a moving body of matter, such as water, as a result of its motion.

❐ kjeldahl nitrogen (kell-doll)

Nitrogen in the form of organic proteins or their decomposition product ammonia, as measured by the Kjeldahl Method.

L

❐ landfill

Facility in which solid waste from municipal and/or industrial sources is disposed; sanitary landfills are those that are operated in accordance with environmental protection standards.

❐ langelier index (li)

An index reflecting the equilibrium pH of a water with respect to calcium and alkalinity. This index is used in stabilizing water to control both corrosion and the deposition of scale.

Langelier index = pH - pHs

where pH = actual pH of the water, and pHs= pH at which the water having the same alkalinity and calcium content is just saturated with calcium carbonate.

❐ large water system

A water system that serves more than 50,000 persons

❐ latency

Time from the first exposure to a chemical until the appearance of a toxic effect.

❐ laundering weir (lawn-der-ing weer)

Sedimentation basin overflow weir. A plate with V-notches along the top to assure a uniform flow rate and avoid short-circuiting.

❐ launders (lawn-ders)

Sedimentation basin and filter discharge channels, consisting of overflow weir plates (in sedimentation basins) and conveying troughs.

❐ lc50

The concentration of a chemical in air or water which is expected to cause death in 50% of test animals living in that air or water.

❐ ld50

The dose of a chemical taken by mouth or absorbed by the skin which is expected to cause death in 50% of the test animals so treated.

❐ leachate

A liquid that results from water collecting contaminants as it trickles through wastes, agricultural pesticides or fertilizers. Leaching may occur in farming areas, feedlots, and landfills, and may result in hazardous substances entering surface water, ground water, or soil.

❐ leaching

The process by which soluble substances are dissolved and transported down through the soil by recharge.

❐ lead service line

A service line made of lead which connects the water main to the building inlet and any lead pigtail, gooseneck or other fitting which is connected to such lead line.

❐ lead(pb)

A heavy metal that is hazardous to health if breathed or swallowed. Its use in gasoline. paints, and plumbing compounds has been sharply restricted or eliminated by federal laws and regulations.

❐ legionella

A genus of bacteria, some species of which have caused a type of pneumonia called Legionnaires Disease.

❐ lesion

A pathological or traumatic discontinuity of tissue or loss of function of a part.

❐ level controls

A float device (or pressure switch) which senses changes in a measured variable and opens or closes a switch in response to that change. In its simplest form, this control might be a floating ball connected mechanically to a switch or valve such as is used to stop water flow into a toilet when the tank is full.

❐ lifetime exposure

Total amount of exposure to a substance that a human would receive in a lifetime (usually assumed to be 70 years).

❐ lindane (lynn-dane)

A pesticide that causes adverse health effects in domestic water supplies and also is toxic to freshwater and marine aquatic life.

❐ linearity (lynn-ee-air-it-ee)

How closely an instrument measures actual values of a variable through its effective range; a measure used to determine the accuracy of an instrument.

❐ linearized multistage model

Derivation of the multistage model, where the data are assumed to be linear at low doses.

❐ littoral zone (lit-or-al)

1) That portion of a body of fresh water extending from the shoreline lakeward to the limit of occupancy of rooted plants.
2) The strip of land along the shoreline between the high and low water levels.

❐ loading

The quantity of a substance entering the environment (soil, water, or air).

❐ loael

Lowest-observed-adverse-effect level; the lowest dose in an experiment which produced an observable adverse effect.

❐ logarithm (log-a-rith-m)

The exponent that indicates the power to which a number must be raised to produce a given number. For example: if B2 = N, the 2 is the logarithm of N (to the base B), or 102 =100 and log10 100 = 2.

M

❐ macroscopic organisms (mack-row-skawp-ick)

Organisms big enough to be seen by the eye without the aid of a microscope.

❐ malignant

Very dangerous or virulent, causing or likely to cause death.

❐ managerial controls

Methods of nonpoint source pollution control that are derived from managerial decisions, such as changes in application times or rates for agrochemicals.

❐ manifold

A large pipe to which a series of smaller pipes are connected. Also called a HEADER.

❐ manometer (man-nah-mut-ter)

An instrument for measuring pressure. Usually, a manometer is a glass tube filled with a liquid that is used to measure the difference in pressure across a flow-measuring device such as an orifice or Venturi meter. The instrument used to measure blood pressure is a type of manometer.

❐ margin of safety (mos)

Maximum amount of exposure producing no measurable effect in animals (or studied humans) divided by the actual amount of human exposure in a population.

❐ mathematical model

Model used during risk assessment to perform extrapolations.

❐ maximum contaminant level (mcl)

The maximum permissible level of a contaminant in water which is delivered to the free flowing outlet of the ultimate user of a public water system, except in the case of turbidity where the maximum permissible level is measured at the point of entry to the distribution system. Contaminants added to the water under circumstances controlled by the user are excluded from this definition, except those contaminants resulting from the corrosion of piping and plumbing caused by water quality.

❐ maximum contaminant level goal (mclg)

The maximum level of a contaminant in drinking water at which no known or anticipated adverse effect on the health of persons would occur, and which allows an adequate margin of safety. Maximum contaminant level goals are non-enforceable health goals.

❐ maximum total trihalomethane potential (mttp)

The maximum concentration of total trihalomethanes produced in a given water containing a disinfectant residual, after 7 days at 25 degrees C or above.

❐ mbas

Methylene - Blue - Active Substances. These substances are used in surfactants or detergents.

❐ measured variable

A characteristic or component part that is sensed and quantified (reduced to a reading of some kind) by a primary element or sensor.

❑ mechanical joint

A flexible device that joins pipes or fittings together by the use of lugs and bolts.

❑ medium-size water system

A water system that serves greater than 3,300 and less than or equal to 50,000 person.

❑ meg

A procedure used for checking the insulation resistance on motors, feeders, buss bar systems, grounds, and branch circuit wiring.

❑ megger (from megohm)

An instrument used for checking the insulation resistance on motors, feeders, buss bar systems, grounds, and branch circuit wiring.

❑ megohm

Meg means one million, so 5 megohms means 5 million ohms. A megger reads in millions of ohms.

❑ meniscus (meh-nis-cuss)

The curved top of a column of liquid (water, oil, mercury) in a small tube. When the liquid wets the sides of the container (as with water), the curve forms a valley. When the confining sides are not wetted (as with mercury), the curve forms a hill or upward bulge.

❑ mesh

One of the openings or spaces in a screen or woven fabric. The value of the mesh is usually given as the number openings per inch. This value does not consider the diameter of the wire or fabric; therefore, the mesh number does not always have a definite relationship to the size of the hole.

❑ mesotrophic (mess-o-tro-rick)

Reservoirs and lakes which contain moderate quantities of nutrients and are moderately productive in terms of aquatic animal and plant life.

❑ metabolism (meh-tab-uh-liz-um)

The sum of the chemical reactions occurring within a cell or a whole organism; includes the energy-releasing breakdown of molecules (catabolism) and the synthesis of new molecules (anabolism).

❑ metabolite

Any product of metabolism, especially a transformed chemical.

❑ metalimnion (met-uh-lim-knee-on)

The middle layer in a thermally stratified lake or reservoir. In this layer there is a rapid decrease in temperature with depth. Also called the THERMOCLINE.

❐ metastatic

Pertaining to the transfer of disease from one organ or part to another not directly connected with it.

❐ methoxychlor (meth-oxy-klor)

A pesticide which causes adverse health effects in domestic water supplies and is also toxic to freshwater and marine aquatic life. The chemical name for methoxychlor is 2,2-bis (P-methoxyphenol)- 1, 1, l - trichloroethane.

❐ methyl orange alkalinity

A measure of the total alkalinity in a water sample. The alkalinity is measured by the amount of standard sulfuric acid required to lower the pH of the water to a pH level of 4.5, as indicated by the change in color of methyl orange from orange to pink. Methyl orange alkalinity is expressed as milligrams per liter equivalent calcium carbonate.

❐ microbial growth (my-krow-bee-ul)

The activity and growth of microorganisms such as bacteria, algae, diatoms, plankton and fungi.

❐ microgram (pg)

One-millionth of a gram (3.5 x 10-8 oz. 0.000000035 oz.).

❐ micrograms per liter (mg/l)

One microgram of a substance dissolved in each liter of water. This unit is equal to parts per billion (ppb) since one liter of water is equal in weight to one billion micrograms.

❐ micron (my-kron)

A unit of length. One millionth of a meter or one thousandth of a millimeter. One micron equals 0.00004 of an inch.

❐ microorganisms (my-crow-or-gan-is-zums)

Living organisms that can be seen individually only with the aid of a microscope.

❐ mil

A unit of length equal to 0.001 of an inch. The diameter of wires and tubing is measured in mils, as is the thickness of plastic sheeting.

❐ milligram (mg)

One-thousandth of a grain (3.5 x 10-1 oz. 0.000035 oz.).

❐ milligrams per liter (mg/l)

A measure of concentration of a dissolved substance. A concentration of one mg/L means that one milligram of a substance is dissolved in each liter of water. For practical purposes, this unit is equal to parts per million (ppm) since one liter of water is equal in weight to one million milligrams. Thus a liter of water containing

10 milligrams of calcium has 10 parts of calcium per one million parts of water, or 10 parts per million (10 ppm).

❑ millimicron (mill-uh-my-kron)

A unit of length equal to 10-3 microns (one thousandth of a micron), 10-6millimeters, or 10-9 meters; correctly called a manometer, nm.

❑ mineralization

The microbial conversion of an element from an organic to an inorganic state.

❑ modeling

Use of mathematical equations to simulate and predict real events and processes.

❑ molar or molarity

A molar solution consists of one gram molecular weight of a compound dissolved in enough water to make one liter of solution. A gram molecular weight is the molecular weight of a compound in grains. For example, the molecular weight of sulfuric acid (H_2SO_4) is 98. A one M solution of sulfuric acid would consist of 98 grains of H_2SO_4 dissolved in enough distilled water to make one liter of solution.

❑ mole

The molecular weight of a substance, usually expressed in grains.

❑ molecule (moll-uh-kule)

The smallest division of a compound that still retains or exhibits all the properties of the substance.

❑ monitoring wells

Wells used to collect ground-water samples for analysis to determine the amount, type, and spread of contaminants In ground water.

❑ monitoring

Measuring concentrations of substances in environmental media or in human or other biological tissues.

❑ monomer (mon-o-mer)

A molecule of low molecular weight capable of reacting with identical or different monomers to form Polymers.

❑ monomictic (mo-no-mick-tick)

Lakes and reservoirs which are relatively deep, do not freeze over during the winter months, and undergo a single stratification and mixing cycle during the year. These lakes and reservoirs usually become destratified during the mixing cycle, usually in the fall of the year.

❑ monovalent

Having a valence of one, such as the cuprous (copper) ion, Cu+.

❐ most probable number (mpn)

MPN is the Most Probable Number of coliform-group organisms per unit volume of sample water. Expressed as the number of organisms per 100 mL off sample water.

❐ motile (mo-till)

Capable of self-propelled movement. A term that is sometimes used to distinguish between certain types of organisms found in water.

❐ motor efficiency

The ratio of energy delivered by a motor to the energy supplied to it during a fixed period or cycle. Motor efficiency ratings will vary depending upon motor manufacturer and usually will range from 88.9 to 90.0 percent.

❐ mtd

Maximum tolerated dose, the dose that an animal species can tolerate for a major portion of its lifetime without significant impairment or toxic effect other than carcinogenicity.

❐ mudballs

Material that is approximately round in shape and varies from pea-sized up to two or more inches in diameter. This material forms in filters and gradually increases in size when not removed by the backwashing process.

❐ mulch

Any substance spread or allowed to remain on the soil surface to conserve soil moisture and shield soil particles from the erosive forces of raindrops and runoff.

❐ multiple use

Use of land for more than one purpose; i.e., grazing of livestock, wildlife production. recreation, watershed, and timber production. Could also apply to use of bodies of water for recreational purposes, fishing, and water supply.

❐ multi-stage model

Mathematical model based on the multi-stage theory of the carcinogenic process, which yields risk estimates either equal to or less than the one-hit model.

❐ multi-stage pump

A pump that has more than one impeller A single-stage pump has one impeller.

❐ municipal sewage

Wastes (mostly liquid) originating from a community; may be composed of domestic wastewaters and/or industrial wastewaters.

❐ mutagen

An agent that causes a permanent genetic change in a cell other than that which occurs during normal genetic recombination.

❐ mutagenicity

The capacity of a chemical or physical agent to cause permanent alteration of the genetic material within living cells.

N

❐ national pollutant discharge

Elimination System permit is the regulatory agency document issued by either a federal or state agency which is designed to control all discharges of pollutants from point sources in U.S. waterways. NPDES permits regulate discharges into navigable waters from all point sources of pollution, including industries, municipal treatment plants, large agricultural feed lots and return Irrigation flows.

❐ national secondary drinking water regulations

Commonly referred to as NSDWRs.

❐ near the first service connection

At one of the 20 percent of all service connections in the entire system that are nearest the water supply treatment facility, as measured by water transport time within the distribution system.

❐ necrosis

Death of cells or tissue.

❐ nematodes

Roundworms, any of which are pathogenic for plants and sometimes animals.

❐ neoplasm

An abnormal growth or tissue, as a tumor.

❐ nephelometric (neff-el-o-met-rick)

A means of measuring turbidity in a sample by using an instrument called a nephelometer. A nephelometer passes light through a sample and the amount of light deflected (usually at a 90- degree angle) is then measured.

❐ nephelometric turbidity unit (ntu)

The unit of measure for turbidity.

❐ neurotoxicity

Exerting a destructive or poisonous effect on nerve tissue.

❐ newton

A force which, when applied to a body having a mass of one kilogram, gives it an acceleration of one meter per second per second.

❐ niosh

The National Institute of Occupational Safety and Health is an organization that tests and approves safety equipment for particular applications. NIOSH is the primary Federal agency engaged in research in the national effort to eliminate on-the-job hazards to the health and safety of working people.

❐ nipdwr

National Interim Primary Drinking Water Regulations.

❐ nitrification inhibitor

A chemical that slows down the conversion of ammonium to nitrate nitrogen.

❐ nitrification

The biochemical transformation of ammonium nitrogen to nitrate nitrogen.

❐ nitrogen fixation

The biological or chemical process by which elemental nitrogen, from the air, is converted to organic or available nitrogen.

❐ nitrogenous (nye-trah-jen-us)

A term used to describe chemical compounds (usually organic) containing nitrogen in combined forms. Proteins and nitrates are nitrogenous compounds.

❐ noael

No-observed-adverse-effect level; the highest dose in an experiment which did not produce an observable adverse effect.

❐ noble metal

Chemically inactive metal (such as gold). A metal that does not corrode easily and is much scarcer (and more valuable) than the so-called useful or base metals.

❐ noel

No-observed-effect level; dose level at which no effects are noted.

❐ nominal diameter

An approximate measurement of the diameter of a pipe. Although the nominal diameter is used to describe the size or diameter of a pipe, it is usually not the exact inside diameter of the pipe.

❐ non-community water system (ncws)

A public water system that is not a community water system. There are two types of NCWSs: transient and non-transient.

❐ non-conventional pollutant

Any pollutant which is not a statutorily listed or which is poorly understood by the scientific community.

❐ non-ionic polymer (non-eye-on-ick)

A polymer that has no net electrical charge.

❐ non-point source

Pollution sources which are diffuse and do not have a single point of origin or are not introduced into a receiving stream from a specific outlet. The pollutants are generally carried off the land by stormwater runoff. The commonly used categories for non-point sources are: agriculture. forestry, urban, mining, construction, dams and channels, land disposal, and saltwater intrusion.

❐ non-potable (non-poe-tuh-bull)

Water that may contain objectionable pollution, contamination, minerals, or infective agents and is considered unsafe and/or unpalatable for drinking.

❐ normal

A normal solution contains one grain equivalent weight of reactant (compound) per. liter of solution. The equivalent weight of an acid is that weight which contains one gram atom of ionizable hydrogen or its chemical equivalent. For example, the equivalent weight of sulfuric acid (H2SO4 is 49 (98 divided by 2 because there are two replaceable hydrogen ions). A one N solution of sulfuric acid would consist of 49 grains of H2SO4 dissolved in enough water to make one liter of solution.

❐ nutrient

Any substance that is assimilated (taken in) by organisms and promotes growth. Nitrogen and phosphorous are nutrients which promote the growth of algae. There are other essential and trace elements which are also considered nutrients.

❐ nutrient pollution

Contamination of water resources by excessive inputs of nutrients; in surface waters, excess algal production is a major concern.

O

❐ odor threshold

The minimum odor of a water sample that can just be detected after successive dilutions with odorless water.

❐ offset (or droop)

The difference between the actual value and the desired value (or set point); characteristic of proportional controllers that do not incorporate reset action.

❐ offstream uses

Water withdrawn from surface or ground water sources for use at another place.

❐ ohm

The unit of electrical resistance. The resistance of a conductor in which one volt produces a current of one ampere.

❑ olfactory fatigue (oh-fak-tore-ee)

A condition in which a person's nose, after exposure to certain odors, is no longer able to detect the odor.

❑ oligotrophic (ah-lig-o-tro-flck)

Reservoirs and lakes which are nutrient poor and contain little aquatic plant or animal life.

❑ one-hit model

Mathematical model based on the biological theory that a single "hit" of some minimum critical amount of a carcinogen at a cellular target- namely, DNA-can initiate an irreversible series of events, eventually leading to a tumor.

❑ operation and maintenance costs

The ongoing, repetitive costs of operating a water system; for example, employee wages and costs for treatment chemicals and periodic equipment repairs.

❑ optimal corrosion control treatment drinking water regulations

The corrosion control treatment that minimizes the lead and copper concentrations at users' taps while insuring that the treatment does not cause the water system to violate any national primary

❑ oral

Of the mouth; through or by the mouth.

❑ organic

Substances that come from animal or plant sources. Organic substances always contain carbon. (Inorganic materials are chemical substances of mineral origin.).

❑ organics

1) A term used to refer to chemical compounds made from carbon molecules. These compounds may be natural materials (such as animal or plant sources) or man made materials (such as synthetic organics).
2) Any form of animal or plant life.

❑ organism

Any form of animal or plant life.

❑ orifice (or-uh-fiss)

An opening (hole) in a plate, wall or partition An orifice flange or plate placed in a pipe consists of a slot or a calibrated circular hole smaller than the pipe diameter. The difference in pressure in the pipe above and at the orifice may be used to determine the flow in the pipe.

❐ oxidation-reduction potential (orp)

The electrical potential required to transfer electrons from one compound or element (the oxidant) to another compound or element (the reductant); used as a qualitative measure of the state of oxidation in water treatment systems.

❐ orthotolidine (or-tho-tol-uh-dine)

Orthotolidine is a calorimetric indicator of chlorine residual. If chlorine is present, a yellow-colored compound is produced. This reagent is no longer approved for chemical analysis.

❐ osha (0-shuh)

The Williams-Steiger Occupational Safety and Health Act of 1970 (OSHA) is a law designed to protect the health and safety of industrial workers and also the operators of water supply systems and treatment plants. OSHA also refers to the federal and state agencies which administrator the OSHA regulations.

❐ osmosis (oz-moe-sis)

The passage of a liquid from a weak solution to a more concentrated solution across a semipermeable membrane. The membrane allows the passage of the solvent (water) but not the dissolved solids (solutes). This process tends to equalize the conditions on either side of the membrane.

❐ overall efficiency pump

The combined efficiency of a pump and motor together. Also called the WIRE-TO-WATER EFFICIENCY.

❐ overdraft

The pumping of water from a groundwater basin or aquifer in excess of the supply flowing into the basin. This pumping results in a depletion or "mining" of the groundwater in the basin.

❐ overflow rate

One of the guidelines for the design of settling tanks and clarifiers in treatment plants. Used by operators to determine if tanks and clarifiers are hydraulically (flow) over- or under loaded. Also called SURFACE LOADING. Overflow Rate (GDP/sq ft) = Flow (GPD)/Surface Area (sq ft)

❐ overturn

The almost spontaneous mixing of all layers of water in a reservoir or lake when the water temperature becomes similar from top to bottom. This may occur in the fall/ winter when the surface waters cool to the same temperature as the bottom waters and also in the spring when the surface waters warms after the ice melts.

❐ oxidation (ox-uh-day-shun)

Oxidation is the addition of oxygen, removal of hydrogen, or the removal of electrons from an element or compound. In the environment, organic matter is oxidized to more stable substances.

❐ oxidation-reduction potential

The electrical potential required to transfer electrons from one compound or element (the oxidant) to another compound or element (the reductant); used as a qualitative measure of the state of oxidation in water treatment systems.

❐ oxidizing agent

Any substance, such as oxygen (O_2) or chlorine (Cl_2), that will readily add (take on) electrons. The opposite is a reducing agent.

❐ ozonation (o-zoe-nay-shun)

The application of ozone to water for disinfection or for taste and odor control.

P

❐ packed tower aeration

A method of treating water to remove volatile organic chemical (VOCs) contaminants. As water is mixed with air, VOCs move from water to air which then passes through carbon filters to trap the contaminants.

❐ palatable (pal-a-ta-ble)

Water at a desirable temperature that is free from objectionable tastes, odors, colors, and turbidity. Pleasing to the senses.

❐ parshall flume

A device used to measure the flow in an open channel. The flume narrows to a throat of fixed dimensions and then expands again. The rate of flow can be calculated by measuring the difference in head (pressure) before and at the throat of the flume.

❐ particle count

The results of a microscopic examination of treated water with a special "particle counter" which classifies suspended particles by number and size.

❐ particulate (par-tick-you-let)

A very small solid suspended in water which can vary widely in size, shape, density, and electrical charge. Colloidal and dispersed particulates are artificially gathered together by the processes of coagulation and flocculation.

❐ partition coefficient

A measure of the extent to which a pesticide is divided between the soil and water phases.

❐ parts per million (ppm)

Parts per million parts, a measurement of concentration on a weight or volume basis. This term is equivalent to milligrams per liter (mg/L) which is the preferred term.

- **pascal**

The pressure or stress of one newton per square meter. (Abbreviated Pa)
1 psi = 6895 Pa = 6.895 kN/sq m = 0.0703 kg/sq cm

- **pathogenic organisms (path-o-jen-ick)**

Organisms, including bacteria, viruses or cysts, capable of causing diseases (typhoid, cholera, dysentery) in a host (such as a person). There are many types of organisms which do NOT cause disease. These organisms are called non-pathogenic.

- **pathogens**

Microorganisms that can cause disease in other organisms or in humans, animals and plants. They may be bacteria, viruses, or parasites and are found in sewage in runoff from animal farms or rural areas populated with domestic and/or wild animals, and in water used for swimming. Fish and shellfish contaminated by pathogens, or the contaminated water itself, can cause serious illnesses.

- **pathology**

The study of disease.

- **percent saturation**

The amount of a substance that is dissolved in a solution compared with the amount that could be dissolved in the solution, expressed as a percent.

- Amount of substance that is dissolved x 100%
- Amount that could be dissolved in solution

- **percolating water (purr-co-lay-ting)**

Water that passes through soil or rocks under the force of gravity.

- **percolation (purr-ko-lay-shun)**

1) The slow seepage of water into and through the ground.
2) The slow passage of water through a filter medium.

- **performance evaluation sample**

A reference sample provided to a laboratory for the purpose of demonstrating that the laboratory can successfully analyze the sample within limits of performance specified by the Agency. The true value of the concentration of the reference material is unknown to the laboratory at the time of the analysis.

- **periphyton (puh-rif-uh-tawn)**

Microscopic plants and animals that are firmly attached to solid surfaces under water such as rocks, logs, pilings and other structures.

- **permeability**

Generally used to refer to the ability of rock or soil to transmit water.

- **permeate (purr-me-ate)**

To penetrate and pass through, as water penetrates and passes through soil and other porous materials.

□ permissible dose

The dose of a chemical that may be received by an individual without the expectation of a significantly harmful result.

□ persistence

The resistance to degradation as measured by the period of time required for complete decomposition of a material.

□ person

An individual, corporation, company, association, partnership; municipality; or State, Federal, or tribal agency.

□ pesticide

Any substance or chemical designed or formulated to kill or control weeds or animal pests.

□ petroleum derivatives

Chemicals formed when gasoline breaks down in contact with ground water.

□ ph (pronounce as separate letters)

pH is an expression of the intensity of the basic or acid condition of a liquid. Mathematically, pH is the logarithm (base 10) of the reciprocal of the hydrogen ion concentration, [H+].

pH= Log (I/[H+])

The pH may range from 0 to 14, where 0 is most acid, 14 most basic, and 7 neutral. Natural waters usually have a pH between 6.5 and 8.5.

□ pharmacokinetics

The dynamic behavior of chemicals inside biological systems; it includes the processes of uptake, distribution, metabolism, and excretion.

□ phenolic compounds (fee-noll-lick)

Organic compounds that are derivatives of benzene.

□ phenolphthalein alkalinity (fee-nol-thay-leen)

The alkalinity in a water sample measured by the amount of standard acid required to lower the pH to a level of 8.3, as indicated by the change in color of phenolphthalein from pink to clear. Phenolphthalein alkalinity is expressed as milligrams per liter equivalent calcium carbonate.

□ photosynthesis (foe-tow-sin-thus-sis)

A process in which organisms, with the aid of chlorophyll (green plant enzyme), convert carbon dioxide and inorganic substances into oxygen and additional plant material, using sunlight for energy. All green plants grow by this process.

□ phytoplankton (flo-tow-pank-ton)

Small, usually microscopic plants (such as algae), found in lakes, reservoirs, and other bodies of water.

❒ pico

A prefix used in the metric system and other scientific systems of measurement which means 10-12 or 0.000000000001.

❒ picocurie (pci)

A measure of radioactivity. One picocurie of radioactivity is equivalent to 0.037 nuclear disintegrations per second.

❒ plan view

A diagram or photo showing a facility as it would appear when looking down on top of it.

❒ plankton

1) Small, usually microscopic, plants (phytoplank- ton) and animals (zooplankton) in aquatic systems.
2) All of the smaller floating, suspended or self-propelled organisms in a body of water.

❒ plug flow

A type of flow that occurs in tanks, basins or reactors when a slug of water moves through a tank without ever dispersing or mixing with the rest of the water flowing through the tank.

❒ plumes

The way polluted water extends downstream from the pollution source (analogous to smoke from a smoke- stack as it drifts downwind in the atmosphere).

❒ point of disinfectant application

The point where disinfectant is applied and water downstream of that point is not subject to recontamination by surface water runoff.

❒ point source

A stationery location or fixed facility from which pollutants are discharged or emitted. Also, any single identifiable source of pollution, e.g., a pipe, ditch, ship, ore pit, factory smokestack.

❒ point-of-entry treatment device

A treatment device applied to the drinking water entering a house or building for the purpose of reducing contaminants in the drinking water distributed throughout the house or building.

❒ point-of-use treatment device

A treatment device applied to a single tap used for the purpose of reducing contaminants in drinking water at that one tap.

❒ pole shader

A copper bar circling the laminated iron core inside the coil of a magnetic starter.

❐ pollutant

Generally, any substance introduced into the environment that adversely affects the usefulness of a resource.

❐ pollution

Generally, the presence of matter or energy whose nature, location or quantity produces undesired environmental effects. Under the Clean Water Act, for example, the term is defined as the man-made or man-induced alteration of the physical, biological, and radiological integrity of water.

❐ polyelectrolyte (polly-ee-leck-tro-lite)

A high- molecular-weight (relatively heavy) substance having points of positive or negative electrical charges that is formed by either natural or man-made processes. Natural polyelectrolytes may be of biological origin or derived from starch products and cellulose derivatives. Man-made polyelectrolytes consist of simple substances that have been made into complex substances of high molecular weight. Used with other chemical coagulants to aid in binding small suspended particles to larger chemical flocs for their removal from water. Often called a polymer.

❐ polymer

A chemical formed by the union of many monomers (a molecule of low molecular weight). Polymers are used with other chemical coagulants to aid in binding small suspended particles to larger chemical flocs for their removal from water. All polyelectrolytes are polymers, but not all polymers are polyelectrolytes.

❐ population at risk

A population subgroup that is more likely to be exposed to a chemical, or is more sensitive to a chemical, than is the general population.

❐ pore

A very small open space in a rock or granular material.

❐ porosity

the capacity of soil of rock to hold water.

❐ positive displacement pump

A type of piston, diaphragm, gear or screw pump that delivers a constant volume with each stroke. Positive displacement pumps are used as chemical solution feeders.

❐ postchlorination

The addition of chlorine to the plant effluent, FOLLOWING plant treatment, for disinfection purposes.

❐ potable water

Water that is safe and satisfactory for drinking and cooking.

❒ potency

Amount of material necessary to produce a given level of a deleterious effect.

❒ potentiation

The effect of one chemical to increase the effect of another chemical.

❒ potentiometric surface

The level to which water will rise in cased wells or other cased excavations into aquifers, measured as feet above mean sea level.

❒ power factor

The ratio of the true power passing through an electric circuit to the product of the voltage and amperage in the circuit. This is a measure of the lag or load of the current with respect to the voltage.

❒ ppb

Parts per billion. Also pg/L or micrograms per liter.

❒ ppm

Parts per million. Also mg/L or milligrams per liter.

❒ prechlorination

The addition of chlorine at the headworks of the plant PRIOR TO other treatment processes mainly for disinfection and control of tastes, odors and aquatic growths. Also applied to aid in coagulation and settling.

❒ precipitate (pre-sip-uh-tate)

1) An insoluble, finely divided substance which is a product of a chemical reaction within a liquid.
2) The separation from solution of an insoluble substance.

❒ precipitation (pre-sip-uh-tay-shun)

1) The process by which atmospheric moisture falls onto a land or water surface as rain, snow, hail, or other forms of moisture.
2) The chemical transformation of a substance in solution into an insoluble form (precipitate).

❒ precision

The ability of an instrument to measure a process variable and to repeatedly obtain the same result. The ability of an instrument to reproduce the same results.

❒ precursor, thm (pre-curse-or)

Natural organic compounds found in all surface and ground waters. These compounds MAY react with halogens (such as chlorine) to form trihalomethanes (try-HAL-o-METH-hanes) (THMs); they MUST be present in order for THMs to form.

▫ prescriptive (pre-skrip-tive)

Water rights which are acquired by diverting water and putting it to use in accordance with specified procedures. These procedures include filing a request to use unused water in a strewn, river or lake with a state agency.

▫ pressure control

A switch which operates on changes in pressure. Usually this is a diaphragm pressing against a spring. When the force on the diaphragm overcomes the spring pressure, the switch is actuated (activated).

▫ pressure head

The vertical distance (in feet) equal to the pressure (in psi) at a specific point. The pressure head is equal to the pressure in psi times 2.31 ft/psi.

▫ prevalence study

An epidemiological study which examines the relationships between diseases and exposures as they exist in a defined population at a particular point in time.

▫ primacy

The responsibility for ensuring that a law is implemented, and the authority to enforce a law and related regulations. A primacy agency has the primary responsibility for administrating and enforcing regulations.

▫ primary element

An instrument which measures (senses) a physical condition or variable of interest. Floats and thermocouples are examples of primary elements.

▫ prime

The action of filling a pump casing with water to remove the air. Most pumps must be primed before startup or they will not pump any water.

▫ prior appropriation

A doctrine of water law that allocates the right to use water on a first-come first-serve basis.

▫ process variable

A physical or chemical quantity which is usually measured and controlled in the operation of a water treatment plant or an industrial plant.

▫ product water

Water that has passed through a water treatment plant. All the treatment processes are completed or finished. This water is the product from the water treatment plant and is ready to be delivered to the consumers. Also called finished water.

▫ profile

A drawing showing elevation plotted against distance, such as the vertical section or side view of a pipeline.

❒ prospective study disease

An epidemiological study which examines the development of disease in a group of persons determined to be presently free of the

❒ prussian blue

A blue paste or liquid (often on a paper like carbon paper) used to show a contact area. Used to determine if gate valve seats fit properly.

❒ psig

Pounds per Square Inch Gage pressure. The pressure within a closed container or pipe measured with a gage in pounds per square inch.

❒ public water system

A system for the provision to the public of piped water for human consumption, If such system has at least fifteen service connections or regularly least 60 days out of the year. Such term includes: 1) any collection, treatment, storage, and distribution facilities under control of the operator of such system and used primarily in connection with such system, and 2) any collection or pretreatment storage facilities not under such control which are used primarily in connection with such system. A public water system is either a "community water system" or a "non-community water system.'

❒ pumping station

Mechanical devices installed in sewer or water systems or other liquid-carrying pipelines that move the liquids to a higher level.

❒ pumping water level

The vertical distance in feet from the centerline of the pump discharge to the level of the free pool while water is being drawn from the pool.

❒ purveyor, water (purr-vay-or)

An agency or person that supplies water (usually potable water).

❒ putrefaction (pew-truh-fack-shun)

Biological decomposition of organic matter, with the production of ill- smelling and tasting products, associated with anaerobic (no oxygen present) conditions.

Q

❒ qualitative

Descriptive of kind, type or direction, as opposed to size, magnitude or degree.

❒ quantitative

Descriptive of size, magnitude or degree.

❒ quicklime

A material that is mostly calcium oxide (CaO) or calcium oxide in natural association with a lesser amount of magnesium oxide. Quicklime is capable of combining with water to form hydrated lime.

R

❒ radial to impeller

Perpendicular to the impeller shaft. Material being pumped flows at a right angle to the impeller.

❒ radical

A group of atoms that is capable of remaining unchanged during a series of chemical reactions. Such combinations (radicals) exist in the molecules of many organic compounds; sulfate (SO_{42}-) is an inorganic radical.

❒ radionuclide

Any man-made or natural element which emits radiation in the form of alpha or beta particles, or as gamma rays.

❒ range

The spread from minimum to maximum values that an instrument is designed to measure. Also see span and effective range.

❒ ranney collector

This water collector is constructed as a dug well from 12 to 16 feet (3.5 to 5 m) in diameter that has been sunk as a caisson near the bank of a river or lake. Screens are driven radially and approximately horizontally from this well into the sand and the gravel deposits underlying the river.

❒ raw water

1) Water in its natural state, prior to any treatment.
2) Usually the water entering the first treatment process of a water treatment plant.

❒ reaeration (re-air-a-shun)

The introduction of air through forced air diffusers into the lower layers of the reservoir. As the air bubbles form and rise through the water, oxygen from the air dissolves into the water and replenishes the dissolved oxygen. The rising bubbles also cause the lower waters to rise to the surface where oxygen from the atmosphere is transferred to the water. This is sometimes called surface reaeration.

❒ reagent (re-a-gent)

A pure chemical substance that is used to make new products or is used in chemical tests to measure, detect, or examine other substances.

❒ recarbonation (re-car-bun-nay-shun)

A process in which carbon dioxide is bubbled into the water being treated to lower the pH. The pH may also be lowered by the addition of acid. Recarbonation is the final stage in the lime-soda ash softening process. This process converts carbonate ions to bicarbonate ions and stabilizes the solution against the precipitation of carbonate compounds.

❒ receiver

A device which indicates the result of a measurement. Most receivers in the water utility field use either a fixed scale and movable indicator (pointer) such as pressure gage or a movable scale and movable indicator like those used on a circular-flow recording chart. Also called an indicator.

❒ receiving waters

All distinct bodies of water that receive runoff or wastewater discharges, such as streams, rivers, ponds, lakes, and estuaries.

❒ receptor

1) In biochemistry: a specialized molecule in a cell that binds a specific chemical with high specificity and high affinity.
2) In exposure assessment: an organism that receives, may receive, or has received environmental exposure to a chemical.

❒ recharge area

Generally, an area that is connected with the underground aquifer(s) by a highly porous soil or rock layer. Water entering a recharge area may travel for miles underground.

❒ recharge rate

The quantity of water per unit time that replenishes or refills an aquifer.

❒ recharge

Process by which rain water (precipitation) seeps into the ground-water system.

❒ recorder

A device that creates a permanent record, on a paper chart or magnetic tape, of the changes of some measured variable.

❒ reducing agent

Any substance, such as base metal (iron) or the sulfide ion (S_2-), that will readily donate (give up) electrons. The opposite is an oxidizing agent.

❒ reduction (re-duck-shun)

Reduction is the addition of hydrogen, removal of oxygen, or the addition of electrons to an element or compound. Under anaerobic conditions (no dissolved oxygen present), sulfur compounds are reduced to odor-producing hydrogen sulfide (H2S and other compounds.

❒ reference

A physical or chemical quantity whose value is known exactly, and thus is used to calibrate or standardize instruments.

❒ reliquifaction (re-lick-we-fack-shun) rem

The unit of dose equivalent from ionizing radiation to the total body or any internal organ or organ system. A millirem (mrem)" is 1/1000 of a rem.

❐ renal

Pertaining to the kidney.

❐ repeat compliance period

Any subsequent compliance period after the initial compliance period.

❐ representative sample

A portion of material or water that is as nearly identical in content and consistency as possible to that in the larger body of material or water being sampled.

❐ reservoir

Any natural or artificial holding area used to store; regulate, or control water.

❐ residual chlorine

The amount of free and/or available chlorine remaining after a given contact time under specified conditions.

❐ residual disinfectant concentration ("c" in ct calculations)

The concentration of disinfectant measured in mg/L in a representative sample of water.

❐ residue

The dry solids remaining after the evaporation of a sample of water or sludge.

❐ respiration

The process in which an organism uses oxygen for its life processes and gives off carbon dioxide.

❐ retrospective study

An epidemiological study which compares diseased persons with non-diseased persons and works back in time to determine exposures.

❐ reverse osmosis (oz-moe-sis)

The application of pressure to a concentrated solution which causes the passage of a liquid from the concentrated solution to a weaker solution across a semipermeable membrane. The membrane allows the passage of the solvent (water) but not the dissolved solids (solutes). The liquid produced is a demineralized water.

❐ reversible effect

An effect which is not permanent, especially adverse effects which diminish when exposure to a toxic chemical is ceased.

❐ rfd (reference dose)

The daily exposure level which, during an entire lifetime of a human, appears to be without appreciable risk on the basis of all facts known at the time.

❑ rill

A small channel eroded into the soil surface by runoff, rills easily can be smoothed out (obliterated) by normal tillage.

❑ riparian rights

A doctrine of state water law under which a land owner is entitled to use the water on or bordering his property, including the right to prevent diversion or misuse of upstream waters. Riparian land is land that borders on surface water.

❑ risk

The potential for realization of unwanted adverse consequences or events.

❑ risk assessment

A qualitative or quantitative evaluation of the environmental and/or health risk resulting from expo- sure to a chemical or physical agent (pollutant); combines exposure assessment results with toxicity assessment results to estimate risk.

❑ risk characterization

Final component of risk assessment that involves integration of the data and analysis involved in hazard evaluation, dose-response evaluation, and human exposure evaluation to determine the likelihood that humans will experience any of the various forms of toxicity associated with a substance.

❑ risk estimate

A description of the probability that organisms exposed to a specified dose of chemical will develop an adverse response (e.g., cancer).

❑ risk factor

Characteristic (e.g., race, sex, age, obesity) or variable (e.g., smoking, occupational exposure level) associated with increased probability of a toxic effect.

❑ risk management

Decisions about whether an assessed risk is sufficiently high to present a public health concern and about the appropriate means for control of a risk judged to be significant.

❑ risk specific dose

The dose associated with a specified risk level.

❑ rodenticide (row-dent-uh-side)

Any substance or chemical used to kill or control rodents.

❑ rotameter (rode-uh-me-ter)

A device used to measure the flow rate of gases and liquids. The gas or liquid being measured flows vertically up a tapered, calibrated tube. Inside the tube is a small ball or bullet-shaped float (it may rotate) that rises or falls depending on the flow rate. The flow rate may be read on a scale behind or on the tube by looking at the middle of the ball or at the widest part or top of the float.

❐ rotor

The rotating part of a machine. The rotor is surrounded by the stationary (non-moving) parts (stator) of the machine.

❐ route of exposure

The avenue by which a chemical comes into contact with an organism (e.g., inhalation, ingestion, dermal contact, injection).

❐ run-off

That part of precipitation, snow melt, or irrigation water that runs off the land into streams or other surface water. It can carry pollutants from the air and land into the receiving waters.

S

❐ sacrificial anode

An easily corroded material deliberately installed in a pipe or tank. The intent of such an installation is to give up (sacrifice) this anode to corrosion while the water supply facilities remain relatively corrosion free.

❐ safe

Condition of exposure under which there is a "practical certainty" that no harm will result in exposed individuals.

❐ safe water

Water that does not contain harmful bacteria, or toxic materials or chemicals. Water may have taste and odor problems, color and certain mineral problems and still be considered safe for drinking.

❐ safe yield

The annual quantity of water that can be taken from a source of supply over a period of years without depleting the source beyond its ability to be replenished naturally in "wet years".

❐ salinity

1)The relative concentration of dissolved salts, usually sodium chloride, in a given water.. 2) A measure of the concentration of dissolved mineral substances in water.

❐ sand filters

Devices that remove some suspended solids from sewage. Air and bacteria decompose additional wastes filtering through the sand so that cleaner water drains from the bed.

❒ sand

Soil particles between 0.05 and 2 .0 mm in diameter.

❒ sanitary sewer

A sewer that transports only wastewaters (from domestic residences and/or industries) to a wastewater treatment plant.

❒ sanitary survey

An on-site review of the water source, facilities, equipment, operation, and maintenance of a public water system for the purpose of evaluating the adequacy of the facilities for producing and distributing safe drinking water.

❒ saprophytes (sap-row-fights)

Organisms living on dead or decaying organic matter. They help natural decomposition of organic matter in water.

❒ saturated zone

The area below the water table where all open spaces are filled with water.

❒ saturator (sat-you-ray-tore)

A device which produces a fluoride solution for the fluoridation process. The device is usually a cylindrical container with granular sodium fluoride on the bottom. Water flows either upward or downward through the sodium fluoride to produce the fluoride solution.

❒ scd (swcd)

Soil Conservation District (also called Soil and Water Conservation District in some areas); a local government entity with in a defined water or soil protection area tat provides assistance to farmers and other local residents in conserving natural resources, especially soil and water.

❒ scfm

Cubic Feet of air per Minute at Standard conditions of temperature, pressure and humidity (O degrees C /14.7 psia /50% relative humidity).

❒ schedule, pipe

A sizing system of arbitrary numbers that specifies the I.D. (inside diameter) and O.D. (outside diameter) for each diameter pipe. This term is used for steel, wrought iron, and some types of plastic pipe.

❒ scs

Soil Conservation Service. An agency of the United States Department of Agriculture that provides technical assistance for resource conservation to farmers, other Federal, state and local agencies, and to local soil conservation districts.

❒ secchi disc (seck-key)

A flat, white disc lowered into the water by a rope until it is just barely visible.

❒ sediment yield

The quantity of sediment arriving at a specific location.

❒ sedimentation

A water treatment process in which solid particles settle out of the water being treated in a large clarifier or sedimentation basin.

❒ seepage

The percolation of water through the soil from unlined channels, ditches, watercourses and water storage facilities.

❒ seize up

Seize up occurs when an engine overheats and a part expands to the point where the engine will not run. Also called "freezing".

❒ semi-confined aquifer

An aquifer that is partially confined by a soil layer (or layers) of low permeability through which recharge and discharge can occur.

❒ sensor

An instrument that measure (senses) a physical condition or variable of interest. Floats and thermocouples are examples of sensors.

❒ septage

The liquid and semisolid contents removed by pumping from a septic tank.

❒ septic (sep-tick)

A condition produced by bacteria when all oxygen supplies are depleted. If severe, bottom deposits and water turn black, give off foul odors, and the water has a greatly increased chlorine demand.

❒ septic system

An onsite system designed to treat and dispose of domestic sewage; a typical septic system consists of a tank that receives wastes from a residence or business. And a system of tile lines or a pit for disposal of the liquid effluent that remains after decomposition of the solids by bacteria in the tank.

❒ sequestration (see-kwes-tray-shun)

A chemical completing (forming or joining together) of metallic cations (such as iron) with certain inorganic compounds, such as phosphate. Sequestration prevents the precipitation of the metals (iron).

❒ service pipe

The pipeline extending from the water main to the building served or to the consumers system.

❒ set point

The position at which the control or controller is set. This is the same as the desired value of the process variable.

❒ sewage

The used water and solids that flow from homes through sewers to a wastewater treatment plant. The preferred term is WASTEWATER.

❒ sewer

An underground system of conduits (pipes and/or tunnels) that collect and transport wastewaters and/or runoff; gravity sewers carry free-flowing water and wastes; pressurized sewers carry pumped wastewaters under pressure.

❒ sewerage system

The network of sewers that carries sewage from point of origin to point of treatment.

❒ shock load

The arrival at a water treatment plant of raw water containing unusual amounts of algae, colloidal matter, color, suspended solids, turbidity, or other pollutants.

❒ short-circuiting

A condition that occurs in tanks or basins when some of the water travels faster than the rest of the flowing water. This is usually undesirable since it may result in shorter contact, reaction, or settling times in comparison with the theoretical (calculated) or presumed detention times.

❒ silt

Soil particles between 0.05 and 0.002 millimeter in approximate diameter.

❒ simulate

To reproduce the action of some process, usually on a smaller scale.

❒ single family structure

A building constructed as a single family residence that is currently used as either a residence or a place of business.

❒ single-stage pump

A pump that has only one impeller. A multi-stage pump has more than one impeller.

❒ sink

A place in the environment where a compound or material collects. See reservoir.

❒ slake

To mix with water with a true chemical combination (hydrolysis) taking place, such as in the slaking of lime.

❒ slope

The slope or inclination of a trench bottom or a trench side wall is the ratio of the vertical distance to the horizontal distance or "rise over run."

❐ slow sand filtration

A process involving passage of raw water through a bed of sand at low velocity (generally less than 0.4 m/h) resulting in substantial particulate removal by physical and biological mechanisms.

❐ sludge (sluj)

The settleable solids separated from water during processing.

❐ slurry (slur-e)

A watery mixture or suspension of insoluble (not dissolved) matter; a thin watery mud or any substance resembling it (such as a grit slurry or a lime slurry).

❐ smcls

Secondary Maximum Contaminant Levels. Secondary MCLs for various water quality indicators are established to protect public welfare.

❐ snarl

Suggested No Adverse Response Level. The concentration of a chemical in water that is expected not to cause an adverse health effect.

❐ soft water

Water having a low concentration of calcium and magnesium ions. According to U.S. Geological Survey guidelines, soft water is water having a hardness of 60 milligrams per liter or less.

❐ software programs

Computer programs; the list of instructions that tell a computer how to perform a given task or tasks.

❐ soil erodibility

A measure of the soil's susceptibility to raindrop impact, runoff and other erosional processes.

❐ soil profile

A vertical section of the earth's highly weathered upper surface often showing several distinct layers, or horizons.

❐ soil structure

The arrangement of soil particles into aggregates.

❐ soil texture

The proportions of soil particles (sand, silt, and clay) in a soil profile.

❐ solder

A metallic compound used to seal the joints between pipes. Until recently, most solder contained 50 percent lead. The use of lead solder containing more than 0.2% lead is now prohibited for pipes carrying potable water.

❐ sole source aquifer

An aquifer that supplies 50 percent or more of the drinking water of an area.

❐ solenoid (so-luh-noid)

A magnetically (electrical coil) operated mechanical device. Solenoids can operate a small valve or a switch.

❐ solution

A liquid mixture of dissolved substances. In a solution it is impossible to see all the separate parts.

❐ sorption

A surface phenomenon which may be either absorption or adsorption, or a combination of the two; often used when the specific mechanism is not known.

❐ sounding tube

A pipe or tube used for measuring the depths of water.

❐ span

The scale or range of values an instrument is designed to measure. Also see range.

❐ specific conductance

A rapid method of estimating the dissolved-solids content of a water supply. The measurement indicates the capacity of a sample of water to carry an electrical current, which is related to the concentration of ionized substances in the water. Also called conductance.

❐ specific gravity

Weight of a particle, substance, or chemical solution in relation to the weight of water. Water has a specific gravity of 1.000 at 4 degrees C (39 degrees F). Particulates in raw water may have a specific gravity of 1.005 to 2.5.

❐ specific yield

The quantity of water that a unit volume of saturated permeable rock or soil will yield when drained by gravity. Specific yield may be expressed as a ratio or as a percentage by volume.

❐ spoil

Excavated material such as soil from the trench of a water main.

❐ spore

The reproductive body of an organism which is capable of giving rise to a new organism either directly or indirectly. A viable (able to live and grow) body regarded as the resting stage of an organism. A spore is usually more resistant to disinfectants and heat than most organisms.

❑ spring

Ground water seeping out of the earth where the water table intersects the ground surface.

❑ spring line

Theoretical center of a pipeline. Also, the guideline for laying a course of bricks.

❑ standard method

See Standard Methods for the Examination of Water and Wastewater.

❑ standard methods for the examination of water and wastewater

A joint publication of the American Public Health Association, American Water Works Association, and the Water Pollution Control Federation which outlines the procedures used to analyze the impurities in water and wastewater.

❑ standard sample

The aliquot of finished drinking water that is examined for the presence of coliform bacteria.

❑ standard solution

A solution in which the exact concentration of a chemical or compound is known.

❑ standard

A physical or chemical quantity whose value is known exactly, and is used to calibrate or standardize instruments.

❑ standardize

To compare with a standard. 1) In wet chemistry, to find out the exact strength of a solution by comparing it with a standard of known strength. 2) To set up an instrument or device to read a standard. This allows you to adjust the instrument so that it reads accurately, or enables you to apply a correction factor to the readings.

❑ starters

Devices used to start up large motors gradually to avoid severe mechanical shock to a driven machine and to prevent disturbance to the electrical lines (causing dimming and flickering of lights).

❑ static head

When water is not moving, the vertical distance (in feet) from a specific point to the water surface is the static head. (The static pressure in psi is the static head in feet times 0.433 psi/ft.) Also see dynamic pressure and static pressure.

❑ static pressure

When water is not moving, the vertical distance (in feet) from a specific point to the water surface is the static head. The static pressure in psi is the static head in feet times 0.433 psi/ft. Also see dynamic pressure and static head.

❒ static water depth

The vertical distance in feet from the centerline of the pump discharge down to the surface level of the free pool while no water is being drawn from the pool or water table.

❒ static water level

1) The elevation or level of the water table in a well when the pump is not operating.
2) The level or elevation to which water would rise in a tube connected to an artesian aquifer, or basin, or conduit under pressure.

❒ stator

That portion of a machine which contains the stationary (non-moving) parts that surround the moving parts (rotor).

❒ sterilization (stare-uh-luh-zay-shun)

The removal or destruction of all microorganisms, including pathogenic and other bacteria, vegetative forms and spores. Compare with disinfection.

❒ stethoscope

An instrument used to magnify sounds and convey them to the ear.

❒ stochastic

Based on the assumption that the actions of a chemical substance results from probabilistic events.

❒ storm sewer

A sewer that collects and transports surface runoff to a discharge point (infiltration basin, receiving stream, treatment plant).

❒ stratification

The formation of separate layers (of temperature, plant, or animal life) in a lake or reservoir. Each layer has similar characteristics such as all water in the layer has the same temperature. Also see thermal stratification.

❒ strip cropping

A crop production system that involves planting alternating strips of row crops and close-growing forage crops; the forage strips intercept and slow runoff from the less protected row crop strips.

❒ subchronic

Of intermediate duration, usually used to describe studies or levels of exposure between 5 and 90 days.

❒ submerged aquatic vegetation (sav)

Aquatic vegetation, such as sea grasses, that cannot withstand excessive drying and therefore live with their leaves at or below the water surface. SAVs provide an important habitat for young fish and other aquatic organisms.

❒ submergence

The distance between the water surface and the media surface in a filter.

❒ suction lift

The NEGATIVE pressure [in feet (meters) of water or inches (centimeters) of mercury vacuum] on the suction side of the pump. The pressure can be measured from the centerline of the pump DOWN TO (lift) the elevation of the hydraulic grade line on the suction side of the pump.

❒ superchlorination (sue-per-klor-uh-nay-shun)

Chlorination with doses that are deliberately selected to produce free or combined residuals so large as to require dechlorination .

❒ supernatant (sue-per-nay-tent)

Liquid removed from settled sludge. Supernatant commonly refers to the liquid between the sludge on the bottom and the water surface of a basin or container.

❒ supersaturated

An unstable condition of a solution (water) in which the solution contains a substance at a concentration greater than the saturation concentration for the substance.

❒ supplier of water

Any person who owns or operates a public water system.

❒ surface loading

One of the guidelines for the design of settling tanks and clarifiers in treatment plants. Used by operators to determine if tanks and clarifiers are hydraulically (flow) over- or underloaded. Also called overflow rate.

❒ surface pump

A mechanism for removing water or waste- water from a sump or wet well.

❒ surface runoff

Precipitation, snow melt, or irrigation in excess of what can infiltrate the soil surface and be stored in small surface depressions; runoff is a major transporter of non-point source pollutants.

❒ surface water

All water naturally open to the atmosphere (rivers, lakes. reservoirs, streams, impoundments, seas, estuaries, etc.) and all springs. wells, or other collectors which are directly influenced by surface water.

❒ surfactant (sir-fac-tent)

Abbreviation for surface-active agent. The active agent in detergents that possesses a high cleaning ability.

❒ surge chamber

A chamber or tank connected to a pipe and located at or near a valve that may quickly open or close or a pump that may suddenly start or stop. When the flow of water in a pipe starts or stops quickly, the surge chamber allows water to flow into or out of the pipe and minimize any sudden positive or negative pressure waves or surges in the pipe.

❒ suspended solids

1) Solids that either float on the surface or are suspended in water or other liquids, and which are largely removable by laboratory filtering.
2) The quantity of material removed from water in a laboratory test, as prescribed in STANDARD METHODS FOR THE EXAMINATION OF WATER AND WASTEWATER.

❒ synergism

An interaction of two or more chemicals which results in an effect that is greater than the sum of their effects taken independently.

❒ system with a single service connection

A system which supplies drinking water to consumers via a single service line.

❒ systemic effects

Effects observed at sites distant from the entry point of a chemical due to its absorption and distribution into the body.

❒ systemic

Relating to whole-body, rather than its individual parts.

T

❒ temperature sensor

A device that opens and closes a switch in response to changes in the temperature. This device might be a metal contact, or a thermocouple that generates minute electrical current proportional to the difference in heat, or a variable resistor whose value changes in response to changes in temperature. Also called a heat sensor.

❒ teratogenesis

The induction of nonhereditary congenital malformations (birth defects) in a developing fetus by exogenous factors acting in the womb; interference with normal embryonic development.

❒ teratogenicity

The capacity of a physical or chemical agent to cause teratogenesis in offspring.

❐ terrace

A broad channel, bench, or embankment constructed across the slope to intercept runoff and detain or channel it to protected outlets, thereby reducing erosion from agricultural areas.

❐ therapeutic index

The ratio of the dose required to produce toxic or lethal effect to dose required to produce nonadverse or therapeutic response.

❐ thermal stratification (strat-uh-fuh-kay-shun)

The formation of layers of different temperatures in a lake or reservoir.

❐ thermocline (thur-moe-kline)

The middle layer in a thermally stratified lake or reservoir. In this layer there is a rapid decrease in temperature with depth. Also called the METALIMNION.

❐ thermocouple

A heat-sensing device made of two conductors of different metals joined at their ends. An electric current is produced when there is a difference in temperature between the ends.

❐ threshold odor

The minimum odor of a water sample that can just be detected after successive dilutions with odorless water. Also called odor threshold

❐ threshold odor number

The greatest dilution of a sample with odor-free water that still yields a just-detectable odor.

❐ threshold

The lowest dose of a chemical at which a specified measurable effect is observed and below which it is not observed.

❐ thrust block

A mass of concrete or similar material appropriately placed around a pipe to prevent movement when the pipe is carrying water. Usually placed at bends and valve structures.

❐ tillage

Plowing, seedbed preparation, and cultivation practices.

❐ time lag

The time required for processes and control systems to respond to a signal or to reach a desired level.

❐ timer

A device for automatically starting or stopping a machine or other device at a given time.

❐ time-weighted average

The average value of a parameter (e.g., concentration of a chemical in air) that varies over time.

❐ tissue

A group of similar cells.

❐ titrate (tie-trate)

To TITRATE a sample, a chemical solution of known strength is added on a drop-by-drop basis until a certain color change, precipitate, or pH change in the sample is observed (end point). Titration is the process of adding the chemical reagent in increments until completion of the reaction, as signaled by the end point.

❐ tncws

See transient non-community water system.

❐ too numerous to count

The total number of bacterial colonies exceeds 200 on a 47-nun diameter membrane filter used for coliform detection.

❐ topography

The arrangement of hills and valleys in a geographic area.

❐ total dissolved phosphorus

Total phosphorus content of material that will pass through a filter of a specific size.

❐ total dissolved solids (tds)

All of the dissolved solids in a water. TDS is measured on a sample of water that has passed through a very fine mesh filter to remove suspended solids. The water passing through the filter is evaporated and the residue represents the dissolved solids.

❐ total dynamic head (tdh)

When a pump is lifting or pumping water, the vertical distance (in feet) from the elevation of the energy grade line on the suction side of the pump to the elevation of the energy grade line on the discharge side of the pump.

❐ total nitrogen

The sum of all nitrogen forms.

❐ total particulate phosphorus

Total phosphorus content of material retained on a filter of a specific size.

❐ total phosphorus

The sum of all phosphorus forms.

❐ total residual chlorine

The amount of available chlorine remaining after a given contact time. The sum of the combined available residual chlorine and the free available residual chlorine.

❐ total trihalomethanes (thms)

The sum of the concentration, in milligrams per liter, of the several trihalomethane compounds, rounded to two significant figures; The sum of the concentration in milligrams per liter of the trihalomethane compounds (trichloromethane [chloroform], dibromochloromethane, bromodichloromethane and tribromomethane [bromoform]), rounded to two significant figures.

❐ totalizer

A device or meter that continuously measures and calculates (adds) total flows in gallons, million gallons, cubic feet, or some other unit of volume measurement.

❐ toxaphene (tox-uh-feen)

A chemical that causes adverse health effects in domestic water supplies and also is toxic to freshwater and marine aquatic life.

❐ toxic (tox-ick)

A substance which is poisonous to an organism.

❐ toxic pollutants

Materials contaminating the environment that cause death, disease. birth defects in organisms that ingest or absorb them. The quantities and length of exposure necessary to cause these effects can vary widely.

❐ toxic substance

A chemical or mixture that may represent an unreasonable risk of injury to health or the environment.

❐ toxicant

A harmful substance or agent that may injure an exposed organism.

❐ toxicity assessment

Characterization of the toxicological properties and effects of a chemical, including all aspects of its absorption, metabolism, excretion and mechanism of action, with special emphasis on establishment of dose response characteristics.

❐ toxicity

The quality or degree of being poisonous or harmful to plant, animal or human life.

❐ toxicology

The science and study of poisons control.

❑ **transducer (trans-due-sir)**

A device which senses some varying condition and converts it to an electrical signal for transmission to some other device (a receiver) for processing or decision making.

❑ **transformation**

Acquisition by a cell of the property of uncontrolled growth.

❑ **transient water system**

A non-community water system that does not serve 25 of the same nonresident persons per day for more than six months per year. Also called a transient non-community water system (TNCWS).

❑ **transmission lines**

Pipelines that transport raw water from its source to a water treatment plant. After treatment, water is usually pumped into pipelines (transmission lines) that are connected to a distribution grid system.

❑ **transmissivity**

The ability of an aquifer to transmit water.

❑ **transpiration (tran-spur-ray-shun)**

The process by which water vapor is released to the atmosphere by living plants.

❑ **treated wastewater**

Wastewater that has been subjected to one or more physical, chemical, and biological processes to reduce its pollution of health hazard.

❑ **tremie (treh-me)**

A device used to place concrete or grout under water.

❑ **trichloroethane (tce) (try-klor-o-eth-hane)**

An organic chemical used as a cleaning solvent that causes adverse health effects in domestic water supplies.

❑ **trihalomethane**

One of a family of organic compounds named as derivatives of methane. THMs are generally the by-product from chlorination of drinking water that contains organic material. The resulting compounds (THMs) are suspected of causing cancer.

❑ **tube settler**

A device that uses bundles of small bore (2 to 3 inches or 50 to 75 mm) tubes installed on an incline as an aid to sedimentation. The tubes may come in a variety of shapes including circular and rectangular. As water rises within the tubes, settling solids fall to the tube surface. As the sludge (from the settled solids) in the tube gains weight, it moves down the tubes and settles to the bottom of the basin for removal by conventional sludge collection means. Tube settlers are sometimes installed in sedimentation basins and clarifiers to improve particle removal.

❐ **tubercle (too-burr-cull)**

A protective crust of corrosion products (rust) which builds up over a pit caused by the loss of metal due to corrosion.

❐ **tuberculation (too-burr-que-lay-shun)**

The development or formation of small mounds of corrosion products (rust) on the inside of iron pipe. These mounds (tubercules) increase the roughness of the inside of the pipe thus increasing resistance to water flow.

❐ **turbid**

Having a cloudy or muddy appearance.

❐ **turbidimeter**

A device that measures the amount of suspended solids in a liquid.

❐ **turbidity (ter-bid-it-tee)**

The cloudy appearance of water caused by the presence of suspended and colloidal matter. In the waterworks field, a turbidity measurement is used to indicate the clarity of water. Technically, turbidity is an optical property of the water based on the amount of light reflected by suspended particles. Turbidity cannot be directly equated to suspended solids because white particles reflect more light than dark-colored particles and many small particles will reflect more light than an equivalent large particle.

U

❐ **uncertainty factor**

A number (equal to or greater than one) used to divide NOAEL or LOAEL values derived from measurements in animals or small groups of humans, in order to estimate a NOAEL value for the whole human population.

❐ **unconfined aquifer**

An aquifer containing water that is not under pressure; the water level in a well is the same as the water table outside the well.

❐ **unit cancer risk**

Estimate of the lifetime risk caused by each unit of exposure in the low exposure region.

❐ **unit hydrograph**

The hydrograph of one inch of storm runoff generated by a rainstorm of fairly uniform intensity within a specific period of time.

❐ unsaturated zone

The area between the land surface and water table in which the pore spaces are only partially filled with water. Also called "zone of aeration".

❐ upper-bound estimate

Estimate not likely to be lower than the true risk.

❐ urban runoff

Storm water from city streets and adjacent domestic or commercial properties that may carry pollutants of various kinds into the sewer systems and/or receiving waters.

❐ user fee

A fee which is collected only from those persons who use a particular service, as opposed to one collected from the public in general. User fees generally vary in proportion to the degree of use of the service.

V

❐ variable costs

Input costs that change as the nature of the production activity of its circumstances change; for example, as production levels vary.

❐ variance

A State with primacy may relieve a public water system from a requirement respecting an MCL by granting a variance if certain conditions exist. These are: 1) the system cannot meet the MCL in spite of the application of best available treatment technology, treatment techniques or other means (taking costs into consideration), due to the characteristics of the raw water sources which are reasonably available to the system, and 2) the variance will not result in an unreasonable public health risk. A system may also be granted a variance from a specified treatment technique if it can show that, due to the nature of the system's raw water source, such treatment is not necessary to public health.

❐ vegetative controls

Non-point source pollution control practices that involve plants (vegetative cover) to reduce erosion and minimize the loss of pollutants.

❐ virus

The smallest form of microorganisms capable of causing disease. Especially, a virus of fecal origin that is infectious to humans by waterborne transmission.

❐ volatile acids

Acids produced during digestion. Fatty acids which are soluble in water and can be steam-distilled at atmospheric pressure. Also called "organic acids." Volatile acids are commonly reported as equivalent to acetic acid.

❐ volatile liquids

Liquids which easily vaporize or evaporate at room temperatures.

❐ volatile solids

Those solids in water or other liquids that are lost on ignition of the dry solids at 550 degrees C.

❐ volatile

Readily vaporizable at a relatively low temperature.

❐ volatilization

Loss of a substance through evaporation.

❐ voltage

The electrical pressure available to cause a flow of current (amperage) when an electrical circuit is closed.

❐ volumetric

A measurement based on the volume of some factor. Volumetric titration is a means of measuring unknown concentrations of water quality indicators in a sample by determining the volume of titrant or liquid reagent needed to complete particular reactions.

❐ vortex

A revolving mass of water which forms a whirlpool. This whirlpool is caused by water flowing out of a small opening in the bottom of a basin or reservoir. A funnel- shaped opening is created downward from the water surface.

W

❐ wastewater treatment plant

A facility that receives wastewaters (and sometimes runoff) from domestic and/or industrial sources, and by a combination of physical, chemical, and biological processes reduces (treats) the wastewaters to less harmful byproducts; known by the acronyms , STP (sewage treatment plant), and POTW.

❐ wastewater

The used water and solids from a community (including used water from industrial processes) that flow to a treatment plant. Storm water, surface water, and ground water infiltration also may be included in the wastewater that enters a wastewater treatment plant. The term -sewage usually refers to household wastes, but this word is being replaced by the term -wastewater.

❐ water budget

A summation of inputs, outputs, and net changes to a particular water resource system over a fixed period. (Also, water balance model).

▫ **water hammer**

The sound like someone hammering on a pipe that occurs when a valve is opened or closed very rapidly. When a valve position is changed quickly, the water pressure in a pipe will increase and decrease back and forth very quickly. This rise and fall in pressures can do serious damage to the system.

▫ **water purveyor (purr-vay-or)**

An agency or person that supplies water (usually potable water).

▫ **water solubility**

The maximum concentration of a chemical compound which can result when it is dissolved in water. If a substance is water soluble, it can very readily disperse through the environment.

▫ **water storage pond**

An impound for liquid wastes, so designated as to accomplish some degree of biochemical treatment of the wastes.

▫ **water supplier**

A person who owns or operates a public water system.

▫ **water table**

The level of ground water. The upper surface of the zone of saturation of groundwater above an impermeable layer of soil or rock (through which water cannot move) as in an unconfined aquifer. This level can be very near the surface of the ground or far below it.

▫ **water treatment lagoon**

An impound for liquid wastes, so designed as to accomplish some degree of biochemical treatment of the wastes.

▫ **water well**

An excavation where the intended use is for the location, acquisition, development, or artificial recharge of groundwater (excluding Sandpoint wells).

▫ **waterborne disease outbreak**

The significant occurrence of acute infectious illness, epidemiologically associated with the ingestion of water from a public water system that is deficient in treatment, as determined by the appropriate local or state agency.

▫ **watershed**

The land area that drains into a stream. An area of land that contributes runoff to one specific delivery point; large watersheds may be composed of several smaller "subsheds", each of which contributes runoff to different locations that ultimately combine at a common delivery point.

□ **watt**

A unit of power equal to one joule per second. The power of a current of one ampere flowing across a potential difference of one volt.

□ **weir (weer)**

1) A wall or plate placed in an open channel and used to measure the flow of water. The depth of the flow over the weir can be used to calculate the flow rate, or a chart or conversion table may be used.
2) A wall or obstruction used to control flow (from settling tanks and clarifiers) to assure uniform flow rate and avoid short-circuiting.

□ **weir diameter (weer)**

Many circular clarifiers have a circular weir within the outside edge of the clarifier. All the water leaving the clarifier flows over this weir. The diameter of the weir is the length of a line from one edge of a weir to the opposite edge and passing through the center of the circle formed by the weir.

□ **weir loading**

A guideline used to determine the length of weir needed on settling tanks and clarifiers in treatment plants. Used by operators to determine if weirs are hydraulically (flow) overloaded.

Weir Loading (GPM/ft) = Flow (GPM)/ Length of Weir (ft)

□ **well**

A bored, drilled, or driven shaft, or a dug hole, whose depth is greater than the largest surface dimension and whose purpose is to reach underground water supplies or oil, or to store or bury fluids below ground.

□ **well field**

Area containing one or more wells that produces usable amount of water.

□ **well monitoring**

The measurement, by on-site instruments or laboratory methods, of the quality of water in a well.

□ **well plug**

A watertight and gastight seal installed in a bore hole or well to prevent movement of fluids.

□ **wet chemistry**

Laboratory procedures used to analyze a sample of water using liquid chemical solutions (wet) instead of, or in addition to, laboratory instruments.

□ **wetlands**

Any number of tidal and nontidal areas characterized by saturated or nearly saturated soils most of the year that form an interface between terrestrial (land-based) and aquatic environments; include freshwater marshes around ponds and channels (rivers and streams), brackish and salt marshes; other common names include swamps and bogs.

❐ **wire-to-water efficiency**

The efficiency of a pump and motor together. Also called the overall efficiency.

❐ **withdrawal**

The process of taking water from a source and conveying it to a place for a particular type of use.

Y

❐ **yield**

The quantity of water (expressed as a rate of flow GPM, GPH, GPD, or total quantity per year) that can be collected for a given use from surface or groundwater sources. The yield may vary with the use proposed, with the plan of development, and also with economic considerations.

Z

❐ **zeta potential**

In coagulation and flocculation procedures, the difference in the electrical charge between the dense layer of ions surrounding the particle and the charge of the bulk of the suspended fluid surrounding this particle. The zeta potential is usually measured in millivolts.

❐ **zone of aeration**

The comparatively dry soil or rock located between the ground surface and the top of the water table.

❐ **zone of saturation**

The soil or rock located below the top of the groundwater table. By definition, the zone of saturation is saturated with water.

❐ **zooplankton (zoe-plank-ton)**

Small, usually microscopic animals(such as protozoans), found in lakes and reservoirs.

Weather Climate Geology

A

❐ anomaly

The difference between the value of a variable (for example temperature) at a given location and its long term average at that location.

C

❐ circulation

The flow or motion of a fluid.

❐ climatology

A quantitative description of climate showing the characteristic values of climate variables over a region. Climate refers to the statistical collection of weather conditions over a specified period of time. Note that the climate taken over different periods of time (30 years, 1000 years) may be different.

❐ composite

An average that is done according to a specific criteria. For example, one could composite the rainfall at a station for all years where the temperature was much above average.

❐ coupled model (or coupled atmosphere-ocean model)

In the context of climate modeling this usually refers to a numerical model which simulates both atmospheric and oceanic motions and temperatures and which takes into account the effects of each component on the other.

❐ cumulonimbus

A cloud type that is dense and vertically developed and is associated with rain (particularly of a convective nature).

J

❐ jet stream

Strong winds concentrated within a narrow zone in the atmosphere. Often used in reference to the axis of maximum mid-latitude westerlies located in the high troposphere.

M

❐ monsoons seasonal winds

They are caused primarily by the greater annual variation in air temperature over large land surfaces compared to ocean surfaces though other factors like land-relief are important.

S

❐ stationary waves

Waves (flow patterns with periodicity in time and/or space) that are fixed relative to Earth.

❐ storm track

The path followed by the center of a low (of atmospheric pressure). In many cases, multiple storms follow the same storm track.

T

❐ teleconnection

A strong statistical relationship between weather in different parts of the globe.

❐ thermocline

As one descends from the surface of the ocean the temperature remains nearly the same as it was at the surface. Soon, however, one encounters a zone in which temperature starts decreasing rapidly with depth. This zone is called the thermocline. The thermocline is important because it can support large scale waves which play a major role in ENSO. In studying the tropical Pacific Ocean, the depth of 20C water ("the 20C isotherm") is often used as a proxy for the depth of the thermocline. Along the equator, the 20C isotherm is typically located at about 50m depth in the eastern pacific, sloping downwards to about 150 m in the western Pacific.

U

❐ upwelling

In ocean dynamics, the upward motion of sub-surface water toward the surface of the ocean. This is often a source of cold, nutrient-rich water. Strong upwelling occurs along the equator where easterly winds are present.

Weather Terminology

A

❐ absolute humidity

A type of humidity that considers the mass of water vapor present per unit volume of space. Also considered as the density of the water vapor. It is usually expressed in grams per cubic meter.

❐ absolute instability

When the lapse rate of a column of air is greater than the dry adiabatic lapse rate. The term absolute is used because this applies whether or not the air is dry or saturated.

❐ absolute temperature scale

A temperature scale with a freezing point of +273°K (Kelvin) and a boiling point of +373°K.

❐ absolute zero

Considered to be the point at which theoretically no molecular activity exists or the temperature at which the volume of a perfect gas vanishes. The value is 0° Kelvin, -273.15° Celsius and -459.67° Fahrenheit.

❐ absorption

The process in which incident radiant energy is retained by a substance. The absorbed radiation is then transformed into molecular energy.

❐ abyssal plain

The flat, gently sloping or nearly level region of the sea floor.

❐ adiabatic process

A thermodynamic change of state in a system in which there is no transfer of heat or mass across the boundaries of the system. In this process, compression will result in warming and expansion will result in cooling.

❐ advection

The horizontal transfer of any property in the atmosphere by the movement of air (wind). Examples include heat and moisture advection.

❐ advection fog

Fog that develops when warm moist air moves over a colder surface, cooling that air to below its dew point.

❑ advisory

Statements that are issued by the National Weather Service for probable weather situations of inconvenience that do not carry the danger of warning criteria, but, if not observed, could lead to hazardous situations. Some examples include snow advisories stating possible slick streets, or fog advisories for patchy fog condition causing temporary restrictions to visibility.

❑ afos

Acronym for Automation of Field Operations and Services. It is the computer system that links National Weather Service offices together for weather data transmission.

❑ air

This is considered the mixture of gases that make up the earth's atmosphere. The principal gases that compose dry air are Nitrogen (N_2) at 78.09%, Oxygen (O_2) at 20.946%, Argon (A) at 0.93%, and Carbon Dioxide (CO_2) at 0.033%. One of the most important constituents of air and most important gases in meteorology is water vapor (H_2O).

❑ air mass

An extensive body of air throughout which the horizontal temperature and moisture characteristics are similar.

❑ air mass thunderstorm

A thunderstorm that is produced by convection within an unstable air mass through an instability mechanism. Such thunderstorms normally occur within a tropical or warm, moist air mass during the summer afternoon as the result of afternoon heating and dissipate soon after sunset. Such thunderstorms are not generally associated with fronts and are less likely to become severe than other types of thunderstorms. However, that does not preclude them from having brief heavy downpours.

❑ air pollution

The soiling of the atmosphere by contaminants to the point that may cause injury to health, property, plant, or animal life, or prevent the use and enjoyment of the outdoors.

❑ air quality standards

The maximum level which will be permitted for a given pollutant. Primary standards are to be sufficiently stringent to protect the public health. Secondary standards must protect the public welfare, including property and aesthetics.

❑ alaskan winds

The downslope air flow that blows through the Alaskan valleys. It is usually given local names, such as Knik, Matanuska, Pruga, Stikine, Taku, Take, Turnagain, or Williwaw.

❐ albedo

The ratio of the amount of radiation reflected from an object's surface compared to the amount that strikes it. This varies according to the texture, color, and expanse of the object's surface and is reported in percentage. Surfaces with high albedo include sand and snow, while low albedo rates include forests and freshly turned earth.

❐ alberta clipper

A fast moving, snow-producing weather system that originates in the lee of the Canadian Rockies. It moves quickly across the northern United States, often bring gusty winds and cold Arctic air.

❐ aleutian low

A semi-permanent, subpolar area of low pressure located in the Gulf of Alaska near the Aleutian Islands. It is a generating area for storms and migratory lows often reach maximum intensity in this area. It is most active during the late fall to late spring. During the summer, it is weaker, retreating towards the North Pole and becoming almost nonexistent. During this time, the North Pacific High pressure system dominates.

❐ altimeter

An instrument used to determine the altitude of an object with respect to a fixed level. The type normally used by meteorologists measures the altitude with respect to sea level pressure.

❐ altimeter setting

The pressure value to which an aircraft altimeter scale is set so that it will indicate the altitude above mean sea level of an aircraft on the ground at the location for which the value was determined.

❐ altitude

In meteorology, the measure of a height of an airborne object in respect to a constant pressure surface or above mean sea level.

❐ altocumulus

Composed of flattened, thick, gray, globular masses, this middle cloud genus is primarily made of water droplets. In the mid-latitudes, cloud bases are usually found between 8,000 and 18,000 feet. A defining characteristic is that it often appears as a wavy billowy layer of cloud, giving it the nickname of "sheep" or "woolpack" clouds. Sometimes confused with cirrocumulus clouds, its elements (individual clouds) have a larger mass and cast a shadow on other elements. It may form several sub-types, such as altocumulus castellanus or altocumulus lenticularis. Virga may also fall from these clouds.

❐ altocumulus castellanus

A middle cloud with vertical development that forms from altocumulus clouds. It is composed primarily of ice crystals in its higher portions and characterized by its turrets, protuberances, or crenelated tops. Its formation indicates instability and turbulence at the altitudes of occurrence.

❑ altostratus

This middle cloud genus is composed of water droplets, and sometimes ice crystals, In the mid-latitudes, cloud bases are generally found between 15,000 and 20,000 feet. White to gray in color, it can create a fibrous veil or sheet, sometimes obscuring the sun or moon. It is a good indicator of precipitation, as it often precedes a storm system. Virga often falls from these clouds.

❑ american meteorological society

An organization whose membership promotes the education and professional advancement of the atmospheric, hydrologic, and oceanographic sciences.

❑ anabatic wind

A wind that is created by air flowing uphill. Valley breezes, produced by local daytime heating, are an example of these winds. The opposite of a katabatic wind.

❑ anemometer

An instrument that measures the speed or force of the wind.

❑ aneroid barometer

An instrument for measuring the atmospheric pressure. It registers the change in the shape of an evacuated metal cell to measure variations on the atmospheric pressure. The aneroid is a thin-walled metal capsule or cell, usually made of phosphor bronze or beryllium copper. The scales on the glass cover measure pressure in both inches and millibars.

❑ anomalous propagation

This refers to the non-standard propagation of a beam of energy, radio or radar, under certain atmospheric conditions, appearing as false (non-precipitation) echoes. May be referred to as A.P.

❑ antarctic

Of or relating to the area around the geographic South Pole, from 90° South to the Antarctic Circle at approximately 66 1/2°South latitude, including the continent of Antarctica. Along the Antarctic Circle, the sun does not set on the day of the summer solstice (approximately December 21st) and does not rise on the day of the winter solstice (approximately June 21st).

❑ antarctic ocean

Although not officially recognized as a separate ocean body, it is commonly applied to those portions of the Atlantic, Pacific, and Indian Oceans that reach the Antarctic continent on their southern extremes.

❑ anticyclone

A relative pressure maximum. An area of pressure that has diverging winds and a rotation opposite to the earth's rotation. This is clockwise the Northern

Hemisphere and counter-clockwise in the Southern Hemisphere. It is the opposite of an area of low pressure, or a cyclone.

❐ anvil

The upper portion of a cumulonimbus cloud that becomes flat and spread-out, sometimes for hundreds of miles downstream from the parent cloud. It may look smooth or fibrous, but in shape, it resembles a blacksmith's anvil. It indicates the mature or decaying stage of a thunderstorm.

❐ aphelion

The point on the earth's orbit that is farthest from the sun. Although the position is part of a 21,000 year cycle, currently it occurs around July, when the earth is about 3 million miles farther from the sun than at perihelion. This term can be applied to any other celestial body in orbit around the sun. It is the opposite of perihelion.

❐ apogee

The point farthest from the earth on the moon's orbit. This term can be applied to any other body orbiting the earth, such as satellites. It is the opposite of perigee.

❐ arctic

Of or relating to the area around the geographic North Pole, from 90° North to the Arctic Circle at approximately 66 1/2 North latitude.

❐ arctic air mass

An air mass that develops around the Arctic, it is characterized by being cold from surface to great heights. The boundary of this air mass is often defined by the Arctic front, a semi-permanent, semi-continuous feature. When this air mass moves from its source region, it may become more shallow in height as it spreads southward.

❐ arctic jet

The jet stream that is situated high in the stratosphere in and around the Arctic or Antarctic Circles. It marks the boundary of polar and arctic air masses.

❐ arctic sea smoke

A type of advection fog that forms primarily over water when cold air passes across warmer waters.

❐ argon (a)

A colorless, odorless inert gas that is the third most abundant constituent of dry air, comprising 0.93% of the total.

❐ arid

A term used for an extremely dry climate. The degree to which a climate lacks effective, life-promoting moisture. It is considered the opposite of humid when speaking of climates.

❐ asos

Acronym for Automated Surface Observing System. This system is a collection of automated weather instruments that collect data. It performs surface based observations from places that do not have a human observer, or that do not have an observer 24 hours a day.

❐ astronomical twilight

The time after nautical twilight has commenced and when the sky is dark enough, away from the sun's location, to allow astronomical work to proceed. It ends when the center of the sun is 18° below the horizon.

❐ atmosphere

The gaseous or air portion of the physical environment that encircles a planet. In the case of the earth, it is held more or less near the surface by the earth's gravitational attraction. The divisions of the atmosphere include the troposphere, the stratosphere, the mesosphere, the ionosphere, and the exosphere.

❐ atmospheric pressure

The pressure exerted by the atmosphere at a given point. Its measurement can be expressed in several ways. One is in millibars. Another is in inches or millimeters of mercury (Hg).

❐ aurora

It is created by the radiant energy emission from the sun and its interaction with the earth's upper atmosphere over the middle and high latitudes. It is seen as a bright display of constantly changing light near the magnetic poles of each hemisphere. In the Northern Hemisphere, it is known as the aurora borealis or Northern Lights, and in the Southern Hemisphere, this phenomena is called the aurora australia.

❐ autumn

The season of the year which occurs as the sun approaches the winter solstice, and characterized by decreasing temperatures in the mid-latitudes. Customarily, this refers to the months of September, October, and November in the North Hemisphere and the months of March, April, and May in the Southern Hemisphere. Astronomically, this is the period between the autumnal equinox and the winter solstice.

❐ avhrr

Acronym for Advanced Very High Resolution Radiometer. It is the main sensor on the U.S. polar orbiting satellites.

❐ aviation weather center

As one of the National Centers for Environmental Prediction, it is the national center for weather information that is used daily by the Federal Aviation Administration, commercial airlines, and private pilots. It is entering a new phase of service, growing to accept global forecasting responsibilities.

❐ awips

Acronym for Advanced Weather Interactive Processing System. It is the computerized system that processes NEXRAD and ASOS data received at National Weather Service Forecast Offices.

❑ azores high

A semi-permanent, subtropical area of high pressure in the North Atlantic Ocean that migrates east and west with varying central pressure. Depending on the season, it has different names. In the Northern Hemispheric winter and early spring, when the Icelandic Low dominates the North Atlantic, it is primarily centered near the Azores Islands. When it is displaced westward, during the summer and fall, the center is located in the western North Atlantic, near Bermuda, and is known as the Bermuda High.

B

❑ backing

A counterclockwise shift in the wind direction in the Northern Hemisphere at a certain location. In the Southern Hemisphere, it is clockwise. This can either happen in the horizontal or the vertical (with height). For example, the wind shifts from the northeast to the north to the northwest. It is the opposite of veering.

❑ backscatter

A radar echo that is reflected, or scattered, at 180 degrees to the direction of the incident wave. Also the scattering of radiant energy into space before it reaches the earth's surface.

❑ ball lightning

A relatively rare form of lightning consisting of a luminous ball, often reddish in color, which moves rapidly along solid objects or remains floating in mid-air.

❑ baroclinity

The state of stratification in a fluid in which surfaces of constant pressure intersect surfaces of constant density. Also known as baroclinicity. An example is the tight temperature gradient along the East Coast of the United States during the winter that gives rise to intense cyclogenesis.

❑ barograph

An instrument that continuously records a barometer's reading of atmospheric pressure.

❑ barometer

An instrument used to measure atmospheric pressure. Two examples are the aneroid barometer and the mercurial barometer.

❑ barometric pressure

The pressure exerted by the atmosphere at a given point. Its measurement can be expressed in several ways. One is in millibars. Another is in inches or millimeters of mercury (Hg).

❐ barotropy

The state of a fluid in which surfaces of constant density or temperature are coincident with surfaces of constant pressure. It is considered zero baroclinity.

❐ barrier winds

Refers to the westerly flow of air along the northern slope of the Brooks Range in northern Alaska that precedes the arrival of colder air from the north.

❐ bathythermograph

A device used to obtain a record of temperature against depth (pressure) in the ocean.

❐ beaufort wind scale

A system of estimating and reporting wind speeds. It is based on the Beaufort Force or Number, which is composed of the wind speed, a descriptive term, and the visible effects upon land objects and/or sea surfaces. The scale was devised by Sir Francis Beaufort (1777-1857), hydrographer to the British Royal Navy.

❐ bellot winds

Refers to the winds in the Canadian Arctic that blow through the narrow Bellot Strait between Somerset Island and the Boothia Peninsula, connecting the Gulf of Boothia and Franklin Strait.

❐ bermuda high

A semi-permanent, subtropical area of high pressure in the North Atlantic Ocean that migrates east and west with varying central pressure. Depending on the season, it has different names. When it is displaced westward, during the Northern Hemispheric summer and fall, the center is located in the western North Atlantic, near Bermuda. In the winter and early spring, it is primarily centered near the Azores Islands.

❐ bernoulli's theorem

A statement of the conservation of energy for a steady, nonviscous, incompressible level flow. It is an inverse relationship in which pressures are least where velocities are greatest. Theorized by Daniel Bernoulli (1700-1782), a Swiss mathematician and physicist.

❐ biosphere

The transition zone between the earth and the atmosphere within which most terrestrial life forms are found. It is considered the outer portion of the geosphere and the inner or lower portion of the atmosphere.

❐ black blizzard

A local term for a violent duststorm on the south-central Great Plains that darkens the sky and casts a pall over the land.

❐ black ice

Thin, new ice on fresh or salt water that appears dark in color because of its transparency. Also refers to thin, transparent ice on road surfaces.

❐ blizzard

A severe weather condition characterized by low temperatures, winds 35 mph or greater, and sufficient falling and/or blowing snow in the air to frequently reduce visibility to 1/4 mile or less for a duration of at least 3 hours. A severe blizzard is characterized by temperatures near or below 10Â°F, winds exceeding 45 mph, and visibility reduced by snow to near zero.

❐ blocking high

The development of a warm ridge or cutoff high aloft at high latitudes which becomes associated with a cold high at the surface, causing a split in the westerly winds. Such a high will move very slowly, tending to move westward during intensification and eastward during dissipation. It prevents the movement of migratory cyclones across its latitudes.

❐ blowing dust

Dust that is raised by the wind to heights of six feet or greater. It is reported as "BLDU" in an observation and on the METAR.

❐ blowing sand

Sand that is raised by the wind to heights of six feet or greater. It is reported as "BLSA" in an observation and on the METAR.

❐ blowing snow

Snow that is raised by the wind to heights of six feet or greater. It is reported as "BLSN" in an observation and on the METAR.

❐ blowing spray

Salt spray that is raised by the wind to heights of six feet or greater. It is reported as "BLPY" in an observation and on the METAR.

❐ blue norther

Refers to a swift-moving cold frontal passage in the southern Great Plains, marked by a dark, blue-black sky with strong wintery winds from the northwest or north and temperatures that may drop 20Â°F to 30Â°F in a few minutes.

❐ boiling point

The temperature at which a liquid changes to a vaporous state. The temperature at which the equilibrium vapor pressure between a liquid and its vapor is equal to the external pressure on the liquid. The boiling point of pure water at standard pressure is 100Â°C or 212Â°F.

❐ boulder wind

A local name referring to an extremely strong downslope wind in the Front Range of the Rocky Mountains near Boulder, Colorado.

❐ boundary layer

The lowest layer of the earth's atmosphere, usually up to 3,300 feet, or one kilometer, from the earth's surface, where the wind is influenced by the friction of the earth's surface and the objects on it.

❐ bow echo

A radar echo signature often associated with severe thunderstorms, especially those that produce wind damage. It is bent outward in a "bow" shape.

❐ boyle's law

States that when the temperature is held constant, the volume of a gas is inversely proportional to its pressure. Therefore, if the pressure increases, the volume decreases and visa versa. For example, if the volume if halved, then the pressure is doubled. If the temperature is held constant, it becomes an isothermal process. Discovered by Robert Boyle (1627-1691), an Irish physicist and chemist and co-founder of the Royal Society.

❐ bright band

A narrow, intense radar echo due to water-covered ice particles at the melting level where reflectivity is at its greatest.

❐ broken

The amount of sky cover for a cloud layer between 5/8ths and 7/8ths, based on the summation layer amount for that layer.

❐ bubble high

A small high that may be created by precipitation and vertical instability associated with thunderstorm activity. A product of downdrafts, it is relatively cold and often has the characteristics of a different air mass. Convergence along the leading edge of a bubble high may help form additional thunderstorms.

❐ buys ballot's law

Describes the relationship of the horizontal wind direction to the pressure distribution. In the Northern Hemisphere, if one stands with one's back to the wind, the pressure on one's left is lower than the pressure on one's right. It is reversed in the Southern Hemisphere. This law was named after the Dutch meteorologist, Buys Ballot, who developed the formula in 1857.

❐ bwer

Acronym for Bounded Weak Echo Region. Refers to radar echo signatures with low reflectivity in the center, surrounded by higher reflectivity. It is associated with strong updrafts and is found in the inflow region of a thunderstorm.

C

❐ calm

Atmospheric conditions devoid of wind or any other air motion. In oceanic terms, it is the apparent absence of motion of the water surface when there is no wind or swell.

❐ calorie

In meteorology, it is the amount of heat required to raise the temperature of one

(1) gram of water one

(2) degree Celsius. It is a unit of heat energy.

❒ cap

Composed of a layer of warmer, dryer air aloft which may suppress or delay the development of thunderstorms. As an air parcel rises, it becomes cooler relative to the ambient, or surrounding, air in the cap and therefore, less buoyant and unable to rise further. Also referred to as a lid.

❒ cape

Acronym for Convective Available Potential Energy. The amount of energy available to create convection, with higher values increasing the possibility for severe weather.

❒ cape verde islands

A group of volcanic islands in the eastern Atlantic Ocean off the coast of West Africa. A Cape Verde hurricane originates near here.

❒ carbon dioxide (co2)

A heavy, colorless gas that is the fourth most abundant constituent of dry air, comprising 0.033% of the total.

❒ catalina eddy

A weak low pressure circulation that may form off the Southern California coast.

❒ ceiling

The lowest cloud layer that is reported as broken or overcast. If the sky is totally obscured, then it is the height of the vertical visibility.

❒ ceiling light

An instrument consisting of a drum and an optical system that projects a narrow vertical beam of light onto a cloud base.

❒ ceilometer

An instrument that is used to measure the angular elevation of a projected light on the base of a cloud. It measures the angle of the cloud base included by the observer (or machine), the ceiling light and the illuminated spot on the cloud.

❒ celestial equator

The projection of the plane of the geographical equator upon the celestial sphere.

❒ celestial sphere

The apparent sphere of infinite radius having the earth as its center. All heavenly bodies (planets, stars, etc.) appear on the "inner surface" of this sphere and the sun moves along the ecliptic.

❑ celsius temperature scale

A temperature scale where water at sea level has a freezing point of 0°C (Celsius) and a boiling point of +100°C. More commonly used in areas that observe the metric system of measurement. Created by Anders Celsius in 1742. In 1948, the Ninth General Conference on Weights and Measures replaced "degree centigrade" with "degree Celsius."

❑ central pressure

The atmospheric pressure at the center of a high or low. It is the highest pressure in a high and the lowest pressure in a low, referring to the sea level pressure of the system on a surface chart.

❑ centrifugal force

The apparent force in a rotating system that deflects masses radially outward from the axis of rotation. This force increases towards the equator and decreases towards the poles.

❑ centripetal force

The force required to keep an object moving in a curved or circular path. It is directed inwards toward the center of the curved path.

❑ charles' law

States that when the pressure is held constant, the volume of a gas varies directly with the temperature. Therefore, if the pressure remains constant, the volume of a gas will increase with the increase of temperature. It was developed by Jacques Charles and is also known as the Charles-Guy-Lussac Law.

❑ chemosphere

A vaguely defined region of the upper atmosphere in which photochemical reactions take place. It includes the top of the stratosphere, all of the mesosphere, and sometimes the lower part of the thermosphere.

❑ cheyenne fog

An upslope fog formed by the westward flow of air from the Missouri River Valley, producing fog on the eastern slopes of the Rockies.

❑ chinook

A type of foehn wind. Refers to the warm downslope wind in the Rocky Mountains that may occur after an intense cold spell when the temperature could rise by 20°F to 40°F in a matter of minutes.

❑ chocolatta north

A West Indian gale that blows from the northwest.

❑ chromosphere

A thin layer of relatively transparent gases above the photosphere of the sun. It is observed best during a total eclipse of the sun.

❑ circulation

The flow or motion of a fluid in or through a given area or volume. In meteorology, it is used to describe the flow of air as it moves around a pressure system in the atmosphere. It describes smaller patterns in semi-permanent pressure systems as well as the relatively permanent global currents of air. In oceanic terms, it is used to describe a water in current flow within a large area, usually a closed circular pattern such as in the North Atlantic.

❑ circulation cells

Large areas of air movement created by the rotation of the earth and the transfer of heat from the equator toward the poles. Circulation is confined to a specific region, such as the tropics, temperate, or polar, that influences the type of weather prevailing there.

❑ cirriform

Clouds composed of small particles, mostly ice crystals. Because the particles are fairly widely dispersed, this usually results in relative transparency and whiteness, often producing a halo phenomena not observed in other clouds forms. These clouds generally have bases above 20,000 feet in the mid-latitudes, and are classified as high clouds. They include all varieties of cirrus, cirrocumulus, and cirrostratus clouds.

❑ cirrocumulus

A cirriform cloud with vertical development, appearing as a thin sheet of small white puffs which give it a rippled effect. It often creates a "mackerel sky", since the ripples may look like fish scales. Sometimes it is confused with altocumulus, however, it has smaller individual masses and does not cast a shadow on other elements. It is also the least common cloud type, often forming from cirrus or cirrostratus, with which it is associated in the sky.

❑ cirrostratus

A cirriform cloud that develops from cirrus spreading out into a thin layer, creating a flat sheetlike appearance. It can give the sky a slightly milky or veiled look. When viewed from the surface of the earth, these ice crystals can create a halo effect around the sun or moon. This cloud is a good precursor of precipitation, indicating it may occur within 12 to 24 hours.

❑ cirrus

One of the three basic cloud forms (the others are cumulus and stratus). It is also one of the three high cloud types. Cirrus are thin, wispy clouds composed of ice crystals and often appear as veil patches or strands. In the mid-latitudes, cloud bases are usually found between 20,000 to 30,000 feet, and it is the highest cloud that forms in the sky, except for the tops, or anvils, of cumulonimbus, which occasionally build to excessive heights.

❑ civil twilight

The time between the moment of sunset, when the sun's apparent upper edge is just at the horizon, until the center of the sun is 6° directly below the horizon.

❐ clear

The state of the sky when no clouds or obscurations are observed or detected from the point of observation.

❐ clear air turbulence

Name given to turbulence that may occur in perfectly clear air without any visual in warning in the form of clouds. It is often found in the vicinity of the jet stream where large shears in the horizontal and vertical are found, although this turbulence is not limited just to jet stream locale. Other areas where it may occur include near mountains, in closed lows aloft, and in regions of wind shear.

❐ clear ice

A glossy, clear, or translucent ice formed by the relatively slow freezing of large supercooled in water droplets. The droplets spread out over an object, such as an aircraft wing's leading edge, prior to complete freezing and forms a sheet of clear ice.

❐ climate

The historical record and description of average daily and in seasonal weather events that help describe a region. Statistics are generally drawn over several decades. The word is derived from the Greek klima, meaning inclination, and reflects the importance early scholars attributed to the sun's influence.

❐ climate analysis center (cac)

The U.S. National Weather Service division that applies new technology and approaches to the analysis, diagnosis, and projection of short term climate fluctuations on a regional and global basis.

❐ climate prediction center (cpc)

A branch of the National Centers for Environmental Prediction,the Center maintains a continuous watch on short-term climate fluctuations and diagnoses and predicts them.

❐ climatology

The study of climate. It includes climatic data, the analysis of the causes of the differences in climate, and the application of climatic data to the solution of specific design or operational problems.

❐ clinometer

An instrument used to measure angles of inclination. Used in conjunction with a ceiling light, it determines cloud height at night, based on the angle of a projected light on the clouds, the observer, and the ceiling light.

❐ closed low

A region of low pressure distinguished by a center of counterclockwise circulation (in the Northern Hemisphere), and is surrounded by one or more isobars or height contours. Closed lows aloft (i.e., above the surface) may become disconnected from the primary westerly flow and thus progress eastward more slowly. It is important to note that a cutoff low is a closed low, but not all closed lows are cutoff lows.

❐ cloud

A visible collection of minute particle matter, such as water droplets and/or ice crystals, in the free air. A cloud forms in the atmosphere as a result of condensation of water vapor. Condensation nuclei, such as in smoke or dust particles, form a surface upon which water vapor can condense.

❐ cloud bank

A well-defined cloud mass that can be observed at a distance. It covers the horizon, but is not directly overhead.

❐ cloudburst

A sudden, heavy rainfall of a showery nature.

❐ coalescence

The merging of two water drops into a single larger drop.

❐ cold

A condition marked by low or decidedly subnormal temperature. The lack of heat.

❐ cold advection

The horizontal movement of colder air into a location. Contrast with warm advection.

❐ cold air funnel

Funnel clouds, usually short-lived, that develop from relatively small showers or thunderstorms when the air aloft is very in cold. Cold air funnels may touch down briefly, but in general are less violent than most other types of tornadoes.

❐ cold core thunderstorms

Thunderstorms formed primarily due to steep lapse rates, especially when very cold air aloft overlies warmer surface air.

❐ cold front

The leading edge of an advancing cold air mass that is under running and displacing the warmer air in its path. Generally, with the passage of a cold front, the temperature and humidity decrease, the pressure rises, and the wind shifts (usually from the southwest to the northwest in the Northern Hemisphere). Precipitation is generally at and/or behind the front, and with a fast-moving system, a squall line may develop ahead of the front.

❐ cold high

A high pressure system that has its coldest temperatures at or near the center of circulation, and horizontally, is thermally barotropic. It is shallow in nature, as circulation decreases with height. Associated with cold Arctic air, it is usually stationary.

❐ cold low

A low pressure system that has its coldest temperatures at or near the center of circulation, and is thermally barotropic with respect to a horizontal plane. Also known as a cold core low. A cut off low is an example, where an isolated pool of colder air is located south of the main westerlies.

❐ cold wave

A rapid fall in temperature within twenty-four hours to temperatures requiring substantially increased protection to agriculture, industry, commerce, and social activities. National Weather Service criteria includes the rate of temperature fall and the minimum to which it falls, depending on the region of the country and time of the in year. The Weather Channel uses the following criteria for a cold wave: a cold spell of two days or more with below normal temperatures in at least fifteen states, with at least five of them more than fifteen degrees below normal.

❐ collada

A strong, steady wind blowing from the north or northwest in the upper part of the Gulf of California and from the northeast in the lower part.

❐ colorado low

A low pressure disturbance that forms in the lee of the Rocky Mountains, usually in southeastern Colorado.

❐ comma cloud

A feature seen on satellite images with a distinctive comma-shape. This is indicative of a synoptic cloud pattern associated with large, well-developed low pressure systems.

❐ condensation

The process by which water vapor undergoes a change in state from a gas to a liquid. It is the opposite physical process of evaporation.

❐ condensation funnel

A funnel-shaped cloud consisting of condensed water drops that has possible rotation.

❐ condensation nuclei

A particle upon which condensation of water vapor occurs. It may be either in a solid or liquid state.

❐ conditional instability

Stable unsaturated air that will result in instability in the event or on the condition that the air becomes saturated. If the air is saturated, it is considered unstable; if air is unsaturated, it is considered stable.

❐ conduction

The transfer of heat through a substance by molecular action or from one substance by being in contact with another.

❐ confluence

A rate at which wind flow comes together along an axis oriented normal to the flow in question. The opposite of diffluence.

❐ constant pressure chart

A chart of a constant pressure surface in which atmospheric pressure is uniform everywhere at any given moment. Elements may include analyses of height above sea level, wind, temperature, and humidity.

❐ constant pressure surface

A surface along which the atmospheric pressure is equal everywhere.

❐ continent

A large land mass rising abruptly from the deep ocean floor, including marginal regions that are shallowly submerged. Continents constitute about one-third of the earth's surface.

❐ continental air mass

An air mass with continental characteristics. It is a secondary characteristic of an air mass classification, signified by the small "c" before the primary characteristic, which is based on source region. For example, cP is an air mass that is continental polar in nature.

❐ continental shelf

The zone around the continents extending from the low-water mark seaward, typically ending in steep slope to the depths of the ocean floor.

❐ contrail

Acronym for CONdensation TRAIL. A cloud-like streamer or trail often seen behind aircraft flying in clear, cold, humid air. A vapor trail is created when the water vapor from the engine exhaust gases are added to the atmosphere.

❐ convection

Motions in a fluid that transport and mix the properties of the fluid. These properties could be heat and/or moisture. When used to imply only upward vertical motion, it is then the opposite of subsidence.

❐ convective condensation level (ccl)

The height at which a parcel of air, if heated sufficiently from below, will rise adiabatically until it is just saturated.

❐ convergence

Wind movement that results in a horizontal net inflow of air into a particular region. Convergent winds at lower levels are associated with upward motion. Contrast with divergence.

❐ cooling degree day

A cooling degree day is given for each degree that the daily mean temperature departs above the baseline of 75 degrees Fahrenheit. It is used to estimate the energy requirements and is an indication of fuel consumption for air conditioning or refrigeration.

❐ coriolis effect

A force per unit mass that arises solely from the earth's rotation, acting as a deflecting force. It is dependent on the latitude and the speed of the moving object. In the Northern Hemisphere, air is deflected to the right of its path, while in the Southern Hemisphere, air is deflected to the left of its path. It is greatest at the poles, North and South, and almost nonexistent at the equator.

❐ coromell

The prevailing evening land breeze which takes place from November to May in the vicinity of La Paz, at the southern tip of Baja California, Mexico.

❐ corona

A pastel halo around the moon or sun created by the diffraction of water droplets. The droplets in the cloud, such as cirrostratus, and the cloud layer itself must be almost perfectly uniform in order for this phenomena to occur. The color display sometimes appears to be iridescent.

❐ corposant

A luminous, sporadic, and often audible, electric discharge. It occurs from objects, especially pointed ones, when the electrical field strength near their surfaces attains a value near 1000 volts per centimeter. It often occurs during stormy weather and might be seen on a ship's mast or yardarm, aircraft, lightning rods, and steeples.

❐ crepuscular rays

Contrasting, alternating bright and dark rays in the sky. Sunlight is scattered by molecules and particles rendering these bright rays visible. Contrast is enhanced by haze, dust, or mist. These rays are more likely to be seen in the late afternoon, as clouds come between the sun and the observer. A similar effect occurs when the sun shines though a break in a layer of clouds.

❐ crystallization

The process of a substance going directly from a vapor form (water vapor) to a solid (ice) at the same temperature, without going through the liquid phase (water). The opposite of sublimation.

❐ cumuliform

Clouds composed of water droplets that exhibit vertical development. The density of the droplets often blocks sunlight, casting shadows on the earth's surface. With increasing vertical height, they are often associated with convection. Bases of these clouds are generally no more than 3,000 feet above the ground, but they can develop past the troposphere in both temperate and tropical latitudes. They are classified as low clouds and include all varieties of cumulus and cumulonimbus. The opposite in type are the horizontal development of stratiform clouds.

❐ cumulonimbus

A vertically developed cumulus cloud, often capped by an anvil-shaped cirriform cloud. Also called a thunderstorm cloud, it is frequently accompanied by heavy showers, lightning, thunder, and sometimes hail, tornadoes or strong, gusty winds.

❐ cumulonimbus mammatus

A portion of a cumulonimbus cloud that appears as a pouch or udder on the under surface of the cloud. Although they do not cause severe weather, they often accompany storms. They may slowly vary in size, since they are an area of negative buoyancy convection, and is associated with severe turbulence in the lower sections of the cloud.

❐ cumulus

One of the three basic cloud forms (the others are cirrus and stratus). It is also one of the two low cloud types. A cloud that develops in a vertical direction from the base (bottom) up. They have flat bases and dome- or cauliflower-shaped upper surfaces. The base of the cloud is often no more than 3,000 feet above the ground, but the top often varies in height. Small, separate cumulus are associated with fair weather (cumulus humilis). With additional heating from the earth's surface, they can grow vertically throughout the day. The top of such a cloud can easily reach 20,000 or more into the troposphere. Under certain atmospheric conditions, these clouds can develop into larger clouds, known as towering cumulus (cumulus congestus), and may produce a rain shower. Further development may create a cumulonimbus.

❐ cumulus congestus

A strongly sprouting cumulus cloud with generally sharp outlines and often with great vertical development. It may occur as tower-like clouds with cauliflower tops. These clouds may produce abundant showers and may develop further into cumulonimbus.

❐ cumulus fractus

Cumulus clouds that appear in irregular fragments, as if they had been shred or torn. Also appears in stratus clouds (called stratus fractus), but not in cirrus clouds.

❐ cumulus humilis

Cumulus clouds with little or no vertical development characterized by a generally flat appearance. Their growth is usually limited by a temperature inversion, which is marked by the unusually uniform height of the clouds. Also called fair-weather cumulus.

❐ cumulus mediocris

Cumulus clouds characterized by moderate vertical development with upper protuberances not very marked in appearance. This cloud does not produce precipitation, but could develop into towering cumulus or cumulonimbus which do.

❐ current

A horizontal movement of water, such as the Gulf Stream off the east coast of North America, or air, such as the jet stream.

❐ cut-off high

A warm high which has become displaced and is on the polarward side of the jet stream. It occurs mostly during the spring and is most frequent over northeastern Siberia, Alaska, and Greenland. It is an example of a blocking high.

❐ cut-off low

A closed cold core low completely removed from the primary westerly flow. Cutoff lows may remain detached from the westerlies for days while exhibiting very little forward (eastward) progress. In some instances, a cutoff low may move to the west, or retrograde, opposite to the prevailing flow. It is important to note that a cutoff low is a closed low, but not all closed lows are cutoff lows.

❐ cyclogenesis

The process that creates a new low pressure system or cyclone, or intensifies a pre-existing one. It is also the first appearance of a trough.

❐ cyclone

An area of closed pressure circulation with rotating and converging winds, the center of which is a relative pressure minimum. The circulation is counterclockwise in the Northern Hemisphere and clockwise in the Southern Hemisphere. Also called a low pressure system and the term used for a tropical cyclone in the Indian Ocean. Other phenomena with cyclonic flow may be referred to by this term, such as dust devils, tornadoes, and tropical and extratropical systems. The opposite of an anticyclone or a high pressure system.

❐ cyclonic flow

Winds that blow in and around a cyclone, that is counterclockwise in the Northern Hemisphere, clockwise in the Southern Hemisphere.

D

❐ daily mean

The average temperature for a day computed by averaging either the hourly readings or, more commonly, the maximum and minimum temperatures.

❐ dalton's law

States that the total pressure exerted by a mixture of gases is equal to the sum of the partial pressures of the gases. Formulated by John Dalton, an English physicist.

❐ data buoys

Buoys placed throughout the Gulf of Mexico and along the Atlantic and Pacific coasts of the United States that relay information on air and water temperature, wind speed, air pressure, and wave conditions via radio signals.

❐ dawn

The first appearance of light in the eastern sky before sunrise. It marks the beginning of morning twilight. The visual display is created by the scattering of light reaching the upper atmosphere prior to the sun's rise to the observer's horizon.

❐ day

Considered a basic unit of time as defined by the earth's motion. It represents the time needed for one complete revolution of the earth about its own axis. Also know as a sidereal day, it is approximately equal to 23 hours, 56 minutes, and 4.09 seconds.

❐ debris cloud

Considered a rotating cloud of debris or dust that is on the ground or near the ground. The debris cloud appearing beneath a thunderstorm will most likely confirm the presence of a tornado.

❐ deepening

Used in describing the history of a low pressure system or an area of cyclonic circulation, it means a decrease in the central pressure of the system. Although it usually describes the action of a pressure system on a constant pressure chart, it also means a surface low is increasing in cyclonic circulation and acquiring more energy.

❐ degree

A measure of temperature difference representing a single division on a temperature scale.

❐ degree day

A measure of the departure of the mean daily temperature from a given standard. That is one degree day for each degree (Fahrenheit or Celsius) of departure above or below the standard during one day.

❐ dense fog advisory

Advisory issued when fog reduces visibility to 1/8 mile or less, creating possible hazardous conditions.

❐ density

The ratio of the mass of a substance to the volume it occupies. In oceanography, it is equivalent to specific gravity and represents the ratio of the weight of a given volume of sea water to that of an equal volume of distilled water at 4.0°C or 39.2°F.

❑ density altitude

The altitude at which a given density is found in the standard atmosphere. Used in aviation, it is computed from the station pressure at takeoff and the virtual temperature at the particular altitude under consideration.

❑ depression

In meteorology, it is another name for an area of low pressure, a low, or trough. It also applies to a stage of tropical cyclone development and is known as a tropical depression to distinguish it from other synoptic features.

❑ derecho

A line of intense, widespread, and fast-moving thunderstorms that moves across a great distance. They are characterized by damaging straight-line winds over hundreds of miles.

❑ dew

Condensation in the form of small water drops that forms on grass and other small objects near the ground when the temperature has fallen to the dew point, generally during the nighttime hours.

❑ dew point

The temperature to which air must be cooled at a constant pressure to become saturated.

❑ diablo winds

Dry winds in the Diablo mountain range in central California that can exceed 60 miles per hour. Similar to the Santa Ana winds, they develop as the wind flows from high pressure over Nevada to lower pressure along the central California coast.

❑ diffluence

A rate at which wind flow spreads apart along an axis oriented normal to the flow in question. The opposite of confluence.

❑ diffraction

The result of light waves interfering with other after passing through a narrow aperture, causing them to bend or spread.

❑ directional shear

The shear created by a rapid change in wind direction with height.

❑ discontinuity

Comparatively large contrast in meteorological elements over a relatively small distance or period of time. In oceanography, it is the abrupt change or jump of a variable at a line or surface.

❑ disturbance

This has several applications. It can apply to a low or cyclone that is small in size and influence. It can also apply to an area that is exhibiting signs of cyclonic development. It may also apply to a stage of tropical cyclone development and is known as a tropical disturbance to distinguish it from other synoptic features.

❒ diurnal

Pertaining to actions or events that occur during a twenty-four hour cycle or recurs every twenty-four hours. Meteorological elements that are measured diurnally include clouds, precipitation, pressure, relative humidity, temperature, and wind.

❒ divergence

Wind movement that results in a horizontal net outflow of air from a particular region. Divergence at lower levels is associated with a downward movement of air from aloft.

❒ dog days

The name given to the very hot summer weather that may persists for four to six weeks between mid-July through early September in the United States. In western Europe, this period may exist from the first week in July to mid-August and is often the period of the greatest frequency of thunder. Named for Sirius, the Dog Star, which lies in conjunction with the sun during this period, it was once believed to intensify the sun's heat during the summer months.

❒ doldrums

Located between 30 degrees North and 30 degrees South latitudes in the vicinity of the equator, this area typically has calm or light and variable winds. Also a nautical term for the equatorial trough.

❒ doppler radar

Weather radar that measures direction and speed of a moving object, such as drops of precipitation, by determining whether atmospheric motion is horizontally toward or away from the radar. Using the Doppler effect, it measures the velocity of particles. Named for J. Christian Doppler, an Austrian physicist, who in 1842 explained why the whistle of an approaching train had a higher pitch than the same whistle when the train was going away.

❒ downburst

A severe localized downdraft from a thunderstorm or shower. This outward burst of cool or colder air creates damaging winds at or near the surface. Sometimes the damage resembles tornadic damage.

❒ downdraft

A sudden descent of cool or cold air to the ground, usually with precipitation, and associated with a thunderstorm or shower.

❒ downpour

A heavy rain.

❒ downslope effect

The warming of an air flow as it descends a hill or mountain slope.

❑ drainage wind

A katabatic wind, it is caused by the cooling of air along the slopes of a mountain.

❑ drifting snow

Snow particles blown from the ground by the wind to a height of less than six feet.

❑ drifts

Normally used when referring to snow or sand particles are deposited behind obstacles or irregularities of the surface or driven into piles by the wind.

❑ drizzle

Slowly falling precipitation in the form of tiny water droplets with diameters less than 0.02 inches or 0.5 millimeters. It falls from stratus clouds and is often associated with low visibility and fog. It is reported as "DZ" in an observation and on the METAR.

❑ dropsonde

A radiosonde dropped with a parachute from an aircraft rather than lifted by a balloon to measure the atmosphere below.

❑ drought

Abnormal dry weather for a specific area that is sufficiently prolonged for the lack of water to cause serious hydrological imbalance.

❑ dry adiabat

The line on a Skew T-Log P chart that depicts the lifting of dry air, or air that is unsaturated. As a parcel rises adiabatically, its pressure decreases and its temperature falls due to the expansion of the air parcel. When an air parcel is unsaturated and rises, then the temperature decreases at a rate of 1°C per 100 meters (5.5°F per 1,000 feet).

❑ dry bulb thermometer

A thermometer used to measure the ambient temperature. The temperature recorded is considered identical to air temperature. One of the two thermometers that make up a psychrometer.

❑ dry line

The boundary between the dry desert air mass of the Southwest U.S. and the moist air mass from the Gulf of Mexico. It usually lies north-south across the central and southern High Plains states during spring and summer. The passage of a dry line results in a sharp decrease in humidity, clearing skies, and a wind shift from southeasterly or south to southwesterly or west. Its presence influences severe weather development in the Great Plains.

❑ dry slot

An area of dry, and usually cloud-free, air that wraps into the southern and eastern sections of a synoptic scale or mesoscale low pressure system. Best seen on a satellite picture, such as a water vapor image.

▫ **dusk**

The period of waning light from the time of sunset to dark.

▫ **dust**

Small particles of earth or other matter suspended in the air. It is reported as "DU" in an observation and for wide spread dust on the METAR.

▫ **dust bowl**

The term given to the area of the Great Plains including Texas, Oklahoma, Kansas, Colorado, and New Mexico that was most greatly affected during the Great Drought of the 1930's.

▫ **dust devil**

A small, rapidly rotating column of wind, made visible by the dust, dirt or debris it picks up. It usually occurs in arid or semi-arid areas and is most likely to develop on clear, dry, hot afternoons in response to surface heating.

▫ **duststorm**

A severe weather condition characterized by strong winds and dust-filled air over a large area. Visibility is reduced to between 5/8ths and 5/16ths statute mile. It is reported as "DS" in an observation and on the METAR.

▫ **d-value**

The deviation of actual altitude along a constant pressure surface from the standard atmosphere altitude of that surface.

▫ **dynamics**

A branch of mechanics that deals with forces and their relations to patterns of motion. In metorology, this relates especially to wind and water patterns.

E

▫ **earthlight (earthshine)**

The faint illumination of the dark part of the moon's disk produced by sunlight reflected onto the moon from the earth's surface and atmosphere.

▫ **earthquake**

A sudden, transient motion or trembling of the earth's crust, resulting from the waves in the earth caused by faulting of the rocks or by volcanic activity.

▫ **easterlies**

Usually applied to the broad patterns of persistent winds with an easterly component, such as the easterly trade winds.

▫ **easterly wave**

An inverted, migratory wave-like disturbance or trough in the tropical region that moves from east to west, generally creating only a shift in winds and rain. The low level convergence and associated convective weather occur on the eastern side of the wave axis. Normally, it moves slower than the atmospheric current in which it is embedded and is considered a weak trough of low pressure. It is often associated with possible tropical cyclone development and is also known as a tropical wave.

❑ echo

The energy return of a radar signal after it has hit the target.

❑ eclipse

The obscuring of one celestial body by another.

❑ ecliptic

The sun's apparent path across the sky that tracks a circle through the celestial sphere.

❑ ecology

The study of the relationships between living organisms and their environment.

❑ eddy

A small disturbance of wind in a large wind flow, which can produce turbulent conditions. They can also be areas of warmer air north of the main westerlies or colder air south of the westerlies. In oceanic circulation, it is a circular movement of water usually formed where currents pass obstructions, between two adjacent currents flowing counter to each other, or along the edge of a permanent current.

❑ el niã'o

The cyclical warming of East Pacific Ocean sea water temperatures off the western coast of South America that can result in significant changes in weather patterns in the United States and elsewhere. This occurs when warm equatorial waters move in and displace the colder waters of the Humbolt Current, cutting off the upwelling process.

❑ electromagnetic radiation

Also called radiation, it is waves of energy propagated though space or through a material media.

❑ electromagnetic spectrum

The band of electromagnetic radiation with components that are separated into their relative wave lengths. The portion of the spectrum that the human eye can detect is called visible light, between the longer infrared waves and the shorter ultraviolet waves. The various types of energy comprising the spectrum are (from longest to shortest) radio, infrared, visible, ultraviolet, x-rays, gamma rays, and cosmic rays.

❑ elevation

The measure of height with respect to a point on the earth's surface above mean sea level.

❐ environment

The sum total of all the external conditions that effect an organism, community, material, or energy.

❐ equator

The geographic circle at 0 degrees latitude on the earth's surface. It is equal distance from the North and South Poles and divides the Northern Hemisphere from the Southern.

❐ equatorial trough

The quasi-continuous area of low pressure between the subtropical high pressure areas in both the Northern and Southern Hemisphere.

❐ equinox

The point at which the ecliptic intersects the celestial equator. Days and nights are most nearly equal in duration. In the Northern Hemisphere, the vernal equinox falls on or about March 20 and the autumnal equinox on or about September 22.

❐ erosion

The movement of soil or rock from one area to another by the action of the sea, running water, moving ice, precipitation, or wind.

❐ evaporation

The physical process by which a liquid, such as water is transformed into a gaseous state, such as water vapor. It is the opposite physical process of condensation.

❐ evapotranspiration

The total amount of water that is transferred from the earth's surface to the atmosphere. It is made up of the evaporation of liquid or solid water plus the transpiration from plants.

❐ exosphere

This region is considered the very outer limits of the earth's atmosphere. Its lower boundary is often called the critical level of escape, where gas atoms are so widely spaced that they rarely collide with one another and have individual orbits. It is estimated to be some 400 plus miles (640 kilometers) above the surface.

❐ extratropical cyclone

Any cyclone not of tropical origin. Generally considered to be a migratory frontal cyclone found in the middle and high latitudes.

❐ eye

The center of a tropical storm or hurricane, characterized by a roughly circular area of light winds and rain-free skies. An eye will usually develop when the maximum sustained wind speeds exceed 78 mph. It can range in size from as small as 5 miles to up to 60 miles, but the average size is 20 miles. In general, when the eye begins to shrink in size, the storm is intensifying.

❑ eye wall

An organized band of convection surrounding the eye, or center, of a tropical cyclone. It contains cumulonimbus clouds, intense rainfall and very strong winds.

F

❑ fahrenheit temperature scale

A temperature scale where water at sea level has a freezing point of +32°F and a boiling point of +212°F. More commonly used in areas that observe the English system of measurement. Created in 1714 by Gabriel Daniel Fahrenheit (1696-1736), a German physicist, who also invented the alcohol and mercury thermometers.

❑ fair

This is a subjective description. Considered as pleasant weather conditions with regard to the time of year and the physical location.

❑ fathom

The common unit of depth in the ocean for countries using the English system of measurement. It is six feet or 1.83 meters. It can also be used in expressing horizontal distance, since 120 fathoms is equal to one cable or nearly on tenth of a nautical mile.

❑ feeder bands

In tropical parlance, the lines or bands of thunderstorms that spiral into and around the center of a tropical system. Also known as outer convective bands, a typical hurricane may have three or more of these bands. They occur in advance of the main rain shield and are usually 40 to 80 miles apart. In thunderstorm development, they are the lines or bands of low level clouds that move or feed into the updraft region of a thunderstorm.

❑ fetch

An area of the water surface over which waves are generated by a wind having a constant direction and speed. Also, it is the name given to the length of the fetch area, measured in the direction of the wind from which the seas are generated. One of the ingredients for lake effect snow is the fetch of the water over which cold air can gain moisture.

❑ few

The amount of sky cover for a cloud layer between 1/8th and 2/8ths, based on the summation layer amount for that layer.

❑ filling

Used in describing the history of a low pressure system or an area of cyclonic circulation, it means an increase in the central pressure of the system. Although it usually describes the action of a pressure system on a constant pressure chart, it also means a surface low is decreasing in cyclonic circulation and losing its characteristics. The opposite of deepening.

❒ firewhirl

A tornado-like rotating column of fire and smoke created by intense heat from a forest fire or volcanic eruption.

❒ first gust

Another name for the initial wind surge observed at the surface as the result of downdrafts forming the leading edge or gust front of a thunderstorm.

❒ flanking line

A line of attached cumulus or towering cumulus clouds of descending height, appearing as stair steps (usually on the southwest side) of the most active part of a supercell.

❒ flash flood

A flood that rises and falls quite rapidly with little or no advance warning, usually as the result of intense rainfall over a relatively small area. Flash floods can be caused by situations such as a sudden excessive rainfall, the failure of a dam, or the thaw of an ice jam.

❒ flood

High water flow or an overflow of rivers or streams from their natural or artificial banks, inundating adjacent low lying areas.

❒ flood plain

Level land that may be submerged by flood waters.

❒ flood stage

The level of a river or stream where overflow onto surrounding areas can occur.

❒ foehn

A warm dry wind on the lee side of a mountain range, whose temperature is increased as the wind descends down the slope. It is created when air flows downhill from a high elevation, raising the temperature by adiabatic compression. Classified as a katabatic wind.

❒ fog

A visible aggregate of minute water droplets suspended in the atmosphere at or near the surface of the earth, reducing horizontal visibility to less than 5/8 statute miles. It is created when the temperature and the dew point of the air have become the same, or nearly the same, and sufficient condensation nuclei are present. It is reported as "FG" in an observation and on the METAR.

❒ fog bank

A fairly well-defined mass of fog observed in the distance. Most commonly seen at sea, over a lake, or along coastal areas.

❐ fogbow

A whitish semicircular arc seen opposite the sun in fog. The outer margin has a reddish tinge, its inner margin has a bluish tinge, and the middle of the band is white. An additional bow with reversed colors sometimes appears inside the first.

❐ forecast

A statement of expected future occurrences. Weather forecasting includes the use of objective models based on certain atmospheric parameters, along with the skill and experience of a meteorologist.

❐ fractus

The elements of cumulus and stratus clouds that appear in irregular fragments, as if they had been shred or torn. Never appears in cirrus clouds. Also known as scud.

❐ freezing drizzle

Drizzle, falling as a liquid, but freezing on impact with the colder ground or other exposed surfaces. It is reported as "FZDZ" in an observation and on the METAR.

❐ freezing fog

Used to describe the phenomena when fog is present and the air temperature is below 0°C. It is reported as "FZFG" in an observation and on the METAR.

❐ freezing point/freeze

The process of changing a liquid to a solid. The temperature at which a liquid solidifies under any given set of conditions. Pure water under atmospheric pressure freezes at 0°C or 32°F. It is the opposite of fusion. In oceanography, the freezing point of water is depressed with increasing salinity.

❐ freezing precipitation

Precipitation that is liquid, but freezes upon impact with a solid surface, such as the ground or other exposed surfaces.

❐ freezing rain

Rain that falls as liquid and freezes upon impact to form a coating of glaze on the colder ground or other exposed surfaces. It is reported as "FZRA" in an observation and on the METAR.

❐ fresh water

Water found rivers, lakes, and rain, that is distinguished from salt water by its appreciable lack of salinity.

❐ friction

In meteorology, it is the turbulent resistance of the earth on the atmosphere. Considered as the resistance of fluids (air and water) to the relative motion of a solid body. The amount is dependent on the size and shape of the body.

❒ friction layer

The thin layer of atmosphere adjacent to the earth's surface. Surface friction is effective in slowing down wind up to approximately 1,500 to 3,000 feet above the ground. Above this level, air tends to flow parallel to the isobars. Wind distribution within this layer is determined by vertical temperature gradient and the physical contours of the underlying surface features.

❒ front

The transition zone or interface between two air masses of different densities, which usually means different temperatures. For example, the area of convergence between warm, moist air and cool, dry air.

❒ frontal passage

It is the passage of a front over a specific point on the surface. It is reflected by the change in dew point and temperature, the shift in wind direction, and the change in atmospheric pressure. Accompanying a passage may be precipitation and clouds. May be referred to as "fropa."

❒ frontogenesis

The birth or creation of a front. This occurs when two adjacent air masses exhibiting different densities and temperatures are brought together by prevailing winds, creating a front. It could happen when either air mass, or both, move over a surface which strengthens their original properties. However, it occurs most often along the eastern coasts of North America and Asia, when the air mass moving out over the ocean has a weak or no distinct boundary. The opposite of frontolysis.

❒ frontolysis

The destruction or dying of a front where the transition zone is losing its contrasting properties. The opposite of frontogenesis.

❒ frost

The covering of ice crystals that forms by direct sublimation on exposed surfaces whose temperature is below freezing.

❒ frozen precipitation

Precipitation that reaches the ground in a frozen state. Examples include snow, snow pellets, snow grains, ice crystals, ice pellets, and hail.

❒ fujita-pearson scale

A scale that classifies the severity of wind damage intensity based on the degree of destruction as it relates to the wind speed as well as path length and path width of the event. It is normally used to identify the most intense damage exhibited by a tornado. Developed by T. Theodore Fujita and Allen Pearson.

❒ funnel cloud

A violent, rotating column of air visibly extending from the base of a towering cumulus or cumulonimbus toward the ground, but not in contact with it. It is reported as "FC" in an observation and on the METAR.

❒ fusion

The change of state from a solid to a liquid at the same temperature. The heat of fusion is the number of gram calories of heat necessary to change one gram of a substance from the solid to the liquid state. It is the opposite of freezing.

G

❒ gale

On the Beaufort Wind Scale, a wind with speeds from 28 to 55 knots (32 to 63 miles per hour). For marine interests, it can be categorized as a moderate gale (28 to 33 knots), a fresh gale (34 to 40 knots), a strong gale (41 to 47 knots), or a whole gale (48 to 55 knots). In 1964, the World Meteorological Organization defined the categories as near gale (28 to 33 knots), gale (34 to 40 knots), strong gale (41 to 47 knots), and storm (48 to 55 knots).

❒ gale warning

A warning for marine interests for impending winds from 28 to 47 knots (32 to 54 miles per hour).

❒ geophysics

The study of the physics or nature of the Earth and its environment. It deals with the composition and physical phenomena of the earth and its liquid and gaseous envelopes. Areas of studies include the atmospheric sciences and meteorology, geology, seismology, and volcanology, and oceanography and related marine sciences, such as hydrology. By extension, it often includes astronomy and the related astro-sciences.

❒ geosphere

Considered the solid portions of the earth, including the hydrosphere and the lithosphere, as opposed to the atmosphere, which lies above it. At their conjunction is the biosphere.

❒ geostationary satellite

An orbiting weather satellite that maintains the same position over the equator during the earth's rotation. Also known as GOES, an acronym for Geostationary Operational Environmental Satellite.

❒ geostrophic wind

A steady horizontal motion of air along straight, parallel isobars or contours in an unchanging pressure or contour field. It is assumed that there is no friction, that the flow is straight with no curvature and there is no divergence or convergence with no vertical acceleration.

❒ glacier winds

Air flow that descends from glaciers, occasionally at a high rate of speed. Caused by the temperature difference between the air in contact with the glacier and the air at the same altitude, it reaches maximum intensity in the early afternoon.

❐ glaze

A smooth clear icy coating of supercooled water droplets that spread out and freeze onto objects on contact. A storm that produces the accretion of glaze is called an ice storm.

❐ gradient wind

A steady horizontal air motion along curved parallel isobars or contours in an unchanging pressure or contour field, assuming there is no friction and no divergence or convergence.

❐ graupel

A form of frozen precipitation consisting of snowflakes or ice crystals and supercooled water droplets frozen together.

❐ gravitation

The mutual attraction between two masses of matter. The rotation of the earth and the atmosphere modifies this attraction to produce the field of gravity.

❐ gravity

The force of attraction of the earth on an object. The direction is downward relative to the earth, and it decreases with elevation or altitude away from the earth's surface.

❐ green flash

A brilliant green coloration of the upper edge of the sun, occasionally seen as the sun's apparent disk is about to set below a clear horizon.

❐ greenhouse effect

The overall warming of the earth's lower atmosphere primarily due to carbon dioxide and water vapor which permit the sun's rays to heat the earth, but then restrict some heat-energy from escaping back into space.

❐ greenwich mean time (gmt)

The name of the twenty-four hour time scale which is used throughout the scientific and military communities. Standard Time begins at Greenwich, England, home of the Royal Observatory which first utilized this method of world time. This is also the Prime Meridian of Longitude. The globe is divided into twenty-four (24) time zones of 15 degrees of arc, or one hour in time apart. To the east of this meridian, time zones are number from 1 to 12 and prefixed with a minus (-), indicting the number of hours to be subtracted to obtain Greenwich Time (GMT). To the west, the time zones are also numbered 1 through 12, but are prefixed with a plus (+), indicating the number of hours to be added to obtain GMT.

❐ ground clutter

A pattern of radar echoes reflecting off fixed ground targets such as buildings or hills near the radar. This may hide or confuse the proper return echo signifying actual precipitation.

❒ ground fog

Fog created when radiational cooling at the earth's surface lowers the temperature of the air near the ground to or below its initial dew point. Primarily takes place at night or early morning.

❒ growing season

Considered the period of the year during which the temperature of cultivated vegetation remains sufficiently high enough to allow plant growth. Usually considered the time period between the last killing frost in the spring and the first killing frost of the autumn. The frost-free growing season is between the first and last occurrence of 32Â°F temperatures in spring and autumn.

❒ gulf stream

The warm, well-defined, swift, relatively narrow ocean current which exists off the east coast of the United States, beginning near Cape Hatteras. The term also applies to the oceanic system of currents that dominate the western and northern Atlantic Ocean: the Florida current, which flows through the Florida Straits between the Florida Keys and Cuba and northwards; the Gulf Stream, which begins around Cape Hatteras and flows northeasterly off the continental slope into the North Atlantic; and the North Atlantic current, which begins around the Grand Banks off Newfoundland and continues east-northeastwards towards the British Isles.

❒ gully washer

A heavy rain shower that occurs suddenly, possibly creating a flash flood.

❒ gust

A sudden significant increase in or rapid fluctuations of wind speed. Peak wind must reach at least 16 knots (18 miles per hour) and the variation between peaks and lulls is at least 10 knots (11.5 miles per hour). The duration is usually less twenty seconds.

❒ gust front

The leading edge of the cool, gusty surface winds produced by thunderstorm downdrafts. Sometimes confused with an outflow boundary.

❒ gustnado

A weak, and usually short-lived, tornado that forms along the gust front of a thunderstorm, appearing as a temporary dust whirl or debris cloud.

H

❒ haboob

Sudanese name for duststorm or sandstorm with strong winds that carry small particles of dirt or sand into the air, particularly severe in areas of drought.

❒ hail

Precipitation that originates in convective clouds, such as cumulonimbus, in the form of balls or irregular pieces of ice, which comes in different shapes and sizes. Hail is considered to have a diameter of 5 millimeter or more; smaller bits of ice are classified as ice pellets, snow pellets, or graupel. Individual lumps are called hailstones. It is reported as "GR" in an observation and on the METAR. Small hail and/or snow pellets is reported as "GS" in an observation and on the METAR.

❐ halo

The ring of light that seems to encircle the sun or moon when veiled by cirrus clouds. To produce this phenomena, the ice crystals must be in a heterogeneous arrangement to refract the sunlight. The most commonly observed is a halo that forms at a 22° radius, although another one at 46° radius may also be seen.

❐ haze

A suspension of fine dust and/or smoke particles in the air. Invisible to the naked eye, the particles reduce visibility by being sufficiently numerous to give the air an opalescent appearance. It is reported as "HZ" in an observation and on the METAR.

❐ heat

A form of energy transferred between two systems by virtue of a difference in temperature. The first law of thermodynamics demonstrated that the heat absorbed by a system may be used by the system to do work or to raise its internal energy.

❐ heat balance

The equilibrium which exists on the average between the radiation received by the earth and atmosphere from the sun and that emitted by the earth and atmosphere. The balance between heat loss (long wave radiation from the earth back into the atmosphere) and heat gain (incoming solar radiation).

❐ heat exhaustion

The effect of excessive heat, particularly when combined with high humidity, on a human being. Signs of heat exhaustion include a general weakness, heavy sweating and clammy skin, dizziness and/or fainting, and muscle cramps.

❐ heat index

The combination of air temperature and humidity that gives a description of how the temperature feels. This is not the actual air temperature.

❐ heat lightning

Lightning that appears as a glowing flash on the horizon. It is actually lightning occurring in distant thunderstorms, just over the horizon and too far away for thunder to be heard.

❐ heat stroke

Introduced to the body by overexposure to high temperatures, particularly when accompanied by high humidity. The signs of heat stroke include when an individual's body temperature is greater than 105°F, the skin is hot and dry, there is a rapid and irregular pulse, perspiration has stopped, and one has lost consciousness.

Seek immediate medical aid. May be called a sun-stroke when caused by direct exposure to the sun.

▫ heat wave

A period of abnormally and uncomfortably hot weather. It could last from several days to several weeks. The Weather Channel uses the following criteria for a heat wave: a minimum of ten states must have 90°F plus temperatures and the temperatures must be at least five degrees above normal in parts of that area for at least two days or more.

▫ heating degree day

One heating degree day is given for each degree that the daily mean temperature is below 65°F. It is used as an indication of fuel consumption.

▫ helicity

A property of a moving fluid, such as air, representing the potential for helical flow (flow that follows a corkscrew pattern). Computed from the vertical wind profile of the lower atmosphere and measured relative to the motion as a storm, it is used to forecast the formation of mesocyclones.

▫ high clouds

A term used to signify cirriform clouds that are composed of ice crystals and generally have bases above 20,000 feet. The main types of high clouds are cirrus,cirrocumulus, and cirrostratus. This altitude applies to the temperate zone. In the polar regions, these clouds may be found at lower altitudes. In the tropics, the defining altitudes for cloud types are generally higher.

▫ high latitudes

The latitude belt roughly between 60° and 90° North and South.

▫ high pressure system

An area of relative pressure maximum that has diverging winds and a rotation opposite to the earth's rotation. This is clockwise the in Northern Hemisphere and counterclockwise in the Southern Hemisphere. It is the opposite of an area of low pressure or a cyclone.

▫ hoarfrost

Another name for frost. A deposit of hoarfrost occurs when air with a dew point below freezing is brought to saturation by cooling.

▫ hook echo

A radar reflectivity pattern observed in a thunderstorm, appearing like a fish hook and indicating favorable conditions for tornadic development. However, hook echoes and tornadoes do not always accompany each other.

▫ horizon

One of several lines or planes used as reference for observation and measurement relative to a given location on the surface of the earth. The geographic horizon, also called the apparent horizon, is the distant line along which earth and sky appear to meet. This is the usual concept of horizon and is used in weather observing. The local horizon is the actual lower boundary of the observed sky or the upper outline of terrestrial objects including nearby natural obstructions, such as mountains.

❐ horse latitudes

Located between 30° North and 30° South in the vicinity of the equator, this area typically has calm or light and variable winds.

❐ hudson bay low

An area of low pressure over or near the Hudson Bay area of Canada that often introduces cold air to the north central and northeast United States.

❐ humboldt current

Also known as the Peru Current, this ocean current flows northward along the western side of South America, offshore Chile and Peru. There is considerable upwelling of the colder subsurface waters due to the prevailing southerly winds. Dominant weather in this area includes coastal fog and low clouds. The presence or lack of this current is a vital part of the meteorological-oceanographic pattern known as El NiÃ±o.

❐ humidity

The amount of water vapor in the air. It is often confused with relative humidity or dew point.

❐ hurricane

The name for a tropical cyclone with sustained winds of 74 miles per hour (65 knots) or greater in the North Atlantic Ocean, Caribbean Sea, Gulf of Mexico, and in the eastern North Pacific Ocean. This same tropical cyclone is known as a typhoon in the western Pacific and a cyclone in the Indian Ocean.

❐ hurricane warning

A formal advisory issued by forecasters at the National Hurricane Center when they have determined that hurricane conditions are expected in a coastal area or group of islands within a 24 hour period. A warning is used to inform the public and marine interests of the storm's location, intensity, and movement.

❐ hurricane watch

A formal advisory issued by forecasters at the National Hurricane Center when they have determined that hurricane conditions are a potential threat to a coastal area or group of islands within a 24 to 36 hour period. A watch is used to inform the public and marine interests of the storm's location, intensity, and movement.

❐ hydrologic cycle

Often called the water cycle, it is the vertical and horizontal transport of water in all its states between the earth, the atmosphere, and the seas.

❐ hydrology

The study of the waters of the earth, especially with relation to the effects of precipitation and evaporation upon the occurrence and character of water in streams, lakes, and on or below the land surface.

❐ hydrometeor

Any any form of atmospheric water vapor, including those blown by the wind off the earth's surface. Liquid or solid water formation that is suspended in the air includes clouds, fog, ice fog, and mist.

1. Drizzle and rain are example of liquid precipitation, while freezing drizzle and freezing rain are examples of freezing precipitation.
2. Solid or frozen precipitation includes ice pellets, hail, snow, snow pellets, snow grains, and ice crystals.
3. Water vapor that evaporates before reaching the ground is virga.
4. Examples of liquid or solid water particles that are lifted off the earth's surface by the wind includes drifting and blowing snow and blowing spray.
5. Dew, frost, rime, and glaze are examples of liquid or solid water deposits on exposed objects.

❐ hydrosphere

Considered as the water portion of the earth's surface. Part of the geosphere.

❐ hygrograph

An instrument that records the hygrometer's measure of water vapor.

❐ hygrometer

An instrument that measures the water vapor content of the atmosphere.

❐ hypothermia

This situation occurs when the core temperature of one's body falls below normal. It is the failure of the body to maintain adequate production of heat under conditions of extreme cold.

I

❐ ice

The solid form of water. It can be found in the atmosphere in the form of ice crystals, snow, ice pellets, and hail for example.

❐ ice crystals

Precipitation in the form of slowly falling, singular or unbranched ice needles, columns, or plates. They make up cirriform clouds, frost, and ice fog. Also, they produce optical phenomena such as halos, coronas, and sun pillars. May be called "diamond dust." It is reported as "IC" in an observation and on the METAR.

▫ ice fog

Fog that is composed of minute ice particles. It occurs in very low temperatures under clear, calm conditions in the polar latitudes and may produce a halo around the sun or moon.

▫ ice jam

An accumulation of broken river ice caught in a narrow channel, frequently producing local flooding. Primarily occurs during a thaw in the late winter or early spring.

▫ ice pellets

Precipitation in the form of transparent or translucent pellets of ice, which are round or irregular in shape. They have a diameter of 0.2 inches (5 mm) or less. They are classified into two types: hard grains of ice consisting of frozen rain drops or largely melted and refrozen snowflakes; pellets of snow encased in a thin layer of ice which have formed from the freezing of droplets intercepted by pellets or water resulting from the partial melting of pellets. It is reported as "PE" in an observation and on the METAR.

▫ ice storm

A severe weather condition characterized by falling freezing precipitation. Such a storm forms a glaze on objects, creating hazardous travel conditions and utility problems.

▫ icelandic low

A semi-permanent, subpolar area of low pressure in the North Atlantic Ocean. Because of its broad area and range of central pressure, it is an area where migratory lows tend to slow down and deepen. It is strongest during a Northern Hemisphere winter and early spring, centered over Iceland and southern Greenland, and is the dominate weather feature in the area. During the summer, it is weaker, less intense, and might divide into two parts, one west of Iceland, the other over the Davis Strait between Greenland and Baffin Island. Then the Azores or Bermuda High becomes the dominate weather feature in the North Atlantic.

▫ icicle

Ice that forms in the shape of a narrow cone hanging point down. It usually forms when liquid water from a sheltered or heated source comes in contact with below-freezing air and freezes more or less rapidly as it flows.

▫ icing

The forming or depositing of ice on an object.

▫ inches of mercury (hg)

The name comes from the use of mercurial barometers which equate the height of a column of mercury with air pressure. One inch of mercury is equivalent to 33.86 millibars or 25.40 millimeters. First devised in 1644 by Evangelista Torricelli (1608-1647), an Italian physicist and mathematician, to explain the fundamental principles of hydromechanics.

A period of abnormally warm weather in mid to late autumn with clear skies and cool nights. A first frost normally precedes this warm spell.

❐ infrared

The long wave, electromagnetic radiation of radiant heat emitted by all hot objects. On the electromagnetic spectrum, it can be found between microwave radiation and visible light. Water vapor, ozone, and carbon dioxide are capable of absorbing or transmitting infrared radiation. May be referred to as IR.

❐ insolation

Solar radiation or heating received at the earth's surface. The name is derived from incoming solar radiation.

❐ instability

The state of equilibrium in which a parcel of air when displaced has a tendency to move further away from its original position. It is the condition of the atmosphere when spontaneous convection and severe weather can occur. Air parcels, when displaced vertically, will accelerate upward, often forming cumulus clouds and possibly thunderstorms.

❐ instrument flight rules (ifr)

Refers to the general weather conditions pilots can expect at the surface and applies to the weather situations at an airport during which a pilot must use instruments to assist take off and landing. IFR conditions for fixed wing aircraft means the minimum cloud ceiling is greater than 500 feet and less than 1,000 feet and/or visibility is greater than 1 mile and less than 3 miles.

❐ instrument shelter

A boxlike structure designed to protect temperature measuring instruments from exposure to direct sunshine, precipitation, and condensation, while at the same time time providing adequate ventilation.

❐ intermountain high

An area of high pressure that occurs during the winter between the Rocky Mountains and the Sierra-Cascade ranges. It blocks the eastward movement of Pacific cyclones. Also called Plateau High or Great Basin High.

❐ international date line

The line of longitude located at 180Â° East or West (with a few local deviations) where the date changes by a day. West of the line it is one day later than east of the line.

❐ intertropical convergence zone (itcz)

An area where the Northern and Southern Hemispheric trade winds converge, usually located between 10 degrees North and South of the equator. It is a broad area of low pressure where both the Coriolis force and the low-level pressure gradient are weak, occasionally allowing tropical disturbances to form. It fluctuates in location, following the sun's rays, so that during the Northern Hemisphere summer, the ITCZ moves northward over the southern North Atlantic and southern Asia.

❐ **inversion**

A departure from the usual increase or decrease of an atmospheric property with altitude. It usually refers to an increase in temperature with increasing altitude, which is a departure from the usual decrease of temperature with height.

❐ **ionosphere**

A complex atmospheric zone of ionized gases that extends between 50 and 400 miles (80 to 640 kilometers) above the earth's surface. It is located between the mesosphere and the exosphere and is included as part of the thermosphere.

❐ **isallobar**

The line of equal change in atmospheric pressure during a certain time period. It marks the change in pressure tendency.

❐ **isobar**

The line drawn on a weather map connecting points of equal barometric pressure.

❐ **isodrosotherm**

The line drawn on a weather map connecting points of equal dew point.

❐ **isohel**

A line drawn through geographic points having equal duration of sunshine or another form of solar radiation during a specified time period.

❐ **isohyet**

The line drawn through geographic points recording equal amounts of rainfall during a given time or for a given of storm.

❐ **isopleth**

A line connecting equal points of value. Also called an isoline.

❐ **isotach**

A line connecting equal wind speeds.

❐ **isotherm**

The line of equal or constant air temperature. If something is isothermal, it is of equal or constant temperature with respect to either time or space.

J

❐ **jet streak**

A region of accelerated wind speed along the axis of a jet stream.

❐ **jet stream**

An area of strong winds that are concentrated in a relatively narrow band in the upper troposphere of the middle latitudes and subtropical regions of the Northern and Southern Hemispheres. Flowing in a semi-continuous band around the globe from west to east, it is caused by the changes in air temperature where the cold polar air moving towards the equator meets the warmer equatorial air moving polarward. It is marked by a concentration of isotherms and strong vertical shear.

K

☐ k index

The measure of thunderstorm potential based on the vertical temperature lapse rate, the moisture content of the lower atmosphere and the vertical extent of the moist layer.

☐ katabatic wind

A wind that is created by air flowing downhill. When this air is warm, it may be called a foehn wind, and regionally it may be known as a Chinook or Santa Ana. When this air is cold or cool, it is called a drainage wind, and regionally it may be known as a mountain breeze or glacier wind. The opposite of an anabatic wind.

☐ katafront

A front where the warm air descends the frontal surface, except in the low layers of the atmosphere.

☐ kelvin temperature scale

A temperature scale with the freezing point of +273°K (Kelvin) and the boiling point of +373° K. It is used primarily for scientific purposes. Also known as the Absolute Temperature Scale. Proposed in 1848 by William T. Kelvin, 1st Baron of Largs (1824-1907), Irish-born Scottish physicist and mathematician.

☐ knot

A nautical unit of speed equal to the velocity at which one nautical mile is traveled in one hour. Used primarily by marine interests and in weather observations. A knot is equivalent to 1.151 statute miles per hour or 1.852 kilometers per hour.

L

☐ lake effect snow

Snow showers that are created when cold dry air passes over a large warmer lake, such as one of the Great Lakes, and picks up moisture and heat.

☐ land breeze

A diurnal coastal breeze that blows offshore, from the land to the sea. It is caused by the temperature difference when the sea surface is warmer than the adjacent land. Predominate during the night, it reaches its maximum about dawn. It blows in the opposite direction of a sea breeze.

☐ landfall

The point at which a tropical cyclone's eye first crosses a land mass.

▫ landspout

A small, weak tornado, which is not formed by a storm-scale rotation. It is generally weaker than a supercell tornado and is not associated with a wall cloud or mesocyclone. It may be observed beneath cumulonimbus or towering cumulus clouds and is the land equivalent of a waterspout.

▫ lapse rate

The change of an atmospheric variable, usually temperature, with height. A steep lapse rate implies a rapid decrease in temperature with height and is a sign of instability.

▫ latent heat

The energy released or absorbed during a change of state.

▫ latitude

The location north or south in reference to the equator, which is designated at zero (0) degrees. Parallel lines that circle the globe both north and south of the equator. The poles are at 90Â° North and South latitude.

▫ lee/leeside/leeward

The side of an object or obstacle, such as a ship's sail, a mountain, or a hill, furthest away from the wind, and therefore, protected from the direct force of the wind. The opposite of windward.

▫ lenticular cloud

A cloud species which has elements resembling smooth lenses or almonds and more or less isolated. These clouds are caused by a wave wind pattern created by the mountains. They are also indicative of down-stream turbulence on the leeward side of a barrier.

▫ level of free convection (lfc)

The level at which a parcel of saturated air becomes warmer than the surrounding air and begins to rise freely. This occurs most readily in a conditionally unstable atmosphere.

▫ lifted index (li)

A measure of atmospheric instability that is obtained by computing the temperature that the air near the ground would have if it were lifted to a higher level and comparing it to the actual temperature at that altitude. Positive values indicate more stable air and negative values indicate instability.

▫ lifting condensation level (lcl)

The height at which a parcel of moist air becomes saturated when it is lifted dry adiabatically.

▫ light waves

That part of the electromagnetic spectrum that contains visible light. The colors, from longest wave length to shortest, are red, orange, yellow, green, blue, indigo, and violet (ROY G. BIV).

❑ **lightning**

A sudden and visible discharge of electricity produced in response to the build up of electrical potential between cloud and ground, between clouds, within a single cloud, or between a cloud and surrounding air.

❑ **line echo wave pattern (lewp)**

A wave-shaped bulge in a line of thunderstorms. It may often be seen as a "S"-shaped radar echo signature and is often associated with severe weather.

❑ **lithometeor**

Atmospheric phenomena which affect the state of the atmosphere. They constitute dry particles that hang suspended in the atmosphere, such as dust, smoke, sand, and haze.

❑ **lithosphere**

The solid, outer portion of the earth's crust coupled to the rigid upper mantle. Part of the geosphere.

❑ **long wave trough**

A wave in the prevailing westerly flow aloft which is characterized by a large length and amplitude. A long wave moves slowly and is persistent. Its position and intensity govern weather patterns over a period of days or weeks.

❑ **longitude**

The location east or west in reference to the Prime Meridian, which is designated as zero (0) degrees longitude. The distance between lines of longitude are greater at the equator and smaller at the higher latitudes, intersecting at the earth's North and South Poles. Time zones are correlated to longitude.

❑ **low clouds**

A term used to signify clouds with bases below 6,000 feet and are of a stratiform or a cumuliform variety. Stratiform clouds include stratus and stratocumulus. Cumuliform clouds include cumulus and cumulonimbus. This altitude applies to the temperate zone. In the polar regions, these clouds may be found at lower altitudes. In the tropics, the defining altitudes for cloud types are generally higher.

❑ **low latitudes**

The latitude belt between 30 and 0 degrees North and South of the equator. Also referred to as the tropical or torrid region.

❑ **low level jet (llj)**

Strong winds that are concentrated in relatively narrow bands in the lower part of the atmosphere. It is often amplified at night. The southerly wind over the US Plains states during spring and summer is a notable example.

❑ **low pressure system**

An area of a relative pressure minimum that has converging winds and rotates in the same direction as the earth. This is counterclockwise in the Northern

Hemisphere and clockwise in the Southern Hemisphere. Also known as an cyclone, it is the opposite of an area of high pressure, or a anticyclone.

❒ lunar eclipse

An eclipse of the moon occurs when the earth is in a direct line between the sun and the moon. The moon does not have any light of its own, instead, it reflects the sun's light. During a lunar eclipse, the moon is in the earth's shadow. It will often look dim and sometimes copper or orange in color.

M

❒ mackerel sky

The name given to cirrocumulus clouds with small vertical extent and composed of ice crystals. The rippled effect gives the appearance of fish scales.

❒ macroburst

A large downburst with an outflow diameter of 2.5 miles (4 kilometers) or larger and damaging winds.

❒ macroscale

The meteorological scale covering an area ranging from the size of a continent to the entire globe.

❒ magnetic poles

Either of the two points on the earth's surface where the magnetic meridians converge. They are not aligned with the geographical poles, but shift and do not lie exactly opposite of the other.

❒ mammatocumulus

An obsolete term for cumulonimbus mammatus, it is a portion of a cumulonimbus cloud that appears as a pouch or udder on the under surface of the cloud. Although they do not cause severe weather, they often accompany storms.

❒ mare's tail

The name given to thin, wispy cirrus clouds composed of ice crystals that appear as veil patches or strands, often resembling a horse's tail.

❒ marginal visual flight rules (mvfr)

Refers to the general weather conditions pilots can expect at the surface. MVFR means Minimum or Marginal Visual Flight Rules. MVFR criteria means a ceiling between 1,000 and 3,000 feet and/or 3 to 5 miles visibility.

❒ maritime air mass

An air mass influenced by the sea. It is a secondary characteristic of an air mass classification, signified by the small "m" before the primary characteristic, which is based on source region. For example, mP is an air mass that is maritime polar in nature. Also known as a marine air mass.

▫ maximum

The greatest value attained by a function, for example, temperature, pressure, or wind speed. The opposite of minimum.

▫ mean sea level

The average height of the sea surface water level. For the United States, it is computed by averaging the levels of all tide stages over a nineteen year period, determined from hourly height readings measured from a fix, predetermined reference level. It is used as a basis for determining elevations, as the reference for all altitudes in upper air measurements, and as the level above which altitude is measured by a pressure altimeter for aviation. Often referred to as MSL.

▫ mean temperature

The average of temperature readings taken over a specified amount of time. Often the average of the maximum and minimum temperatures.

▫ measured ceiling

A ceiling classification applied when the ceiling value has been determined by an instrument, such as a ceilometer or ceiling light, or by the known heights of unobscured portions of objects, other than natural landmarks, near the runway.

▫ melting level

The altitude at which ice crystals and snow flakes melt as they descend through the atmosphere.

▫ melting point

The temperature at which a solid substance undergoes fusion, changing from a solid to a liquid state. Contrast with freezing point.

▫ mercurial barometer

An instrument used for measuring the change in atmospheric pressure. It uses a long glass tube, open at one end and closed at the other. After first filling the open end with mercury, it is then temporarily sealed and placed into a cistern of mercury. A nearly perfect vacuum is established at the closed end after the mercury descends. The height of the column of mercury in the tube is a measurement of air pressure. As atmospheric pressure increases, the mercury is forced from the cistern up the tube; when the atmospheric pressure decreases, the mercury flows back into the cistern. Measurement is taken in inches of mercury. Although mercurial barometers are very accurate, practicality has led observers to use aneroid barometers. First used by Evangelista Torricelli (1608-1647), an Italian physicist and mathematician, to explain the fundamental principles of hydromechanics.

▫ meridional flow

Atmospheric circulation in which the north and south, or meridional, component of motion is unusually pronounced. This weakens the zonal flow.

❐ mesocyclone

A area of rotation of storm size that may often be found on the southwest part of a supercell. Its circulation can be larger than the tornado that may develop within it, but not necessarily. Originally a radar term for a rotation signature that met certain criteria, it is best seen on Doppler radar.

❐ mesohigh

A small, concentrated area of high pressure that may be created by the cold outflow and rain-cooled air from thunderstorms. It often forms a pseudo cold front or squall line on its leading edges.

❐ mesolow

A small scale low pressure center, ranging from the size of an individual thunderstorm to many tens of miles.

❐ mesoscale

The scale of meteorological phenomena that range in size from several kilometers to around 100 kilometers. This includes MCCs, MCSs, and squall lines. Smaller phenomena are classified as microscale while larger are classified as synoptic-scale.

❐ mesoscale convective complex (mcc)

A large mesoscale convective system (MCS) which is about the size of the state of Ohio or Iowa and lasts at least 6 hours. Generally forming during the afternoon and evening, the complex normally reaches its peak intensity at night when heavy rainfall and flooding become the primary threat. Severe weather may occur at anytime.

❐ mesoscale convective system (mcs)

A large organized convective weather system comprised of a number of individual thunderstorms. It normally persists for several hours and may be rounded or linear in shape. This term is often used to describe a cluster of thunderstorms that does not meet the criteria of a mesoscale convective complex (MCC).

❐ mesosphere

The layer of the atmosphere located between the stratosphere and the ionosphere, where temperatures drop rapidly with increasing height. It extends between 31 and 50 miles (17 to 80 kilometers) above the earth's surface.

❐ metar

Acronym for METeorological Aerodrome Report. It is the primary observation code used in the United States to satisfy requirements for reporting surface meteorological data. Minimum reporting requirements includes wind, visibility, runway visual range, present weather, sky condition, temperature, dew point, and altimeter setting.

❐ meteorology/meteorologist

The science and study of the atmosphere and atmospheric phenomena. Various areas of meteorology include agricultural, applied, astrometerology, aviation, dynamic, hydrometeorology, operational, and synoptic, to name a few. A scientist who studies the atmosphere and atmospheric phenomena.

❐ microbarograph

A instrument designed to continuously record a barometer's reading of very small changes in atmospheric pressure.

❐ microburst

A severe localized wind blasting down from a thunderstorm. It covers an area less than 2.5 miles (4 kilometers) in diameter and is of short duration, usually less than 5 minutes.

❐ microscale

The smallest scale of meteorological phenomena that range in size from a few centimeters to a few kilometers. Larger phenomena are classified as mesoscale. It also refers to small scale meteorological phenomena with life spans of less than a few minutes that affect very small areas and are strongly influenced by local conditions of temperature and terrain.

❐ middle clouds

A term used to signify clouds with bases between 6,000 and 18,000 feet. At the higher altitudes, they may also have some ice crystals, but they are composed mainly of water droplets. Altocumulus, altostratus, and nimbostratus are the main types of middle clouds. This altitude applies to the temperate zone. In the polar regions, these clouds may be found at lower altitudes. In the tropics, the defining altitudes for cloud types are generally higher.

❐ middle latitudes

The latitude belt roughly between 35 and 65 degrees North and South. May be referred to as the temperate region.

❐ millibar (mb)

The standard unit of measurement for atmospheric pressure used by the National Weather Service. One millibar is equivalent to 100 newtons per square meter. Standard surface pressure is 1,013.2 millibars.

❐ minimum

The least value attained by a function, for example, temperature, pressure, or wind speed. The opposite of maximum.

❐ mist

A collection of microscopic water droplets suspended in the atmosphere. It does not reduce visibility as much as fog and is often confused with drizzle.

❐ mixed layer

It is the upper portion of the boundary layer in which air is thoroughly mixed by convection. In oceanography, it is the layer of the water that is mixed through wave action or thermohaline convection.

❑ mixed precipitation

Any of the following combinations of freezing and frozen precipitation: snow and sleet, snow and freezing rain, or sleet alone. Rain may also be present.

❑ moist adiabat

The line on a Skew T-Log P chart that depicts the change in temperature of saturated air as it rises and undergoes cooling due to adiabatic expansion. As saturated air rises, the temperature changes at a rate of 0.55Â°C per 100 meters (2-3Â°F per 1,000 feet).

❑ moisture

Refers to the water vapor content in the atmosphere, or the total water, liquid, solid or vapor, in a given volume of air.

❑ monsoon

The seasonal shift of winds created by the great annual temperature variation that occurs over large land areas in contrast with associated ocean surfaces. The monsoon is associated primarily with the moisture and copious rains that arrive with the southwest flow across southern India. The name is derived from the word mausim, Arabic for season. This pattern is most evident on the southern and eastern sides of Asia, although it does occur elsewhere, such as in the southwestern United States.

❑ mountain breeze

A katabatic wind, it is formed at night by the radiational cooling along mountainsides. As the slopes become colder than the surrounding atmosphere, the lower levels of air cool and drain to the lowest point of the terrain. It may reach several hundred feet in depth, and extreme cases, attain speeds of 50 knots or greater. It blows in the opposite direction of a valley breeze.

❑ mountain wave

A wave in the atmosphere caused by a barrier, such as a mountain. Sometimes it is marked by lenticular clouds to the lee side of mountain barriers. May be called a standing wave or a lee wave.

❑ mud slide

Fast moving soil, rocks and water that flow down mountain slopes and canyons during a heavy a downpour of rain.

❑ muggy

A subjective term for warm and excessively humid weather.

❑ multicell storm

A thunderstorm made up of two or more single-cell storms.

❑ multiple vortex tornado

A tornado which has two or more condensation funnels or debris clouds, often rotating around a common center.

N

❑ nadir

The point on any given observer's celestial sphere diametrically opposite of one's zenith.

❑ national centers for environmental prediction (ncep)

As part of the National Weather Service, the centers provide timely, accurate, and continually improving worldwide forecast guidance products. Some of the centers include the Aviation Weather Center, the Climate Prediction Center, the Storm Prediction Center, and the Tropical Prediction Center.

❑ national climatic data center (ncdc)

The agency that archives climatic data from the National Oceanic and Atmospheric Administration as well as other climatological organizations. For further information, contact the NCDC, located in Asheville, North Carolina.

❑ national hurricane center (nhc)

A branch of the Tropical Prediction Center, it is the office of the National Weather Service that is responsible for tracking and forecasting tropical cyclones over the North Atlantic, Caribbean, Gulf of Mexico, and the Eastern Pacific.

❑ national meteorological center (nmc)

Now incorporated into the National Centers for Environmental Prediction, it was the division of the National Weather Service that produced, processed, handled, and distributed meteorological and oceanographic information to users throughout the Northern Hemisphere, specifically U.S. governmental organizations.

❑ national oceanic and atmospheric administration (noaa)

A branch of the U.S. Department of Commerce, it is the parent organization of the National Weather Service. It promotes global environmental stewardship, emphasizing atmospheric and marine resources.

❑ national severe storms forecast center (nssfc)

As of October 1995, the responsibilities of this Center were divided into two branches, the Storm Prediction Center and the Aviation Weather Center.

❑ national severe storms laboratory (nssl)

A branch of the National Oceanic and Atmospheric Administration, it provides accurate and timely forecasts and warnings of hazardous weather events, especially flash floods, hail, lightning, tornadoes, and other severe wind storms.

❑ national weather association (nwa)

An organization whose membership promotes excellence in operational meteorology and related activities, recognizing the professional as well as the volunteer.

❐ national weather service (nws)

A primary branch of the National Oceanic and Atmospheric Administration, it is responsible for all aspects of observing and forecasting atmospheric conditions and their consequences, including severe weather and flood warnings.

❐ nautical mile

A unit of length used in marine navigation that is equal to a minute of arc of a great circle on a sphere. One international nautical mile is equivalent to 1,852 meters or 1.151 statue miles.

❐ nautical twilight

The time after civil twilight, when the brighter stars used for celestial navigation have appeared and the horizon may still be seen. It ends when the center of the sun is 12 degrees below the horizon, and it is too difficult to perceive the horizon, preventing accurate sighting of stars.

❐ neap tide

A tide of decreased range, which occurs about every two weeks when the moon is at one quarter or three-quarters full.

❐ negative vorticity advection

The advection of lower values of vorticity into an area.

❐ nephelococcygia

A term applied when people find familiar objects within the shape of a cloud.

❐ newhall winds

The local name for winds blowing downward from desert uplands through the Newhall Pass southward into the San Fernando Valley, north of Los Angeles.

❐ newton

The unit of force giving a mass of about one kilogram (2.205 pounds) an acceleration of about one meter (1 yard) per second per second.

❐ nexrad

Acronym for NEXT Generation Weather RADAR. A network of advanced Doppler radars implemented in the United States between 1992 and 1996, it detects the location and intensity of precipitation out to a range of 143 miles from the radar site. NEXRAD Doppler radar is highly sensitive and can detect precipitation from very light rain and snow up to the strongest thunderstorms with accuracy and detail. However, sometimes the radar's extreme sensitivity will cause ground clutter and other non-precipitation echoes to be displayed in the vicinity of the radar site.

❐ night

The period of the day between dusk and dawn.

❑ nimbostratus

This cloud exhibits a combination of rain or snow, and sometimes the base of the cloud cannot be seen because of the heaviness of precipitation. They are generally associated with fall and winter conditions, but can occur during any season.

❑ nitrogen (N_2)

A colorless, tasteless, odorless gas that is the most abundant constituent of dry air. It comprises 78.09%.

❑ noctilucent clouds

Rarely seen clouds of tiny ice particles that form approximately 75 to 90 kilometers above the earth's surface. They have been seen only during twilight (dusk and dawn) during the summer months in the higher latitudes. They may appear bright against a dark night sky, with a blue-silver color or orange-red.

❑ nocturnal thunderstorms

Thunderstorms which develop after sunset. They are often associated with the strengthening of the low level jet and are most common over the Plains states. They also occur over warm water and may be associated with the seaward extent of the overnight land breeze.

❑ nor'easter

A cyclonic storm occurring off the east coast of North America. These winter weather events are notorious for producing heavy snow, rain, and tremendous waves that crash onto Atlantic beaches, often causing beach erosion and structural damage. Wind gusts associated with these storms can exceed hurricane force in intensity. A nor'easter gets its name from the continuously strong northeasterly winds blowing in from the ocean ahead of the storm and over the coastal areas.

❑ normal

The recognized standard value of a meteorological element as it has been averaged in a given location over a fixed number of years. Normals are concerned with the distribution of data within limits of common occurrence. The parameters may include temperatures (high, low, and deviation), pressure, precipitation (rain, snow, etc.), winds (speed and direction), thunderstorms, amount of clouds, percent relative humidity, etc.

❑ north pacific high

A semi-permanent, subtropical area of high pressure in the North Pacific Ocean. It is strongest in the Northern Hemispheric summer and is displaced towards the equator during the winter when the Aleutian Low becomes more dominate.

❑ nowcast

A short-term weather forecast for expected conditions in the next few hours.

❑ numerical forecasting

The use of numerical models, such as the fundamental equations of hydrodynamics subjected to observed initial conditions, to forecast the weather.

These models are run on high-speed computers at the National Centers for Environmental Prediction.

O

❑ obscuration

Any phenomena in the atmosphere, excluding precipitation, that reduces horizontal visibility. According to the National Weather Service, some of the obstructions to visibility include blowing and widespread dust, fog (including freezing fog and patchy fog), haze, mist, sand and blowing sand, smoke, blowing spray, and volcanic ash. It is reported as "X" in an observation and on the METAR.

❑ observation

In meteorology, the evaluation of one or more meteorological elements, such as temperature, pressure, or wind, that describe the state of the atmosphere, either at the earth's surface or aloft. An observer is one who records the evaluations of the meteorological elements.

❑ occluded front

Also known as an occlusion, it is a complex front formed when a cold front overtakes a warm front. It develops when three thermally different air masses conflict. The type of frontal boundary they create depends on the manner in which they meet.

❑ ocean

The intercommunicating body of salt water occupying the depressions of the earth's surface, or one of its major primary subdivisions, bounded by the continents, or the equator, and other imaginary lines. A sea is subdivision of an ocean.

❑ oceanography

The study of the ocean, embracing and integrating all knowledge pertaining to the ocean's physical boundaries, the chemistry and physics of sea water, and marine biology.

❑ omega block

A warm high aloft which has become displaced and is on the polarward side of the jet stream. It frequently occurs in the late winter and early spring in the Northern Hemisphere. The name comes from its resemblance to the Greek letter, Omega, when analyzed on upper air charts.

❑ opaque

A condition where a material, such as a cloud, blocks the passage of radiant energy, especially light. Opaque sky cover refers to the amount of sky cover that completely hides all that might be above it.

❑ orographic lifting

Where the flow of air is forced up and over barriers such as highlands or mountains. Moist air being forced aloft begins to cool, consequently condensation forms, and rain or snow begins to fall. By the time the air reaches the leeward side of the barrier, it sinks and warms, resulting in decreasing relative humidity, cessation of precipitation, and the dissipation of clouds. May be called an orographic uplift.

❐ outflow

Also referred to as an outflow boundary, it is the outward flow of air from a system, such as a thunderstorm. It is the result of cold downdrafts and its passage includes a wind shift and temperature drop.

❐ overcast

The amount of sky cover for a cloud layer that is 8/8ths, based on the summation layer amount for that layer.

❐ overrunning

This occurs when a relatively warm air mass is forced above a cooler air mass of greater density. Weather generally associated with this event includes cloudiness, cool temperatures, and steady precipitation.

❐ oxygen (O_2)

A colorless, tasteless, odorless gas that is the second most abundant constituent of dry air, comprising 20.946%.

❐ ozone (O_3)

A nearly colorless gas and a form of oxygen (O2). It is composed of an oxygen molecule made up of three oxygen atoms instead of two.

❐ ozone layer

An atmospheric layer that contains a high proportion of oxygen that exists as ozone. It acts as a filtering mechanism against incoming ultraviolet radiation. It is located between the troposphere and the stratosphere, around 9.5 to 12.5 miles (15 to 20 kilometers) above the earth's surface.

P

❐ palmer drought index

A long-term meteorological drought severity index produced by the NOAA/ USDA (Department of Agriculture) Joint Agricultural Weather Facility. The index depicts prolonged times, as in months or years, of abnormal dryness or wetness. It responds slowly, changing little from week to week, and reflects long-term moisture runoff, recharge, and deep percolation, as well as evapotranspiration.

❐ palouser

A strong, dangerous, katabatic wind that descends from the mountains into the Palouse River valley in northern Idaho and eastern Washington. May be called a Cow-Killer.

❑ parcel

A volume of air small enough to contain uniform distribution of its meteorological properties and large enough to remain relatively self-contained and respond to all meteorological processes.

❑ parhelion

The scientific name for sun dogs. Either of two colored luminous spots that appear at roughly 22 degrees on both sides of the sun at the same elevation. They are caused by the refraction of sunlight passing through ice crystals. They are most commonly seen during winter in the middle latitudes and are exclusively associated with cirriform clouds. They are also known as mock suns.

❑ partial obscuration

Denotes that 1/8th or more of the sky, but not all of the sky, is hidden by any surface-based phenomena in the atmosphere, excluding precipitation. It often reduces horizontal visibility but not the vertical. It is reported as "X" in an observation and on the METAR.

❑ partly cloudy

The state of the weather when the clouds are conspicuously present, but do not completely dull the sky or the day at any moment. The National Weather Service does not have an amount of sky cover for this condition.

❑ pascal

The unit of pressure produced when one newton acts on about one square meter.

❑ pascal's law

When an external pressure is applied to any confined fluid at rest, the pressure is increased at every point in the fluid by the amount of external pressure applied. It means that the pressure of the atmosphere is exerted not only downward on the surface of an object, but also in all directions against a surface which is exposed to the atmosphere. Formulated by Blaise Pascal (1623-1662), a French mathematician, theologian, and physicist.

❑ peak gust

The highest instantaneous wind speed observed or recorded.

❑ perigee

The point nearest the earth on the moon's orbit. This term can be applied to any other body orbiting the earth, such as satellites. It is the opposite of apogee.

❑ perihelion

The point of the earth's orbit that is nearest to the sun. Although the position is part of a 21,000 year cycle, currently it occurs around January, when the earth is about 3 million miles closer to the sun than at aphelion. This term can be applied to any other celestial body in orbit around the sun. It is the opposite of aphelion.

❐ photometer

Any of a number of atmospheric phenomena which appear as luminous patterns in the sky. They do not directly cause adverse weather. They include halos, coronas. rainbows, and fogbows.

❐ photosphere

The intensely bright portion of the sun visible to the unaided eye; the "surface" of the sun. Reaching temperatures estimated at about 11,000Â°F, it is the portion of the sun's atmosphere which emits continuous electromagnetic radiation.

❐ pilot balloon

A small balloon whose ascent is used to determine the direction and speed of low level atmospheric winds. Also known as a pibal.

❐ pilot report

A report of in-flight weather by an aircraft pilot or crew member. Often referred to as a PIREP.

❐ plan position indicator

Also known as a PPI Scope, it is a radar indicator scope displaying range and azimuth of targets in polar coordinates.

❐ plow/plough wind

The spreading downdraft and strong straight-line winds preceding a thunderstorm. So named in the American Midwest because of its ability to flatten tall grasses as it passes.

❐ polar air mass

An air mass that forms over a high latitude region. Continental polar air (cP) is formed over cold surface regions and is typically very stable with low moisture. Maritime polar air (mP), produced over warmer waters, is less stable with high moisture.

❐ polar front

A semi-continuous, semi-permanent boundary between polar air masses and tropical air masses. An integral part of an early meteorological theory known as the Polar Front Theory.

❐ polar jet

Marked by a concentration of isotherms and strong vertical shear, this jet is the boundary between the polar air and the subtropical air. It often divides into two branches, the north and the south, and marks the high speed core of the prevailing westerlies. It is associated with the location and motion of the high and low pressure areas of the middle latitudes, and therefore, is variable in position, elevation, and wind speed. Its position tends to migrate south in the Northern Hemispheric winter and north in the summer, and its core winds increase during the winter and become less strong in the summer.

❐ polar-orbiting satellite

A satellite whose orbit passes over both of the earth's between poles.

❐ poles/polar

The poles are the geographic point at 90 degrees latitude North and South on the earth's surface. They are equal distance from the equator. The polar region is considered to be that area between 60Â° and 90Â° latitude, both North and South.

❐ pollutant

Particles, gases, or liquid aerosols in the atmosphere which have an undesirable effect on humans or their surroundings. Something unfavorable to health and life that has been added to the environment.

❐ positive vorticity advection

The advection of higher values of vorticity into an area. It is also known as cyclonic vorticity.

❐ pounds per square inch (psi)

A unit for measuring pressure. One PSI equals the pressure resulting from a force of one pound force acting over an area of one square inch.

❐ precipitation

Any and all forms of water, liquid or solid, that falls from clouds and reaches the ground. This includes drizzle, freezing drizzle, freezing rain, hail, ice crystals, ice pellets, rain, snow, snow pellets, and snow grains. The amount of fall is usually expressed in inches of liquid water depth of the substance that has fallen at a given point over a specified time period.

❐ pre-frontal squall line

A line of thunderstorms that precedes an advancing cold front.

❐ pre-frontal trough

An elongated area of relatively low pressure preceding a cold front that is usually associated with a shift in wind direction.

❐ pressure

The force per unit area exerted by the weight of the atmosphere above a point on or above the earth's surface.

❐ pressure altimeter

An aneroid barometer calibrated to indicate altitude in feet instead of units of pressure. It is read accurately only in a standard atmosphere and when the correct altimeter setting is used.

❐ pressure altitude

The altitude in standard atmosphere at which a given pressure will be observed. It is the indicated altitude of a pressure altimeter at an altitude setting of 29.92

inches of mercury, and is therefore the indicated altitude above the 29.92 constant pressure surface.

❑ pressure change

The net difference between the barometric pressure at the beginning and ending of a specified interval of time, usually the three hour period preceding an observation.

❑ pressure characteristic

The pattern of the pressure change during the specified period of time, usually the three hour period preceding an observation. This is recorded in three categories: falling, rising, or steady.

❑ pressure gradient

The amount of pressure change that occurs over a fixed distance at a fixed altitude.

❑ pressure jump

A sudden increase in the observed atmospheric pressure or station pressure.

❑ pressure tendency

The pressure characteristic and amount of pressure change during a specified time period, usually the three hour period preceding the observation.

❑ prevailing visibility

It is considered representative of visibility conditions at the observation station. It is the greatest distance that can be seen throughout at least half the horizon circle, but not necessarily continuous.

❑ prevailing wind

A wind that blows from one direction more frequently than any other during a given period, such as a day, month, season, or year.

❑ profiler

A type of Doppler radar that typically measures both wind speed and direction from the surface to 55,000 feet in the atmosphere.

❑ prognostic chart

A chart of forecast predictions that may include pressure, fronts. precipitation, temperature, and other meteorological elements. Also known as a prog.

❑ psychrometer

An instrument used to measure water vapor content of the atmosphere. It consists of two thermometers, a wet bulb and dry bulb. May also be referred to as a sling psychrometer.

❑ pulse

A very short duration of time. In regard to a radar, it is a brief burst of a electromagnetic radiation emitted by the radar.

Q

❐ quantitative precipitation forecast (qpf)

A forecast of rainfall, snowfall or liquid equivalent of snowfall.

❐ quasi-stationary front

A front which is nearly stationary or moves very little since the last synoptic position. Also known as a stationary front.

R

❐ radar

Acronym for radio detection and ranging. An electronic instrument used to detect distant objects and measure their range by how they scatter or reflect radio energy. Precipitation and clouds are detected by measuring the strength of the electromagnetic signal reflected back.

❐ radarsonde observation

An upper air observation used to determine winds and other meteorological data, by tracking the range, elevation, and azimuth of a radar target carried aloft. A type of rawinsonde.

❐ radial velocity

A type of velocity that expresses motion toward or away from a given location. In Doppler radar, it is the component of motion that is parallel to the radar beam.

❐ radiation

The process by which energy is propagated through any medium by virtue of the wave motion of that medium. Electromagnetic radiation, which emits heat and light, is one form. Sound waves are another.

❐ radiation fog

Fog that is created when radiational cooling at the earth's surface lowers the temperature of the air near the ground to or below its dew point. Formation is best when there is a shallow surface layer of relatively moist air beneath a drier layer, clear skies, and light surface winds. This primarily occurs during the night or early morning.

❐ radiational cooling

The cooling of the earth's surface and the adjacent air. Although it occurs primarily at night, it happens when the earth's surface suffers a net loss of heat due to outgoing radiation.

❐ radiosonde

An instrument attached to a weather balloon used to measure pressure, temperature, humidity, and winds aloft. Observations are made when the radiosonde is aloft and emits radio signals as it ascends.

❒ rain

Precipitation in the form of liquid water droplets greater than 0.5 mm. If widely scattered, the drop size may be smaller. It is reported as "R" in an observation and on the METAR. The intensity of rain is based on rate of fall. "Very light" (R--) means that the scattered drops do not completely wet a surface. "Light" (R-) means it is greater than a trace and up to 0.10 inch an hour. "Moderate" (R) means the rate of fall is between 0.11 to 0.30 inch per hour. "Heavy" (R+) means over 0.30 inch per hour.

❒ rain forest

A forest which grows in a region of heavy annual precipitation. There are two major types, tropical and temperate.

❒ rain gauge

An instrument used to measure the amount of rain that has fallen. Measurement is done in hundredths of inches (0.01").

❒ rain shadow

Also referred to as a precipitation shadow, it is the region on the lee side of a mountain or similar barrier where the precipitation is less than on the windward side.

❒ rainbow

A luminous arc featuring all colors of the visible light spectrum (red, orange, yellow, green, blue, indigo, and violet). It is created by refraction, total reflection, and the dispersion of light. It is visible when the sun is shining through air containing water spray or raindrops, which occurs during or immediately after a rain shower. The bow is always observed in the opposite side of the sky from the sun.

❒ rainfall

The amount of precipitation of any type, primarily liquid. It is usually the amount that is measured by a rain gauge.

❒ range resolution

The ability of radar to distinguish between targets on the same azimuth but at different ranges.

❒ rawinsonde

An upper air observation that evaluates the winds, temperature, relative humidity, and pressure aloft by means of a balloon-attached radiosonde that is tracked by a radar or radio direction-finder. It is a radiosonde observation combined with a winds-aloft observation, called a rawin.

❒ reconnaissance (recco) code

An aircraft weather reconnaissance code that has come to refer primarily to in-flight tropical weather observations, but actually signifies any detailed weather observation or investigation from an aircraft in flight.

❒ **reflectivity**

A measure of the process by which a surface can turn back a portion of incident radiation into the medium through which the radiation approached. It also refers to the degree by which precipitation is able to reflect a radar beam.

❒ **refraction**

The bending of light or radar beam as it passes through a zone of contrasting properties, such as atmospheric density, water vapor, or temperature.

❒ **relative humidity**

A type of humidity that considers the ratio of the actual vapor pressure of the air to the saturation vapor pressure. It is usually expressed in percentage.

❒ **relative vorticity**

The sum of the rotation of an air parcel about the axis of the pressure system and the rotation of the parcel about its own axis.

❒ **resolution**

In relation to radar, it is the ability to read two distinct targets separately. The clearer the resolution, the nearer the two objects can be to each other and still be distinguishable.

❒ **retrogression**

In meteorology, it is the movement of a weather system in a direction opposite to the direction of the basic flow in which it is embedded. Often used in reference to a long wave trough or other macroscale feature. For example, a long wave trough that may move slightly westward when the "normal" movement and flow is eastward.

❒ **ridge**

An elongated area of high atmospheric pressure that is associated with an area of maximum anticyclonic circulation. The opposite of a trough.

❒ **rime**

The rapid freezing of supercooled water droplets as they touch an exposed object, forming a white opaque granular deposit of ice. It is one of the results of an ice storm, and when formed on aircraft it is called rime icing.

❒ **rip current**

It is formed by a strong surface water movement, or current, of a short duration that flows seaward from the shore. The return flow is piled up onshore by the incoming waves and wind. It is localized, of narrow width, and its position relative to the beach can change as the wave condition changes. Therefore, the higher the waves, the stronger the current.

❒ **rocketsonde**

A type of radiosonde that is shot into the atmosphere by a rocket, allowing it to collect data during its parachute descent from a higher position in the atmosphere than a balloon could reach.

❒ roll cloud

A relatively rare, low-level, horizontal, tube-shaped cloud. Although they are associated with a thunderstorm, they are completely detached from the base of the cumulonimbus cloud.

❒ rotation

The spinning of a body, such as the earth, about its axis.

❒ rotor cloud

An altocumulus cloud formation that can be found in the lee of a mountain or similar barrier. The air rotates around a horizontal axis, creating turbulence. Altocumulus lenticularis is an example.

❒ runway visual range (rvr)

It is the maximum distance at which the runway, or the specified lights or markers delineating it, can be seen from a position above a specified point on its center line. This value is normally determined by visibility sensors located alongside and higher than the center line of the runway. RVR is calculated from visibility, ambient light level, and runway light intensity.

❒ saffir-simpson damage-potential scale

Developed in the early 1970s by Herbert Saffir, a consulting engineer, and Robert Simpson, then Director of the National Hurricane Center, it is a measure of hurricane intensity on a scale of 1 to 5. The scale categorizes potential damage based on barometric pressure, wind speeds, and surge.

S

❒ salinity

A measure of the quantity of dissolved salts in sea water. The total amount of dissolved solids in sea water in parts per thousand by weight.

❒ salt water

The water of the ocean, distinguished from fresh water by its appreciable salinity.

❒ sand

Loose particles of hard, broken rock or minerals. In observing, sand is reported when particles of sand are raised to sufficient height that reduces visibility. It is reported as "SA" in an observation and on the METAR.

❒ sandstorm

A strong wind carrying sand particles through the air. They are low level occurences, usually only ten feet in height to not more than fifty feet above the surface. Due to the frequent winds created by surface heating, they are most predominate during the day and die out in the night. Visibility is reduced to between

5/8ths and 6/16ths statute mile, and if less than 5/16ths, then the storm is considered a heavy sandstorm. It is reported as "SS" in an observation and on the METAR.

❐ santa ana winds

The hot, dry winds, generally from the east, that funnel through the Santa Ana river valley south of the San Gabriel and San Bernadino Mountains in southern California, including the Los Angeles basin. Classified as katabatic, it occurs most often during the winter and it is an example of a foehn wind.

❐ sargasso sea

An area of the North Atlantic Ocean between Bermuda and the Azores. It is in the middle of the North Atlantic oceanic gyre, with converging surface waters. Consequently, it has less biological features than any other region of the ocean because the lack of mixing with more nutrient-rich waters.

❐ satellite

Any object that orbits a celestial body, such as a moon. However, the term is often used in reference to the manufactured objects that orbit the earth, either in a geostationary or a polar manner. Some of the information that is gathered by weather satellites, such as GOES9, includes upper air temperatures and humidity, recording the temperatures of cloud tops, land, and ocean, monitoring the movement of clouds to determine upper level wind speeds, tracing the movement of water vapor, monitoring the sun and solar activity, and relaying data from weather instruments around the world.

❐ satellite images

Images taken by a weather satellite that reveal information, such as the flow of water vapor, the movement of frontal system, and the development of a tropical system. Looping individual images aids meteorologists in forecasting. One way a picture can be taken is as a visible shot, that is best during times of visible light (daylight). Another way is as an IR (infrared) shot, that reveals cloud temperatures and can be used day or night.

❐ saturate

To treat or charge something to the point where no more can be absorbed, dissolved, or retained. In meteorology, it is used when discussing the amount of water vapor in a volume of air.

❐ saturation point

The point when the water vapor in the atmosphere is at its maximum level for the existing temperature.

❐ scattered

The amount of sky cover for a cloud layer between 3/8ths and 4/8ths, based on the summation layer amount for that layer.

❐ scattering

The process by which small particles suspended in the air diffuse a portion of the incident radiation in all directions. This is a primary reason for colors, such as blue skies, rainbows, and orange sunsets. When working with radars, this often refers to the more or less random changes in direction of radio energy.

❐ scud

Low fragments of clouds, usually stratus fractus, that are unattached and below a layer of higher clouds, either nimbostratus or cumulonimbus. They are often along and behind cold fronts and gust fronts, being associated with cool moist air, such as an outflow from a thunderstorm. When observed from a distance, they are sometimes mistaken for tornadoes.

❐ sea breeze

A diurnal coastal breeze that blows onshore, from the sea to the land. It is caused by the temperature difference when the surface of the land is warmer than the adjacent body of water. Predominate during the day, it reaches its maximum early to mid afternoon. It blows in the opposite direction of a land breeze.

❐ sea breeze front

A coastal phenomena, it is restricted to large bodies of water and their immediate coast lines. This is usually the landward extent of the sea breeze. Due to the imbalance of heating between land and water, a region of maximum upward motion or convergence occurs by mid-afternoon in the summer some 10 to 15 miles inland. Air mass thunderstorms or a line of towering cumulus clouds with showers can form along the front. At the beach, there are blue skies and a light breeze. This often occurs along the coast of the Gulf of Mexico and Florida's east coast.

❐ sea fog

A type of advection fog which forms in warm moist air cooled to saturation as the air moves across cold water.

❐ sea ice

Ice that is formed by the freezing of sea water. It forms first as small crystals, thickens into sludge, and coagulates into sheet ice, pancake ice, or ice floes of various shapes and sizes.

❐ sea level

The height or level of the sea surface at any time. It is used as a reference for elevations above and below.

❐ sea level pressure

The atmospheric pressure at mean sea level, usually determined from the observed station pressure.

❐ sea mile

A unit of length distinguished from a nautical mile. One sea mile is equivalent to 1,000 fathoms (6,000 feet).

❐ sea spray

Sometimes called salt spray, it is the drops of sea water (salt water) blown from the top of a wave.

❐ sea surface temperature (sst)

The temperature of the water's surface. It is measured using buoy and ship data, infrared satellite imagery, and coastal observations.

❐ season

A division of the year according to some regularly recurring phenomena, usually astronomical or climatic. For example, in the Northern Hemisphere, winter is said to begin on the winter solstice and end on the vernal equinox when spring begins, covering the months of December, January, and February. In the tropics, there is the dry and the rainy season, depending on the amount of precipitation.

❐ semi-permanent pressure systems

A relatively stable, stationary pressure-and-wind system where the pressure is predominately high or low with the changing season. They are not of a transitory nature, like migratory lows that develop from temperature and density differences.

❐ severe thunderstorm

A thunderstorm with winds measuring 50 knots (58 mph) or greater, 3/4 inch hail or larger, or tornadoes. Severe thunderstorms may also produce torrential rain and frequent lightning.

❐ severe weather

Generally, any destructive weather event, but usually applies to localized storms, such as blizzards, intense thunderstorms, or tornadoes.

❐ shear

It is the rate of change over a short duration. In wind shear, it can refer to the frequent change in wind speed within a short distance. It can occur vertically or horizontally. Directional shear is a frequent change in direction within a short distance, which can also occur vertically or horizontally. When used in reference to Doppler radar, it describes the change in radial velocity over short distances horizontally.

❐ shear line

A line of maximum horizontal wind shear. A narrow zone across which there is an abrupt change in the horizontal wind component parallel to it.

❐ short wave

A progressive wave of smaller amplitude, wave length, and duration than a long wave. It moves in the same direction as the basic current in which it is embedded and may induce upward vertical motion ahead of it. They are more numerous than long waves and often disappear with height in the atmosphere.

❒ showalter stability index

A measure of the local static stability of the atmosphere. It is determined by lifting an air parcel to 500 millibars and then comparing its temperature to that of the environment. If the parcel is colder than its new environment, then the atmosphere is more stable. If the parcel is warmer than its new environment, then the atmosphere is unstable and the potential for thunderstorm development and severe weather increases.

❒ shower

Precipitation from a convective cloud that is characterized by its sudden beginning and ending, changes in intensity, and rapid changes in the appearance of the sky. It occurs in the form of rain (SHRA), snow (SHSN), or ice (SHPE). It is reported as "SH" in an observation and on the METAR.

❒ siberian express

A fierce, cold flow of air that originates in Siberia, then moves into Alaska and northern Canada before moving southward into the United States.

❒ siberian high

The semi-permanent high pressure area that forms over Siberia during the winter. The average central pressure exceeds 1030 millibars from late November to early March. It is characterized by clear, dry weather. Over southern Asia, the predominate surface wind is northeasterly, just the opposite of the predominate summer winds which bring the monsoon.

❒ sidereal time

The measure of time as defined by the diurnal motion of the vernal equinox. A sidereal day is equivalent to one complete rotation of the earth relative to the equinox, which is 23 hours, 56 minutes, and 4.091 seconds. A sidereal year is the interval required for the earth to make one absolute revolution around the sun, which is 365 days, 6 hours, 9 minutes, and 9.5 seconds.

❒ skew t-log p diagram

A thermodynamic diagram, using the temperature and the logarithm of pressure as coordinates. It is used to evaluate and forecast air parcel properties. Some values that can be determined are the Convective Condensation Level (CCL), the Lifting Condensation Level (LCL), and the Level of Free Convection (LFC).

❒ sky

The vault-like apparent surface against which all aerial objects are seen from the earth.

❒ sky cover

The amount of the celestial dome that is hidden by clouds and/or obscurations.

❒ sleet

Also known as ice pellets, it is winter precipitation in the form of small bits or pellets of ice that rebound after striking the ground or any other hard surface. It is reported as "PE" in an observation and on the METAR.

☐ slush

Snow or ice on the ground that has been reduced to a softy watery mixture by rain and/or warm temperatures.

☐ small craft advisory

An advisory issued for marine interests, especially for operators of small boats or other vessels. Conditions include wind speeds between 20 knots (23 mph) and 34 knots (39 mph).

☐ smoke

Small particles produced by combustion that are suspended in the air. A transition to haze may occur when the smoke particles have traveled great distance (25 to 100 miles or more), and when the larger particles have settled out. The remaining particles become widely scattered through the atmosphere. It is reported as "FU" in an observation and on the METAR.

☐ snow

Frozen precipitation in the form of white or translucent ice crystals in complex branched hexagonal form. It most often falls from stratiform clouds, but can fall as snow showers from cumuliform ones. It usually appears clustered into snowflakes. It is reported as "SN" in an observation and on the METAR.

☐ snow advisory

A statement or advisory issued when snow is expected to create hazardous travel conditions. It warns of less severe weather conditions than a winter storm.

☐ snow banner

A plume of snow blown off a mountain crest, resembling smoke blowing from a volcano.

☐ snow blindness

Temporary blindness or impaired vision that results from bright sunlight reflected off the snow surface. The medical term is niphablepsia.

☐ snow cover

The areal extent of ground covered by the snow. It is usually expressed as a percent of the total area of a given region.

☐ snow creep

A continuous, extremely slow, downhill movement of a layer of snow.

☐ snow crust

The crisp, almost icy, surface on fallen snow, usually formed by the slight melting and refreezing of the surface snow.

☐ snow depth

The actual depth of snow on the ground at any instant during a storm, or after any single snowstorm or series of storms.

❒ snow devil

A small, rotating wind that picks up loose snow instead of dirt (like a dust devil) or water (like a waterspout). Formed mechanically by the convergence of local air currents. May be called a snowspout.

❒ snow eater

Any warm down slope wind, or foehn, that blows over snowy terrain and melts the snow.

❒ snow flurry/flurries

Light showers of snow, generally very brief without any measurable accumulation. May be reported as "SHSN--" in an observation and on the METAR.

❒ snow garland

Snow appearing as a beautiful long thick rope draped on trees, fences and other objects. Formed by the surface tension of thin films of water bonding individual snow crystals.

❒ snow grains

Frozen precipitation in the form of very small, white, opaque grains of ice. The solid equivalent of drizzle. It is reported as "SG" in an observation and on the METAR.

❒ snow level

The elevation in mountainous terrain where the precipitation changes from rain to snow, depending on the temperature structure of the associated air mass.

❒ snow line

The lowest elevation area of a perennial snow field on high terrain, such as a mountain range.

❒ snow pellets

Frozen precipitation in the form of white, round or conical opaque grains of ice. Their diameter ranges from 0.08 to 0.2 inch (2 to 5 mm). They are easily crushed and generally break up after rebounding from a hard surface, unlike hail. Sometimes it is called small or soft hail. It is reported as "GS" in an observation and on the METAR.

❒ snow roller

The product of moist, cohesive snow that when initiated by wind rolls across the landscape, gathering snow until it can no longer move. It is shaped like a rolled sleeping bag, some reaching four feet across and seven feet in diameter.

❒ snow shower

Frozen precipitation in the form of snow, characterized by its sudden beginning and ending. It is reported as "SHSN" in an observation and on the METAR.

❒ snow squall

A heavy snow shower accompanied by sudden strong winds, or a squall.

❐ snowburn

A burn of the skin, like a sunburn, but caused by the sun's rays reflected off the snow surface.

❐ snowfall

The rate at which snow falls, usually expressed in inches of snow depth over a six hour period.

❐ snowflakes

An ice crystal or an aggregate of ice crystals which fall from clouds.

❐ snowpack

The amount of annual accumulation of snow at higher elevations.

❐ solar day

The complete rotation of the earth in relation to the sun. Although it varies, an average has determined a mean solar day of 24 hours. It is universally used for civil purposes.

❐ solar eclipse

An eclipse of the sun occurs when the moon is in a direct line between the sun and the earth, casting some of the earth's surface in its shadow. The moon's disk shaped outline appears to cover the sun's brighter surface, or photosphere. That part of the earth that is directly in the moon's shadow will see a total eclipse of the sun, while the areas around it will see a partial eclipse.

❐ solstice

The point at which the sun is the furthest on the ecliptic from the celestial equator. The point at which sun is at maximum distance from the equator and days and nights are most unequal in duration. The Tropic of Cancer and the Tropic of Capricorn are those parallels of latitude which lies directly beneath a solstice. In the Northern Hemisphere, the winter solstice falls on or about December 21 and the summer solstice on or about June 21.

❐ sounding

A plot of the atmosphere, using data rom upper air or radiosonde observations. Usually confined to a vertical profile of the temperatures, dew points, and winds above a fixed location.

❐ southern oscillation

A periodic reversal of the pressure pattern across the tropical Pacific Ocean during El NiÑo events. It is represents the distribution of temperature and pressure over an oceanic area.

❐ specific humidity

The ratio of the density of the water vapor to the density of the air, a mix of dry air and water vapor. It is expressed in grams per gram or in grams per kilograms.

The specific humidity of an air parcel remains constant unless water vapor is added to or taken from the parcel.

❑ spring

The season of the year which occurs as the sun approaches the summer solstice, and characterized by increasing temperatures in the mid-latitudes. Customarily, this refers to the months of March, April, and May in the North Hemisphere, and the months of September, October, and November in the Southern Hemisphere. Astronomically, this is the period between the vernal equinox and the summer solstice.

❑ spring tide

A tide of increased range, which occurs about every two weeks when the moon is new or full.

❑ squall

A sudden onset of strong winds with speeds increasing to at least 16 knots (18 miles per hour) and sustained at 22 or more knots (25 miles per hour) for at least one minute. The intensity and duration is longer than that of a gust. It is reported as "SQ"s in an observation and on the METAR.

❑ squall line

A narrow band or line of active thunderstorms that is not associated with a cold front. It may form from an outflow boundary or the leading edge of a mesohigh.

❑ st. elmo's fire

A luminous, and often audible, electric discharge that is sporadic in nature. It occurs from objects, especially pointed ones, when the electrical field strength near their surfaces attains a value near 1000 volts per centimeter. It often occurs during stormy weather and might be seen on a ship's mast or yardarm, aircraft, lightning rods, and steeples. Also known as corposant or corona discharge.

❑ stable/stability

Occurs when a rising air parcel becomes denser than the surrounding air. It will then return to its original position. When the density of the air parcel remains the same as the surrounding air after being lifted, it is also considered stable, since it does not have the tendency to rise or sink further.

❑ stagnation area

An area that has a combination of stable stratification, weak horizontal wind speed, and little, if any, significant precipitation. It is usually associated with an area of high pressure.

❑ standard atmosphere

A standard atmosphere has been defined by the International Civil Aeronautical Organization (ICAO). It assumes a mean sea level temperature of 15°C a standard sea level pressure of 1,013.25 millibars or 29.92 inches of mercury, and a temperature lapse rate of 0.65°C per 100 meters up to 11 kilometers in the atmosphere.

❐ standard surface pressure

The measurement of one atmosphere of pressure under standard conditions. It is equivalent to 1,013.25 millibars, 29.92 inches of mercury, 760 millimeters of mercury, 14.7 pounds per square inch, or 1.033 grams per square centimeter.

❐ standing cloud

Any type of isolated cloud, generally formed over peaks or ridges of mountainous areas, that appears stationary or standing over the terrain.

❐ standing wave

An atmospheric wave that is stationary with respect to the medium in which it is embedded.

❐ station elevation

The vertical distance above mean sea level that is the reference level for all current measurements of atmospheric pressure at that station.

❐ station pressure

The atmospheric pressure with respect to the station elevation.

❐ stationary front

A front which is nearly stationary or moves very little since the last synoptic position. May be known as a quasi-stationary front.

❐ steam fog

A type of advection fog that is produced by evaporation when cool air passes over a warm wet surface and the fog rises, giving the appearance of steam. Also called sea smoke when it occurs over the ocean.

❐ storm

An individual low pressure disturbance, complete with winds, clouds, and precipitation. The name is associated with destructive or unpleasant weather. Storm-scale refers to disturbances the size of individual thunderstorms.

❐ storm prediction center (spc)

A branch of the National Centers for Environmental Prediction, the Center monitors and forecasts severe and non-severe thunderstorms, tornadoes, and other hazardous weather phenomena across the United States. Formerly known as the Severe Local Storms (SELS) unit of the National Severe Storms Forecast Center.

❐ storm tracks

The path or tracks generally followed by a cyclonic disturbance.

❐ storm winds

On the Beaufort Wind Scale, a wind with speeds from 56 to 63 knots (64 to 72 miles per hour).

❒ straight-line winds

Any surface wind that is not associated with rotation. An example is the first gust from a thunderstorm, as opposed to tornadic winds.

❒ stratiform

Clouds composed of water droplets that exhibit no or have very little vertical development. The density of the droplets often blocks sunlight, casting shadows on the earth's surface. Bases of these clouds are generally no more than 6,000 feet above the ground. They are classified as low clouds, and include all varieties of stratus and stratocumulus. The opposite in type are the vertical development of cumuliform clouds.

❒ stratocumulus

A low cloud composed of layers or patches of cloud elements. It can form from cumulus clouds becoming more stratiformed and often appears as regularly arranged elements that may be tessellated, rounded, or roll-shaped with relatively flat tops and bases. It is light or dark gray in color, depending on the size of the water droplets and the amount of sunlight that is passing through them.

❒ stratopause

The boundary zone or transition layer between the stratosphere and the mesosphere. Characterized by a decrease in temperature with increasing altitude.

❒ stratosphere

The layer of the atmosphere located between the troposphere and the mesosphere, characterized by a slight temperature increase and absence of clouds. It extends between 11 and 31 miles (17 to 50 kilometers) above the earth's surface. It is the location of the earth's ozone layer.

❒ stratus

One of the three basic cloud forms (the others are cirrus and cumulus. It is also one of the two low cloud types. It is a sheet like cloud that does not exhibit individual elements, and is, perhaps, the most common of all low clouds. Thick and gray, it is seen in low, uniform layers and rarely extends higher than 5,000 feet above the earth's surface. A veil of stratus may give the sky a hazy appearance. Fog may form from a stratus cloud that touches the ground. Although it can produce drizzle or snow, it rarely produces heavy precipitation. Clouds producing heavy precipitation may exist above a layer of stratus.

❒ stratus fractus

Stratus clouds that appear in irregular fragments, as if they had been shred or torn. Also appears in cumulus clouds (called cumulus fractus), but not in cirrus clouds.

❒ sublimation

The process of a solid (ice) changing directly into a gas (water vapor), or water vapor changing directly into ice, at the same temperature, without ever going through the liquid state (water). The opposite of crystallization.

▫ subpolar

The region bordering the polar region, between 50° and 70° North and South latitude. This is generally an area of semi-permanent low pressure that exists and where the Aleutian and Icelandic Lows may be found. However, a dome of high pressure may form over the cold continental surfaces during the winter, for example, the North American High and the Siberian High.

▫ subrefraction

Less than normal bending of light or a radar beam as it passes through a zone of contrasting properties, such as atmospheric density, water vapor, or temperature.

▫ subsidence

A sinking or downward motion of air, often seen in anticyclones. It is most prevalent when there is colder, denser air aloft. It is often used to imply the opposite of atmospheric convection.

▫ subtropical

The region between the tropical and temperate regions, an area between 35° and 40° North and South latitude. This is generally an area of semi-permanent high pressure that exists and is where the Azores and North Pacific Highs may be found.

▫ subtropical air

An air mass that forms over the subtropical region. The air is typically warm with a high moisture content due to the low evaporative process.

▫ subtropical jet

Marked by a concentration of isotherms and vertical shear, this jet is the boundary between the subtropical air and the tropical air. It is found approximately between 25° and 35° North latitude and usually above an altitude of 40,000 feet. Its position tends to migrate south in the Northern Hemispheric winter and north in the summer.

▫ summation layer amount

The amount of sky cover for each layer is given in eighths of sky cover attributable to clouds or obscurations. The summation amount for any given layer is equal to the sum of the sky cover for the layer being evaluated plus the sky cover for all lower layers, including partial obscuration. A summation amount for a layer can not exceed 8/8ths.

▫ summer

Astronomically, this is the period between the summer solstice and the autumnal equinox. It is characterized as having the warmest temperatures of the year, except in some tropical regions. Customarily, this refers to the months of June, July, and August in the North Hemisphere, and the months of December, January, and February in the Southern Hemisphere.

▫ sun dog

Either of two colored luminous spots that appear at roughly 22° on both sides of the sun at the same elevation. They are caused by the refraction of sunlight passing through ice crystals. They are most commonly seen during winter in the middle latitudes and are exclusively associated with cirriform clouds. The scientific name for sun dogs is parhelion and they are also known as mock suns.

❑ sun pillar

Horizontal ice crystals in the form of plates, which occur in clouds and ice fog near the earth's surface, reflect sunlight into vertical sun pillars for a spectacular display.

❑ sunrise

The daily appearance of the sun on the eastern horizon as a result of the earth's rotation. In the United States, it is considered as that instant when the upper edge of the sun appears on the sea level horizon. In Great Britain, the center of the sun's disk is used instead. Time of sunrise is calculated for mean sea level.

❑ sunset

The daily disappearance of the sun below the western horizon as a result of the earth's rotation. In the United States, it is considered as that instant when the upper edge of the sun just disappears below the sea level horizon. In Great Britain, the center of the sun's disk is used instead. Time of sunset is calculated for mean sea level.

❑ supercell

A severe thunderstorm characterized by a rotating, long-lived, intense updraft. Although not very common, they produce a relatively large amount of severe weather, in particular, extremely large hail, damaging straight-line winds, and practically all violent tornadoes.

❑ supercooling

The reduction of the temperature of any liquid below the melting point of that substance's solid phase. Cooling a substance beyond its nominal freezing point. Supercooled water is water that remains in a liquid state when it is at a temperature that is well below freezing. The smaller and purer the water droplets, the more likely they can become supercooled.

❑ superrefraction

Greater than normal bending of light or radar beam as it passes through a zone of contrasting properties, such as atmospheric density, water vapor, or temperature.

❑ surface boundary layer

The lowest layer of the earth's atmosphere, usually up to 3,300 feet, or one kilometer, from the earth's surface, where the wind is influenced by the friction of the earth's surface and the objects on it.

❑ surge

The increase in sea water height from the level that would normally occur were there no storm. Although the most dramatic surges are associated with hurricanes, even smaller low pressure systems can cause a slight increase in the sea level if the wind and fetch is just right. It is estimated by subtracting the normal astronomic tide from the observed storm tide.

❒ swell

Ocean waves that have traveled out of their generating area. Swell characteristically exhibits a more regular and longer period and has flatter wave crests than waves within their fetch.

❒ synoptic chart

Any map or chart that depicts meteorological or atmospheric conditions over a large area at any given time.

❒ synoptic scale

The size of migratory high and low pressure systems in the lower troposphere that cover a horizontal area of several hundred miles or more.

❒ syzygy

The points in the moon's orbit about the earth at which the moon is new or full.

T

❒ teleconnections

Information used by forecasters to determine what the weather might be elsewhere when compared with past weather conditions at the same degree of longitude.

❒ temperate climate

Climates with distinct winter and summer seasons, typical of regions found between the Tropics of Cancer and Capricorn and the Arctic and Antarctic Circles. Considered the climate of the middle latitudes.

❒ temperature

The measure of molecular motion or the degree of heat of a substance. It is measured on an arbitrary scale from absolute zero, where the molecules theoretically stop moving. It is also the degree of hotness or coldness. In surface observations, it refers primarily to the free air or ambient temperature close to the surface of the earth.

❒ terminal doppler weather radar (tdwr)

Doppler radar installed at major airports throughout the United States to detect microbursts.

❒ terrestrial radiation

Long wave radiation that is emitted by the earth back into the atmosphere. Most of it is absorbed by the water vapor in the atmosphere, while less than ten percent is radiated directly into space.

❐ texas norther

Local name in the south-central Great Plains for strong winter winds blowing north or northwest following a sharp cold front with dropping temperatures. Marked by a dark, blue-black sky.

❐ thaw

A warm spell of weather when ice and snow melt. To free something from the binding action of ice by warming it to a temperature above the melting point of ice.

❐ theodolite

An optical instrument used to track the motion of a pilot balloon, or pibal, by measuring the elevation and azimuth angles.

❐ thermal low

Also known as heat low, it is an area of low pressure due to the high temperatures caused by intensive heating at the surface. It tends to remain stationary over its source area, with weak cyclonic circulation. There are no fronts associated with it. An example is the low that develops over southwestern United States and northwestern Mexico during the summer months.

❐ thermocline

A vertical negative temperature gradient in some layer of a body of water which is appreciably greater than the gradients above and below it. In the ocean, this may be seasonal, due to the heating of the surface water in the summer, or permanent.

❐ thermodynamics

Study of the processes that involve the transformation of heat into mechanical work, of mechanical work into heat, or the flow of heat from a hotter body to a colder body.

❐ thermograph

Essentially, a self-recording thermometer. A thermometer that continuously records the temperature on a chart.

❐ thermohaline

In oceanography, it pertains to when both temperature and salinity act together. An example is thermohaline circulation which is vertical circulation induced by surface cooling, which causes convective overturning and consequent mixing.

❐ thermometer

An instrument used for measuring temperature. The different scales used in meteorology are Celsius, Fahrenheit, and Kelvin or Absolute.

❐ thermosphere

A thermal classification, it is the layer of the atmosphere located between the mesosphere and outer space. It is a region of steadily increasing temperature with altitude, and includes all of the exosphere and most, if not all, of the ionosphere.

❐ thickness

The thickness of a layer in the atmosphere is proportional to the mean temperature of that whole layer. The layer most often used in meteorology is between 1000 and 500 millibars. There can be different temperature profiles in the lowest layer of the atmosphere with the same 1000-500 millibar thickness value, depending on what is happening above that lowest layer. For example, if the lower levels are warming but higher levels are cooling, the overall mean temperature, the thickness, could remain the same. Likewise, on a sunny day, the amount of incoming solar radiation, affects the temperature right at the earth's surface, without necessarily having much effect on the thickness of the whole layer.

❐ thunder

The sound emitted by rapidly expanding gases along the channel of a lightning discharge. Over three-quarters of lightning's electrical discharge is used in heating the gases in the atmosphere in and immediately around the visible channel. Temperatures can rise to over 10,000 Â°C in microseconds, resulting in a violent pressure wave, composed of compression and rarefaction. The rumble of thunder is created as one's ear catches other parts of the discharge, the part of the lightning flash nearest registering first, then the parts further away.

❐ thunder snow

A wintertime thunderstorm from which falls snow instead of rain. Violent updrafts and at or below freezing temperatures throughout the atmosphere, from surface to high aloft, discourage the melting of snow and ice into rain. Intense snowfall rates often occur during these situations.

❐ thunderstorm

Produced by a cumulonimbus cloud, it is a microscale event of relatively short duration characterized by thunder, lightning, gusty surface winds, turbulence, hail, icing, precipitation, moderate to extreme up and downdrafts, and under the most severe conditions, tornadoes.

❐ tide

The periodic rising and falling of the earth's oceans and atmosphere. It is the result of the tide-producing forces of the moon and the sun acting on the rotating earth. This propagates a wave through the atmosphere and along the surface of the earth's waters.

❐ tilt

The inclination to the vertical of a significant feature of the pressure pattern or of the field of moisture or temperature. For example, midlatitide troughs tend to display a westward tilt with altitude through the troposphere.

❐ tiros

A series of television infrared observation satellites that demonstrated the feasibility and capability of observing the cloud cover and weather patterns of earth from space. An experimental program, it was the first spaceborne system that allowed meteorologists to acquire information that was immediately put to use in an operational setting. The first U.S. weather satellite, TIROS I, was launched on April 1, 1960, and TIROS X, the last of the series, was launched on July 2, 1965.

❑ tornado

A violently rotating column of air in contact with and extending between a convective cloud and the surface of the earth. It is the most destructive of all storm-scale atmospheric phenomena. They can occur anywhere in the world given the right conditions, but are most frequent in the United States in an area bounded by the Rockies on the west and the Appalachians in the east.

❑ tornado alley

A geographic corridor in the United States which stretches north from Texas to Nebraska and Iowa. In terms of sheer numbers, this section of the United States receives more tornadoes than any other.

❑ towering cumulus

Another name for cumulus congestus, it is a rapidly growing cumulus or an individual dome-shaped clouds whose height exceeds its width. Its distinctive cauliflower top often mean showers below, but lacking the characteristic anvil of a cumulonimbus, it is not a thunderstorm.

❑ trace

Generally, an unmeasurable or insignificant quantity. A precipitation amount of less than 0.005 inch.

❑ trade winds

Two belts of prevailing winds that blow easterly from the subtropical high pressure centers towards the equatorial trough. Primarily lower level winds, they are characterized by their great consistency of direction. In the Northern Hemisphere, the trades blow from the northeast, and in the Southern Hemisphere, the trades blow from the southeast.

❑ trajectory

The curve that a body, such as a celestial object, describes in space. This applies to air parcel movement also.

❑ translucent

Not transparent, but clear enough to allow light to pass through.

❑ transmissometer

An electronic instrument system which provides a continuous record of the atmospheric transmission between two fixed points. By showing the transmissivity of light through the atmosphere, the horizontal visibility may be determined.

❑ transparent

A condition where a material is clear enough not to block the passage of radiant energy, especially light.

❐ transpiration

The process by which water in plants is transferred as water vapor to the atmosphere.

❐ triple point

The point at which any three atmospheric boundaries meet. It is most often used to refer to the point of occlusion of an extratropical cyclone where the cold, warm, and occluded fronts meet. Cyclogenesis may occur at a triple point. It is also the condition of temperature and pressure under which the gaseous, liquid, and solid forms of a substance can exist in equilibrium.

❐ tropic of cancer

The most northern point on the earth where the sun is directly overhead, located at approximately 23.5 degrees North latitude.

❐ tropic of capricorn

The most southern point on the earth where the sun is directly overhead, located at approximately 23.5 degrees South latitude.

❐ tropical air mass

An air mass that forms in the tropics or subtropics over the low latitudes. Maritime tropical air is produced over oceans and is warm and humid, while continental tropical air is formed over arid regions and is very hot and dry.

❐ tropical cyclone

A warm core low pressure system which develops over tropical, and sometimes subtropical, waters, and has an organized circulation. Depending on sustained surface winds, the system is classified as a tropical disturbance, a tropical depression, a tropical storm, or a hurricane or typhoon.

❐ tropical depression

A tropical cyclone in which the maximum sustained surface winds are 38 miles per hour (33 knots) or less. Characteristically having one or more closed isobars, it may form slowly from a tropical disturbance or an easterly wave which has continued to organize.

❐ tropical disturbance

An area of organized convection, originating in the tropics and occasionally the subtropics, that maintains its identity for 24 hours or more. It is often the first developmental stage of any subsequent tropical depression, tropical storm, or hurricane.

❐ tropical prediction center (tpc)

A division of the National Centers for Environmental Prediction, the Center issues watches, warnings, forecasts, and analyses of hazardous weather conditions in the tropics for both domestic and international communities. The National Hurricane Center is a branch.

❒ tropical storm

A tropical cyclone in which the maximum sustained surface winds are from 39 miles per hour (34 knots) to 73 miles per hour (63 knots). At this point, the system is given a name to identify and track it.

❒ tropical wave

Another name for an easterly wave, it is an area of relatively low pressure moving westward through the trade wind easterlies. Generally, it is associated with extensive cloudiness and showers, and may be associated with possible tropical cyclone development.

❒ tropics/tropical

The region of the earth located between the Tropic of Cancer, at 23.5 degrees North latitude, and the Tropic of Capricorn, at 23.5 degrees South latitude. It encompasses the equatorial region, an area of high temperatures and considerable precipitiation during part of the year.

❒ tropopause

The boundary zone or transition layer between the troposphere and the stratosphere. This is characterized by little or no increase or decrease in temperature or change in lapse rate with increasing altitude.

❒ troposphere

The lowest layer of the atmosphere located between the earth's surface to approximately 11 miles (17 kilometers) into the atmosphere. Characterized by clouds and weather, temperature generally decreases with increasing altitude.

❒ trough

An elongated area of low atmospheric pressure that is associated with an area of minimum cyclonic circulation. The opposite of a ridge.

❒ tsunami

An ocean wave with a long period that is formed by an underwater earthquake or landslide, or volcanic eruption. It may travel unnoticed across the ocean for thousands of miles from its point of origin and builds up to great heights over shallower water. Also known as a seismic sea wave, and incorrectly, as a tidal wave.

❒ tule fog

Ground fog in the central valley of California and the leading cause of weather-related casualties in that state. It forms at night and in the early morning when the ground cools, lowering the air temperature near the ground to or below its initial dew point.

❒ turbulence

The irregular and instantaneous motions of air which is made up of a number of small of eddies that travel in the general air current. Atmospheric turbulence is caused by random fluctuations in the wind flow. It can be caused by thermal or convective currents, differences in terrain and wind speed, along a frontal zone, or variation in temperature and pressure.

- **twilight**

Often called dusk, it is the evening period of waning light from the time of sunset to dark. The time of increasing light in the morning is called dawn. Twilight ends in the evening or begins in the morning at a specific time and can be categorized into three areas of decreasing light. Civil twilight is the time in the evening when car headlights need to be turned on to be seen by other drivers. Nautical twilight is when the bright stars used by navigators have appeared and the horizon may still be seen. Astronomical twilight is when the sunlight is still shining on the higher levels of the atmosphere, yet it is dark enough for astronomical work to begin. During dawn, the reverse order occurs until full daylight.

- **twister**

A slang term used in the United States for a tornado.

- **typhoon**

The name for a tropical cyclone with sustained winds of 74 miles per hour (65 knots) or greater in the western North Pacific Ocean. This same tropical cyclone is known as a hurricane in the eastern North Pacific and North Atlantic Ocean, and as a cyclone in the Indian Ocean.

U

- **ultraviolet**

Electromagnetic radiation that has a wavelength shorter than visible light and longer than x-rays. Although it accounts for only 4 to 5 percent of the total energy of insolation, it is responsible for many complex photochemical reactions, such as fluorescence and the formation of ozone.

- **undercast**

In aviation, it is an opaque cloud layer viewed from an observation point above the layer. From the ground, it would be considered an overcast.

- **united states weather bureau**

The official name of the National Weather Service prior to 1970.

- **universal time coordinate**

One of several names for the twenty-four hour time which is used throughout the scientific and military communities.

- **university corporation for atmospheric research (ucar)**

A non-profit university membership consortium which carries out programs to benefit atmospheric, oceanic, and related sciences around the globe. Among other activities, UCAR operates the National Center for Atmospheric Research (NCAR) with National Science Foundation sponsorship.

❑ unstable/ instability

Occurs when a rising air parcel becomes less dense than the surrounding air. Since its temperature will not cool as rapidly as the surrounding environment, it will continue to rise on its own.

❑ updraft

A small scale current of air with vertical motion. If there is enough moisture, then it may condense, forming a cumulus cloud, the first step towards thunderstorm development.

❑ upper air/upper level

The portion of the atmosphere which is above the lower troposphere. It is generally applied to the levels above 850 millibars. Therefore, upper level lows and highs, troughs, winds, observations, and charts all apply to atmospheric phenomena above the surface.

❑ upslope effect

The cooling of an air flow as it ascends a hill or mountain slope. If there is enough moisture and the air is stable, stratiform clouds and precipitation may form. If the air is unstable, there might be an increased chance of thunderstorm development.

❑ upslope fog

Fog that forms when warm, moist surface air is forced up a slope by the wind. It is adiabatically cooled to below its initial dew point, which means the air cools by expansion as it rises. It forms best where there is a gradual slope, and it can become quite deep, requiring considerable time to dissipate.

❑ upwelling

The process by which water rises from a lower to a higher depth, usually as a result of divergence and offshore currents. It influences climate by bringing colder, more nutrient-rich water to the surface. A vital factor of the El Niño event.

❑ upwind

The direction from which the wind is blowing. Also the windward side of an object. The opposite of the downwind or leeward side.

V

❑ valley breeze

An anabatic wind, it is formed during the day by the heating of the valley floor. As the ground becomes warmer than the surrounding atmosphere, the lower levels of air heat and rise, flowing up the mountainsides. It blows in the opposite direction of a mountain breeze.

❐ vapor pressure

The pressure exerted by the molecules of a given vapor. In meteorology, it is considered as the part of total atmospheric pressure due to the water vapor content. It is independent of other gases or vapors.

❐ vapor trail

A cloudlike streamer or trail often seen behind aircraft flying in clear, cold, humid air. A vapor trail is created when the water vapor from the engine exhaust gases are added to the atmosphere. Also called a contrail, for condensation trail.

❐ variable ceiling

Occurs when the height of a ceiling layer increases and decreases rapidly, the ascribed height is the average of all the varying values.

❐ veering

A clockwise shift in the wind direction in the Northern Hemisphere at a certain location. In the Southern Hemisphere, it is counterclockwise. This can either happen horizontally or vertically (with height). For example, the wind shifts from the north to the northeast to the east. It is the opposite of backing.

❐ vernal equinox

Taking place in the Northern Hemispheric spring, it is the point at which the ecliptic intersects the celestial equator. Days and nights are most nearly equal in duration. It falls on or about March 20 and is considered the beginning of spring in the Northern Hemisphere and autumn in the Southern Hemisphere. It is the astronomical opposite of the autumnal equinox.

❐ vertical temperature profile

A series of temperature measurements taken at various levels in the atmosphere that show the thermal structure of the atmosphere over a specific location. Obtained through a rawinsonde sounding or comparable method, and exhibited in a skew t-log p diagram.

❐ vertical visibility

The distance an observer can see vertically into an undefined ceiling, or the height corresponding to the top of a ceiling light projector beam, or the height at which a ceiling balloon disappears during the presence of an indefinite ceiling.

❐ vertical wind profile

A series of wind direction and wind speed measurements taken at various levels in the atmosphere that show the wind structure of the atmosphere over a specific location. Obtained through a rawinsonde sounding or comparable method, and exhibited in a skew t-log p diagram.

❐ virga

Streaks or wisps of precipitation, such as water or ice particles, that fall from clouds but evaporate before reaching the ground. From a distance, the event sometimes may be mistaken for a funnel cloud or tornado. Typically, it may fall from altocumulus, altostratus, or high based cumuonimbus.

▫ visibility

A measure of the opacity of the atmosphere, and therefore, the greatest distance one can see prominent objects with normal eyesight. The National Weather Service has various terms for visibility. Surface visibility is the prevailing visibility determined from the usual point of observation. Prevailing visibility is considered representative of visibility conditions at the station. Sector visibility is the visibility in a specified direction that represents at least a 45 degree arc of the horizon circle. Tower visibility is the prevailing visibility determined from the airport traffic control tower (ATCT) at stations that also report surface visibility.

▫ visible light

The portion of the electromagnetic spectrum that can be detected by the human eye. It travels at the same speed as all other radiation, that is at 186,000 mile per second. It has a wave length longer than ultraviolet light and shorter than x-rays.

▫ visual flight rules (vfr)

Refers to the general weather conditions pilots can expect at the surface. VFR criteria means a ceiling greater than 3,000 feet and greater than 5 miles.

▫ vortex

Any circular or rotary flow in the atmosphere that possesses vorticity.

▫ vorticity

The measurement of the rotation of a small air parcel. It has vorticity when the parcel spins as it moves along its path. Although the axis of the rotation can extend in any direction, meteorologists are primarily concerned with the rotational motion about an axis that is perpendicular to the earth's surface. If it does not spin, it is said to have zero vorticity. In the Northern Hemisphere, the vorticity is positive when the parcel has a counterclockwise, or cyclonic, rotation. It is negative when the parcel has clockwise, or anticyclonic, rotation.

▫ vorticity maximum

A center of vorticity, or the maximum of the vorticity field of a fluid.

W

▫ walker circulation

A deep east-west overturning in the atmosphere normally confined to within about 20 degrees latitude of the equator extending from low-levels to near the tropopause. One side of the Walker Circulation is associated with rising motion, clouds, and rain while the opposite side with sinking motion, generally fair weather and little or no rain. This circulation is large in spatial extend normally extending anywhere from way around the world or completely around the world (thus either 1 or 2 upward and 1 or 2 downward branches to the circulation). The rising and sinking branches are climatologically anchored to specific geographical locations, but these locations are seasonally dependent. In winter the the rising branch is centered near the "maritime continent" of greater Indonesia while in summer it is centered near southern India. The sinking branches are generally found in the eastern equatorial Pacific in winter and summer. Sub-branches of rising motion are generally tied to equatorial areas of South America and Africa in summer.

❐ wall cloud

An abrupt lowering of a cloud from its parent cloud base, a cumulonimbus or supercell, with no visible precipitation underneath. Forming in the area of a thunderstorm updraft, or inflow area, it exhibits rapid upward movement and cyclonic rotation. It often develops before strong or violent tornadoes.

❐ warm

To have or give out heat to a moderate or adequate degree. A subjective term for temperatures between cold and hot. In meteorology, an air parcel that is warm is only so in relation to another parcel.

❐ warm advection

The horizontal movement of warmer air into a location.

❐ warm front

The leading edge of an advancing warm air mass that is replacing a retreating relatively colder air mass. Generally, with the passage of a warm front, the temperature and humidity increase, the pressure rises, and although the wind shifts (usually from the southwest to the northwest in the Northern Hemisphere), it is not as pronounced as with a cold frontal passage. Precipitation, in the form of rain, snow, or drizzle, is generally found ahead of the surface front, as well as convective showers and thunderstorms. Fog is common in the cold air ahead of the front. Although clearing usually occurs after passage, some conditions may produced fog in the warm air.

❐ warm high

A high pressure system that has its warmest temperatures at or near the center of circulation. Contrast with a cold high.

❐ warm low

A low pressure system that has its warmest temperatures at or near the center of circulation. Also referred to as a warm core low.

❐ warning

A forecast issued when severe weather has developed, is already occurring and reported, or is detected on radar. Warnings state a particular hazard or imminent danger, such as tornadoes, severe thunderstorms, flash and river floods, winter storms, heavy snows, etc.

❐ wasatch winds

Strong winds blowing easterly out of the Wasatch Mountains in Utah, sometimes reaching speeds greater than 75 miles per hour.

❐ watch

A forecast issued well in advance of a severe weather event to alert the public of the possibility of a particular hazard, such as tornadoes, severe thunderstorms, flash and river floods, winter storms, or heavy snows.

❐ water

Refers to the chemical compound, H_2O, as well as its liquid form. At atmospheric temperatures and pressures, it can exist in all three phases: solid (ice), liquid (water), and gaseous (water vapor). It is a vital, life-sustaining part of life on earth.

❐ water cycle

The vertical and horizontal transport of water in all its states between the earth, the atmosphere, and the seas.

❐ water vapor (H_2O)

Water in gaseous form. It is one of the most import constituents of the atmosphere. Due to its molecular content, air containing water vapor is lighter than dry air. This contributes to the reason why moist air has a tendency to rise.

❐ waterspout

A small, weak tornado, which is not formed by a storm-scale rotation. It is generally weaker than a supercell tornado and is not associated with a wall cloud or mesocyclone. It may be observed beneath cumulonimbus or towering cumulus clouds and is the water equivalent of a landspout.

❐ wave cyclone

A cyclone which forms and moves along a front. The circulation around the cyclone's center produces a wavelike deformation on the front. May also be call a migratory cyclone or low.

❐ wave length

The least distance between particles moving in the same phase of oscillation of a wave. In oceanography, it is the horizontal distance between the highest parts of two successive wave crests above the still water level, separated by a trough that is below the still water level, and it is measured in meters.

❐ wave(s)

In general, any pattern with some roughly identifiable periodicity in time and/ or space. It is also considered as a disturbance that moves through or over the surface of the medium with speed dependent on the properties of the medium. In meteorology, this applies to atmospheric waves, such as long waves and short waves. In oceanography, this applies to waves generated by mechanical means, such as currents, turbidity, and the wind.

❐ weather

The state of the atmosphere at a specific time and with respect to its effect on life and human activities. It is the short term variations of the atmosphere, as opposed to the long term, or climatic, changes. It is often referred to in terms of brightness, cloudiness, humidity, precipitation, temperature, visibility, and wind.

❐ weather surveillance radar (wsr-88d)

The newest generation of Doppler radars, the 1988 Doppler weather radar. The radar units, with help from a set of computers, show very detailed images of precipitation and other phenomena, including air motions within a storm.

❐ weather vane

Originally used as a wind vane, it is an instrument that indicates the wind direction. The name developed based on observations on what kind of weather occurred with certain wind directions. Creative designs often adorn the tops of barns and houses.

❐ weathering

The decay and breakup of rocks on the earth's surface by natural chemical and mechanical processes. The mechanical action includes large changes of temperature, extreme temperatures, frost, or the impact of wind borne sand or water. Chemical action includes the chemical reactions between atmospheric constituents in a moist environments or in rain water. Biological agents are mainly fungi which attack organic material.

❐ wedge

Primarily refers to an elongated area of shallow high pressure at the earth's surface. It is generally associated with cold air east of the Rockies or Appalachians. It is another name for a ridge, ridge line, or ridge axis. Contrast with a trough. Wedge is also a slang term for a large, wide tornado with a wedge-like shape.

❐ west virginia high

An area of stagnant high pressure located over West Virginia during Indian Summer.

❐ westerlies

Usually applied to the broad patterns of persistent winds with a westerly component. It is the dominant persistent atmospheric motion, centered over the midlatitudes of each hemisphere. Near the earth's surface, the westerlies extend from approximately 35 to 65 degrees latitude, while in the upper levels they extend further polarward and equatorward.

❐ wet bulb depression

Dependent on the temperature and the humidity of the air, it is the difference between the dry bulb and the wet bulb readings.

❐ wet bulb thermometer

A thermometer used to measure the lowest temperature in the ambient atmosphere in its natural state by evaporating water from a wet muslin-covered bulb of a thermometer. The wet bulb temperature is used to compute dew point and relative humidity. One of the two thermometers that make up a psychrometer.

❑ whirlwind

A small-scale, rapidly rotating column of wind, formed thermally and most likely to develop on clear, dry, hot afternoons. Often called a dust devil when visible by the dust, dirt or debris it picks up. Also slang for a landspout or a tornado.

❑ whiteout

When visibility is near zero due to blizzard conditions or occurs on sunless days when clouds and surface snow seem to blend, erasing the horizon and creating a completely white vista.

❑ wind

Air that flows in relation to the earth's surface, generally horizontally. There are four areas of wind that are measured: direction, speed, character (gusts and squalls), and shifts. Surface winds are measured by wind vanes and anemometers, while upper level winds are detected through pilot balloons, rawin, or aircraft reports.

❑ wind chill index

The calculation of temperature that takes into consideration the effects of wind and temperature on the human body. Describes the average loss of body heat and how the temperature feels. This is not the actual air temperature.

❑ wind direction

The direction from which the wind is blowing. For example, an easterly wind is blowing from the east, not toward the east. It is reported with reference to true north, or 360 degrees on the compass, and expressed to the nearest 10 degrees, or to one of the 16 points of the compass (N, NE, WNW, etc.).

❑ wind shear

The rate of wind speed or direction change with distance. Vertical wind shear is the rate of change of the wind with respect to altitude. Horizontal wind shear is the rate of change on a horizontal plane.

❑ wind shift

The term applied to a change in wind direction of 45 degrees or more, which takes place in less than 15 minutes. It may the result of a frontal passage, from katabatic winds, sea breezes, or thunderstorms, and in some instances, the change may be gradual or abrupt.

❑ wind speed

The rate of the motion of the air on a unit of time. It can be measured in a number of ways. In observing, it is measured in knots, or nautical miles per hour. The unit most often used in the United States is miles per hour.

❑ wind vane

An instrument that indicates the wind direction. The end of the vane which offers the greatest resistance to the motion of the air moves to the downwind position.

❐ wind wave

An ocean or lake wave resulting from the action of wind on the water's surface. After it leaves its fetch area, it is considered a swell.

❐ windward

The direction from which the wind is blowing. Also the upwind side of an object. The opposite of the downwind or leeward side.

❐ winter

Astronomically, this is the period between the winter solstice and the vernal equinox. It is characterized as having the coldest temperatures of the year, when the sun is primarily over the opposite hemisphere. Customarily, this refers to the months of December, January, and February in the North Hemisphere, and the months of June, July, and August in the Southern Hemisphere.

❐ winter storm

Any one of several storm systems that develop during the late fall to early spring and deposit wintry precipitation, such as snow, freezing rain, or ice.

❐ world meteorological organization (wmo)

From weather prediction to air pollution research, climate change related activities, ozone layer depletion studies and tropical storm forecasting, the World Meteorological Organization coordinates global scientific activity to allow increasingly prompt and accurate weather.

X

❐ x-rays

The portion of the electromagnetic spectrum that has a very short wave length. It has a wave length longer than gamma rays, yet shorter than visible light. X-rays can penetrate various thicknesses of all solids, and when absorbed by a gas, can result in ionization.

Y

❐ year

The interval required for the earth to complete one revolution around the sun. A sidereal year, which is the time it take for the earth to make one absolute revolution around the sun, is 365 days, 6 hours, 9 minutes, and 9.5 seconds. The calendar year begins at 12 o'clock midnight local time on the night of December 31st-January 1st. Currently, the Gregorian calendar of 365 days is used, with 366 days every four years, a leap year. The tropical year, also called the mean solar year, is dependent on the seasons. It is the interval between two consecutive returns of the sun to the vernal equinox. In 1900, that took 365 days, 5 hours, 48 minutes, and 46 seconds, and it is decreasing at the rate of 0.53 second per century.

❐ yellow snow

Snow that is given golden, or yellow, appearance by the presence of pine or cypress pollen in it.

Z

❐ zenith

The point which is elevated 90 degrees from all points on a given observer's astronomical horizon. The point on any given observer's celestial sphere that lies directly above him. The opposite of nadir.

❐ zodiac

The position of the sun during the course of the year as it appears to move though successive constellations. Also, the band where the ecliptic runs centrally through the celestial sphere and contains the sun, the moon, and all the planets except Venus and Pluto.

❐ zonal flow

The flow of air along a latitudinal component of existing flow, normally from west to east.

❐ zonal index

The measure of the strength of the westerly winds of the middle latitudes. It is expressed as the horizontal pressure difference between 35 degrees and 55 degrees latitude, or as the corresponding geostrophic wind.

❐ zulu time

One of several names for the twenty-four hour time which is used throughout the scientific and military communities.